Matrix Mathematics

Using a modern matrix-based approach, this rigorous second course in linear algebra helps upper-level undergraduates in mathematics, data science, and the physical sciences transition from basic theory to advanced topics and applications. Its clarity of exposition together with many illustrations, 900+ exercises, and 350 conceptual and numerical examples aid the student's understanding. Concise chapters promote a focused progression through essential ideas. Topics are derived and discussed in detail, including the singular value decomposition, Jordan canonical form, spectral theorem, QR factorization, normal matrices, Hermitian matrices, and positive definite matrices. Each chapter ends with a bullet list summarizing important concepts. New to this edition are chapters on matrix norms and positive matrices, many new sections on topics including interpolation and LU factorization, 300+ additional problems, many new examples, and color-enhanced figures. Prerequisites include a first course in linear algebra and basic calculus sequence. Instructor resources are available.

Stephan Ramon Garcia is W. M. Keck Distinguished Service Professor and Chair of the Department of Mathematics and Statistics at Pomona College. He is the author of five books and over 100 research articles in operator theory, complex analysis, matrix analysis, number theory, discrete geometry, and combinatorics. He has served on the editorial boards of the *Proceedings of the American Mathematical Society, Notices of the American Mathematical Society, Involve,* and *The American Mathematical Monthly.* He received six teaching awards from three different institutions and is a fellow of the American Mathematical Society, which has awarded him the inaugural Dolciani Prize for Excellence in Research.

Roger A. Horn was Professor and Chair of the Department of Mathematical Sciences at the Johns Hopkins University and then Research Professor of Mathematics at the University of Utah until his retirement in 2015. His publications include *Matrix Analysis* (2nd edition, Cambridge, 2012) and *Topics in Matrix Analysis* (both written with Charles R. Johnson, Cambridge, 1991), as well as more than 100 research articles in matrix analysis, statistics, health services research, complex variables, probability, differential geometry, and analytic number theory. He was the editor of *The American Mathematical Monthly* and has served on the editorial boards of the *SIAM Journal of Matrix Analysis, Linear Algebra and Its Applications,* and the *Electronic Journal of Linear Algebra.*

CAMBRIDGE MATHEMATICAL TEXTBOOKS

Cambridge Mathematical Textbooks is a program of undergraduate and beginning graduate-level textbooks for core courses, new courses, and interdisciplinary courses in pure and applied mathematics. These texts provide motivation with plenty of exercises of varying difficulty, interesting examples, modern applications, and unique approaches to the material.

Advisory Board

Paul T. Allen, *Lewis & Clark College*
Melody Chan, *Brown University*
Teena Gerhardt, *Michigan State University*
Illya Hicks, *Rice University*
Greg Lawler, *University of Chicago*
John M. Lee, *University of Washington*
Lawrence Washington, *University of Maryland, College Park*
Talithia Williams, *Harvey Mudd College*

A complete list of books in the series can be found at www.cambridge.org/mathematics

Recent titles include the following:

Chance, Strategy, and Choice: An Introduction to the Mathematics of Games and Elections, S. B. Smith
Set Theory: A First Course, D. W. Cunningham
Chaotic Dynamics: Fractals, Tilings, and Substitutions, G. R. Goodson
A Second Course in Linear Algebra, S. R. Garcia & R. A. Horn
Introduction to Experimental Mathematics, S. Eilers & R. Johansen
Exploring Mathematics: An Engaging Introduction to Proof, J. Meier & D. Smith
A First Course in Analysis, J. B. Conway
Introduction to Probability, D. F. Anderson, T. Seppäläinen & B. Valkó
Linear Algebra, E. S. Meckes & M. W. Meckes
A Short Course in Differential Topology, B. I. Dundas
Abstract Algebra with Applications, A. Terras
Complex Analysis, D. E. Marshall
Abstract Algebra: A Comprehensive Introduction, J. W. Lawrence & F. A. Zorzitto
An Invitation to Combinatorics, S. Shahriari
Modern Mathematical Logic, Joseph Mileti

Matrix Mathematics
A Second Course in Linear Algebra

SECOND EDITION

STEPHAN RAMON GARCIA
Pomona College

ROGER A. HORN
Tampa, Florida

Shaftesbury Road, Cambridge CB2 8EA, United Kingdom

One Liberty Plaza, 20th Floor, New York, NY 10006, USA

477 Williamstown Road, Port Melbourne, VIC 3207, Australia

314–321, 3rd Floor, Plot 3, Splendor Forum, Jasola District Centre, New Delhi – 110025, India

103 Penang Road, #05–06/07, Visioncrest Commercial, Singapore 238467

Cambridge University Press is part of Cambridge University Press & Assessment, a department of the University of Cambridge.

We share the University's mission to contribute to society through the pursuit of education, learning and research at the highest international levels of excellence.

www.cambridge.org
Information on this title: www.cambridge.org/highereducation/isbn/9781108837101

DOI: 10.1017/9781108938426

First published 2017
Second edition published 2023

Printed in the United Kingdom by TJ Books Limited, Padstow, Cornwall

A catalogue record for this publication is available from the British Library.

Library of Congress Cataloging-in-Publication Data
Names: Garcia, Stephan Ramon, author. | Horn, Roger A., author.
Title: Matrix mathematics : a second course in linear algebra /
Stephan Ramon Garcia, Pomona College, California, Roger A. Horn, Tampa, Florida.
Other titles: Second course in linear algebra
Description: Second edition. | Cambridge, United Kingdom ; New York, NY,
USA : Cambridge University Press, 2023. | Series: Cambridge mathematical
textbooks | Includes bibliographical references and index.
Identifiers: LCCN 2022048148 | ISBN 9781108837101 (hardback) |
ISBN 9781108938426 (ebook)
Subjects: LCSH: Algebras, Linear – Textbooks.
Classification: LCC QA184.2 .G37 2023 | DDC 512/.5–dc23/eng20230327
LC record available at https://lccn.loc.gov/2022048148

ISBN 978-1-108-83710-1 Hardback

Additional resources for this publication at www.cambridge.org/garcia-horn.

To our families:

Gizem, Reyhan, and Altay
Susan;
Craig, Cori, Cole, and Carson;
Howard, Heidi, Archer, and Ella Ceres

"A broad coverage of more advanced topics, rich set of exercises, and thorough index make this stylish book an excellent choice for a second course in linear algebra."

Nick Higham, *University of Manchester*

"This textbook thoroughly covers all the material you'd expect in a Linear Algebra course plus modern methods and applications. These include topics like the Fourier transform, eigenvalue adjustments, stochastic matrices, interlacing, power method and more. With 20 chapters of such material, this text would be great for a multi-part course and a reference book that all mathematicians should have."

Deanna Needell, *University of California, Los Angeles*

"The original edition of Garcia and Horn's *Second Course in Linear Algebra* was well-written, well-organized, and contained several interesting topics that students should see – but rarely do in first-semester linear algebra – such as the singular value decomposition, Gershgorin circles, Cauchy's interlacing theorem, and Sylvester's inertia theorem. This new edition also has all of this, together with useful new material on matrix norms. Any student with the opportunity to take a second course on linear algebra would be lucky to have this book."

Craig Larson, *Virginia Commonwealth University*

"An extremely versatile Linear Algebra textbook that allows numerous combinations of topics for a traditional course or a more modern and applications-oriented class. Each chapter contains the exact amount of information, presented in a very easy-to-read style, and a plethora of interesting exercises to help the students deepen their knowledge and understanding of the material."

Maria Isabel Bueno Cachadina, *University of California, Santa Barbara*

"This is an excellent textbook. The topics flow nicely from one chapter to the next and the explanations are very clearly presented. The material can be used for a good second course in Linear Algebra by appropriately choosing the chapters to use. Several options are possible. The breadth of subjects presented makes this book a valuable resource."

Daniel B. Szyld, *Temple University and President of the International Linear Algebra Society*

"With a careful selection of topics and a deft balance between theory and applications, the authors have created a perfect textbook for a second course on Linear Algebra. The exposition is clear and lively. Rigorous proofs are supplemented by a rich variety of examples, figures, and problems."

Rajendra Bhatia, *Ashoka University*

"The authors have provided a contemporary, methodical, and clear approach to a broad and comprehensive collection of core topics in matrix theory. They include a wealth of illustrative examples and accompanying exercises to re-enforce the concepts in each chapter. One unique aspect of this book is the inclusion of a large number of concepts that arise in many interesting applications that do not typically appear in other books. I expect this text will be a compelling reference for active researchers and instructors in this subject area."

Shaun Fallat, *University of Regina*

"It starts from scratch, but manages to cover an amazing variety of topics, of which quite a few cannot be found in standard textbooks. All matrices in the book are over complex numbers, and the connections to physics, statistics, and engineering are regularly highlighted. Compared with the first edition, two new chapters and 300 new problems have been added, as well as many new conceptual examples. Altogether, this is a truly impressive book."

Claus Scheiderer, *University of Konstanz*

Contents

Preface for the Second Edition

New to this Edition

This is the second edition of *A Second Course in Linear Algebra*. The new title reflects an approach to advanced linear algebra that emphasizes matrix factorizations and algorithms. The second edition includes:

• New chapters on Matrix Norms and Positive Matrices.

• Revisions that incorporate classroom experience with students.

• New sections on Interpolation, Orthogonal polynomials, Gaussian quadrature, *LU* factorization, unitary equivalence and bidiagonal matrices, induced matrix norms, iterative algorithms such as the power and point Jacobi methods, and Perron-Frobenius theory.

• Color-enhanced figures.

• More than 300 new problems and many new conceptual and numerical examples.

• A comprehensive solution manual available to instructors.

Target Readership

Matrix mathematics and linear algebra are increasingly relevant in a world focused on the acquisition and analysis of data. Consequently, this book is intended for students of pure and applied mathematics, computer science, economics, engineering, mathematical biology, operations research, physics, and statistics. We assume that the reader has completed a lower-division calculus sequence and a first course in linear algebra. Analysis is not a prerequisite for this book.

Key Features of the Book

• Block matrices are employed systematically.

• Matrix factorizations and unitary transformations are emphasized.

• More than 350 examples illustrate concepts introduced in the text.

- Topics for a one-semester course can be selected in many ways to match the needs and interests of the class.

- Reviews of complex numbers, polynomials, and basic linear algebra are included.

- More than 90 figures illustrate the geometric foundations of linear algebra.

- Special topics include polynomial interpolation, orthogonal polynomials, Gaussian quadrature, matrix norms, Perron–Frobenius theory, and the Google matrix.

- Every chapter includes problems, more than 900 in total.

- Notes at the end of chapters provide sources of additional information.

- Each chapter ends with a bullet list of important concepts.

- Symbols used in the book are listed in a table of notation, with page references.

- A comprehensive index helps readers locate concepts and definitions. More than 2,000 entries enhance the value of the book as a reference.

- Concise, direct presentation and language level are suitable for an international audience.

Coverage of the Book

Matrices and vector spaces in this book are over the complex field. The use of complex scalars is essential to the study of eigenvalues, even for real matrices, and is consistent with modern numerical linear algebra software. Moreover, it is aligned with applications in physics (complex wave functions and Hermitian matrices in quantum mechanics), electrical engineering (analysis of circuits and signals in which both phase and amplitude are important), statistics (time series and characteristic functions), and computer science (fast Fourier transforms, convergent matrices in iterative algorithms, and quantum computing).

While studying linear algebra with this book, students can observe and practice good mathematical communication skills. These skills include how to state (and read) a theorem carefully; how to choose (and use) hypotheses; how to prove a statement by induction, by contradiction, or by contraposition; how to improve a theorem by weakening its hypotheses or strengthening its conclusions; how to use counterexamples; and how to write a cogent solution to a problem.

The following topics in the book are useful in applications of linear algebra, but fall outside the realm of linear transformations and similarity, so they may be absent from textbooks that adopt an abstract operator approach:

- Gershgorin's disk theorem on eigenvalue location

- The pivot-column decomposition and full-rank factorizations

- Commutants and trace-zero matrices (Shoda's theorem)

- QR, bidiagonal, triangular, and Cholesky factorizations

- Discrete Fourier transforms

- Circulant matrices

- Eigenvalue adjustments and the Google matrix

- Nonnegative matrices (Markov matrices) and positive matrices (Perron's theorem)

- The singular value and compact singular value decompositions

- Low-rank approximations to a data matrix

- Generalized inverses (Moore–Penrose inverses)

- Positive semidefinite matrices

- Schur complements

- Hadamard (entrywise) and Kronecker (tensor) products

- The Schur product theorem

- Matrix norms and the spectral radius

- Error bounds for eigenvalue computations (Bauer–Fike theorem)

- Convergent matrices, power-bounded matrices, and iterative algorithms

- Least-squares and minimum-norm solutions

- Complex symmetric matrices

- Inertia of normal matrices

- Eigenvalue and singular-value interlacing

- Inequalities among eigenvalues, singular values, and diagonal entries

Structure of the Book

A comprehensive list of symbols and notation (with page references) follows the Preface.

Chapter 1 reviews complex and real vector spaces, with numerous examples. The essential concepts of linear independence, linear dependence, and spanning lists are introduced, and the book's first matrix factorization emerges: the pivot-column decomposition.

Chapter 2 focuses on bases, dimension, and change-of-basis matrices. Matrix similarity arises as the relation between the representations of a linear transformation with respect to two bases. Lagrange interpolation provides examples of bases in vector spaces of polynomials. We observe instability with interpolation at equally spaced nodes and better behavior with interpolation at Chebyshev nodes. Integration of a polynomial interpolation leads to Simpson's rule and other Newton–Cotes quadrature formulae.

The block-matrix paradigm used throughout the book is introduced in Chapter 3. Block-matrix notation is useful in thinking about and communicating mathematical concepts. It focuses attention on the main ideas, instead of a quagmire of symbols and subscripts. Block matrices are central to the logic and coding of modern numerical algorithms. Row and column partitions are essential to the representation of a matrix product as the sum of outer products. Block Gaussian elimination leads to the Schur complement and determinant formulae for bordered matrices. Kronecker (tensor) products are a special topic at the end of this chapter.

Rank is the core concept in Chapter 4, which begins with the rank-nullity theorem and the subspace-intersection theorem. Block-matrix methods and full-rank factorizations are employed to present the basic rank inequalities for matrix sums and products. We present an algorithm to obtain LU factorizations that makes use of the outer-product representation for a matrix product. Shoda's theorem about matrix commutators and trace-zero matrices is a special topic in the final section.

Chapters 5 and 6 review geometry in the Euclidean plane and use it to motivate axioms for inner product spaces and normed linear spaces. Topics include orthogonal vectors, orthogonal projections, orthonormal bases, orthogonalization, the Riesz representation theorem, adjoints, and applications of the theory to Fourier series. Orthogonal polynomials and Gaussian quadrature are the special topic in Chapter 6.

Chapter 7 introduces unitary matrices, which are used in modern computational algorithms because they are easy to invert and exhibit superior stability properties in numerical calculations. In this chapter, we use unitary matrices to construct the QR factorization and a unitary similarity to upper Hessenberg form.

Chapter 8 discusses orthogonal projections, best approximations, least-squares solutions of linear systems, and the use of QR factorizations to solve the normal equations.

Chapter 9 introduces eigenvalues, eigenvectors, and geometric multiplicity. We show that an $n \times n$ real or complex matrix has at least one and not more than n distinct eigenvalues, and use Gershgorin's disk theorem to identify a region in the complex plane that contains them.

Chapter 10 deals with the characteristic polynomial and algebraic multiplicity. We develop criteria for diagonalizability and define primary matrix functions of a diagonalizable matrix. Topics include Fibonacci numbers, the eigenvalues of AB and BA, commutants, and simultaneous diagonalization.

Chapter 11 features Schur's triangularization theorem and a related result for a commuting family. Schur's theorem is used to prove the Cayley–Hamilton theorem: each square matrix is annihilated by its characteristic polynomial. The latter result motivates introduction of the minimal polynomial and a study of its properties. We prove Sylvester's theorem on linear matrix equations and use it to show that every square matrix is similar to a block diagonal matrix with unispectral diagonal blocks. The special topic in this chapter discusses perturbations of the Google matrix that facilitate computation of website rankings.

Chapter 12 builds on the preceding chapter to show that every square matrix is similar to a Jordan matrix that is unique up to permutation of its direct summands.

We discuss several applications of the Jordan canonical form in Chapter 13. They include systems of linear differential equations, an analysis of the Jordan structures of AB and BA, convergent and power-bounded matrices, a limit theorem for stochastic matrices that have positive entries, similarity of a matrix to its transpose, and similarity of a matrix to its complex conjugate.

Chapter 14 is about normal matrices: matrices that commute with their conjugate transpose. The spectral theorem says that a matrix is normal if and only if it is unitarily diagonalizable. Hermitian, skew-Hermitian, unitary, real-orthogonal, real-symmetric, and circulant matrices are all normal.

Positive semidefinite matrices are the subject of Chapter 15. These matrices arise in statistics (correlation matrices and the normal equations), mechanics (kinetic and potential energy in a vibrating system), and geometry (ellipsoids). Topics include matrix square roots, simultaneous diagonalization of quadratic forms, Cholesky factorization, and Hadamard and Kronecker products.

The principal result in Chapter 16 is the singular value decomposition, which is at the heart of many modern numerical algorithms in statistics, control theory, approximation, image compression, and data analysis. Topics include the compact singular value decomposition and polar decompositions, with special attention to uniqueness of these factorizations. A special topic is unitary equivalence of a complex matrix to an upper bidiagonal matrix.

In Chapter 17, the singular value decomposition is used to compress an image or data matrix. Other applications of the singular value decomposition discussed are the generalized inverse (Moore–Penrose inverse) of a matrix; inequalities between singular values and eigenvalues; the spectral norm of a matrix; perturbation bounds for linear systems and eigenvalue problems; and canonical forms for matrices that are complex symmetric or idempotent.

Chapter 18 investigates eigenvalue interlacing phenomena for Hermitian matrices that are bordered or are subjected to an additive perturbation. Related results include an interlacing theorem for singular values, a determinant criterion for positive definiteness, and inequalities that link the eigenvalues and diagonal entries of a Hermitian matrix. We prove Sylvester's inertia theorem for Hermitian matrices and a generalized inertia theorem for normal matrices.

Norms and matrix norms are the topics in Chapter 19. Eight examples of matrix norms are presented, with a systematic account of inequalities between pairs of them. Many iterative algorithms require that a particular matrix has spectral radius less than 1, which would be the case if some matrix norm of that matrix is less than 1. We analyze two iterative algorithms: the point Jacobi method to solve a linear system and the power method to find a dominant eigenpair. Facts about matrix norms are used to prove Gelfand's formula for the spectral radius.

Chapter 20 is devoted to Perron's theorem about the dominant eigenpair and limits of powers of a real square matrix with positive entries. This result has been used in diverse fields such as economic modeling, team ranking, population dynamics, genetics, and city planning. We prove it using facts about matrix norms and the Jordan canonical form.

Four short appendices review notation and concepts for complex numbers, polynomials, basic linear algebra, and mathematical induction. The appendices are provided for reference, and readers can consult them as needed.

The appendices are followed by a short list of references and an extensive index.

Acknowledgments

We thank Zachary Glassman for producing many of our illustrations and for answering our LaTeX questions. The cute animal illustrations in Figures 2.1, 2.2, 7.2, 9.1, 9.4, 9.5, 9.6, A.2, A.4, and A.5 were produced with the wonderful Ti*k*Zlings package of samcarter.

We thank Zhongshan Li, Dennis Merino, Russ Merris, and Fuzhen Zhang for their comments on evolving manuscripts for this book.

We thank the students who attended the first author's advanced linear algebra courses at Pomona College during fall 2014, fall 2015, spring 2019, and fall 2022. In particular, we thank Ahmed Al Fares, Andreas Biekert, Arsum Chaudhary, Andi Chen, Wanning Chen, Alex Cloud, Bill DeRose, Jacob Fiksel, Logan Gilbert, Sheridan Grant, Adam He, David Khatami, Cheng Wai Koo, Bo Li, Shiyue Li, Samantha Morrison, Nathanael Roy, Michael Someck, Sallie Walecka, Angie Wang, Summer Will, Wentao Yuan, and Alan Zhou for their comments.

Special thanks to Zoë Batterman, Christopher Donnay, Gordon Elnagar, Ciaran Evans, Elizabeth Sarapata, Adam Starr, Adam Waterbury, and Chris Wang for their eagle-eyed reading of the text.

S. R. Garcia thanks the National Science Foundation for support in the form of grants DMS-2054002 and DMS-1800123.

Notation

$\in, \notin$	is / is not an element of
$\subseteq$	is a subset of
$\varnothing$	the empty set
$\cup$	union
$\cap$	intersection
$\times$	Cartesian product
$f : X \to Y$	f is a function from X into Y
$\Longrightarrow$	implies
$\Longleftarrow$	is implied by
$\Longleftrightarrow$	if and only if
$\approx$	approximately equal
$x \mapsto y$	implicit definition of a function that maps x to y
$\mathbb{N} = \{1, 2, 3, \ldots\}$	the set of natural numbers
$\mathbb{Z} = \{\ldots, -2, -1, 0, 1, 2, \ldots\}$	the set of integers
$\mathbb{R}$	the set of real numbers
$\mathbb{C}$	the set of complex numbers
$\mathbb{F}$	field of scalars ($\mathbb{F} = \mathbb{R}$ or $\mathbb{C}$)
$\operatorname{Re} z$	real part of the complex number z (p. 417)
$\operatorname{Im} z$	imaginary part of the complex number z (p. 417)
$|z|$	modulus of the complex number z (p. 420)
$\arg z$	argument of the complex number z (p. 420)
$[a, b]$	a real interval that includes its endpoints a, b
$\mathcal{U}, \mathcal{V}, \mathcal{W}$	vector spaces
$\mathscr{U}, \mathscr{V}, \mathscr{W}$	subsets of vector spaces
$a, b, c, \ldots$	scalars
$\mathbf{a}, \mathbf{b}, \mathbf{c}, \ldots$	(column) vectors
$A, B, C, \ldots$	matrices
$\mathsf{M}_{m \times n}(\mathbb{F})$	the set of $m \times n$ matrices with entries in $\mathbb{F}$
$\mathsf{M}_n(\mathbb{F})$	the set of $n \times n$ matrices with entries in $\mathbb{F}$
$\mathsf{M}_{m \times n}$	the set of $m \times n$ matrices with entries in $\mathbb{C}$
M_n	the set of $n \times n$ matrices with entries in $\mathbb{C}$
$\cong$	an equivalence relation (p. 51)
$\deg p$	degree of a polynomial p (p. 428)
δ_{ij}	Kronecker delta (p. 434)
$(\mathbf{x})_i$	ith entry of a vector $\mathbf{x}$ (p. 439)
I_n	$n \times n$ identity matrix (p. 434)

I	identity matrix (size inferred from context) (p. 434)
$\operatorname{diag}(x_1, x_2, \ldots, x_n)$	diagonal matrix with diagonal entries $x_1, x_2, \ldots, x_n$ (p. 435)
$A^0 = I$	convention for zeroth power of a matrix (p. 437)
A^{T}	transpose of A (p. 437)
$A^{-\mathsf{T}}$	inverse of A^{T} (p. 437)
$\overline{A}$	conjugate of A (p. 437)
A^*	conjugate transpose (adjoint) of A (p. 437)
A^{-*}	inverse of A^* (p. 437)
$\operatorname{tr} A$	trace of A (p. 438)
$\det A$	determinant of A (p. 441)
$\operatorname{adj} A$	adjugate of A (p. 442)
$\operatorname{sgn} \sigma$	sign of a permutation σ (p. 443)
$\mathcal{P}_n$	set of complex polynomials of degree at most n (p. 4)
$\mathcal{P}_n(\mathbb{R})$	set of real polynomials of degree at most n (p. 4)
$\mathcal{P}$	set of all complex polynomials (p. 4)
$C_{\mathbb{F}}[a, b]$	set of continuous $\mathbb{F}$-valued functions on $[a, b]$ (p. 4)
$C[a, b]$	set of continuous $\mathbb{C}$-valued functions on $[a, b]$ (p. 4)
$\operatorname{null} A$	null space of a matrix A (p. 5)
$\operatorname{col} A$	column space of a matrix A (p. 6)
$\operatorname{row} A$	row space of a matrix A (p. 6)
$\mathcal{P}_{\text{even}}$	set of even complex polynomials (p. 7)
$\mathcal{P}_{\text{odd}}$	set of odd complex polynomials (p. 7)
$A\mathcal{U}$	A acting on a subspace $\mathcal{U}$ (p. 6)
$\operatorname{span} \mathscr{S}$	span of a subset $\mathscr{S}$ of a vector space (p. 8)
$\mathbf{e}$	all-ones vector (p. 10)
$\mathcal{U} \cap \mathcal{W}$	intersection of subspaces $\mathcal{U}$ and $\mathcal{W}$ (p. 11)
$\mathcal{U} + \mathcal{W}$	sum of subspaces $\mathcal{U}$ and $\mathcal{W}$ (p. 12)
$\mathcal{U} \oplus \mathcal{W}$	direct sum of subspaces $\mathcal{U}$ and $\mathcal{W}$ (p. 12)
$\mathbf{v}_1, \mathbf{v}_2, \ldots, \widehat{\mathbf{v}_j}, \ldots, \mathbf{v}_r$	list of vectors with $\mathbf{v}_j$ omitted (p. 17)
$\mathbf{e}_1, \mathbf{e}_2, \ldots, \mathbf{e}_n$	standard basis for $\mathbb{F}^n$ (p. 28)
E_{ij}	matrix with (i, j) entry 1 and all others 0 (p. 28)
$\dim \mathcal{V}$	dimension of $\mathcal{V}$ (p. 28)
$[\mathbf{v}]_\beta$	coordinate vector of $\mathbf{v}$ with respect to a basis β (p. 33)
$\mathcal{L}(\mathcal{V}, \mathcal{W})$	set of linear transformations from $\mathcal{V}$ to $\mathcal{W}$ (p. 35)
$\mathcal{L}(\mathcal{V})$	set of linear transformations from $\mathcal{V}$ to itself (p. 35)
$\ker T$	kernel of T (p. 36)
$\operatorname{ran} T$	range of T (p. 36)
I	identity linear transformation (p. 38)
$\operatorname{rank} A$	rank of a matrix A (p. 31)
$\operatorname{nullity} A$	nullity of a matrix A (p. 74)
$\star$	unspecified matrix entry (p. 63)
$A \oplus B$	direct sum of matrices A and B (p. 64)
$A \otimes B$	Kronecker product of matrices A and B (p. 68)
$\operatorname{vec} A$	vector of stacked columns of a matrix A (p. 69)

$[A, B]$	commutator of matrices A and B (p. 85)
$\langle \mathbf{x}, \mathbf{y} \rangle$	inner product of vectors $\mathbf{x}$ and $\mathbf{y}$ (p. 98)
$\langle A, B \rangle_\mathrm{F}$	Frobenius inner product of matrices A and B (p. 99)
$\perp$	orthogonal (p. 100)
$\lVert \mathbf{x} \rVert$	norm of a vector $\mathbf{x}$ (p. 101)
$\lVert \mathbf{x} \rVert_2$	Euclidean norm of a vector $\mathbf{x}$ (p. 101)
$\lVert A \rVert_\mathrm{F}$	Frobenius norm of a matrix A (p. 101)
$\lVert \mathbf{x} \rVert_1$	ℓ^1 norm (absolute sum norm) of a vector $\mathbf{x}$ (p. 106)
$\lVert \mathbf{x} \rVert_\infty$	ℓ^∞ norm (max norm) of a vector $\mathbf{x}$ (p. 106)
${}_\gamma[T]_\beta$	matrix representation of $T \in \mathcal{L}(\mathcal{V}, \mathcal{W})$ with respect to bases β and γ (p. 118)
$f(x^+)$	one-sided limit from the right (p. 126)
$f(x^-)$	one-sided limit from the left (p. 126)
$P_\mathbf{v}$	projection onto a unit vector $\mathbf{v}$ (p. 138)
F_n	$n \times n$ Fourier matrix (p. 140)
$\mathcal{U}^\perp$	orthogonal complement of a set $\mathcal{U}$ (p. 158)
$P_\mathcal{U}$	orthogonal projection onto a subspace $\mathcal{U}$ (p. 163)
$d(\mathbf{v}, \mathcal{U})$	distance from $\mathbf{v}$ to $\mathcal{U}$ (p. 167)
$G(\mathbf{u}_1, \mathbf{u}_2, \ldots, \mathbf{u}_n)$	Gram matrix (p. 170)
$g(\mathbf{u}_1, \mathbf{u}_2, \ldots, \mathbf{u}_n)$	Gram determinant (p. 170)
$\mathrm{spec}\, A$	spectrum of A (p. 186)
$\mathcal{E}_\lambda(A)$	eigenspace of A for eigenvalue λ (p. 189)
$\mathscr{G}_k(A)$	kth Gershgorin disk of A (p. 190)
$\mathscr{G}(A)$	Gershgorin region of A (p. 190)
$R'_k(A)$	kth deleted absolute row sum of A (p. 190)
$R_k(A)$	kth absolute row sum of A (p. 190)
$p_A(z)$	characteristic polynomial of A (p. 202)
$\mathscr{F}'$	commutant of a set of matrices $\mathscr{F}$ (p. 213)
e^A	matrix exponential of A (p. 213)
$m_A(z)$	minimal polynomial of A (p. 226)
C_p	companion matrix of the polynomial p (p. 227)
$J_k(\lambda)$	$k \times k$ Jordan block with eigenvalue λ (p. 241)
J_k	$k \times k$ nilpotent Jordan block (p. 241)
$w_1, w_1, \ldots, w_q$	Weyr characteristic of a matrix (p. 248)
$\rho(A)$	spectral radius of A (p. 259)
$\mathcal{G}_\lambda(A)$	generalized eigenspace of A for eigenvalue λ (p. 253)
$\Delta(A)$	defect from normality of A (p. 282)
H_n	$n \times n$ Hartley matrix (p. 286)
R_n	real part of F_n (p. 286)
T_n	imaginary part of F_n (p. 286)
$A \circ B$	Hadamard product of A and B (p. 315)
$A \bullet B$	Jordan product of A and B (p. 320)
$\sigma_{\max}(A)$	maximum singular value (p. 346)
$\lVert A \rVert_2$	spectral norm of a matrix A (p. 347)
$\sigma_{\min}(A)$	minimum singular value (p. 350)
$\sigma_1(A), \sigma_2(A), \ldots$	singular values of A (p. 350)

$A^\dagger$	pseudoinverse of A (p. 353)
$\kappa_2(A)$	spectral condition number of A (p. 358)
$\mathscr{W}(A)$	numerical range of A (p. 384)
$\|\mathbf{x}\|_{[k]}$	k-norm of a vector $\mathbf{x}$ (p. 391)
$\|\mathbf{x}\|_S$	$\|S\mathbf{x}\|$, in which S is invertible (p. 391)
$\|A\|_1$	ℓ_1 norm of a matrix A (p. 391)
$\|A\|_\infty$	max norm of a matrix A (p. 392)
$N_{n\text{-max}}(A)$	n-max norm of a matrix A (p. 392)
$N_{\text{col max}}(A)$	column max norm of a matrix A (p. 392)
$N_S(A)$	$\|SAS^{-1}\|$ (p. 393)
$N_1(A)$	maximum absolute column-sum norm (p. 394)
$N_\infty(A)$	maximum absolute row-sum norm (p. 395)
$A_{(k)}$	$k \times k$ leading principal submatrix of A (p. 436)
S_n	nth statement in an induction (p. 447)
$|A|$	modulus of A (p. 334)
$|A|$	$A = [a_{ij}] \implies |A| = [|a_{ij}|]$ (Chapter 20 only)
$|\mathbf{x}|$	$\mathbf{x} = [x_i] \implies |\mathbf{x}| = [|x_i|]$ (Chapter 20 only)

1   Vector Spaces

Many types of mathematical objects can be added and scaled: vectors in the plane, real-valued functions on a given real interval, polynomials, and real or complex matrices. Through long experience with these and other examples, mathematicians have identified a short list of essential features (axioms) that define a consistent and inclusive mathematical framework known as a vector space.

The theory of vector spaces and linear transformations provides a conceptual framework and vocabulary for linear mathematical models of diverse phenomena. Even inherently non-linear physical theories may be well approximated for a broad range of applications by linear theories, whose natural setting is in real or complex vector spaces.

Examples of vector spaces include the two-dimensional real plane (the setting for plane analytic geometry and two-dimensional Newtonian mechanics) and three-dimensional real Euclidean space (the setting for solid analytic geometry, classical electromagnetism, and analytical dynamics). Other kinds of vector spaces abound in science and engineering. For example, standard mathematical models in quantum mechanics, electrical circuits, and signal processing use complex vector spaces. Many scientific theories exploit the formalism of vector spaces, which supplies powerful mathematical tools that are based only on the axioms for a vector space and their logical consequences, not on the details of a particular application.

In this chapter, we provide formal definitions of real and complex vector spaces, and many examples. Among the important concepts introduced are linear combinations, span, linear independence, and linear dependence.

1.1   What Is a Vector Space?

A vector space comprises four things that work together in harmony:

(a) A field $\mathbb{F}$ of *scalars*, which in this book is either $\mathbb{C}$ (complex numbers) or $\mathbb{R}$ (real numbers).

(b) A set $\mathcal{V}$ of objects called *vectors*.

(c) An operation of *vector addition* that takes any pair of vectors $\mathbf{u}, \mathbf{v} \in \mathcal{V}$ and assigns to them a vector in $\mathcal{V}$ denoted by $\mathbf{u} + \mathbf{v}$ (their *sum*).

(d) An operation of *scalar multiplication* that takes any scalar $c \in \mathbb{F}$ and any vector $\mathbf{u} \in \mathcal{V}$ and assigns to them a vector in $\mathcal{V}$ denoted by $c\mathbf{u}$.

Definition 1.1.1 Let $\mathbb{F} = \mathbb{R}$ or $\mathbb{C}$. Then $\mathcal{V}$ is a *vector space over the field* $\mathbb{F}$ (alternatively, $\mathcal{V}$ is an $\mathbb{F}$-*vector space*) if the scalars $\mathbb{F}$, the vectors $\mathcal{V}$, and the operations of vector addition and scalar multiplication satisfy the following axioms:

(1) There is a unique element $\mathbf{0} \in \mathcal{V}$ that is the additive identity element for vector addition, that is, $\mathbf{0} + \mathbf{u} = \mathbf{u}$ for all $\mathbf{u} \in \mathcal{V}$. The vector $\mathbf{0}$ is the *zero vector*.

(2) Vector addition is commutative: $\mathbf{u} + \mathbf{v} = \mathbf{v} + \mathbf{u}$ for all $\mathbf{u}, \mathbf{v} \in \mathcal{V}$.

(3) Vector addition is associative: $\mathbf{u} + (\mathbf{v} + \mathbf{w}) = (\mathbf{u} + \mathbf{v}) + \mathbf{w}$ for all $\mathbf{u}, \mathbf{v}, \mathbf{w} \in \mathcal{V}$.

(4) Additive inverses exist and are unique: for each $\mathbf{u} \in \mathcal{V}$, there is a unique $-\mathbf{u} \in \mathcal{V}$ such that $\mathbf{u} + (-\mathbf{u}) = \mathbf{0}$.

(5) The number 1 is the identity element for scalar multiplication: $1\mathbf{u} = \mathbf{u}$ for all $\mathbf{u} \in \mathcal{V}$.

(6) Multiplication in $\mathbb{F}$ and scalar multiplication are compatible: $a(b\mathbf{u}) = (ab)\mathbf{u}$ for all $a, b \in \mathbb{F}$ and all $\mathbf{u} \in \mathcal{V}$.

(7) Scalar multiplication distributes over vector addition: $c(\mathbf{u} + \mathbf{v}) = c\mathbf{u} + c\mathbf{v}$ for all $c \in \mathbb{F}$ and all $\mathbf{u}, \mathbf{v} \in \mathcal{V}$.

(8) Addition in $\mathbb{F}$ distributes over scalar multiplication: $(a + b)\mathbf{u} = a\mathbf{u} + b\mathbf{u}$ for all $a, b \in \mathbb{F}$ and all $\mathbf{u} \in \mathcal{V}$.

A vector space over $\mathbb{R}$ is a *real vector space*; a vector space over $\mathbb{C}$ is a *complex vector space*. To help distinguish vectors from scalars, we often denote vectors (elements of the set $\mathcal{V}$) by boldface lowercase letters such as $\mathbf{a}$, $\mathbf{b}$, $\mathbf{u}$, and $\mathbf{v}$. In particular, this distinguishes the scalar 0 from the vector $\mathbf{0}$.

We often need to derive a conclusion from the fact that a vector $c\mathbf{u}$ is the zero vector, so we should look carefully at how that can happen.

Theorem 1.1.2 *Let $\mathcal{V}$ be an $\mathbb{F}$-vector space, let $c \in \mathbb{F}$, and let $\mathbf{u} \in \mathcal{V}$. The following statements are equivalent:*

(a) $c = 0$ *or* $\mathbf{u} = \mathbf{0}$.

(b) $c\mathbf{u} = \mathbf{0}$.

Proof (a) $\Rightarrow$ (b) First suppose that $\mathbf{u} = \mathbf{0}$. Then

$$
\begin{aligned}
c\mathbf{0} &= c\mathbf{0} + \mathbf{0} && \text{Axiom (1)} \\
&= c\mathbf{0} + \big(c\mathbf{0} + (-(c\mathbf{0}))\big) && \text{Axiom (4)} \\
&= (c\mathbf{0} + c\mathbf{0}) + (-(c\mathbf{0})) && \text{Axiom (3)} \\
&= c(\mathbf{0} + \mathbf{0}) + (-(c\mathbf{0})) && \text{Axiom (7)} \\
&= c\mathbf{0} + (-(c\mathbf{0})) && \text{Axiom (1)} \\
&= \mathbf{0} && \text{Axiom (4).}
\end{aligned}
$$

In particular, observe that

$$c\mathbf{0} = \mathbf{0} \text{ for any } c \in \mathbb{F}. \tag{1.1.3}$$

Now let $c = 0$ and compute

$$
\begin{aligned}
0\mathbf{u} &= 0\mathbf{u} + \mathbf{0} && \text{Axiom (1)} \\
&= 0\mathbf{u} + \big(0\mathbf{u} + (-(0\mathbf{u}))\big) && \text{Axiom (4)} \\
&= (0\mathbf{u} + 0\mathbf{u}) + (-(0\mathbf{u})) && \text{Axiom (3)} \\
&= (0 + 0)\mathbf{u} + (-(0\mathbf{u})) && \text{Axiom (8)} \\
&= 0\mathbf{u} + (-(0\mathbf{u})) && \\
&= \mathbf{0} && \text{Axiom (4).}
\end{aligned}
$$

(b) $\Rightarrow$ (a) Suppose that $c\mathbf{u} = \mathbf{0}$. If $c = 0$, we are done. If $c \neq 0$, then

$$\begin{aligned}
\mathbf{u} &= 1\mathbf{u} & \text{Axiom (5)} \\
&= (c^{-1}c)\mathbf{u} \\
&= c^{-1}(c\mathbf{u}) & \text{Axiom (6)} \\
&= c^{-1}\mathbf{0} \\
&= \mathbf{0} & \text{by (1.1.3).} \qquad \square
\end{aligned}$$

Corollary 1.1.4 *Let* $\mathcal{V}$ *be an* $\mathbb{F}$-*vector space. Then* $(-1)\mathbf{u} = -\mathbf{u}$ *for every* $\mathbf{u} \in \mathcal{V}$.

Proof Let $\mathbf{u} \in \mathcal{V}$. We must show that $(-1)\mathbf{u} + \mathbf{u} = \mathbf{0}$. Use the vector-space Axioms (5) and (8), together with the preceding theorem, and compute

$$\begin{aligned}
(-1)\mathbf{u} + \mathbf{u} &= (-1)\mathbf{u} + 1\mathbf{u} & \text{Axiom (5)} \\
&= (-1 + 1)\mathbf{u} & \text{Axiom (8)} \\
&= 0\mathbf{u} \\
&= \mathbf{0} & \text{Theorem 1.1.2.} \qquad \square
\end{aligned}$$

Addition in a vector space is an operation on only two vectors. We define addition of three or more vectors via a sequence of two-vector additions. We can define $\mathbf{u} + \mathbf{v} + \mathbf{w}$ to be

$$(\mathbf{u} + \mathbf{v}) + \mathbf{w} \quad \text{or} \quad \mathbf{u} + (\mathbf{v} + \mathbf{w})$$

because Axiom (4) (associativity) says that these expressions are equal. We can define $\mathbf{u} + \mathbf{v} + \mathbf{w} + \mathbf{x}$ in a similar fashion (insert suitable parentheses) to be

$$(\mathbf{u} + \mathbf{v}) + (\mathbf{w} + \mathbf{x}), \quad \mathbf{u} + (\mathbf{v} + (\mathbf{w} + \mathbf{x})), \quad \text{or} \quad (\mathbf{u} + (\mathbf{v} + \mathbf{w})) + \mathbf{x}$$

because we can prove that these expressions are equal. For example, two applications of Axiom (4) show that

$$(\mathbf{u} + (\mathbf{v} + \mathbf{w})) + \mathbf{x} = ((\mathbf{u} + \mathbf{v}) + \mathbf{w}) + \mathbf{x} = (\mathbf{u} + \mathbf{v}) + (\mathbf{w} + \mathbf{x}).$$

If $n \geq 3$, we define $\mathbf{v}_1 + \mathbf{v}_2 + \cdots + \mathbf{v}_n$ (a finite sum) via any sequence of two-vector additions obtained by insertion of suitable parentheses. It follows from Axiom (4) that the sum obtained does not depend on how the parentheses are inserted.

1.2 Examples of Vector Spaces

Axiom (1) ensures that every vector space contains a zero vector, so a vector space cannot be empty. However, the axioms for a vector space permit $\mathcal{V}$ to contain only the zero vector. Such a vector space is not interesting, and we often need to exclude it when formulating theorems.

Definition 1.2.1 Let $\mathcal{V}$ be an $\mathbb{F}$-vector space. If $\mathcal{V} = \{\mathbf{0}\}$, then $\mathcal{V}$ is a *zero vector space*; if $\mathcal{V} \neq \{\mathbf{0}\}$, then $\mathcal{V}$ is a *nonzero vector space*.

In each of the following examples, we describe the elements of the set $\mathcal{V}$ (the *vectors*), the zero vector, and the operations of scalar multiplication and vector addition. In this book, the field $\mathbb{F}$ is always either $\mathbb{C}$ or $\mathbb{R}$.

Example 1.2.2 Let $V = \mathbb{F}^n$, the set of $n \times 1$ matrices (column vectors) with entries from $\mathbb{F}$. For typographical convenience, we often write $\mathbf{u} = [u_i]$ or $\mathbf{u} = [u_1 \ u_2 \ \ldots \ u_n]^\mathsf{T}$ instead of[1]

$$\mathbf{u} = \begin{bmatrix} u_1 \\ u_2 \\ \vdots \\ u_n \end{bmatrix} \in \mathbb{F}^n \quad \text{for } u_1, u_2, \ldots, u_n \in \mathbb{F}.$$

Vector addition of $\mathbf{u} = [u_i]$ and $\mathbf{v} = [v_i]$ is defined by $\mathbf{u} + \mathbf{v} = [u_i + v_i]$, and scalar multiplication by elements of $\mathbb{F}$ is defined by $c\mathbf{u} = [cu_i]$; we refer to these as *entrywise operations*. The zero vector in $\mathbb{F}^n$ is $\mathbf{0}_n = [0 \ 0 \ \ldots \ 0]^\mathsf{T}$. We often omit the subscript from a zero vector when its size can be inferred from context.

Example 1.2.3 Let $V \in \mathsf{M}_{1 \times n}(\mathbb{F})$, the set of $1 \times n$ matrices (row vectors) with entries from $\mathbb{F}$. Vector addition and scalar multiplication are defined entrywise, as in the preceding example. The zero vector is the row vector $\mathbf{0}^\mathsf{T}$.

Example 1.2.4 Let $V = \mathsf{M}_{m \times n}(\mathbb{F})$, the set of $m \times n$ matrices with entries from $\mathbb{F}$. Vector addition and scalar multiplication are defined entrywise, as in the preceding two examples. The zero vector in $\mathsf{M}_{m \times n}(\mathbb{F})$ is the matrix $0_{m \times n} \in \mathsf{M}_{m \times n}(\mathbb{F})$, all entries of which are zero. We often omit the subscripts from a zero matrix if its size can be inferred from context.

Example 1.2.5 Let $V = \mathcal{P}_n$, the set of polynomials of degree at most n with complex coefficients. The set of polynomials of degree at most n with real coefficients is denoted by $\mathcal{P}_n(\mathbb{R})$. Addition of polynomials is defined by adding the coefficients of corresponding monomials. For example, if $p(z) = iz^2 - 5$ and $q(z) = -7z^2 + 3z + 2$ in $\mathcal{P}_2$, then $(p+q)(z) = (i-7)z^2 + 3z - 3$. Scalar multiplication of a polynomial by a scalar c is defined by multiplying each coefficient by c. For example, $(4p)(z) = 4iz^2 - 20$. The zero vector in $\mathcal{P}_n$ is the zero polynomial; see Appendix B.1.

Example 1.2.6 Let $V = \mathcal{P}$, the set of all polynomials with complex coefficients. The operations of vector addition and scalar multiplication are the same as in the preceding example, and the zero vector in $\mathcal{P}$ is again the zero polynomial.

Example 1.2.7 Let V and W be vector spaces over the same field $\mathbb{F}$. The *Cartesian product* $V \times W$ is the set of all ordered pairs $(\mathbf{v}, \mathbf{w})$ in which $\mathbf{v} \in V$ and $\mathbf{w} \in W$. Vector addition of $(\mathbf{v}_1, \mathbf{w}_1)$ and $(\mathbf{v}_2, \mathbf{w}_2)$ is defined by $(\mathbf{v}_1, \mathbf{w}_1) + (\mathbf{v}_2, \mathbf{w}_2) = (\mathbf{v}_1 + \mathbf{v}_2, \mathbf{w}_1 + \mathbf{w}_2)$, in which $\mathbf{v}_1 + \mathbf{v}_2$ and $\mathbf{w}_1 + \mathbf{w}_2$ denote the results of vector addition operations in V and W, respectively. Scalar multiplication by elements of $\mathbb{F}$ is defined by $c(\mathbf{v}, \mathbf{w}) = (c\mathbf{v}, c\mathbf{w})$, in which $c\mathbf{v}$ and $c\mathbf{w}$ denote the results of scalar multiplication operations in V and W, respectively. The zero vector in $V \times W$ is $(\mathbf{0}, \mathbf{0})$, which employs the respective zero vectors in V and W. The elements of the vector spaces V and W can come from different sets, but it is essential that both vector spaces be over the same field.

Example 1.2.8 Let $V = C_\mathbb{F}[a, b]$, the set of continuous $\mathbb{F}$-valued functions on an interval $[a, b] \subset \mathbb{R}$ with $a < b$. If the field designator is absent, it is understood that $\mathbb{F} = \mathbb{C}$, that is, $C[0, 1]$ is $C_\mathbb{C}[0, 1]$. The operations of vector addition and scalar multiplication are defined

---

[1] If you encounter an unfamiliar symbol in this book, consult the Notation section for a cross-reference to a definition. The Notation section entry for the symbol A^T identifies it as the transpose of a matrix A and points to Appendix C.2.

pointwise. If $f, g \in C_{\mathbb{F}}[a, b]$, then $f + g$ is the $\mathbb{F}$-valued function on $[a, b]$ defined by $(f+g)(t) = f(t) + g(t)$ for each $t \in [a, b]$. If $c \in \mathbb{F}$, the $\mathbb{F}$-valued function cf is defined by $(cf)(t) = cf(t)$ for each $t \in [a, b]$. A theorem from calculus ensures that $f + g$ and cf are continuous if f and g are continuous, so sums and scalar multiples of elements of $C_{\mathbb{F}}[a, b]$ are in $C_{\mathbb{F}}[a, b]$. The zero vector in $C_{\mathbb{F}}[a, b]$ is the *zero function*, which takes the value zero at every point in $[a, b]$.

Example 1.2.9 Let V be the set of all infinite sequences $\mathbf{u} = (u_1, u_2, \ldots)$, in which each $u_i \in \mathbb{F}$ and $u_i \neq 0$ for only finitely many values of the index i. The operations of vector addition and scalar multiplication are defined entrywise. The zero vector in V is the *zero infinite sequence* $\mathbf{0} = (0, 0, \ldots)$. We say that V is the $\mathbb{F}$-vector space of *finitely nonzero sequences*.

1.3 Subspaces

Definition 1.3.1 A *subspace* of an $\mathbb{F}$-vector space V is a subset $\mathcal{U} \subseteq V$ that is an $\mathbb{F}$-vector space with the same vector addition and scalar multiplication operations as in V.

A subspace is nonempty; it is a vector space, so it contains a zero vector.

Example 1.3.2 If V is an $\mathbb{F}$-vector space, then $\{\mathbf{0}\}$ and V itself are subspaces of V.

To show that a subset $\mathcal{U}$ of an $\mathbb{F}$-vector space V is a subspace, we do not need to verify the vector-space Axioms (2)–(3) and (5)–(8) because they are automatically satisfied; we say that $\mathcal{U}$ *inherits* these properties from V. However, we must show the following:

(a) Sums and scalar multiples of elements of $\mathcal{U}$ are in $\mathcal{U}$ (that is, $\mathcal{U}$ is *closed under vector addition and scalar multiplication*).

(b) $\mathcal{U}$ contains the zero vector of V.

(c) $\mathcal{U}$ contains an additive inverse for each of its elements.

The following theorem describes a streamlined way to verify these conditions.

Theorem 1.3.3 *Let V be an $\mathbb{F}$-vector space and let $\mathcal{U}$ be a nonempty subset of V. Then $\mathcal{U}$ is a subspace of V if and only if $c\mathbf{u} + \mathbf{v} \in \mathcal{U}$ for all $\mathbf{u}, \mathbf{v} \in \mathcal{U}$ and all $c \in \mathbb{F}$.*

Proof If $\mathcal{U}$ is a subspace, $\mathbf{u}, \mathbf{v} \in \mathcal{U}$, and $c \in \mathbb{F}$, then $c\mathbf{u} + \mathbf{v} \in \mathcal{U}$ because a subspace is closed under scalar multiplication and vector addition.

Conversely, suppose that $c\mathbf{u} + \mathbf{v} \in \mathcal{U}$ for all $\mathbf{u}, \mathbf{v} \in \mathcal{U}$ and every $c \in \mathbb{F}$. We must verify the properties (a), (b), and (c) in the preceding list.

(a) We have $c\mathbf{u} = c\mathbf{u} + \mathbf{0} \in \mathcal{U}$ and $\mathbf{u} + \mathbf{v} = 1\mathbf{u} + \mathbf{v} \in \mathcal{U}$ for all $c \in \mathbb{F}$ and all $\mathbf{u}, \mathbf{v} \in \mathcal{U}$.

(b) Let $\mathbf{u} \in \mathcal{U}$. Corollary 1.1.4 ensures that $(-1)\mathbf{u}$ is the additive inverse of $\mathbf{u}$, so $\mathbf{0} = (-1)\mathbf{u} + \mathbf{u} \in \mathcal{U}$.

(c) Since $(-1)\mathbf{u} = (-1)\mathbf{u} + \mathbf{0}$, it follows that the additive inverse of $\mathbf{u}$ is in $\mathcal{U}$. $\qquad\square$

The following examples use the criterion in the preceding theorem to verify that a certain subset of a vector space is a subspace.

Example 1.3.4 Let $A \in \mathsf{M}_{m \times n}(\mathbb{F})$. The *null space* of A is

$$\text{null } A = \{\mathbf{x} \in \mathbb{F}^n : A\mathbf{x} = \mathbf{0}\} \subseteq \mathbb{F}^n. \tag{1.3.5}$$

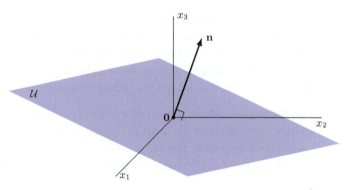

Figure 1.1 A plane $\mathcal{U}$ through the origin is a subspace of $\mathbb{R}^3$.

Since $A\mathbf{0}_n = \mathbf{0}_m$, the zero vector of $\mathbb{F}^n$ is in null A, which is therefore not empty. If $\mathbf{x}, \mathbf{y} \in$ null A, then $A\mathbf{x} = \mathbf{0}$ and $A\mathbf{y} = \mathbf{0}$. For any $c \in \mathbb{F}$, we have $A(c\mathbf{x} + \mathbf{y}) = cA\mathbf{x} + A\mathbf{y} = c\mathbf{0} + \mathbf{0} = \mathbf{0}$, so $c\mathbf{x} + \mathbf{y} \in$ null A. The preceding theorem ensures that null A is a subspace of $\mathbb{F}^n$.

Example 1.3.6 Let $\mathbf{n} = [a\ b\ c]^\mathsf{T} \in \mathbb{R}^3$ be nonzero and refer to Figure 1.1. The set $\mathcal{U} = \{\mathbf{x} = [x_1\ x_2\ x_3]^\mathsf{T} \in \mathbb{R}^3 : ax_1 + bx_2 + cx_3 = 0\}$ is the plane in $\mathbb{R}^3$ that contains the zero vector and has normal vector $\mathbf{n}$. Since $\mathbf{n}^\mathsf{T} \in \mathsf{M}_{1\times 3}(\mathbb{R})$ and $\mathbf{n}^\mathsf{T}\mathbf{x} = ax_1 + bx_2 + cx_3$, it follows that $\mathcal{U} =$ null $\mathbf{n}^\mathsf{T}$. The preceding example ensures that $\mathcal{U}$ is a subspace of $\mathbb{R}^3$.

Example 1.3.7 Let $A \in \mathsf{M}_{m\times n}(\mathbb{F})$. The *column space* of A is

$$\text{col}\,A = \{A\mathbf{x} : \mathbf{x} \in \mathbb{F}^n\} \subseteq \mathbb{F}^m. \tag{1.3.8}$$

Since $A\mathbf{0}_n = \mathbf{0}_m$, the zero vector of $\mathbb{F}^m$ is in col A, which is therefore not empty. If $\mathbf{u}, \mathbf{v} \in$ col A, then there are $\mathbf{x}, \mathbf{y} \in \mathbb{F}^n$ such that $\mathbf{u} = A\mathbf{x}$ and $\mathbf{v} = A\mathbf{y}$. For any $c \in \mathbb{F}$, we have $c\mathbf{u} + \mathbf{v} = cA\mathbf{x} + A\mathbf{y} = A(c\mathbf{x}+\mathbf{y})$, so $c\mathbf{u}+\mathbf{v} \in$ col A. Theorem 1.3.3 ensures that col A is a subspace of $\mathbb{F}^m$.

Example 1.3.9 Let $A \in \mathsf{M}_{m\times n}(\mathbb{F})$. The *row space* of A is

$$\text{row}\,A = \{\mathbf{x}^\mathsf{T}A : \mathbf{x} \in \mathbb{F}^m\} \subseteq \mathsf{M}_{1\times n}(\mathbb{F}). \tag{1.3.10}$$

Arguments similar to those in the preceding example show that row A is a subspace of $\mathsf{M}_{1\times n}(\mathbb{F})$. The row vector $\mathbf{x}^\mathsf{T}A$ and the column vector $A^\mathsf{T}\mathbf{x}$ are transposes of each other. This one-to-one correspondence between the elements of row A and col A^T permits us to deduce properties of one of these subspaces from properties of the other.

Example 1.3.11 Let $\mathcal{V} = \mathsf{M}_{m\times n}(\mathbb{F})$ and let $\mathcal{U} \subseteq \mathcal{V}$ be the subset of matrices whose last row has only zero entries. The zero matrix is in $\mathcal{U}$. Sums and scalar multiples of elements of $\mathcal{U}$ have zero last row. It follows that $\mathcal{U}$ is a subspace of $\mathcal{V}$.

Example 1.3.12 Let $A \in \mathsf{M}_m(\mathbb{F})$ and let $\mathcal{U}$ be a subspace of $\mathsf{M}_{m\times n}(\mathbb{F})$. We claim that

$$A\mathcal{U} = \{AX : X \in \mathcal{U}\}$$

is a subspace of $\mathsf{M}_{m\times n}(\mathbb{F})$. Since $0 \in \mathcal{U}$, we have $0 = A0 \in A\mathcal{U}$, which is therefore not empty. Moreover, $cAX + AY = A(cX + Y) \in A\mathcal{U}$ for any scalar c and any $X, Y \in \mathcal{U}$. Theorem 1.3.3 ensures that $A\mathcal{U}$ is a subspace of $\mathsf{M}_{m\times n}(\mathbb{F})$. For example, $A\mathsf{M}_{m\times 1}(\mathbb{F}) = $ col A.

The next four examples involve subspaces whose elements are polynomials.

Example 1.3.13 $\mathcal{P}_5$ is a subspace of $\mathcal{P}$; see Examples 1.2.5 and 1.2.6. Sums and scalar multiples of polynomials of degree 5 or less are in $\mathcal{P}_5$.

Example 1.3.14 $\mathcal{P}_5(\mathbb{R})$ is a subset of $\mathcal{P}_5$, but it is not a subspace. For example, the scalar 1 is in $\mathcal{P}_5(\mathbb{R})$, but $i1 = i \notin \mathcal{P}_5(\mathbb{R})$. The issue here is that the scalars for the vector space $\mathcal{P}_5(\mathbb{R})$ are real numbers and the scalars for the vector space $\mathcal{P}_5$ are complex numbers. A subspace and the vector space that contains it must use the same field of scalars.

Example 1.3.15 A polynomial $p \in \mathcal{P}$ is *even* if $p(-z) = p(z)$ for all z. We denote the set of even polynomials by $\mathcal{P}_{\text{even}}$. A polynomial p is *odd* if $p(-z) = -p(z)$ for all z. We denote the set of odd polynomials by $\mathcal{P}_{\text{odd}}$. For example, $p(z) = 2 + 3z^2$ is even and $p(z) = 5z + 4z^3$ is odd. Constant polynomials are even; the zero polynomial is both even and odd. Each of $\mathcal{P}_{\text{even}}$ and $\mathcal{P}_{\text{odd}}$ is a subspace of $\mathcal{P}$.

Example 1.3.16 The complex vector space $\mathcal{P}$ is a subspace of the complex vector space $C[a, b]$. Every polynomial is a continuous function, and $cp + q \in \mathcal{P}$ whenever $p, q \in \mathcal{P}$ and $c \in \mathbb{C}$. Theorem 1.3.3 ensures that $\mathcal{P}$ is a subspace of $C[a, b]$.

1.4 Linear Combinations, Lists, and Span

The basic operations in an $\mathbb{F}$-vector space permit us to multiply vectors by scalars and then add them. For example, in the real vector space $\mathbb{R}^2$, consider the vectors $\mathbf{u}$, $\mathbf{v}$, $\mathbf{w}$, and $\mathbf{z}$ illustrated in Figure 1.2. A computation reveals that

$$7\mathbf{u} - 5\mathbf{v} + \mathbf{w} = 7\begin{bmatrix} 1 \\ 1 \end{bmatrix} - 5\begin{bmatrix} 1 \\ 2 \end{bmatrix} + \begin{bmatrix} -1 \\ 2 \end{bmatrix} = \begin{bmatrix} 1 \\ -1 \end{bmatrix} = \mathbf{z}, \tag{1.4.1}$$

so $\mathbf{z}$ is a sum of scalar multiples of the vectors $\mathbf{u}$, $\mathbf{v}$, and $\mathbf{w}$. We also have

$$-\mathbf{u} + \mathbf{v} - \mathbf{w} = -\begin{bmatrix} 1 \\ 1 \end{bmatrix} + \begin{bmatrix} 1 \\ 2 \end{bmatrix} - \begin{bmatrix} -1 \\ 2 \end{bmatrix} = \begin{bmatrix} 1 \\ -1 \end{bmatrix} = \mathbf{z}, \tag{1.4.2}$$

which expresses $\mathbf{z}$ in two different ways as a sum of scalar multiples of $\mathbf{u}$, $\mathbf{v}$, and $\mathbf{w}$, respectively. The following definition provides vocabulary to describe computations like these.

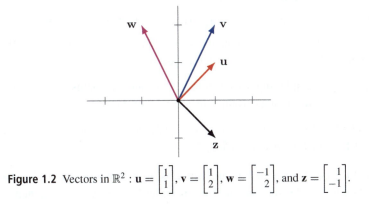

Figure 1.2 Vectors in $\mathbb{R}^2$: $\mathbf{u} = \begin{bmatrix} 1 \\ 1 \end{bmatrix}$, $\mathbf{v} = \begin{bmatrix} 1 \\ 2 \end{bmatrix}$, $\mathbf{w} = \begin{bmatrix} -1 \\ 2 \end{bmatrix}$, and $\mathbf{z} = \begin{bmatrix} 1 \\ -1 \end{bmatrix}$.

Definition 1.4.3 Let $\mathcal{U}$ be a nonempty subset of an $\mathbb{F}$-vector space $\mathcal{V}$. A *linear combination* of elements of $\mathcal{U}$ is an expression of the form

$$c_1\mathbf{v}_1 + c_2\mathbf{v}_2 + \cdots + c_r\mathbf{v}_r, \tag{1.4.4}$$

in which r is a positive integer, $\mathbf{v}_1, \mathbf{v}_2, \ldots, \mathbf{v}_r \in \mathcal{U}$, and $c_1, c_2, \ldots, c_r \in \mathbb{F}$. A linear combination (1.4.4) is *trivial* if $c_1 = c_2 = \cdots = c_r = 0$; otherwise, it is *nontrivial*.

A linear combination is, by definition, a sum of finitely many scalar multiples of vectors. For example, (1.4.1) shows that the vector $\mathbf{z}$ in Figure 1.2 is a linear combination of $\mathbf{u}$, $\mathbf{v}$, and $\mathbf{w}$; (1.4.2) expresses $\mathbf{z}$ as a different linear combination of these vectors.

Example 1.4.5 Every element of $\mathcal{P}$ is a linear combination of $1, z, z^2, \ldots$.

Definition 1.4.6 A *list* of vectors in an $\mathbb{F}$-vector space $\mathcal{V}$ is a nonempty finite sequence $\beta = \mathbf{v}_1, \mathbf{v}_2, \ldots, \mathbf{v}_r$ of vectors in $\mathcal{V}$. The vectors $\mathbf{v}_1, \mathbf{v}_2, \ldots, \mathbf{v}_r$ are the *elements* of the list β. A *nonzero list* has at least one nonzero element. We often denote a list by a lowercase Greek letter such as β or γ.

A subtle, but important, point is that a vector can appear more than once in a list. For example, $\beta = z, z^2, z^2, z^2, z^3$ is a list of five vectors in $\mathcal{P}_3$. However, the set of vectors in the list β is $\{z, z^2, z^3\}$. Sets do not have multiplicities; see Appendix C.1. Accounting for the multiplicities of scalars and vectors is often important in linear algebra; lists help us do this.

A second important point is that order matters in a list. For example, $\beta = z, z^2, z^3$ and $\gamma = z^3, z^2, z$ are different lists of vectors in $\mathcal{P}_3$.

Definition 1.4.7 Let $\mathcal{U}$ be a subset of an $\mathbb{F}$-vector space $\mathcal{V}$. If $\mathcal{U} \neq \varnothing$, then span $\mathcal{U}$ is the set all of linear combinations of elements of $\mathcal{U}$; we define span $\varnothing = \{\mathbf{0}\}$. If $\beta = \mathbf{v}_1, \mathbf{v}_2, \ldots, \mathbf{v}_r$ is a list of vectors in $\mathcal{V}$, we define span $\beta = $ span$\{\mathbf{v}_1, \mathbf{v}_2, \ldots, \mathbf{v}_r\}$, that is, the span of a list is the span of the set of vectors in the list.

Suppose that a list of vectors β is obtained from a list γ by reordering its elements. The commutativity of vector addition ensures that span $\beta = $ span γ.

Example 1.4.8 If $\mathbf{u} \in \mathcal{V}$, then Theorem 1.3.3 ensures that span$\{\mathbf{u}\} = \{c\mathbf{u} : c \in \mathbb{F}\}$ is a subspace of $\mathcal{V}$. In particular, span$\{\mathbf{0}\} = \{\mathbf{0}\}$.

Example 1.4.9 Let $A = [\mathbf{a}_1 \ \mathbf{a}_2 \ \ldots \ \mathbf{a}_n] \in \mathsf{M}_{m \times n}(\mathbb{F})$ (see (C.2.2) or (3.1.1) for this presentation of a matrix, partitioned according to its columns) and consider the list $\beta = \mathbf{a}_1, \mathbf{a}_2, \ldots, \mathbf{a}_n$ of vectors in the $\mathbb{F}$-vector space $\mathbb{F}^m$. Then

$$\text{span } \beta = \{x_1\mathbf{a}_1 + x_2\mathbf{a}_2 + \cdots + x_n\mathbf{a}_n : x_1, x_2, \ldots, x_n \in \mathbb{F}\} = \{A\mathbf{x} : \mathbf{x} \in \mathbb{F}^n\} = \text{col } A,$$

that is, the span of the columns of a matrix is its column space. A vector $\mathbf{y} \in \mathbb{F}^m$ is in the span of the columns of A if and only if $\mathbf{y} = A\mathbf{x}$ for some $\mathbf{x} \in \mathbb{F}^n$.

The preceding example suggests a powerful result: an inclusion of column spaces is equivalent to the existence of a certain matrix factorization.

Theorem 1.4.10 *Let* $Y = [\mathbf{y}_1 \ \mathbf{y}_2 \ \ldots \ \mathbf{y}_p] \in \mathsf{M}_{m \times p}(\mathbb{F})$ *and let* $A \in \mathsf{M}_{m \times n}(\mathbb{F})$. *Then* col $Y \subseteq$ col A *if and only if* $Y = AX$ *for some* $X \in \mathsf{M}_{n \times p}(\mathbb{F})$.

Proof If col $Y \subseteq$ col A, then each column of Y is in the column space of A. Example 1.4.9 ensures that each $\mathbf{y}_j = A\mathbf{x}_j$ for some $\mathbf{x}_j \in \mathbb{F}^n$. If we let $X = [\mathbf{x}_1 \ \mathbf{x}_2 \ \ldots \ \mathbf{x}_p]$, then

$$Y = [\mathbf{y}_1 \ \mathbf{y}_2 \ \ldots \ \mathbf{y}_p] = [A\mathbf{x}_1 \ A\mathbf{x}_2 \ \ldots \ A\mathbf{x}_p] = A[\mathbf{x}_1 \ \mathbf{x}_2 \ \ldots \ \mathbf{x}_p] = AX.$$

Conversely, if $Y = AX, X \in \mathsf{M}_{n \times p}(\mathbb{F})$, and $\mathbf{u} \in \mathrm{col}\, Y$, then Example 1.4.9 tells us that $\mathbf{u} = Y\mathbf{v}$ for some $\mathbf{v} \in \mathbb{F}^p$. Consequently, $\mathbf{u} = Y\mathbf{v} = AX\mathbf{v} = A(X\mathbf{v}) \in \mathrm{col}\, A$, so $\mathrm{col}\, Y \subseteq \mathrm{col}\, A$. $\qquad \square$

Example 1.4.11 Let

$$A = \begin{bmatrix} \mathbf{a}_1^\mathsf{T} \\ \vdots \\ \mathbf{a}_m^\mathsf{T} \end{bmatrix} \in \mathsf{M}_{m \times n}(\mathbb{F})$$

(see (C.2.3) or (3.1.11) for this presentation of a matrix, partitioned according to its rows). If $\beta = \mathbf{a}_1^\mathsf{T}, \mathbf{a}_2^\mathsf{T}, \ldots, \mathbf{a}_m^\mathsf{T}$, then

$$\mathrm{span}\, \beta = \{x_1 \mathbf{a}_1^\mathsf{T} + x_2 \mathbf{a}_2^\mathsf{T} + \cdots + x_m \mathbf{a}_m^\mathsf{T} : x_1, x_2, \ldots, x_m \in \mathbb{F}\} = \{\mathbf{x}^\mathsf{T} A : \mathbf{x} \in \mathbb{F}^m\} = \mathrm{row}\, A,$$

that is, the span of the rows of A is its row space. It can also be thought of as (the transpose of) the span of the columns of A^T.

Example 1.4.12 Consider the list $\beta = z, z^3$ of elements of $\mathcal{P}_3$. Then $\mathrm{span}\, \beta = \{c_1 z + c_2 z^3 : c_1, c_2 \in \mathbb{C}\}$ is a subspace of $\mathcal{P}_3$ because it is nonempty and

$$c(a_1 z + a_2 z^3) + (b_1 z + b_2 z^3) = (ca_1 + b_1)z + (ca_2 + b_2)z^3$$

is a linear combination of vectors in the list β for all $c, a_1, a_2, b_1, b_2 \in \mathbb{C}$.

The span of a subset of a vector space is always a subspace.

Theorem 1.4.13 *Let $\mathcal{U}$ be a subset of an $\mathbb{F}$-vector space V.*

(a) *span $\mathcal{U}$ is a subspace of V.*

(b) *$\mathcal{U} \subseteq \mathrm{span}\, \mathcal{U}$.*

(c) *$\mathcal{U} = \mathrm{span}\, \mathcal{U}$ if and only if $\mathcal{U}$ is a subspace of V.*

(d) *span (span $\mathcal{U}$) = span $\mathcal{U}$.*

Proof First suppose that $\mathcal{U} = \varnothing$. Then Definition 1.4.7 says that $\mathrm{span}\, \varnothing = \{\mathbf{0}\}$, which is a subspace of V. The empty set is a subset of every set, so $\varnothing \subset \{\mathbf{0}\} = \mathrm{span}\, \varnothing$. Both implications in (c) are vacuous. For (d), we have $\mathrm{span}\, (\mathrm{span}\, \varnothing) = \mathrm{span}\{\mathbf{0}\} = \{\mathbf{0}\} = \mathrm{span}\, \varnothing$; see Example 1.4.8.

Now suppose that $\mathcal{U} \neq \varnothing$. If $\mathbf{u}, \mathbf{v} \in \mathrm{span}\, \mathcal{U}$ and $c \in \mathbb{F}$, then each of $\mathbf{u}, \mathbf{v}, c\mathbf{u}$, and $c\mathbf{u} + \mathbf{v}$ is a linear combination of elements of $\mathcal{U}$, so each is in $\mathrm{span}\, \mathcal{U}$. Theorem 1.3.3 ensures that $\mathrm{span}\, \mathcal{U}$ is a subspace of V. The assertion in (b) follows from the fact that $1\mathbf{u} = \mathbf{u}$ is an element of $\mathrm{span}\, \mathcal{U}$ for each $\mathbf{u} \in \mathcal{U}$. To prove the two implications in (c), first suppose that $\mathcal{U} = \mathrm{span}\, \mathcal{U}$. Then (a) ensures that $\mathcal{U}$ is a subspace of V. Conversely, if $\mathcal{U}$ is a subspace of V, then it is closed under vector addition and scalar multiplication, so $\mathrm{span}\, \mathcal{U} \subseteq \mathcal{U}$. The containment $\mathcal{U} \subseteq \mathrm{span}\, \mathcal{U}$ in (b) ensures that $\mathcal{U} = \mathrm{span}\, \mathcal{U}$. The assertion in (d) follows from (a) and (c). $\qquad \square$

Theorem 1.4.14 *Let $\mathcal{U}$ and $\mathcal{W}$ be subsets of an $\mathbb{F}$-vector space V. If $\mathcal{U} \subseteq \mathcal{W}$, then $\mathrm{span}\, \mathcal{U} \subseteq \mathrm{span}\, \mathcal{W}$.*

Proof If $\mathcal{U} = \varnothing$, then $\mathrm{span}\, \mathcal{U} = \{\mathbf{0}\} \subseteq \mathrm{span}\, \mathcal{W}$. If $\mathcal{U} \neq \varnothing$, then every linear combination of elements of $\mathcal{U}$ is a linear combination of elements of $\mathcal{W}$. $\qquad \square$

Example 1.4.15 Let $\mathscr{U} = \{1, z - 2z^2, z^2 + 5z^3, z^3, 1 + 4z^2\}$. We claim that span $\mathscr{U} = \mathcal{P}_3$. To verify this, observe that

$$1 = 1,$$
$$z = (z - 2z^2) + 2(z^2 + 5z^3) - 10z^3,$$
$$z^2 = (z^2 + 5z^3) - 5z^3, \text{ and}$$
$$z^3 = z^3.$$

Thus, $\{1, z, z^2, z^3\} \subseteq$ span $\mathscr{U} \subseteq \mathcal{P}_3$. Now invoke the two preceding theorems, compute

$$\mathcal{P}_3 = \text{span}\{1, z, z^2, z^3\} \subseteq \text{span}(\text{span}\,\mathscr{U}) = \text{span}\,\mathscr{U} \subseteq \mathcal{P}_3,$$

and conclude that span $\mathscr{U} = \mathcal{P}_3$.

Definition 1.4.16 Let V be an $\mathbb{F}$-vector space. Let $\mathscr{U}$ be a subset of V and let β be a list of vectors in V. Then $\mathscr{U}$ *spans* V ($\mathscr{U}$ is a *spanning set* for V) if span $\mathscr{U} = V$. The list β *spans* V (β is a *spanning list*) if span $\beta = V$.

It is convenient to say that "$\mathbf{v}_1, \mathbf{v}_2, \ldots, \mathbf{v}_r$ span V" rather than "the list of vectors $\mathbf{v}_1, \mathbf{v}_2, \ldots, \mathbf{v}_r$ spans V." If a list of vectors in $\mathbb{F}^m$ is the list of columns of a matrix $A \in \mathsf{M}_{m \times n}(\mathbb{F})$, it is also convenient to say "the columns of A span V rather than "the list of columns of A span V."

Example 1.4.17 Let $B \in \mathsf{M}_{n \times p}(\mathbb{F})$. The columns of B span $\mathbb{F}^n$ if and only if col $B = \mathbb{F}^n$.

Example 1.4.18 Each of the sets $\{1, z, z^2, z^3\}$ and $\{1, z - 2z^2, z^2 + 5z^3, z^3, 1 + 4z^2\}$ spans $\mathcal{P}_3$; see Example 1.4.15.

Example 1.4.19 Let $A = [\mathbf{a}_1 \ \mathbf{a}_2 \ \ldots \ \mathbf{a}_n] \in \mathsf{M}_n(\mathbb{F})$ be invertible, let $\mathbf{y} \in \mathbb{F}^n$, and let $A^{-1}\mathbf{y} = [x_1 \ x_2 \ \ldots \ x_n]^\mathsf{T}$. Then $\mathbf{y} = A(A^{-1}\mathbf{y}) = x_1\mathbf{a}_1 + x_2\mathbf{a}_2 + \cdots + x_n\mathbf{a}_n$ is a linear combination of the columns of A. We conclude that if $A \in \mathsf{M}_n(\mathbb{F})$ is invertible, then its columns span $\mathbb{F}^n$.

Example 1.4.20 The identity matrix I_n is invertible and its columns are

$$\mathbf{e}_1 = \begin{bmatrix} 1 \\ 0 \\ \vdots \\ 0 \end{bmatrix}, \qquad \mathbf{e}_2 = \begin{bmatrix} 0 \\ 1 \\ \vdots \\ 0 \end{bmatrix}, \ldots, \qquad \mathbf{e}_n = \begin{bmatrix} 0 \\ 0 \\ \vdots \\ 1 \end{bmatrix}. \tag{1.4.21}$$

Consequently, span$\{\mathbf{e}_1, \mathbf{e}_2, \ldots, \mathbf{e}_n\} = \mathbb{F}^n$. Any $\mathbf{u} = [u_i] \in \mathbb{F}^n$ can be expressed as $\mathbf{u} = u_1\mathbf{e}_1 + u_2\mathbf{e}_2 + \cdots + u_n\mathbf{e}_n$. For example, the *all-ones vector* $\mathbf{e} \in \mathbb{F}^n$ can be expressed as $\mathbf{e} = \mathbf{e}_1 + \mathbf{e}_2 + \cdots + \mathbf{e}_n = [1 \ 1 \ \ldots \ 1]^\mathsf{T}$.

Example 1.4.22 Consider the vectors $\mathbf{u}$ and $\mathbf{v}$ in Figure 1.2, and let $A = [\mathbf{u} \ \mathbf{v}] = \begin{bmatrix} 1 & 1 \\ 1 & 2 \end{bmatrix}$. A computation using (C.2.7) reveals that $A^{-1} = \begin{bmatrix} 2 & -1 \\ -1 & 1 \end{bmatrix}$, so A is invertible. The preceding example ensures that span$\{\mathbf{u}, \mathbf{v}\} = \mathbb{R}^2$, so each vector in $\mathbb{R}^2$ is a linear combination of $\mathbf{u}$ and $\mathbf{v}$. Equivalently, the system of linear equations $A\mathbf{x} = \mathbf{y}$ is consistent for each $\mathbf{y} \in \mathbb{R}^2$.

Example 1.4.23 Let $A \in \mathsf{M}_{m \times n}(\mathbb{F})$ and $B \in \mathsf{M}_{n \times p}(\mathbb{F})$. If the columns of B span $\mathbb{F}^n$, then Example 1.4.19 ensures that

$$\text{col } AB = \{AB\mathbf{x} : \mathbf{x} \in \mathbb{F}^p\} = \{A\mathbf{y} : \mathbf{y} \in \mathbb{F}^n\} = \text{col } A.$$

Thus, $\operatorname{col} AB = \operatorname{col} A$ if $B \in \mathsf{M}_n(\mathbb{F})$ is invertible. If $C \in \mathsf{M}_m(\mathbb{F})$ is invertible, then so is C^T. Then $\operatorname{col} A^\mathsf{T} = \operatorname{col}(A^\mathsf{T} C^\mathsf{T}) = \operatorname{col}(CA)^\mathsf{T}$, which means that $\operatorname{row} CA = \operatorname{row} A$ if C is invertible.

Example 1.4.24 Let $\mathcal{V}$ and $\mathcal{W}$ be $\mathbb{F}$-vector spaces. Suppose that $\{\mathbf{v}_1, \mathbf{v}_2, \ldots, \mathbf{v}_m\}$ is a spanning set for $\mathcal{V}$ and $\{\mathbf{w}_1, \mathbf{w}_2, \ldots, \mathbf{w}_n\}$ is a spanning set for $\mathcal{W}$. If $(\mathbf{v}, \mathbf{w}) \in \mathcal{V} \times \mathcal{W}$, there are scalars c_i and d_j such that

$$(\mathbf{v}, \mathbf{w}) = (\mathbf{v}, \mathbf{0}) + (\mathbf{0}, \mathbf{w}) = \left(\sum_{i=1}^{m} c_i \mathbf{v}_i, \mathbf{0} \right) + \left(\mathbf{0}, \sum_{j=1}^{n} d_j \mathbf{w}_j \right) = \sum_{i=1}^{m} c_i(\mathbf{v}_i, \mathbf{0}) + \sum_{j=1}^{n} d_j(\mathbf{0}, \mathbf{w}_j).$$

Therefore, the $m + n$ vectors $(\mathbf{v}_1, \mathbf{0}), \ldots, (\mathbf{v}_m, \mathbf{0}), (\mathbf{0}, \mathbf{w}_1), \ldots, (\mathbf{0}, \mathbf{w}_n)$ span $\mathcal{V} \times \mathcal{W}$.

1.5 Intersections, Sums, and Direct Sums of Subspaces

In this section, we discuss several ways to combine given subspaces to form new subspaces.

Theorem 1.5.1 *Let $\mathcal{U}$ and $\mathcal{W}$ be subspaces of an $\mathbb{F}$-vector space $\mathcal{V}$. Their intersection $\mathcal{U} \cap \mathcal{W} = \{\mathbf{v} : \mathbf{v} \in \mathcal{U} \text{ and } \mathbf{v} \in \mathcal{W}\}$ is a subspace of $\mathcal{V}$.*

Proof The zero vector is in both $\mathcal{U}$ and $\mathcal{W}$, so it is in $\mathcal{U} \cap \mathcal{W}$. Thus, $\mathcal{U} \cap \mathcal{W}$ is nonempty. If $\mathbf{u}, \mathbf{v} \in \mathcal{U}$ and $\mathbf{u}, \mathbf{v} \in \mathcal{W}$, then for any $c \in \mathbb{F}$, the vector $c\mathbf{u} + \mathbf{v}$ is in both $\mathcal{U}$ and $\mathcal{W}$ since they are subspaces. Consequently, $c\mathbf{u} + \mathbf{v} \in \mathcal{U} \cap \mathcal{W}$, so Theorem 1.3.3 ensures that $\mathcal{U} \cap \mathcal{W}$ is a subspace of $\mathcal{V}$. $\qquad\square$

Figure 1.3 illustrates the intersection of a pair of two-dimensional subspaces in $\mathbb{R}^3$. The following examples show that the union of subspaces need not be a subspace.

Example 1.5.2 In the real vector space $\mathbb{R}^2$, the union of the subspaces $\mathcal{X} = \operatorname{span}\{\mathbf{e}_1\}$ and $\mathcal{Y} = \operatorname{span}\{\mathbf{e}_2\}$ is not a subspace of $\mathbb{R}^2$ since $\mathbf{e}_1 + \mathbf{e}_2 \notin \mathcal{X} \cup \mathcal{Y}$; see Figure 1.4.

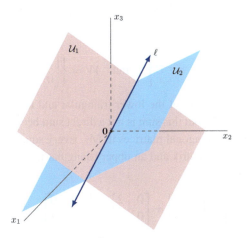

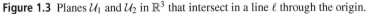

Figure 1.3 Planes $\mathcal{U}_1$ and $\mathcal{U}_2$ in $\mathbb{R}^3$ that intersect in a line ℓ through the origin.

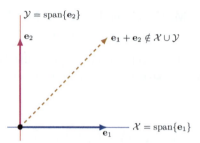

Figure 1.4 The union of two subspaces in $\mathbb{R}^2$ need not be a subspace of $\mathbb{R}^2$; see Example 1.5.2.

Example 1.5.3 In $\mathcal{P}$, the union $\mathcal{P}_{\text{even}} \cup \mathcal{P}_{\text{odd}}$ is the set of polynomials that are even or odd. However, $z^2 \in \mathcal{P}_{\text{even}}$ and $z \in \mathcal{P}_{\text{odd}}$, but $z + z^2 \notin \mathcal{P}_{\text{even}} \cup \mathcal{P}_{\text{odd}}$, so $\mathcal{P}_{\text{even}} \cup \mathcal{P}_{\text{odd}}$ is not a subspace of $\mathcal{P}$.

The span of the union of subspaces is a subspace since the span of any set is a subspace.

Definition 1.5.4 Let $\mathcal{U}$ and $\mathcal{W}$ be subspaces of an $\mathbb{F}$-vector space $\mathcal{V}$. The *sum* of $\mathcal{U}$ and $\mathcal{W}$ is the subspace span $(\mathcal{U} \cup \mathcal{W})$. It is denoted by $\mathcal{U} + \mathcal{W}$.

In the preceding definition, span $(\mathcal{U} \cup \mathcal{W})$ consists of all linear combinations of vectors that are in $\mathcal{U}$ or in $\mathcal{W}$. Since both $\mathcal{U}$ and $\mathcal{W}$ are closed under vector addition, it follows that

$$\mathcal{U} + \mathcal{W} = \text{span}\,(\mathcal{U} \cup \mathcal{W}) = \{\mathbf{u} + \mathbf{w} : \mathbf{u} \in \mathcal{U} \text{ and } \mathbf{w} \in \mathcal{W}\}.$$

Example 1.5.5 In the notation of Example 1.5.2, we have $\mathcal{X} + \mathcal{Y} = \mathbb{R}^2$ since $\begin{bmatrix} x \\ y \end{bmatrix} = x\mathbf{e}_1 + y\mathbf{e}_2$ for all $x, y \in \mathbb{R}$.

Example 1.5.6 We have $\mathcal{P}_{\text{even}} + \mathcal{P}_{\text{odd}} = \mathcal{P}$. For example, $z^5 + iz^4 + z - 2 = (iz^4 - 2) + (z^5 + z)$ is a sum of a vector in $\mathcal{P}_{\text{even}}$ and a vector in $\mathcal{P}_{\text{odd}}$.

In the two preceding examples, the respective pairs of subspaces have an important special property that we identify in the following definition.

Definition 1.5.7 Let $\mathcal{U}$ and $\mathcal{W}$ be subspaces of an $\mathbb{F}$-vector space $\mathcal{V}$. If $\mathcal{U} \cap \mathcal{W} = \{\mathbf{0}\}$, then the sum of $\mathcal{U}$ and $\mathcal{W}$ is a *direct sum*. It is denoted by $\mathcal{U} \oplus \mathcal{W}$.

Example 1.5.8 Let

$$\mathcal{U} = \left\{ \begin{bmatrix} a & 0 \\ b & c \end{bmatrix} \in \mathsf{M}_2 : a, b, c \in \mathbb{C} \right\} \quad \text{and} \quad \mathcal{W} = \left\{ \begin{bmatrix} x & y \\ 0 & z \end{bmatrix} \in \mathsf{M}_2 : x, y, z \in \mathbb{C} \right\}$$

be the subspaces of M_2 consisting of the lower triangular and upper triangular matrices, respectively. Then $\mathcal{U} + \mathcal{W} = \mathsf{M}_2$, but this sum is not a direct sum because $\mathcal{U} \cap \mathcal{W} \neq \{\mathbf{0}\}$; instead, $\mathcal{U} \cap \mathcal{W}$ is the subspace of all diagonal matrices in M_2. Every square matrix can be expressed as a sum of a lower triangular matrix and an upper triangular matrix, but the summands need not be unique. For example,

$$\begin{bmatrix} 1 & 2 \\ 3 & 4 \end{bmatrix} = \begin{bmatrix} 1 & 0 \\ 3 & 4 \end{bmatrix} + \begin{bmatrix} 0 & 2 \\ 0 & 0 \end{bmatrix} = \begin{bmatrix} -1 & 0 \\ 3 & 3 \end{bmatrix} + \begin{bmatrix} 2 & 2 \\ 0 & 1 \end{bmatrix}.$$

Direct sums are important because any vector in a direct sum has a unique representation with respect to the direct summands.

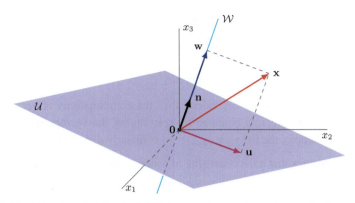

Figure 1.5 $\mathcal{U} + \mathcal{W} = \mathcal{U} \oplus \mathcal{W}$, in which $\mathcal{U}$ is the plane $ax_1 + bx_2 + cx_3 = 0$ with normal vector $\mathbf{n} = [a\ b\ c]^\mathsf{T}$, and $\mathcal{W} = \mathrm{span}\{\mathbf{n}\}$. Theorem 1.5.9 ensures that each $\mathbf{x} \in \mathcal{U} + \mathcal{W}$ has a unique representation $\mathbf{x} = \mathbf{u} + \mathbf{w}$, in which $\mathbf{u} \in \mathcal{U}$ and $\mathbf{w} \in \mathcal{W}$.

Theorem 1.5.9 *Let $\mathcal{U}$ and $\mathcal{W}$ be subspaces of an $\mathbb{F}$-vector space $\mathcal{V}$ and suppose that $\mathcal{U} \cap \mathcal{W} = \{\mathbf{0}\}$. Then each vector in $\mathcal{U} \oplus \mathcal{W}$ is uniquely expressible as a sum of a vector in $\mathcal{U}$ and a vector in $\mathcal{W}$.*

Proof Suppose that $\mathbf{u}_1, \mathbf{u}_2 \in \mathcal{U}$, $\mathbf{w}_1, \mathbf{w}_2 \in \mathcal{W}$, and $\mathbf{u}_1 + \mathbf{w}_1 = \mathbf{u}_2 + \mathbf{w}_2$. Then $\mathbf{u}_1 - \mathbf{u}_2 = \mathbf{w}_2 - \mathbf{w}_1 \in \mathcal{U} \cap \mathcal{W}$. But $\mathcal{U} \cap \mathcal{W} = \{\mathbf{0}\}$, so $\mathbf{u}_1 - \mathbf{u}_2 = \mathbf{0}$ and $\mathbf{w}_2 - \mathbf{w}_1 = \mathbf{0}$. Thus, $\mathbf{u}_1 = \mathbf{u}_2$ and $\mathbf{w}_1 = \mathbf{w}_2$. $\square$

Example 1.5.10 Let $\mathcal{U}$ be a plane through the origin in $\mathbb{R}^3$ with nonzero normal vector $\mathbf{n} = [a\ b\ c]^\mathsf{T}$, and let $\mathcal{W} = \mathrm{span}\{\mathbf{n}\}$; see Figure 1.5. If $\mathbf{x} \in \mathcal{U} \cap \mathcal{W}$, then $\mathbf{x} = d\mathbf{n}$ for some $d \in \mathbb{R}$, and $\mathbf{n}^\mathsf{T}\mathbf{x} = 0$ since $\mathcal{U} = \mathrm{null}\ \mathbf{n}^\mathsf{T}$; see Example 1.3.6. But $\mathbf{n}^\mathsf{T}\mathbf{x} = d\mathbf{n}^\mathsf{T}\mathbf{n} = d(a^2 + b^2 + c^2) = 0$ if and only if $d = 0$, so $\mathbf{x} = \mathbf{0}$. We conclude that each $\mathbf{x} \in \mathcal{U} + \mathcal{W}$ is a unique linear combination $\mathbf{x} = \mathbf{u} + \mathbf{w}$, in which $\mathbf{u} \in \mathcal{U}$ and $\mathbf{w} \in \mathcal{W}$. In fact, each $\mathbf{x} \in \mathbb{R}^3$ can be represented in this way; see Theorem 2.2.11.

Example 1.5.11 Although $\mathcal{P}_4 = \mathrm{span}\{1, z, z^2, z^3\} + \mathrm{span}\{z^3, z^4\}$, this sum is not a direct sum since $\mathrm{span}\{1, z, z^2, z^3\} \cap \mathrm{span}\{z^3, z^4\} = \mathrm{span}\{z^3\} \neq \{0\}$.

Example 1.5.12 We have $\mathcal{P} = \mathcal{P}_{\mathrm{even}} \oplus \mathcal{P}_{\mathrm{odd}}$ since the only polynomial that is both even and odd is the zero polynomial; see Example 1.3.15.

Example 1.5.13 M_2 is the direct sum of the subspace of strictly lower triangular matrices and the subspace of upper triangular matrices.

1.6 Linear Dependence and Linear Independence

Equations (1.4.1) and (1.4.2) link the vectors in Figure 1.2. We use them to compute

$$\mathbf{0} = \mathbf{z} - \mathbf{z} = (7\mathbf{u} - 5\mathbf{v} + \mathbf{w}) - (-\mathbf{u} + \mathbf{v} - \mathbf{w}) = 8\mathbf{u} - 6\mathbf{v} + 2\mathbf{w}, \qquad (1.6.1)$$

which expresses the zero vector as a nontrivial linear combination of $\mathbf{u}$, $\mathbf{v}$, and $\mathbf{w}$. The following definition says that these three vectors are linearly dependent.

Definition 1.6.2 A list of vectors $\beta = \mathbf{v}_1, \mathbf{v}_2, \ldots, \mathbf{v}_r$ in an $\mathbb{F}$-vector space $\mathcal{V}$ is *linearly dependent* if there are scalars $c_1, c_2, \ldots, c_r \in \mathbb{F}$, not all zero, such that $c_1\mathbf{v}_1 + c_2\mathbf{v}_2 + \cdots + c_r\mathbf{v}_r = \mathbf{0}$.

Here are some facts about linear dependence:

- If $\beta = \mathbf{v}_1, \mathbf{v}_2, \ldots, \mathbf{v}_r$ is linearly dependent, then the commutativity of vector addition ensures that any list of r vectors obtained by rearranging the vectors in β is also linearly dependent. For example, the list $\mathbf{v}_r, \mathbf{v}_{r-1}, \ldots, \mathbf{v}_1$ is linearly dependent.

- Theorem 1.1.2 ensures that a list consisting of a single vector $\mathbf{v}$ is linearly dependent if and only if $\mathbf{v} = \mathbf{0}$.

- A list of two vectors is linearly dependent if and only if one of the vectors is a scalar multiple of the other. If $c_1\mathbf{v}_1 + c_2\mathbf{v}_2 = \mathbf{0}$ and $c_1 \neq 0$, then $\mathbf{v}_1 = -c_1^{-1}c_2\mathbf{v}_2$; if $c_2 \neq 0$, then $\mathbf{v}_2 = -c_2^{-1}c_1\mathbf{v}_1$. Conversely, if $\mathbf{v}_1 = c\mathbf{v}_2$, then $1\mathbf{v}_1 + (-c)\mathbf{v}_2 = \mathbf{0}$ is a nontrivial linear combination.

- A list of vectors is linearly dependent if and only if one of the vectors is a linear combination of the others. If $c_1\mathbf{v}_1 + c_2\mathbf{v}_2 + \cdots + c_r\mathbf{v}_r = \mathbf{0}$ and $c_j \neq 0$, then $\mathbf{v}_j = -c_j^{-1}\sum_{i \neq j} c_i\mathbf{v}_i$. Conversely, if $\mathbf{v}_j = \sum_{i \neq j} d_i\mathbf{v}_i$, then $1\mathbf{v}_j + \sum_{i \neq j}(-d_i)\mathbf{v}_i = \mathbf{0}$ is a nontrivial linear combination.

- Any list of vectors that includes the zero vector is linearly dependent. For example, if the list is $\mathbf{v}_1, \mathbf{v}_2, \ldots, \mathbf{v}_r$ and $\mathbf{v}_r = \mathbf{0}$, then $0\mathbf{v}_1 + 0\mathbf{v}_2 + \cdots + 0\mathbf{v}_{r-1} + 1\mathbf{v}_r = \mathbf{0}$ is a nontrivial linear combination.

- Any list of vectors that includes the same vector twice is linearly dependent. For example, if $\mathbf{v}_{r-1} = \mathbf{v}_r$ in the list $\mathbf{v}_1, \mathbf{v}_2, \ldots, \mathbf{v}_{r-1}, \mathbf{v}_r$, then $0\mathbf{v}_1 + 0\mathbf{v}_2 + \cdots + 0\mathbf{v}_{r-2} + 1\mathbf{v}_{r-1} + (-1)\mathbf{v}_r = \mathbf{0}$ is a nontrivial linear combination.

Example 1.6.3 Equation (1.6.1) permits us to express each of the vectors involved as a linear combination of the others: $\mathbf{u} = \frac{3}{4}\mathbf{v} - \frac{1}{4}\mathbf{w}$, $\mathbf{v} = \frac{4}{3}\mathbf{u} + \frac{1}{3}\mathbf{w}$, and $\mathbf{w} = 3\mathbf{v} - 4\mathbf{u}$.

If a linearly dependent list is lengthened by appending additional vectors, then it remains linearly dependent.

Theorem 1.6.4 *If a list of vectors $\mathbf{v}_1, \mathbf{v}_2, \ldots, \mathbf{v}_r$ in an $\mathbb{F}$-vector space $\mathcal{V}$ is linearly dependent, then the list $\mathbf{v}_1, \mathbf{v}_2, \ldots, \mathbf{v}_r, \mathbf{v}$ is linearly dependent for any $\mathbf{v} \in \mathcal{V}$.*

Proof If $\sum_{i=1}^{r} c_i\mathbf{v}_i = \mathbf{0}$ and some $c_j \neq \mathbf{0}$, then $\sum_{i=1}^{r} c_i\mathbf{v}_i + 0\mathbf{v} = \mathbf{0}$ is a nontrivial linear combination. $\square$

Example 1.6.5 The identity (1.6.1) shows that the list $\mathbf{u}, \mathbf{v}, \mathbf{w}$ of vectors in Figure 1.2 is linearly dependent. The preceding theorem ensures that the list $\mathbf{u}, \mathbf{v}, \mathbf{w}, \mathbf{z}$ is also linearly dependent.

The opposite of linear dependence is linear independence.

Definition 1.6.6 A list of vectors $\beta = \mathbf{v}_1, \mathbf{v}_2, \ldots, \mathbf{v}_r$ in an $\mathbb{F}$-vector space $\mathcal{V}$ is *linearly independent* if it is not linearly dependent. That is, β is linearly independent if the only scalars $c_1, c_2, \ldots, c_r \in \mathbb{F}$ such that $c_1\mathbf{v}_1 + c_2\mathbf{v}_2 + \cdots + c_r\mathbf{v}_r = \mathbf{0}$ are $c_1 = c_2 = \cdots = c_r = 0$.

It is convenient to say that "$\mathbf{v}_1, \mathbf{v}_2, \ldots, \mathbf{v}_r$ are linearly independent" rather than "the list of vectors $\mathbf{v}_1, \mathbf{v}_2, \ldots, \mathbf{v}_r$ is linearly independent." If a list of vectors in $\mathbb{F}^m$ is the list of columns of a matrix $A \in \mathsf{M}_{m \times n}(\mathbb{F})$, it is also convenient to say "the columns of A are linearly independent" rather than "the list of columns of A is linearly independent." We observe similar conventions for linear dependence and for the rows of a matrix.

Unlike lists, which are nonempty and finite by definition, sets can be empty or can contain infinitely many distinct vectors. We define linear independence and linear dependence for sets of vectors as follows:

Definition 1.6.7 A subset $\mathscr{S}$ of a vector space V is *linearly independent* if every list of distinct vectors in $\mathscr{S}$ is linearly independent; $\mathscr{S}$ is *linearly dependent* if some list of distinct vectors in $\mathscr{S}$ is linearly dependent.

Here are some facts about linear independence:

- Whether a list is linearly independent does not depend on how it is ordered.

- A list consisting of a single vector $\mathbf{v}$ is linearly independent if and only if $\mathbf{v} \neq \mathbf{0}$.

- A list of two vectors is linearly independent if and only if neither vector in the list is a scalar multiple of the other.

- A list of vectors is linearly independent if and only if no vector in the list can be expressed as a linear combination of the others.

The following example shows that the linear independence of a list of n vectors in $\mathbb{F}^m$ can be formulated as a statement about the null space of a matrix in $\mathsf{M}_{m \times n}(\mathbb{F})$.

Example 1.6.8 A list $\beta = \mathbf{a}_1, \mathbf{a}_2, \ldots, \mathbf{a}_n$ of vectors in $\mathbb{F}^m$ is linearly independent if and only if the only scalars $x_1, x_2, \ldots, x_n \in \mathbb{F}$ such that

$$x_1 \mathbf{a}_1 + x_2 \mathbf{a}_2 + \cdots + x_n \mathbf{a}_n = \mathbf{0} \tag{1.6.9}$$

are $x_1 = x_2 = \cdots = x_n = 0$. Let $A = [\mathbf{a}_1\ \mathbf{a}_2\ \ldots\ \mathbf{a}_n]$ and $\mathbf{x} = [x_1\ x_2\ \ldots\ x_n]^\mathsf{T}$. The linear combination in (1.6.9) is $A\mathbf{x}$, so β is linearly independent if and only if $\mathbf{x} = \mathbf{0}$ is the only vector in $\mathbb{F}^n$ such that $A\mathbf{x} = \mathbf{0}$. That is, β is linearly independent if and only if null $A = \{\mathbf{0}\}$.

Example 1.6.10 Let $A \in \mathsf{M}_{m \times n}(\mathbb{F})$ and $B \in \mathsf{M}_{p \times m}(\mathbb{F})$. If the columns of B are linearly independent, then $BA\mathbf{x} = \mathbf{0}$ if and only if $A\mathbf{x} = \mathbf{0}$, that is, null BA = null A. In particular, null BA = null A if $B \in \mathsf{M}_m(\mathbb{F})$ is invertible.

Example 1.6.11 Let $A \in \mathsf{M}_{m \times n}(\mathbb{F})$ and let β be the list of its rows. Then β is linearly independent if and only if $\mathbf{x} = \mathbf{0}$ is the only vector in $\mathbb{F}^m$ such that $\mathbf{x}^\mathsf{T} A = \mathbf{0}^\mathsf{T}$. That is, β is linearly independent if and only if null $A^\mathsf{T} = \{\mathbf{0}\}$.

Example 1.6.12 In $\mathcal{P}$, the vectors $1, z, z^2, \ldots, z^n$ are linearly independent. A linear combination $c_0 + c_1 z + \cdots + c_n z^n$ is the zero polynomial if and only if $c_0 = c_1 = \cdots = c_n = 0$; see Appendix B.4.

Example 1.6.13 In $C[-\pi, \pi]$, the vectors $1, e^{it}, e^{2it}, \ldots, e^{nit}$ are linearly independent for each $n = 1, 2, \ldots$; see Appendix A.5. This follows from P.2.55; see also P.6.10 and Theorem 6.1.10.

Example 1.6.14 Let $A = [\mathbf{a}_1\ \mathbf{a}_2\ \ldots\ \mathbf{a}_n] \in \mathsf{M}_n(\mathbb{F})$ be invertible and suppose that $x_1, x_2, \ldots, x_n \in \mathbb{F}$ are scalars such that $x_1 \mathbf{a}_1 + x_2 \mathbf{a}_2 + \cdots + x_n \mathbf{a}_n = \mathbf{0}$. Let $\mathbf{x} = [x_1\ x_2\ \ldots\ x_n]^\mathsf{T} \in \mathbb{F}^n$. Then $A\mathbf{x} = x_1 \mathbf{a}_1 + x_2 \mathbf{a}_2 + \cdots + x_n \mathbf{a}_n = \mathbf{0}$ and hence $\mathbf{x} = A^{-1}(A\mathbf{x}) = A^{-1}\mathbf{0} = \mathbf{0}$. It follows that $x_1 = x_2 = \cdots = x_n = 0$, so $\mathbf{a}_1, \mathbf{a}_2, \ldots, \mathbf{a}_n$ are linearly independent. We conclude that if $A \in \mathsf{M}_n(\mathbb{F})$ is invertible, then its columns are linearly independent in $\mathbb{F}^n$.

Example 1.6.15 The identity matrix I_n is invertible. Consequently, its columns $\mathbf{e}_1, \mathbf{e}_2, \ldots, \mathbf{e}_n$ (see (1.4.21)) are linearly independent in $\mathbb{F}^n$. One can see this directly by observing that $[u_1\ u_2\ \ldots\ u_n]^\mathsf{T} = u_1\mathbf{e}_1 + u_2\mathbf{e}_2 + \cdots + u_n\mathbf{e}_n = \mathbf{0}$ if and only if $u_1 = u_2 = \cdots = u_n = 0$.

Example 1.6.16 Let V and W be $\mathbb{F}$-vector spaces. Suppose that the list $\beta = \mathbf{v}_1, \mathbf{v}_2, \ldots, \mathbf{v}_m$ of vectors in V is linearly independent and the list $\gamma = \mathbf{w}_1, \mathbf{w}_2, \ldots, \mathbf{w}_n$ of vectors in W is linearly independent. If there are scalars c_i and d_j such that

$$(\mathbf{0}, \mathbf{0}) = \sum_{i=1}^m c_i(\mathbf{v}_i, \mathbf{0}) + \sum_{j=1}^n d_j(\mathbf{0}, \mathbf{w}_j) = \left(\sum_{i=1}^m c_i\mathbf{v}_i, \sum_{j=1}^n d_j\mathbf{w}_j \right),$$

then $c_1 = c_2 = \cdots = c_m = 0$ and $d_1 = d_2 = \cdots = d_n = 0$. Therefore, the $m + n$ vectors $(\mathbf{v}_1, \mathbf{0}), \ldots (\mathbf{v}_m, \mathbf{0}), (\mathbf{0}, \mathbf{w}_1), \ldots, (\mathbf{0}, \mathbf{w}_n)$ in the Cartesian product $V \times W$ are linearly independent.

Example 1.6.17 Complex conjugation[2] in $\mathbb{C}^n$ preserves linear independence, that is, a list $\beta = \mathbf{u}_1, \mathbf{u}_2, \ldots, \mathbf{u}_r \in \mathbb{C}^n$ is linearly independent if and only if the list $\overline{\beta} = \overline{\mathbf{u}_1}, \overline{\mathbf{u}_2}, \ldots, \overline{\mathbf{u}_r}$ is linearly independent. Suppose that $\mathbf{u}_1, \mathbf{u}_2, \ldots, \mathbf{u}_r$ are linearly independent and $c_1\overline{\mathbf{u}_1} + c_2\overline{\mathbf{u}_2} + \cdots + c_r\overline{\mathbf{u}_r} = \mathbf{0}$. Then $\overline{c_1}\mathbf{u}_1 + \overline{c_2}\mathbf{u}_2 + \cdots + \overline{c_r}\mathbf{u}_r = \overline{\mathbf{0}} = \mathbf{0}$, so each $\overline{c_i} = 0$ and hence each $c_i = 0$. Thus, $\overline{\beta}$ is linearly independent. If $\overline{\beta}$ is linearly independent, the preceding argument confirms that $\beta = \overline{\overline{\beta}}$ is linearly independent.

The most important property of a linearly independent list of vectors is that it provides a representation of each vector in its span as a linear combination with unique coefficients.

Theorem 1.6.18 *Let $\beta = \mathbf{v}_1, \mathbf{v}_2, \ldots, \mathbf{v}_r$ be a linearly independent list of vectors in an $\mathbb{F}$-vector space V. Then*

$$a_1\mathbf{v}_1 + a_2\mathbf{v}_2 + \cdots + a_r\mathbf{v}_r = b_1\mathbf{v}_1 + b_2\mathbf{v}_2 + \cdots + b_r\mathbf{v}_r \qquad (1.6.19)$$

if and only if $a_j = b_j$ for each $j = 1, 2, \ldots, r$.

Proof The identity (1.6.19) is equivalent to

$$(a_1 - b_1)\mathbf{v}_1 + (a_2 - b_2)\mathbf{v}_2 + \cdots + (a_r - b_r)\mathbf{v}_r = \mathbf{0}. \qquad (1.6.20)$$

Since β is linearly independent, (1.6.20) is satisfied if and only if $a_j - b_j = 0$ for each $j = 1, 2, \ldots, r$. $\qquad \square$

It is convenient to summarize related results that we use many times.

Theorem 1.6.21 *Let $A \in \mathsf{M}_{m \times n}(\mathbb{F})$ and $B \in \mathsf{M}_{m \times p}(\mathbb{F})$.*

(a) $\operatorname{col} B \subseteq \operatorname{col} A$ *if and only if there is an $X \in \mathsf{M}_{n \times p}(\mathbb{F})$ such that $B = AX$.*

(b) *If B has linearly independent columns, $X \in \mathsf{M}_{n \times p}(\mathbb{F})$, and $AX = B$, then X has linearly independent columns.*

(c) *If A has linearly independent columns, $X, Y \in \mathsf{M}_{n \times p}(\mathbb{F})$, and $AX = AY$, then $X = Y$.*

---

[2] If you encounter an unfamiliar term in this book, look it up in the index. The index entry for complex conjugation of a matrix points to Appendix C.2.

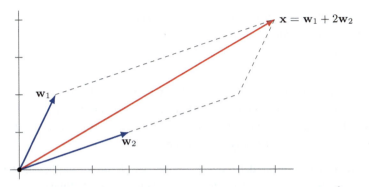

Figure 1.6 Linear combination of linearly independent vectors in $\mathbb{R}^2$.

Proof (a) This is Theorem 1.4.10.

(b) If $\mathbf{v} \in \text{null}\,X$, then $B\mathbf{v} = AX\mathbf{v} = A\mathbf{0} = \mathbf{0}$, which implies that $\mathbf{v} = \mathbf{0}$ since B has linearly independent columns. We conclude that $\text{null}\,X = \{\mathbf{0}\}$ (Example 1.6.8), so X has linearly independent columns.

(c) If $AX = AY$, then $A(X - Y) = 0$, so every column of $X - Y$ is in $\text{null}\,A$. However, $\text{null}\,A = \{\mathbf{0}\}$ since A has linearly independent columns. Therefore, every column of $X - Y$ is $\mathbf{0}$ and $X = Y$.   □

Example 1.6.22 In Figure 1.6 the list $\mathbf{w}_1, \mathbf{w}_2$ is linearly independent and $\mathbf{x} = \mathbf{w}_1 + 2\mathbf{w}_2$. Theorem 1.6.18 ensures that this is the only way to write $\mathbf{x}$ as a linear combination of $\mathbf{w}_1$ and $\mathbf{w}_2$. We encountered a different situation in Figure 1.2, where we were able to express the vector $\mathbf{z}$ as two different linear combinations of the vectors in the linearly dependent list $\mathbf{u}, \mathbf{v}, \mathbf{w}$; see (1.4.1) and (1.4.2).

It is convenient to have a notation for the list obtained by omitting a vector from a given list.

Definition 1.6.23 Let $r \geq 2$ and let $\beta = \mathbf{v}_1, \mathbf{v}_2, \ldots, \mathbf{v}_r$ be a list of vectors in an $\mathbb{F}$-vector space $\mathcal{V}$. If $j \in \{1, 2, \ldots, r\}$, the list of $r - 1$ vectors obtained by omitting $\mathbf{v}_j$ from β is denoted by $\mathbf{v}_1, \mathbf{v}_2, \ldots, \widehat{\mathbf{v}}_j, \ldots, \mathbf{v}_r$.

Example 1.6.24 If $\beta = \mathbf{v}_1, \mathbf{v}_2, \mathbf{v}_3, \mathbf{v}_4$, then $\mathbf{v}_1, \mathbf{v}_2, \widehat{\mathbf{v}}_3, \mathbf{v}_4$ is the list $\mathbf{v}_1, \mathbf{v}_2, \mathbf{v}_4$.

If a linearly independent list is shortened by removing some (but not all) of its elements, then it remains linearly independent.

Theorem 1.6.25 *Let* $r \geq 2$ *and suppose that a list* $\mathbf{v}_1, \mathbf{v}_2, \ldots, \mathbf{v}_r$ *of vectors in an* $\mathbb{F}$-*vector space* $\mathcal{V}$ *is linearly independent. Then the list* $\mathbf{v}_1, \mathbf{v}_2, \ldots, \widehat{\mathbf{v}}_j, \ldots, \mathbf{v}_r$ *is linearly independent for any* $j \in \{1, 2, \ldots, r\}$.

Proof The claim is equivalent to the assertion that if $\mathbf{v}_1, \mathbf{v}_2, \ldots, \widehat{\mathbf{v}}_j, \ldots, \mathbf{v}_r$ is linearly dependent, then $\mathbf{v}_1, \mathbf{v}_2, \ldots, \mathbf{v}_r$ is linearly dependent. Theorem 1.6.4 ensures that this is the case. □

Theorem 1.6.26 *Let* $\beta = \mathbf{v}_1, \mathbf{v}_2, \ldots, \mathbf{v}_r$ *be a list of vectors in a nonzero* $\mathbb{F}$-*vector space* $\mathcal{V}$.

(a) *Suppose that* β *is linearly independent and does not span* $\mathcal{V}$. *If* $\mathbf{v} \in \mathcal{V}$ *and* $\mathbf{v} \notin \text{span}\,\beta$, *then the list* $\mathbf{v}_1, \mathbf{v}_2, \ldots, \mathbf{v}_r, \mathbf{v}$ *is linearly independent.*

(b) *Suppose that β is linearly dependent and* $\operatorname{span}\beta = \mathcal{V}$. *If* $c_1\mathbf{v}_1 + c_2\mathbf{v}_2 + \cdots + c_r\mathbf{v}_r = \mathbf{0}$ *is a nontrivial linear combination and* $j \in \{1, 2, \ldots, r\}$ *is any index such that* $c_j \neq 0$, *then* $\mathbf{v}_1, \mathbf{v}_2, \ldots, \widehat{\mathbf{v}}_j, \ldots, \mathbf{v}_r$ *spans* $\mathcal{V}$.

Proof (a) Suppose that $c_1\mathbf{v}_1 + c_2\mathbf{v}_2 + \cdots + c_r\mathbf{v}_r + c\mathbf{v} = \mathbf{0}$. If $c \neq 0$, then $\mathbf{v} = -c^{-1}\sum_{i=1}^{r} c_i\mathbf{v}_i \in$ span$\{\mathbf{v}_1, \mathbf{v}_2, \ldots, \mathbf{v}_r\}$, which is a contradiction. Thus, $c = 0$, and hence $c_1\mathbf{v}_1 + c_2\mathbf{v}_2 + \cdots + c_r\mathbf{v}_r = \mathbf{0}$. The linear independence of β implies that $c_1 = c_2 = \cdots = c_r = 0$. We conclude that the list $\mathbf{v}_1, \mathbf{v}_2, \ldots, \mathbf{v}_r, \mathbf{v}$ is linearly independent.

(b) If $r = 1$, then the linear dependence of the list $\beta = \mathbf{v}_1$ implies that $\mathbf{v}_1 = \mathbf{0}$, so $\mathcal{V} = \operatorname{span}\beta = \{\mathbf{0}\}$, which is a contradiction. Thus, $r \geq 2$. Since $c_j \neq 0$, we have $\mathbf{v}_j = -c_j^{-1}\sum_{i\neq j} c_i\mathbf{v}_i$. This identity can be used to eliminate $\mathbf{v}_j$ from any linear combination in which it appears, so any vector that is a linear combination of the r vectors in the list β (namely, every vector in $\mathcal{V}$) is also a linear combination of the $r - 1$ vectors in the list $\mathbf{v}_1, \mathbf{v}_2, \ldots, \widehat{\mathbf{v}}_j, \ldots, \mathbf{v}_r$. $\qquad\square$

1.7 The Pivot Column Decomposition

If we are given a linearly dependent nonzero list of vectors in an $\mathbb{F}$-vector space, then Theorem 1.6.26.b says that one of its elements can be omitted to obtain a shorter list with the same span. The following theorem uses Theorem 1.6.26.a to construct a shorter list that has the same span and is linearly independent.

Theorem 1.7.1 *Let* $\beta = \mathbf{v}_1, \mathbf{v}_2, \ldots, \mathbf{v}_p$ *be a nonzero list of vectors in an* $\mathbb{F}$-*vector space. There is an* $s \in \{1, 2, \ldots, p\}$ *and unique indices* $j_1, j_2, \ldots, j_s$ *such that:*

(a) $1 \leq j_1 < j_2 < \cdots < j_s \leq p$.

(b) $\gamma = \mathbf{v}_{j_1}, \mathbf{v}_{j_2}, \ldots, \mathbf{v}_{j_s}$ *is linearly independent.*

(c) $\operatorname{span}\gamma = \operatorname{span}\beta$.

(d) *If* $j < j_1$, *then* $\mathbf{v}_j = \mathbf{0}$.

(e) *If* $s > 1$, $2 \leq k \leq s$, *and* $j_{k-1} < j < j_k$, *then* $\mathbf{v}_j \in \operatorname{span}\{\mathbf{v}_{j_1}, \mathbf{v}_{j_2}, \ldots, \mathbf{v}_{j_{k-1}}\}$.

Proof Since β is nonzero, we may let $j_1 = \min\{j : \mathbf{v}_j \neq \mathbf{0}\}$. Then $\beta_1 = \mathbf{v}_{j_1}$ is linearly independent since $\mathbf{v}_{j_1} \neq \mathbf{0}$. If $\operatorname{span}\beta_1 = \operatorname{span}\beta$, then let $s = 1$ and stop. If not, let $j_2 = \min\{j > j_1 : \mathbf{v}_j \notin \operatorname{span}\beta_1\}$. Theorem 1.6.26.a ensures that $\beta_2 = \mathbf{v}_{j_1}, \mathbf{v}_{j_2}$ is linearly independent. If $\operatorname{span}\beta_2 = \operatorname{span}\beta$, then let $s = 2$ and stop. If not, continue this left-to-right vector selection process until it terminates with a linearly independent list $\beta_s = \mathbf{v}_{j_1}, \mathbf{v}_{j_2}, \ldots, \mathbf{v}_{j_s}$ such that $\operatorname{span}\beta_s = \operatorname{span}\beta$. $\qquad\square$

Definition 1.7.2 The indices $j_1, j_2, \ldots, j_s$ in Theorem 1.7.1 are the *pivot indices* of the list $\beta = \mathbf{v}_1, \mathbf{v}_2, \ldots, \mathbf{v}_p$. The vectors $\mathbf{v}_{j_1}, \mathbf{v}_{j_2}, \ldots, \mathbf{v}_{j_s}$ are the *pivot vectors* of β.

Example 1.7.3 Let β be the list of columns of

$$A = \begin{bmatrix} 0 & 5 & 10 & 0 & 1 & 13 & 19 & 2 & 13 \\ 0 & 4 & 8 & 0 & 2 & 14 & 20 & 1 & 11 \\ 0 & 3 & 6 & 0 & 3 & 15 & 21 & 4 & 21 \\ 0 & 2 & 4 & 0 & 4 & 16 & 22 & 3 & 19 \\ 0 & 1 & 2 & 0 & 5 & 17 & 23 & 5 & 26 \end{bmatrix} = [\mathbf{a}_1 \ \mathbf{a}_2 \ \ldots \ \mathbf{a}_9] \in \mathsf{M}_{5\times 9},$$

so span $\beta = \operatorname{col} A$. Follow the steps in the proof of the preceding theorem to construct a linearly independent spanning list of columns. The first nonzero column is $\mathbf{a}_2$, so $\beta_1 = \mathbf{a}_2$. The columns $\mathbf{a}_3 = 2\mathbf{a}_2$ and $\mathbf{a}_4 = \mathbf{0}$ are in span β_1 but $\mathbf{a}_5$ is not, so $\beta_2 = \mathbf{a}_2, \mathbf{a}_5$. The columns $\mathbf{a}_6 = 2\mathbf{a}_2 + 3\mathbf{a}_5$ and $\mathbf{a}_7 = 3\mathbf{a}_2 + 4\mathbf{a}_5$ are in span β_2 but $\mathbf{a}_8$ is not, so $\beta_3 = \mathbf{a}_2, \mathbf{a}_5, \mathbf{a}_8$. Finally, $\mathbf{a}_9 = \mathbf{a}_2 + 2\mathbf{a}_5 + 3\mathbf{a}_8$ is in span β_3. The list $\beta_3 = \mathbf{a}_2, \mathbf{a}_5, \mathbf{a}_8$ spans $\operatorname{col} A$ and Theorem 1.6.26.a ensures that it is linearly independent. The pivot vectors are $\mathbf{a}_2, \mathbf{a}_5, \mathbf{a}_8$ and the pivot indices are $j_1 = 2, j_2 = 5$, and $j_3 = 8$.

Definition 1.7.4 The *pivot columns* and *pivot indices* of $A = [\mathbf{a}_1 \ \mathbf{a}_2 \ \ldots \ \mathbf{a}_n] \in \mathsf{M}_{m\times n}(\mathbb{F})$ are the respective pivot vectors and pivot indices of the list $\mathbf{a}_1, \mathbf{a}_2, \ldots, \mathbf{a}_n$.

The pivot columns of a nonzero matrix A are linearly independent and span its column space. These two properties lead to a factorization that reveals linear dependencies among the columns of A.

Theorem 1.7.5 (Pivot Column Decomposition) *Let $A = [\mathbf{a}_1 \ \mathbf{a}_2 \ \ldots \ \mathbf{a}_n] \in \mathsf{M}_{m\times n}(\mathbb{F})$ be nonzero, let $j_1 < j_2 < \cdots < j_s$ be its pivot indices, and let $P = [\mathbf{a}_{j_1} \ \mathbf{a}_{j_2} \ \ldots \ \mathbf{a}_{j_s}] \in \mathsf{M}_{m\times s}(\mathbb{F})$.*

(a) *The s columns of P are linearly independent, $1 \leq s \leq n$, and $\operatorname{col} P = \operatorname{col} A$.*

(b) *There is a unique $R = [\mathbf{r}_1 \ \mathbf{r}_2 \ \ldots \ \mathbf{r}_n] \in \mathsf{M}_{s\times n}(\mathbb{F})$ such that $A = PR$.*

(c) *$\mathbf{r}_{j_k} = \mathbf{e}_k \in \mathbb{F}^s$ for each $k = 1, 2, \ldots, s$.*

(d) *If $j < j_1$, then $\mathbf{r}_j = \mathbf{0}$.*

(e) *If $s > 1$, $2 \leq k \leq s$, and $j_{k-1} < j < j_k$, then $\mathbf{r}_j \in \operatorname{span}\{\mathbf{e}_1, \mathbf{e}_2, \ldots, \mathbf{e}_{j_{k-1}}\}$.*

(f) *The rows of R are linearly independent and $\operatorname{null} A = \operatorname{null} R$.*

Proof (a) See Theorem 1.7.1 and its parts (b) and (c).

(b) See (a) and (c) of Theorem 1.6.21.

(c) Since $A = PR$, we have $\mathbf{a}_{j_k} = [\mathbf{a}_{j_1} \ \mathbf{a}_{j_2} \ \ldots \ \mathbf{a}_{j_s}]\mathbf{r}_{j_k}$ for each $k = 1, 2, \ldots, s$. Linear independence of the columns of P ensures that $\mathbf{r}_{j_k} = \mathbf{e}_k$.

(d) and (e) We have $\mathbf{a}_j = [\mathbf{a}_{j_1} \ \mathbf{a}_{j_2} \ \ldots \ \mathbf{a}_{j_s}]\mathbf{r}_j$ for each $j = 1, 2, \ldots, n$, and the assertions follow from (d) and (e) of Theorem 1.7.1.

(f) Let $\mathbf{x} = [x_1 \ x_2 \ \ldots \ x_s]^\mathsf{T}$. For each $k = 1, 2, \ldots, s$, we have $\mathbf{r}_{j_k} = \mathbf{e}_k$, so the entry of $\mathbf{x}^\mathsf{T} R$ in position j_k is x_k. If $\mathbf{x}^\mathsf{T} R = \mathbf{0}^\mathsf{T}$, then $x_1 = x_2 = \cdots = x_s = 0$ and hence $\mathbf{x} = \mathbf{0}$. This ensures that R has linearly independent rows. Now let $\mathbf{y} \in \mathbb{F}^n$. If $A\mathbf{y} = \mathbf{0}$, then $P(R\mathbf{y}) = \mathbf{0}$ and the linear independence of the columns of P ensures that $R\mathbf{y} = \mathbf{0}$; this means that $\operatorname{null} A \subseteq \operatorname{null} R$. Conversely, if $R\mathbf{y} = \mathbf{0}$, then $A\mathbf{y} = P(R\mathbf{y}) = \mathbf{0}$, so $\operatorname{null} R \subseteq \operatorname{null} A$. $\qquad\square$

Definition 1.7.6 The *pivot column decomposition* of a nonzero $A \in \mathsf{M}_{m \times n}(\mathbb{F})$ is the unique factorization $A = PR$ described in Theorem 1.7.5.

Example 1.7.7 Our analysis of the 5×9 matrix A in Example 1.7.3 shows that its pivot column decomposition is $A = PR$, in which $P = [\mathbf{a}_2 \ \mathbf{a}_5 \ \mathbf{a}_8] \in \mathsf{M}_{5 \times 3}$ and

$$R = \begin{bmatrix} 0 & 1 & 2 & 0 & 0 & 2 & 3 & 0 & 1 \\ 0 & 0 & 0 & 0 & 1 & 3 & 4 & 0 & 2 \\ 0 & 0 & 0 & 0 & 0 & 0 & 0 & 1 & 3 \end{bmatrix} = [\mathbf{r}_1 \ \mathbf{r}_2 \ \dots \ \mathbf{r}_9].$$

Observe that the submatrix of R consisting of its pivot columns (2, 5, and 8) is a 3×3 identity matrix, and the rows of R exhibit the staircase pattern asserted in part (e) of the preceding theorem.

Vectors in the null space of R (and hence in the null space of A) can be identified by inspecting the non-pivot columns of R and using the identity $A = PR$. For example, inspection of column 3 of R reveals that $\mathbf{a}_3 = P\mathbf{r}_3 = 2\mathbf{a}_{j_1} = 2\mathbf{a}_2$. Thus, $\mathbf{a}_3 - 2\mathbf{a}_2 = A(\mathbf{e}_3 - 2\mathbf{e}_2) = \mathbf{0}$, so $\mathbf{w}_3 = \mathbf{e}_3 - 2\mathbf{e}_2 \in \text{null}\, A$. Inspection of column 6 of R reveals that $\mathbf{a}_6 = P\mathbf{r}_6 = 2\mathbf{a}_{j_1} + 3\mathbf{a}_{j_2}$. Thus,

$$\mathbf{a}_6 - 2\mathbf{a}_{j_1} - 3\mathbf{a}_{j_2} = \mathbf{a}_6 - 2\mathbf{a}_2 - 3\mathbf{a}_5 = A(\mathbf{e}_6 - 2\mathbf{e}_2 - 3\mathbf{e}_5) = \mathbf{0},$$

and hence $\mathbf{w}_6 = \mathbf{e}_6 - 2\mathbf{e}_2 - 3\mathbf{e}_5 \in \text{null}\, A$. A similar inspection of the remaining four non-pivot columns of R identifies the columns of

$$W = [\mathbf{w}_1 \ \mathbf{w}_3 \ \mathbf{w}_4 \ \mathbf{w}_6 \ \mathbf{w}_7 \ \mathbf{w}_9] = \begin{bmatrix} 1 & 0 & 0 & 0 & 0 & 0 \\ 0 & -2 & 0 & -2 & -3 & -1 \\ 0 & 1 & 0 & 0 & 0 & 0 \\ 0 & 0 & 1 & 0 & 0 & 0 \\ 0 & 0 & 0 & -3 & -4 & -2 \\ 0 & 0 & 0 & 1 & 0 & 0 \\ 0 & 0 & 0 & 0 & 1 & 0 \\ 0 & 0 & 0 & 0 & 0 & -3 \\ 0 & 0 & 0 & 0 & 0 & 1 \end{bmatrix} \qquad (1.7.8)$$

as linearly independent vectors in $\text{null}\, A$. See P.1.14.

Example 1.7.9 What is the equation of the plane in $\mathbb{R}^3$ through $\mathbf{0}$ that contains $\mathbf{u} = [1 \ 2 \ 3]^\mathsf{T}$ and $\mathbf{v} = [4 \ 5 \ 6]^\mathsf{T}$? Example 1.3.6 tells us that if $A = [\mathbf{u} \ \mathbf{v}]$ and $\mathbf{n} \in \text{null}\, A^\mathsf{T}$, then $\mathbf{n}^\mathsf{T}\mathbf{x} = \mathbf{0}$ is the equation we seek. Factor $A^\mathsf{T} = PR$ as in Theorem 1.7.5 and obtain

$$P = \begin{bmatrix} 1 & 2 \\ 4 & 5 \end{bmatrix} \qquad \text{and} \qquad R = \begin{bmatrix} 1 & 0 & -1 \\ 0 & 1 & 2 \end{bmatrix}.$$

Then $\mathbf{n} = [1 \ -2 \ 1]^\mathsf{T} \in \text{null}\, R = \text{null}\, A^\mathsf{T}$, so $x_1 - 2x_2 + x_3 = 0$ is the desired equation.

1.8 Problems

P.1.1 The vectors $[1 \ 2 \ 3]^\mathsf{T}$ and $[3 \ 2 \ 1]^\mathsf{T}$ span a plane through the origin in $\mathbb{R}^3$. Use the method in Example 1.7.9 to find the equation of the plane.

P.1.2 Is $[1 \ 0 \ 1]^\mathsf{T} \in \mathbb{R}^3$ in the span of $[-1 \ 3 \ 2]^\mathsf{T}$ and $[1 \ 1 \ -2]^\mathsf{T}$? Why?

P.1.3 Is $[7 \ 8 \ 9]^\mathsf{T} \in \mathbb{R}^3$ in the span of $[1 \ 2 \ 3]^\mathsf{T}$ and $[4 \ 5 \ 6]^\mathsf{T}$? Why? Is $[10 \ 11 \ 12]^\mathsf{T}$?

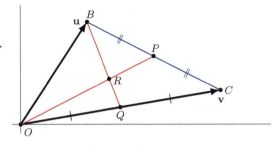

Figure 1.7 The plane triangle OBC.

P.1.4 Choose any three pairs of vectors in Figure 1.2. Show that each pair is linearly independent and spans $\mathbb{R}^2$. *Hint*: Examples 1.4.22 and 1.6.14.

P.1.5 Show that $\text{span}\{1, z^2, z^4, \ldots\} = \mathcal{P}_{\text{even}}$ and $\text{span}\{z, z^3, z^5, \ldots\} = \mathcal{P}_{\text{odd}}$.

P.1.6 Let $\mathcal{V} = \{p \in \mathcal{P} : p(iz) = p(z)\}$. (a) Show that $\mathcal{V}$ is a subspace of $\mathcal{P}$. (b) Show that $\text{span}\{1, z^4, z^8, \ldots\} = \mathcal{V}$.

P.1.7 Let OBC be a plane triangle determined by linearly independent vectors $\mathbf{u}$ and $\mathbf{v}$; see Figure 1.7. Let P and Q be the respective midpoints of sides BC and OC. (a) Show that the medians from O and B are the line segments $OP = \{\frac{t}{2}\mathbf{u} + \frac{t}{2}\mathbf{v} : 0 \le t \le 1\}$ and $BQ = \{(1 - s)\mathbf{u} + \frac{s}{2}\mathbf{v} : 0 \le s \le 1\}$. (b) For what values of t and s do these line segments intersect? How does linear independence play a role here? Let R denote the point of intersection. (c) Show that the ratios of lengths $|BR| : |RQ|$ and $|OR| : |RP|$ are both $2 : 1$. (d) Where do OP and BQ intersect the median from C?

P.1.8 Consider the following lists of vectors in $\mathbb{R}^3$:

$$\beta = \begin{bmatrix} 1 \\ 2 \\ 3 \end{bmatrix}, \begin{bmatrix} 4 \\ s \\ 6 \end{bmatrix}, \begin{bmatrix} 7 \\ 8 \\ 9 \end{bmatrix} \quad \text{and} \quad \gamma = \begin{bmatrix} 4 \\ 2 \\ 6 \end{bmatrix}, \begin{bmatrix} 1 \\ t \\ -1 \end{bmatrix}, \begin{bmatrix} 2 \\ -2 \\ 3 \end{bmatrix}.$$

For what values of s is β linearly dependent? For what values of t is γ linearly dependent?

P.1.9 The matrix

$$A = \begin{bmatrix} 3 & 1 & \boxed{?} & 4 & \boxed{?} \\ 5 & 3 & \boxed{?} & 5 & \boxed{?} \\ 0 & 4 & \boxed{?} & -1 & \boxed{?} \end{bmatrix}$$

is factored $A = PR$, as in Theorem 1.7.5. If

$$R = \begin{bmatrix} 1 & 0 & -\frac{3}{4} & 0 & -\frac{45}{8} \\ 0 & 1 & \frac{1}{4} & 0 & \frac{7}{8} \\ 0 & 0 & 0 & 1 & \frac{9}{2} \end{bmatrix},$$

what are the missing entries of A?

P.1.10 Let $B \in \mathsf{M}_n$. (a) Show that $\{A \in \mathsf{M}_n : \text{tr}\, BA = 0\}$ is a subspace of M_n. (b) When is $\{A \in \mathsf{M}_n : \det BA = 0\}$ a subspace of M_n? Why?

P.1.11 Show that the set of $n \times n$ symmetric matrices is a subspace of M_n but the set of $n \times n$ Hermitian matrices is not. *Hint*: M_n is a complex vector space.

P.1.12 Show that the set of $n \times n$ Hermitian matrices is a real vector space.

P.1.13 Let

$$A = \begin{bmatrix} 1 & 2 & 3 \\ 4 & 5 & 6 \\ 7 & 8 & 9 \end{bmatrix}. \tag{1.8.1}$$

(a) Choose two of the following and show that each is a factorization $A = XY$, in which the columns of X and the rows of Y are each linearly independent.

$$\begin{bmatrix} 1 & 2 \\ 4 & 5 \\ 7 & 8 \end{bmatrix}\begin{bmatrix} 1 & 0 & -1 \\ 0 & 1 & 2 \end{bmatrix}, \begin{bmatrix} 2 & 3 \\ 5 & 6 \\ 8 & 9 \end{bmatrix}\begin{bmatrix} 2 & 1 & 0 \\ -1 & 0 & 1 \end{bmatrix}, \begin{bmatrix} 1 & 3 \\ 4 & 6 \\ 7 & 9 \end{bmatrix}\begin{bmatrix} 1 & \frac{1}{2} & 0 \\ 0 & \frac{1}{2} & 1 \end{bmatrix}, \begin{bmatrix} -1 & 3 \\ -1 & 9 \\ -1 & 15 \end{bmatrix}\begin{bmatrix} \frac{1}{2} & -\frac{1}{2} & -\frac{3}{2} \\ \frac{1}{2} & \frac{1}{2} & \frac{1}{2} \end{bmatrix}.$$

(b) Which of these factorizations is the one described in Theorem 1.7.5? Why?

P.1.14 Verify the assertion in Example 1.7.7 that $A = PR$. Explain how columns 1, 3, 5, and 6 of the matrix W in (1.7.8) are obtained from non-pivot columns of R. Why are the columns of W linearly independent?

P.1.15 In the spirit of the examples in Section 1.2, explain how $\mathbb{C}^n$ is a vector space over $\mathbb{R}$. Is $\mathbb{R}^n$ a vector space over $\mathbb{C}$?

P.1.16 Prove that the only subspaces of the real vector space $\mathbb{R}^1$ are $\{0\}$ and $\mathbb{R}^1$.

P.1.17 Prove that the only subspaces of the real vector space $\mathbb{R}^2$ are $\{0\}$, $\mathbb{R}^2$, and span$\{\mathbf{v}\}$ for any nonzero $\mathbf{v} \in \mathbb{R}^2$.

P.1.18 Let $\mathcal{V}$ be the set of real 2×2 matrices of the form $\mathbf{v} = \begin{bmatrix} 1 & v \\ 0 & 1 \end{bmatrix}$. Define vector addition by $\mathbf{v} + \mathbf{w} = \begin{bmatrix} 1 & v \\ 0 & 1 \end{bmatrix}\begin{bmatrix} 1 & w \\ 0 & 1 \end{bmatrix}$ (ordinary matrix multiplication) and scalar multiplication by $c\mathbf{v} = \begin{bmatrix} 1 & cv \\ 0 & 1 \end{bmatrix}$. Show that $\mathcal{V}$ together with these two operations is a real vector space. What is the zero vector in $\mathcal{V}$?

P.1.19 Consider the Cartesian product $\mathbb{R}^2 \times \mathbb{R}$ of real vector spaces, consisting of ordered pairs $(\mathbf{v}, w)$ in which $\mathbf{v} \in \mathbb{R}^2$ is a real vector and $w \in \mathbb{R}$ is a real scalar. (a) What are the operations of vector addition and scalar multiplication by real scalars in $\mathbb{R}^2 \times \mathbb{R}$? (b) What is the zero vector in $\mathbb{R}^2 \times \mathbb{R}$? (c) Show that $\mathbb{R}^2 \times \mathbb{R}$ is a vector space. (d) Discuss how $\mathbb{R}^3$ is related to $\mathbb{R}^2 \times \mathbb{R}$.

P.1.20 Show that the intersection of any (possibly infinite) collection of subspaces of an $\mathbb{F}$-vector space is a subspace.

P.1.21 Let $\mathcal{U}$ be a subset of an $\mathbb{F}$-vector space $\mathcal{V}$. Show that span $\mathcal{U}$ is the intersection of all the subspaces of $\mathcal{V}$ that contain $\mathcal{U}$. *Hint*: Consider the cases $\mathcal{U} \neq \varnothing$ and $\mathcal{U} = \varnothing$.

P.1.22 Let $\mathcal{U}$ and $\mathcal{W}$ be subspaces of an $\mathbb{F}$-vector space $\mathcal{V}$. Prove that $\mathcal{U} \cup \mathcal{W}$ is a subspace of $\mathcal{V}$ if and only if $\mathcal{U} \subseteq \mathcal{W}$ or $\mathcal{W} \subseteq \mathcal{U}$.

P.1.23 Let $\mathcal{U}$ be a subspace of an $\mathbb{F}$-vector space. (a) Prove that $\mathcal{U} + \{\mathbf{0}\} = \mathcal{U}$ and $\mathcal{U} \cap \{\mathbf{0}\} = \{\mathbf{0}\}$. (b) Prove that $\mathcal{U} + \mathcal{U} = \mathcal{U}$ and $\mathcal{U} \cap \mathcal{U} = \mathcal{U}$.

P.1.24 Let $\mathcal{V}$ be an $\mathbb{F}$-vector space and suppose that $\mathcal{V} = \text{span}\{\mathbf{v}_1, \mathbf{v}_2, \ldots, \mathbf{v}_n\}$. Suppose that $\mathcal{W}$ is a nonzero subspace of $\mathcal{V}$. Do there exist indices $j_1, j_2, \ldots, j_r$ such that $\mathcal{W} = \text{span}\{\mathbf{v}_{j_1}, \mathbf{v}_{j_2}, \ldots, \mathbf{v}_{j_r}\}$? Why or why not?

P.1.25 Let $\mathcal{U}, \mathcal{V}, \mathcal{W}$ be subspaces of an $\mathbb{F}$-vector space. (a) Must $(\mathcal{U} + \mathcal{V}) + \mathcal{W} = \mathcal{U} + (\mathcal{V} + \mathcal{W})$? Why? (b) Must $(\mathcal{U} \cap \mathcal{V}) \cap \mathcal{W} = \mathcal{U} \cap (\mathcal{V} \cap \mathcal{W})$? Why?

P.1.26 Let $\mathcal{U}, \mathcal{V}, \mathcal{W}$ be subspaces of an $\mathbb{F}$-vector space. Must $(\mathcal{U} + \mathcal{V}) \cap \mathcal{W} = (\mathcal{U} + \mathcal{W}) \cap (\mathcal{V} + \mathcal{W})$? Why?

P.1.27   (a) Show that $\operatorname{col} A = \mathbb{R}^3$ for

$$A = [\mathbf{a}_1\ \mathbf{a}_2\ \mathbf{a}_3\ \mathbf{a}_4] = \begin{bmatrix} 1 & 1 & 1 & 0 \\ 0 & 1 & 1 & 1 \\ 0 & 0 & 1 & 1 \end{bmatrix} \in \mathsf{M}_n(\mathbb{R}).$$

(b) Find $i, j, k \in \{1, 2, 3, 4\}$ such that $\operatorname{span}\{\mathbf{a}_i, \mathbf{a}_j, \mathbf{a}_k\} = \mathbb{R}^3$. (c) Find distinct $i, j, k \in \{1, 2, 3, 4\}$ such that $\operatorname{span}\{\mathbf{a}_i, \mathbf{a}_j, \mathbf{a}_k\} \neq \mathbb{R}^3$.

P.1.28   (a) Show that $\operatorname{col} A = \mathbb{R}^3$ for

$$A = [\mathbf{a}_1\ \mathbf{a}_2\ \mathbf{a}_3\ \mathbf{a}_4] = \begin{bmatrix} 1 & 1 & 1 & 1 \\ 0 & 1 & 1 & 0 \\ 0 & 0 & 1 & 1 \end{bmatrix} \in \mathsf{M}_n(\mathbb{R}).$$

(b) Show that $\operatorname{span}\{\mathbf{a}_i, \mathbf{a}_j, \mathbf{a}_k\} = \mathbb{R}^3$ for all distinct $i, j, k \in \{1, 2, 3, 4\}$.

P.1.29   Let $\mathbf{u} \in \mathbb{F}^m$ and $\mathbf{v} \in \mathbb{F}^n$ be nonzero and let $A = \mathbf{u}\mathbf{v}^\mathsf{T}$. (a) What is the size of A? (b) Show that $\operatorname{col} A = \operatorname{span}\{\mathbf{u}\}$. (c) Find the factors P and R in the pivot column decomposition $A = PR$. (d) Find $n - 1$ linearly independent vectors in $\operatorname{null} A$. *Hint*: Example 1.7.7.

P.1.30   Let $n \geq 2$ and let $A, B \in \mathsf{M}_n$. Which of the following subsets of M_n is a subspace of the complex vector space M_n? Why? (a) All invertible matrices. (b) All noninvertible matrices. (c) All A such that $A^2 = 0$. (d) All matrices whose first column is zero. (e) All lower triangular matrices. (f) All $X \in \mathsf{M}_n$ such that $AX + X^\mathsf{T}B = 0$. (g) $\{AXB : X \in \mathsf{M}_n\}$.

P.1.31   (a) Let $A = \begin{bmatrix} a & b \\ c & d \end{bmatrix}$. Compute the entries of $A^2 - (\operatorname{tr} A)\, A + (\det A)\, I$. (b) Is there a $B \in \mathsf{M}_2$ such that $\beta = I, B, B^2$ is linearly independent? If yes, exhibit one. If no, explain why. (c) Is there a $C \in \mathsf{M}_3$ such that $\beta = I, C, C^2$ is linearly independent? If yes, exhibit one. If no, explain why.

P.1.32   (a) Give an example of a nonzero $A \in \mathsf{M}_2(\mathbb{F})$ such that $\operatorname{col} A = \operatorname{null} A$. (b) Give an example of $A \in \mathsf{M}_2(\mathbb{F})$ such that $\operatorname{col} A \oplus \operatorname{null} A = \mathbb{F}^2$.

P.1.33   Suppose that $A \in \mathsf{M}_2$ is invertible. Prove that $\operatorname{span}\{\ldots, A^{-2}, A^{-1}, I, A, A^2, \ldots\} \neq \mathsf{M}_2$. *Hint*: Consider $\operatorname{span}\{I, A, A^2, \ldots\}$, $\operatorname{span}\{I, A^{-1}, A^{-2}, \ldots\}$, and P.1.31.

P.1.34   Let $\mathcal{V}$ be a real vector space and suppose that the list $\beta = \mathbf{u}, \mathbf{v}, \mathbf{w}$ of vectors in $\mathcal{V}$ is linearly independent. Show that the list $\gamma = \mathbf{u} - \mathbf{v}, \mathbf{v} - \mathbf{w}, \mathbf{w} + \mathbf{u}$ is linearly independent. What about the list $\delta = \mathbf{u} + \mathbf{v}, \mathbf{v} + \mathbf{w}, \mathbf{w} + \mathbf{u}$?

P.1.35   Let $\mathcal{V}$ be an $\mathbb{F}$-vector space, let $\mathbf{w}_1, \mathbf{w}_2, \ldots, \mathbf{w}_r \in \mathcal{V}$, and suppose that at least one $\mathbf{w}_j$ is nonzero. Explain why $\operatorname{span}\{\mathbf{w}_1, \mathbf{w}_2, \ldots, \mathbf{w}_r\} = \operatorname{span}\{\mathbf{w}_i : i = 1, 2, \ldots, r \text{ and } \mathbf{w}_i \neq \mathbf{0}\}$.

P.1.36   Prove that $\mathcal{U} = \{p \in \mathcal{P}_3 : p(0) = 0\}$ is a subspace of $\mathcal{P}_3$ and show that $\mathcal{U} = \operatorname{span}\{z, z^2, z^3\}$.

P.1.37   Let $\beta = p_1, p_2, \ldots, p_k \in \mathcal{P}$ be a list of nonzero polynomials such that for each $i \neq j$ the degrees of p_i and p_j are different. Prove that β is linearly independent.

P.1.38   State the converse of Theorem 1.6.4. Is it true or false? Give a proof or a counter-example.

P.1.39   In M_n, let $\mathcal{U}$ denote the set of strictly lower triangular matrices and let $\mathcal{W}$ denote the set of strictly upper triangular matrices. (a) Show that $\mathcal{U}$ and $\mathcal{W}$ are subspaces of M_n. (b) What is $\mathcal{U} + \mathcal{W}$? Is this sum a direct sum? Why?

P.1.40 Let $\mathcal{U}$ denote the set of symmetric matrices in M_n and let $\mathcal{V}$ denote the set of skew-symmetric matrices in M_n. Show that (a) $\mathcal{U}$ and $\mathcal{V}$ are subspaces of M_n, and (b) $\mathsf{M}_n = \mathcal{U} \oplus \mathcal{V}$. Hint: $A = \frac{1}{2}(A + A^\mathsf{T}) + \frac{1}{2}(A - A^\mathsf{T})$.

P.1.41 Let $f \in C[-1, 1]$. Then f is *even* if $f(-x) = f(x)$ for all $x \in [-1, 1]$, and f is *odd* if $f(-x) = -f(x)$ for all $x \in [-1, 1]$. Let $\mathcal{U}$ denote the set of even functions in $C[-1, 1]$ and let $\mathcal{V}$ denote the set of odd functions in $C[-1, 1]$. Show that (a) $\mathcal{U}$ and $\mathcal{V}$ are subspaces of $C[-1, 1]$, and (b) $C[-1, 1] = \mathcal{U} \oplus \mathcal{V}$. Hint: $f(x) = \frac{1}{2}(f(x) + f(-x)) + \frac{1}{2}(f(x) - f(-x))$.

P.1.42 Which of the following are subspaces of $C[0, 1]$. Why? (a) $\{f \in C[0, 1] : \int_0^1 f(t)\,dt = 0\}$. (b) $\{f \in C[0, 1] : \int_0^1 f(t)\,dt = 1\}$. (c) $\{f \in C[0, 1] : \int_0^1 f^2(t)\,dt = 0\}$. (d) $\{f \in C[0, 1] : \int_0^1 \sin(\pi t)f(t)\,dt = 0\}$.

P.1.43 Let $\mathcal{V}$ denote the set of twice-differentiable functions in $C[0, 1]$ such that $f''(t) + e^{t^2}f'(t) - (\cos t)f(t) = 0$ for all $t \in [0, 1]$. Show that $\mathcal{V}$ is a subspace of $C[0, 1]$.

P.1.44 Let $\mathbf{v}_1, \mathbf{v}_2, \ldots, \mathbf{v}_r \in \mathbb{C}^n$ and let $A \in \mathsf{M}_n$ be invertible. Prove that $A\mathbf{v}_1, A\mathbf{v}_2, \ldots, A\mathbf{v}_r$ are linearly independent if and only if $\mathbf{v}_1, \mathbf{v}_2, \ldots, \mathbf{v}_r$ are linearly independent.

P.1.45 Suppose that $\mathbf{v}_1, \mathbf{v}_2, \ldots, \mathbf{v}_r \in \mathbb{C}^n$ are linearly independent and $A \in \mathsf{M}_n$ is not invertible. If $1 \leq r \leq n - 1$, show by example that $A\mathbf{v}_1, A\mathbf{v}_2, \ldots, A\mathbf{v}_r$ can be linearly independent.

P.1.46 Show that $\{t^k : k = 0, 1, 2, \ldots\} \subset C_\mathbb{R}[0, 1]$ is linearly independent.

P.1.47 In $\mathbb{R}^3$, a line through the origin is the span of a nonzero vector. If $\mathbf{n}_1 = [a\ b\ c]^\mathsf{T}$ and $\mathbf{n}_2 = [d\ e\ f]^\mathsf{T}$ are nonzero normal vectors for the respective planes $\mathcal{U}_1$ and $\mathcal{U}_2$ in Figure 1.3, show that the line ℓ (their intersection) is the span of the vector $\mathbf{v} = [bf - ce\quad cd - af\quad ae - bd]^\mathsf{T}$.

P.1.48 Suppose that $\mathbf{u} = [u_1\ u_2\ u_3]^\mathsf{T}$ and $\mathbf{v} = [v_1\ v_2\ v_3]^\mathsf{T}$ are linearly independent vectors in $\mathbb{R}^3$, and let $\mathbf{n} = [a\ b\ c]^\mathsf{T}$, in which $a = u_2v_3 - v_2u_3$, $b = v_1u_3 - u_1v_3$, and $c = u_1v_2 - v_1u_2$. Show that the plane $\operatorname{span}\{\mathbf{u}, \mathbf{v}\}$ in $\mathbb{R}^3$ is the same as the plane $\{[x_1\ x_2\ x_3]^\mathsf{T} \in \mathbb{R}^3 : ax_1 + bx_2 + cx_3 = 0\}$.

1.9 Notes

A *field* is a mathematical construct whose axioms capture the essential features of ordinary arithmetic operations with real or complex numbers. Examples of other fields include the rational numbers (ratios of integers), complex algebraic numbers (zeros of polynomials with integer coefficients), real rational functions, and the integers modulo a prime number. The only fields we deal with in this book are the real and complex numbers. For information about general fields, see [DF04].

Linear independence of a list of vectors is a fundamental concept in linear algebra. Independence of events and random variables is a fundamental concept in probability and statistics. Beware: these two notions of "independence" have nothing to do with each other.

1.10 Some Important Concepts

- Subspaces of a vector space.

- Column space, row space, and null space of a matrix.

- Linear combinations and span.
- Linear independence and linear dependence of a list of vectors.
- How to select a linearly independent list of vectors from a nonzero list.
- The pivot column decomposition of a matrix.

2   Bases and Similarity

Linearly independent lists of vectors that span a vector space are of special importance. They provide a bridge between the abstract world of vector spaces and the concrete world of matrices. They permit us to define the dimension of a vector space and motivate the concept of matrix similarity.

2.1 What Is a Basis?

Definition 2.1.1 Let V be an $\mathbb{F}$-vector space and let n be a positive integer. A list of vectors $\beta = \mathbf{v}_1, \mathbf{v}_2, \ldots, \mathbf{v}_n$ in V is a *basis* for V if span $\beta = V$ and β is linearly independent.

By definition, a basis is a finite list of vectors.

Theorem 1.7.1 ensures that the column space of any $A \in \mathsf{M}_{m \times n}(\mathbb{F})$ has a basis comprised of the pivot columns of A. The following example shows that every column of A may be needed to construct a basis for col A.

Example 2.1.2 Let V be the real vector space $\mathbb{R}^2$ and consider the list $\beta = \begin{bmatrix} 2 \\ 1 \end{bmatrix}, \begin{bmatrix} 1 \\ 1 \end{bmatrix}$. The span of β consists of all vectors of the form

$$x_1 \begin{bmatrix} 2 \\ 1 \end{bmatrix} + x_2 \begin{bmatrix} 1 \\ 1 \end{bmatrix} = \begin{bmatrix} 2 & 1 \\ 1 & 1 \end{bmatrix} \begin{bmatrix} x_1 \\ x_2 \end{bmatrix} = A\mathbf{x},$$

in which $A = \begin{bmatrix} 2 & 1 \\ 1 & 1 \end{bmatrix}$ and $\mathbf{x} = \begin{bmatrix} x_1 \\ x_2 \end{bmatrix} \in \mathbb{R}^2$. The columns of A are the vectors in the list β, which spans the column space of A. A calculation reveals that $A^{-1} = \begin{bmatrix} 1 & -1 \\ -1 & 2 \end{bmatrix}$. For any $\mathbf{y} = \begin{bmatrix} y_1 \\ y_2 \end{bmatrix} \in \mathbb{R}^2$, we have $\mathbf{y} = A(A^{-1}\mathbf{y})$, in which $A^{-1}\mathbf{y} = \begin{bmatrix} y_1 - y_2 \\ -y_1 + 2y_2 \end{bmatrix}$. This shows that span $\beta = \mathbb{R}^2$. If $A\mathbf{x} = \mathbf{0}$, then $\mathbf{x} = A^{-1}(A\mathbf{x}) = A^{-1}\mathbf{0} = \mathbf{0}$, so null $A = \{\mathbf{0}\}$ and β is linearly independent. Thus, β is a basis for $\mathbb{R}^2$.

The preceding example is a special case of the following theorem. For its converse, see Theorem 2.3.13.

Theorem 2.1.3 *If $A \in \mathsf{M}_n(\mathbb{F})$ is invertible, then the list of its columns is a basis for $\mathbb{F}^n$.*

Proof The columns of any invertible matrix in $\mathsf{M}_n(\mathbb{F})$ span $\mathbb{F}^n$ and are linearly independent; see Examples 1.4.19 and 1.6.14.   □

Let β be a basis for an $\mathbb{F}$-vector space V. Each vector in V is a linear combination of the elements of β since span $\beta = V$. The coefficients in that linear combination are uniquely determined since β is linearly independent; see Theorem 1.6.18. However, a vector space can have many different bases. Our next task is to investigate how they are related.

Lemma 2.1.4 (Replacement Lemma) *Let V be a nonzero $\mathbb{F}$-vector space and let r be a positive integer. Suppose that $\beta = \mathbf{u}_1, \mathbf{u}_2, \ldots, \mathbf{u}_r$ spans V. Let $\mathbf{v} \in V$ be nonzero and let*

$$\mathbf{v} = \sum_{i=1}^{r} c_i \mathbf{u}_i. \tag{2.1.5}$$

(a) $c_j \neq 0$ *for some $j \in \{1, 2, \ldots, r\}$.*

(b) *If $c_j \neq 0$, then the list*

$$\mathbf{v}, \mathbf{u}_1, \mathbf{u}_2 \ldots, \widehat{\mathbf{u}}_j, \ldots, \mathbf{u}_r \tag{2.1.6}$$

(the element $\mathbf{u}_j$ is omitted) spans V.

(c) *If β is a basis for V and $c_j \neq 0$, then (2.1.6) is a basis for V.*

(d) *Suppose that β is a basis for V and $r \geq 2$. If $\mathbf{v} \notin \operatorname{span}\{\mathbf{u}_1, \mathbf{u}_2, \ldots, \mathbf{u}_k\}$ for some $k \in \{1, 2, \ldots, r - 1\}$, then*

$$\mathbf{v}, \mathbf{u}_1, \ldots, \mathbf{u}_k, \mathbf{u}_{k+1}, \ldots, \widehat{\mathbf{u}}_j, \ldots, \mathbf{u}_r \tag{2.1.7}$$

is a basis for V for some $j \in \{k + 1, k + 2, \ldots, r\}$.

Proof (a) If each $c_i = 0$, then $\mathbf{v} = \sum_{i=1}^{r} c_i \mathbf{u}_i = \mathbf{0}$, which is a contradiction.

(b) The list $\mathbf{v}, \mathbf{u}_1, \mathbf{u}_2, \ldots, \mathbf{u}_r$ is linearly dependent because $\mathbf{v}$ is a linear combination of $\mathbf{u}_1, \mathbf{u}_2, \ldots, \mathbf{u}_r$. Therefore, the assertion follows from Theorem 1.6.26.b.

(c) We must show that the list (2.1.6) is linearly independent. Suppose that $c\mathbf{v} + \sum_{i \neq j} b_i \mathbf{u}_i = \mathbf{0}$. If $c \neq 0$, then

$$\mathbf{v} = -c^{-1} \sum_{i \neq j} b_i \mathbf{u}_i, \tag{2.1.8}$$

which is different from the representation (2.1.5), in which $c_j \neq 0$. Having two different representations of $\mathbf{v}$ as a linear combination of elements of β contradicts Theorem 1.6.18, so $c = 0$. Consequently, $\sum_{i \neq j} b_i \mathbf{u}_i = \mathbf{0}$ and the linear independence of β ensures that each $b_i = 0$. Thus, the list (2.1.6) is linearly independent.

(d) Because $\mathbf{v} \notin \operatorname{span}\{\mathbf{u}_1, \mathbf{u}_2, \ldots, \mathbf{u}_k\}$, we have $c_j \neq 0$ in the representation (2.1.5) for some $j \in \{k + 1, k + 2, \ldots, r\}$. The assertion now follows from (c). □

Example 2.1.9 The list $\beta = 1, z, z^2$ is a basis for $\mathcal{P}_2$. Then $p(z) = z^2 - z$ is a linear combination of the form (2.1.5) in which $c_1 = 0$, $c_2 = -1$, and $c_3 = 1$. Let us verify the assertion in Lemma 2.1.4 that we may replace either z or z^2 with $p(z)$ since c_2 and c_3 are nonzero. The list $1, z, z^2 - z$ is a basis for $\mathcal{P}_2$ (since $z^2 = (z^2 - z) + z$) and $1, z^2 - z, z^2$ is also a basis for $\mathcal{P}_2$ (since $z = z^2 - (z^2 - z)$). But $z^2 - z, z, z^2$ is not a basis for $\mathcal{P}_2$ because its span does not contain 1. This does not contradict Lemma 2.1.4 since $c_1 = 0$.

The next theorem shows that the number of elements in a basis for V is an upper bound for the number of elements in any linearly independent list of vectors in V.

Theorem 2.1.10 *Let V be an $\mathbb{F}$-vector space and let r and n be positive integers. Suppose that $\beta = \mathbf{u}_1, \mathbf{u}_2, \ldots, \mathbf{u}_n$ is a basis for V and $\gamma = \mathbf{v}_1, \mathbf{v}_2, \ldots, \mathbf{v}_r$ is linearly independent.*

(a) $r \leq n$.

(b) *If $r = n$, then γ is a basis for V.*

Proof Suppose that $r \geq n$. For each $k = 1, 2, \ldots, n$, we claim that there are indices $i_1, i_2, \ldots, i_{n-k} \in \{1, 2, \ldots, n\}$ such that $\gamma_k = \mathbf{v}_k, \mathbf{v}_{k-1}, \ldots, \mathbf{v}_1, \mathbf{u}_{i_1}, \mathbf{u}_{i_2}, \ldots, \mathbf{u}_{i_{n-k}}$ is a basis for $\mathcal{V}$. We proceed by induction. The base case $k = 1$ follows from (a) and (c) of Lemma 2.1.4. The induction step follows from Lemma 2.1.4.d because the linear independence of γ ensures that $\mathbf{v}_{k+1} \notin \mathrm{span}\{\mathbf{v}_1, \mathbf{v}_2, \ldots, \mathbf{v}_k\}$. The case $k = n$ implies that $\gamma_n = \mathbf{v}_n, \mathbf{v}_{n-1}, \ldots, \mathbf{v}_1$ is a basis for $\mathcal{V}$. If $r = n$, this proves (b). If $r > n$, then $\mathbf{v}_{n+1} \in \mathrm{span}\{\mathbf{v}_n, \mathbf{v}_{n-1}, \ldots, \mathbf{v}_1\}$, which contradicts the linear independence of γ. We conclude that $r > n$ is impossible, which proves (a).   □

The following corollary is of fundamental importance, and it permits us to define the dimension of a vector space in the next section.

Corollary 2.1.11 *Let m and n be positive integers. If $\mathbf{v}_1, \mathbf{v}_2, \ldots, \mathbf{v}_m$ and $\mathbf{w}_1, \mathbf{w}_2, \ldots, \mathbf{w}_n$ are bases of an $\mathbb{F}$-vector space $\mathcal{V}$, then $m = n$.*

Proof The preceding theorem ensures that $m \leq n$ and $n \leq m$.   □

2.2 Dimension

Definition 2.2.1 Let $\mathcal{V}$ be an $\mathbb{F}$-vector space and let n be a positive integer. If there is a list $\mathbf{v}_1, \mathbf{v}_2, \ldots, \mathbf{v}_n$ of vectors that is a basis for $\mathcal{V}$, then $\mathcal{V}$ is *n-dimensional* ($\mathcal{V}$ has *dimension n*). The zero vector space has *dimension zero*. If $\mathcal{V}$ has dimension n for some nonnegative integer n, then $\mathcal{V}$ is *finite dimensional*; otherwise, $\mathcal{V}$ is *infinite dimensional*. If $\mathcal{V}$ is finite dimensional, its dimension is denoted by $\dim \mathcal{V}$.

Example 2.2.2 The list $\beta = \mathbf{e}_1, \mathbf{e}_2, \ldots, \mathbf{e}_n$ of vectors in (1.4.21) is linearly independent and its span is $\mathbb{F}^n$. The basis β is the *standard basis* for $\mathbb{F}^n$. There are n vectors in β, so $\dim \mathbb{F}^n = n$.

Example 2.2.3 In the $\mathbb{F}$-vector space $\mathsf{M}_{m \times n}(\mathbb{F})$, consider the matrices E_{pq} for $1 \leq p \leq m$ and $1 \leq q \leq n$, defined as follows: the (i, j) entry of E_{pq} is 1 if $(i, j) = (p, q)$; it is 0 otherwise. The mn matrices E_{pq} (arranged in a list in any desired order) are a basis, so $\dim \mathsf{M}_{m \times n}(\mathbb{F}) = mn$. For example, if $m = 2$ and $n = 3$, then $\dim \mathsf{M}_{2 \times 3}(\mathbb{F}) = 6$ and

$$E_{11} = \begin{bmatrix} 1 & 0 & 0 \\ 0 & 0 & 0 \end{bmatrix}, \qquad E_{12} = \begin{bmatrix} 0 & 1 & 0 \\ 0 & 0 & 0 \end{bmatrix}, \qquad E_{13} = \begin{bmatrix} 0 & 0 & 1 \\ 0 & 0 & 0 \end{bmatrix},$$

$$E_{21} = \begin{bmatrix} 0 & 0 & 0 \\ 1 & 0 & 0 \end{bmatrix}, \qquad E_{22} = \begin{bmatrix} 0 & 0 & 0 \\ 0 & 1 & 0 \end{bmatrix}, \qquad E_{23} = \begin{bmatrix} 0 & 0 & 0 \\ 0 & 0 & 1 \end{bmatrix}.$$

Example 2.2.4 Let $\mathcal{V}$ and $\mathcal{W}$ be $\mathbb{F}$-vector spaces with respective bases $\beta = \mathbf{v}_1, \mathbf{v}_2, \ldots, \mathbf{v}_m$ and $\gamma = \mathbf{w}_1, \mathbf{w}_2, \ldots, \mathbf{w}_n$. In Example 1.4.24, we found that the list $\alpha = (\mathbf{v}_1, \mathbf{0}), (\mathbf{v}_2, \mathbf{0}), \ldots, (\mathbf{v}_m, \mathbf{0}),$ $(\mathbf{0}, \mathbf{w}_1), (\mathbf{0}, \mathbf{w}_2), \ldots, (\mathbf{0}, \mathbf{w}_n)$ spans the Cartesian product $\mathcal{V} \times \mathcal{W}$, and in Example 1.6.16 we found that α is linearly independent. Therefore, α is a basis for $\mathcal{V} \times \mathcal{W}$, which has dimension $m + n$.

Example 2.2.5 In Section 2.7, we show how to construct many different bases for $\mathcal{P}_n$, which has dimension $n + 1$. A different basis is associated with each set of $n + 1$ distinct points in $\mathbb{C}$.

Example 2.2.6 In Example 1.6.12 we saw that $1, z, z^2, \ldots, z^n$ are linearly independent in $\mathcal{P}$ for each $n = 1, 2, \ldots$. Theorem 2.1.10 says that if $\mathcal{P}$ is finite dimensional, then $\dim \mathcal{P} \geq n$ for each $n = 1, 2, \ldots$. Since this is impossible, $\mathcal{P}$ is infinite dimensional.

Example 2.2.7 In the vector space $\mathcal{V}$ of finitely nonzero sequences (see Example 1.2.9), consider the vectors $\mathbf{v}_k$ that have a 1 in position k and zeros elsewhere. For each $n = 1, 2, \ldots$, the vectors $\mathbf{v}_1, \mathbf{v}_2, \ldots, \mathbf{v}_n$ are linearly independent, so $\mathcal{V}$ is infinite dimensional.

The following theorem says two things about a nonzero $\mathbb{F}$-vector space $\mathcal{V}$: (a) any finite set that spans $\mathcal{V}$ contains the elements of a basis; and (b) if $\mathcal{V}$ is finite dimensional, then any linearly independent list of vectors can be extended to a basis.

Theorem 2.2.8 *Let $\mathcal{V}$ be a nonzero $\mathbb{F}$-vector space, let r be a positive integer, and let $\mathbf{v}_1, \mathbf{v}_2, \ldots, \mathbf{v}_r \in \mathcal{V}$.*

(a) *If $\operatorname{span}\{\mathbf{v}_1, \mathbf{v}_2, \ldots, \mathbf{v}_r\} = \mathcal{V}$, then $\mathcal{V}$ is finite dimensional, $n = \dim \mathcal{V} \leq r$, and there are indices $i_1, i_2, \ldots, i_n \in \{1, 2, \ldots, r\}$ such that the list $\mathbf{v}_{i_1}, \mathbf{v}_{i_2}, \ldots, \mathbf{v}_{i_n}$ is a basis for $\mathcal{V}$.*

(b) *Suppose that $\mathcal{V}$ is finite dimensional and $\dim \mathcal{V} = n > r$. If $\mathbf{v}_1, \mathbf{v}_2, \ldots, \mathbf{v}_r$ are linearly independent, then there are $n - r$ vectors $\mathbf{w}_1, \mathbf{w}_2, \ldots, \mathbf{w}_{n-r} \in \mathcal{V}$ such that $\mathbf{v}_1, \mathbf{v}_2, \ldots, \mathbf{v}_r, \mathbf{w}_1, \mathbf{w}_2, \ldots, \mathbf{w}_{n-r}$ is a basis for $\mathcal{V}$.*

Proof (a) We proved this (and more) in Theorem 1.7.1.

(b) The hypothesis is that $\mathbf{v}_1, \mathbf{v}_2, \ldots, \mathbf{v}_r \in \mathcal{V}$ are linearly independent and $r < n = \dim \mathcal{V}$. Since $\mathbf{v}_1, \mathbf{v}_2, \ldots, \mathbf{v}_r$ do not span $\mathcal{V}$ (Corollary 2.1.11), Theorem 1.6.26.a ensures there is a $\mathbf{v} \in \mathcal{V}$ such that $\mathbf{v}_1, \mathbf{v}_2, \ldots, \mathbf{v}_r, \mathbf{v}$ is linearly independent. If this longer list spans $\mathcal{V}$, it is a basis. If not, invoke Theorem 1.6.26.a again. Theorem 2.1.10 ensures that this process terminates in at most $n - r$ steps. $\square$

A basis is simultaneously a maximal linearly independent list and a minimal spanning list.

Corollary 2.2.9 *Let n be a positive integer, let $\mathcal{V}$ be an n-dimensional $\mathbb{F}$-vector space, and let $\beta = \mathbf{v}_1, \mathbf{v}_2, \ldots, \mathbf{v}_n$ be a list of vectors in $\mathcal{V}$.*

(a) *If β spans $\mathcal{V}$, then it is a basis.*

(b) *If β is linearly independent, then it is a basis.*

Proof (a) If β is not linearly independent, then the preceding theorem ensures that a strictly shorter list is a basis. This contradicts the assumption that $\dim \mathcal{V} = n$.

(b) If β does not span $\mathcal{V}$, then the preceding theorem ensures that a strictly longer list is a basis, which contradicts the assumption that $\dim \mathcal{V} = n$. $\square$

Example 2.2.10 Let $A \in \mathsf{M}_n(\mathbb{F})$ and let β be the list of columns of A (vectors in $\mathbb{F}^n$). The two parts of Corollary 2.2.9 ensure that $\operatorname{col} A = \mathbb{F}^n$ if and only if $\operatorname{null} A = \{\mathbf{0}\}$. That is, the linear system $A\mathbf{x} = \mathbf{b}$ has a solution for each $\mathbf{b} \in \mathbb{F}^n$ if and only if the only solution to $A\mathbf{x} = \mathbf{0}$ is $\mathbf{x} = \mathbf{0}$.

Theorem 2.2.11 *Let $\mathcal{U}$ be a subspace of an n-dimensional $\mathbb{F}$-vector space $\mathcal{V}$. Then $\mathcal{U}$ is finite dimensional and $\dim \mathcal{U} \leq n$, with equality if and only if $\mathcal{U} = \mathcal{V}$.*

Proof If $\mathcal{U} = \{\mathbf{0}\}$, then $\dim \mathcal{U} = 0$ and there is nothing to prove, so we may assume that $\mathcal{U} \neq \{\mathbf{0}\}$. Let $\mathbf{v}_1 \in \mathcal{U}$ be nonzero. If $\mathrm{span}\{\mathbf{v}_1\} = \mathcal{U}$, then $\dim \mathcal{U} = 1$. If $\mathrm{span}\{\mathbf{v}_1\} \neq \mathcal{U}$, Theorem 1.6.26.a ensures that there is a $\mathbf{v}_2 \in \mathcal{U}$ such that the list $\mathbf{v}_1, \mathbf{v}_2$ is linearly independent. If $\mathrm{span}\{\mathbf{v}_1, \mathbf{v}_2\} = \mathcal{U}$, then $\dim \mathcal{U} = 2$; if $\mathrm{span}\{\mathbf{v}_1, \mathbf{v}_2\} \neq \mathcal{U}$, Theorem 1.6.26.a ensures that there is a $\mathbf{v}_3 \in \mathcal{U}$ such that the list $\mathbf{v}_1, \mathbf{v}_2, \mathbf{v}_3$ is linearly independent. Repeat until a linearly independent spanning list is obtained. Since no linearly independent list of vectors in $\mathcal{V}$ contains more than n elements (Theorem 2.1.10), this process terminates in $r \leq n$ steps with a linearly independent list of vectors $\mathbf{v}_1, \mathbf{v}_2, \ldots, \mathbf{v}_r$ whose span is $\mathcal{U}$. Thus, $r = \dim \mathcal{U} \leq n$ with equality only if $\mathbf{v}_1, \mathbf{v}_2, \ldots, \mathbf{v}_n$ is a basis for $\mathcal{V}$ (Theorem 2.1.10 again), in which case $\mathcal{U} = \mathcal{V}$. $\qquad\square$

The preceding theorem ensures that any pair of subspaces $\mathcal{U}$ and $\mathcal{W}$ of a finite-dimensional vector space $\mathcal{V}$ has a finite-dimensional sum $\mathcal{U} + \mathcal{W}$ and a finite-dimensional intersection $\mathcal{U} \cap \mathcal{W}$, since each is a subspace of $\mathcal{V}$.

An initial application of Theorem 2.2.11 yields a result about left and right inverses of a square matrix. A matrix $B \in \mathsf{M}_n(\mathbb{F})$ is a *left inverse* (respectively, *right inverse*) of $A \in \mathsf{M}_n(\mathbb{F})$ if $BA = I$ (respectively, $AB = I$). A square matrix need not have a left inverse, but if it does, then that left inverse is also a right inverse. If a square matrix has a right inverse, then that right inverse is also a left inverse. We now show how these remarkable facts (Theorem 2.2.13) follow from the finite dimensionality of $\mathsf{M}_n(\mathbb{F})$.

Lemma 2.2.12 *If $A \in \mathsf{M}_n(\mathbb{F})$, then there is an integer $k \geq 0$ and a polynomial q such that $q(0) = 1$ and $A^k q(A) = 0$.*

Proof Since $\dim \mathsf{M}_n(\mathbb{F}) = n^2$ (Example 2.2.3), the list $I, A, A^2, \ldots, A^{n^2}$ of $n^2 + 1$ matrices is linearly dependent. Thus, there are scalars $c_0, c_1, c_2, \ldots, c_{n^2}$, not all zero, such that $c_0 I + c_1 A + c_2 A^2 + \cdots + c_{n^2} A^{n^2} = 0$. Let $p(z) = c_0 + c_1 z + c_2 z^2 + \cdots + c_{n^2} z^{n^2}$, so $p(A) = 0$ (see Appendix B.5). Let $k = \min\{j : c_j \neq 0\}$ and let $q(z) = p(z)/(c_k z^k)$. Then $q(z) = 1 + \frac{c_{k+1}}{c_k} z + \cdots$, so $q(0) = 1$ and $A^k q(A) = p(A)/c_k = 0$. $\qquad\square$

Theorem 2.2.13 (Left and Right Inverses) *Let $A, B \in \mathsf{M}_n(\mathbb{F})$. Then $AB = I$ if and only if $BA = I$.*

Proof Suppose that $AB = I$. We claim that $A^k B^k = I$ for each $k = 0, 1, 2 \ldots$. Proceed by induction. In the base case $k = 0$, we have $A^0 B^0 = II = I$ (see (C.2.8) for the convention $A^0 = I$). Suppose that $k \geq 1$ and $A^{k-1} B^{k-1} = I$. Then $I = AB = A(I)B = A(A^{k-1} B^{k-1})B = A^k B^k$, which completes the induction.

The preceding lemma ensures that there is a nonnegative integer k and a polynomial q such that $q(0) = 1$ and $B^k q(B) = 0$. Then $0 = B^k q(B) = A^k B^k q(B) = I q(B) = q(B)$. Since $q(0) = 1$ and $q(B) = 0$, the degree of q is at least 1. Write $q(z) = c_m z^m + c_{m-1} z^{m-1} + \cdots + c_1 z + 1$, in which $m \geq 1$ and $c_m \neq 0$. Since $AB = I$,

$$
\begin{aligned}
0 &= A q(B) \\
&= A(c_m B^m + c_{m-1} B^{m-1} + \cdots + c_1 B + I) \\
&= c_m A B^m + c_{m-1} A B^{m-1} + \cdots + c_1 AB + A \\
&= c_m (AB) B^{m-1} + c_{m-1}(AB) B^{m-2} + \cdots + c_1 (AB) + A \\
&= c_m B^{m-1} + c_{m-1} B^{m-2} + \cdots + c_1 I + A,
\end{aligned}
$$

which shows that $A = -c_m B^{m-1} - c_{m-1} B^{m-2} - \cdots - c_1 I$. Therefore, A commutes with B, and $I = AB = BA$.

If $BA = I$, interchange A and B in the proof. $\qquad\square$

Finite dimensionality and squareness are essential assumptions in the preceding theorem; see P.2.36 and P.2.37. The uniqueness of the inverse of a square matrix is addressed in the following corollary. See P.2.38 for the non-square non-unique case.

Corollary 2.2.14 (Uniqueness of Inverse) *Let $A, B, C \in \mathsf{M}_n(\mathbb{F})$. If $AB = I$ or $BA = I$ and either $AC = I$ or $CA = I$, then $B = C$.*

Proof Suppose that $AC = I$ and $BA = I$. The preceding theorem ensures that $CA = I$ and $AB = I$. Then $B = IB = (CA)B = C(AB) = CI = C$. Similar arguments prove the remaining three cases. $\qquad\square$

2.3 Full-Rank Factorizations

The column space and row space of a matrix are finite-dimensional subspaces that play important roles in matrix mathematics. In this section, we discover how their dimensions are related.

Theorem 2.3.1 *Let $A \in \mathsf{M}_{m\times n}(\mathbb{F})$. Then*

$$\dim \operatorname{col} A = \dim \operatorname{col} A^{\mathsf{T}} = \dim \operatorname{row} A \leq \min\{m, n\}. \tag{2.3.2}$$

Proof Since the columns of A^{T} are the transposes of the rows of A, it suffices to prove the first equality in (2.3.2). If $A = 0$, then its row and column spaces are zero dimensional. If A is nonzero, then Theorem 1.7.5 ensures that there is an $s \in \{1, 2, \ldots, n\}$ and matrices $P \in \mathsf{M}_{m\times s}(\mathbb{F})$ and $R \in \mathsf{M}_{s\times n}(\mathbb{F})$ such that $A = PR$, $\operatorname{col} A = \operatorname{col} P$, and P has linearly independent columns.

The s columns of P span the column space of A and are linearly independent, so they are a basis for $\operatorname{col} A$. Thus, $\dim \operatorname{col} A = s$. We have $A^{\mathsf{T}} = R^{\mathsf{T}} P^{\mathsf{T}}$, so $\operatorname{col} A^{\mathsf{T}} \subseteq \operatorname{col} R^{\mathsf{T}}$. The s columns of R^{T} span the column space of R^{T}, so $\dim \operatorname{col} R^{\mathsf{T}} \leq s = \dim \operatorname{col} A$ (Theorem 2.2.8.a). Theorem 2.2.11 ensures that $\dim \operatorname{col} A^{\mathsf{T}} \leq \dim \operatorname{col} R^{\mathsf{T}} \leq \dim \operatorname{col} A$, and hence

$$\dim \operatorname{col} A^{\mathsf{T}} \leq \dim \operatorname{col} A. \tag{2.3.3}$$

Apply (2.3.3) to A^{T} and find that

$$\dim \operatorname{col} A = \dim \operatorname{col}(A^{\mathsf{T}})^{\mathsf{T}} \leq \dim \operatorname{col} A^{\mathsf{T}}. \tag{2.3.4}$$

Combine (2.3.3) and (2.3.4) and obtain $\dim \operatorname{col} A \leq \dim \operatorname{col} A^{\mathsf{T}} \leq \dim \operatorname{col} A$, so $\dim \operatorname{col} A = \dim \operatorname{col} A^{\mathsf{T}}$. Finally, $\operatorname{col} A \subseteq \mathbb{F}^m$, $\dim \mathbb{F}^m = m$, $\operatorname{col} A^{\mathsf{T}} \subseteq \mathbb{F}^n$, and $\dim \mathbb{F}^n = n$, so Theorem 2.2.11 ensures that $\dim \operatorname{col} A \leq \min\{m, n\}$. $\qquad\square$

Definition 2.3.5 Let $A \in \mathsf{M}_{m\times n}(\mathbb{F})$. The *row rank* of A is the dimension of $\operatorname{row} A$, the *column rank* of A is the dimension of $\operatorname{col} A$, and the *rank* of A is the common value of its row rank and column rank. We say that A has *full row rank* if $\operatorname{rank} A = m$; it has *full column rank* if $\operatorname{rank} A = n$; it has *full rank* if $\operatorname{rank} A = \min\{m, n\}$.

Example 2.3.6 Consider

$$A = \begin{bmatrix} 1 & 1 \\ 0 & 0 \\ 0 & 1 \end{bmatrix}, \qquad B = \begin{bmatrix} 2 & 0 & 1 \\ 1 & 1 & 0 \end{bmatrix}, \quad \text{and} \quad C = \begin{bmatrix} 1 & 1 & 1 \\ 1 & 1 & 1 \end{bmatrix}.$$

Then A and B have full rank but C does not. In each case, inspection verifies that row rank equals column rank.

With the help of Theorem 2.3.1 and the vocabulary of Definition 2.3.5, we can restate and strengthen Theorem 1.7.5 to obtain a basic matrix factorization.

Theorem 2.3.7 (Full-Rank Factorization) *Let $A \in M_{m \times n}(\mathbb{F})$ be nonzero, let $r = \operatorname{rank} A$, and let the columns of $X \in M_{m \times r}(\mathbb{F})$ be a basis for* col A. *Then there is a unique $Y \in M_{r \times n}(\mathbb{F})$ such that $A = XY$. Moreover,* rank $Y = r$, *the rows of Y are a basis for* row A, *and* null $A =$ null Y.

Proof Since the columns of X span the column space of A, Theorem 1.4.10 ensures that there is some $Y \in M_{r \times n}(\mathbb{F})$ such that $A = XY$. The columns of X are linearly independent, so Theorem 1.6.21 tells us that Y is unique. Since $A^\mathsf{T} = Y^\mathsf{T} X^\mathsf{T}$, we have col $A^\mathsf{T} \subseteq$ col Y^T. Theorems 2.3.1 and 2.2.11 ensure that $r = \operatorname{rank} A^\mathsf{T} \leq \operatorname{rank} Y^\mathsf{T} \leq r$. We conclude that rank $Y^\mathsf{T} = r$, so rank $Y = r$ and the columns of Y^T are a basis for col A^T, that is, the rows of Y are a basis for row A. To show that null $A =$ null Y, observe that the columns of X are linearly independent, so $A\mathbf{u} = X(Y\mathbf{u}) = \mathbf{0}$ if and only if $Y\mathbf{u} = \mathbf{0}$. $\square$

Definition 2.3.8 The factorization $A = XY$ described in the preceding theorem is a *full-rank factorization* of A.

Theorem 1.7.1 provides one choice for the factor X whose columns are a basis for col A, but there are many others; see P.1.13.

Example 2.3.9 A full-rank factorization is not unique. For example, P.1.13 provides four different full-rank factorizations of the matrix (1.8.1). Here are two more:

$$\begin{bmatrix} 5 & 8 \\ 23 & 38 \\ 41 & 68 \end{bmatrix} \begin{bmatrix} 1 & 6 & 11 \\ -\frac{1}{2} & -\frac{7}{2} & -\frac{13}{2} \end{bmatrix} \quad \text{and} \quad \begin{bmatrix} 14 & -19 \\ 32 & -43 \\ 50 & -67 \end{bmatrix} \begin{bmatrix} \frac{11}{2} & \frac{3}{2} & -\frac{5}{2} \\ 4 & 1 & -2 \end{bmatrix}.$$

Example 2.3.10 Suppose that a nonzero $A \in M_n$ is idempotent (that is, $A^2 = A$) and rank $A = r$. If $A = XY$ is a full-rank factorization, then X and Y^T are $n \times r$, each has full rank, and $XYXY = A^2 = A = XY$. Write this identity as $X(YX - I_r)Y = 0$. Since X has linearly independent columns, it follows that $(YX - I_r)Y = 0$. Since Y has linearly independent rows, it follows that $YX - I_r = 0$ and hence $YX = I_r$. Then

$$\operatorname{tr} A = \operatorname{tr} XY = \operatorname{tr} YX = \operatorname{tr} I_r = r = \operatorname{rank} A. \tag{2.3.11}$$

If $A = 0$, which is idempotent, then we also have tr $A =$ rank A. Thus, the rank of any idempotent matrix equals its trace. If A is idempotent and tr $A = n$, then rank $A = n$, so A is invertible and $A = A^{-1}A^2 = A^{-1}A = I$.

Example 2.3.12 The preceding example leads to another proof of Theorem 2.2.13. Suppose that $A, B \in M_n$ and $AB = I$. Then $(BA)^2 = B(AB)A = BIA = BA$, so BA is idempotent. Since tr $BA =$ tr $AB =$ tr $I = n$, Example 2.3.10 ensures that $BA = I$.

We use full-rank factorizations to prove rank inequalities in Chapter 3 and to compute eigenvalues of low-rank matrices in Chapter 10.

The final theorem in this section provides the converse of Theorem 2.1.3.

Theorem 2.3.13 *Let* $\mathbf{a}_1, \mathbf{a}_2, \ldots, \mathbf{a}_n \in \mathbb{F}^n$, *let* $\beta = \mathbf{a}_1, \mathbf{a}_2, \ldots, \mathbf{a}_n$, *and let* $A = [\mathbf{a}_1 \ \mathbf{a}_2 \ \ldots \ \mathbf{a}_n]$. *The following are equivalent.*

(a) β *is a basis for* $\mathbb{F}^n$.

(b) A *is invertible.*

(c) $\operatorname{rank} A = n$.

Proof (a) $\Rightarrow$ (b) Since $\operatorname{span} \beta = \mathbb{F}^n$, for each $j = 1, 2, \ldots, n$ there is a $\mathbf{b}_j \in \mathbb{F}^n$ such that $A\mathbf{b}_j = \mathbf{e}_j$. Let $B = [\mathbf{b}_1 \ \mathbf{b}_2 \ \ldots \ \mathbf{b}_n]$. Then

$$AB = A[\mathbf{b}_1 \ \mathbf{b}_2 \ \ldots \ \mathbf{b}_n] = [A\mathbf{b}_1 \ A\mathbf{b}_2 \ \ldots \ A\mathbf{b}_n] = [\mathbf{e}_1 \ \mathbf{e}_2 \ \ldots \ \mathbf{e}_n] = I.$$

Theorem 2.2.13 ensures that $BA = I$, so $B = A^{-1}$.

(b) $\Rightarrow$ (c) Theorem 2.1.3 ensures that β is a basis for $\mathbb{F}^n$. Since β is a basis for $\operatorname{col} A$, we conclude that $\dim \operatorname{col} A = n$.

(c) $\Rightarrow$ (a) Since $\operatorname{col} A$ is a subspace of $\mathbb{F}^n$ and $\dim \operatorname{col} A = \dim \mathbb{F}^n = n$, Theorem 2.2.11 implies that $\operatorname{col} A = \mathbb{F}^n$. Therefore, β spans $\mathbb{F}^n$ and Corollary 2.2.9.a ensures that it is a basis for $\mathbb{F}^n$. □

2.4 Coordinate Vectors and Matrix Representations of Linear Transformations

Definition 2.4.1 Let $\beta = \mathbf{v}_1, \mathbf{v}_2, \ldots, \mathbf{v}_n$ be a basis for a finite-dimensional $\mathbb{F}$-vector space $\mathcal{V}$. Write any vector $\mathbf{u} \in \mathcal{V}$ as a linear combination

$$\mathbf{u} = c_1 \mathbf{v}_1 + c_2 \mathbf{v}_2 + \cdots + c_n \mathbf{v}_n, \tag{2.4.2}$$

in which the coefficients are uniquely determined (Theorem 1.6.18). The β-*coordinate vector* of $\mathbf{u}$ is

$$[\mathbf{u}]_\beta = \begin{bmatrix} c_1 \\ c_2 \\ \vdots \\ c_n \end{bmatrix}. \tag{2.4.3}$$

The scalars $c_1, c_2, \ldots, c_n$ in (2.4.2) are the *coordinates of* $\mathbf{u}$ with respect to the basis β. The function $\mathbf{u} \mapsto [\mathbf{u}]_\beta$ from $\mathcal{V}$ to $\mathbb{F}^n$ is the β-*basis representation function.*

Example 2.4.4 Let $\mathcal{V} = \mathcal{P}_2$ and consider the basis $\beta = f_1, f_2, f_3$, in which $f_1 = 1, f_2 = 2z - 1$, and $f_3 = 6z^2 - 6z + 1$. A calculation reveals that $1 = f_1, z = \frac{1}{2}f_1 + \frac{1}{2}f_2$, and $z^2 = \frac{1}{3}f_1 + \frac{1}{2}f_2 + \frac{1}{6}f_3$, so the β-coordinate vector of $p(z) = c_0 1 + c_1 z + c_2 z^2$ is

$$[p]_\beta = \frac{1}{6} \begin{bmatrix} 6c_0 + 3c_1 + 2c_2 \\ 3c_1 + 3c_2 \\ c_2 \end{bmatrix}.$$

Example 2.4.5 Let $V = \mathbb{R}^2$ and consider the basis $\beta = \mathbf{v}_1, \mathbf{v}_2$, in which $\mathbf{v}_1 = \begin{bmatrix} 2 \\ 1 \end{bmatrix}$ and $\mathbf{v}_2 = \begin{bmatrix} 1 \\ 1 \end{bmatrix}$. Then $\mathbf{e}_1 = \mathbf{v}_1 - \mathbf{v}_2$ and $\mathbf{e}_2 = -\mathbf{v}_1 + 2\mathbf{v}_2$. If $\mathbf{y} = y_1\mathbf{e}_1 + y_2\mathbf{e}_2$, then

$$\mathbf{y} = y_1(\mathbf{v}_1 - \mathbf{v}_2) + y_2(-\mathbf{v}_1 + 2\mathbf{v}_2) = (y_1 - y_2)\mathbf{v}_1 + (-y_1 + 2y_2)\mathbf{v}_2.$$

Thus, the β-coordinate vector of $\mathbf{y}$ is

$$[\mathbf{y}]_\beta = \begin{bmatrix} y_1 - y_2 \\ -y_1 + 2y_2 \end{bmatrix} = \begin{bmatrix} 1 & -1 \\ -1 & 2 \end{bmatrix}\begin{bmatrix} y_1 \\ y_2 \end{bmatrix} = V^{-1}\mathbf{y},$$

in which $V = \begin{bmatrix} 2 & 1 \\ 1 & 1 \end{bmatrix} = [\mathbf{v}_1 \ \mathbf{v}_2]$.

The approach in the preceding example can be employed to obtain the coordinate vector of any vector in $\mathbb{F}^n$ with respect to any basis.

Theorem 2.4.6 *Let $\beta = \mathbf{v}_1, \mathbf{v}_2, \ldots, \mathbf{v}_n$ be a basis for $\mathbb{F}^n$ and let $V = [\mathbf{v}_1 \ \mathbf{v}_2 \ \cdots \ \mathbf{v}_n] \in \mathsf{M}_n(\mathbb{F})$. For any $\mathbf{x} \in \mathbb{F}^n$, its β-coordinate vector is $[\mathbf{x}]_\beta = V^{-1}\mathbf{x}$.*

Proof If $\mathbf{x} = x_1\mathbf{v}_1 + x_2\mathbf{v}_2 + \cdots + x_n\mathbf{v}_n$, then $[\mathbf{x}]_\beta = [x_1 \ x_2 \ \cdots \ x_n]^\mathsf{T}$ and $V[\mathbf{x}]_\beta = \mathbf{x}$. Thus, $[\mathbf{x}]_\beta = V^{-1}\mathbf{x}$. $\qquad\qquad\square$

The β-basis representation function (2.4.3) provides a one-to-one correspondence between vectors in the n-dimensional $\mathbb{F}$-vector space V and vectors in $\mathbb{F}^n$. Theorem 1.6.18 ensures that this correspondence is one to one. It is onto because, for any given column vector on the right side of (2.4.3), the vector $\mathbf{u}$ defined by (2.4.2) satisfies the identity (2.4.3). In addition, the β-representation function has the following important property. If $\mathbf{u}, \mathbf{w} \in V$,

$$\mathbf{u} = a_1\mathbf{v}_1 + a_2\mathbf{v}_2 + \cdots + a_n\mathbf{v}_n, \qquad \mathbf{w} = b_1\mathbf{v}_1 + b_2\mathbf{v}_2 + \cdots + b_n\mathbf{v}_n,$$

and $c \in \mathbb{F}$, then

$$c\mathbf{u} + \mathbf{w} = (ca_1 + b_1)\mathbf{v}_1 + (ca_2 + b_2)\mathbf{v}_2 + \cdots + (ca_n + b_n)\mathbf{v}_n.$$

Consequently,

$$[c\mathbf{u} + \mathbf{w}]_\beta = \begin{bmatrix} ca_1 + b_1 \\ ca_2 + b_1 \\ \vdots \\ ca_n + b_n \end{bmatrix} = c\begin{bmatrix} a_1 \\ a_2 \\ \vdots \\ a_n \end{bmatrix} + \begin{bmatrix} b_1 \\ b_2 \\ \vdots \\ b_n \end{bmatrix} = c[\mathbf{u}]_\beta + [\mathbf{w}]_\beta. \qquad (2.4.7)$$

This identity says something subtle and important. The addition and scalar multiplication operations on its left side are operations in the $\mathbb{F}$-vector space V; the addition and scalar multiplication operations on its right side are operations in the $\mathbb{F}$-vector space $\mathbb{F}^n$. The β-basis representation function links these two pairs of operations in a manner that respects linear algebraic operations in V and in $\mathbb{F}^n$. Informally, we conclude that any n-dimensional $\mathbb{F}$-vector space V is fundamentally the same as $\mathbb{F}^n$; formally, we say that V and $\mathbb{F}^n$ are *isomorphic*.

Theorem 2.4.8 *Let V be an n-dimensional $\mathbb{F}$-vector space, let β be a basis for V, and let $\gamma = \mathbf{u}_1, \mathbf{u}_2, \ldots, \mathbf{u}_k$ be a list of vectors in V. Then γ is linearly independent if and only if $[\mathbf{u}_1]_\beta, [\mathbf{u}_2]_\beta, \ldots, [\mathbf{u}_k]_\beta$ is linearly independent.*

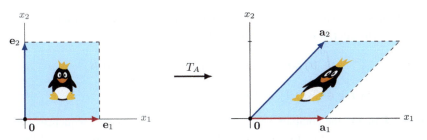

Figure 2.1 The linear transformation $T_A : \mathbb{R}^2 \to \mathbb{R}^2$ induced by the matrix A in Example 2.4.12.

Proof Let $c_1, c_2, \ldots, c_k \in \mathbb{F}$ and use (2.4.7) to write

$$\left[\sum_{i=1}^{k} c_i \mathbf{u}_i \right]_\beta = \sum_{i=1}^{k} c_i [\mathbf{u}_i]_\beta.$$

The left side is zero if and only if the right side is zero. Therefore, γ is linearly independent if and only if $[\mathbf{u}_1]_\beta, [\mathbf{u}_2]_\beta, \ldots, [\mathbf{u}_k]_\beta$ is linearly independent. $\qquad\square$

Definition 2.4.9 Let V and W be vector spaces over the same field $\mathbb{F}$. A function $T : V \to W$ is a *linear transformation* if $T(c\mathbf{u} + \mathbf{v}) = cT\mathbf{u} + T\mathbf{v}$ for all $\mathbf{u}, \mathbf{v} \in V$ and all $c \in \mathbb{F}$. If $V = W$, a linear transformation $T : V \to V$ is sometimes called a *linear operator* (or just an *operator*). The set of linear transformations from V to W is denoted by $\mathfrak{L}(V, W)$. If $V = W$, this is abbreviated by $\mathfrak{L}(V, V) = \mathfrak{L}(V)$.

For notational convenience (and by analogy with the conventional notation for matrix-vector products), $T(\mathbf{v})$ is usually written $T\mathbf{v}$.

A linear transformation $T : V \to W$ takes the zero vector of V to the zero vector of W.

Lemma 2.4.10 *If $T \in \mathfrak{L}(V, W)$, then $T\mathbf{0} = \mathbf{0}$.*

Proof $T\mathbf{0} = T(0\mathbf{0}) = 0(T\mathbf{0}) = \mathbf{0}$. $\qquad\square$

Example 2.4.11 For a given basis β of an n-dimensional $\mathbb{F}$-vector space V, the function $T\mathbf{v} = [\mathbf{v}]_\beta$ is a linear transformation from V to the $\mathbb{F}$-vector space $\mathbb{F}^n$.

Example 2.4.12 For a given $A \in \mathsf{M}_{m \times n}(\mathbb{F})$, properties of matrix arithmetic ensure that the function $T_A : \mathbb{F}^n \to \mathbb{F}^m$ defined by $T_A \mathbf{x} = A\mathbf{x}$ is a linear transformation. For example, $A = \begin{bmatrix} 1 & 1 \\ 0 & 1 \end{bmatrix} = [\mathbf{a}_1 \ \mathbf{a}_2] \in \mathsf{M}_2(\mathbb{R})$ induces the linear transformation $T_A : \mathbb{R}^2 \to \mathbb{R}^2$ depicted in Figure 2.1.

Definition 2.4.13 The linear transformation T_A defined in the preceding example is the *linear transformation induced by A*.

Example 2.4.14 On the complex vector space $\mathcal{P}$, the function $T : \mathcal{P} \to \mathcal{P}$ defined by $Tp = p'$ (differentiation) is a linear operator. This is because the derivative of a polynomial is a polynomial and $(cp + q)' = cp' + q'$ for any $c \in \mathbb{C}$ and any $p, q \in \mathcal{P}$.

Example 2.4.15 The function $T : C_\mathbb{R}[0, 1] \to C_\mathbb{R}[0, 1]$ defined by

$$(Tf)(t) = \int_0^t f(s)\, ds$$

is a linear operator. This is because the indicated integral of a continuous function is continuous (even better, it is differentiable) and

$$\int_0^t (cf(s) + g(s))\, ds = c \int_0^t f(s)\, ds + \int_0^t g(s)\, ds$$

for any $c \in \mathbb{R}$, any $f, g \in C_\mathbb{R}[0,1]$, and any $t \in [0,1]$.

Example 2.4.16 On the complex vector space $\mathcal{V}$ of finitely nonzero sequences (see Examples 1.2.9 and 2.2.7), define the *right shift* $T(x_1, x_2, \ldots) = (0, x_1, x_2, \ldots)$ and the *left shift* $S(x_1, x_2, x_3, \ldots) = (x_2, x_3, \ldots)$. A computation reveals that T and S are linear operators. See P.2.36 for some interesting properties of these operators.

Let $\mathcal{V}$ and $\mathcal{W}$ be vector spaces over the same field $\mathbb{F}$ and let $T \in \mathcal{L}(\mathcal{V}, \mathcal{W})$. The *kernel* and *range* of T are $\ker T = \{ \mathbf{v} \in \mathcal{V} : T\mathbf{v} = \mathbf{0} \}$ and $\operatorname{ran} T = \{ T\mathbf{v} : \mathbf{v} \in \mathcal{V} \}$, respectively. The same arguments used in Examples 1.3.4 and 1.3.7 to show that the null space and column space of a matrix are subspaces also show that $\ker T$ is a subspace of $\mathcal{V}$ and $\operatorname{ran} T$ is a subspace of $\mathcal{W}$. A convenient way to show that a subset of a vector space is a subspace is to identify it as the kernel or range of a linear transformation; see P.2.17.

Theorem 2.4.17 *Let $\mathcal{V}$ and $\mathcal{W}$ be vector spaces over $\mathbb{F}$. Then $T \in \mathcal{L}(\mathcal{V}, \mathcal{W})$ is one to one if and only if $\ker T = \{\mathbf{0}\}$.*

Proof Lemma 2.4.10 says that $T\mathbf{0} = \mathbf{0}$. If T is one to one and $T\mathbf{x} = \mathbf{0}$, then $\mathbf{x} = \mathbf{0}$ and hence $\ker T = \{\mathbf{0}\}$. Now suppose that $\ker T = \{\mathbf{0}\}$. If $T\mathbf{x} = T\mathbf{y}$, then $\mathbf{0} = T\mathbf{x} - T\mathbf{y} = T(\mathbf{x} - \mathbf{y})$, which says that $\mathbf{x} - \mathbf{y} \in \ker T$. Consequently, $\mathbf{x} - \mathbf{y} = \mathbf{0}$ and hence $\mathbf{x} = \mathbf{y}$. $\qquad\square$

A noteworthy fact about a linear transformation on a finite-dimensional vector space is that if its action on a basis is known, then its action on every vector is determined. The following example illustrates the principle, which is formalized in Theorem 2.4.19.

Example 2.4.18 Consider the basis $\beta = 1, z, z^2$ of $\mathcal{P}_2$ and the linear transformation $T : \mathcal{P}_2 \to \mathcal{P}_1$ defined by $Tp = p'$ (differentiation). Then $T1 = 0, Tz = 1$, and $Tz^2 = 2z$. Consequently, for any $p(z) = c_2 z^2 + c_1 z + c_0 \in \mathcal{P}_2$,

$$Tp = T(c_2 z^2 + c_1 z + c_0) = c_2 Tz^2 + c_1 Tz + c_0 T1 = c_2(2z) + c_1.$$

Theorem 2.4.19 *Let $\mathcal{V}$ and $\mathcal{W}$ be $\mathbb{F}$-vector spaces. Suppose that $\mathcal{V}$ is finite dimensional and nonzero. Let $\beta = \mathbf{v}_1, \mathbf{v}_2, \ldots, \mathbf{v}_n$ be a basis for $\mathcal{V}$ and let $T \in \mathcal{L}(\mathcal{V}, \mathcal{W})$. Then $\operatorname{ran} T = \operatorname{span}\{T\mathbf{v}_1, T\mathbf{v}_2, \ldots, T\mathbf{v}_n\}$. In particular, $\operatorname{ran} T$ is finite dimensional and $\dim \operatorname{ran} T \leq n$.*

Proof Compute

$$\begin{aligned}
\operatorname{ran} T &= \{ T\mathbf{v} : \mathbf{v} \in \mathcal{V} \} \\
&= \{ T(c_1 \mathbf{v}_1 + c_2 \mathbf{v}_2 + \cdots + c_n \mathbf{v}_n) : c_1, c_2, \ldots, c_n \in \mathbb{F} \} \\
&= \{ c_1 T\mathbf{v}_1 + c_2 T\mathbf{v}_2 + \cdots + c_n T\mathbf{v}_n : c_1, c_2, \ldots, c_n \in \mathbb{F} \} \\
&= \operatorname{span}\{ T\mathbf{v}_1, T\mathbf{v}_2, \ldots, T\mathbf{v}_n \}.
\end{aligned}$$

Theorem 2.2.8.a ensures that $\operatorname{ran} T$ is finite dimensional and $\dim \operatorname{ran} T \leq n$. $\qquad\square$

If $\mathcal{V}$ and $\mathcal{W}$ are nonzero finite-dimensional $\mathbb{F}$-vector spaces, the preceding theorem can be further refined. Let $\beta = \mathbf{v}_1, \mathbf{v}_2, \ldots, \mathbf{v}_n$ be a basis for $\mathcal{V}$, let $\gamma = \mathbf{w}_1, \mathbf{w}_2, \ldots, \mathbf{w}_m$ be a basis for $\mathcal{W}$, and let $T \in \mathcal{L}(\mathcal{V}, \mathcal{W})$. Express $\mathbf{v} \in \mathcal{V}$ as $\mathbf{v} = c_1 \mathbf{v}_1 + c_2 \mathbf{v}_2 + \cdots + c_n \mathbf{v}_n$, that is, $[\mathbf{v}]_\beta = [c_1 \ c_2 \ \ldots \ c_n]^\mathsf{T}$. Then $T\mathbf{v} = c_1 T\mathbf{v}_1 + c_2 T\mathbf{v}_2 + \cdots + c_n T\mathbf{v}_n$, so

$$[T\mathbf{v}]_\gamma = c_1 [T\mathbf{v}_1]_\gamma + c_2 [T\mathbf{v}_2]_\gamma + \cdots + c_n [T\mathbf{v}_n]_\gamma. \tag{2.4.20}$$

Now define the matrix

$$_\gamma[T]_\beta = \big[[T\mathbf{v}_1]_\gamma \ [T\mathbf{v}_2]_\gamma \ \ldots \ [T\mathbf{v}_n]_\gamma \big] \in \mathsf{M}_{m \times n}(\mathbb{F}), \tag{2.4.21}$$

whose jth column is the γ-coordinate vector of $T\mathbf{v}_j$. We can rewrite (2.4.20) as

$$[T\mathbf{v}]_\gamma = {}_\gamma[T]_\beta [\mathbf{v}]_\beta. \tag{2.4.22}$$

Once we have fixed a basis β of $\mathcal{V}$ and a basis γ of $\mathcal{W}$, we can determine $T\mathbf{v}$ in two steps. First, compute the matrix $_\gamma[T]_\beta$ in (2.4.21); this needs to be done only once. Then, for each $\mathbf{v}$ of interest, compute its β-coordinate vector $[\mathbf{v}]_\beta$ and calculate the matrix-vector product in (2.4.22) to determine the γ-coordinate vector $[T\mathbf{v}]_\gamma$, which expresses $T\mathbf{v}$ as a linear combination of the vectors in γ.

Definition 2.4.23 Let $\mathcal{V}$ and $\mathcal{W}$ be finite-dimensional $\mathbb{F}$-vector spaces. Let $\beta = \mathbf{v}_1, \mathbf{v}_2, \ldots, \mathbf{v}_n$ be a basis for $\mathcal{V}$, let $\gamma = \mathbf{w}_2, \mathbf{w}_2, \ldots, \mathbf{w}_m$ be a basis for $\mathcal{W}$, and let $T \in \mathcal{L}(\mathcal{V}, \mathcal{W})$. Then

$$_\gamma[T]_\beta = \big[[T\mathbf{v}_1]_\gamma \ [T\mathbf{v}_2]_\gamma \ \ldots \ [T\mathbf{v}_n]_\gamma \big] \in \mathsf{M}_{m \times n}(\mathbb{F})$$

is the *matrix of T with respect to the bases β and γ*; more succinctly, we refer to it as the *β-γ matrix representation of T*.

Example 2.4.24 With T and β as in Example 2.4.18, consider the basis $\gamma = 1, z$ of $\mathcal{P}_1$. Then

$$[T1]_\gamma = [0]_\gamma = \begin{bmatrix} 0 \\ 0 \end{bmatrix}, \qquad [Tz]_\gamma = [1]_\gamma = \begin{bmatrix} 1 \\ 0 \end{bmatrix}, \qquad [Tz^2]_\gamma = [2z]_\gamma = \begin{bmatrix} 0 \\ 2 \end{bmatrix},$$

and $_\gamma[T]_\beta = \begin{bmatrix} 0 & 1 & 0 \\ 0 & 0 & 2 \end{bmatrix}$. If $[p]_\beta = [c_0 \ c_1 \ c_2]^\mathsf{T}$, then

$$[Tp]_\gamma = {}_\gamma[T]_\beta [p]_\beta = \begin{bmatrix} 0 & 1 & 0 \\ 0 & 0 & 2 \end{bmatrix} \begin{bmatrix} c_0 \\ c_1 \\ c_2 \end{bmatrix} = \begin{bmatrix} c_1 \\ 2c_2 \end{bmatrix},$$

so $Tp = c_1 1 + (2c_2)z = c_1 + 2c_2 z$, which agrees with our computation in Example 2.4.18.

Example 2.4.25 Let $T : \mathcal{P}_2 \to \mathbb{C}$ be the linear transformation defined by $Tp = p(2)$ (evaluate $p(z)$ at $z = 2$). Consider the basis $\beta = 1, z, z^2$ for $\mathcal{P}_2$ and the basis $\gamma = 1$ for the one-dimensional vector space $\mathbb{C}$. Then $[T1]_\gamma = 1$, $[Tz]_\gamma = 2$, and $[Tz^2]_\gamma = 4$, so $_\gamma[T]_\beta = [1 \ 2 \ 4]$. If $p(z) = c_0 + c_1 z + c_2 z^2$, then

$$[Tp]_\gamma = {}_\gamma[T]_\beta [p]_\beta = [1 \ 2 \ 4] \begin{bmatrix} c_0 \\ c_1 \\ c_2 \end{bmatrix} = c_0 + 2c_1 + 4c_2.$$

This is a special case of Theorem 6.4.4.

2.5 Change of Basis

The identity (2.4.22) contains a wealth of information. Consider the special case in which $\mathcal{W} = \mathcal{V}$ is n-dimensional and $n \geq 1$. Suppose that $\beta = \mathbf{v}_1, \mathbf{v}_2, \ldots, \mathbf{v}_n$ and $\gamma = \mathbf{w}_1, \mathbf{w}_2, \ldots, \mathbf{w}_n$ are bases for $\mathcal{V}$. The *identity linear transformation* $I \in \mathcal{L}(\mathcal{V})$ is defined by $I\mathbf{v} = \mathbf{v}$ for all $\mathbf{v} \in \mathcal{V}$. The β-β matrix representation of I is

$$\begin{aligned}
_\beta[I]_\beta &= \begin{bmatrix} [I\mathbf{v}_1]_\beta & [I\mathbf{v}_2]_\beta & \cdots & [I\mathbf{v}_n]_\beta \end{bmatrix} \\
&= \begin{bmatrix} [\mathbf{v}_1]_\beta & [\mathbf{v}_2]_\beta & \cdots & [\mathbf{v}_n]_\beta \end{bmatrix} \\
&= \begin{bmatrix} \mathbf{e}_1 & \mathbf{e}_2 & \cdots & \mathbf{e}_n \end{bmatrix} \\
&= I_n \in \mathsf{M}_n(\mathbb{F}),
\end{aligned}$$

the $n \times n$ identity matrix. What can we say about

$$_\gamma[I]_\beta = \begin{bmatrix} [\mathbf{v}_1]_\gamma & [\mathbf{v}_2]_\gamma & \cdots & [\mathbf{v}_n]_\gamma \end{bmatrix} \quad \text{and} \quad _\beta[I]_\gamma = \begin{bmatrix} [\mathbf{w}_1]_\beta & [\mathbf{w}_2]_\beta & \cdots & [\mathbf{w}_n]_\beta \end{bmatrix}? \quad (2.5.1)$$

For any $\mathbf{v} \in \mathcal{V}$, use (2.4.22) and compute

$$I_n[\mathbf{v}]_\gamma = [\mathbf{v}]_\gamma = [I\mathbf{v}]_\gamma = {}_\gamma[I]_\beta [\mathbf{v}]_\beta = {}_\gamma[I]_\beta [I\mathbf{v}]_\beta = {}_\gamma[I]_\beta {}_\beta[I]_\gamma [\mathbf{v}]_\gamma.$$

This identity ensures that the jth column of $_\gamma[I]_{\beta\beta}[I]_\gamma$ is

$$_\gamma[I]_\beta {}_\beta[I]_\gamma \, \mathbf{e}_j = {}_\gamma[I]_\beta {}_\beta[I]_\gamma [\mathbf{v}_j]_\gamma = [\mathbf{v}_j]_\gamma = \mathbf{e}_j,$$

which is the jth column of I_n. Therefore,

$$I_n = {}_\gamma[I]_\beta {}_\beta[I]_\gamma. \qquad (2.5.2)$$

The identity (2.5.2) and Theorem 2.2.13 tell us that $_\gamma[I]_\beta$ is invertible and $_\beta[I]_\gamma$ is its inverse.

Definition 2.5.3 The matrix $_\gamma[I]_\beta$ defined in (2.5.1) is the β-γ *change-of-basis* matrix. It describes how to represent each vector in the basis β as a linear combination of the vectors in the basis γ.

Example 2.5.4 Figure 2.2 illustrates the standard basis $\beta = \mathbf{e}_1, \mathbf{e}_2$ in $\mathbb{R}^2$ and the basis $\gamma = \mathbf{w}_1, \mathbf{w}_2$ in $\mathbb{R}^2$, in which $[\mathbf{w}_1]_\beta = \begin{bmatrix} 1 \\ 2 \end{bmatrix}$ and $[\mathbf{w}_2]_\beta = \begin{bmatrix} 3 \\ 1 \end{bmatrix}$. A computation reveals that $[\mathbf{e}_1]_\gamma = \begin{bmatrix} -1/5 \\ 2/5 \end{bmatrix}$ and $[\mathbf{e}_2]_\gamma = \begin{bmatrix} 3/5 \\ -1/5 \end{bmatrix}$. We have $_\beta[I]_\gamma = \begin{bmatrix} [\mathbf{w}_1]_\beta & [\mathbf{w}_2]_\beta \end{bmatrix} = \begin{bmatrix} 1 & 3 \\ 2 & 1 \end{bmatrix}$ and

$$_\beta[I]_\gamma^{-1} = \begin{bmatrix} -\frac{1}{5} & \frac{3}{5} \\ \frac{2}{5} & -\frac{1}{5} \end{bmatrix},$$

which is consistent with the identity

$$_\gamma[I]_\beta = \begin{bmatrix} [\mathbf{e}_1]_\gamma & [\mathbf{e}_2]_\gamma \end{bmatrix} = \begin{bmatrix} -\frac{1}{5} & \frac{3}{5} \\ \frac{2}{5} & -\frac{1}{5} \end{bmatrix}.$$

Example 2.5.5 Let $\mathbf{v}_1 = \begin{bmatrix} 1 \\ 1 \end{bmatrix}$, $\mathbf{v}_2 = \begin{bmatrix} 1 \\ -1 \end{bmatrix}$, and $\mathbf{w}_1 = \begin{bmatrix} 1 \\ 2 \end{bmatrix}$, $\mathbf{w}_2 = \begin{bmatrix} 1 \\ 3 \end{bmatrix}$. Each of the lists $\beta = \mathbf{v}_1, \mathbf{v}_2$ and $\gamma = \mathbf{w}_1, \mathbf{w}_2$ is a basis for $\mathbb{R}^2$. We can compute the columns of $_\gamma[I]_\beta = \begin{bmatrix} [\mathbf{v}_1]_\gamma & [\mathbf{v}_2]_\gamma \end{bmatrix}$ by solving some linear equations. For example, the entries of $[\mathbf{v}_1]_\gamma = \begin{bmatrix} a \\ b \end{bmatrix}$ are the coefficients in the representation

$$\mathbf{v}_1 = \begin{bmatrix} 1 \\ 1 \end{bmatrix} = a \begin{bmatrix} 1 \\ 2 \end{bmatrix} + b \begin{bmatrix} 1 \\ 3 \end{bmatrix} = \begin{bmatrix} a+b \\ 2a+3b \end{bmatrix}$$

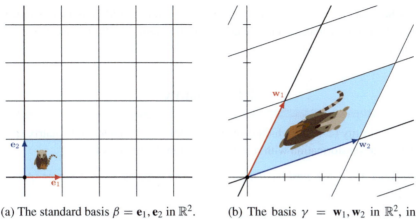

(a) The standard basis $\beta = \mathbf{e}_1, \mathbf{e}_2$ in $\mathbb{R}^2$.   (b) The basis $\gamma = \mathbf{w}_1, \mathbf{w}_2$ in $\mathbb{R}^2$, in which $\mathbf{w}_1 = [1\ 2]^{\mathsf{T}}$ and $\mathbf{w}_2 = [3\ 1]^{\mathsf{T}}$.

Figure 2.2 Relationship between two bases in $\mathbb{R}^2$.

of $\mathbf{v}_1$ as a linear combination of $\mathbf{w}_1$ and $\mathbf{w}_2$. The solution is $a = 2$ and $b = -1$, so $[\mathbf{v}_1]_\gamma = \begin{bmatrix} 2 \\ -1 \end{bmatrix}$. Solving the linear equation $\mathbf{v}_2 = a\mathbf{w}_1 + b\mathbf{w}_2$ provides the second column of $_\gamma[I]_\beta$, which is $[\mathbf{v}_2]_\gamma = \begin{bmatrix} 4 \\ -3 \end{bmatrix}$. Thus,

$$_\gamma[I]_\beta = \begin{bmatrix} 2 & 4 \\ -1 & -3 \end{bmatrix} \quad \text{and} \quad _\beta[I]_\gamma = {_\gamma[I]_\beta^{-1}} = \begin{bmatrix} \frac{3}{2} & 2 \\ -\frac{1}{2} & -1 \end{bmatrix}. \tag{2.5.6}$$

Here is an alternative approach. Let $V = [\mathbf{v}_1\ \mathbf{v}_2]$ and $W = [\mathbf{w}_1\ \mathbf{w}_2]$. Let $W^{-1}V = X = [\mathbf{x}_1\ \mathbf{x}_2] = [x_{ij}]$, so $V = WX$. Then $\mathbf{v}_1 = W\mathbf{x}_1 = x_{11}\mathbf{w}_1 + x_{21}\mathbf{w}_2$ represents $\mathbf{v}_1$ as a linear combination of the basis vectors in γ, so $[\mathbf{v}_1]_\gamma = [W\mathbf{x}_1]_\gamma = \mathbf{x}_1$. In the same way, we find that $[\mathbf{v}_2]_\gamma = \mathbf{x}_2$. Thus,

$$_\gamma[I]_\beta = \big[[\mathbf{v}_1]_\gamma\ [\mathbf{v}_2]_\gamma\big] = [\mathbf{x}_1\ \mathbf{x}_2] = X = W^{-1}V.$$

As a check, compute

$$W^{-1}V = \begin{bmatrix} 3 & -1 \\ -2 & 1 \end{bmatrix}\begin{bmatrix} 1 & 1 \\ 1 & -1 \end{bmatrix} = \begin{bmatrix} 2 & 4 \\ -1 & -3 \end{bmatrix},$$

which is consistent with (2.5.6).

Theorem 2.5.7 *Let n be a positive integer, let V be an n-dimensional $\mathbb{F}$-vector space, and let $\beta = \mathbf{v}_1, \mathbf{v}_2, \ldots, \mathbf{v}_n$ be a basis for V.*

(a) *Let $\gamma = \mathbf{w}_1, \mathbf{w}_2, \ldots, \mathbf{w}_n$ be a basis for V. The change-of-basis matrix $_\gamma[I]_\beta \in \mathsf{M}_n(\mathbb{F})$ is invertible and its inverse is $_\beta[I]_\gamma$.*

(b) *If $S \in \mathsf{M}_n(\mathbb{F})$ is invertible, then there is a basis γ for V such that $S = {_\gamma[I]_\beta}$.*

Proof (a) We proved the first assertion in our discussion of (2.5.2).

(b) Let $S^{-1} = [\mathbf{s}_1\ \mathbf{s}_2\ \ldots\ \mathbf{s}_n]$ and define $\mathbf{w}_1, \mathbf{w}_2, \ldots, \mathbf{w}_n \in V$ by $[\mathbf{w}_j]_\beta = \mathbf{s}_j$. Let $\gamma = \mathbf{w}_1, \mathbf{w}_2, \ldots, \mathbf{w}_n$. If $c_1\mathbf{w}_1 + c_2\mathbf{w}_2 + \cdots + c_n\mathbf{w}_n = \mathbf{0}$, then

$$\mathbf{0} = \left[\sum_{j=1}^{n} c_j\mathbf{w}_j\right]_\beta = \sum_{j=1}^{n} c_j[\mathbf{w}_j]_\beta = \sum_{j=1}^{n} c_j\mathbf{s}_j,$$

which implies that $c_1 = c_2 = \cdots = c_n = 0$ since the columns of S^{-1} are linearly independent. We conclude that γ is linearly independent, and Theorem 2.1.10 ensures that it is a basis for $\mathcal{V}$. Then $S^{-1} = {}_\beta[I]_\gamma$ and $S = {}_\gamma[I]_\beta$. $\qquad\square$

2.6 Similarity

Consider the identity (2.4.22), which gives us the tools to understand how matrix representations of a linear transformation with respect to different bases are related.

Theorem 2.6.1 *Let n be a positive integer, let $\mathcal{V}$ be an n-dimensional $\mathbb{F}$-vector space, and let $T \in \mathcal{L}(\mathcal{V})$.*

(a) *Let β and γ be bases for $\mathcal{V}$, and let $S = {}_\gamma[I]_\beta$. Then S is invertible and*

$$
{}_\gamma[T]_\gamma = {}_\gamma[I]_\beta\,{}_\beta[T]_\beta\,{}_\beta[I]_\gamma = S\,{}_\beta[T]_\beta\,S^{-1}. \tag{2.6.2}
$$

(b) *Let $S \in \mathsf{M}_n(\mathbb{F})$ be invertible and let β be a basis for $\mathcal{V}$. Then there is a basis γ for $\mathcal{V}$ such that ${}_\gamma[T]_\gamma = S\,{}_\beta[T]_\beta\,S^{-1}$.*

Proof (a) Theorem 2.3.13 ensures that S is invertible. For each $\mathbf{v} \in \mathcal{V}$,

$$
\begin{aligned}
{}_\gamma[T]_\gamma[\mathbf{v}]_\gamma &= [T\mathbf{v}]_\gamma = [I(T\mathbf{v})]_\gamma = {}_\gamma[I]_\beta\,[T\mathbf{v}]_\beta \\
&= {}_\gamma[I]_\beta\,{}_\beta[T]_\beta\,[\mathbf{v}]_\beta = {}_\gamma[I]_\beta\,{}_\beta[T]_\beta\,[I\mathbf{v}]_\beta \\
&= {}_\gamma[I]_\beta\,{}_\beta[T]_\beta\,{}_\beta[I]_\gamma\,[\mathbf{v}]_\gamma,
\end{aligned}
$$

which verifies (2.6.2); see P.2.8.

(b) Theorem 2.5.7 ensures that there is a basis γ for $\mathcal{V}$ such that $S = {}_\gamma[I]_\beta$, so $S^{-1} = {}_\beta[I]_\gamma$ and the assertion follows from (2.6.2). $\qquad\square$

Example 2.6.3 Let $\beta = \mathbf{e}_1, \mathbf{e}_2$ be the standard basis for $\mathbb{R}^2$ and let the basis $\gamma = \mathbf{w}_1, \mathbf{w}_2$ be as in Example 2.5.5. Then

$$
{}_\beta[I]_\gamma = \big[[\mathbf{w}_1]_\beta \ [\mathbf{w}_2]_\beta\big] = \begin{bmatrix} 1 & 1 \\ 2 & 3 \end{bmatrix} \quad \text{and} \quad {}_\gamma[I]_\beta = {}_\beta[I]_\gamma^{-1} = \begin{bmatrix} 3 & -1 \\ -2 & 1 \end{bmatrix}.
$$

Theorem 2.4.19 says that a linear transformation on $\mathbb{R}^2$ is uniquely determined by its action on a basis. Let $T : \mathbb{R}^2 \to \mathbb{R}^2$ be the linear transformation such that $T\mathbf{e}_1 = 2\mathbf{e}_1$ and $T\mathbf{e}_2 = 3\mathbf{e}_2$, so

$$
{}_\beta[T]_\beta = \begin{bmatrix} 2 & 0 \\ 0 & 3 \end{bmatrix}.
$$

We have

$$
\begin{aligned}
T\mathbf{w}_1 &= T(\mathbf{e}_1 + 2\mathbf{e}_2) = T\mathbf{e}_1 + 2T\mathbf{e}_2 = 2\mathbf{e}_1 + 6\mathbf{e}_2, \\
T\mathbf{w}_2 &= T(\mathbf{e}_1 + 3\mathbf{e}_2) = T\mathbf{e}_1 + 3T\mathbf{e}_2 = 2\mathbf{e}_1 + 9\mathbf{e}_2,
\end{aligned}
$$

so

$$
{}_\beta[T]_\gamma = \big[[T\mathbf{y}_1]_\beta \ [T\mathbf{y}_2]_\beta\big] = \begin{bmatrix} 2 & 2 \\ 6 & 9 \end{bmatrix}.
$$

The preceding theorem ensures that

$$\gamma[T]_\gamma = \gamma[I]_{\beta\beta}[T]_{\beta\beta}[I]_\gamma = \begin{bmatrix} 3 & -1 \\ -2 & 1 \end{bmatrix}\begin{bmatrix} 2 & 0 \\ 0 & 3 \end{bmatrix}\begin{bmatrix} 1 & 1 \\ 2 & 3 \end{bmatrix} = \begin{bmatrix} 0 & -3 \\ 2 & 5 \end{bmatrix}.$$

Definition 2.6.4 Let $A, B \in M_n(\mathbb{F})$. Then A and B are *similar over* $\mathbb{F}$ if there is an invertible $S \in M_n(\mathbb{F})$ such that $A = SBS^{-1}$.

If $A = SBS^{-1}$ and there is a need to emphasize the role of S, we say that A *is similar to* B *via the similarity matrix* S.

Corollary 2.6.5 *Let* $A, B \in M_n(\mathbb{F})$. *The following are equivalent:*

(a) *A and B are similar over* $\mathbb{F}$.

(b) *There is an n-dimensional $\mathbb{F}$-vector space V, bases β and γ for V, and a $T \in \mathcal{L}(V)$ such that $A = \beta[T]_\beta$ and $B = \gamma[T]_\gamma$.*

Proof (a) $\Rightarrow$ (b) Let $S \in M_n(\mathbb{F})$ be an invertible matrix such that $A = SBS^{-1}$. Let $V = \mathbb{F}^n$ and let T be the linear operator $T_A : \mathbb{F}^n \to \mathbb{F}^n$ induced by A (see Definition 2.4.13). Let β be the standard basis for $\mathbb{F}^n$ and let γ be the list of columns of S; Theorem 2.1.3 ensures that γ is a basis. Then $\beta[T]_\beta = A$ and $\beta[I]_\gamma = S$, so

$$SBS^{-1} = A = \beta[T]_\beta = \beta[I]_\gamma \, \gamma[T]_\gamma \, \gamma[I]_\beta = S_\gamma[T]_\gamma S^{-1}.$$

Consequently, $SBS^{-1} = S_\gamma[T]_\gamma S^{-1}$, which implies that $B = \gamma[T]_\gamma$.

(b) $\Rightarrow$ (a) This implication is Theorem 2.6.1.a. $\square$

Example 2.6.6 Let $T : \mathcal{P}_2 \to \mathcal{P}_2$ be the linear transformation $Tp = p'$ (differentiation), and consider the bases

$$\beta = 1, z, z^2, \qquad \gamma = 1, 2z - 1, 6z^2 + z + 1, \quad \text{and} \quad \alpha = z + 1, 1 - z, 3z^2 + z.$$

Then

$$\beta[T]_\beta = \begin{bmatrix} 0 & 1 & 0 \\ 0 & 0 & 2 \\ 0 & 0 & 0 \end{bmatrix}, \quad \gamma[T]_\gamma = \begin{bmatrix} 0 & 2 & 7 \\ 0 & 0 & 6 \\ 0 & 0 & 0 \end{bmatrix}, \quad \text{and} \quad \alpha[T]_\alpha = \begin{bmatrix} \frac{1}{2} & -\frac{1}{2} & \frac{7}{2} \\ \frac{1}{2} & -\frac{1}{2} & -\frac{5}{2} \\ 0 & 0 & 0 \end{bmatrix}. \quad (2.6.7)$$

The preceding corollary ensures that these matrices are pairwise similar, and its proof shows that the similarities can be accomplished with change-of-basis matrices. For example,

$$\beta[I]_\alpha = \begin{bmatrix} 1 & 1 & 0 \\ 1 & -1 & 1 \\ 0 & 0 & 3 \end{bmatrix} \quad \text{and} \quad \alpha[I]_\beta = \begin{bmatrix} \frac{1}{2} & \frac{1}{2} & -\frac{1}{6} \\ \frac{1}{2} & -\frac{1}{2} & \frac{1}{6} \\ 0 & 0 & \frac{1}{3} \end{bmatrix}.$$

A calculation reveals that $\beta[I]_\alpha \, \alpha[I]_\beta = I$ and

$$\begin{bmatrix} 1 & 1 & 0 \\ 1 & -1 & 1 \\ 0 & 0 & 3 \end{bmatrix}\begin{bmatrix} \frac{1}{2} & -\frac{1}{2} & \frac{7}{2} \\ \frac{1}{2} & -\frac{1}{2} & -\frac{5}{2} \\ 0 & 0 & 0 \end{bmatrix}\begin{bmatrix} \frac{1}{2} & \frac{1}{2} & -\frac{1}{6} \\ \frac{1}{2} & -\frac{1}{2} & \frac{1}{6} \\ 0 & 0 & \frac{1}{3} \end{bmatrix} = \begin{bmatrix} 0 & 1 & 0 \\ 0 & 0 & 2 \\ 0 & 0 & 0 \end{bmatrix}.$$

Notice that the three pairwise similar matrices (2.6.7) have the same trace, determinant, and rank. This is not an accident.

Theorem 2.6.8 *Let $A, B \in \mathsf{M}_n$. If A is similar to B, then* $\operatorname{tr} A = \operatorname{tr} B$, $\det A = \det B$, *and* $\operatorname{rank} A = \operatorname{rank} B$.

Proof Let $S \in \mathsf{M}_n$ be invertible and such that $A = SBS^{-1}$. Then (C.2.9) ensures that
$$\operatorname{tr} A = \operatorname{tr}\big(S(BS^{-1})\big) = \operatorname{tr}\big((BS^{-1})S\big) = \operatorname{tr} B.$$
The product rule for determinants ensures that
$$\det A = \det SBS^{-1} = (\det S)(\det B)(\det(S^{-1}))$$
$$= (\det S)(\det S)^{-1}(\det B) = \det B.$$
Example 1.4.23 ensures that $\operatorname{col} SBS^{-1} = \operatorname{col} SB$ and $\operatorname{row} SB = \operatorname{row} B$. Now use Theorem 2.3.1 to compute
$$\operatorname{rank} A = \operatorname{rank} SBS^{-1} = \dim \operatorname{col} SBS^{-1} = \dim \operatorname{col} SB$$
$$= \dim \operatorname{row} SB = \dim \operatorname{row} B = \operatorname{rank} B. \qquad \square$$

The following theorem includes some important facts about polynomials and similarity.

Theorem 2.6.9 *Let $A, B, X \in \mathsf{M}_n(\mathbb{F})$, let $S \in \mathsf{M}_n$ be invertible, and let p be a polynomial.*

(a) *If $AX = XB$, then $p(A)X = Xp(B)$.*

(b) *If $A = SBS^{-1}$, then $p(A) = Sp(B)S^{-1}$.*

(c) *If $A = SBS^{-1}$, then $A - \lambda I = S(B - \lambda I)S^{-1}$ for every $\lambda \in \mathbb{F}$.*

(d) *If there is a $\lambda \in \mathbb{F}$ such that $A - \lambda I = S(B - \lambda I)S^{-1}$, then $A = SBS^{-1}$.*

(e) *If A is similar to B, then $\operatorname{rank}(A - \lambda I)^k = \operatorname{rank}(B - \lambda I)^k$ for all $\lambda \in \mathbb{F}$ and each $k = 1, 2, \ldots$.*

Proof (a) We first use induction to prove that $A^j X = XB^j$ for $j = 0, 1, 2, \ldots$. The base case $j = 0$ is $IX = XI$, which is true. For the induction step, suppose that $A^j X = XB^j$ for some j. Then $A^{j+1}X = AA^j X = AXB^j = XBB^j = XB^{j+1}$, which completes the induction.

Let $p(z) = c_k z^k + \cdots + c_1 z + c_0$. Then
$$p(A)X = (c_k A^k + \cdots + c_1 A + c_0 I)X$$
$$= c_k(A^k X) + \cdots + c_1(AX) + c_0 X$$
$$= c_k(XB^k) + \cdots + c_1(XB) + c_0 X$$
$$= X(c_k B^k + \cdots + c_1 B + c_0 I)$$
$$= Xp(B).$$

(b) If $A = SBS^{-1}$, then $AS = SB$. Part (a) ensures that $p(A)S = Sp(B)$ and hence $p(A) = Sp(B)S^{-1}$.

(c) This follows from (b) with $p(z) = z - \lambda$.

(d) If there is a $\lambda \in \mathbb{F}$ such that $A - \lambda I = S(B - \lambda I)S^{-1}$, then $A - \lambda I = SBS^{-1} - \lambda SS^{-1} = SBS^{-1} - \lambda I$, so $A = SBS^{-1}$.

(e) Parts (b) and (c) ensure that $(A - \lambda I)^k$ is similar to $(B - \lambda I)^k$, and Theorem 2.6.8 says that their ranks are equal. $\qquad \square$

Example 2.6.10 Let
$$A = \begin{bmatrix} 1 & 1 & 0 \\ 0 & 1 & 1 \\ 0 & 0 & 1 \end{bmatrix} \quad \text{and} \quad B = \begin{bmatrix} 1 & 1 & 0 \\ 0 & 1 & 0 \\ 0 & 0 & 1 \end{bmatrix}.$$

Then $\operatorname{tr} A = \operatorname{tr} B = 3$, $\det A = \det B = 1$, and $\operatorname{rank} A = \operatorname{rank} B = 3$. However, $\operatorname{rank}(A - I) = 2$ and $\operatorname{rank}(B - I) = 1$, so Theorem 2.6.9.c and Theorem 2.6.8 tell us that A is not similar to B. We have a lot more to say about similarity in Chapter 12.

Similarity satisfies the following properties:

(a) A is similar to A for all $A \in M_n(\mathbb{F})$. $\hspace{2cm}$ *Reflexive*

(b) A is similar to B if and only if B is similar to A. $\hspace{2cm}$ *Symmetric*

(c) If A is similar to B and B is similar to C, then A is similar to C. $\hspace{1cm}$ *Transitive*

To verify these properties, consider

(a) $A = IAI^{-1}$.

(b) If $A = SBS^{-1}$, then $B = (S^{-1})A(S^{-1})^{-1}$.

(c) If $A = SBS^{-1}$ and $B = RCR^{-1}$, then $A = (SR)C(SR)^{-1}$.

Definition 2.6.11 A relation between pairs of matrices is an *equivalence relation* if it is reflexive, symmetric, and transitive.

Similarity is the first of several equivalence relations to be introduced in this book.

Similar matrices represent (with respect to possibly different bases) the same linear operator, so they can be expected to share many important properties. Some of these shared properties are: rank, determinant, trace, eigenvalues, characteristic polynomial, minimal polynomial, and Jordan canonical form. We have a lot to look forward to as we study these properties in the following chapters.

2.7 Polynomial Bases and Lagrange Interpolation

For any distinct $x_1, x_2 \in \mathbb{R}$, and any $y_1, y_2 \in \mathbb{R}$, there is a line $y = mx + b$ that passes through the points (x_1, y_1) and (x_2, y_2). That is, we can find a real polynomial of degree at most one that takes any given real values y_1 and y_2 at the points x_1 and x_2. In this section, we investigate the problem of finding a complex polynomial of degree at most $n - 1$ (that is, an element of $\mathcal{P}_{n-1}$) that takes given values at n distinct points in $\mathbb{C}$. The concepts of span, linear independence, and basis in the vector space $\mathcal{P}_{n-1}$ play key roles in our investigation.

Example 2.7.1 Let $z_1, z_2 \in \mathbb{C}$. If $z_1 \neq z_2$, then

$$\ell_1(z) = \frac{z - z_2}{z_1 - z_2} \quad \text{and} \quad \ell_2(z) = \frac{z - z_1}{z_2 - z_1} \tag{2.7.2}$$

are in $\mathcal{P}_1$. They have been crafted so that $\ell_1(z_1) = \ell_2(z_2) = 1$ and $\ell_1(z_2) = \ell_2(z_1) = 0$, which we write succinctly as $\ell_j(z_k) = \delta_{jk}$ for $j, k \in \{1, 2\}$; see (C.2.5) for the Kronecker delta notation. For any $w_1, w_2 \in \mathbb{C}$, the linear combination $p = w_1\ell_1 + w_2\ell_2$ is in $\mathcal{P}_1$ and satisfies

$$p(z_1) = w_1\ell_1(z_1) + w_2\ell_2(z_1) = w_1\delta_{11} + w_2\delta_{21} = w_1, \quad \text{and}$$
$$p(z_2) = w_1\ell_1(z_2) + w_2\ell_2(z_2) = w_1\delta_{12} + w_2\delta_{22} = w_2.$$

For any $f \in \mathcal{P}_1$, the linear combination $g = f(z_1)\ell_1 + f(z_2)\ell_2$ of the polynomials (2.7.2) satisfies

$$g(z_k) = f(z_1)\ell_1(z_k) + f(z_2)\ell_2(z_k) = f(z_1)\delta_{1k} + f(z_2)\delta_{2k} = f(z_k)$$

for $k = 1, 2$. The identity theorem for polynomials (Appendix B.4) ensures that $f = g$. That is, $\beta = \ell_1, \ell_2$ spans the vector space $\mathcal{P}_1$ and

$$f = f(z_1)\ell_1 + f(z_2)\ell_2 \qquad (2.7.3)$$

for all $f \in \mathcal{P}_1$. If a linear combination $h = a_1\ell_1 + a_2\ell_2$ is the zero polynomial, then equation (2.7.3) ensures that $a_1 = h(z_1) = 0$ and $a_2 = h(z_2) = 0$, that is, β is linearly independent. We conclude that β is a basis for $\mathcal{P}_1$.

In the following theorem, we build on the ideas in the preceding example. We construct, for each list of n distinct complex numbers $z_1, z_2, \ldots, z_n$, a basis $\beta = \ell_1, \ell_2, \ldots, \ell_n$ for $\mathcal{P}_{n-1}$ such that $\ell_j(z_k) = \delta_{jk}$ for all $j, k \in \{1, 2, \ldots, n\}$. This basis provides a solution to the following interpolation problem: find a polynomial of degree at most $n - 1$ that takes given real or complex values at n given distinct real or complex points.

Theorem 2.7.4 (Lagrange Interpolation) *Let $n \geq 1$ and let $z_1, z_2, \ldots, z_n \in \mathbb{C}$ be distinct. If $n = 1$, let $\ell_1(z) = 1$. If $n \geq 2$, let*

$$\ell_j(z) = \prod_{\substack{1 \leq i \leq n \\ i \neq j}} \frac{z - z_i}{z_j - z_i} \qquad (2.7.5)$$

for each $j = 1, 2, \ldots, n$, and let $\beta = \ell_1, \ell_2, \ldots, \ell_n$.

(a) *β is a basis for $\mathcal{P}_{n-1}$, and each $f \in \mathcal{P}_{n-1}$ has the unique representation*

$$f = f(z_1)\ell_1 + f(z_2)\ell_2 + \cdots + f(z_n)\ell_n. \qquad (2.7.6)$$

(b) *For each choice of $w_1, w_2, \ldots, w_n \in \mathbb{C}$, there is a unique $p \in \mathcal{P}_{n-1}$ such that $p(z_k) = w_k$ for $k = 1, 2, \ldots, n$. Moreover, p has the unique representation*

$$p = w_1\ell_1 + w_2\ell_2 + \cdots + w_n\ell_n \qquad (2.7.7)$$

as a linear combination of elements of β.

(c) *If $z_1, z_2, \ldots, z_n, w_1, w_2, \ldots, w_n \in \mathbb{R}$, then the polynomials in (2.7.5) and (2.7.7) are in $\mathcal{P}_{n-1}(\mathbb{R})$.*

Proof The definition (2.7.5) says that the polynomials

$$\ell_j(z) = \frac{(z - z_1) \cdots (z - z_{j-1})}{(z_j - z_1) \cdots (z_j - z_{j-1})} \frac{(z - z_{j+1}) \cdots (z - z_n)}{(z_j - z_{j+1}) \cdots (z_j - z_n)} \quad \text{for } j = 1, 2, \ldots, n,$$

are well defined since $z_1, z_2, \ldots, z_n$ are distinct. These polynomials have degree $n - 1$, and a computation reveals their most important property: $\ell_j(z_k) = \delta_{jk}$ for $j, k \in \{1, 2, \ldots, n\}$.

(a) For any complex-valued function f on $\mathbb{C}$,

$$f_n(z) = f(z_1)\ell_1(z) + f(z_2)\ell_2(z) + \cdots + f(z_n)\ell_n(z)$$

is a linear combination of elements of $\mathcal{P}_{n-1}$, so it is a polynomial of degree at most $n - 1$ that satisfies

$$f_n(z_k) = \sum_{j=1}^{n} f(z_j)\ell_j(z_k) = \sum_{j=1}^{n} f(z_j)\delta_{jk} = f(z_k) \quad \text{for } k = 1, 2, \ldots, n.$$

If $f \in \mathcal{P}_{n-1}$, the identity theorem for polynomials (Appendix B.4) ensures that $f = f_n$, which shows that $\beta = \ell_1, \ell_2, \ldots, \ell_n$ spans $\mathcal{P}_{n-1}$. If a linear combination $h = a_1\ell_1 + a_2\ell_2 + \cdots + a_n\ell_n$

is the zero polynomial, then (2.7.6) ensures that $a_k = h(z_k) = 0$ for each $k = 1, 2, \ldots, n$. We conclude that the spanning list β is linearly independent and hence it is a basis for $\mathcal{P}_{n-1}$.

(b) Let p be the polynomial defined by (2.7.7). Then $p(z_k) = w_k$ for each $k = 1, 2, \ldots, n$, and (a) ensures that

$$p = w_1 \ell_1 + w_2 \ell_2 + \cdots + w_n \ell_n = p(z_1)\ell_1 + p(z_2)\ell_2 + \cdots + p(z_n)\ell_n$$

is the unique representation of p as a linear combination of the elements of β.

(c) The polynomials ℓ_j in (2.7.5) have real coefficients if $z_1, z_2, \ldots, z_n$ are real, and the linear combination in (2.7.7) has real coefficients if $w_1, w_2, \ldots, w_n$ are real. $\square$

Definition 2.7.8 The polynomials ℓ_j defined in (2.7.5) are the *Lagrange basis polynomials* corresponding to the distinct *nodes* $z_1, z_2, \ldots, z_n \in \mathbb{C}$. The list $\beta = \ell_1, \ell_2, \ldots, \ell_n$ is the *Lagrange basis* for $\mathcal{P}_{n-1}$ corresponding to the distinct nodes $z_1, z_2, \ldots, z_n \in \mathbb{C}$. The equation (2.7.7) is the *Lagrange interpolation formula*.

So long as the nodes $z_1, z_2, \ldots, z_n$ are distinct, they may be chosen in any way we like, and any such choice is associated with a basis $\beta = \ell_1, \ell_2, \ldots, \ell_n$ for $\mathcal{P}_{n-1}$. Lagrange interpolation is used to approximate functions and design numerical integration rules. Other applications include scaling and sampling digital images, data compression, and cryptographic sharing schemes (see P.2.62 for an example).

If the equations (2.7.5) are expanded in powers of z, each element of the Lagrange basis $\beta = \ell_1, \ell_2, \ldots, \ell_n$ is expressed as a linear combination of elements of the basis $\gamma = z^{n-1}, z^{n-2}, \ldots, z, 1$. For example, if $n = 3$, a calculation reveals that

$$\ell_1(z) = \frac{z^2 - (z_2 + z_3)z + z_2 z_3}{(z_1 - z_2)(z_1 - z_3)},$$

$$\ell_2(z) = \frac{z^2 - (z_1 + z_3)z + z_1 z_3}{(z_2 - z_1)(z_2 - z_3)}, \quad \text{and} \qquad (2.7.9)$$

$$\ell_3(z) = \frac{z^2 - (z_1 + z_2)z + z_1 z_2}{(z_3 - z_1)(z_3 - z_2)}.$$

These identities say that the β-γ change-of-basis matrix for $n = 3$ is the (necessarily invertible) matrix

$$_\gamma[I]_\beta = \begin{bmatrix} \dfrac{1}{(z_1 - z_2)(z_1 - z_3)} & \dfrac{1}{(z_2 - z_1)(z_2 - z_3)} & \dfrac{1}{(z_3 - z_1)(z_3 - z_2)} \\[2ex] \dfrac{-(z_2 + z_3)}{(z_1 - z_2)(z_1 - z_3)} & \dfrac{-(z_1 + z_3)}{(z_2 - z_1)(z_2 - z_3)} & \dfrac{-(z_1 + z_2)}{(z_3 - z_1)(z_3 - z_2)} \\[2ex] \dfrac{z_2 z_3}{(z_1 - z_2)(z_1 - z_3)} & \dfrac{z_1 z_3}{(z_2 - z_1)(z_2 - z_3)} & \dfrac{z_1 z_2}{(z_3 - z_1)(z_3 - z_2)} \end{bmatrix}. \qquad (2.7.10)$$

Conversely, each element of the basis γ is a linear combination of elements of the basis β. If we apply (2.7.6) to the polynomials in the basis $\gamma = z^2, z, 1$, we obtain

$$\begin{aligned} z^2 &= z_1^2 \ell_1(z) &+& z_2^2 \ell_2(z) &+& z_3^2 \ell_3(z), \\ z &= z_1 \ell_1(z) &+& z_2 \ell_2(z) &+& z_3 \ell_3(z), \quad \text{and} \\ 1 &= \ell_1(z) &+& \ell_2(z) &+& \ell_3(z). \end{aligned}$$

These identities say that the γ-β change-of-basis matrix is the (necessarily invertible) 3×3 Vandermonde matrix

$$_\beta[I]_\gamma = \begin{bmatrix} z_1^2 & z_1 & 1 \\ z_2^2 & z_2 & 1 \\ z_3^2 & z_3 & 1 \end{bmatrix} ; \tag{2.7.11}$$

see Appendix C.2. Its inverse is the β-γ change-of-basis matrix (2.7.10).

Example 2.7.12 Consider the change-of-basis matrices (2.7.10) and (2.7.11). The $(1, 1)$ entry of $_\beta[I]_\gamma {}_\gamma[I]_\beta$ is

$$\frac{z_1^2 - (z_2 + z_3)z_1 + z_2 z_3}{(z_1 - z_2)(z_1 - z_3)}.$$

With the help of the first equation in (2.7.9) (or via direct calculation), we identify this expression as $\ell_1(z_1) = 1$. The $(2, 3)$ entry of $_\beta[I]_\gamma {}_\gamma[I]_\beta$ is

$$\frac{z_2^2 - (z_1 + z_2)z_2 + z_1 z_2}{(z_3 - z_1)(z_3 - z_2)},$$

which is $\ell_3(z_2) = 0$. The pattern revealed here is that $_\beta[I]_\gamma {}_\gamma[I]_\beta = [\ell_j(z_i)] = [\delta_{ji}] = I$, which is consistent with Theorem 2.5.7.a; see P.2.47.

The following special case of (2.7.9) leads to Simpson's rule for numerical integration.

Example 2.7.13 The Lagrange basis polynomials corresponding to the nodes $z_1 = -1$, $z_2 = 0$, and $z_3 = 1$ are

$$\ell_1(z) = \frac{z(z-1)}{2} = \frac{1}{2}z^2 - \frac{1}{2}z,$$

$$\ell_2(z) = \frac{(z+1)(z-1)}{-1} = -z^2 + 1, \quad \text{and}$$

$$\ell_3(z) = \frac{z(z+1)}{2} = \frac{1}{2}z^2 + \frac{1}{2}z.$$

If f is a given function on a finite real interval $[a, b]$, one might try to approximate it with the interpolating polynomial

$$f_n = f(x_1)\ell_1 + f(x_2)\ell_2 + \cdots + f(x_n)\ell_n \in \mathcal{P}_{n-1}, \tag{2.7.14}$$

in which $\beta = \ell_1, \ell_2, \ldots, \ell_n$ is a Lagrange basis associated with some choice of nodes $a \leq x_1 < x_2 < \cdots < x_n \leq b$ in $[a, b]$. A good approximation to a smooth function can often be obtained over a short interval. For example, Figure 2.3 shows a graph of $f(x) = \ln(x + 2)$ and an interpolating polynomial (2.7.14) with three equally spaced nodes in the interval $[-1, 1]$.

Even though a continuous real-valued function on a bounded real interval can be uniformly approximated by a polynomial (see P.8.28), trying to do this with an interpolating polynomial (2.7.14) with equally spaced nodes can be spectacularly unsuccessful. In Figure 2.4, the Lagrange interpolating polynomials oscillate wildly and fail to give good approximations on the whole interval. Increasing the number of nodes can make the oscillations even worse, but a strategic choice of unequally spaced nodes can deliver much better results. In Figure 2.5, the nodes of the interpolating polynomial are at

$$x_k = \cos\left(\frac{2k-1}{30}\pi\right) \quad \text{for } k = 1, 2, \ldots, 15, \tag{2.7.15}$$

which are the zeros of a Chebyshev polynomial of degree 15; see Section 6.11.

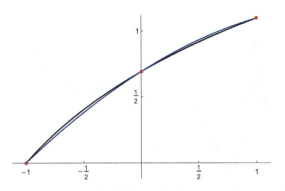

Figure 2.3 Graphs of $f(x) = \ln(x + 2)$ (black) and a three-point Lagrange interpolation polynomial (blue).

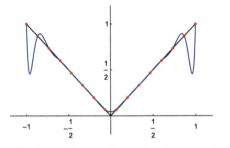

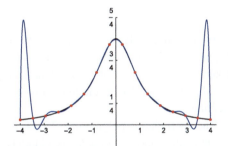

(a) Graphs of $f(x) = |x|$ and a sixteen-point Lagrange interpolation polynomial (blue).

(b) Graphs of $f(x) = \frac{1}{1+x^2}$ and a sixteen-point Lagrange interpolation polynomial (blue).

Figure 2.4 Lagrange interpolating polynomials using equally spaced nodes may fail to give good approximations on the whole interval.

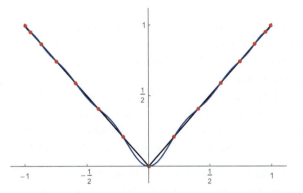

Figure 2.5 Graphs of $f(x) = |x|$ and a fifteen-point Lagrange interpolation polynomial (blue) using the Chebyshev nodes (2.7.15).

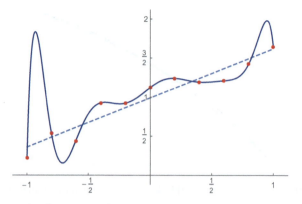

Figure 2.6 Eleven points in an approximately linear arrangement (red), their Lagrange interpolating polynomial (blue), and the best-fit line (dashed) determined by the method of least squares (see Section 8.5).

Interpolating polynomials are poorly suited for modeling data and discerning trends. In Figure 2.6, eleven points in an approximately linear arrangement are plotted, along with a Lagrange interpolating polynomial that interpolates them. It is useless in identifying the linear trend that is revealed in the least-squares linear model (dashed line) fitted to the data (see Section 8.5).

The interpolating polynomial (2.7.14) can be used to approximate a definite integral. If we integrate the approximation f_n, we obtain an approximation to the integral of f:

$$\int_a^b f(x)\,dx \approx \int_a^b f_n(x)\,dx = \int_a^b \left(\sum_{k=1}^n f(x_k)\ell_k(x) \right) dx$$

$$= \sum_{k=1}^n f(x_k) \underbrace{\int_a^b \ell_k(x)\,dx}_{w_k} = \sum_{k=1}^n w_k f(x_k). \qquad (2.7.16)$$

The integrals of the Lagrange basis polynomials (the weights w_k) depend on the interval and the nodes, but not on f; they can be computed, tabulated, and used repeatedly for any desired f. A numerical integration formula (2.7.16) is an *n-point rule*; if the nodes x_k are equally spaced, it is an *n-point Newton–Cotes rule*.

Theorem 2.7.4.a ensures that $f_n = f$ if $f \in \mathcal{P}_{n-1}$, so an n-point rule gives the exact value of the integral if $f \in \mathcal{P}_{n-1}$, regardless of the choice of nodes. However, if we are careful, we can do much better: there is a choice of unequally spaced nodes for which the n-point rule (2.7.16) gives the exact value of the integral for all $f \in \mathcal{P}_{2n-1}$; see Theorem 6.9.6.

Example 2.7.17 Consider the real interval $[-1, 1]$ and the nodes $x_1 = -1, x_2 = 0, x_3 = 1$. The corresponding Lagrange basis polynomials are displayed in Example 2.7.13. The weights are

$$w_1 = \int_{-1}^1 \ell_1(x)\,dx = \frac{1}{3}, \quad w_2 = \int_{-1}^1 \ell_2(x)\,dx = \frac{4}{3}, \quad \text{and} \quad w_3 = \int_{-1}^1 \ell_3(x)\,dx = \frac{1}{3}.$$

The corresponding 3-point Newton–Cotes rule

$$\int_{-1}^{1} f(x)\,dx \approx \frac{1}{3}f(-1) + \frac{4}{3}f(0) + \frac{1}{3}f(1) \tag{2.7.18}$$

is a special case of *Simpson's rule*; see P.2.60. For example, if $f(x) = \ln(x+2)$, then

$$\int_{-1}^{1} \ln(x+2)\,dx \approx \frac{1}{3}\ln 1 + \frac{4}{3}\ln 2 + \frac{1}{3}\ln 3 = \frac{1}{3}\ln 48 = 1.2904\ldots,$$

which differs from the exact value $3\ln 3 - 2 = 1.2958\ldots$ by less than 0.42%.

2.8 Problems

P.2.1 Verify the assertions in Example 2.3.9.

P.2.2 Let $A \in M_n$ with $a_{ij} = 1$ if $i = j$ or $i + j = n$, and $a_{ij} = 0$ otherwise. Compute rank A and find a basis for col A.

P.2.3 Let $A \in M_n$ with $a_{ij} = 1$ if $i + j$ is even and $a_{ij} = 0$ if $i + j$ is odd. Compute rank A and find a basis for col A.

P.2.4 Let $A \in M_n$ with $a_{ij} = 1$ if $1 \in \{i,j\}$ or $n \in \{i,j\}$. Compute rank A and find a basis for col A.

P.2.5 Let $\mathbf{x} \in \mathbb{F}^m$ and $\mathbf{y} \in \mathbb{F}^n$ be nonzero. Compute rank $\mathbf{xy}^\mathsf{T}$.

P.2.6 Let $A = \begin{bmatrix} a & b \\ c & d \end{bmatrix} \in M_2$ and suppose that dim row $A = 1$. If $[a\ b] = \lambda[c\ d]$, how are the columns $\begin{bmatrix} a \\ c \end{bmatrix}$ and $\begin{bmatrix} b \\ d \end{bmatrix}$ related?

P.2.7 Let $A = [a_{ij}] \in M_n$. Show that the entries of A are determined by the action of A on the standard basis via the identity $a_{ij} = \mathbf{e}_i^* A \mathbf{e}_j$. *Hint*: $A = I^* A I$, in which $I = [\mathbf{e}_1\ \mathbf{e}_2\ \ldots\ \mathbf{e}_n]$.

P.2.8 Let $A, B \in M_{m \times n}(\mathbb{F})$. Show that $A\mathbf{x} = B\mathbf{x}$ for every $\mathbf{x} \in \mathbb{F}^n$ if and only if $A = B$. *Hint*: Use the preceding problem.

P.2.9 Let $\mathbf{u} = [1\ 0\ 1]^\mathsf{T}$, $\mathbf{v}_1 = [0\ 1+i\ 1-i]^\mathsf{T}$, $\mathbf{v}_2 = [2i\ 1\ 0]^\mathsf{T}$, and $\mathbf{v}_3 = [2\ -1\ 1]^\mathsf{T}$. Let $\beta = \mathbf{v}_1, \mathbf{v}_2, \mathbf{v}_3$. Show that β is a basis for $\mathbb{C}^3$ and compute $[\mathbf{u}]_\beta$.

P.2.10 State and prove the assertions in Corollary 2.2.14 that are not addressed in the proof.

P.2.11 Compute rank A, in which

$$A = \begin{bmatrix} 0 & 3 & 1 & 5 & 0 & 5 & 1 & 4 & 4 & 0 \\ 2 & 3 & 0 & 0 & 1 & 3 & 3 & 2 & 1 & 5 \\ 1 & 4 & 0 & 1 & 5 & 3 & 3 & 0 & 0 & 0 \\ 2 & 4 & 5 & 0 & 2 & 0 & 2 & 0 & 4 & 3 \end{bmatrix} \in M_{4 \times 10}.$$

Hint: Find four columns whose span includes $\mathbf{e}_1, \mathbf{e}_2, \mathbf{e}_3, \mathbf{e}_4$.

P.2.12 Which of $A = \begin{bmatrix} 1 & 1 \\ 0 & 1 \end{bmatrix}$, $B = \begin{bmatrix} 1 & 1 \\ 1 & 1 \end{bmatrix}$, $C = \begin{bmatrix} 1 & 1 \\ 0 & 0 \end{bmatrix}$, and $D = \begin{bmatrix} 1 & 0 \\ 0 & 0 \end{bmatrix}$ are similar to each other?

P.2.13 Let $A \in M_n$. (a) Show that A is similar to I if and only if $A = I$. (b) Show that A is similar to 0 if and only if $A = 0$. (c) Show that A is similar to λI for some scalar λ if and only if $A = \lambda I$.

P.2.14 Show that $A = \begin{bmatrix} 0 & 0 \\ 0 & 0 \end{bmatrix}$ and $B = \begin{bmatrix} 0 & 1 \\ 0 & 0 \end{bmatrix}$ have the same trace and determinant but are not similar.

P.2.15 Let $A = [a_{ij}] \in \mathsf{M}_n$ and suppose that

$$a_{ij} = \begin{cases} 1 & \text{if } j = i+1, \\ 0 & \text{otherwise.} \end{cases}$$

What are the values of $\operatorname{rank}(A^k)$ for $k = 0, 1, 2, \ldots$?

P.2.16 Let $\beta = \mathbf{u}_1, \mathbf{u}_2, \ldots, \mathbf{u}_n$ be a basis for a nonzero $\mathbb{F}$-vector space $\mathcal{V}$. (a) If any vector in $\mathcal{V}$ is appended to β, explain why the resulting list still spans $\mathcal{V}$ but is not linearly independent. (b) If any vector in β is omitted, explain why the resulting list is still linearly independent but no longer spans $\mathcal{V}$.

P.2.17 Review Example 1.3.15. Show that each set $\mathcal{P}_{\text{even}}$ and $\mathcal{P}_{\text{odd}}$ is the kernel of a linear operator on the vector space $\mathcal{P}$. *Hint*: P.1.41.

P.2.18 Let $\mathcal{V}$ be an n-dimensional $\mathbb{F}$-vector space with $n \geq 2$ and let $\beta = \mathbf{v}_1, \mathbf{v}_2, \ldots, \mathbf{v}_r$ be a list of vectors in $\mathcal{V}$ with $1 \leq r < n$. Show that β does not span $\mathcal{V}$.

P.2.19 Let $\mathbf{x}_1, \mathbf{x}_2, \ldots, \mathbf{x}_k \in \mathbb{R}^n$. Then $\mathbf{x}_1, \mathbf{x}_2, \ldots, \mathbf{x}_k$ are real vectors in $\mathbb{C}^n$. Show that $\mathbf{x}_1, \mathbf{x}_2, \ldots, \mathbf{x}_k$ are linearly independent in $\mathbb{R}^n$ if and only if they are linearly independent in $\mathbb{C}^n$.

P.2.20 Let $A \in \mathsf{M}_{m \times n}(\mathbb{R})$. Definition 2.3.5 suggests that the rank of A (considered as an element of $\mathsf{M}_{m \times n}(\mathbb{F})$) might depend on whether $\mathbb{F} = \mathbb{R}$ or $\mathbb{F} = \mathbb{C}$. Show that $\operatorname{rank} A = r$ (considered as an element of $\mathsf{M}_{m \times n}(\mathbb{R})$) if and only if $\operatorname{rank} A = r$ (considered as an element of $\mathsf{M}_{m \times n}(\mathbb{C})$).

P.2.21 Let $A \in \mathsf{M}_{m \times n}(\mathbb{F})$ and let $\mathbf{x}_1, \mathbf{x}_2, \ldots, \mathbf{x}_k \in \mathbb{F}^n$. (a) If $\mathbf{x}_1, \mathbf{x}_2, \ldots, \mathbf{x}_k$ are linearly independent and $\operatorname{rank} A = n$, show that $A\mathbf{x}_1, A\mathbf{x}_2, \ldots, A\mathbf{x}_k \in \mathbb{F}^m$ are linearly independent. (b) If $A\mathbf{x}_1, A\mathbf{x}_2, \ldots, A\mathbf{x}_k \in \mathbb{F}^m$ are linearly independent, show that $\mathbf{x}_1, \mathbf{x}_2, \ldots, \mathbf{x}_k$ are linearly independent. (c) Although the hypothesis that A has full column rank is not needed in (b), show by example that it cannot be omitted in (a).

P.2.22 Let $A \in \mathsf{M}_{m \times n}$ Show that (a) $\operatorname{rank} A = \operatorname{rank} \overline{A}$, and (b) $\operatorname{rank} A = \operatorname{rank} A^*$.

P.2.23 Let $A \in \mathsf{M}_n$ be invertible. Why is A^T invertible? Show that $(A^\mathsf{T})^{-1} = (A^{-1})^\mathsf{T}$.

P.2.24 Let $P \in \mathsf{M}_n$ be idempotent. Show that $P = 0$ if and only if $\operatorname{tr} P = 0$.

P.2.25 If $A = XY$ is a full-rank factorization, show that $A = (XB)(B^{-1}Y)$ is a full-rank factorization for any invertible $B \in \mathsf{M}_{\operatorname{rank} A}$.

P.2.26 Show that each of the following is a full-rank factorization of (1.8.1):

$$\begin{bmatrix} 2 & 3 \\ 5 & 6 \\ 8 & 9 \end{bmatrix} \begin{bmatrix} 2 & 1 & 0 \\ -1 & 0 & 1 \end{bmatrix}, \quad \begin{bmatrix} 1 & 3 \\ 4 & 6 \\ 7 & 9 \end{bmatrix} \begin{bmatrix} 1 & \frac{1}{2} & 0 \\ 0 & \frac{1}{2} & 1 \end{bmatrix}, \quad \text{and} \quad \begin{bmatrix} -1 & 3 \\ -1 & 9 \\ -1 & 15 \end{bmatrix} \begin{bmatrix} \frac{1}{2} & -\frac{1}{2} & -\frac{3}{2} \\ \frac{1}{2} & \frac{1}{2} & \frac{1}{2} \end{bmatrix}.$$

P.2.27 If $A \in \mathsf{M}_{m \times n}$ has an invertible $k \times k$ submatrix B, show that $\operatorname{rank} A \geq k$. *Hint*: If the entries of B lie in distinct columns $j_1, j_2, \ldots, j_k$ of A, show that those columns are linearly independent.

P.2.28 If $A \in \mathsf{M}_{m \times n}$ is nonzero and $\operatorname{rank} A = r$, show that: (a) A has an invertible $r \times r$ submatrix, and (b) no square submatrix of A of size greater than r is invertible. (c) Conclude that $\operatorname{rank} A$ is the size of the largest invertible submatrix of A.

P.2.29 Let $\mathcal{V} \subseteq \mathsf{M}_n$ denote the subspace of all skew-symmetric matrices $(A = -A^\mathsf{T})$. Compute $\dim \mathcal{V}$ and find a basis for $\mathcal{V}$.

P.2.30 What is the dimension of the real vector space of Hermitian matrices in $\mathsf{M}_n(\mathbb{C})$? See P.1.12.

P.2.31 Let

$$A = \begin{bmatrix} 0 & 1 & 0 \\ 0 & 0 & 1 \\ 0 & 0 & 0 \end{bmatrix} \quad \text{and} \quad B = \begin{bmatrix} 0 & 1 & 0 \\ 0 & 0 & 1 \\ 1 & 0 & 0 \end{bmatrix}.$$

(a) Let $V \subseteq M_3$ denote the span of the set of all sums and products of nonnegative powers of A and A^T (including $A^0 = I$). Compute dim V and find a basis for V. (b) Do the same for B.

P.2.32 Let $\mathcal{U}$, $\mathcal{V}$, and $\mathcal{W}$ be $\mathbb{F}$-vector spaces. Let $S \in \mathcal{L}(\mathcal{U}, \mathcal{V})$ and $T \in \mathcal{L}(\mathcal{V}, \mathcal{W})$. Define $TS : \mathcal{U} \to \mathcal{W}$ by $(TS)\mathbf{u} = T(S\mathbf{u})$. Show that $TS \in \mathcal{L}(\mathcal{U}, \mathcal{W})$.

P.2.33 Let $A \in M_n$. Show that $T : M_n \to M_n$ defined by $T(A) = AX - XA$ is a linear transformation.

P.2.34 Verify the three matrix representations of the differentiation operator in Example 2.6.6 and the similarity of $_\beta[T]_\beta$ and $_\alpha[T]_\alpha$.

P.2.35 Let $n \geq 1$, and let $\beta = \mathbf{v}_1, \mathbf{v}_2, \ldots, \mathbf{v}_n$ and $\gamma = \mathbf{w}_1, \mathbf{w}_2, \ldots, \mathbf{w}_n$ be bases of $\mathbb{F}^n$. Define the $n \times n$ matrices $V = [\mathbf{v}_1 \ \mathbf{v}_2 \ \cdots \ \mathbf{v}_n]$ and $W = [\mathbf{w}_1 \ \mathbf{w}_2 \ \cdots \ \mathbf{w}_n]$. Let $S = {}_\gamma[I]_\beta$ be the β-γ change-of-basis matrix. (a) Explain why $V = WS$ and deduce that $S = W^{-1}V$. See Example 2.5.5. (b) Why is $_\beta[I]_\gamma = V^{-1}W$?

P.2.36 Let V be the complex vector space of finitely nonzero sequences. Let T and S be the right-shift and left-shift operators defined in Example 2.4.16. Show that $ST = I$ but $TS \neq I$. Does this contradict Theorem 2.2.13?

P.2.37 Let

$$A = \begin{bmatrix} 1 & 0 & 0 \\ 0 & 1 & 0 \end{bmatrix} \quad \text{and} \quad B = \begin{bmatrix} 1 & 0 \\ 0 & 1 \\ 0 & 0 \end{bmatrix}.$$

Show that $AB = I$ but $BA \neq I$. Does this contradict Theorem 2.2.13?

P.2.38 Show by example that $AB = I$ and $AC = I$ need not imply $B = C$ if the matrices involved are not square. Does this contradict Corollary 2.2.14?

P.2.39 Show that: (a) $\begin{bmatrix} 0 & 1 \\ 0 & 0 \end{bmatrix}$ is similar to $\begin{bmatrix} 0 & c \\ 0 & 0 \end{bmatrix}$ for all nonzero $c \in \mathbb{C}$. *Hint*: Consider $S = \begin{bmatrix} 1 & 0 \\ 0 & c \end{bmatrix}$. (b) $\begin{bmatrix} 1 & 1 \\ 0 & 1 \end{bmatrix}$ is similar to $\begin{bmatrix} 1 & c \\ 0 & 1 \end{bmatrix}$ for all nonzero $c \in \mathbb{C}$.

P.2.40 Prove that the following define equivalence relations $\cong$ on M_n. (a) $A \cong B$ if null $A =$ null B. (b) $A \cong B$ if $A = XB$, in which $X \in M_n$ is invertible. (c) $A \cong B$ if $A = XBY$, in which $X, Y \in M_n$ are invertible. (d) $A \cong B$ if $A = X^*BX$, in which $X \in M_n$ is invertible. (e) $A \cong B$ if $A = UBU^*$, in which $U^*U = I$.

P.2.41 What is the rank of

$$A = \begin{bmatrix} 1 & n+1 & 2n+1 & \cdots & (n-1)n+1 \\ 2 & n+2 & 2n+2 & \cdots & (n-1)n+2 \\ \vdots & \vdots & \vdots & \ddots & \vdots \\ n & 2n & 3n & \cdots & n^2 \end{bmatrix}?$$

Find a basis for col A.

P.2.42 Let $A \in M_n(\mathbb{R})$ and $B \in M_n(\mathbb{C})$. Let $B = X + iY$, in which $X, Y \in M_n(\mathbb{R})$. If $AB = I$, show that $AX = XA = I$ and $Y = 0$. Deduce that a real square matrix has a complex inverse if and only if it has a real inverse.

P.2.43 Explain why dim $\mathbb{C}^n = 2n$ if $\mathbb{C}^n$ is considered as a real vector space.

P.2.44 Let $z_1 = 1 + i$ and $z_2 = -1 + 2i$. Find polynomials $\ell_1, \ell_2 \in \mathcal{P}_1$ such that $\ell_j(z_k) = \delta_{jk}$, for $j, k \in \{1, 2\}$. Verify that $1 = \ell_1(z) + \ell_2(z)$ and $z = z_1\ell_1(z) + z_2\ell_2(z)$.

P.2.45 Let $z_1, z_2, \ldots, z_n \in \mathbb{C}$ be distinct. Define $\ell(z) = (z - z_1)(z - z_2) \cdots (z - z_n)$. Show that the Lagrange interpolation formula (2.7.7) can be expressed as

$$p(z) = \sum_{k=1}^{n} w_k \frac{\ell(z)}{(z - z_k)\ell'(z_k)}.$$

P.2.46 Let $f(x) = x^2$ and $n = 3$. Calculate the Lagrange basis $\beta = \ell_1, \ell_2, \ell_3$ and compute the values of $f(z_1)\ell_1(z) + f(z_2)\ell_2(z) + f(z_3)\ell_3(z)$ if the nodes are: (a) $z_1 = 1$, $z_2 = 2$, and $z_3 = 3$. (b) $z_1 = 1$, $z_2 = 3$, and $z_3 = 5$. (c) $z_1 = -2$, $z_2 = 0$, and $z_3 = 1$.

P.2.47 Use the change-of-basis matrices (2.7.10) and (2.7.11) to compute the $(2, 2)$ and $(3, 2)$ entries of $_\beta[I]_\gamma \,_\gamma[I]_\beta$.

P.2.48 Let $n = 3$, let $\beta = \ell_1, \ell_2, \ell_3$ be the Lagrange basis for the nodes $-1, 0, 1$, and let $\gamma = z^2, z, 1$. Use (2.7.10) and (2.7.11) to compute the entries of (a) $_\beta[I]_\gamma$, (b) $_\gamma[I]_\beta$, (c) $_\beta[I]_\gamma \,_\gamma[I]_\beta$, and (d) $_\gamma[I]_\beta \,_\beta[I]_\gamma$.

P.2.49 Let $a, b, c, d \in \mathbb{C}$. Show that there is a $p \in \mathcal{P}_2$ such that $p(-1) = a$, $p(0) = b$, $p(1) = c$, and $p(2) = d$ if and only if $a - 3b + 3c - d = 0$.

P.2.50 Let $n \geq 2$ and let $\ell_1, \ell_2, \ldots, \ell_n$ be the Lagrange basis for $\mathcal{P}_{n-1}$ corresponding to distinct nodes $z_1, z_2, \ldots, z_n \in \mathbb{C}$. (a) For any $a \in \mathbb{C}$ and each $r = 1, 2, \ldots, n - 1$, show that $(a - z)^r = \sum_{k=1}^{n} (a - z_k)^r \ell_k(z)$. (b) Show that $\sum_{k=1}^{n} (z - z_k)^r \ell_k(z) = 0$ for each $r = 1, 2, \ldots, n - 1$.

P.2.51 Let $\beta = \ell_1, \ell_2, \ldots, \ell_n$ be the Lagrange basis for $\mathcal{P}_{n-1}$ corresponding to distinct nodes $z_1, z_2, \ldots, z_n \in \mathbb{C}$, and let $\gamma = z^{n-1}, z^{n-2}, \ldots, z, 1$. Prove that

$$\begin{array}{rcl}
z^{n-1} & = & z_1^{n-1}\ell_1(z) + z_2^{n-1}\ell_2(z) + \cdots + z_n^{n-1}\ell_n(z) \\
z^{n-2} & = & z_1^{n-2}\ell_1(z) + z_2^{n-2}\ell_2(z) + \cdots + z_n^{n-2}\ell_n(z) \\
& \vdots & \\
z & = & z_1\ell_1(z) + z_2\ell_2(z) + \cdots + z_n\ell_n(z) \\
1 & = & \ell_1(z) + \ell_2(z) + \cdots + \ell_n(z).
\end{array}$$

Explain why the γ-β change-of-basis matrix $_\beta[I]_\gamma$ is

$$V(z_1, z_2, \ldots, z_n) = \begin{bmatrix} z_1^{n-1} & z_1^{n-2} & \cdots & z_1 & 1 \\ z_2^{n-1} & z_2^{n-2} & \cdots & z_2 & 1 \\ \vdots & \vdots & \ddots & \vdots & \vdots \\ z_n^{n-1} & z_n^{n-2} & \cdots & z_n & 1 \end{bmatrix}. \tag{2.8.1}$$

Why must $V(z_1, z_2, \ldots, z_n)$ be invertible (no computations, please!) if $z_1, z_2, \ldots, z_n$ are distinct? The matrix (2.8.1) is an n-by-n Vandermonde matrix.

P.2.52 In the notation of the preceding problem, if $\delta = 1, z, z^2, \ldots, z^{n-1}$ and $\alpha = \ell_n, \ell_{n-1}, \ldots, \ell_1$, what are $_\beta[I]_\delta$, $_\alpha[I]_\gamma$, and $_\alpha[I]_\delta$? These three matrices are alternative versions of the Vandermonde matrix (2.8.1).

P.2.53 Let $z_1, z_2, \ldots, z_n \in \mathbb{C}$ be distinct. Provide details for the following steps to evaluate the determinant of the Vandermonde matrix (2.8.1):

(a) $p(z) = \det V(z_1, z_2, \ldots, z_{n-1}, z) \in \mathcal{P}_{n-1}$ and $p(z_1) = p(z_2) = \cdots = p(z_{n-1}) = 0$.

(b) $p(z) = c(z - z_1)(z - z_2) \cdots (z - z_{n-1})$, in which the scalar c depends on $z_1, z_2, \ldots, z_{n-1}$.

(c) The coefficient of the term z^{n-1} in p is $(-1)^{n+1} \det V(z_1, z_2, \ldots, z_{n-1})$. *Hint:* Laplace expansion along the last row.

(d) $\det V(z_1, z_2, \ldots, z_n) = (z_n - z_1)(z_n - z_2) \cdots (z_n - z_{n-1}) \det V(z_1, z_2, \ldots, z_{n-1})$.

(e) $\det V(z_1, z_2) = z_1 - z_2$.

(f) $\det V(z_1, z_2, \ldots, z_n) = \displaystyle\prod_{1 \le i < j \le n} (z_i - z_j)$. *Hint:* Induction.

P.2.54 Consider the Vandermonde matrix

associated with the nodes $-1, 0, 1$; see (2.7.11). Use Example 2.7.13 to determine its inverse. Check your answer.

P.2.55 If functions $f_1, f_2, \ldots, f_n \in C_{\mathbb{R}}[-1, 1]$ are at least $n - 1$ times differentiable on $(-1, 1)$, one may be able to prove that they are linearly independent in the following way: If $a_1, a_2, \ldots, a_n \in \mathbb{R}$ and if $h = a_1 f_1 + a_2 f_2 + \cdots + a_n f_n \in C_{\mathbb{R}}[-1, 1]$ is the zero function, then each derivative $h', h'', \ldots, h^{(n-1)}$ is the zero function on $(-1, 1)$. For any $x_0 \in (-1, 1)$, we have a homogeneous system of linear equations

$$
\begin{array}{ccccccc}
a_1 f_1(x_0) & + & a_2 f_2(x_0) & + & \cdots & + & a_n f_n(x_0) & = & 0 \\
a_1 f_1'(x_0) & + & a_2 f_2'(x_0) & + & \cdots & + & a_n f_n'(x_0) & = & 0 \\
\vdots & & \vdots & & \vdots & \ddots & \vdots & & \vdots \\
a_1 f_1^{(n-1)}(x_0) & + & a_2 f_2^{(n-1)}(x_0) & + & \cdots & + & a_n f_n^{(n-1)}(x_0) & = & 0.
\end{array}
\tag{2.8.2}
$$

If

$$
\det
\begin{bmatrix}
f_1(x_0) & f_2(x_0) & \cdots & f_n(x_0) \\
f_1'(x_0) & f_2'(x_0) & \cdots & f_n'(x_0) \\
\vdots & \vdots & \ddots & \vdots \\
f_1^{(n-1)}(x_0) & f_2^{(n-1)}(x_0) & \cdots & f_n^{(n-1)}(x_0)
\end{bmatrix}
\tag{2.8.3}
$$

is nonzero, then $a_1 = a_2 = \cdots = a_n = 0$ and it follows that $f_1, f_2, \ldots, f_n$ are linearly independent. Use this method with $x_0 = 0$ to show the linear independence of:

(a) $1, x, x^2, \ldots, x^n$,

(b) $e^{\lambda_1 x}, e^{\lambda_2 x}, \ldots, e^{\lambda_n x}$ if $\lambda_1, \lambda_2, \ldots, \lambda_n \in \mathbb{R}$ are distinct, and

(c) $\ln(x + 2), \cos x, e^x$.

Remark: The determinant (2.8.3) is the *Wronskian* of the functions $f_1, f_2, \ldots, f_n$ at the point x_0. Nonvanishing of the Wronskian at a point is a sufficient condition for linear independence. The following problem addresses the converse implication.

P.2.56 Consider $f_1(x) = x^2$ and $f_2(x) = x|x|$. Show that (a) their Wronskian (2.8.3) is zero for all $x_0 \in [-1, 1]$, and (b) $f_1, f_2 \in C_{\mathbb{R}}[-1, 1]$ are linearly independent.

P.2.57 Why must Simpson's rule (2.7.18) give the exact value of $\int_{-1}^{1} f(x)\, dx$ if $f \in \mathcal{P}_2$? Show that it also gives the exact value if $f \in \mathcal{P}_3$.

P.2.58 Use the method in P.2.55 to show that the Lagrange basis polynomials in Example 2.7.13 are linearly independent.

P.2.59 In the n-point rule (2.7.16), show that $w_1 + w_2 + \cdots + w_n = b - a$.

P.2.60 *Simpson's rule* approximates $\int_a^b f(x)\,dx$ in the following way: Choose an odd $n \geq 3$, let $h = (b-a)/(n-1)$, and let $x_j = a + (j-1)h$ for $j = 1, 2, \ldots, n$. In each subinterval $[x_j, x_{j+2}]$, for $j = 1, 3, 5, \ldots, n-2$, let $\ell_j, \ell_{j+1}, \ell_{j+2}$ be the Lagrange basis corresponding to the nodes x_j, x_{j+1}, x_{j+2} (see Example 2.7.13). Approximate f on $[x_j, x_{j+2}]$ by its Lagrange interpolating polynomial, integrate the approximation

$$\int_{x_j}^{x_{j+2}} f(x)\,dx \approx \sum_{k=0}^{2} f(x_{j+k}) \int_{x_j}^{x_{j+2}} \ell_{j+k}(x)\,dx,$$

and add. Show that the resulting approximation to the definite integral is

$$\int_a^b f(x)\,dx = \sum_{j=1,3,5,\ldots,n-2} \int_{x_j}^{x_{j+2}} f(x)\,dx$$

$$\approx \frac{h}{3}\big(f(a) + 4f(x_2) + 2f(x_3) + 4f(x_4) + \cdots + 2f(x_{n-2}) + 4f(x_{n-1}) + f(b)\big).$$

This approximation is a *composite Simpson rule*; (2.7.18) illustrates the case $n = 3$.

P.2.61 Continue with the notation of the preceding problem. Show that the Simpson rule with $[a,b] = [-1,1]$ and $n = 5$ is

$$\int_{-1}^{1} f(x)\,dx \approx \frac{1}{6}\big(f(-1) + 4f(-\tfrac{1}{2}) + 2f(0) + 4f(\tfrac{1}{2}) + f(1)\big).$$

For $f(x) = \ln(x+2)$, show that the value of this approximation is $\frac{1}{6}\ln(151{,}875/64) = 1.2953\ldots$, which differs from the exact value by less than 0.04%.

P.2.62 Eight business partners $P_1, P_2, \ldots, P_8$ keep emergency cash in a bank account that has a secret passcode a_0. The partners agree that any three of them acting together (but no fewer) may obtain and use the passcode. Here is the system they use: Their banker generates, in confidence, two random integers a_1, a_2 and eight distinct random integers $n_1, n_2, \ldots, n_8$, forms the secret polynomial $p(x) = a_2 x^2 + a_1 x + a_0$, and computes the eight values $w_i = p(n_i)$. The banker gives partner P_i the pair of integers (n_i, w_i) for $i = 1, 2, \ldots, 8$. If three distinct partners P_i, P_j, P_k agree to obtain the passcode, they use the nodes n_i, n_j, n_k to compute a Lagrange basis $\beta = \ell_1, \ell_2, \ell_3$ of $\mathcal{P}_2$. They then use Theorem 2.7.4.a to compute the secret polynomial p, which reveals the passcode.

    (a) Show that the passcode is $w_i \ell_1(0) + w_j \ell_2(0) + w_k \ell_3(0)$.

    (b) Suppose that the following pairs are distributed to the partners:

        (1, 1443),   (2, 1786),   (3, 2263),   (4, 2874),   (5, 3619),   (6, 4498),
        (7, 5511),   (8, 6658).

        Choose any three pairs and find the passcode.

    (c) Describe a system such that any m of q partners could obtain the passcode, but no fewer could do so.

P.2.63 Discover a formula for $f(n) = 1 + 2 + \cdots + n$ as follows: You suspect that f is a polynomial of degree two, say $f(x) = ax^2 + bx + c$. Evaluate $(n, f(n))$ for three values of n of your choosing. Use these data to compute a Lagrange basis $\beta = \ell_1, \ell_2, \ell_3$ for $\mathcal{P}_2$ and use (2.7.7) to represent f. See Appendix D.1 for a proof that $f(n)$ is the correct value for the sum, once you have the formula.

P.2.64 Use the method in the preceding problem to discover a formula for $g(n) = 1^2 + 2^2 + 3^2 + \cdots + n^2$. Use induction to prove that your formula is correct.

2.9 Notes

The broad topics of interpolation and approximation are discussed in depth in [Dav63]. This is a good place to read about the mysterious oscillations observed with Lagrange interpolation at many equally spaced nodes (Figure 2.4) and the role of Chebyshev polynomials in obtaining better uniform approximations (Figure 2.5).

For numerical stability, it is important that all the weights w_k in an n-point rule (2.7.16) be positive. This is the case for n-point Newton–Cotes rules with $n \le 6$, but for $n \ge 7$ some weights are negative. In Section 6.9, we discuss a class of n-point rules for which the weights are positive for all $n = 2, 3, \ldots$.

2.10 Some Important Concepts

- Basis for a vector space.

- Dimension of a vector space.

- $\operatorname{rank} A = \operatorname{rank} A^\mathsf{T} = \dim \operatorname{col} A = \dim \operatorname{row} A$.

- Full-rank factorizations.

- Functions that are linear transformations.

- The action of a linear transformation on a basis determines its action on all vectors.

- β-γ matrix representation of a linear transformation.

- Change-of-basis matrices are invertible, and vice versa.

- Similarity of matrices.

- Two matrices are similar if and only if each is a matrix representation of the same linear transformation.

- Lagrange interpolation.

- Every choice of n distinct nodes determines a Lagrange basis for the vector space of polynomials of degree at most $n - 1$.

3   Block Matrices

A matrix is not just an array of scalars. It can be thought of as an array of submatrices in many different ways. We begin by regarding a matrix as an array of columns and we explore some implications of this viewpoint for matrix products and Cramer's rule. We turn to arrays of rows, which lead to additional insights for matrix products. We discuss determinants of block matrices, block versions of elementary matrices, and Cauchy's formula for the determinant of a bordered matrix. Finally, we introduce the Kronecker product, which provides a way to construct block matrices with a special structure.

3.1   Row and Column Partitions

Let $\mathbf{b}_1, \mathbf{b}_2, \ldots, \mathbf{b}_n \in \mathbb{F}^r$ be the columns of $B \in \mathsf{M}_{r \times n}(\mathbb{F})$. Then the presentation

$$B = [\mathbf{b}_1 \ \mathbf{b}_2 \ \ldots \ \mathbf{b}_n] \tag{3.1.1}$$

is *partitioned according to its columns.* For any $\mathbf{x} = [x_1 \ x_2 \ \ldots \ x_n]^\mathsf{T} \in \mathbb{F}^n$,

$$B\mathbf{x} = [\mathbf{b}_1 \ \mathbf{b}_2 \ \ldots \ \mathbf{b}_n] \begin{bmatrix} x_1 \\ x_2 \\ \vdots \\ x_n \end{bmatrix} = x_1 \mathbf{b}_1 + x_2 \mathbf{b}_2 + \cdots + x_n \mathbf{b}_n \tag{3.1.2}$$

is a linear combination of the columns of B. The coefficients are the entries of $\mathbf{x}$.

Let $A \in \mathsf{M}_{m \times r}(\mathbb{F})$. Use the same column partition of B to write $AB \in \mathsf{M}_{m \times n}(\mathbb{F})$ as

$$AB = [A\mathbf{b}_1 \ A\mathbf{b}_2 \ \ldots \ A\mathbf{b}_n]. \tag{3.1.3}$$

This presentation partitions AB according to its columns, each of which is a linear combination of the columns of A. The coefficients are the entries of the corresponding column of B.

The identity (3.1.2) is plausible; it is a formal product of a row object and a column object. To prove that it is correct, we must verify that corresponding entries of the left and right sides are the same. Let $B = [b_{ij}] \in \mathsf{M}_{r \times n}(\mathbb{F})$, so the columns of B are the vectors

$$\mathbf{b}_j = \begin{bmatrix} b_{1j} \\ b_{2j} \\ \vdots \\ b_{rj} \end{bmatrix} \quad \text{for } j = 1, 2, \ldots, n.$$

For any $i \in \{1, 2, \ldots, r\}$, the ith entry of $B\mathbf{x}$ is $\sum_{k=1}^n b_{ik} x_k$. This equals the ith entry of $x_1 \mathbf{b}_1 + x_2 \mathbf{b}_2 + \cdots + x_n \mathbf{b}_n$, which is $x_1 b_{i1} + x_2 b_{i2} + \cdots + x_n b_{in}$.

Example 3.1.4 Let

$$A = \begin{bmatrix} 1 & 2 \\ 3 & 4 \end{bmatrix}, \qquad B = \begin{bmatrix} 4 & 5 & 2 \\ 6 & 7 & 1 \end{bmatrix}, \tag{3.1.5}$$

$\mathbf{b}_1 = \begin{bmatrix} 4 \\ 6 \end{bmatrix}, \mathbf{b}_2 = \begin{bmatrix} 5 \\ 7 \end{bmatrix}$, and $\mathbf{b}_3 = \begin{bmatrix} 2 \\ 1 \end{bmatrix}$. Then (3.1.3) in this case is

$$AB = \begin{bmatrix} \begin{bmatrix} 1 & 2 \\ 3 & 4 \end{bmatrix}\begin{bmatrix} 4 \\ 6 \end{bmatrix} & \begin{bmatrix} 1 & 2 \\ 3 & 4 \end{bmatrix}\begin{bmatrix} 5 \\ 7 \end{bmatrix} & \begin{bmatrix} 1 & 2 \\ 3 & 4 \end{bmatrix}\begin{bmatrix} 2 \\ 1 \end{bmatrix} \end{bmatrix} = \begin{bmatrix} 16 & 19 & 4 \\ 36 & 43 & 10 \end{bmatrix}.$$

Example 3.1.6 The identity (3.1.2) permits us to construct a matrix that maps a given basis $\mathbf{x}_1, \mathbf{x}_2, \ldots, \mathbf{x}_n$ for $\mathbb{F}^n$ to another given basis $\mathbf{y}_1, \mathbf{y}_2, \ldots, \mathbf{y}_n$ for $\mathbb{F}^n$. The matrices $X = [\mathbf{x}_1 \ \mathbf{x}_2 \ \ldots \ \mathbf{x}_n]$ and $Y = [\mathbf{y}_1 \ \mathbf{y}_2 \ \ldots \ \mathbf{y}_n]$ are invertible and satisfy $X\mathbf{e}_i = \mathbf{x}_i$, $X^{-1}\mathbf{x}_i = \mathbf{e}_i$, and $Y\mathbf{e}_i = \mathbf{y}_i$ for $i = 1, 2, \ldots, n$. Thus, $YX^{-1}\mathbf{x}_i = Y\mathbf{e}_i = \mathbf{y}_i$ for $i = 1, 2, \ldots, n$.

Example 3.1.7 The identity (3.1.3) provides a short proof of the fact that $A \in \mathsf{M}_n$ is invertible if $A\mathbf{x} = \mathbf{y}$ is consistent for each $\mathbf{y} \in \mathbb{F}^n$. By hypothesis, there exist $\mathbf{b}_1, \mathbf{b}_2, \ldots, \mathbf{b}_n \in \mathbb{F}^n$ such that $A\mathbf{b}_i = \mathbf{e}_i$ for $i = 1, 2, \ldots, n$. Let $B = [\mathbf{b}_1 \ \mathbf{b}_2 \ \ldots \ \mathbf{b}_n]$. Then $AB = A[\mathbf{b}_1 \ \mathbf{b}_2 \ \ldots \ \mathbf{b}_n] = [A\mathbf{b}_1 \ A\mathbf{b}_2 \ \ldots \ A\mathbf{b}_n] = [\mathbf{e}_1 \ \mathbf{e}_2 \ \ldots \ \mathbf{e}_n] = I$, so Theorem 2.2.13 ensures that $B = A^{-1}$.

Column partitions permit us to give a short proof of Cramer's rule. Although conceptually elegant, Cramer's rule is not recommended for use in numerical algorithms.

Theorem 3.1.8 (Cramer's Rule) *Let* $A = [\mathbf{a}_1 \ \mathbf{a}_2 \ \ldots \ \mathbf{a}_n] \in \mathsf{M}_n(\mathbb{F})$ *be invertible, let* $\mathbf{y} \in \mathbb{F}^n$, *and let*

$$A_i(\mathbf{y}) = [\mathbf{a}_1 \ \ldots \ \mathbf{a}_{i-1} \ \mathbf{y} \ \mathbf{a}_{i+1} \ \ldots \ \mathbf{a}_n] \in \mathsf{M}_n(\mathbb{F})$$

denote the matrix obtained by replacing the ith column of A with $\mathbf{y}$. *Let* $\mathbf{x} = [x_1 \ x_2 \ \ldots \ x_n]^\mathsf{T}$, *in which*

$$x_1 = \frac{\det A_1(\mathbf{y})}{\det A}, \qquad x_2 = \frac{\det A_2(\mathbf{y})}{\det A}, \ldots \qquad x_n = \frac{\det A_n(\mathbf{y})}{\det A}. \tag{3.1.9}$$

Then $\mathbf{x}$ *is the unique solution to* $A\mathbf{x} = \mathbf{y}$.

Proof Let $\mathbf{x} = [x_i] \in \mathbb{F}^n$ be the unique solution to $A\mathbf{x} = \mathbf{y}$; see Appendix C.3. For $i = 1, 2, \ldots, n$, let

$$X_i = [\mathbf{e}_1 \ \ldots \ \mathbf{e}_{i-1} \ \mathbf{x} \ \mathbf{e}_{i+1} \ \ldots \ \mathbf{e}_n] \in \mathsf{M}_n(\mathbb{F})$$

denote the matrix obtained by replacing the ith column of I_n with $\mathbf{x}$. Perform a Laplace expansion along the ith row of X_i and obtain $\det X_i = x_i \det I_{n-1} = x_i$. Since $A\mathbf{x} = \mathbf{y}$ and $A\mathbf{e}_j = \mathbf{a}_j$ for each $j = 1, 2, \ldots, n$, we have

$$\begin{aligned} X_i &= [\mathbf{e}_1 \ \ldots \ \mathbf{e}_{i-1} \ \mathbf{x} \ \mathbf{e}_{i+1} \ \ldots \ \mathbf{e}_n] \\ &= [A^{-1}\mathbf{a}_1 \ \ldots \ A^{-1}\mathbf{a}_{i-1} \ A^{-1}\mathbf{y} \ A^{-1}\mathbf{a}_{i+1} \ \ldots \ A^{-1}\mathbf{a}_n] \\ &= A^{-1}[\mathbf{a}_1 \ \ldots \ \mathbf{a}_{i-1} \ \mathbf{y} \ \mathbf{a}_{i+1} \ \ldots \ \mathbf{a}_n] \\ &= A^{-1}A_i(\mathbf{y}). \end{aligned}$$

Then for each $i = 1, 2, \ldots, n$,

$$x_i = \det X_i = \det\big(A^{-1}A_i(\mathbf{y})\big) = \det(A^{-1})\det A_i(\mathbf{y}) = \frac{\det A_i(\mathbf{y})}{\det A}. \qquad \square$$

Example 3.1.10 Let

$$A = \begin{bmatrix} 1 & 2 & 3 \\ 8 & 9 & 4 \\ 7 & 6 & 5 \end{bmatrix}, \qquad \mathbf{x} = \begin{bmatrix} x_1 \\ x_2 \\ x_3 \end{bmatrix}, \quad \text{and} \quad \mathbf{y} = \begin{bmatrix} 2 \\ 2 \\ 2 \end{bmatrix}.$$

Since $\det A = -48$ is nonzero, $A\mathbf{x} = \mathbf{y}$ has a unique solution. To form A_1, A_2, and A_3, replace the first, second, and third columns of A by $\mathbf{y}$, respectively:

$$A_1(\mathbf{y}) = \begin{bmatrix} 2 & 2 & 3 \\ 2 & 9 & 4 \\ 2 & 6 & 5 \end{bmatrix}, \qquad A_2(\mathbf{y}) = \begin{bmatrix} 1 & 2 & 3 \\ 8 & 2 & 4 \\ 7 & 2 & 5 \end{bmatrix}, \quad \text{and} \quad A_3(\mathbf{y}) = \begin{bmatrix} 1 & 2 & 2 \\ 8 & 9 & 2 \\ 7 & 6 & 2 \end{bmatrix}.$$

Then

$$x_1 = \frac{\det A_1(\mathbf{y})}{\det A} = \frac{20}{-48} = -\frac{5}{12},$$

$$x_2 = \frac{\det A_2(\mathbf{y})}{\det A} = \frac{-16}{-48} = \frac{1}{3}, \quad \text{and}$$

$$x_3 = \frac{\det A_3(\mathbf{y})}{\det A} = \frac{-28}{-48} = \frac{7}{12}.$$

What we have done for columns, we can also do for rows. Let $\mathbf{a}_1, \mathbf{a}_2, \ldots, \mathbf{a}_m \in \mathbb{F}^r$ and let

$$A = \begin{bmatrix} \mathbf{a}_1^\mathsf{T} \\ \mathbf{a}_2^\mathsf{T} \\ \vdots \\ \mathbf{a}_m^\mathsf{T} \end{bmatrix} \in \mathsf{M}_{m \times r}(\mathbb{F}). \tag{3.1.11}$$

For any $\mathbf{x} \in \mathbb{F}^m$,

$$\mathbf{x}^\mathsf{T} A = [x_1 \; x_2 \; \ldots \; x_m] \begin{bmatrix} \mathbf{a}_1^\mathsf{T} \\ \mathbf{a}_2^\mathsf{T} \\ \vdots \\ \mathbf{a}_m^\mathsf{T} \end{bmatrix} = x_1 \mathbf{a}_1^\mathsf{T} + x_2 \mathbf{a}_2^\mathsf{T} + \cdots + x_m \mathbf{a}_m^\mathsf{T} \tag{3.1.12}$$

and

$$AB = \begin{bmatrix} \mathbf{a}_1^\mathsf{T} B \\ \mathbf{a}_2^\mathsf{T} B \\ \vdots \\ \mathbf{a}_m^\mathsf{T} B \end{bmatrix} \in \mathsf{M}_{m \times n}(\mathbb{F}). \tag{3.1.13}$$

These presentations partition A and AB *according to their rows* and make it clear that each row of AB is a linear combination of the rows of B. The coefficients in that linear combination are the entries of the corresponding row of A.

Example 3.1.14 With A and B as in (3.1.5), $\mathbf{a}_1^\mathsf{T} = [1 \; 2]$ and $\mathbf{a}_2^\mathsf{T} = [3 \; 4]$. In this case (3.1.13) is

$$AB = \begin{bmatrix} [1 \; 2] \begin{bmatrix} 4 & 5 & 2 \\ 6 & 7 & 1 \end{bmatrix} \\ [3 \; 4] \begin{bmatrix} 4 & 5 & 2 \\ 6 & 7 & 1 \end{bmatrix} \end{bmatrix} = \begin{bmatrix} 16 & 19 & 4 \\ 36 & 43 & 10 \end{bmatrix}.$$

If we use the row partition of A in (3.1.11) and the column partition of B in (3.1.1), we obtain the presentation

$$AB = \begin{bmatrix} \mathbf{a}_1^{\mathsf{T}} \\ \vdots \\ \mathbf{a}_m^{\mathsf{T}} \end{bmatrix} [\mathbf{b}_1 \ \mathbf{b}_2 \ \dots \ \mathbf{b}_n] = \begin{bmatrix} \mathbf{a}_1^{\mathsf{T}}\mathbf{b}_1 & \mathbf{a}_1^{\mathsf{T}}\mathbf{b}_2 & \cdots & \mathbf{a}_1^{\mathsf{T}}\mathbf{b}_n \\ \mathbf{a}_2^{\mathsf{T}}\mathbf{b}_1 & \mathbf{a}_2^{\mathsf{T}}\mathbf{b}_2 & \cdots & \mathbf{a}_2^{\mathsf{T}}\mathbf{b}_n \\ \vdots & \vdots & \ddots & \vdots \\ \mathbf{a}_m^{\mathsf{T}}\mathbf{b}_1 & \mathbf{a}_m^{\mathsf{T}}\mathbf{b}_2 & \cdots & \mathbf{a}_m^{\mathsf{T}}\mathbf{b}_n \end{bmatrix} \in \mathsf{M}_{m \times n}(\mathbb{F}) \qquad (3.1.15)$$

for the matrix product AB. The scalars $\mathbf{a}_i^{\mathsf{T}}\mathbf{b}_j$ in (3.1.15) are often referred to as *inner products*, though that term is strictly correct only for real matrices (the inner product of $\mathbf{b}_j$ and $\mathbf{a}_i$ in $\mathbb{C}^n$ is $\mathbf{a}_i^*\mathbf{b}_j$; see Example 5.3.3).

Example 3.1.16 With the matrices A and B in (3.1.5), the identity (3.1.15) is

$$AB = \begin{bmatrix} [1 \ 2]\begin{bmatrix}4\\6\end{bmatrix} & [1 \ 2]\begin{bmatrix}5\\7\end{bmatrix} & [1 \ 2]\begin{bmatrix}2\\1\end{bmatrix} \\ [3 \ 4]\begin{bmatrix}4\\6\end{bmatrix} & [3 \ 4]\begin{bmatrix}5\\7\end{bmatrix} & [3 \ 4]\begin{bmatrix}2\\1\end{bmatrix} \end{bmatrix} = \begin{bmatrix} 16 & 19 & 4 \\ 36 & 43 & 10 \end{bmatrix}.$$

We now turn to a final way to present a matrix product. Let $A \in \mathsf{M}_{m \times r}$ and $B \in \mathsf{M}_{r \times n}$. Partition $A = [\mathbf{a}_1 \ \mathbf{a}_2 \ \dots \ \mathbf{a}_r]$ according to its columns, and partition B according to its rows, so

$$B = \begin{bmatrix} \mathbf{b}_1^{\mathsf{T}} \\ \mathbf{b}_2^{\mathsf{T}} \\ \vdots \\ \mathbf{b}_r^{\mathsf{T}} \end{bmatrix}.$$

Then

$$AB = [\mathbf{a}_1 \ \mathbf{a}_2 \ \dots \ \mathbf{a}_r] \begin{bmatrix} \mathbf{b}_1^{\mathsf{T}} \\ \vdots \\ \mathbf{b}_r^{\mathsf{T}} \end{bmatrix} = \mathbf{a}_1\mathbf{b}_1^{\mathsf{T}} + \mathbf{a}_2\mathbf{b}_2^{\mathsf{T}} + \cdots + \mathbf{a}_r\mathbf{b}_r^{\mathsf{T}} \qquad (3.1.17)$$

is presented as a sum of $m \times n$ matrices, each of which has rank at most one. The summands in (3.1.17) are *outer products*. The outer-product presentation (3.1.17) of a matrix product is a key part of an algorithm to construct some basic matrix factorizations; see Sections 4.3 and 15.4.

Example 3.1.18 With the matrices A and B in (3.1.5), the identity (3.1.17) is

$$AB = \begin{bmatrix}1\\3\end{bmatrix}[4 \ 5 \ 2] + \begin{bmatrix}2\\4\end{bmatrix}[6 \ 7 \ 1] = \begin{bmatrix} 4 & 5 & 2 \\ 12 & 15 & 6 \end{bmatrix} + \begin{bmatrix} 12 & 14 & 2 \\ 24 & 28 & 4 \end{bmatrix} = \begin{bmatrix} 16 & 19 & 4 \\ 36 & 43 & 10 \end{bmatrix}.$$

Let $A = [\mathbf{a}_1 \ \mathbf{a}_2 \ \dots \ \mathbf{a}_r] \in \mathsf{M}_{n \times r}$ and $B = [\mathbf{b}_1 \ \mathbf{b}_2 \ \dots \ \mathbf{b}_m] \in \mathsf{M}_{n \times m}$. If each column of A is a linear combination of the columns of B, then (3.1.2) says that there are $\mathbf{x}_1, \mathbf{x}_2, \dots, \mathbf{x}_r \in \mathbb{C}^m$ such that $\mathbf{a}_i = B\mathbf{x}_i$ for $i = 1, 2, \dots, r$. If we let $X = [\mathbf{x}_1 \ \mathbf{x}_2 \ \dots \ \mathbf{x}_r] \in \mathsf{M}_{m \times r}$, the identity

$$A = BX \qquad (3.1.19)$$

is a compact way to say that each column of A is a linear combination of the columns of B, that is, $\text{col } A \subseteq \text{col } B$; see Theorem 1.4.10.

It can be useful to group together some adjacent columns and present $B \in \mathsf{M}_{n \times r}$ in the partitioned form $B = [B_1 \ B_2 \ \ldots \ B_k]$ in which $B_j \in \mathsf{M}_{n \times r_j}$ for $j = 1, 2, \ldots, k$, and $r_1 + r_2 + \cdots + r_k = r$. If $A \in \mathsf{M}_{m \times n}$, then $AB_j \in \mathsf{M}_{m \times r_j}$ and

$$AB = [AB_1 \ AB_2 \ \ldots \ AB_k]. \tag{3.1.20}$$

An analogous partition according to groups of rows is also possible.

Example 3.1.21 Partition the matrix B in (3.1.5) as

$$B = [B_1 \ B_2], \qquad B_1 = \begin{bmatrix} 4 & 5 \\ 6 & 7 \end{bmatrix}, \quad \text{and} \quad B_2 = \begin{bmatrix} 2 \\ 1 \end{bmatrix}.$$

Then

$$AB = [AB_1 \ AB_2] = \begin{bmatrix} \begin{bmatrix} 1 & 2 \\ 3 & 4 \end{bmatrix}\begin{bmatrix} 4 & 5 \\ 6 & 7 \end{bmatrix} & \begin{bmatrix} 1 & 2 \\ 3 & 4 \end{bmatrix}\begin{bmatrix} 2 \\ 1 \end{bmatrix} \end{bmatrix} = \begin{bmatrix} 16 & 19 & 4 \\ 36 & 43 & 10 \end{bmatrix}.$$

Theorem 2.2.8.b says that any linearly independent list in a finite-dimensional $\mathbb{F}$-vector space can be extended to a basis. The following theorem is a block-matrix version of this basic fact.

Theorem 3.1.22 *Suppose that* $1 \leq k < n$, $U, V \in \mathsf{M}_{n \times k}(\mathbb{F})$, *and* $\operatorname{rank} U = \operatorname{rank} V = k$.

(a) *There is a* $W \in \mathsf{M}_{n \times (n-k)}(\mathbb{F})$ *such that* $[V \ W]$ *is invertible.*

(b) *There is an invertible* $C \in \mathsf{M}_n(\mathbb{F})$ *such that* $U = CV$.

Proof (a) Since the columns of $V = [\mathbf{v}_1 \ \mathbf{v}_2 \ \ldots \ \mathbf{v}_k]$ are linearly independent, Theorem 2.2.8.b ensures that there are $n - k$ additional vectors such that $\beta = \mathbf{v}_1, \mathbf{v}_2, \ldots, \mathbf{v}_k, \mathbf{w}_1, \mathbf{w}_2, \ldots, \mathbf{w}_{n-k}$ is a basis for $\mathbb{F}^n$. Let $W = [\mathbf{w}_1 \ \mathbf{w}_2 \ \ldots \ \mathbf{w}_{n-k}]$. Theorem 2.3.13 says that $[V \ W]$ is invertible.

(b) Choose $W, Z \in \mathsf{M}_{n \times (n-k)}(\mathbb{F})$ such that $F = [V \ W]$ and $G = [U \ Z]$ are invertible. Then $I_n = F^{-1}F = [F^{-1}V \ F^{-1}W]$ and $I_n = G^{-1}G = [G^{-1}U \ G^{-1}Z]$, so $U = (GF^{-1})V$ and $C = GF^{-1}$ is invertible. $\qquad\qquad\square$

Example 3.1.23 Each $A \in \mathsf{M}_n(\mathbb{F})$ can be factored as $A = BDC$, in which $B, C \in \mathsf{M}_n(\mathbb{F})$ are invertible and

$$D = \operatorname{diag}(\underbrace{1, 1, \ldots, 1}_{\operatorname{rank} A}, \underbrace{0, 0, \ldots, 0}_{n - \operatorname{rank} A}).$$

Why? If $A = 0$, choose $D = 0$ and $B = C = I$. If A is invertible, choose $B = A$ and $D = C = I$. If $r = \operatorname{rank} A$ and $1 \leq r \leq n - 1$, let $A = XY$ be a full-rank factorization and let $B = [X \ U]$ and $C^\mathsf{T} = [Y^\mathsf{T} \ V]$ be $n \times n$ and invertible. Then $A = XI_rY = BDC$.

Example 3.1.24 The "side-by-side" method for matrix inversion can be justified with a block-matrix calculation. If $A \in \mathsf{M}_n$ is invertible, then its reduced row echelon form is I. Let $R = E_k E_{k-1} \cdots E_1$ be the product of elementary matrices $E_1, E_2, \ldots, E_k$ that encode row operations that row reduce A to I. Then $RA = I$ and $R = A^{-1}$ (Theorem 2.2.13). Thus, $R[A \ I] = [RA \ R] = [I \ A^{-1}]$, so reducing the block matrix $[A \ I]$ to its reduced row echelon form reveals A^{-1}.

Our final observation about row and column partitions is that they provide a convenient way to show that the determinant is a linear function of each column (respectively, each row) of a square matrix.

Theorem 3.1.25 *Let* $A = [\mathbf{a}_1\ \mathbf{a}_2\ \ldots\ \mathbf{a}_n] \in \mathsf{M}_n$, *let* $j \in \{1, 2, \ldots, n\}$, *and partition* $A = [A_1\ \mathbf{a}_j\ A_2]$, *in which* $A_1 = [\mathbf{a}_1\ \ldots\ \mathbf{a}_{j-1}]$ *(not present if $j = 1$) and* $A_2 = [\mathbf{a}_{j+1}\ \ldots\ \mathbf{a}_n]$ *(not present if $j = n$). For* $\mathbf{x}_1, \mathbf{x}_2, \ldots, \mathbf{x}_m \in \mathbb{C}^n$ *and* $c_1, c_2, \ldots, c_m \in \mathbb{C}$, *we have*

$$\det\left[A_1\ \sum_{k=1}^{m} c_k \mathbf{x}_k\ A_2\right] = \sum_{k=1}^{m} c_k \det[A_1\ \mathbf{x}_k\ A_2]. \tag{3.1.26}$$

Proof For $k = 1, 2, \ldots, m$, let $\mathbf{x}_k = [x_{1k}\ x_{2k}\ \ldots\ x_{nk}]^\mathsf{T}$ and evaluate the left side of (3.1.26) with a Laplace expansion along column j; see (C.4.1). We have

$$\det\left[A_1\ \sum_{k=1}^{m} c_k \mathbf{x}_k\ A_2\right] = \sum_{i=1}^{n} (-1)^{i+j}\left(\sum_{k=1}^{m} c_k x_{ik}\right) \det A_{ij}$$

$$= \sum_{k=1}^{m} c_k \sum_{i=1}^{n} (-1)^{i+j} x_{ik} \det A_{ij}$$

$$= \sum_{k=1}^{m} c_k \det [A_1\ \mathbf{x}_k\ A_2],$$

in which $\det A_{ij}$ is the (i, j) minor of A. $\square$

Since $\det A = \det A^\mathsf{T}$, there is an analogous result for rows.

3.2 Block Partitions and Direct Sums

Consider the 2×2 block partitions

$$A = \begin{bmatrix} A_{11} & A_{12} \\ A_{21} & A_{22} \end{bmatrix} \in \mathsf{M}_{m \times n}(\mathbb{F}) \quad \text{and} \quad B = \begin{bmatrix} B_{11} & B_{12} \\ B_{21} & B_{22} \end{bmatrix} \in \mathsf{M}_{p \times q}(\mathbb{F}), \tag{3.2.1}$$

in which each A_{ij} is $m_i \times n_j$, each B_{ij} is $p_i \times q_j$,

$$m_1 + m_2 = m, \qquad n_1 + n_2 = n, \qquad p_1 + p_2 = p, \quad \text{and} \quad q_1 + q_2 = q.$$

To form $A + B$ we must have $m = p$ and $n = q$. To use block-matrix operations to compute

$$A + B = \begin{bmatrix} A_{11} + B_{11} & A_{12} + B_{12} \\ A_{21} + B_{21} & A_{22} + B_{22} \end{bmatrix}$$

we must have $m_i = p_i$ and $n_i = q_i$ for $i = 1, 2$. If these conditions are satisfied, then the partitions (3.2.1) are *conformal for addition* (or just *conformal*).

To form AB we must have $n = p$, in which case AB is $m \times q$. To use block-matrix operations to compute

$$AB = \begin{bmatrix} A_{11} & A_{12} \\ A_{21} & A_{22} \end{bmatrix}\begin{bmatrix} B_{11} & B_{12} \\ B_{21} & B_{22} \end{bmatrix} = \begin{bmatrix} A_{11}B_{11} + A_{12}B_{21} & A_{11}B_{12} + A_{12}B_{22} \\ A_{21}B_{11} + A_{22}B_{21} & A_{21}B_{12} + A_{22}B_{22} \end{bmatrix}$$

we must have $n_i = p_i$ for each $i = 1, 2$. If these conditions are satisfied, then the partitions (3.2.1) are *conformal for multiplication* (or just *conformal*).

Partitioned matrices (3.2.1) are examples of *block matrices*; the submatrices A_{ij} are the *blocks*. For notational simplicity we have described 2×2 block matrices, but all of these ideas carry over to other conformally structured block matrices. For example,

$$[A_1\ A_2]\begin{bmatrix} B_1 \\ B_2 \end{bmatrix} = A_1 B_1 + A_2 B_2 \tag{3.2.2}$$

and

$$\begin{bmatrix} A_1 \\ A_2 \end{bmatrix} [B_1 \ B_2] = \begin{bmatrix} A_1 B_1 & A_1 B_2 \\ A_2 B_1 & A_2 B_2 \end{bmatrix} \tag{3.2.3}$$

if the respective partitions are conformal for multiplication.

A *block $m \times n$ matrix* is an array

$$A = \begin{bmatrix} A_{11} & A_{12} & \cdots & A_{1n} \\ A_{21} & A_{22} & \cdots & A_{2n} \\ \vdots & \vdots & \ddots & \vdots \\ A_{m1} & A_{m2} & \cdots & A_{mn} \end{bmatrix},$$

in which $A_{ij} \in \mathsf{M}_{p_i \times q_j}(\mathbb{F})$ for $i = 1, 2, \dots, m$ and $j = 1, 2, \dots, n$. If $A_{ij} = 0$ whenever $i > j$, then A is *block upper triangular*. If $A_{ij} = 0$ whenever $i < j$, then A is *block lower triangular*. If $A_{ij} = 0$ whenever $i \neq j$, then A is *block diagonal*.

Example 3.2.4 If $A \in \mathsf{M}_m$ and $B \in \mathsf{M}_n$ are invertible, then

$$\begin{bmatrix} A & 0 \\ 0 & B \end{bmatrix} \begin{bmatrix} A^{-1} & 0 \\ 0 & B^{-1} \end{bmatrix} = \begin{bmatrix} AA^{-1} & 0 \\ 0 & BB^{-1} \end{bmatrix} = \begin{bmatrix} I_m & 0 \\ 0 & I_n \end{bmatrix} = I.$$

This computation and Theorem 2.2.13 show that $\begin{bmatrix} A & 0 \\ 0 & B \end{bmatrix}^{-1} = \begin{bmatrix} A^{-1} & 0 \\ 0 & B^{-1} \end{bmatrix}$.

Example 3.2.5 Let $X \in \mathsf{M}_{m \times n}$ and consider the block triangular matrix

$$\begin{bmatrix} I_m & X \\ 0 & I_n \end{bmatrix} \in \mathsf{M}_{m+n}.$$

Then

$$\begin{bmatrix} I & -X \\ 0 & I \end{bmatrix} \begin{bmatrix} I & X \\ 0 & I \end{bmatrix} = \begin{bmatrix} I & X - X \\ 0 & I \end{bmatrix} = \begin{bmatrix} I & 0 \\ 0 & I \end{bmatrix} = I,$$

and hence

$$\begin{bmatrix} I & X \\ 0 & I \end{bmatrix}^{-1} = \begin{bmatrix} I & -X \\ 0 & I \end{bmatrix}. \tag{3.2.6}$$

The following example presents a generalization of (3.2.6).

Example 3.2.7 Consider a 2×2 block triangular matrix

$$\begin{bmatrix} Y & X \\ 0 & Z \end{bmatrix} = \begin{bmatrix} Y & 0 \\ 0 & Z \end{bmatrix} \begin{bmatrix} I & Y^{-1}X \\ 0 & I \end{bmatrix},$$

in which $Y \in \mathsf{M}_n$ and $Z \in \mathsf{M}_m$ are invertible. Then we can use (3.2.6) to compute

$$\begin{bmatrix} Y & X \\ 0 & Z \end{bmatrix}^{-1} = \begin{bmatrix} I & Y^{-1}X \\ 0 & I \end{bmatrix}^{-1} \begin{bmatrix} Y & 0 \\ 0 & Z \end{bmatrix}^{-1} = \begin{bmatrix} I & -Y^{-1}X \\ 0 & I \end{bmatrix} \begin{bmatrix} Y^{-1} & 0 \\ 0 & Z^{-1} \end{bmatrix}$$

$$= \begin{bmatrix} Y^{-1} & -Y^{-1}XZ^{-1} \\ 0 & Z^{-1} \end{bmatrix}. \tag{3.2.8}$$

The identity (3.2.8) implies a useful fact about triangular and block triangular matrices.

Theorem 3.2.9 *Let $n \geq 2$ and suppose that an $n \times n$ block matrix $A = [A_{ij}]$ is block upper (respectively, lower) triangular and has invertible diagonal blocks. Then A is invertible, its inverse is block upper (respectively, lower) triangular, and the diagonal blocks of A^{-1} are $A_{11}^{-1}, A_{22}^{-1}, \dots, A_{nn}^{-1}$, in that order.*

Proof Suppose that A is $n \times n$ block upper triangular. We proceed by induction on n, the number of blocks on the diagonal of A. In the base case $n = 2$, the identity (3.2.8) ensures that

$$\begin{bmatrix} A_{11} & A_{12} \\ 0 & A_{22} \end{bmatrix}^{-1} = \begin{bmatrix} A_{11}^{-1} & \star \\ 0 & A_{22}^{-1} \end{bmatrix},$$

which is block upper triangular and has the asserted diagonal blocks. The symbol $\star$ indicates an entry or block whose value is not relevant to the argument. For the induction step, let $n \geq 3$ and suppose that every $n \times n$ block upper triangular matrix with invertible diagonal blocks has an inverse that is block upper triangular and has the asserted diagonal blocks. Let

$$A = \begin{bmatrix} B & \star \\ 0 & C \end{bmatrix},$$

in which B is an $n \times n$ block upper triangular matrix. It follows from (3.2.8) that

$$A^{-1} = \begin{bmatrix} B^{-1} & \star \\ 0 & C^{-1} \end{bmatrix}.$$

The induction hypothesis ensures that B^{-1} is block upper triangular and has the asserted diagonal blocks. Thus, A^{-1} is block upper triangular and has the asserted diagonal blocks.

If A is block lower triangular, the assertion follows from considering A^{T}. $\qquad\square$

If $A \in \mathsf{M}_n(\mathbb{F})$ is upper triangular and all its diagonal entries are nonzero, the preceding theorem says that A is invertible, its inverse is upper triangular, and the diagonal entries of A^{-1} are the reciprocals of the diagonal entries of A.

The identity (3.2.6) leads to a basic similarity of a 2×2 block upper triangular matrix that we will use to prove a block-diagonalization theorem (Theorem 11.4.3).

Theorem 3.2.10 *Let $B \in \mathsf{M}_m$, let $C, X \in \mathsf{M}_{m \times n}$, and let $D \in \mathsf{M}_n$. Then*

$$\begin{bmatrix} B & C \\ 0 & D \end{bmatrix} \quad \text{is similar to} \quad \begin{bmatrix} B & C + XD - BX \\ 0 & D \end{bmatrix}. \tag{3.2.11}$$

Proof Use (3.2.6) and compute the similarity

$$\begin{bmatrix} I & X \\ 0 & I \end{bmatrix} \begin{bmatrix} B & C \\ 0 & D \end{bmatrix} \begin{bmatrix} I & -X \\ 0 & I \end{bmatrix} = \begin{bmatrix} B & C + XD - BX \\ 0 & D \end{bmatrix}. \qquad\square$$

Example 3.2.12 The transpose and conjugate transpose operate on block matrices as follows:

$$\begin{bmatrix} A_{11} & A_{12} & \cdots & A_{1n} \\ A_{21} & A_{22} & \cdots & A_{2n} \\ \vdots & \vdots & \ddots & \vdots \\ A_{m1} & A_{m2} & \cdots & A_{mn} \end{bmatrix}^{\mathsf{T}} = \begin{bmatrix} A_{11}^{\mathsf{T}} & A_{21}^{\mathsf{T}} & \cdots & A_{m1}^{\mathsf{T}} \\ A_{12}^{\mathsf{T}} & A_{22}^{\mathsf{T}} & \cdots & A_{m2}^{\mathsf{T}} \\ \vdots & \vdots & \ddots & \vdots \\ A_{1n}^{\mathsf{T}} & A_{2n}^{\mathsf{T}} & \cdots & A_{mn}^{\mathsf{T}} \end{bmatrix}$$

and

$$\begin{bmatrix} A_{11} & A_{12} & \cdots & A_{1n} \\ A_{21} & A_{22} & \cdots & A_{2n} \\ \vdots & \vdots & \ddots & \vdots \\ A_{m1} & A_{m2} & \cdots & A_{mn} \end{bmatrix}^{*} = \begin{bmatrix} A_{11}^{*} & A_{21}^{*} & \cdots & A_{m1}^{*} \\ A_{12}^{*} & A_{22}^{*} & \cdots & A_{m2}^{*} \\ \vdots & \vdots & \ddots & \vdots \\ A_{1n}^{*} & A_{2n}^{*} & \cdots & A_{mn}^{*} \end{bmatrix}.$$

Example 3.2.13 The product of conformally partitioned block lower (respectively, upper) triangular matrices is block lower (respectively, upper) triangular, and the diagonal blocks of the product are products of the factors' corresponding diagonal blocks. For example,

$$\begin{bmatrix} A_{11} & 0 & 0 \\ A_{21} & A_{22} & 0 \\ A_{31} & A_{32} & A_{33} \end{bmatrix} \begin{bmatrix} B_{11} & 0 & 0 \\ B_{21} & B_{22} & 0 \\ B_{31} & B_{32} & B_{33} \end{bmatrix} = \begin{bmatrix} A_{11}B_{11} & 0 & 0 \\ \star & A_{22}B_{22} & 0 \\ \star & \star & A_{33}B_{33} \end{bmatrix}.$$

In particular, the product of lower (respectively, upper) triangular matrices is lower (respectively, upper) triangular and the diagonal entries of the product are products of the factors' corresponding diagonal entries. For example,

$$\begin{bmatrix} a_{11} & a_{12} & a_{13} \\ 0 & a_{22} & a_{23} \\ 0 & 0 & a_{33} \end{bmatrix} \begin{bmatrix} b_{11} & b_{12} & b_{13} \\ 0 & b_{22} & b_{23} \\ 0 & 0 & b_{33} \end{bmatrix} = \begin{bmatrix} a_{11}b_{11} & \star & \star \\ 0 & a_{22}b_{22} & \star \\ 0 & 0 & a_{33}b_{33} \end{bmatrix}.$$

A *direct sum* of square matrices is a block matrix that is *block diagonal*, that is, every off-diagonal block is zero:

$$A \oplus B = \begin{bmatrix} A & 0 \\ 0 & B \end{bmatrix} \quad \text{and} \quad A_{11} \oplus A_{22} \oplus \cdots \oplus A_{kk} = \begin{bmatrix} A_{11} & & \\ & \ddots & \\ & & A_{kk} \end{bmatrix}; \qquad (3.2.14)$$

we often suppress zero blocks for visual clarity. The square matrices $A_{11}, A_{22},\ldots, A_{kk}$ in (3.2.14) are *direct summands*. Our convention is that if a direct summand has size 0, then it is omitted from the direct sum. For any scalars $\lambda_1, \lambda_2, \ldots, \lambda_n$, the direct sum $[\lambda_1] \oplus [\lambda_2] \oplus \cdots \oplus [\lambda_n]$ of 1×1 matrices is often written as

$$\text{diag}(\lambda_1, \lambda_2, \ldots, \lambda_n) = \begin{bmatrix} \lambda_1 & & \\ & \ddots & \\ & & \lambda_n \end{bmatrix} \in \mathsf{M}_n.$$

Linear combinations and products of conformally partitioned direct sums all act blockwise; so do powers and polynomials. Let A, B, C, D be matrices of appropriate sizes, let a, b be scalars, let k be an integer, and let p be a polynomial. The following relations can be verified with computations.

(a) $a(A \oplus B) + b(C \oplus D) = (aA + bC) \oplus (aB + bD)$.

(b) $(A \oplus B)(C \oplus D) = AC \oplus BD$.

(c) $(A \oplus B)^k = A^k \oplus B^k$.

(d) $p(A \oplus B) = p(A) \oplus p(B)$.

The following fact about left and right multiplication by diagonal matrices comes up frequently in matrix manipulations. Let $A \in \mathsf{M}_{m \times n}$ be presented according to its entries, columns, and rows as

$$A = [a_{ij}] = [\mathbf{c}_1 \ \mathbf{c}_2 \ \ldots \ \mathbf{c}_n] = \begin{bmatrix} \mathbf{r}_1^\mathsf{T} \\ \vdots \\ \mathbf{r}_m^\mathsf{T} \end{bmatrix} \in \mathsf{M}_{m \times n}.$$

Let $\Lambda = \text{diag}(\lambda_1, \lambda_2, \ldots, \lambda_n)$ and $M = \text{diag}(\mu_1, \mu_2, \ldots, \mu_m)$. Then

$$A\Lambda = [\lambda_j a_{ij}] = [\lambda_1 \mathbf{c}_1 \ \lambda_2 \mathbf{c}_2 \ \ldots \ \lambda_n \mathbf{c}_n] \qquad (3.2.15)$$

and

$$MA = [\mu_i a_{ij}] = \begin{bmatrix} \mu_1 \mathbf{r}_1^\mathsf{T} \\ \vdots \\ \mu_m \mathbf{r}_m^\mathsf{T} \end{bmatrix}. \tag{3.2.16}$$

Thus, right multiplication of A by a diagonal matrix Λ multiplies the columns of A by the corresponding diagonal entries of Λ. Left multiplication of A by a diagonal matrix M multiplies the rows of A by the corresponding diagonal entries of M.

We say that $\lambda_1, \lambda_2, \dots, \lambda_k$ are *distinct* if $\lambda_i \neq \lambda_j$ whenever $i \neq j$. The following lemma about commuting block matrices is at the heart of several important results.

Lemma 3.2.17 *Let $\lambda_1, \lambda_2, \dots, \lambda_k$ be scalars and let $\Lambda = \lambda_1 I_{n_1} \oplus \lambda_2 I_{n_2} \oplus \cdots \oplus \lambda_k I_{n_k} \in \mathsf{M}_n$. Partition the rows and columns of the $k \times k$ block matrix $A = [A_{ij}] \in \mathsf{M}_n$ conformally with Λ, so each $A_{ij} \in \mathsf{M}_{n_i \times n_j}$.*

(a) *If A is block diagonal, then $A\Lambda = \Lambda A$.*

(b) *If $A\Lambda = \Lambda A$ and $\lambda_1, \lambda_2, \dots, \lambda_k$ are distinct, then A is block diagonal.*

Proof Use (3.2.15) and (3.2.16) to compute

$$A\Lambda = \begin{bmatrix} A_{11} & \cdots & A_{1k} \\ \vdots & \ddots & \vdots \\ A_{k1} & \cdots & A_{kk} \end{bmatrix} \begin{bmatrix} \lambda_1 I_{n_1} & & \\ & \ddots & \\ & & \lambda_k I_{n_k} \end{bmatrix} = \begin{bmatrix} \lambda_1 A_{11} & \cdots & \lambda_k A_{1k} \\ \vdots & \ddots & \vdots \\ \lambda_1 A_{k1} & \cdots & \lambda_k A_{kk} \end{bmatrix}$$

and

$$\Lambda A = \begin{bmatrix} \lambda_1 I_{n_1} & & \\ & \ddots & \\ & & \lambda_k I_{n_k} \end{bmatrix} \begin{bmatrix} A_{11} & \cdots & A_{1k} \\ \vdots & \ddots & \vdots \\ A_{k1} & \cdots & A_{kk} \end{bmatrix} = \begin{bmatrix} \lambda_1 A_{11} & \cdots & \lambda_1 A_{1k} \\ \vdots & \ddots & \vdots \\ \lambda_k A_{k1} & \cdots & \lambda_k A_{kk} \end{bmatrix}.$$

If A is block diagonal, then $A_{ij} = 0$ for $i \neq j$ and the preceding computation confirms that $A\Lambda = \Lambda A$. If $A\Lambda = \Lambda A$, then $\lambda_j A_{ij} = \lambda_i A_{ij}$ and hence $(\lambda_i - \lambda_j) A_{ij} = 0$ for $1 \leq i, j \leq k$. If $\lambda_i - \lambda_j \neq 0$, then $A_{ij} = 0$. Thus, every off-diagonal block is a zero matrix if $\lambda_1, \lambda_2, \dots, \lambda_k$ are distinct. $\square$

Example 3.2.18 The hypothesis in Lemma 3.2.17.b that $\lambda_1, \lambda_2, \dots, \lambda_k$ are distinct is critical. Indeed, if $\lambda_1 = \lambda_2 = \cdots = \lambda_k = 1$, then $\Lambda = I$ commutes with every $A \in \mathsf{M}_k$. In this case, $A\Lambda = \Lambda A$ does not imply that A is a block diagonal matrix partitioned conformally with Λ.

3.3 Determinants of Block Matrices

The determinant of a diagonal or triangular matrix is the product of its main diagonal entries. This observation can be generalized to block diagonal or block triangular matrices. The key idea is a determinant identity for the direct sum of a square matrix and an identity matrix.

Lemma 3.3.1 *Let $A \in \mathsf{M}_m$ and let $n \in \{1, 2, \dots\}$. Then $\det \begin{bmatrix} I_n & 0 \\ 0 & A \end{bmatrix} = \det A = \det \begin{bmatrix} A & 0 \\ 0 & I_n \end{bmatrix}$.*

Proof Let $B_n = \begin{bmatrix} I_n & 0 \\ 0 & A \end{bmatrix}$ and $C_n = \begin{bmatrix} A & 0 \\ 0 & I_n \end{bmatrix}$, let S_n be the statement that $\det B_n = \det A = \det C_n$, and proceed by induction. The base case $n = 1$ follows by Laplace expansion by minors along the first and last rows of B_1 and C_1, respectively. For the inductive step, assume that S_n is true and observe that

$$\det B_{n+1} = \det \begin{bmatrix} I_{n+1} & 0 \\ 0 & A \end{bmatrix} = \det \begin{bmatrix} 1 & 0 & 0 \\ 0 & I_n & 0 \\ 0 & 0 & A \end{bmatrix} = \det \begin{bmatrix} 1 & 0 \\ 0 & B_n \end{bmatrix}, \qquad (3.3.2)$$

and

$$\det C_{n+1} = \det \begin{bmatrix} A & 0 \\ 0 & I_{n+1} \end{bmatrix} = \det \begin{bmatrix} A & 0 & 0 \\ 0 & I_n & 0 \\ 0 & 0 & 1 \end{bmatrix} = \det \begin{bmatrix} C_n & 0 \\ 0 & 1 \end{bmatrix}. \qquad (3.3.3)$$

In a Laplace expansion by minors along the first row (see Appendix C.4) of the final determinant in (3.3.2), only one summand has a nonzero coefficient:

$$\det \begin{bmatrix} 1 & 0 \\ 0 & B_n \end{bmatrix} = 1 \cdot (-1)^{1+1} \det B_n = \det B_n.$$

In a Laplace expansion by minors along the last row of the final determinant in (3.3.3), only one summand has a nonzero coefficient:

$$\det \begin{bmatrix} C_n & 0 \\ 0 & 1 \end{bmatrix} = 1 \cdot (-1)^{(n+1)+(n+1)} \det C_n = \det C_n.$$

The induction hypothesis ensures that $\det B_n = \det A = \det C_n$, so we conclude that $\det B_{n+1} = \det A = \det C_{n+1}$. This completes the induction. $\qquad \square$

Theorem 3.3.4 *Let $A \in \mathsf{M}_r$, $B \in \mathsf{M}_{r \times (n-r)}$, and $D \in \mathsf{M}_{n-r}$. Then $\det \begin{bmatrix} A & B \\ 0 & D \end{bmatrix} = (\det A)(\det D)$.*

Proof Write

$$\begin{bmatrix} A & B \\ 0 & D \end{bmatrix} = \begin{bmatrix} I_r & 0 \\ 0 & D \end{bmatrix} \begin{bmatrix} I_r & B \\ 0 & I_{n-r} \end{bmatrix} \begin{bmatrix} A & 0 \\ 0 & I_{n-r} \end{bmatrix}$$

and use the product rule for determinants:

$$\det \begin{bmatrix} A & B \\ 0 & D \end{bmatrix} = \left(\det \begin{bmatrix} I_r & 0 \\ 0 & D \end{bmatrix} \right) \left(\det \begin{bmatrix} I_r & B \\ 0 & I_{n-r} \end{bmatrix} \right) \left(\det \begin{bmatrix} A & 0 \\ 0 & I_{n-r} \end{bmatrix} \right).$$

The preceding lemma ensures that the first and last factors in this product are $\det D$ and $\det A$, respectively. The matrix in the middle factor is an upper triangular matrix with every main diagonal entry equal to 1, so its determinant is 1. $\qquad \square$

Block Gaussian elimination illustrates the utility of block-matrix manipulations. Let

$$M = \begin{bmatrix} A & B \\ C & D \end{bmatrix} \in \mathsf{M}_n, \quad \text{in which } A \in \mathsf{M}_r \text{ and } D \in \mathsf{M}_{n-r}. \qquad (3.3.5)$$

The block row operations

$$\begin{bmatrix} I_r & 0 \\ -CA^{-1} & I_{n-r} \end{bmatrix} \begin{bmatrix} A & B \\ C & D \end{bmatrix} = \begin{bmatrix} A & B \\ 0 & D - CA^{-1}B \end{bmatrix} \quad \text{if } A \text{ is invertible}, \qquad (3.3.6)$$

or

$$\begin{bmatrix} I_r & -BD^{-1} \\ 0 & I_{n-r} \end{bmatrix} \begin{bmatrix} A & B \\ C & D \end{bmatrix} = \begin{bmatrix} A - BD^{-1}C & 0 \\ C & D \end{bmatrix} \quad \text{if } D \text{ is invertible}, \qquad (3.3.7)$$

reduce M to block triangular form by multiplying it on the left with matrices that have determinant 1. They involve block versions of Type III elementary matrices; see p. 440.

Apply Theorem 3.3.4 and the product rule for determinants to (3.3.6) to see that

$$\det\begin{bmatrix} A & B \\ C & D \end{bmatrix} = \det\begin{bmatrix} A & B \\ 0 & D - CA^{-1}B \end{bmatrix} = (\det A)\det(D - CA^{-1}B). \tag{3.3.8}$$

The expression

$$M/A = D - CA^{-1}B \tag{3.3.9}$$

is the *Schur complement of A in M* and the identity

$$\det M = \det\begin{bmatrix} A & B \\ C & D \end{bmatrix} = (\det A)\det(M/A) \tag{3.3.10}$$

is the *Schur determinant formula*. It permits us to evaluate the determinant of a large block matrix by computing the determinants of two smaller matrices, provided that a certain leading principal submatrix is invertible. The Schur complement of D in M can be defined in a similar manner starting with (3.3.7); see P.3.6.

Theorem 3.3.11 *Let $A \in \mathsf{M}_k$ be invertible, suppose that $1 \le k < n$, and let $M = \begin{bmatrix} A & B \\ C & D \end{bmatrix} \in \mathsf{M}_n$. Then M is invertible if and only if $M/A = D - CA^{-1}B$ is invertible.*

Proof Since $\det A \ne 0$, it follows that $\det M = (\det A)\det(M/A) \ne 0$ if and only if $\det(M/A) \ne 0$. □

Definition 3.3.12 If $A \in \mathsf{M}_n(\mathbb{F})$, $\mathbf{x}, \mathbf{y} \in \mathbb{F}^n$, and $c \in \mathbb{F}$, then

$$\begin{bmatrix} c & \mathbf{x}^\mathsf{T} \\ \mathbf{y} & A \end{bmatrix}, \quad \begin{bmatrix} \mathbf{x}^\mathsf{T} & c \\ A & \mathbf{y} \end{bmatrix}, \quad \begin{bmatrix} A & \mathbf{x} \\ \mathbf{y}^\mathsf{T} & c \end{bmatrix}, \quad \text{and} \quad \begin{bmatrix} \mathbf{x} & A \\ c & \mathbf{y}^\mathsf{T} \end{bmatrix}$$

are *bordered matrices*. They are obtained from A by *bordering*.

Example 3.3.13 In (3.3.5), let $r = 1$, let a be a nonzero scalar, and let $\mathbf{b}, \mathbf{c} \in \mathbb{C}^{n-1}$. Then (3.3.8) is the *reduction formula*

$$\det M = \det\begin{bmatrix} a & \mathbf{b}^\mathsf{T} \\ \mathbf{c} & D \end{bmatrix} = a\,\det\left(D - \frac{1}{a}\mathbf{c}\mathbf{b}^\mathsf{T}\right) = a\,\det(M/a) \tag{3.3.14}$$

for the determinant of a bordered matrix. It ensures that M is invertible if and only if $M/a = D - \frac{1}{a}\mathbf{c}\mathbf{b}^\mathsf{T}$ is invertible. This principle plays a key role in our proof of Theorem 4.3.5.

Example 3.3.15 In (3.3.5), let $r = n - 1$, let $A \in \mathsf{M}_{n-1}$ be invertible, and let $\mathbf{b}, \mathbf{c} \in \mathbb{C}^{n-1}$. Then (3.3.8) is the *Cauchy expansion*

$$\det M = \det\begin{bmatrix} A & \mathbf{b} \\ \mathbf{c}^\mathsf{T} & d \end{bmatrix}$$

$$= (d - \mathbf{c}^\mathsf{T} A^{-1}\mathbf{b})\det A \tag{3.3.16}$$

$$= d\det A - \mathbf{c}^\mathsf{T}(\operatorname{adj} A)\mathbf{b} \tag{3.3.17}$$

for the determinant of a bordered matrix. It ensures that M is invertible if and only if $M/A = d - \mathbf{c}^\mathsf{T} A^{-1}\mathbf{b} \ne 0$. The formulation (3.3.17) is valid even if A is not invertible; see (C.4.3).

We care about algorithms that transform matrices into triangular or block triangular form because some computational problems are easier to solve after such a transformation. For example, consider a system of linear equations $Ax = y$, in which A has the block triangular form

$$A = \begin{bmatrix} A_{11} & A_{12} \\ 0 & A_{22} \end{bmatrix}$$

and the blocks A_{11} and A_{22} are square. Partition $\mathbf{x} = \begin{bmatrix} \mathbf{x}_1 \\ \mathbf{x}_2 \end{bmatrix}$ and $\mathbf{y} = \begin{bmatrix} \mathbf{y}_1 \\ \mathbf{y}_2 \end{bmatrix}$ conformally with A and write the system as a pair of smaller systems:

$$A_{11}\mathbf{x}_1 = \mathbf{y}_1 - A_{12}\mathbf{x}_2$$
$$A_{22}\mathbf{x}_2 = \mathbf{y}_2.$$

First solve $A_{22}\mathbf{x}_2 = \mathbf{y}_2$ for $\mathbf{x}_2$, and then solve $A_{11}\mathbf{x}_1 = \mathbf{y}_1 - A_{12}\mathbf{x}_2$ for $\mathbf{x}_1$. This is an example of *block backward substitution*.

3.4 Kronecker Products

Block matrices are central to a matrix "product" that finds applications in physics, signal processing, digital imaging, linear matrix equations, and many areas of pure mathematics.

Definition 3.4.1 Let $A = [a_{ij}] \in \mathsf{M}_{m \times n}$ and $B \in \mathsf{M}_{p \times q}$. The *Kronecker product* of A and B is the block matrix

$$A \otimes B = \begin{bmatrix} a_{11}B & a_{12}B & \cdots & a_{1n}B \\ a_{21}B & a_{22}B & \cdots & a_{2n}B \\ \vdots & \vdots & \ddots & \vdots \\ a_{m1}B & a_{m2}B & \cdots & a_{mn}B \end{bmatrix} \in \mathsf{M}_{mp \times nq}. \tag{3.4.2}$$

The Kronecker product is also called the *tensor product*. In the special case $n = q = 1$, the definition (3.4.2) says that the Kronecker product of $\mathbf{x} = [x_i] \in \mathbb{C}^m$ and $\mathbf{y} \in \mathbb{C}^p$ is

$$\mathbf{x} \otimes \mathbf{y} = \begin{bmatrix} x_1\mathbf{y} \\ x_2\mathbf{y} \\ \vdots \\ x_m\mathbf{y} \end{bmatrix} \in \mathbb{C}^{mp}.$$

Example 3.4.3 Using the matrices A and B in (3.1.5),

$$A \otimes B = \begin{bmatrix} B & 2B \\ 3B & 4B \end{bmatrix} = \left[\begin{array}{ccc|ccc} 4 & 5 & 2 & 8 & 10 & 4 \\ 6 & 7 & 1 & 12 & 14 & 2 \\ \hline 12 & 15 & 6 & 16 & 20 & 8 \\ 18 & 21 & 3 & 24 & 28 & 4 \end{array} \right] \in \mathsf{M}_{4 \times 6}$$

and

$$B \otimes A = \begin{bmatrix} 4A & 5A & 2A \\ 6A & 7A & A \end{bmatrix} = \left[\begin{array}{cc|cc|cc} 4 & 8 & 5 & 10 & 2 & 4 \\ 12 & 16 & 15 & 20 & 6 & 8 \\ \hline 6 & 12 & 7 & 14 & 1 & 2 \\ 18 & 24 & 21 & 28 & 3 & 4 \end{array} \right] \in \mathsf{M}_{4 \times 6}.$$

Although $A \otimes B \neq B \otimes A$, these two matrices have the same size and contain the same entries.

The Kronecker product satisfies many identities that one would like a product to obey:

$$c(A \otimes B) = (cA) \otimes B = A \otimes (cB) \tag{3.4.4}$$

$$(A + B) \otimes C = A \otimes C + B \otimes C \tag{3.4.5}$$

$$A \otimes (B + C) = A \otimes B + A \otimes C \tag{3.4.6}$$

$$(A \otimes B) \otimes C = A \otimes (B \otimes C) \tag{3.4.7}$$

$$(A \otimes B)^\mathsf{T} = A^\mathsf{T} \otimes B^\mathsf{T} \tag{3.4.8}$$

$$\overline{A \otimes B} = \overline{A} \otimes \overline{B} \tag{3.4.9}$$

$$(A \otimes B)^* = A^* \otimes B^* \tag{3.4.10}$$

$$I_m \otimes I_n = I_{mn}. \tag{3.4.11}$$

There is a relationship between the ordinary product and the Kronecker product.

Theorem 3.4.12 (Mixed Product Property) *Let $A \in \mathsf{M}_{m \times n}$, $B \in \mathsf{M}_{p \times q}$, $C \in \mathsf{M}_{n \times r}$, and $D \in \mathsf{M}_{q \times s}$. Then*

$$(A \otimes B)(C \otimes D) = (AC) \otimes (BD) \in \mathsf{M}_{mp \times rs}. \tag{3.4.13}$$

Proof If $A = [a_{ij}]$ and $C = [c_{ij}]$, then $A \otimes B = [a_{ij}B]$ and $C \otimes D = [c_{ij}D]$. The (i,j) block of $(A \otimes B)(C \otimes D)$ is

$$\sum_{k=1}^{n} (a_{ik}B)(c_{kj}D) = \left(\sum_{k=1}^{n} a_{ik}c_{kj} \right) BD = (AC)_{ij}BD,$$

in which $(AC)_{ij}$ denotes the (i,j) entry of AC. This identity shows that the (i,j) block of $(A \otimes B)(C \otimes D)$ is the (i,j) block of $(AC) \otimes (BD)$. $\qquad\square$

Corollary 3.4.14 *If $A \in \mathsf{M}_m$ and $B \in \mathsf{M}_n$ are invertible, then $A \otimes B$ is invertible and $(A \otimes B)^{-1} = A^{-1} \otimes B^{-1}$.*

Proof $(A \otimes B)(A^{-1} \otimes B^{-1}) = (AA^{-1}) \otimes (BB^{-1}) = I_m \otimes I_n = I_{mn}. \qquad\square$

The following definition introduces a way to convert a matrix to a vector that is compatible with Kronecker-product operations.

Definition 3.4.15 Let $X = [\mathbf{x}_1 \; \mathbf{x}_2 \; \ldots \; \mathbf{x}_n] \in \mathsf{M}_{m \times n}$. The operator vec $: \mathsf{M}_{m \times n} \to \mathbb{C}^{mn}$ is defined by

$$\mathrm{vec}\, X = \begin{bmatrix} \mathbf{x}_1 \\ \mathbf{x}_2 \\ \vdots \\ \mathbf{x}_n \end{bmatrix} \in \mathbb{C}^{mn},$$

that is, vec stacks the columns of X vertically.

Theorem 3.4.16 *Let $A \in \mathsf{M}_{m \times n}$, $X \in \mathsf{M}_{n \times p}$, and $B \in \mathsf{M}_{p \times q}$. Then* $\mathrm{vec}\,(AXB) = (B^{\mathsf{T}} \otimes A)\,\mathrm{vec}\,X$.

Proof Let $B = [b_{ij}] = [\mathbf{b}_1 \ \mathbf{b}_2 \ \cdots \ \mathbf{b}_q]$ and $X = [\mathbf{x}_1 \ \mathbf{x}_2 \ \cdots \ \mathbf{x}_p]$. The kth column of AXB is

$$A X \mathbf{b}_k = A \sum_{i=1}^{p} b_{ik} \mathbf{x}_i = [b_{1k}A \ b_{2k}A \ \cdots \ b_{pk}A]\,\mathrm{vec}\,X = (\mathbf{b}_k^{\mathsf{T}} \otimes A)\,\mathrm{vec}\,X.$$

Stack these vectors vertically and obtain

$$\mathrm{vec}\,AXB = \begin{bmatrix} \mathbf{b}_1^{\mathsf{T}} \otimes A \\ \mathbf{b}_2^{\mathsf{T}} \otimes A \\ \vdots \\ \mathbf{b}_q^{\mathsf{T}} \otimes A \end{bmatrix} \mathrm{vec}\,X = (B^{\mathsf{T}} \otimes A)\,\mathrm{vec}\,X. \qquad \square$$

3.5 Problems

P.3.1 Let $X = [X_1 \ X_2] \in \mathsf{M}_{m \times n}$, in which $X_1 \in \mathsf{M}_{m \times n_1}$, $X_2 \in \mathsf{M}_{m \times n_2}$, and $n_1 + n_2 = n$. Compute $X^{\mathsf{T}}X$ and XX^{T}.

P.3.2 Let

$$A = \begin{bmatrix} -1 & 2 & 3 \\ 2 & -3 & 1 \\ 3 & 1 & -2 \end{bmatrix} = \begin{bmatrix} -1 & 2 & 3 \\ 2 & -3 & 1 \\ 3 & 1 & -2 \end{bmatrix} \quad \text{and} \quad B = \begin{bmatrix} 1 & 3 & -2 \\ 3 & -2 & 1 \\ -2 & 1 & 3 \end{bmatrix}.$$

(a) Compute AB in two different ways by partitioning B conformally with each presentation of A and then performing block-matrix multiplication. (b) Compute AB in the standard manner, and verify that all three answers agree.

P.3.3 Let

$$A = \begin{bmatrix} a_{11} & a_{12} & a_{13} & a_{14} & a_{15} \\ a_{21} & a_{22} & a_{23} & a_{24} & a_{25} \\ a_{31} & a_{32} & a_{33} & a_{34} & a_{35} \\ a_{41} & a_{42} & a_{43} & a_{44} & a_{45} \end{bmatrix} \quad \text{and} \quad B = \begin{bmatrix} b_{11} & b_{12} & b_{13} \\ b_{21} & b_{22} & b_{23} \\ b_{31} & b_{32} & b_{33} \\ b_{41} & b_{42} & b_{43} \\ b_{51} & b_{52} & b_{53} \end{bmatrix}.$$

How should B be partitioned to compute AB with block matrices?

P.3.4 Let $A \in \mathsf{M}_{m \times n}$ and $B = \mathsf{M}_{n \times m}$. Show that

$$C = \begin{bmatrix} I - BA & B \\ 2A - ABA & AB - I \end{bmatrix}$$

is an involution and $\mathrm{tr}\,C = n - m$.

P.3.5 Verify that the matrices on the right sides of (3.1.15) and (3.1.17) have the same entries.

P.3.6 Let $M \in \mathsf{M}_n$ be the 2×2 block matrix (3.3.5). If D is invertible, then the *Schur complement of D in M* is $M/D = A - BD^{-1}C$. Show that $\det M = (\det D)\det(M/D)$.

P.3.7 (a) Partition

$$A = \begin{bmatrix} 2 & 2 & 3 \\ 2 & 9 & 7 \\ 4 & -3 & 8 \end{bmatrix} \tag{3.5.1}$$

as a 2×2 block matrix $A = [A_{ij}]$, in which $A_{11} = [2]$ is 1×1. Verify that the reduced form (3.3.6) obtained by block Gaussian elimination is

$$\begin{bmatrix} 2 & 2 & 3 \\ 0 & 7 & 4 \\ 0 & -7 & 2 \end{bmatrix}. \tag{3.5.2}$$

(b) Now perform standard row-wise Gaussian elimination on (3.5.1) to zero out all the entries in the first column below the $(1, 1)$ entry. Verify that you obtain the same reduced form (3.5.2).

P.3.8 Let $A, B, C, D \in M_n$ and let $M = \begin{bmatrix} A & B \\ C & D \end{bmatrix}$. (a) Suppose that A is invertible. If A commutes with B, show that $\det M = \det(DA - CB)$; if A commutes with C, show that $\det M = \det(AD - CB)$. (b) Suppose that D is invertible. If D commutes with C, show that $\det M = \det(AD - BC)$; if D commutes with B, show that $\det M = \det(DA - BC)$. (c) What can you say if $n = 1$?

P.3.9 Let $A, B \in M_n$, let $M = \begin{bmatrix} A & B \\ B & A \end{bmatrix}$, and let $S = \begin{bmatrix} I_n & I_n \\ I_n & -I_n \end{bmatrix}$. Show that $\det M = \det(A + B) \det(A - B)$. *Hint*: Compute SMS^{-1}.

P.3.10 Let $A, B \in M_n$, let $M = \begin{bmatrix} A & B \\ B & -A \end{bmatrix}$, and let $S = \begin{bmatrix} I_n & 0 \\ 0 & iI_n \end{bmatrix}$. Show that $\det M = (-1)^n \det(A + iB) \det(A - iB)$. *Hint*: Compute SMS.

P.3.11 Let $A, B \in M_n$, let $M = \begin{bmatrix} A & -B \\ B & A \end{bmatrix}$, and let $S = \begin{bmatrix} I_n & iI_n \\ 0 & I_n \end{bmatrix}$. Show that $\det M = \det(A + iB) \det(A - iB)$. *Hint*: Compute SMS^{-1}.

P.3.12 Compute

$$\det \begin{bmatrix} 0 & 0 & 0 & 1 & 4 & 6 \\ 0 & 0 & 0 & 0 & 2 & 5 \\ 0 & 0 & 0 & 0 & 0 & 3 \\ 3 & 5 & 6 & 0 & 0 & 0 \\ 0 & 2 & 4 & 0 & 0 & 0 \\ 0 & 0 & 1 & 0 & 0 & 0 \end{bmatrix}.$$

P.3.13 Suppose that $a, b, c, d \in \mathbb{R}$ and $A\mathbf{x} = \mathbf{y}$, in which

$$A = \begin{bmatrix} 1 & 0 & 0 & 0 & 0 & 0 & 4 \\ a & 2 & 0 & 0 & 0 & 0 & 7 \\ c & b & 1 & 0 & 0 & 0 & 4 \\ a & d & a & 2 & 0 & 0 & 7 \\ d & b & c & b & 1 & 0 & 4 \\ c & a & c & d & a & 2 & 7 \\ a & c & b & d & c & b & 1 \end{bmatrix}, \quad \mathbf{x} = \begin{bmatrix} x_1 \\ x_2 \\ x_3 \\ x_4 \\ x_5 \\ x_6 \\ 4 \end{bmatrix} \quad \text{and} \quad \mathbf{y} = \begin{bmatrix} 0 \\ 0 \\ 0 \\ 0 \\ 0 \\ 0 \\ 1 \end{bmatrix}.$$

What are the exact numerical values of $\det A$ and x_4? Your answers should not include a, b, c, d.

P.3.14 Let $M = \begin{bmatrix} A & B \\ 0 & D \end{bmatrix}$, in which A and D are square. Let p be a polynomial. Prove that $p(M) = \begin{bmatrix} p(A) & \star \\ 0 & p(D) \end{bmatrix}$.

P.3.15 A 2×2 block matrix $M = \begin{bmatrix} A & -B \\ B & A \end{bmatrix} \in M_{2n}(\mathbb{R})$, in which each block is $n \times n$, is a *matrix of complex type*. Let $J_{2n} = \begin{bmatrix} 0 & I_n \\ -I_n & 0 \end{bmatrix}$. Show that $M \in M_{2n}(\mathbb{R})$ is a matrix of complex type if and only if J_{2n} commutes with M.

P.3.16 Suppose that $M, N \in M_{2n}(\mathbb{R})$ are matrices of complex type. Show that $M + N$ and MN are matrices of complex type. If M is invertible, show that M^{-1} is also a matrix of complex type. *Hint*: Use the criterion in the preceding problem.

P.3.17 A 2×2 block matrix $M = \begin{bmatrix} A & B \\ B & A \end{bmatrix} \in M_{2n}$, in which each block is $n \times n$, is *block centrosymmetric*. Let $L_{2n} = \begin{bmatrix} 0 & I_n \\ I_n & 0 \end{bmatrix}$. Show that M is block centrosymmetric if and only if L_{2n} commutes with M.

P.3.18 Suppose that $M, N \in \mathsf{M}_{2n}$ are block centrosymmetric. Show that $M + N$ and MN are block centrosymmetric. If M is invertible, show that M^{-1} is also block centrosymmetric. *Hint*: Use the criterion in the preceding problem.

P.3.19 If an invertible matrix M is partitioned as a 2×2 block matrix as in (3.3.5), there is a conformally partitioned presentation of its inverse:

$$M^{-1} = \begin{bmatrix} (A - BD^{-1}C)^{-1} & -A^{-1}B(D - CA^{-1}B)^{-1} \\ -D^{-1}C(A - BD^{-1}C)^{-1} & (D - CA^{-1}B)^{-1} \end{bmatrix}, \tag{3.5.3}$$

provided that all the indicated inverses exist. (a) Verify (3.5.3) and show that

$$M^{-1} = \begin{bmatrix} A^{-1} & 0 \\ 0 & D^{-1} \end{bmatrix} \begin{bmatrix} A & -B \\ -C & D \end{bmatrix} \begin{bmatrix} (M/D)^{-1} & 0 \\ 0 & (M/A)^{-1} \end{bmatrix}. \tag{3.5.4}$$

(b) Derive the identity $\begin{bmatrix} I_k & 0 \\ X & I_{n-k} \end{bmatrix}^{-1} = \begin{bmatrix} I_k & 0 \\ -X & I_{n-k} \end{bmatrix}$ from (3.5.3). (c) If all its blocks are 1×1 matrices, show that (3.5.3) reduces to $M^{-1} = \begin{bmatrix} a & b \\ c & d \end{bmatrix}^{-1} = \frac{1}{\det M} \begin{bmatrix} d & -b \\ -c & a \end{bmatrix}$.

P.3.20 Suppose that $A \in \mathsf{M}_{n \times m}$ and $B \in \mathsf{M}_{m \times n}$. Use (3.3.10), P.3.6, and the block matrix $\begin{bmatrix} I_n & -A \\ B & I_m \end{bmatrix}$ to derive the *Sylvester determinant identity*

$$\det(I_n + AB) = \det(I_m + BA), \tag{3.5.5}$$

which relates the determinant of an $n \times n$ matrix to the determinant of an $m \times m$ matrix.

P.3.21 Let $\mathbf{u}, \mathbf{v} \in \mathbb{C}^n$ and let $z \in \mathbb{C}$. (a) Show that $\det(I + z\mathbf{u}\mathbf{v}^\mathsf{T}) = 1 + z\mathbf{v}^\mathsf{T}\mathbf{u}$. (b) If $A \in \mathsf{M}_n$ is invertible, show that

$$\det(A + z\mathbf{u}\mathbf{v}^\mathsf{T}) = \det A + z(\det A)(\mathbf{v}^\mathsf{T}A^{-1}\mathbf{u}). \tag{3.5.6}$$

(c) Can you show that $\det(A + z\mathbf{u}\mathbf{v}^\mathsf{T}) = \det A + z\mathbf{v}^\mathsf{T}(\mathrm{adj}\,A)\mathbf{u}$ for all $A \in \mathsf{M}_n$?

P.3.22 Let $A \in \mathsf{M}_n$ be invertible and let $\mathbf{u}, \mathbf{v} \in \mathbb{C}^n$. (a) Show that $\mathrm{rank}(A + \mathbf{u}\mathbf{v}^\mathsf{T}) = n$ if and only if $\mathbf{v}^\mathsf{T}A^{-1}\mathbf{u} \neq -1$. (b) If $\mathbf{v}^\mathsf{T}A^{-1}\mathbf{u} = -1$, what can you say about $\mathrm{rank}(A + \mathbf{u}\mathbf{v}^\mathsf{T})$?

P.3.23 Use (3.1.2) and (3.1.3) (and their notation) to verify the associative identity $(AB)\mathbf{x} = A(B\mathbf{x})$. *Hint*: $[A\mathbf{b}_1 \ A\mathbf{b}_2 \ \ldots \ A\mathbf{b}_n]\mathbf{x} = \sum_i x_i A\mathbf{b}_i = A(\sum_i x_i \mathbf{b}_i)$.

P.3.24 Suppose that $A \in \mathsf{M}_{m \times r}$, $B \in \mathsf{M}_{r \times n}$, and $C \in \mathsf{M}_{n \times p}$. Use the preceding problem to verify the associative identity $(AB)C = A(BC)$. *Hint*: Let $\mathbf{x}$ be a column of C.

P.3.25 Let $A, B \in \mathsf{M}_{m \times n}$. Suppose that $r = \mathrm{rank}\,A \geq 1$ and let $s = \mathrm{rank}\,B$. Let $A = XY$ be a full-rank factorization. Show that $\mathrm{col}\,A \subseteq \mathrm{col}\,B$ if and only if $B = [X\ X_2]Z$, in which $Z \in \mathsf{M}_{s \times n}$ has full rank, $X_2 \in \mathsf{M}_{m \times (s-r)}$, and $\mathrm{rank}[X\ X_2] = s$.

P.3.26 Let $A, B \in \mathsf{M}_{m \times n}$ and suppose that $r = \mathrm{rank}\,A \geq 1$. Show that the following are equivalent:

(a) $\mathrm{col}\,A = \mathrm{col}\,B$.

(b) There are full-rank matrices $X \in \mathsf{M}_{m \times r}$ and $Y, Z \in \mathsf{M}_{r \times n}$ such that $A = XY$ and $B = XZ$.

(c) There is an invertible $S \in \mathsf{M}_n$ such that $B = AS$.

Hint: If $A = XY$ is a full-rank factorization, then $A = [X\ 0_{n \times (n-r)}]W$, in which $W = \begin{bmatrix} Y \\ Y_2 \end{bmatrix} \in \mathsf{M}_n$ is invertible.

P.3.27 Let $A, C \in \mathsf{M}_m$ and $B, D \in \mathsf{M}_n$. Show that there is an invertible $Z \in \mathsf{M}_{m+n}$ such that $A \oplus B = (C \oplus D)Z$ if and only if there are invertible $X \in \mathsf{M}_m$ and $Y \in \mathsf{M}_n$ such that $A = CX$ and $B = DY$.

P.3.28 Suppose that $A \in \mathsf{M}_{n \times p}$ and $B = \mathsf{M}_{n \times q}$ have full column rank. If $\operatorname{col} A = \operatorname{col} B$, show that $p = q$. How is this related to Corollary 2.1.11? *Hint:* $A = BX$ and $B = AY$.

P.3.29 Let $A, B \in \mathsf{M}_{m \times n}$ and suppose that there is an $X \in \mathsf{M}_n$ such that $A = BX$. If $\operatorname{rank} A = n$, show that $\operatorname{rank} B = n$ and X is invertible. What can you say about the converse?

P.3.30 Let $A = \begin{bmatrix} a & b \\ c & d \end{bmatrix}$ and $B = [e\ f]$. Verify that $A \otimes B$ and $B \otimes A$ are not equal, but they have the same size and the same sets of entries.

P.3.31 Compute $A \otimes I_2 + I_2 \otimes B$, in which $A = \begin{bmatrix} 1 & 2 \\ 3 & 4 \end{bmatrix}$ and $B = \begin{bmatrix} 5 & 6 \\ 7 & 8 \end{bmatrix}$.

P.3.32 If $A \otimes B = 0$, show that $A = 0$ or $B = 0$. Is this true for the ordinary matrix product AB?

P.3.33 Choose two of the Kronecker product identities (3.4.4)–(3.4.10) and prove them.

P.3.34 Explain why $(A \otimes B \otimes C)(D \otimes E \otimes F) = (AD) \otimes (BE) \otimes (CF)$ if all the matrices have appropriate sizes.

P.3.35 If $A, B, C, D, R, S \in \mathsf{M}_n$, R and S are invertible, $A = RBR^{-1}$, and $C = SDS^{-1}$, show that $A \otimes C = (R \otimes S)(B \otimes D)(R \otimes S)^{-1}$.

P.3.36 Let $A = [a_{ij}] \in \mathsf{M}_m$ and $B = [b_{ij}] \in \mathsf{M}_n$ be upper triangular matrices. Prove the following: (a) $A \otimes B \in \mathsf{M}_{mn}$ is an upper triangular matrix whose mn diagonal entries are (in some order) the mn scalars $a_{ii}b_{jj}$, for $i = 1, 2, \ldots, m$ and $j = 1, 2, \ldots, n$. (b) $A \otimes I_n + I_m \otimes B$ is an upper triangular matrix whose mn diagonal entries are (in some order) the mn scalars $a_{ii} + b_{jj}$, for $i = 1, 2, \ldots, m$ and $j = 1, 2, \ldots, n$.

P.3.37 Let $A \in \mathsf{M}_m$ and $B \in \mathsf{M}_n$. Show that $A \otimes I_n$ commutes with $I_m \otimes B$.

P.3.38 Let $\mathbf{a} = [0\ 1\ 2]^\mathsf{T}$ and $\mathbf{b} = [3\ 4\ 5]^\mathsf{T}$. (a) Compute $\mathbf{a}\mathbf{b}^\mathsf{T}$ and $\mathbf{a}^\mathsf{T}\mathbf{b}$. (b) Compute $\mathbf{b}\mathbf{a}^\mathsf{T}$ and $\mathbf{b}^\mathsf{T}\mathbf{a}$. (c) Compute $\mathbf{a} \otimes \mathbf{b}^\mathsf{T}$, $\mathbf{b} \otimes \mathbf{a}^\mathsf{T}$, $\mathbf{a} \otimes \mathbf{b}$, and $\mathbf{b} \otimes \mathbf{a}$. (d) Compute $\operatorname{vec}(\mathbf{a}\mathbf{b}^\mathsf{T})$ and $\operatorname{vec}(\mathbf{b}\mathbf{a}^\mathsf{T})$. (e) Discuss.

3.6 Notes

In the identities in P.3.8, the hypotheses that A and D are invertible can be omitted; see [HJ13, (0.8.5.13)]. For some historical comments about the Kronecker product and more of its properties, see [HJ94, Chapter 4].

3.7 Some Important Concepts

- Row and column partitions of a matrix.

- Conformal partitions for addition and multiplication of matrices.

- Matrix multiplication: by rows, by columns, via inner products, via outer products.

- Block-diagonal matrices.

- Schur complement and the determinant of a 2×2 block matrix.

- Block versions of elementary matrices.

- Determinant of a bordered matrix (reduction formula and Cauchy expansion).

- Kronecker product and its properties.

Rank, Triangular Factorizations, and Row Equivalence

In this chapter, we collect some important facts about matrices: the rank-nullity theorem; the intersection and sum of column spaces; rank inequalities for sums and products of matrices; the LU factorization and solutions of linear systems; row equivalence, the pivot column decomposition, and the reduced row echelon form. In a final capstone section, we use linear dependence, the trace, block matrices, induction, and similarity to characterize matrices that are commutators. Throughout the chapter, we emphasize the block-matrix methods introduced in Chapter 3.

4.1 The Rank-Nullity Theorem and Subspace Intersection

Figure 4.1 illustrates the null space and column space of a matrix $A \in M_{m \times n}(\mathbb{F})$. The following theorem reveals how the dimensions of these subspaces are related. The dimension of $\operatorname{col} A$ is the rank of A, and the dimension of $\operatorname{null} A$ is the *nullity* of A.

Theorem 4.1.1 (Rank-Nullity Theorem) *Let $A \in M_{m \times n}(\mathbb{F})$. Then*

$$\operatorname{rank} A + \operatorname{nullity} A = n, \tag{4.1.2}$$

that is, the sum of the rank and nullity of A is the number of columns of A.

Proof Let $d = \operatorname{nullity} A$. To show that $\operatorname{rank} A = n - d$, we consider three cases.

(a) If $d = n$, then $A = 0$ and $\operatorname{rank} A = 0$.

(b) If $d = 0$, then the columns of A are linearly independent (see Example 1.6.8), so $\operatorname{rank} A = n$.

(c) If $1 \leq d \leq n - 1$, let the columns of $V \in M_{n \times d}(\mathbb{F})$ be a basis for $\operatorname{null} A$. Theorem 3.1.22 ensures that there is a $W \in M_{n \times (n-d)}(\mathbb{F})$ such that $U = [V \ W] \in M_n(\mathbb{F})$ is invertible. Example 1.4.23 ensures that $\operatorname{col} A = \operatorname{col} AU$ so it suffices to show that $\operatorname{rank} AU = n - d$. Observe that

$$\operatorname{col} A = \operatorname{col} AU = \operatorname{col}(A[V \ W]) = \operatorname{col}[AV \ AW] = \operatorname{col}[0 \ AW] = \operatorname{col} AW$$

and $AW \in M_{m \times (n-d)}(\mathbb{F})$. Suppose that $\mathbf{y} \in \mathbb{F}^{n-d}$ and $AW\mathbf{y} = \mathbf{0}$. Then $W\mathbf{y} \in \operatorname{null} A$, so there is an $\mathbf{x} \in \mathbb{F}^d$ such that $W\mathbf{y} = V\mathbf{x}$. Let $\mathbf{z} = \begin{bmatrix} \mathbf{x} \\ -\mathbf{y} \end{bmatrix} \in \mathbb{F}^n$ and compute

$$U\mathbf{z} = \begin{bmatrix} V & W \end{bmatrix} \begin{bmatrix} \mathbf{x} \\ -\mathbf{y} \end{bmatrix} = V\mathbf{x} - W\mathbf{y} = \mathbf{0}.$$

Since U is invertible, it follows that $\mathbf{z} = \mathbf{0}$, and hence $\mathbf{y} = \mathbf{0}$. Therefore, $\operatorname{null} AW = \{\mathbf{0}\}$, and (b) ensures that $\operatorname{rank} A = \operatorname{rank} AW = n - d$. $\qquad \square$

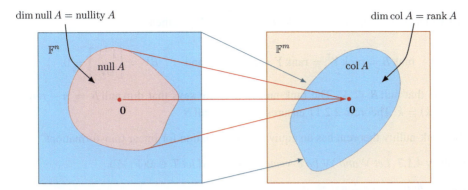

dim null A = nullity A

dim col A = rank A

$\mathbb{F}^n$

null A

$\mathbb{F}^m$

col A

0

0

Figure 4.1 For $A \in \mathsf{M}_{m \times n}(\mathbb{F})$, the column space of A need not equal $\mathbb{F}^m$; this occurs if and only if A has linearly independent rows. The null space of A need not be $\{\mathbf{0}\}$; this occurs if and only if A has linearly independent columns. The rank-nullity theorem says that rank A + nullity $A = n$.

For a square matrix $A \in \mathsf{M}_n(\mathbb{F})$, (4.1.2) implies that null $A = \{\mathbf{0}\}$ if and only if col $A = \mathbb{F}^n$. This is equivalent to a familiar fact about systems of linear equations: for each $\mathbf{y} \in \mathbb{F}^n$, the linear system $A\mathbf{x} = \mathbf{y}$ has a solution $\mathbf{x} \in \mathbb{F}^n$ if and only if $A\mathbf{x} = \mathbf{0}$ has only the trivial solution.

Example 4.1.3 In Example 1.7.7, we obtained six linearly independent vectors (the columns of (1.7.8)) in the null space of the 5×9 matrix A in Example 1.7.3. Since rank $A = 3$, the rank-nullity theorem ensures that these six vectors are a basis for null A.

Example 4.1.4 Let $A \in \mathsf{M}_{m \times n}(\mathbb{F})$. Figure 4.1 illustrates the linear transformation $T_A : \mathbb{F}^n \to \mathbb{F}^m$ induced by A. We know that rank $A = $ rank A^T (Theorem 2.3.1) and the rank-nullity theorem ensures that nullity $A^\mathsf{T} = m - $ rank A^T. The following statements are equivalent:

- T_A is onto.
- col $A = \mathbb{F}^m$.
- rank $A = m$.
- rank $A^\mathsf{T} = m$.

- nullity $A^\mathsf{T} = 0$.
- A^T has linearly independent columns.
- A has linearly independent rows.

Example 4.1.5 The rank-nullity theorem provides another proof of Theorem 2.2.13 (left and right inverses). If $A, B \in \mathsf{M}_n(\mathbb{F})$ and $AB = I$, then $A(B\mathbf{x}) = (AB)\mathbf{x} = \mathbf{x}$ for every $\mathbf{x} \in \mathbb{F}^n$. Therefore, col $A = \mathbb{F}^n$ and rank $A = n$. Theorem 4.1.1 ensures that nullity $A = 0$. Compute $A(I - BA) = A - (AB)A = A - IA = A - A = 0$. Since null $A = \{\mathbf{0}\}$, every column of $I - BA$ is the zero vector. Therefore, $BA = I$.

Example 4.1.6 Let $\mathcal{V}$ be a k-dimensional subspace of $\mathbb{F}^n$. We claim that there are $A, B \in \mathsf{M}_n(\mathbb{F})$ such that $\mathcal{V} = $ col A and $\mathcal{V} = $ null B. If $k = 0$, we may take $A = 0$ and $B = I$. If $k = n$, we may take $A = I$ and $B = 0$. Suppose that $1 \leq k \leq n - 1$ and let the columns of $V \in \mathsf{M}_{n \times k}(\mathbb{F})$ be a basis for $\mathcal{V}$. If $A = [V\ 0] \in \mathsf{M}_n(\mathbb{F})$, then col $A = $ col $V = \mathcal{V}$. Theorem 3.1.22 ensures that there is a $W \in \mathsf{M}_{n \times (n-k)}(\mathbb{F})$ such that $C = [V\ W]$ is invertible. Partition $C^{-\mathsf{T}} = [X\ Y]$, in which $Y \in \mathsf{M}_{n \times (n-k)}(\mathbb{F})$. Then rank $Y = n - k$ and

$$\begin{bmatrix} I_k & 0 \\ 0 & I_{n-k} \end{bmatrix} = I_n = C^{-1}C = \begin{bmatrix} X^\mathsf{T} \\ Y^\mathsf{T} \end{bmatrix} \begin{bmatrix} V & W \end{bmatrix} = \begin{bmatrix} X^\mathsf{T}V & X^\mathsf{T}W \\ Y^\mathsf{T}V & Y^\mathsf{T}W \end{bmatrix}.$$

In particular, $Y^\mathsf{T} V = 0_{(n-k)\times k}$. If we let $B = \begin{bmatrix} Y^\mathsf{T} \\ 0 \end{bmatrix} \in \mathsf{M}_n(\mathbb{F})$, then

$$\operatorname{rank} B = \operatorname{rank} Y^\mathsf{T} = \operatorname{rank} Y = n - k \qquad \text{and} \qquad BV = \begin{bmatrix} Y^\mathsf{T} V \\ 0V \end{bmatrix} = 0.$$

It follows that $\operatorname{null} B \subseteq V$. The rank-nullity theorem says that $\dim \operatorname{null} B = n - \operatorname{rank} B = n - (n-k) = k$. Theorem 2.2.11 ensures that $V = \operatorname{null} B$.

The rank-nullity theorem has an equivalent formulation for linear transformations.

Corollary 4.1.7 *Let V and W be $\mathbb{F}$-vector spaces and let $T \in \mathcal{L}(V, W)$.*

(a) *If V is finite dimensional, then*

$$\dim \operatorname{ran} T + \dim \ker T = \dim V. \tag{4.1.8}$$

(b) *If V and W are finite dimensional and $\dim V = \dim W$, then T is one to one if and only if T is onto.*

(c) *If T is one to one and onto, then $T^{-1} \in \mathcal{L}(W, V)$.*

Proof (a) Let $n = \dim V$ and $d = \dim \ker T$. If $n = 0$ or $d = n$, there is nothing to prove, so we may assume that $0 \le d < n$. Let $\beta = \mathbf{v}_1, \mathbf{v}_2, \ldots, \mathbf{v}_n$ be a basis for V. Theorem 2.4.19 ensures that $\operatorname{ran} T$ is finite dimensional and $\dim \operatorname{ran} T \le n$. Let $m = \dim \operatorname{ran} T$, let $\gamma = \mathbf{w}_1, \mathbf{w}_2, \ldots, \mathbf{w}_m$ be a basis for $\operatorname{ran} T$, and let $A = {}_\gamma[T]_\beta \in \mathsf{M}_{m\times n}(\mathbb{F})$. Since $[T\mathbf{v}]_\gamma = A[\mathbf{v}]_\beta$ for every $\mathbf{v} \in V$ (see (2.4.22)), it follows that $\mathbf{v} \in \ker T$ if and only if $[\mathbf{v}]_\beta \in \operatorname{null} A$, and $\mathbf{w} \in \operatorname{ran} T$ if and only if $[\mathbf{w}]_\gamma \in \operatorname{col} A$. Since the β-basis representation function preserves linear independence (Theorem 2.4.8), it follows that $\dim \operatorname{ran} T = \dim \operatorname{col} A = \operatorname{rank} A$ and $\dim \ker T = \operatorname{nullity} A$. Therefore, the identities (4.1.8) and (4.1.2) are equivalent.

(b) Let $n = \dim V = \dim W$. Theorem 2.2.11 says that $\operatorname{ran} T = W$ (that is, T is onto) if and only if $\dim \operatorname{ran} T = n$, and (4.1.8) ensures that $\dim \operatorname{ran} T = n$ if and only if $\dim \ker T = 0$. Theorem 2.4.17 says that $\dim \ker T = 0$ if and only if T is one to one.

(c) If T is one to one and onto, then $T^{-1}\mathbf{w}_1 = \mathbf{v}_1$ and $T^{-1}\mathbf{w}_2 = \mathbf{v}_2$ if and only if $T\mathbf{v}_1 = \mathbf{w}_1$ and $T\mathbf{v}_2 = \mathbf{w}_2$; see Appendix C.1. If $c \in \mathbb{F}$, then $T(c\mathbf{v}_1 + \mathbf{v}_2) = cT\mathbf{v}_1 + T\mathbf{v}_2 = c\mathbf{w}_1 + \mathbf{w}_2$. Therefore, $T^{-1}(c\mathbf{w}_1 + \mathbf{w}_2) = c\mathbf{v}_1 + \mathbf{v}_2 = cT^{-1}\mathbf{w}_1 + T^{-1}\mathbf{w}_2$, so T^{-1} is a linear transformation. $\square$

Block matrices can help us analyze sums and intersections of subspaces; see Figure 4.2. For example, if $B \in \mathsf{M}_{m\times p}(\mathbb{F})$ and $C \in \mathsf{M}_{m\times q}(\mathbb{F})$, the sum of their column spaces is

$$\operatorname{col} B + \operatorname{col} C = \{B\mathbf{x} + C\mathbf{y} : \mathbf{x} \in \mathbb{F}^p \text{ and } \mathbf{y} \in \mathbb{F}^q\} = \left\{ [B\ C]\begin{bmatrix} \mathbf{x} \\ \mathbf{y} \end{bmatrix} : \mathbf{x} \in \mathbb{F}^p, \mathbf{y} \in \mathbb{F}^q \right\}$$

$$= \{[B\ C]\mathbf{z} : \mathbf{z} \in \mathbb{F}^{p+q}\} = \operatorname{col}[B\ C].$$

If $B \in \mathsf{M}_{p\times n}(\mathbb{F})$ and $C \in \mathsf{M}_{q\times n}(\mathbb{F})$, the intersection of their null spaces is

$$\operatorname{null} B \cap \operatorname{null} C = \{\mathbf{x} \in \mathbb{F}^n : B\mathbf{x} = 0 \text{ and } C\mathbf{x} = 0\} = \left\{ \mathbf{x} \in \mathbb{F}^n : \begin{bmatrix} B\mathbf{x} \\ C\mathbf{x} \end{bmatrix} = 0 \right\}$$

$$= \left\{ \mathbf{x} \in \mathbb{F}^n : \begin{bmatrix} B \\ C \end{bmatrix}\mathbf{x} = 0 \right\} = \operatorname{null}\begin{bmatrix} B \\ C \end{bmatrix}.$$

In the following theorem, we study the subspaces $\operatorname{col}[B\ C]$, $\operatorname{null}[B\ C]$, and $\operatorname{col} B \cap \operatorname{col} C$.

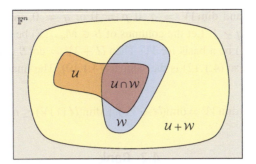

Figure 4.2 Venn-diagrammatic illustration of the intersection and sum of subspaces $\mathcal{U}$ and $\mathcal{W}$ of $\mathbb{F}^n$.

Theorem 4.1.9 *Let $B \in \mathsf{M}_{m \times p}(\mathbb{F})$ and $C \in \mathsf{M}_{m \times q}(\mathbb{F})$ have full column rank. Then*

$$\operatorname{rank}[B\ C] + \dim(\operatorname{col} B \cap \operatorname{col} C) = p + q. \tag{4.1.10}$$

Proof The rank-nullity theorem says that $\operatorname{rank}[B\ C] + \operatorname{nullity}[B\ C] = p + q$, so it suffices to show that $\operatorname{nullity}[B\ C] = \dim(\operatorname{col} B \cap \operatorname{col} C)$. Let $d = \dim(\operatorname{col} B \cap \operatorname{col} C)$ and let $\mathbf{z} \in \operatorname{null}[B\ C]$. Partition $\mathbf{z} = \begin{bmatrix} \mathbf{x} \\ \mathbf{y} \end{bmatrix}$, in which $\mathbf{x} \in \mathbb{F}^p$ and $\mathbf{y} \in \mathbb{F}^q$. Then

$$[B\ C]\mathbf{z} = [B\ C]\begin{bmatrix} \mathbf{x} \\ \mathbf{y} \end{bmatrix} = B\mathbf{x} + C\mathbf{y} = \mathbf{0}, \qquad \text{so} \qquad B\mathbf{x} = C(-\mathbf{y}) \in \operatorname{col} B \cap \operatorname{col} C.$$

If $d = 0$, then $B\mathbf{x} = C(-\mathbf{y}) = \mathbf{0}$. Since B and C have full column rank, it follows that $\mathbf{x}$ and $\mathbf{y}$ are zero vectors, $\mathbf{z} = \mathbf{0}$, $\operatorname{null}[B\ C] = \{\mathbf{0}\}$, and $\operatorname{nullity}[B\ C] = 0$.

If $d \geq 1$, let the columns of $Z \in \mathsf{M}_{m \times d}(\mathbb{F})$ be a basis for $\operatorname{col} B \cap \operatorname{col} C$. Then there is a unique $\mathbf{v} \in \mathbb{F}^d$ such that $B\mathbf{x} = C(-\mathbf{y}) = Z\mathbf{v}$. It follows from the inclusions $\operatorname{col} Z \subseteq \operatorname{col} B$ and $\operatorname{col} Z \subseteq \operatorname{col} C$ that there are $X \in \mathsf{M}_{p \times d}(\mathbb{F})$ and $Y \in \mathsf{M}_{q \times d}(\mathbb{F})$ such that $Z = BX$ and $Z = CY$. Since Z, B, and C have linearly independent columns, Theorem 1.6.21 ensures that X and Y are unique and have linearly independent columns. Then $B\mathbf{x} = Z\mathbf{v} = BX\mathbf{v}$ and $C(-\mathbf{y}) = Z\mathbf{v} = CY\mathbf{v}$, which imply that $\mathbf{x} = X\mathbf{v}$ and $\mathbf{y} = -Y\mathbf{v}$. Thus, every vector in $\operatorname{null}[B\ C]$ can be represented as

$$\mathbf{z} = \begin{bmatrix} \mathbf{x} \\ \mathbf{y} \end{bmatrix} = \begin{bmatrix} X\mathbf{v} \\ -Y\mathbf{v} \end{bmatrix} = \begin{bmatrix} X \\ -Y \end{bmatrix}\mathbf{v},$$

which shows that the columns of $\begin{bmatrix} X \\ -Y \end{bmatrix}$ span $\operatorname{null}[B\ C]$. If $\begin{bmatrix} X \\ -Y \end{bmatrix}\mathbf{v} = \mathbf{0}$, then $X\mathbf{v} = \mathbf{0}$. The linear independence of the columns of X ensures that $\mathbf{v} = \mathbf{0}$. We conclude that the d columns of $\begin{bmatrix} X \\ -Y \end{bmatrix}$ are a basis for $\operatorname{null}[B\ C]$. $\qquad\qquad\square$

An important consequence of the preceding theorem is a sufficient condition for two subspaces to have a nonzero intersection. This condition plays a key role in our study of eigenvalue interlacing; see Theorem 18.7.1.

Theorem 4.1.11 (Subspace Intersection) *Let $\mathcal{U}$ and $\mathcal{W}$ be subspaces of $\mathbb{F}^n$. Then*

$$\dim(\mathcal{U} + \mathcal{W}) + \dim(\mathcal{U} \cap \mathcal{W}) = \dim\mathcal{U} + \dim\mathcal{W}. \tag{4.1.12}$$

If $k \geq 1$ and $\dim\mathcal{U} + \dim\mathcal{W} \geq n + k$, then $\mathcal{U} \cap \mathcal{W}$ contains k linearly independent vectors.

Proof Let $\dim \mathcal{U} = p$ and $\dim \mathcal{W} = q$. If $p = 0$ or $q = 0$, there is nothing to prove. Suppose that $p \geq 1$ and $q \geq 1$. Let the columns of $B \in \mathsf{M}_{n \times p}(\mathbb{F})$ be a basis for $\mathcal{U}$ and let the columns of $C \in \mathsf{M}_{n \times q}(\mathbb{F})$ be a basis for $\mathcal{W}$. Then $\mathcal{U} + \mathcal{W} = \operatorname{col} B + \operatorname{col} C = \operatorname{col}[B \ C]$ and $\mathcal{U} \cap \mathcal{W} = \operatorname{col} B \cap \operatorname{col} C$, so (4.1.12) is the same as (4.1.10). The final assertion follows from a computation:

$$n + k \leq \dim \mathcal{U} + \dim \mathcal{W} = \dim(\mathcal{U} + \mathcal{W}) + \dim(\mathcal{U} \cap \mathcal{W}) \leq n + \dim(\mathcal{U} \cap \mathcal{W}). \qquad \square$$

4.2 Rank

If $A \in \mathsf{M}_{m \times n}$ has full column rank, the rank-nullity theorem (Theorem 4.1.1) ensures that $\operatorname{null} A = \{\mathbf{0}\}$. This observation is the foundation of the following theorem about matrix products that preserve rank. Products with this property play an important role in reductions of matrices to various standard forms.

Theorem 4.2.1 *Let $A \in \mathsf{M}_{m \times n}(\mathbb{F})$. If $X \in \mathsf{M}_{p \times m}(\mathbb{F})$ has full column rank and $Y \in \mathsf{M}_{n \times q}(\mathbb{F})$ has full row rank, then*

$$\operatorname{rank} A = \operatorname{rank} XAY. \tag{4.2.2}$$

In particular, (4.2.2) is valid if $X \in \mathsf{M}_m(\mathbb{F})$ and $Y \in \mathsf{M}_n(\mathbb{F})$ are invertible.

Proof First consider $XA \in \mathsf{M}_{p \times n}(\mathbb{F})$. Since X has full column rank, the rank-nullity theorem ensures that $\operatorname{null} X = \{\mathbf{0}\}$. If $\mathbf{u} \in \mathbb{F}^n$, then $(XA)\mathbf{u} = X(A\mathbf{u}) = \mathbf{0}$ if and only if $A\mathbf{u} = \mathbf{0}$. Thus, $\operatorname{null} XA = \operatorname{null} A$ and the rank-nullity theorem tells us that $\operatorname{rank} XA = \operatorname{rank} A$ since XA and A have the same number of columns. To analyze $AY \in \mathsf{M}_{m \times q}(\mathbb{F})$, consider its transpose (which has full column rank), and apply Theorem 2.3.1 twice:

$$\operatorname{rank} AY = \operatorname{rank}(AY)^{\mathsf{T}} = \operatorname{rank}(Y^{\mathsf{T}} A^{\mathsf{T}}) = \operatorname{rank} A^{\mathsf{T}} = \operatorname{rank} A. \qquad \square$$

Example 4.2.3 If the columns of $A = [\mathbf{a}_1 \ \mathbf{a}_2 \ \dots \ \mathbf{a}_n] \in \mathsf{M}_{m \times n}$ are linearly independent and $X \in \mathsf{M}_m$ is invertible, then $\operatorname{rank} XA = \operatorname{rank} A = n$. This means that the columns of $XA = [X\mathbf{a}_1 \ X\mathbf{a}_2 \ \dots \ X\mathbf{a}_n]$ are linearly independent.

The following theorem provides an upper bound on the rank of a product and a lower bound on the rank of an augmented matrix.

Theorem 4.2.4 *Let $A \in \mathsf{M}_{m \times k}$, $B \in \mathsf{M}_{k \times n}$, and $C \in \mathsf{M}_{m \times p}$. Then*

$$\operatorname{rank} AB \leq \min\{\operatorname{rank} A, \operatorname{rank} B\}, \tag{4.2.5}$$

and $\operatorname{rank} A = \operatorname{rank} AB$ if and only if $\operatorname{col} A = \operatorname{col} AB$. Also,

$$\max\{\operatorname{rank} A, \operatorname{rank} C\} \leq \operatorname{rank}[A \ C], \tag{4.2.6}$$

and $\operatorname{rank} A = \operatorname{rank}[A \ C]$ if and only if $\operatorname{col} C \subseteq \operatorname{col} A$.

Proof Since $\operatorname{col} AB \subseteq \operatorname{col} A$, Theorem 2.2.11 ensures that $\operatorname{rank} AB = \dim \operatorname{col} AB \leq \dim \operatorname{col} A = \operatorname{rank} A$, with equality if and only if $\operatorname{col} A = \operatorname{col} AB$. Apply this inequality to the transpose of AB and obtain

$$\operatorname{rank} AB = \operatorname{rank}(AB)^{\mathsf{T}} = \operatorname{rank}(B^{\mathsf{T}} A^{\mathsf{T}}) \leq \operatorname{rank} B^{\mathsf{T}} = \operatorname{rank} B.$$

Finally, $\operatorname{col} A \subseteq \operatorname{col} A + \operatorname{col} C = \operatorname{col}[A \ C]$, so Theorem 2.2.11 ensures that

$$\operatorname{rank} A = \dim \operatorname{col} A \leq \dim \operatorname{col}[A \ C] = \operatorname{rank}[A \ C],$$

with equality if and only if $\operatorname{col} A = \operatorname{col} A + \operatorname{col} C$, which occurs if and only if $\operatorname{col} C \subseteq \operatorname{col} A$. $\square$

The following theorems about rank exploit the identity (3.2.2). We begin with a lower bound on the rank of a product.

Theorem 4.2.7 *Let $A \in \mathsf{M}_{m \times k}$ and $B \in \mathsf{M}_{k \times n}$. Then*

$$\operatorname{rank} A + \operatorname{rank} B - k \leq \operatorname{rank} AB. \tag{4.2.8}$$

Proof First suppose that $AB = 0$. If $A = 0$, then (4.2.8) is equivalent to the (correct) inequality $\operatorname{rank} B \leq k$. If $\operatorname{rank} A \geq 1$, then every column of B is in $\operatorname{null} A$, so the rank-nullity theorem tells us that

$$\operatorname{rank} B = \dim \operatorname{col} B \leq \operatorname{nullity} A = k - \operatorname{rank} A.$$

Therefore, $\operatorname{rank} A + \operatorname{rank} B \leq k$ and the assertion is proved in the case $AB = 0$.

Now suppose that $\operatorname{rank} AB = r \geq 1$ and let $AB = XY$ be a full-rank factorization. Let

$$C = [A \ X] \in \mathsf{M}_{m \times (k+r)} \quad \text{and} \quad D = \begin{bmatrix} B \\ -Y \end{bmatrix} \in \mathsf{M}_{(k+r) \times n}.$$

Then $CD = AB - XY = 0$, so (4.2.6) and the preceding case ensure that

$$\operatorname{rank} A + \operatorname{rank} B \leq \operatorname{rank} C + \operatorname{rank} D \leq k + r = k + \operatorname{rank} AB. \qquad \square$$

The next theorem bounds the rank of a sum from above and below.

Theorem 4.2.9 *Let $A, B \in \mathsf{M}_{m \times n}$. Then*

$$|\operatorname{rank} A - \operatorname{rank} B| \leq \operatorname{rank}(A + B) \leq \operatorname{rank} A + \operatorname{rank} B. \tag{4.2.10}$$

Proof The asserted inequalities are valid if $A = 0$ or $B = 0$, so we assume that $\operatorname{rank} A = r \geq 1$ and $\operatorname{rank} B = s \geq 1$. Let $A = X_1 Y_1$ and $B = X_2 Y_2$ be full-rank factorizations, and let

$$C = [X_1 \ X_2] \in \mathsf{M}_{m \times (r+s)} \quad \text{and} \quad D = \begin{bmatrix} Y_1 \\ Y_2 \end{bmatrix} \in \mathsf{M}_{(r+s) \times n}.$$

Then $A + B = X_1 Y_1 + X_2 Y_2 = CD$, so (4.2.5) and (2.3.2) ensure that

$$\operatorname{rank}(A + B) = \operatorname{rank} CD \leq \min\{\operatorname{rank} C, \operatorname{rank} D\} \leq \min\{r + s, r + s\} = \operatorname{rank} A + \operatorname{rank} B,$$

which is the upper bound in (4.2.10). Now use this upper bound to compute

$$\operatorname{rank} A = \operatorname{rank}(A + B + (-B)) \leq \operatorname{rank}(A + B) + \operatorname{rank} B$$

and

$$\operatorname{rank} B = \operatorname{rank}(A + B + (-A)) \leq \operatorname{rank}(A + B) + \operatorname{rank} A.$$

Combine these inequalities to get $\pm (\operatorname{rank} A - \operatorname{rank} B) \leq \operatorname{rank}(A + B)$, which is the lower bound in (4.2.10). $\square$

4.3 *LU* Factorization

A triangular system of linear equations can be solved by forward or backward substitution. In this section, we develop a strategy for reducing a linear system to a pair of triangular systems.

Example 4.3.1 A lower triangular system, such as

$$\begin{bmatrix} 1 & 0 \\ 2 & 1 \end{bmatrix} \begin{bmatrix} w_1 \\ w_2 \end{bmatrix} = \begin{bmatrix} 14 \\ 64 \end{bmatrix},$$

can be solved by forward substitution (solve the equations from the top down). Substitute the solution $w_1 = 14$ of the first equation into the second and obtain $w_2 = 64 - 2w_1 = 64 - 28 = 36$. An upper triangular system, such as

$$\begin{bmatrix} 4 & 5 \\ 0 & 6 \end{bmatrix} \begin{bmatrix} x_1 \\ x_2 \end{bmatrix} = \begin{bmatrix} 14 \\ 36 \end{bmatrix},$$

can be solved by backward substitution (solve the equations from the bottom up). Substitute the solution $x_2 = 6$ of the second equation into the first and obtain $4x_1 = 14 - 5x_2 = 14 - 30 = -16$, so $x_1 = -4$.

Now consider the linear system $A\mathbf{x} = \mathbf{y}$, in which $\begin{bmatrix} 4 & 5 \\ 8 & 16 \end{bmatrix}$. Since

$$A = \begin{bmatrix} 4 & 5 \\ 8 & 16 \end{bmatrix} = \underbrace{\begin{bmatrix} 1 & 0 \\ 2 & 1 \end{bmatrix}}_{L} \underbrace{\begin{bmatrix} 4 & 5 \\ 0 & 6 \end{bmatrix}}_{U}, \tag{4.3.2}$$

we can obtain a solution in two steps. First use forward substitution to solve the lower triangular system $L\mathbf{w} = \mathbf{y}$ for $\mathbf{w}$, and then use backward substitution to solve the upper triangular system $U\mathbf{x} = \mathbf{w}$ for $\mathbf{x}$. This works if we have a factorization $A = LU$, but how can we get one? The key observation is that the first row of U in (4.3.2) is the same as the first row of A, the first column of L is $\frac{1}{4}$ times the first column of A, and 4 is the $(1, 1)$ entry of A. We lean heavily on this observation in our proof of the next theorem.

Definition 4.3.3 A *unit lower triangular matrix* is a lower triangular matrix with all its main diagonal entries 1. A matrix A has an *LU-factorization* if $A = LU$, in which L is unit lower triangular and U is upper triangular.

Definition 4.3.4 If $\mathbf{v} = \begin{bmatrix} a \\ \mathbf{c} \end{bmatrix} \in \mathbb{C}^k$ and $a \neq 0$, we *unitize* $\mathbf{v}$ by forming $a^{-1}\mathbf{v} = \begin{bmatrix} 1 \\ a^{-1}\mathbf{c} \end{bmatrix}$.

Our interest in *LU* factorizations is in the context of solving systems of linear equations, so it is natural to assume that the matrix to be factored is invertible.

Theorem 4.3.5 (*LU* factorization) *Let* $n \geq 2$ *and suppose that* $A \in \mathsf{M}_n$ *is invertible.*

(a) *A has an LU factorization if and only if every leading principal submatrix of A is invertible.*

(b) *If* $A = L_1 U_1$ *and* $A = L_2 U_2$ *are LU factorizations, then* $L_1 = L_2$ *and* $U_1 = U_2$.

Proof Let $X_{(k)}$ denote the $k \times k$ leading principal submatrix of a matrix X; see Figure C.6.

(a) ($\Rightarrow$) Suppose that $A = LU$ is an LU factorization and let $k \in \{1, 2, \ldots, n-1\}$. We must show that $A_{(k)}$ is invertible. Partition

$$A = \begin{bmatrix} A_{(k)} & A_{12} \\ A_{21} & A_{22} \end{bmatrix}, \quad L = \begin{bmatrix} L_{(k)} & 0 \\ L_{21} & L_{22} \end{bmatrix}, \quad \text{and} \quad U = \begin{bmatrix} U_{(k)} & U_{12} \\ 0 & U_{22} \end{bmatrix} \tag{4.3.6}$$

conformally. Theorem 3.3.4 ensures that

$$\det A = (\det L)(\det U) = (\det L_{(k)})(\det L_{22})(\det U_{(k)})(\det U_{22}).$$

Since $\det A \neq 0$, it follows that $L_{(k)}$ and $U_{(k)}$ are invertible. Therefore, $A_{(k)} = L_{(k)} U_{(k)}$ is invertible. If $k = n$, then $A_{(n)} = A$ is invertible by assumption.

($\Leftarrow$) Suppose that $\det A_{(k)} \neq 0$ for each $k = 1, 2, \ldots, n$. Partition $A = \begin{bmatrix} a & \mathbf{b}^{\mathsf{T}} \\ \mathbf{c} & D \end{bmatrix}$, in which $a \in \mathbb{C}$, $\mathbf{b}, \mathbf{c} \in \mathbb{C}^{n-1}$, and $D \in \mathsf{M}_{n-1}$. Since $a = A_{(1)} \neq 0$, we can use the unitized first column and the first row of A to form the outer product

$$a^{-1} \begin{bmatrix} a \\ \mathbf{c} \end{bmatrix} \begin{bmatrix} a & \mathbf{b}^{\mathsf{T}} \end{bmatrix} = \begin{bmatrix} 1 \\ a^{-1}\mathbf{c} \end{bmatrix} \begin{bmatrix} a & \mathbf{b}^{\mathsf{T}} \end{bmatrix} = \begin{bmatrix} a & \mathbf{b}^{\mathsf{T}} \\ \mathbf{c} & a^{-1}\mathbf{c}\mathbf{b}^{\mathsf{T}} \end{bmatrix} \tag{4.3.7}$$

and use it to reduce A to a matrix that is bordered with a row and column of zeros:

$$A^{[1]} = A - a^{-1} \begin{bmatrix} a \\ \mathbf{c} \end{bmatrix} \begin{bmatrix} a & \mathbf{b}^{\mathsf{T}} \end{bmatrix} = \begin{bmatrix} 0 & \mathbf{0}^{\mathsf{T}} \\ \mathbf{0} & D - a^{-1}\mathbf{c}\mathbf{b}^{\mathsf{T}} \end{bmatrix} = \begin{bmatrix} 0 & \mathbf{0}^{\mathsf{T}} \\ \mathbf{0} & A/a \end{bmatrix}. \tag{4.3.8}$$

The block $A/a = D - a^{-1}\mathbf{c}\mathbf{b}^{\mathsf{T}}$ is the Schur complement of a in A; see (3.3.9) and (3.3.14).

Let $k \in \{1, 2, \ldots, n-1\}$ and let $\mathbf{v}_{(k)} \in \mathbb{C}^k$ denote the vector whose entries are the first k entries of a longer vector $\mathbf{v}$. Inspection of (4.3.7) shows that the leading principal submatrix of A of size $k + 1$ is

$$A_{(k+1)} = \begin{bmatrix} a & \mathbf{b}_{(k)}^{\mathsf{T}} \\ \mathbf{c}_{(k)} & D_{(k)} \end{bmatrix}.$$

It is invertible and $a \neq 0$, so Theorem 3.3.11 ensures that the Schur complement $A_{(k+1)}/a = D_{(k)} - \frac{1}{a}\mathbf{c}_{(k)}\mathbf{b}_{(k)}^{\mathsf{T}}$ is also invertible. Notice that $A_{(k+1)}/a$ is the $k \times k$ leading principal submatrix of $D - \frac{1}{a}\mathbf{c}\mathbf{b}^{\mathsf{T}} = A/a$. Therefore, the $k \times k$ leading principal submatrix of A/a is invertible for each $k = 1, 2, \ldots, n-1$. This means that the invertibility hypotheses on the original matrix A are inherited by the trailing principal submatrix of the reduced matrix $A^{[1]}$. Consequently, we can reduce $A^{[1]}$ by subtracting the outer product of its unitized second column and second row, thereby obtaining a matrix that is bordered with two rows and columns of zeros. Its trailing principal submatrix of size $n - 2$ again inherits the property of having invertible principal submatrices. After n reduction steps, we arrive at a zero matrix and obtain a representation of A as a sum of outer products of vectors with increasing numbers of initial zero entries. The identity (3.1.17) recognizes this sum as a matrix product, which is an LU factorization of A. The computations in the following example illustrate the sequence of reduction steps.

(b) If $A = L_1 U_1$ and $A = L_2 U_2$ are LU factorizations, then the factors are invertible and $L_1^{-1} L_2 = U_1 U_2^{-1}$. We know that L_1^{-1} is unit lower triangular (Theorem 3.2.9). Therefore, $L_1^{-1} L_2$ is unit lower triangular and equal to $U_1 U_2^{-1}$, which is upper triangular (Example 3.2.13). We conclude that $L_1^{-1} L_2$ is diagonal, so $L_1^{-1} L_2 = I = U_1 U_2^{-1}$. Thus, $L_1 = L_2$ and $U_1 = U_2$. $\quad\square$

Example 4.3.9 The determinants of the leading principal submatrices of

$$A = \begin{bmatrix} 1 & 5 & 6 & 7 \\ -1 & -3 & 2 & 2 \\ 2 & 16 & 39 & 51 \\ -4 & -10 & -2 & -39 \end{bmatrix}$$

are $\det A_{(1)} = 1$, $\det A_{(2)} = 2$, $\det A_{(3)} = 6$, and $\det A = 24$. We can use the reduction algorithm in the proof of the preceding theorem to obtain the LU factorization of A. The first reduction uses the outer product of the unitized first column and row of A:

$$A^{[1]} = A - \begin{bmatrix} 1 \\ -1 \\ 2 \\ -4 \end{bmatrix} \begin{bmatrix} 1 & 5 & 6 & 7 \end{bmatrix} = \begin{bmatrix} 0 & 0 & 0 & 0 \\ 0 & 2 & 8 & 9 \\ 0 & 6 & 27 & 37 \\ 0 & 10 & 22 & -11 \end{bmatrix}.$$

The second reduction uses the outer product of the unitized second column and row of $A^{[1]}$:

$$A^{[2]} = A^{[1]} - \frac{1}{2} \begin{bmatrix} 0 \\ 2 \\ 6 \\ 10 \end{bmatrix} \begin{bmatrix} 0 & 2 & 8 & 9 \end{bmatrix} = \begin{bmatrix} 0 & 0 & 0 & 0 \\ 0 & 0 & 0 & 0 \\ 0 & 0 & 3 & 10 \\ 0 & 0 & -18 & -56 \end{bmatrix}.$$

Now subtract the outer product of the unitized third column and row of $A^{[2]}$:

$$A^{[3]} = A^{[2]} - \frac{1}{3} \begin{bmatrix} 0 \\ 0 \\ 3 \\ -18 \end{bmatrix} \begin{bmatrix} 0 & 0 & 3 & 10 \end{bmatrix} = \begin{bmatrix} 0 & 0 & 0 & 0 \\ 0 & 0 & 0 & 0 \\ 0 & 0 & 0 & 0 \\ 0 & 0 & 0 & 4 \end{bmatrix}.$$

The final reduction uses the unitized last column and row of $A^{[3]}$:

$$A^{[4]} = A^{[3]} - \frac{1}{4} \begin{bmatrix} 0 \\ 0 \\ 0 \\ 4 \end{bmatrix} \begin{bmatrix} 0 & 0 & 0 & 4 \end{bmatrix} = \begin{bmatrix} 0 & 0 & 0 & 0 \\ 0 & 0 & 0 & 0 \\ 0 & 0 & 0 & 0 \\ 0 & 0 & 0 & 0 \end{bmatrix}.$$

The factors L and U are obtained with the assistance of (3.1.17) from the outer products in the four reductions:

$$A = \underbrace{\begin{bmatrix} 1 & 0 & 0 & 0 \\ -1 & 1 & 0 & 0 \\ 2 & 3 & 1 & 0 \\ -4 & 5 & -6 & 1 \end{bmatrix}}_{L} \underbrace{\begin{bmatrix} 1 & 5 & 6 & 7 \\ 0 & 2 & 8 & 9 \\ 0 & 0 & 3 & 10 \\ 0 & 0 & 0 & 4 \end{bmatrix}}_{U}.$$

Example 4.3.10 An invertible matrix need not have an LU factorization. For example, if the invertible matrix $A = \begin{bmatrix} 0 & 1 \\ 1 & 1 \end{bmatrix} = [a_{ij}] \in \mathsf{M}_2$ has an LU factorization $A = LU$ with $L = [\ell_{ij}]$ and $U = [u_{ij}]$, then $0 = a_{11} = \ell_{11}u_{11}$ and $1 = a_{12}a_{21} = \ell_{11}u_{12}\ell_{21}u_{11} = (\ell_{11}u_{11})u_{12}\ell_{21} = a_{11}u_{12}\ell_{21} = 0$, which is impossible. Notice that $A_{(1)}$ is not invertible, so Theorem 4.3.5 does not apply.

Example 4.3.11 A non-invertible matrix can have an LU factorization, which need not be unique. For example,

$$A = \begin{bmatrix} 1 & 1 & 1 \\ 1 & 1 & 1 \\ 1 & 1 & 1 \end{bmatrix} = \begin{bmatrix} 1 & 0 & 0 \\ 1 & 1 & 0 \\ 1 & 1 & 1 \end{bmatrix} \begin{bmatrix} 1 & 1 & 1 \\ 0 & 0 & 0 \\ 0 & 0 & 0 \end{bmatrix} = \begin{bmatrix} 1 & 0 & 0 \\ 1 & 1 & 0 \\ 1 & 0 & 1 \end{bmatrix} \begin{bmatrix} 1 & 1 & 1 \\ 0 & 0 & 0 \\ 0 & 0 & 0 \end{bmatrix}.$$

4.4 Row Equivalence

If A and B are matrices of the same size, when is there an invertible matrix C such that $A = CB$? Theorem 4.2.1 ensures that $\operatorname{rank} A = \operatorname{rank} CB = \operatorname{rank} B$, so it is necessary that A and B have the same rank (equivalently, the same nullity). The example

$$A = \begin{bmatrix} 1 & 0 \\ 0 & 0 \end{bmatrix}, \qquad B = \begin{bmatrix} 0 & 1 \\ 0 & 0 \end{bmatrix}, \quad \text{and} \quad CB = \begin{bmatrix} 0 & \star \\ 0 & \star \end{bmatrix}$$

shows that equal rank is not sufficient. The following theorem identifies necessary and sufficient conditions and generalizes Theorem 3.1.22, which considers the special case $\operatorname{null} A = \operatorname{null} B = \{\mathbf{0}\}$.

Theorem 4.4.1 *Let $A, B \in \mathsf{M}_{m \times n}(\mathbb{F})$. The following are equivalent:*

(a) $\operatorname{null} A = \operatorname{null} B$.

(b) $B = CA$ *for some invertible $C \in \mathsf{M}_m(\mathbb{F})$.*

(c) $\operatorname{row} A = \operatorname{row} B$.

Each of the preceding statements is equivalent to the following if A and B are nonzero:

(d) *The pivot column decompositions of A and B are $A = PR$ and $B = QR$, in which P and Q are the respective pivot matrices of A and B.*

Proof Let $r = \operatorname{rank} A$. If $r = 0$, then $A = 0$ and each statement implies that $B = 0$ and there is nothing to prove. The case $r = n$ is Theorem 3.1.22. Thus, we may suppose that $1 \le r < n$.

(a) $\Rightarrow$ (b) The rank-nullity theorem tells us that $\dim \operatorname{null} A = n - r$, so $\dim \operatorname{null} B = n - r$ as well and $\operatorname{rank} B = r$. Let the columns of $X \in \mathsf{M}_{n \times (n-r)}(\mathbb{F})$ be a basis for $\operatorname{null} A$. Theorem 3.1.22 ensures that there is a $Y \in \mathsf{M}_{n \times r}(\mathbb{F})$ such that $Z = [X \ Y] \in \mathsf{M}_n(\mathbb{F})$ is invertible. Then

$$r = \operatorname{rank} A = \operatorname{rank} AZ = \operatorname{rank}[AX \ AY] = \operatorname{rank}[0 \ AY] = \operatorname{rank} AY$$

and

$$r = \operatorname{rank} B = \operatorname{rank} BZ = \operatorname{rank}[BX \ BY] = \operatorname{rank}[0 \ BY] = \operatorname{rank} BY.$$

Theorem 3.1.22 ensures that there is an invertible $C \in \mathsf{M}_m(\mathbb{F})$ such that $BY = CAY$, that is, $(B - CA)Y = 0$. Since $BX = AX = 0$, we also have $(B - CA)X = 0$, and hence

$$(B - CA)Z = [\,(B - CA)X \quad (B - CA)Y\,] = [0 \ 0] = 0.$$

It follows that $B - CA = 0$ since Z is invertible.

(b) $\Rightarrow$ (c) Since $B = CA$, every row of B is a linear combination of rows of A and hence $\operatorname{row} B \subseteq \operatorname{row} A$. But $A = C^{-1}B$, so the same reasoning shows that $\operatorname{row} A \subseteq \operatorname{row} B$. We conclude that $\operatorname{row} A = \operatorname{row} B$.

(c) $\Rightarrow$ (a) Let the columns of $X \in M_{n \times r}(\mathbb{F})$ be a basis for $\mathrm{col}\, A^\mathsf{T} = \mathrm{col}\, B^\mathsf{T}$. Theorem 2.3.7 ensures that there are full-rank matrices $Y, Z \in M_{r \times m}(\mathbb{F})$ such that $A^\mathsf{T} = XY$ and $B^\mathsf{T} = XZ$. Then $A = Y^\mathsf{T} X^\mathsf{T}$, $B = Z^\mathsf{T} X^\mathsf{T}$, and $\mathrm{null}\, A = \mathrm{null}\, X^\mathsf{T} = \mathrm{null}\, B$.

(b) $\Rightarrow$ (d) If $A \neq 0$, let $A = PR$ be its pivot column decomposition and let $B = CA$, in which C is invertible. Theorem 4.2.1 ensures that the left-to-right column selection algorithm in Theorem 1.7.5 produces the same pivot indices for A and CA, so CP is the pivot matrix of B. Therefore, $B = CA = CPR = (CP)R$ is the pivot column decomposition of B.

(d) $\Rightarrow$ (c) If A and B are nonzero and if $A = PR$ and $B = QR$ are their respective pivot column decompositions, then these are full-rank factorizations and Theorem 2.3.7 ensures that $\mathrm{row}\, A = \mathrm{row}\, R = \mathrm{row}\, B$. $\qquad\square$

Definition 4.4.2 $A, B \in M_{m \times n}(\mathbb{F})$ are *row equivalent* if $\mathrm{row}\, A = \mathrm{row}\, B$; they are *left equivalent* if there is an invertible $C \in M_m(\mathbb{F})$ such that $A = CB$.

The preceding theorem says that A and B are row equivalent if and only if they are left equivalent, and it provides a computational test for either equivalence: If A and B are nonzero, compute their (unique) pivot column decompositions $A = PR_1$ and $B = QR_2$ and see if $R_1 = R_2$. The proof of Theorem 1.7.5 contains an algorithm to compute these factors by manipulating columns and using their linear dependencies. An alternative way to compute them manipulates rows instead, and makes essential use of left equivalence, as we now explain.

Let $A \in M_{m \times n}(\mathbb{F})$ and suppose that $\mathrm{rank}\, A = r \geq 1$. If $A = PR$ is its pivot column decomposition, then Theorem 3.1.22 says there is a $W \in M_{m \times (m-r)}(\mathbb{F})$ such that $G = [P \ W] \in M_m(\mathbb{F})$ is invertible. Then

$$G^{-1}P = \begin{bmatrix} I_r \\ 0 \end{bmatrix} \quad \text{and} \quad G^{-1}A = G^{-1}PR = \begin{bmatrix} I_r \\ 0 \end{bmatrix} R = \begin{bmatrix} R \\ 0 \end{bmatrix} = E \in M_{m \times n}(\mathbb{F}),$$

so E is left equivalent to A. The structure of R (see Theorem 1.7.5) ensures that E is in reduced row echelon form; see the list on p. 440. The reduced row echelon form of A can be computed via a sequence of left multiplications by elementary matrices (invertible matrices of three types), which means that the factor R in the pivot column decomposition of A (which is E with the bottom rows of zeros removed) can be determined via row operations that make no use of column dependencies.

Example 4.4.3 Let A be the matrix in (P.1.13). We compute its reduced row echelon form E with a sequence of row reductions (blockwise and entrywise Gaussian elimination acting on bolded entries) that make use of (3.3.6) and (3.3.7):

$$\begin{bmatrix} 1 & 0 & 0 \\ -4 & 1 & 0 \\ -7 & 0 & 1 \end{bmatrix} \begin{bmatrix} 1 & 2 & 3 \\ \mathbf{4} & 5 & 6 \\ \mathbf{7} & 8 & 9 \end{bmatrix} = \begin{bmatrix} 1 & 2 & 3 \\ 0 & -3 & -6 \\ 0 & -6 & -12 \end{bmatrix}$$

$$\begin{bmatrix} 1 & 0 & 0 \\ 0 & -\frac{1}{3} & 0 \\ 0 & 0 & 1 \end{bmatrix} \begin{bmatrix} 1 & 2 & 3 \\ 0 & \mathbf{-3} & \mathbf{-6} \\ 0 & -6 & -12 \end{bmatrix} = \begin{bmatrix} 1 & 2 & 3 \\ 0 & 1 & 2 \\ 0 & -6 & -12 \end{bmatrix}$$

$$\begin{bmatrix} 1 & 0 & 0 \\ 0 & 1 & 0 \\ 0 & 6 & 1 \end{bmatrix} \begin{bmatrix} 1 & 2 & 3 \\ 0 & 1 & 2 \\ 0 & \mathbf{-6} & \mathbf{-12} \end{bmatrix} = \begin{bmatrix} 1 & 2 & 3 \\ 0 & 1 & 2 \\ 0 & 0 & 0 \end{bmatrix}$$

$$\begin{bmatrix} 1 & -2 & 0 \\ 0 & 1 & 0 \\ 0 & 0 & 1 \end{bmatrix} \begin{bmatrix} 1 & \mathbf{2} & 3 \\ 0 & 1 & 2 \\ 0 & 0 & 0 \end{bmatrix} = \begin{bmatrix} 1 & 0 & -1 \\ 0 & 1 & 2 \\ 0 & 0 & 0 \end{bmatrix} = E.$$

Now discard the bottom row of zeros and obtain

$$R = \begin{bmatrix} 1 & 0 & -1 \\ 0 & 1 & 2 \end{bmatrix}. \tag{4.4.4}$$

If we partition $A = [\mathbf{a}_1\ \mathbf{a}_2\ \mathbf{a}_3]$ and let $P = [\mathbf{p}_1\ \mathbf{p}_2]$, then (4.4.4) and the identity $A = PR$ tell us that

$$[\mathbf{a}_1\ \mathbf{a}_2\ \mathbf{a}_3] = [\mathbf{p}_1\ \mathbf{p}_2] \begin{bmatrix} 1 & 0 & -1 \\ 0 & 1 & 2 \end{bmatrix} = [\mathbf{p}_1\ \mathbf{p}_2\ -\mathbf{p}_1+2\mathbf{p}_2].$$

Therefore, $P = [\mathbf{a}_1\ \mathbf{a}_2]$. This demonstrates how both factors of the pivot column decomposition of A can be computed with a sequence of elementary row operations.

The uniqueness of the reduced row echelon form of a matrix A follows from the uniqueness of its pivot column decomposition. If A had two different reduced row echelon forms, it would have two different pivot column decompositions, which it does not. Since R is determined uniquely by A, and E is determined uniquely by R, no matter how we perform a sequence of row operations on A, when we arrive at a matrix in reduced row echelon form, it must be E.

In the preceding example, the product of the matrices that encode the row reductions is

$$G^{-1} = \begin{bmatrix} 1 & -2 & 0 \\ 0 & 1 & 0 \\ 0 & 0 & 1 \end{bmatrix} \begin{bmatrix} 1 & 0 & 0 \\ 0 & 1 & 0 \\ 0 & 6 & 1 \end{bmatrix} \begin{bmatrix} 1 & 0 & 0 \\ 0 & -\frac{1}{3} & 0 \\ 0 & 0 & 1 \end{bmatrix} \begin{bmatrix} 1 & 0 & 0 \\ -4 & 1 & 0 \\ -7 & 0 & 1 \end{bmatrix} = \begin{bmatrix} -\frac{5}{3} & \frac{2}{3} & 0 \\ \frac{4}{3} & -\frac{1}{3} & 0 \\ 1 & -2 & 1 \end{bmatrix}. \tag{4.4.5}$$

Its inverse is

$$G = \begin{bmatrix} 1 & 2 & 0 \\ 4 & 5 & 0 \\ 7 & 8 & 1 \end{bmatrix} = [P \star],$$

which reveals the columns of the factor P in the pivot column decomposition:

$$A = \begin{bmatrix} 1 & 2 \\ 4 & 5 \\ 7 & 8 \end{bmatrix} \begin{bmatrix} 1 & 0 & -1 \\ 0 & 1 & 2 \end{bmatrix} = PR.$$

4.5 Commutators and Shoda's Theorem

If T and S are linear operators on a vector space V, their *commutator* is the linear operator $[T, S] = TS - ST$. If $A, B \in \mathsf{M}_n$, their *commutator* is the matrix $[A, B] = AB - BA$. The commutator of operators is zero if and only if the operators commute, so if we want to understand how noncommuting matrices or operators are related, we might begin by studying their commutator.

Example 4.5.1 In one-dimensional quantum mechanics, one considers a vector space V of suitably differentiable complex-valued functions of x (position) and t (time). The *position operator* T is defined by $(Tf)(x, t) = xf(x, t)$, and the *momentum operator* S is defined by $(Sf)(x, t) = -i\hbar \frac{\partial}{\partial x} f(x, t)$, in which $\hbar$ is a constant. Compute

$$TSf = x\left(-i\hbar \frac{\partial f}{\partial x}\right) = -i\hbar x \frac{\partial f}{\partial x} \quad \text{and} \quad STf = -i\hbar \frac{\partial}{\partial x}(xf) = -i\hbar x \frac{\partial f}{\partial x} - i\hbar f.$$

We have $(TS - ST)f = i\hbar f$ for all $f \in \mathcal{V}$, that is, $TS - ST = i\hbar I$. This commutator identity implies the *Heisenberg uncertainty principle*, which says that precise simultaneous measurement of position and momentum is impossible in a one-dimensional quantum mechanical system.

The commutator of the position and momentum operators is a nonzero scalar multiple of the identity, but this cannot happen for matrices. If $A, B \in \mathsf{M}_n$, then (C.2.9) ensures that

$$\mathrm{tr}(AB - BA) = \mathrm{tr}\,AB - \mathrm{tr}\,BA = \mathrm{tr}\,AB - \mathrm{tr}\,AB = 0. \tag{4.5.2}$$

However, $\mathrm{tr}(cI_n) = nc \neq 0$ if $c \neq 0$. Consequently, a nonzero scalar matrix cannot be a commutator.

How can we decide if a given matrix is a commutator? It must have trace zero, but is this necessary condition also sufficient? The following lemma is the first step toward proving that it is.

Lemma 4.5.3 *Let $n \geq 2$ and let $A \in \mathsf{M}_n(\mathbb{F})$. The list $\mathbf{x}, A\mathbf{x}$ is linearly dependent for all $\mathbf{x} \in \mathbb{F}^n$ if and only if A is a scalar matrix.*

Proof If $A = cI_n$ for some $c \in \mathbb{F}$, then the list $\mathbf{x}, A\mathbf{x}$ equals $\mathbf{x}, c\mathbf{x}$, which is linearly dependent for all $\mathbf{x} \in \mathbb{F}^n$. Conversely, suppose that $\mathbf{x}, A\mathbf{x}$ is linearly dependent for all $\mathbf{x} \in \mathbb{F}^n$. Since $\mathbf{e}_i, A\mathbf{e}_i$ is linearly dependent for each $i = 1, 2, \ldots, n$, there are scalars $a_1, a_2, \ldots, a_n$ such that each $A\mathbf{e}_i = a_i\mathbf{e}_i$. Consequently, A is a diagonal matrix. If $\mathbf{e} \in \mathbb{F}^n$ is the all-ones vector, then the linear dependence of $\mathbf{e}, A\mathbf{e}$ implies that there is a scalar λ such that $A\mathbf{e} = \lambda\mathbf{e}$. Then

$$\sum_{i=1}^{n} \lambda\mathbf{e}_i = \lambda\mathbf{e} = A\mathbf{e} = A\sum_{i=1}^{n}\mathbf{e}_i = \sum_{i=1}^{n}A\mathbf{e}_i = \sum_{i=1}^{n}a_i\mathbf{e}_i.$$

The linear independence of $\mathbf{e}_1, \mathbf{e}_2, \ldots, \mathbf{e}_n$ ensures that $a_i = \lambda$ for each $i = 1, 2, \ldots, n$. Therefore, $A = \lambda I$. $\square$

The next step is to show that any nonscalar matrix is similar to a matrix with at least one zero diagonal entry.

Lemma 4.5.4 *Let $n \geq 2$ and let $A \in \mathsf{M}_n(\mathbb{F})$. If A is not a scalar matrix, then it is similar over $\mathbb{F}$ to a matrix that has a zero entry in its $(1, 1)$ position.*

Proof The preceding lemma ensures that there is an $\mathbf{x} \in \mathbb{F}^n$ such that the list $\mathbf{x}, A\mathbf{x}$ is linearly independent. If $n = 2$, let $S = [\mathbf{x}\ A\mathbf{x}]$. If $n > 2$, invoke Theorem 3.1.22 and choose $B \in \mathsf{M}_{n \times (n-2)}$ such that $S = [\mathbf{x}\ A\mathbf{x}\ B]$ is invertible. Let $S^{-*} = [\mathbf{y}\ Y]$, in which $Y \in \mathsf{M}_{n \times (n-1)}$. The $(1, 2)$ entry of $I_n = S^{-1}S$ is zero; this entry is $\mathbf{y}^*A\mathbf{x}$. Therefore,

$$S^{-1}AS = \begin{bmatrix} \mathbf{y}^* \\ Y^* \end{bmatrix} A[\mathbf{x}\ A\mathbf{x}\ B] = \begin{bmatrix} \mathbf{y}^*A\mathbf{x} & \star \\ \star & A_1 \end{bmatrix} = \begin{bmatrix} 0 & \star \\ \star & A_1 \end{bmatrix}, \tag{4.5.5}$$

which has a zero entry in its $(1, 1)$ position. $\square$

Because $\mathrm{tr}\,A = \mathrm{tr}(S^{-1}AS) = 0 + \mathrm{tr}\,A_1$, the matrix A_1 in (4.5.5) has the same trace as A, so our construction suggests the induction in the following argument.

Theorem 4.5.6 *Let $A \in \mathsf{M}_n(\mathbb{F})$. Then A is similar over $\mathbb{F}$ to a matrix, each of whose diagonal entries is $\frac{1}{n}\mathrm{tr}\,A$.*

Proof If $n = 1$, then there is nothing to prove, so assume that $n \geq 2$. The matrix $B = A - (\frac{1}{n}\mathrm{tr}\,A)I_n$ has trace zero. If we can show that B is similar over $\mathbb{F}$ to a matrix C

with zero diagonal entries, then Theorem 2.6.9.d ensures that $A = B + (\frac{1}{n} \operatorname{tr} A)I_n$ is similar over $\mathbb{F}$ to $C + (\frac{1}{n} \operatorname{tr} A)I_n$, which is a matrix of the asserted form. The preceding lemma ensures that B is similar over $\mathbb{F}$ to a matrix of the form $\begin{bmatrix} 0 & \star \\ \star & B_1 \end{bmatrix}$, in which $B_1 \in \mathsf{M}_{n-1}(\mathbb{F})$ and $\operatorname{tr} B_1 = 0$. We proceed by induction. Let P_k be the statement that B is similar over $\mathbb{F}$ to a matrix of the form

$$\begin{bmatrix} C_k & \star \\ \star & B_k \end{bmatrix}, \quad \text{in which } C_k \in \mathsf{M}_k(\mathbb{F}), \ B_k \in \mathsf{M}_{n-k}(\mathbb{F}), \ \operatorname{tr} B_k = 0,$$

and C_k has zero diagonal entries. We have established the base case P_1. Suppose that $k < n-1$ and P_k is true. If B_k is a scalar matrix, then $B_k = 0$ and the assertion is proved. If B_k is not a scalar matrix, then there is an invertible $S_k \in \mathsf{M}_{n-k}(\mathbb{F})$ such that $S_k^{-1} B_k S_k$ has a zero entry in position $(1, 1)$. Then

$$\begin{bmatrix} I_k & 0 \\ 0 & S_k^{-1} \end{bmatrix} \begin{bmatrix} C_k & \star \\ \star & B_k \end{bmatrix} \begin{bmatrix} I_k & 0 \\ 0 & S_k \end{bmatrix} = \begin{bmatrix} C_k & \star \\ \star & S_k^{-1} B_k S_k \end{bmatrix} = \begin{bmatrix} C_{k+1} & \star \\ \star & B_{k+1} \end{bmatrix},$$

in which $C_{k+1} \in \mathsf{M}_{k+1}(\mathbb{F})$ has zero diagonal entries, $B_{k+1} \in \mathsf{M}_{n-k-1}(\mathbb{F})$, and $\operatorname{tr} B_{k+1} = 0$. This shows that P_k implies P_{k+1}, which completes the induction. Therefore, P_{n-1} is true and B is similar over $\mathbb{F}$ to a matrix of the form $\begin{bmatrix} C_{n-1} & \star \\ \star & b \end{bmatrix}$, in which $C_{n-1} \in \mathsf{M}_{n-1}(\mathbb{F})$ has zero diagonal entries. Thus, $0 = \operatorname{tr} B = \operatorname{tr} C_{n-1} + b = b$, which shows that B is similar to a matrix with zero diagonal. $\qquad\square$

We can now prove the following characterization of commutators.

Theorem 4.5.7 (Shoda) *Let $A \in \mathsf{M}_n(\mathbb{F})$. Then A is a commutator of matrices in $\mathsf{M}_n(\mathbb{F})$ if and only if $\operatorname{tr} A = 0$.*

Proof The necessity of the trace condition is established in (4.5.2), so we consider only its sufficiency. Suppose that $\operatorname{tr} A = 0$. The preceding corollary ensures that there is an invertible $S \in \mathsf{M}_n(\mathbb{F})$ such that $S^{-1}AS = B = [b_{ij}]$ has zero diagonal entries. Let $X = \operatorname{diag}(1, 2, \ldots, n)$ and let $Y = [y_{ij}]$. Then $XY - YX = [iy_{ij}] - [jy_{ij}] = [(i-j)y_{ij}]$. If we let

$$y_{ij} = \begin{cases} (i-j)^{-1} b_{ij} & \text{if } i \neq j, \\ 1 & \text{if } i = j, \end{cases}$$

then $XY - YX = B$, and A is the commutator of SXS^{-1} and SYS^{-1}. $\qquad\square$

4.6 Problems

P.4.1   Let $A \in \mathsf{M}_{m \times n}(\mathbb{F})$ and see Figure 4.1. Prove that $\operatorname{col} A = \mathbb{F}^m$ if and only if the rows of A are linearly independent.

P.4.2   Use LU factorization to solve the linear system $A\mathbf{x} = \mathbf{y}$, in which

$$A = \begin{bmatrix} 2 & 1 & 3 \\ 4 & 5 & 6 \\ 7 & 8 & 9 \end{bmatrix} \quad \text{and} \quad \mathbf{y} = \begin{bmatrix} 4 \\ 2 \\ 1 \end{bmatrix}.$$

P.4.3 Use LU factorization to solve the linear system $A\mathbf{x} = \mathbf{y}$, in which

$$A = \begin{bmatrix} 1 & -1 & 1 \\ 0 & 1 & 1 \\ 1 & 1 & 1 \end{bmatrix} \quad \text{and} \quad \mathbf{y} = \begin{bmatrix} 10 \\ 9 \\ 8 \end{bmatrix}.$$

P.4.4 Suppose that $A \in M_n$ and $A = LU$, in which $L \in M_n$ is unit lower triangular and $U = [u_{ij}] \in M_n$ is upper triangular. Show that $\det A = u_{11}u_{22} \cdots u_{nn}$.

P.4.5 Find LU factorizations of

$$A = \begin{bmatrix} 1 & 2 & 3 \\ 4 & 5 & 6 \\ 7 & 8 & 10 \end{bmatrix} \quad \text{and} \quad B = \begin{bmatrix} 1 & 1 & 1 & 1 \\ 1 & 2 & 2 & 2 \\ 1 & 2 & 3 & 3 \\ 1 & 2 & 3 & 4 \end{bmatrix}.$$

Use your factorizations to compute $\det A$ and $\det B$.

P.4.6 Show that neither LU factorization in Example 4.3.11 can be obtained with the reduction algorithm in Theorem 4.3.5. What goes wrong?

P.4.7 Let A be the matrix in (1.8.1) and let $\mathbf{y} = [1\ 1\ 1]^\mathsf{T}$. Use (4.4.4) and (4.4.5) to show that all solutions of the linear system $A\mathbf{x} = \mathbf{y}$ are given by $\mathbf{x} = [-1\ 1\ 0]^\mathsf{T} + t[1\ -2\ 0]^\mathsf{T}$ with $t \in \mathbb{R}$. *Hint*: Consider $G^{-1}A\mathbf{x} = G^{-1}\mathbf{y}$, in which G^{-1} is the matrix in (4.4.5) that encodes the composition of all the row reductions. Why this choice?

P.4.8 If $A \in M_n(\mathbb{F})$ and $A = B(I_s \oplus 0_{n-s})C$, in which $B, C \in M_n(\mathbb{F})$ are invertible, show that $s = \operatorname{rank} A$.

P.4.9 If $A \in M_n$ is to be represented as $A = XY^\mathsf{T}$ for some $X, Y \in M_{n \times r}$, explain why r cannot be smaller than $\operatorname{rank} A$.

P.4.10 For $X \in M_{m \times n}$, let $\nu(X)$ denote the nullity of X. (a) Show that $\nu(X^\mathsf{T}) = \nu(X) + m - n$. (b) Let $A \in M_{m \times k}$ and $B \in M_{k \times n}$. Show that the rank inequality (4.2.8) is equivalent to

$$\nu(AB) \leq \nu(A) + \nu(B). \tag{4.6.1}$$

(c) If $A, B \in M_n$, show that the rank inequality (4.2.5) is equivalent to

$$\max\{\nu(A), \nu(B)\} \leq \nu(AB). \tag{4.6.2}$$

The inequalities (4.6.1) and (4.6.2) are known as *Sylvester's law of nullity*. (d) Consider $A = [0\ 1]$ and $B = [1\ 0]^\mathsf{T}$. Show that the inequality (4.6.2) need not be valid for matrices that are not square.

P.4.11 Let V be the infinite-dimensional $\mathbb{F}$-vector space of finitely nonzero sequences; see Example 1.2.9. Define $T : V \to V$ by $T(v_1, v_2, v_3, \dots) = (v_1, 2v_2, 3v_3, \dots)$. Show that $T \in \mathcal{L}(V)$ and $T^{-1} \in \mathcal{L}(V)$.

P.4.12 Let $A \in M_{m \times n}$ and suppose that $B \in M_{m \times n}$ is obtained by changing the value of exactly one entry of A. Show that $\operatorname{rank} B \in \{(\operatorname{rank} A) - 1, \operatorname{rank} A, (\operatorname{rank} A) + 1\}$. Give examples to illustrate all three possibilities.

P.4.13 Let V be a nonzero n-dimensional $\mathbb{F}$-vector space and let $T \in \mathcal{L}(V, \mathbb{F})$ be a nonzero linear transformation. Why is $\dim \ker T = n - 1$?

P.4.14 Let $V = \operatorname{span}\{AB - BA : A, B \in M_n\}$. (a) Show that the function $\operatorname{tr} : M_n \to \mathbb{C}$ is a linear transformation. (b) Why is $\dim \ker \operatorname{tr} = n^2 - 1$? (c) Prove that $\dim V \leq n^2 - 1$. (d) Let $E_{ij} = \mathbf{e}_i\mathbf{e}_j^\mathsf{T}$, every entry of which is zero except for a 1 in the (i, j) position. Show that $E_{ij}E_{k\ell} = \delta_{jk}E_{i\ell}$ for $1 \leq i, j, k, \ell \leq n$. (e) Find a basis for V and show that $V = \ker \operatorname{tr}$. *Hint*: Work out the case $n = 2$ first.

P.4.15 Let $\mathcal{U}$ and $\mathcal{W}$ be subspaces of $\mathbb{F}^n$. Show that $\mathcal{U} + \mathcal{W}$ is a direct sum if and only if $\dim(\mathcal{U} + \mathcal{W}) = \dim \mathcal{U} + \dim \mathcal{W}$.

P.4.16 Let $\mathcal{U}$, $\mathcal{V}$, and $\mathcal{W}$ be subspaces of $\mathbb{F}^n$. (a) Show that $\dim(\mathcal{U} \cap \mathcal{V} \cap \mathcal{W}) \geq \dim \mathcal{U} + \dim \mathcal{V} + \dim \mathcal{W} - 2n$. (b) If $\dim \mathcal{U} + \dim \mathcal{V} + \dim \mathcal{W} \geq 2n + k$, show that $\mathcal{U} \cap \mathcal{V} \cap \mathcal{W}$ contains k linearly independent vectors.

P.4.17 If every leading principal submatrix of $A \in \mathsf{M}_n$ is invertible, show that $A = LDU$, in which L and U^T are square and unit lower triangular, D is diagonal and invertible, and the factors L, D, and U are unique.

P.4.18 (a) Let $A = \begin{bmatrix} a & b \\ c & d \end{bmatrix}$ and $a \neq 0$. Show that A has an LU factorization in which $L = \begin{bmatrix} 1 & 0 \\ a^{-1}c & 1 \end{bmatrix}$ and $U = \begin{bmatrix} a & b \\ 0 & A/a \end{bmatrix}$. What does this say if $\det A = 0$? (b) Let $A \in \mathsf{M}_n$. If $\det A_{(k)} \neq 0$ for each $k = 1, 2, \ldots, n-1$ and $\det A = 0$, show that A has an LU factorization in which the last row of U is zero.

P.4.19 Use LU factorization to find all solutions of the linear system

$$\begin{bmatrix} 1 & 2 & 3 \\ 4 & 5 & 6 \\ 7 & 8 & 9 \end{bmatrix} \begin{bmatrix} x_1 \\ x_2 \\ x_3 \end{bmatrix} = \begin{bmatrix} 10 \\ 9 \\ 8 \end{bmatrix}.$$

P.4.20 (a) Attempt an LU factorization of

$$A = \begin{bmatrix} 1 & 1 & 2 \\ 1 & 1 & 3 \\ 4 & 5 & 6 \end{bmatrix}$$

and show that it fails after the first reduction step. (b) Interchange the second and third rows of A, and compute an LU factorization. (c) Interchange the second and third columns of A, and compute an LU factorization. (d) What are the effects of these interchanges in the context of solving a linear system $A\mathbf{x} = \mathbf{y}$?

P.4.21 A non-square matrix can have an LU factorization. (a) Verify that

$$A = \begin{bmatrix} 2 & 1 & 3 \\ 4 & 5 & 6 \\ 7 & 8 & 9 \\ 10 & 11 & 12 \end{bmatrix} = \begin{bmatrix} 1 & 0 & 0 \\ 2 & 1 & 0 \\ \frac{7}{2} & \frac{3}{2} & 1 \\ 5 & 2 & 2 \end{bmatrix} \begin{bmatrix} 2 & 1 & 3 \\ 0 & 3 & 0 \\ 0 & 0 & -\frac{3}{2} \end{bmatrix} = LU.$$

(b) Use the factorization in (a) to obtain the LU factorization

$$A^\mathsf{T} = \begin{bmatrix} 1 & 0 & 0 \\ \frac{1}{2} & 1 & 0 \\ \frac{3}{2} & 0 & 1 \end{bmatrix} \begin{bmatrix} 2 & 4 & 7 & 10 \\ 0 & 3 & \frac{9}{2} & 6 \\ 0 & 0 & -\frac{3}{2} & -3 \end{bmatrix}.$$

(c) Use the LU factorizations in (a) and (b) to show that $A^\mathsf{T}\mathbf{x} = \mathbf{y}$ has a solution for every $\mathbf{y} \in \mathbb{C}^3$, but $A\mathbf{z} = \mathbf{w} = [w_i] \in \mathbb{C}^4$ has a solution if and only if $w_2 - 2w_3 + w_4 = 0$. *Hint:* If $L\mathbf{u} = \mathbf{w}$ has a solution $\mathbf{u} = [u_i] \in \mathbb{C}^3$, why are u_1, u_2, and u_3 uniquely determined and $5u_1 + 2u_2 + 2u_3 = w_4$? (d) Verify that $[0\ 1\ -2\ 1]^\mathsf{T} \in \mathrm{null}\,A^\mathsf{T}$. This is not an accident; see P.8.29.

P.4.22 Suppose that $A \in \mathsf{M}_{m \times n}$ and let $p = \min\{m, n\}$. If every leading principal submatrix of A is invertible, show that A has an LU factorization in which L is $m \times p$ and U is $p \times n$. *Hint*: Consider $[X \ Y] = [LU \ Y] = L[U \ L^{-1}Y]$ and $\begin{bmatrix} X \\ Y \end{bmatrix} = \begin{bmatrix} LU \\ Y \end{bmatrix} = \begin{bmatrix} L \\ YU^{-1} \end{bmatrix} U$.

P.4.23 (a) Verify the LU factorizations

$$A = \begin{bmatrix} 1 & 2 \\ 3 & 4 \\ 5 & 6 \end{bmatrix} = \begin{bmatrix} 1 & 0 \\ 3 & 1 \\ 5 & 2 \end{bmatrix} \begin{bmatrix} 1 & 2 \\ 0 & -2 \end{bmatrix} \quad \text{and} \quad B = \begin{bmatrix} 1 & 2 & 3 \\ 4 & 5 & 6 \end{bmatrix} = \begin{bmatrix} 1 & 0 \\ 4 & 1 \end{bmatrix} \begin{bmatrix} 1 & 2 & 3 \\ 0 & -3 & -6 \end{bmatrix}.$$

(b) Use these factorizations to show that $Bx = y$ has a solution for every $y \in \mathbb{C}^2$, while $Ax = y = [y_i] \in \mathbb{C}^3$ has a solution if and only if $y_1 - 2y_2 + y_3 = 0$. (c) Verify that $[1 \ -2 \ 1]^\mathsf{T} \in \text{null } A^\mathsf{T}$.

P.4.24 Let $n \geq 2$ and let $A \in \mathsf{M}_n$ be the symmetric tridiagonal matrix with entries 2 on the main diagonal and -1 on the superdiagonal and subdiagonal. (a) Compute the LU factorization of A for $n = 5$. (b) Verify (multiply the factors) that $A = LU$, in which

$$L = \begin{bmatrix} 1 & & & & \\ -\frac{1}{2} & 1 & & & \\ & -\frac{2}{3} & \ddots & & \\ & & \ddots & 1 & \\ & & & -\frac{n-1}{n} & 1 \end{bmatrix} \quad \text{and} \quad U = \begin{bmatrix} 2 & -1 & & & \\ & \frac{3}{2} & -1 & & \\ & & \ddots & \ddots & \\ & & & \frac{n}{n-1} & -1 \\ & & & & \frac{n+1}{n} \end{bmatrix}$$

are $n \times n$ and bidiagonal. (c) Show that $\det A_{(k)} = k + 1$ for each $k = 1, 2, \ldots, n$.

P.4.25 Let $A, L, U \in \mathsf{M}_n$. Suppose that $A = LU$, and that

$$A = \begin{bmatrix} A_{11} & A_{12} \\ A_{21} & A_{22} \end{bmatrix}, \quad L = \begin{bmatrix} L_{11} & 0 \\ L_{21} & L_{22} \end{bmatrix}, \quad \text{and} \quad U = \begin{bmatrix} U_{11} & U_{12} \\ 0 & U_{22} \end{bmatrix}$$

are partitioned conformally, with $A_{11}, L_{11}, U_{11} \in \mathsf{M}_k$. Prove that

$$\text{rank}\begin{bmatrix} A_{11} & A_{12} \end{bmatrix} + \text{rank}\begin{bmatrix} A_{11} \\ A_{21} \end{bmatrix} \leq \text{rank } A_{11} + k. \tag{4.6.3}$$

Hint: Show that $[A_{11} \ A_{12}] = L_{11}[U_{11} \ U_{12}]$ and $\text{rank}[A_{11} \ A_{12}] \leq \text{rank } L_{11}$. Then show that $\text{rank}\begin{bmatrix} A_{11} \\ A_{21} \end{bmatrix} \leq \text{rank } U_{11}$ and $\text{rank } L_{11} + \text{rank } U_{11} - k \leq \text{rank } L_{11}U_{11} = \text{rank } A_{11}$.

P.4.26 If $A \in \mathsf{M}_n$ and $A = LU$, in which $L, U^\mathsf{T} \in \mathsf{M}_n$ are lower triangular, show that the inequality (4.6.3) is satisfied for each $k = 1, 2, \ldots, n$. (a) Is (4.6.3) satisfied if A_{11} is invertible? (b) Is (4.6.3) satisfied if $k = 1$ and $A = \begin{bmatrix} 0 & 0 \\ 0 & 1 \end{bmatrix}$, $A = \begin{bmatrix} 0 & 1 \\ 0 & 0 \end{bmatrix}$, or $A = \begin{bmatrix} 0 & 1 \\ 1 & 0 \end{bmatrix}$? Find factors L and U in these cases, if possible. (c) Is (4.6.3) satisfied if $k = 1, 2, 3$ and A is the matrix in Example 4.3.11?

P.4.27 Let $A, B \in \mathsf{M}_n(\mathbb{F})$. Show that $\text{rank } AB + n = \text{rank}\begin{bmatrix} 0 & A \\ B & I \end{bmatrix}$. *Hint*: Consider $\begin{bmatrix} I & -A \\ 0 & I \end{bmatrix} \begin{bmatrix} 0 & A \\ B & I \end{bmatrix} \begin{bmatrix} I & 0 \\ -B & I \end{bmatrix}$.

P.4.28 Let $A \in M_{m \times n}$ and $B \in M_{n \times p}$. (a) Show that (4.1.2) is a special case of the identity

$$\text{rank } AB + \dim(\text{null } A \cap \text{col } B) = \text{rank } B. \qquad (4.6.4)$$

(b) Prove (4.6.4). Proceed as in the proof of (4.1.2). *Hint*: Let $d = \dim(\text{null } A \cap \text{col } B)$. What happens if $d = 0$? $d = \text{rank } B$? If $1 \leq d < \text{rank } B$, let the columns of V be a basis for null $A \cap \text{col } B$ and let the columns of $U = [V\ W]$ be a basis for col B. Then col $AB = \text{col } AU$.

P.4.29 Let $A \in M_{m \times n}$, $B \in M_{n \times p}$, and $C \in M_{p \times q}$. (a) Show that null $A \cap \text{col } BC \subseteq \text{null } A \cap$ col B. (b) Use (4.6.4) to prove the *Frobenius rank inequality*

$$\text{rank } AB + \text{rank } BC \leq \text{rank } B + \text{rank } ABC.$$

Hint: rank $ABC = \text{rank } BC - \dim(\text{null } A \cap \text{col } BC)$. (c) Show that (4.2.5) and (4.2.8) are special cases of the Frobenius rank inequality.

P.4.30 Let $A, B \in M_n$. (a) Is rank $AB = \text{rank } BA$? (b) Is rank $AA^\mathsf{T} = \text{rank } A^\mathsf{T}A$? Why?

P.4.31 Let $\mathcal{V}$ be a complex vector space. Let $T : M_n \to \mathcal{V}$ be a linear transformation such that $T(XY) = T(YX)$ for all $X, Y \in M_n$. Show that $T(A) = (\frac{1}{n} \text{tr } A)T(I_n)$ for all $A \in M_n$ and dim ker $T = n^2 - 1$. *Hint*: $A = (A - (\frac{1}{n} \text{tr } A)I_n) + (\frac{1}{n} \text{tr } A)I_n$.

P.4.32 Let $\Phi : M_n \to \mathbb{C}$ be a linear transformation. Show that $\Phi = \text{tr}$ if and only if $\Phi(I_n) = n$ and $\Phi(XY) = \Phi(YX)$ for all $X, Y \in M_n$.

P.4.33 In the proof of Theorem 4.5.7, why may X be taken to be any diagonal matrix with distinct diagonal entries? For example, why does $X = \text{diag}(n, n - 1, \ldots, 2, 1)$ work?

P.4.34 Let $A = \begin{bmatrix} 1 & 2 \\ 3 & -1 \end{bmatrix}$. (a) Use the construction in the proof of Lemma 4.5.4 to find a matrix that is similar to A and has both diagonal entries equal to zero. (b) Use the construction in the proof of Theorem 4.5.7 to find matrices X and Y such that $A = XY - YX$. (c) Find another pair of matrices Z and W such that $A = ZW - WZ$.

P.4.35 For each of the following matrices, compute its reduced row echelon form, its pivot column decomposition, and bases for its column space and null space.

(a) $\begin{bmatrix} -1 & -2 & 2 & 3 & 3 \\ 3 & 6 & 1 & 5 & -1 \\ 2 & 4 & -1 & 0 & 2 \end{bmatrix}$
(b) $\begin{bmatrix} 1 & -1 & 2 & 1 & 3 \\ 1 & 0 & 1 & 1 & 1 \\ 3 & 2 & 2 & 3 & 2 \end{bmatrix}$

P.4.36 Let

$$A = \begin{bmatrix} 1 & 2 & 3 & ? & ? & 6 \\ 6 & 5 & 4 & ? & ? & 1 \\ 2 & 3 & 4 & ? & ? & 1 \\ 1 & 6 & 5 & ? & ? & 2 \end{bmatrix} \quad \text{and} \quad E = \begin{bmatrix} 1 & 0 & 0 & 0 & 0 & 0 \\ 0 & 1 & 0 & -1 & -2 & 0 \\ 0 & 0 & 1 & 2 & 3 & 0 \\ 0 & 0 & 0 & 0 & 0 & 1 \end{bmatrix},$$

in which E is the reduced row echelon form of A. Find the missing columns of A and the pivot column decomposition of A.

P.4.37 Let $A = PR$ be a pivot column decomposition, in which

$$A = \begin{bmatrix} 4 & 8 & 1 & 6 & 1 & 0 & 9 & 1 \\ 1 & 2 & 0 & 1 & 1 & 1 & 5 & 0 \\ 2 & \boxed{?} & 1 & \boxed{?} & 6 & 1 & 8 & 1 \\ 1 & 2 & 2 & 5 & 1 & 0 & \boxed{?} & 1 \\ 0 & 0 & 1 & 2 & 2 & 1 & \boxed{?} & 6 \\ 1 & 2 & 0 & 1 & 1 & 2 & 8 & 1 \end{bmatrix} \quad \text{and} \quad R = \begin{bmatrix} 1 & 2 & 0 & 1 & 0 & 0 & 2 & 0 \\ 0 & 0 & 1 & 2 & 0 & 0 & 1 & 0 \\ 0 & 0 & 0 & 0 & 1 & 0 & 0 & 0 \\ 0 & 0 & 0 & 0 & 0 & 1 & 3 & 0 \\ 0 & 0 & 0 & 0 & 0 & 0 & 0 & 1 \end{bmatrix}.$$

Find P and the missing entries of A.

P.4.38 Suppose that $A \in \mathsf{M}_{6\times 4}$ is row equivalent to

$$B = \begin{bmatrix} 1 & 0 & 0 & 1 \\ 0 & 2 & 3 & 2 \\ 0 & 0 & 0 & 4 \\ 0 & 0 & 0 & 0 \\ 0 & 0 & 0 & 0 \\ 0 & 0 & 0 & 0 \end{bmatrix}.$$

(a) Find a basis for $\operatorname{null} A$. (b) What is $\dim \operatorname{null} A^{\mathsf{T}}$?

P.4.39 Find a matrix with constant diagonal that is similar to $\begin{bmatrix} 1 & 1 \\ 1 & -1 \end{bmatrix}$.

P.4.40 Express $\begin{bmatrix} 1 & 1 \\ 1 & -1 \end{bmatrix}$ as a commutator.

P.4.41 Express $\begin{bmatrix} 1 & 0 \\ 0 & -1 \end{bmatrix}$ as a commutator.

P.4.42 Express $\operatorname{diag}(1, i, -1, -i)$ as a commutator.

P.4.43 Let $\mathcal{C} = \{A \in \mathsf{M}_n : A \text{ is a commutator}\}$. Show that $\mathcal{C}$ is a subspace of M_n.

4.7 Notes

The rank of a matrix can be characterized in many different ways: the number of leading ones in its reduced row echelon form (Appendix C.3); the dimension of its column space (Definition 2.3.5); the parameter s in its pivot column decomposition (Theorem 1.7.5); the dimension of its row space (Theorem 2.3.1); via the dimension of its null space (Theorem 4.1.1); the maximum number of its linearly independent columns (Corollary 2.2.9.b); the maximum size of an invertible submatrix (P.2.28); and the number of its nonzero singular values (Theorem 16.1.4).

There is a lot more to be said about LU factorizations and their variants; see [HJ13, Section 3.5] and [GVL13, Chapter 3]. For example, the rows of any square matrix can be permuted in such a way that the resulting matrix has an LU factorization. This fact is exploited in commercial software to solve systems of linear equations, since permuting the rows of A and $\mathbf{y}$ in a linear system $A\mathbf{x} = \mathbf{y}$ is equivalent to reordering the linear equations. In 1997, while he was an undergraduate at University of Connecticut, Pavel Okunev proved that the necessary condition in P.4.25 is also sufficient. That is, $A = LU$ for some lower triangular matrices L and U^{T} if and only if (4.6.3) is satisfied for each $k = 1, 2, \ldots, n$.

Lemma 4.5.4 and Theorem 4.5.6 describe special structures that can be achieved for the diagonal of a matrix that is similar to a given $A \in \mathsf{M}_n$. Other results of this type are: If A is not a scalar matrix, then it is similar to a matrix whose diagonal entries are $0, 0, \ldots, 0, \mathrm{tr}\, A$. If $A \neq 0$, then it is similar to a matrix whose diagonal entries are all nonzero. For an exposition of these results; see [Pra14, Section 15].

4.8 Some Important Concepts

- The rank-nullity theorem.

- Dimension of the intersection of subspaces.

- Rank of a sum and product of matrices.

- *LU* factorization.

- Row equivalence.

- Trace-zero matrices and commutators.

5   Inner Products and Norms

Many abstract concepts that make linear algebra a powerful mathematical tool have their roots in plane geometry, so we begin the study of inner product spaces with a review of basic properties of lengths and angles in the real two-dimensional plane $\mathbb{R}^2$. Guided by these geometrical properties, we formulate axioms for inner products and norms, which provide generalized notions of length (norm) and perpendicularity (orthogonality) in abstract vector spaces.

5.1 The Pythagorean Theorem and the Law of Cosines

Given the lengths of two orthogonal line segments that form the sides of a right triangle in the real Euclidean plane, the classical Pythagorean theorem describes how to find the length of its hypotenuse.

Theorem 5.1.1 (Classical Pythagorean Theorem) *If a and b are the lengths of two sides of a right triangle $\mathcal{T}$, and if c is the length of its hypotenuse, then $a^2 + b^2 = c^2$.*

Proof Construct a square with side c and place four copies of $\mathcal{T}$ around it to make a larger square with side $a + b$; see Figure 5.1. The area of the larger square equals the area of the smaller square plus four times the area of $\mathcal{T}$, so $(a + b)^2 = c^2 + 4\left(\frac{1}{2}ab\right)$, and hence $a^2 + 2ab + b^2 = c^2 + 2ab$. We conclude that $a^2 + b^2 = c^2$. $\square$

    The length of any side of a plane triangle (it need not be a right triangle) is determined by the lengths of its other two sides and the cosine of the angle between them. This is a consequence of the Pythagorean theorem.

Theorem 5.1.2 (Law of Cosines) *Let a and b be the lengths of two sides of a plane triangle, let θ be the angle between the two sides, and let c be the length of the third side. Then $a^2 + b^2 - 2ab\cos\theta = c^2$. If $\theta = \pi/2$ (a right angle), then $a^2 + b^2 = c^2$.*

Proof See Figure 5.2. The Pythagorean theorem ensures that

$$c^2 = (a - b\cos\theta)^2 + (b\sin\theta)^2 = a^2 - 2ab\cos\theta + b^2\cos^2\theta + b^2\sin^2\theta$$

$$= a^2 - 2ab\cos\theta + b^2(\cos^2\theta + \sin^2\theta) = a^2 - 2ab\cos\theta + b^2.$$

If $\theta = \pi/2$, then $\cos\theta = 0$; the triangle is a right triangle and the law of cosines reduces to the classical Pythagorean theorem. $\square$

    The law of cosines implies a familiar fact about plane triangles: the length of one side is not greater than the sum of the lengths of the other two sides.

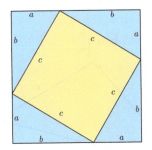

Figure 5.1 Proof of the classical Pythagorean theorem.

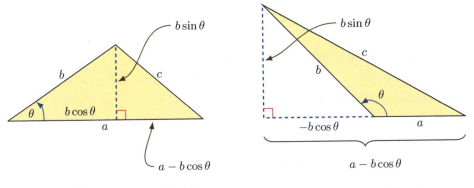

(a) $\cos\theta$ is positive if $0 < \theta < \frac{\pi}{2}$.   (b) $\cos\theta$ is negative if $\frac{\pi}{2} < \theta < \pi$.

Figure 5.2 Proof of the law of cosines.

Corollary 5.1.3 (Triangle Inequality) *Let a, b, and c be the lengths of the sides of a plane triangle. Then*

$$c \leq a + b. \tag{5.1.4}$$

Proof Let θ be the angle between the sides whose lengths are a and b. Since $-\cos\theta \leq 1$, Theorem 5.1.2 tells us that $c^2 = a^2 - 2ab\cos\theta + b^2 \leq a^2 + 2ab + b^2 = (a+b)^2$. Thus, $c \leq a + b$. ☐

5.2 Angles and Lengths in the Plane

Consider the triangle in Figure 5.3 whose vertices are given by $\mathbf{0}$ and the real Cartesian coordinate vectors $\mathbf{a} = \begin{bmatrix} a_1 \\ a_2 \end{bmatrix}$ and $\mathbf{b} = \begin{bmatrix} b_1 \\ b_2 \end{bmatrix}$. If we place

$$\mathbf{c} = \mathbf{a} - \mathbf{b} = \begin{bmatrix} a_1 - b_1 \\ a_2 - b_2 \end{bmatrix}$$

so that its initial point is at $\mathbf{b}$, it forms the third side of a triangle. Inspired by the classical Pythagorean theorem, we introduce the notation

$$\|\mathbf{a}\| = \sqrt{a_1^2 + a_2^2} \tag{5.2.1}$$

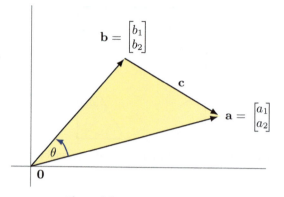

Figure 5.3 Angles and vectors.

to indicate the (Euclidean) length of the vector **a**. Then

$$\|\mathbf{a}\|^2 = a_1^2 + a_2^2, \qquad \|\mathbf{b}\|^2 = b_1^2 + b_2^2, \qquad \text{and} \quad \|\mathbf{c}\|^2 = (a_1 - b_1)^2 + (a_2 - b_2)^2.$$

The law of cosines tells us that

$$\|\mathbf{a}\|^2 + \|\mathbf{b}\|^2 - 2\|\mathbf{a}\|\|\mathbf{b}\|\cos\theta = \|\mathbf{c}\|^2.$$

Thus,

$$a_1^2 + a_2^2 + b_1^2 + b_2^2 - 2\|\mathbf{a}\|\|\mathbf{b}\|\cos\theta = (a_1 - b_1)^2 + (a_2 - b_2)^2$$
$$= a_1^2 - 2a_1b_1 + b_1^2 + a_2^2 - 2a_2b_2 + b_2^2,$$

and therefore,

$$a_1b_1 + a_2b_2 = \|\mathbf{a}\|\|\mathbf{b}\|\cos\theta. \tag{5.2.2}$$

The left side of (5.2.2) is the *dot product* of **a** and **b**; we denote it by

$$\mathbf{a} \cdot \mathbf{b} = a_1b_1 + a_2b_2. \tag{5.2.3}$$

The Euclidean length and the dot product are related by (5.2.1), which we can write as

$$\|\mathbf{a}\| = \sqrt{\mathbf{a} \cdot \mathbf{a}}. \tag{5.2.4}$$

The identity (5.2.2) is

$$\mathbf{a} \cdot \mathbf{b} = \|\mathbf{a}\|\|\mathbf{b}\|\cos\theta, \tag{5.2.5}$$

in which θ is the angle (see Figure 5.4) between **a** and **b**; $\theta = \pi/2$ (a right angle) if and only if **a** and **b** are orthogonal, in which case $\mathbf{a} \cdot \mathbf{b} = 0$.

Because $|\cos\theta| \leq 1$, the identity (5.2.5) implies that

$$|\mathbf{a} \cdot \mathbf{b}| \leq \|\mathbf{a}\|\|\mathbf{b}\|. \tag{5.2.6}$$

After a few computations, one verifies that the dot product satisfies the following properties for $\mathbf{a}, \mathbf{b}, \mathbf{c} \in \mathbb{R}^2$ and $c \in \mathbb{R}$:

(a) $\mathbf{a} \cdot \mathbf{a}$ is real and nonnegative   *Nonnegativity*

(b) $\mathbf{a} \cdot \mathbf{a} = 0$ if and only if $\mathbf{a} = \mathbf{0}$ *Positivity*

(c) $(\mathbf{a} + \mathbf{b}) \cdot \mathbf{c} = \mathbf{a} \cdot \mathbf{c} + \mathbf{b} \cdot \mathbf{c}$ *Additivity*

(d) $(c\mathbf{a}) \cdot \mathbf{b} = c(\mathbf{a} \cdot \mathbf{b})$ *Homogeneity*

(e) $\mathbf{a} \cdot \mathbf{b} = \mathbf{b} \cdot \mathbf{a}$ *Symmetry*

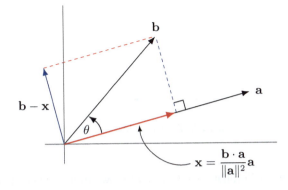

Figure 5.4 The orthogonal projection of one vector onto another.

This list of properties suggests that the first position in the dot product enjoys a favored status, but it does not. The symmetry property ensures that the dot product is additive and homogeneous in both positions:

$$\mathbf{a} \cdot (\mathbf{b} + \mathbf{c}) = (\mathbf{b} + \mathbf{c}) \cdot \mathbf{a} = \mathbf{b} \cdot \mathbf{a} + \mathbf{c} \cdot \mathbf{a} = \mathbf{a} \cdot \mathbf{b} + \mathbf{a} \cdot \mathbf{c}$$

and

$$\mathbf{a} \cdot (c\mathbf{b}) = (c\mathbf{b}) \cdot \mathbf{a} = c(\mathbf{b} \cdot \mathbf{a}) = c(\mathbf{a} \cdot \mathbf{b}).$$

In Figure 5.4,

$$\mathbf{x} = (\|\mathbf{b}\| \cos \theta) \frac{\mathbf{a}}{\|\mathbf{a}\|} = \frac{\mathbf{b} \cdot \mathbf{a}}{\|\mathbf{a}\|^2} \mathbf{a} \qquad (5.2.7)$$

is the *projection of* **b** *onto* **a**. Then $\mathbf{b} - \mathbf{x} = \mathbf{b} - \frac{\mathbf{b} \cdot \mathbf{a}}{\|\mathbf{a}\|^2} \mathbf{a}$ is orthogonal to **a** (and hence also to **x**) because

$$(\mathbf{b} - \mathbf{x}) \cdot \mathbf{a} = \left(\mathbf{b} - \frac{\mathbf{b} \cdot \mathbf{a}}{\|\mathbf{a}\|^2} \mathbf{a} \right) \cdot \mathbf{a} = \mathbf{b} \cdot \mathbf{a} - \frac{\mathbf{b} \cdot \mathbf{a}}{\|\mathbf{a}\|^2} (\mathbf{a} \cdot \mathbf{a})$$

$$= \mathbf{b} \cdot \mathbf{a} - \mathbf{b} \cdot \mathbf{a} = 0.$$

Thus, $\mathbf{b} = \mathbf{x} + (\mathbf{b} - \mathbf{x})$ expresses **b** as the sum of two vectors, one parallel to **a** and one orthogonal to **a**.

From the properties of the dot product, we deduce that the Euclidean length function has the following properties for $\mathbf{a}, \mathbf{b} \in \mathbb{R}^2$ and $c \in \mathbb{R}$:

(a) $\|\mathbf{a}\|$ is real and nonnegative.   *Nonnegativity*

(b) $\|\mathbf{a}\| = 0$ if and only if $\mathbf{a} = \mathbf{0}$.   *Positivity*

(c) $\|c\mathbf{a}\| = |c| \|\mathbf{a}\|$.   *Homogeneity*

(d) $\|\mathbf{a} + \mathbf{b}\| \leq \|\mathbf{a}\| + \|\mathbf{b}\|$.   *Triangle Inequality*

(e) $\|\mathbf{a} + \mathbf{b}\|^2 + \|\mathbf{a} - \mathbf{b}\|^2 = 2\|\mathbf{a}\|^2 + 2\|\mathbf{b}\|^2$.   *Parallelogram Identity*

Nonnegativity and positivity follow directly from the corresponding properties of the dot product. Homogeneity follows from the homogeneity and symmetry of the dot product:

$$\|c\mathbf{a}\|^2 = c\mathbf{a} \cdot c\mathbf{a} = c(\mathbf{a} \cdot c\mathbf{a}) = c(c\mathbf{a} \cdot \mathbf{a}) = c^2(\mathbf{a} \cdot \mathbf{a}) = |c|^2(\mathbf{a} \cdot \mathbf{a}) = |c|^2 \|\mathbf{a}\|^2.$$

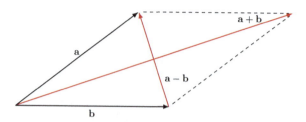

Figure 5.5 The parallelogram identity.

The triangle inequality is Corollary 5.1.3, which follows from the law of cosines. The parallelogram identity (see Figure 5.5) says that the sum of the squares of the lengths of the two diagonals of a plane parallelogram equals the sum of the squares of the lengths of its four sides. This follows from the additivity of the dot product in both positions:

$$\|\mathbf{a} + \mathbf{b}\|^2 + \|\mathbf{a} - \mathbf{b}\|^2 = (\mathbf{a} + \mathbf{b}) \cdot (\mathbf{a} + \mathbf{b}) + (\mathbf{a} - \mathbf{b}) \cdot (\mathbf{a} - \mathbf{b})$$
$$= (\mathbf{a} \cdot \mathbf{a} + \mathbf{a} \cdot \mathbf{b} + \mathbf{b} \cdot \mathbf{a} + \mathbf{b} \cdot \mathbf{b})$$
$$+ (\mathbf{a} \cdot \mathbf{a} - \mathbf{a} \cdot \mathbf{b} - \mathbf{b} \cdot \mathbf{a} + \mathbf{b} \cdot \mathbf{b})$$
$$= 2(\mathbf{a} \cdot \mathbf{a} + \mathbf{b} \cdot \mathbf{b})$$
$$= 2(\|\mathbf{a}\|^2 + \|\mathbf{b}\|^2).$$

5.3 Inner Products

Guided by our experience with plane geometry and the dot product, we make the following definition.

Definition 5.3.1 Let $\mathbb{F} = \mathbb{R}$ or $\mathbb{C}$. An *inner product* on an $\mathbb{F}$-vector space $\mathcal{V}$ is a function $\langle \cdot, \cdot \rangle : \mathcal{V} \times \mathcal{V} \to \mathbb{F}$ that satisfies the following axioms for any $\mathbf{u}, \mathbf{v}, \mathbf{w} \in \mathcal{V}$ and $c \in \mathbb{F}$:

(a) $\langle \mathbf{v}, \mathbf{v} \rangle$ is real and nonnegative.   *Nonnegativity*

(b) $\langle \mathbf{v}, \mathbf{v} \rangle = 0$ if and only if $\mathbf{v} = \mathbf{0}$.   *Positivity*

(c) $\langle \mathbf{u} + \mathbf{v}, \mathbf{w} \rangle = \langle \mathbf{u}, \mathbf{w} \rangle + \langle \mathbf{v}, \mathbf{w} \rangle$.   *Additivity*

(d) $\langle c\mathbf{u}, \mathbf{v} \rangle = c\langle \mathbf{u}, \mathbf{v} \rangle$.   *Homogeneity*

(e) $\langle \mathbf{u}, \mathbf{v} \rangle = \overline{\langle \mathbf{v}, \mathbf{u} \rangle}$.   *Conjugate Symmetry*

The nonnegativity, positivity, additivity, and homogeneity axioms reflect familiar properties of the dot product on $\mathbb{R}^2$; the conjugate-symmetry axiom (sometimes called the *Hermitian axiom*) looks like the symmetry of the dot product if $\mathbb{F} = \mathbb{R}$, but it is different if $\mathbb{F} = \mathbb{C}$. It ensures that

$$\langle a\mathbf{v}, a\mathbf{v} \rangle = a\langle \mathbf{v}, a\mathbf{v} \rangle = a\overline{\langle a\mathbf{v}, \mathbf{v} \rangle} = a(\overline{a}\overline{\langle \mathbf{v}, \mathbf{v} \rangle}) = a\overline{a}\langle \mathbf{v}, \mathbf{v} \rangle = |a|^2 \langle \mathbf{v}, \mathbf{v} \rangle,$$

in agreement with (a).

The additivity, homogeneity, and conjugate-symmetry axioms ensure that

$$\langle a\mathbf{u} + b\mathbf{v}, \mathbf{w} \rangle = \langle a\mathbf{u}, \mathbf{w} \rangle + \langle b\mathbf{v}, \mathbf{w} \rangle = a\langle \mathbf{u}, \mathbf{w} \rangle + b\langle \mathbf{v}, \mathbf{w} \rangle,$$

so an inner product is *linear* in its first position. However,

$$\langle \mathbf{u}, a\mathbf{v} + b\mathbf{w} \rangle = \overline{\langle a\mathbf{v} + b\mathbf{w}, \mathbf{u} \rangle} = \overline{\langle a\mathbf{v}, \mathbf{u} \rangle + \langle b\mathbf{w}, \mathbf{u} \rangle}$$

$$= \overline{\langle a\mathbf{v}, \mathbf{u} \rangle} + \overline{\langle b\mathbf{w}, \mathbf{u} \rangle} = \overline{a}\overline{\langle \mathbf{v}, \mathbf{u} \rangle} + \overline{b}\overline{\langle \mathbf{w}, \mathbf{u} \rangle}$$

$$= \overline{a}\langle \mathbf{u}, \mathbf{v} \rangle + \overline{b}\langle \mathbf{u}, \mathbf{w} \rangle.$$

If $\mathbb{F} = \mathbb{C}$, we can summarize the preceding computations as "the inner product is *conjugate linear* in its second position." If $\mathbb{F} = \mathbb{R}$, then $a = \overline{a}$ and $b = \overline{b}$, so the inner product is linear in its second position. Since an inner product on a complex vector space is linear in its first position and conjugate linear in its second position, one says that it is *sesquilinear* (one-and-a-half linear). An inner product on a real vector space is *bilinear* (twice linear).

Definition 5.3.2 Let $\mathbb{F} = \mathbb{R}$ or $\mathbb{C}$. An *inner product space* is an $\mathbb{F}$-vector space $\mathcal{V}$, together with an inner product $\langle \cdot, \cdot \rangle : \mathcal{V} \times \mathcal{V} \to \mathbb{F}$. We say that $\mathcal{V}$ is an $\mathbb{F}$-*inner product space* or that $\mathcal{V}$ is an *inner product space over* $\mathbb{F}$.

Some examples of inner product spaces are in the following examples.

Example 5.3.3 Consider $\mathcal{V} = \mathbb{F}^n$ as a vector space over $\mathbb{F}$. For $\mathbf{u} = [u_i], \mathbf{v} = [v_i] \in \mathcal{V}$, let

$$\langle \mathbf{u}, \mathbf{v} \rangle = \mathbf{v}^* \mathbf{u} = \sum_{i=1}^{n} u_i \overline{v_i}.$$

This is the *standard inner product* on $\mathbb{F}^n$. If $\mathbb{F} = \mathbb{R}$, then $\langle \mathbf{u}, \mathbf{v} \rangle = \mathbf{v}^{\mathsf{T}}\mathbf{u}$. If $\mathbb{F} = \mathbb{R}$ and $n = 2$, then $\langle \mathbf{u}, \mathbf{v} \rangle = \mathbf{u} \cdot \mathbf{v}$ is the dot product on $\mathbb{R}^2$. If $n = 1$, the "vectors" in $\mathcal{V} = \mathbb{F}$ are scalars and $\langle c, d \rangle = c\overline{d}$.

Example 5.3.4 Let $\mathcal{V} = \mathcal{P}_n$ be the complex vector space of polynomials of degree at most n. Fix a finite, nonempty real interval $[a, b]$, and define

$$\langle p, q \rangle = \int_a^b p(t)\overline{q(t)}\,dt.$$

This is known as the L^2 *inner product* on $\mathcal{P}_n$ over the interval $[a, b]$. Verification of the nonnegativity, additivity, homogeneity, and conjugate-symmetry axioms is straightforward. Verification of the positivity axiom requires some analysis and algebra. If p is a polynomial such that

$$\langle p, p \rangle = \int_a^b p(t)\overline{p(t)}\,dt = \int_a^b |p(t)|^2\,dt = 0,$$

one can use properties of the integral and the continuity of the nonnegative function $|p|$ to show that $p(t) = 0$ for all $t \in [a, b]$. It follows that p is the zero polynomial, which is the zero element of the vector space $\mathcal{P}_n$.

Example 5.3.5 Let $\mathbb{F} = \mathbb{R}$ or $\mathbb{C}$. Let $\mathcal{V} = \mathsf{M}_{m \times n}(\mathbb{F})$ be the $\mathbb{F}$-vector space of $m \times n$ matrices over $\mathbb{F}$, let $A = [a_{ij}] \in \mathcal{V}$, let $B = [b_{ij}] \in \mathcal{V}$, and define the *Frobenius inner product*

$$\langle A, B \rangle_{\mathrm{F}} = \mathrm{tr}\, B^* A. \tag{5.3.6}$$

Let $B^* A = [c_{ij}] \in \mathsf{M}_n(\mathbb{F})$ and compute

$$\langle A, B \rangle_{\mathrm{F}} = \mathrm{tr}\, B^* A = \sum_{j=1}^{n} c_{jj} = \sum_{j=1}^{n} \left(\sum_{i=1}^{m} \overline{b_{ij}} a_{ij} \right) = \sum_{i,j} a_{ij}\overline{b_{ij}}.$$

Since

$$\operatorname{tr} A^*A = \sum_{i,j} |a_{ij}|^2 \geq 0, \tag{5.3.7}$$

we see that $\operatorname{tr} A^*A = 0$ if and only if $A = 0$. Conjugate symmetry follows from the fact that $\operatorname{tr} X^* = \overline{\operatorname{tr} X}$ for any $X \in \mathsf{M}_n$. Compute

$$\langle A, B \rangle_F = \operatorname{tr} B^*A = \operatorname{tr}(A^*B)^* = \overline{\operatorname{tr} A^*B} = \overline{\langle B, A \rangle_F}.$$

If $n = 1$, then $\mathcal{V} = \mathbb{F}^m$ and the Frobenius inner product is the standard inner product on $\mathbb{F}^m$.

In the preceding examples, the vector spaces are finite dimensional. In the following examples, they are not. Infinite-dimensional inner product spaces play important roles in physics (quantum mechanics), aeronautics (model approximation), and engineering (signal analysis), as well as in mathematics itself.

Example 5.3.8 Let $\mathcal{V} = C_\mathbb{F}[a, b]$ be the $\mathbb{F}$-vector space of continuous $\mathbb{F}$-valued functions on the finite, nonempty real interval $[a, b]$. For any $f, g \in \mathcal{V}$, define

$$\langle f, g \rangle = \int_a^b f(t) \overline{g(t)} \, dt. \tag{5.3.9}$$

This is the L^2 *inner product* on $C_\mathbb{F}[a, b]$. Verification of the nonnegativity, additivity, homogeneity, and conjugate-symmetry axioms is straightforward. Positivity follows in the same manner as in Example 5.3.4.

Example 5.3.10 If $\langle \cdot, \cdot \rangle$ is an inner product on $\mathcal{V}$ and if c is a positive real scalar, then $c \langle \cdot, \cdot \rangle$ is also an inner product on $\mathcal{V}$. For example, the L^2 inner product (5.3.9) on $[-\pi, \pi]$ often appears in the modified form

$$\langle f, g \rangle = \frac{1}{\pi} \int_{-\pi}^{\pi} f(t) \overline{g(t)} \, dt \tag{5.3.11}$$

in the study of Fourier series; see Section 6.8.

Example 5.3.12 Let $\mathcal{V}$ be the complex vector space of finitely nonzero sequences $\mathbf{v} = (v_1, v_2, \dots)$; see Example 1.2.9. For any $\mathbf{u}, \mathbf{v} \in \mathcal{V}$, define $\langle \mathbf{u}, \mathbf{v} \rangle = \sum_{i=1}^{\infty} u_i \overline{v_i}$. The indicated sum involves only finitely many nonzero summands since each vector has only finitely many nonzero entries. Verification of the nonnegativity, additivity, homogeneity, and conjugate-symmetry axioms is straightforward. To verify positivity, observe that if $\mathbf{u} \in \mathcal{V}$ and $0 = \langle \mathbf{u}, \mathbf{u} \rangle = \sum_{i=1}^{\infty} u_i \overline{u_i} = \sum_{i=1}^{\infty} |u_i|^2$, then each $u_i = 0$, so $\mathbf{u}$ is the zero vector in $\mathcal{V}$.

Based on our experience with orthogonal lines and dot products of real vectors in the plane, we define orthogonality in an inner product space.

Definition 5.3.13 Let $\mathcal{V}$ be an inner product space with inner product $\langle \cdot, \cdot \rangle$. Then $\mathbf{u}, \mathbf{v} \in \mathcal{V}$ are *orthogonal* if $\langle \mathbf{u}, \mathbf{v} \rangle = 0$. If $\mathbf{u}, \mathbf{v} \in \mathcal{V}$ are orthogonal, we write $\mathbf{u} \perp \mathbf{v}$. Nonempty subsets $\mathscr{S}_1, \mathscr{S}_2 \subseteq \mathcal{V}$ are *orthogonal* if $\mathbf{u} \perp \mathbf{v}$ for every $\mathbf{u} \in \mathscr{S}_1$ and $\mathbf{v} \in \mathscr{S}_2$.

Three important properties of orthogonality follow from the definition and the axioms for an inner product.

Theorem 5.3.14 *Let $\mathcal{V}$ be an inner product space with inner product $\langle \cdot, \cdot \rangle$.*

(a) *$\mathbf{u} \perp \mathbf{v}$ if and only if $\mathbf{v} \perp \mathbf{u}$.*

(b) *$\mathbf{0} \perp \mathbf{u}$ for every $\mathbf{u} \in \mathcal{V}$.*

(c) *If $\mathbf{v} \perp \mathbf{u}$ for every $\mathbf{u} \in \mathcal{V}$, then $\mathbf{v} = \mathbf{0}$.*

Proof

(a) $\langle \mathbf{u}, \mathbf{v} \rangle = \overline{\langle \mathbf{v}, \mathbf{u} \rangle}$, so $\langle \mathbf{u}, \mathbf{v} \rangle = 0$ if and only if $\langle \mathbf{v}, \mathbf{u} \rangle = 0$.

(b) $\langle \mathbf{0}, \mathbf{u} \rangle = \langle 0\mathbf{0}, \mathbf{u} \rangle = 0\langle \mathbf{0}, \mathbf{u} \rangle = 0$ for every $\mathbf{u} \in \mathcal{V}$.

(c) If $\mathbf{v} \perp \mathbf{u}$ for every $\mathbf{u} \in \mathcal{V}$, then $\langle \mathbf{v}, \mathbf{v} \rangle = 0$ and the positivity axiom implies that $\mathbf{v} = \mathbf{0}$. $\quad\square$

Corollary 5.3.15 *Let $\mathcal{V}$ be an inner product space and let $\mathbf{v}, \mathbf{w} \in \mathcal{V}$. If $\langle \mathbf{u}, \mathbf{v} \rangle = \langle \mathbf{u}, \mathbf{w} \rangle$ for all $\mathbf{u} \in \mathcal{V}$, then $\mathbf{v} = \mathbf{w}$.*

Proof If $\langle \mathbf{u}, \mathbf{v} \rangle = \langle \mathbf{u}, \mathbf{w} \rangle$ for all $\mathbf{u} \in \mathcal{V}$, then $\langle \mathbf{u}, \mathbf{v} - \mathbf{w} \rangle = 0$ for all $\mathbf{u}$. Part (c) of the preceding theorem ensures that $\mathbf{v} - \mathbf{w} = \mathbf{0}$. $\quad\square$

5.4 The Norm Derived from an Inner Product

By analogy with the dot product and Euclidean length in the plane, a generalized length can be defined in any inner product space.

Definition 5.4.1 Let $\mathcal{V}$ be an inner product space with inner product $\langle \cdot, \cdot \rangle$. The function $\| \cdot \| : \mathcal{V} \to [0, \infty)$ defined by

$$\|\mathbf{v}\| = \sqrt{\langle \mathbf{v}, \mathbf{v} \rangle} \tag{5.4.2}$$

is *the norm derived from the inner product* $\langle \cdot, \cdot \rangle$. For brevity, we refer to (5.4.2) as *the norm on $\mathcal{V}$.*

The definition ensures that $\|\mathbf{v}\| \geq 0$ for all $\mathbf{v} \in \mathcal{V}$.

Example 5.4.3 The norm on $\mathcal{V} = \mathbb{F}^n$ derived from the standard inner product is the *Euclidean norm*

$$\|\mathbf{u}\|_2 = \left(\mathbf{u}^* \mathbf{u} \right)^{1/2} = \left(\sum_{i=1}^{n} |u_i|^2 \right)^{1/2} \quad \text{for } \mathbf{u} = [u_i] \in \mathbb{F}^n. \tag{5.4.4}$$

The Euclidean norm is also called the ℓ_2 *norm*.

Example 5.4.5 The norm on $\mathcal{V} = \mathsf{M}_{m \times n}(\mathbb{F})$ derived from the Frobenius inner product is the *Frobenius norm*

$$\|A\|_{\mathrm{F}}^2 = \langle A, A \rangle_{\mathrm{F}} = \operatorname{tr} A^* A = \sum_{i=1}^{m} \sum_{j=1}^{n} |a_{ij}|^2 \quad \text{for } A = [a_{ij}] \in \mathsf{M}_{m \times n}.$$

The Frobenius norm is sometimes called the *Schur norm* or the *Hilbert–Schmidt norm*. If $n = 1$, the Frobenius norm is the Euclidean norm on $\mathbb{F}^m$.

Example 5.4.6 The norm derived from the L^2 inner product on $C[a, b]$ (see Example 5.3.8) is *the L^2 norm*

$$\|f\| = \left(\int_a^b |f(t)|^2 \, dt \right)^{1/2}. \tag{5.4.7}$$

If we think of the integral as a limit of Riemann sums, there is a natural analogy between (5.4.7) and (5.4.4).

Example 5.4.8 Consider the complex inner product space $V = C[-\pi, \pi]$ with the L^2 inner product and norm. The functions $\cos t$ and $\sin t$ are in V, so

$$\|\sin t\|^2 = \int_{-\pi}^{\pi} \sin^2 t \, dt = \frac{1}{2} \int_{-\pi}^{\pi} (1 - \cos 2t) \, dt = \frac{1}{2} \left(t - \frac{1}{2} \sin 2t \right) \Big|_{-\pi}^{\pi} = \pi$$

and

$$\langle \sin t, \cos t \rangle = \int_{-\pi}^{\pi} \sin t \, \cos t \, dt = \frac{1}{2} \int_{-\pi}^{\pi} \sin 2t \, dt = -\frac{1}{4} \cos 2t \Big|_{-\pi}^{\pi} = 0.$$

Thus, with respect to the L^2 norm and inner product on $C[-\pi, \pi]$, the function $\sin t$ has norm $\sqrt{\pi}$ and is orthogonal to $\cos t$.

The derived norm (5.4.2) satisfies many of the properties of Euclidean length in the plane.

Theorem 5.4.9 *Let V be an $\mathbb{F}$-inner product space with inner product $\langle \cdot, \cdot \rangle$ and derived norm $\| \cdot \|$. Let $\mathbf{u}, \mathbf{v} \in V$ and $c \in \mathbb{F}$.*

(a) $\|\mathbf{u}\|$ *is real and nonnegative.*   *Nonnegativity*

(b) $\|\mathbf{u}\| = 0$ *if and only if $\mathbf{u} = \mathbf{0}$.*   *Positivity*

(c) $\|c\mathbf{u}\| = |c| \|\mathbf{u}\|$.   *Homogeneity*

(d) *If $\langle \mathbf{u}, \mathbf{v} \rangle = 0$, then $\|\mathbf{u} + \mathbf{v}\|^2 = \|\mathbf{u}\|^2 + \|\mathbf{v}\|^2$.*   *Pythagorean Theorem*

(e) $\|\mathbf{u} + \mathbf{v}\|^2 + \|\mathbf{u} - \mathbf{v}\|^2 = 2\|\mathbf{u}\|^2 + 2\|\mathbf{v}\|^2$.   *Parallelogram Identity*

Proof (a) Nonnegativity is built in to Definition 5.4.1.

(b) Positivity follows from the positivity of the inner product. If $\|\mathbf{u}\| = 0$, then $\langle \mathbf{u}, \mathbf{u} \rangle = 0$, which implies that $\mathbf{u} = \mathbf{0}$.

(c) Homogeneity follows from homogeneity and conjugate symmetry of the inner product:

$$\|c\mathbf{u}\| = (\langle c\mathbf{u}, c\mathbf{u} \rangle)^{1/2} = (c\bar{c} \langle \mathbf{u}, \mathbf{u} \rangle)^{1/2} = (|c|^2 \langle \mathbf{u}, \mathbf{u} \rangle)^{1/2} = |c| \langle \mathbf{u}, \mathbf{u} \rangle^{1/2} = |c| \|\mathbf{u}\|.$$

(d) Compute

$$\|\mathbf{u} + \mathbf{v}\|^2 = \langle \mathbf{u} + \mathbf{v}, \mathbf{u} + \mathbf{v} \rangle = \langle \mathbf{u}, \mathbf{u} \rangle + \langle \mathbf{u}, \mathbf{v} \rangle + \langle \mathbf{v}, \mathbf{u} \rangle + \langle \mathbf{v}, \mathbf{v} \rangle$$
$$= \langle \mathbf{u}, \mathbf{u} \rangle + 0 + 0 + \langle \mathbf{v}, \mathbf{v} \rangle = \|\mathbf{u}\|^2 + \|\mathbf{v}\|^2.$$

(e) Use additivity in both positions of the inner product to compute

$$\|\mathbf{u} + \mathbf{v}\|^2 + \|\mathbf{u} - \mathbf{v}\|^2 = \langle \mathbf{u} + \mathbf{v}, \mathbf{u} + \mathbf{v} \rangle + \langle \mathbf{u} - \mathbf{v}, \mathbf{u} - \mathbf{v} \rangle$$
$$= \langle \mathbf{u}, \mathbf{u} \rangle + \langle \mathbf{u}, \mathbf{v} \rangle + \langle \mathbf{v}, \mathbf{u} \rangle + \langle \mathbf{v}, \mathbf{v} \rangle$$
$$+ \langle \mathbf{u}, \mathbf{u} \rangle - \langle \mathbf{u}, \mathbf{v} \rangle - \langle \mathbf{v}, \mathbf{u} \rangle + \langle \mathbf{v}, \mathbf{v} \rangle$$
$$= 2(\langle \mathbf{u}, \mathbf{u} \rangle + \langle \mathbf{v}, \mathbf{v} \rangle)$$
$$= 2(\|\mathbf{u}\|^2 + \|\mathbf{v}\|^2). \qquad \square$$

Definition 5.4.10 Let V be an inner product space with derived norm $\| \cdot \|$. Then $\mathbf{u} \in V$ is a *unit vector* if $\|\mathbf{u}\| = 1$.

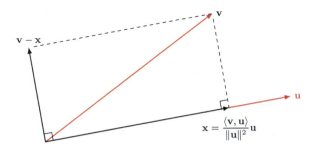

Figure 5.6 The orthogonal projection of one vector onto another.

Any nonzero $\mathbf{u} \in \mathcal{V}$ can be *normalized* to create a unit vector $\mathbf{u}/\|\mathbf{u}\|$ that is proportional to $\mathbf{u}$:

$$\left\| \frac{\mathbf{u}}{\|\mathbf{u}\|} \right\| = \frac{\|\mathbf{u}\|}{\|\mathbf{u}\|} = 1 \quad \text{if} \quad \mathbf{u} \neq \mathbf{0}. \tag{5.4.11}$$

In an inner product space $\mathcal{V}$, a generalization of (5.2.7) defines the projection of one vector onto another. Let $\mathbf{u} \in \mathcal{V}$ be nonzero and form

$$\mathbf{x} = \frac{\langle \mathbf{v}, \mathbf{u} \rangle}{\|\mathbf{u}\|^2} \mathbf{u} = \left\langle \mathbf{v}, \frac{\mathbf{u}}{\|\mathbf{u}\|} \right\rangle \frac{\mathbf{u}}{\|\mathbf{u}\|}, \tag{5.4.12}$$

which is the inner product of $\mathbf{v}$ with the unit vector in the direction of $\mathbf{u}$, times the unit vector in the direction of $\mathbf{u}$. This is *the projection of* $\mathbf{v}$ *onto* $\mathbf{u}$; see Figure 5.6. Since

$$\langle \mathbf{v} - \mathbf{x}, \mathbf{u} \rangle = \langle \mathbf{v}, \mathbf{u} \rangle - \langle \mathbf{x}, \mathbf{u} \rangle = \langle \mathbf{v}, \mathbf{u} \rangle - \left\langle \frac{\langle \mathbf{v}, \mathbf{u} \rangle}{\|\mathbf{u}\|^2} \mathbf{u}, \mathbf{u} \right\rangle$$

$$= \langle \mathbf{v}, \mathbf{u} \rangle - \langle \mathbf{v}, \mathbf{u} \rangle \frac{\langle \mathbf{u}, \mathbf{u} \rangle}{\|\mathbf{u}\|^2} = \langle \mathbf{v}, \mathbf{u} \rangle - \langle \mathbf{v}, \mathbf{u} \rangle = 0,$$

$\mathbf{v} - \mathbf{x}$ is orthogonal to $\mathbf{u}$ (and hence also to $\mathbf{x}$). Consequently,

$$\mathbf{v} = \mathbf{x} + (\mathbf{v} - \mathbf{x}) \tag{5.4.13}$$

expresses $\mathbf{v}$ as a sum of orthogonal vectors, one proportional to $\mathbf{u}$ and the other orthogonal to $\mathbf{u}$.

An important inequality that generalizes (5.2.6) is valid in any inner product space.

Theorem 5.4.14 (Cauchy–Schwarz Inequality) *Let $\mathcal{V}$ be an inner product space with inner product $\langle \cdot, \cdot \rangle$ and derived norm $\| \cdot \|$. Then*

$$|\langle \mathbf{u}, \mathbf{v} \rangle| \leq \|\mathbf{u}\| \|\mathbf{v}\| \tag{5.4.15}$$

for all $\mathbf{u}, \mathbf{v} \in \mathcal{V}$ with equality if and only if $\mathbf{u}$ and $\mathbf{v}$ are linearly dependent, that is, if and only if one of them is a scalar multiple of the other.

Proof If $\mathbf{u} = \mathbf{0}$ or $\mathbf{v} = \mathbf{0}$ there is nothing to prove: both sides of (5.4.15) are zero and $\mathbf{u}, \mathbf{v}$ are linearly dependent. Thus, we may assume that $\mathbf{u}$ and $\mathbf{v}$ are nonzero. Define $\mathbf{x}$ as in (5.4.12)

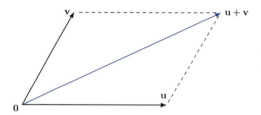

Figure 5.7 The triangle inequality is $\|\mathbf{u} + \mathbf{v}\| \leq \|\mathbf{u}\| + \|\mathbf{v}\|$.

and write $\mathbf{v} = \mathbf{x} + (\mathbf{v} - \mathbf{x})$ as in (5.4.13). Since $\mathbf{x}$ and $\mathbf{v} - \mathbf{x}$ are orthogonal, Theorem 5.4.9.d ensures that

$$\|\mathbf{v}\|^2 = \|\mathbf{x}\|^2 + \|\mathbf{v} - \mathbf{x}\|^2$$

$$\geq \|\mathbf{x}\|^2 = \left\| \frac{\langle \mathbf{v}, \mathbf{u} \rangle}{\|\mathbf{u}\|^2} \mathbf{u} \right\|^2 \tag{5.4.16}$$

$$= \frac{|\langle \mathbf{v}, \mathbf{u} \rangle|^2}{\|\mathbf{u}\|^4} \|\mathbf{u}\|^2 = \frac{|\langle \mathbf{v}, \mathbf{u} \rangle|^2}{\|\mathbf{u}\|^2},$$

so $\|\mathbf{u}\|^2 \|\mathbf{v}\|^2 \geq |\langle \mathbf{v}, \mathbf{u} \rangle|^2 = |\overline{\langle \mathbf{u}, \mathbf{v} \rangle}|^2 = |\langle \mathbf{u}, \mathbf{v} \rangle|^2$. Consequently, $\|\mathbf{u}\| \|\mathbf{v}\| \geq |\langle \mathbf{u}, \mathbf{v} \rangle|$.

If (5.4.16) is an equality, then $\mathbf{v} - \mathbf{x} = \mathbf{0}$ and $\mathbf{v} = \mathbf{x} = \langle \mathbf{v}, \mathbf{u} \rangle \|\mathbf{u}\|^{-2} \mathbf{u}$, so $\mathbf{v}$ and $\mathbf{u}$ are linearly dependent. Conversely, if $\mathbf{v}$ and $\mathbf{u}$ are linearly dependent, then $\mathbf{v} = c\mathbf{u}$ for some nonzero scalar c, in which case

$$\mathbf{x} = \frac{\langle \mathbf{v}, \mathbf{u} \rangle}{\|\mathbf{u}\|^2} \mathbf{u} = \frac{\langle c\mathbf{u}, \mathbf{u} \rangle}{\|\mathbf{u}\|^2} \mathbf{u} = c\mathbf{u} = \mathbf{v},$$

and hence (5.4.16) is an equality. $\square$

Example 5.4.17 Let $\lambda_1, \lambda_2, \ldots, \lambda_n \in \mathbb{C}$. Consider the all-ones vector $\mathbf{e} \in \mathbb{C}^n$, the vector $\mathbf{u} = [\lambda_i] \in \mathbb{C}^n$, and the standard inner product on $\mathbb{C}^n$. The Cauchy–Schwarz inequality ensures that

$$\left| \sum_{i=1}^n \lambda_i \right|^2 = |\langle \mathbf{u}, \mathbf{e} \rangle|^2 \leq \|\mathbf{u}\|_2^2 \|\mathbf{e}\|_2^2 = \left(\sum_{i=1}^n |\lambda_i|^2 \right) \left(\sum_{i=1}^n 1^2 \right) = n \sum_{i=1}^n |\lambda_i|^2$$

with equality if and only if $\mathbf{e}$ and $\mathbf{u}$ are linearly dependent, that is, if and only if $\mathbf{u} = c\mathbf{e}$ for some $c \in \mathbb{C}$. Thus, $|\sum_{i=1}^n \lambda_i| \leq \sqrt{n} (\sum_{i=1}^n |\lambda_i|^2)^{1/2}$, with equality if and only if $\lambda_1 = \lambda_2 = \cdots = \lambda_n$.

Example 5.4.18 Now consider $\mathbf{p} = [p_i] \in \mathbb{R}^n$ with nonnegative entries that sum to 1, that is, $\mathbf{e}^\mathsf{T} \mathbf{p} = 1$. The Cauchy–Schwarz inequality ensures that

$$\left| \sum_{i=1}^n p_i \lambda_i \right|^2 = \left| \sum_{i=1}^n \sqrt{p_i}(\sqrt{p_i}\lambda_i) \right|^2 \leq \left(\sum_{i=1}^n p_i \right) \left(\sum_{i=1}^n p_i |\lambda_i|^2 \right) = \sum_{i=1}^n p_i |\lambda_i|^2,$$

with equality if and only if $\mathbf{p}$ and $\mathbf{u} = [\lambda_i] \in \mathbb{C}^n$ are linearly dependent.

In addition to the basic properties listed in Theorem 5.4.9, a derived norm also satisfies an inequality that generalizes (5.1.4); see Figure 5.7.

Corollary 5.4.19 (Triangle Inequality for a Derived Norm) *Let V be an inner product space and let $\mathbf{u}, \mathbf{v} \in V$. Then*

$$\|\mathbf{u} + \mathbf{v}\| \leq \|\mathbf{u}\| + \|\mathbf{v}\|, \tag{5.4.20}$$

with equality if and only if one of the vectors is a real nonnegative scalar multiple of the other.

Proof We invoke the additivity and conjugate symmetry of the inner product, together with the Cauchy–Schwarz inequality (5.4.15):

$$\begin{aligned}
\|\mathbf{u} + \mathbf{v}\|^2 &= \langle \mathbf{u} + \mathbf{v}, \mathbf{u} + \mathbf{v} \rangle \\
&= \langle \mathbf{u}, \mathbf{u} \rangle + \langle \mathbf{u}, \mathbf{v} \rangle + \langle \mathbf{v}, \mathbf{u} \rangle + \langle \mathbf{v}, \mathbf{v} \rangle \\
&= \langle \mathbf{u}, \mathbf{u} \rangle + \langle \mathbf{u}, \mathbf{v} \rangle + \overline{\langle \mathbf{u}, \mathbf{v} \rangle} + \langle \mathbf{v}, \mathbf{v} \rangle \\
&= \|\mathbf{u}\|^2 + 2\operatorname{Re}\langle \mathbf{u}, \mathbf{v} \rangle + \|\mathbf{v}\|^2 \\
&\leq \|\mathbf{u}\|^2 + 2|\langle \mathbf{u}, \mathbf{v} \rangle| + \|\mathbf{v}\|^2 & (5.4.21) \\
&\leq \|\mathbf{u}\|^2 + 2\|\mathbf{u}\|\|\mathbf{v}\| + \|\mathbf{v}\|^2 & (5.4.22) \\
&= \left(\|\mathbf{u}\| + \|\mathbf{v}\| \right)^2.
\end{aligned}$$

We conclude that $\|\mathbf{u} + \mathbf{v}\| \leq \|\mathbf{u}\| + \|\mathbf{v}\|$, with equality if and only if the inequalities (5.4.22) and (5.4.21) are equalities. Equality in the Cauchy–Schwarz inequality (5.4.22) occurs if and only if there is a scalar c such that $\mathbf{u} = c\mathbf{v}$ or $\mathbf{v} = c\mathbf{u}$. Equality in (5.4.21) occurs if and only if c is real and nonnegative. $\qquad\square$

In an inner product space, the norm is determined by the inner product since $\|\mathbf{v}\|^2 = \langle \mathbf{v}, \mathbf{v} \rangle$. The following result shows that the inner product is determined by the norm.

Theorem 5.4.23 (Polarization Identities) *Let V be an $\mathbb{F}$-inner product space and let $\mathbf{u}, \mathbf{v} \in V$.*

(a) *If $\mathbb{F} = \mathbb{R}$, then*

$$\langle \mathbf{u}, \mathbf{v} \rangle = \frac{1}{4}\left(\|\mathbf{u} + \mathbf{v}\|^2 - \|\mathbf{u} - \mathbf{v}\|^2 \right). \tag{5.4.24}$$

(b) *If $\mathbb{F} = \mathbb{C}$, then*

$$\langle \mathbf{u}, \mathbf{v} \rangle = \frac{1}{4}\left(\|\mathbf{u} + \mathbf{v}\|^2 - \|\mathbf{u} - \mathbf{v}\|^2 + i\|\mathbf{u} + i\mathbf{v}\|^2 - i\|\mathbf{u} - i\mathbf{v}\|^2 \right). \tag{5.4.25}$$

Proof (a) If $\mathbb{F} = \mathbb{R}$, then

$$\begin{aligned}
\|\mathbf{u} + \mathbf{v}\|^2 - \|\mathbf{u} - \mathbf{v}\|^2 &= (\|\mathbf{u}\|^2 + 2\langle \mathbf{u}, \mathbf{v} \rangle + \|\mathbf{v}\|^2) - (\|\mathbf{u}\|^2 + 2\langle \mathbf{u}, -\mathbf{v} \rangle + \| -\mathbf{v}\|^2) \\
&= 2\langle \mathbf{u}, \mathbf{v} \rangle + \|\mathbf{v}\|^2 + 2\langle \mathbf{u}, \mathbf{v} \rangle - \|\mathbf{v}\|^2 \\
&= 4\langle \mathbf{u}, \mathbf{v} \rangle.
\end{aligned}$$

(b) If $\mathbb{F} = \mathbb{C}$, then

$$\begin{aligned}
\|\mathbf{u} + \mathbf{v}\|^2 - \|\mathbf{u} - \mathbf{v}\|^2 &= (\|\mathbf{u}\|^2 + 2\operatorname{Re}\langle \mathbf{u}, \mathbf{v} \rangle + \|\mathbf{v}\|^2) - (\|\mathbf{u}\|^2 + 2\operatorname{Re}\langle \mathbf{u}, -\mathbf{v} \rangle + \| -\mathbf{v}\|^2) \\
&= 2\operatorname{Re}\langle \mathbf{u}, \mathbf{v} \rangle + \|\mathbf{v}\|^2 + 2\operatorname{Re}\langle \mathbf{u}, \mathbf{v} \rangle - \|\mathbf{v}\|^2 \\
&= 4\operatorname{Re}\langle \mathbf{u}, \mathbf{v} \rangle
\end{aligned}$$

and

$$\|\mathbf{u} + i\mathbf{v}\|^2 - \|\mathbf{u} - i\mathbf{v}\|^2 = (\|\mathbf{u}\|^2 + 2\operatorname{Re}\langle\mathbf{u}, i\mathbf{v}\rangle + \|i\mathbf{v}\|^2)$$
$$- (\|\mathbf{u}\|^2 + 2\operatorname{Re}\langle\mathbf{u}, -i\mathbf{v}\rangle + \|-i\mathbf{v}\|^2)$$
$$= -2\operatorname{Re} i\langle\mathbf{u}, \mathbf{v}\rangle + \|\mathbf{v}\|^2 - 2\operatorname{Re} i\langle\mathbf{u}, \mathbf{v}\rangle - \|\mathbf{v}\|^2$$
$$= -4\operatorname{Re} i\langle\mathbf{u}, \mathbf{v}\rangle$$
$$= 4\operatorname{Im}\langle\mathbf{u}, \mathbf{v}\rangle.$$

Therefore, the right side of (5.4.25) is

$$\frac{1}{4}(4\operatorname{Re}\langle\mathbf{u}, \mathbf{v}\rangle + 4i\operatorname{Im}\langle\mathbf{u}, \mathbf{v}\rangle) = \operatorname{Re}\langle\mathbf{u}, \mathbf{v}\rangle + i\operatorname{Im}\langle\mathbf{u}, \mathbf{v}\rangle = \langle\mathbf{u}, \mathbf{v}\rangle. \qquad \square$$

5.5 Normed Vector Spaces

In the preceding section, we showed how a length function (norm) on a vector space can be derived from an inner product. We now introduce other length functions that are useful in applications, but might not be derived from an inner product.

Definition 5.5.1 A *norm* on an $\mathbb{F}$-vector space $\mathcal{V}$ is a function $\|\cdot\| : \mathcal{V} \to [0, \infty)$ that has the following properties for any $\mathbf{u}, \mathbf{v} \in \mathcal{V}$ and $c \in \mathbb{F}$:

(a) $\|\mathbf{u}\|$ is real and nonnegative.   *Nonnegativity*

(b) $\|\mathbf{u}\| = 0$ if and only if $\mathbf{u} = 0$.   *Positivity*

(c) $\|c\mathbf{u}\| = |c|\,\|\mathbf{u}\|$.   *Homogeneity*

(d) $\|\mathbf{u} + \mathbf{v}\| \le \|\mathbf{u}\| + \|\mathbf{v}\|$.   *Triangle Inequality*

In the following three examples, $\mathcal{V}$ is the $\mathbb{F}$-vector space $\mathbb{F}^n$, $\mathbf{u} = [u_i] \in \mathbb{F}^n$, $\mathbf{v} = [v_i] \in \mathbb{F}^n$, and $\mathbf{e}_1$ and $\mathbf{e}_2$ are the first two standard basis vectors in $\mathbb{F}^n$.

Example 5.5.2 The function $\|\mathbf{u}\|_1 = |u_1| + |u_2| + \cdots + |u_n|$ is the ℓ_1 *norm* (or *absolute sum norm*) on $\mathbb{F}^n$. It satisfies the nonnegativity, positivity, and homogeneity axioms in the preceding definition of a norm. To verify the triangle inequality, we invoke the triangle inequality for the modulus function on $\mathbb{F}$ and compute

$$\|\mathbf{u} + \mathbf{v}\|_1 = |u_1 + v_1| + |u_2 + v_2| + \cdots + |u_n + v_n|$$
$$\le |u_1| + |v_1| + |u_2| + |v_2| + \cdots + |u_n| + |v_n|$$
$$= \|\mathbf{u}\|_1 + \|\mathbf{v}\|_1.$$

Since $\|\mathbf{e}_1 + \mathbf{e}_2\|_1^2 + \|\mathbf{e}_1 - \mathbf{e}_2\|_1^2 = 8 > 4 = 2\|\mathbf{e}_1\|_1^2 + 2\|\mathbf{e}_2\|_1^2$, the ℓ_1 norm does not satisfy the parallelogram identity; see Theorem 5.4.9.e. Consequently, it is not derived from an inner product.

Example 5.5.3 The function $\|\mathbf{u}\|_\infty = \max\{|u_i| : 1 \le i \le n\}$ is the ℓ_∞ *norm* (or *max norm*) on $\mathbb{F}^n$. Verification of the nonnegativity, positivity, and homogeneity axioms is straightforward. To verify the triangle inequality, let k be any index such that $\|\mathbf{u} + \mathbf{v}\|_\infty = |u_k + v_k|$. Use the triangle inequality for the modulus function on $\mathbb{F}$ to compute

$$\|\mathbf{u} + \mathbf{v}\|_\infty = |u_k + v_k| \le |u_k| + |v_k| \le \|\mathbf{u}\|_\infty + \|\mathbf{v}\|_\infty.$$

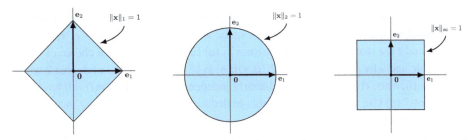

(a) Unit ball for $\|\cdot\|_1$ on $\mathbb{R}^2$.   (b) Unit ball for $\|\cdot\|_2$ on $\mathbb{R}^2$.   (c) Unit ball for $\|\cdot\|_\infty$ on $\mathbb{R}^2$.

Figure 5.8 Unit balls for the ℓ_1, ℓ_2, and ℓ_∞ norms on $\mathbb{R}^2$.

The computation $\|\mathbf{e}_1 + \mathbf{e}_2\|_\infty^2 + \|\mathbf{e}_1 - \mathbf{e}_2\|_\infty^2 = 2 < 4 = 2\|\mathbf{e}_1\|_\infty^2 + 2\|\mathbf{e}_2\|_\infty^2$ shows that the ℓ_∞ norm does not satisfy the parallelogram identity. Hence it is not derived from an inner product.

Definition 5.5.4 A *normed vector space* is a real or complex vector space V, equipped with a norm $\|\cdot\| : V \to [0, \infty)$. The *unit ball* of a normed space is $\{\mathbf{v} \in V : \|\mathbf{v}\| \le 1\}$.

The unit balls for the ℓ_1, ℓ_2 (Euclidean), and ℓ_∞ norms on $\mathbb{R}^2$ are illustrated in Figure 5.8.

Definition 5.5.5 Let V be a normed vector space with norm $\|\cdot\|$. Then $\mathbf{v} \in V$ is a *unit vector* if $\|\mathbf{v}\| = 1$.

Any nonzero vector $\mathbf{v}$ in a normed vector space can be scaled to create a unit vector $\mathbf{v}/\|\mathbf{v}\|$:

$$\text{If } \mathbf{v} \neq \mathbf{0}, \text{ then } \left\| \frac{\mathbf{v}}{\|\mathbf{v}\|} \right\| = \frac{\|\mathbf{v}\|}{\|\mathbf{v}\|} = 1.$$

The process of scaling a nonzero vector to a unit vector is called *normalization*.

5.6 Problems

P.5.1 Let $\mathscr{P}$ be the parallelogram in the first quadrant of $\mathbb{R}^2$ whose vertices are $\mathbf{0}$, $\mathbf{v}$, $\mathbf{w}$, and $\mathbf{v} + \mathbf{w}$. Let $A = [\mathbf{v}\ \mathbf{w}] \in M_2(\mathbb{R})$. Show that the area of $\mathscr{P}$ is $\sqrt{\|\mathbf{v}\|^2\|\mathbf{w}\|^2 - \langle \mathbf{v}, \mathbf{w}\rangle^2} = |\det A|$. What does Theorem 5.4.14 say about $\mathscr{P}$ if $\det A = 0$?

P.5.2 Let V be a real inner product space and let $\mathbf{x}, \mathbf{y} \in V$. Show that $\langle \mathbf{x}, \mathbf{y}\rangle = 0$ if and only if $\|\mathbf{x} + \mathbf{y}\| = \|\mathbf{x} - \mathbf{y}\|$.

P.5.3 Let V be a complex inner product space and let $\mathbf{x}, \mathbf{y} \in V$. (a) If $\langle \mathbf{x}, \mathbf{y}\rangle = 0$, show that $\|\mathbf{x}+\mathbf{y}\| = \|\mathbf{x}-\mathbf{y}\|$. (b) What can you say about the converse? Consider the case $\mathbf{y} = i\mathbf{x}$.

P.5.4 Provide details for the following alternative proof of the Cauchy–Schwarz inequality in a real inner product space V. For nonzero vectors $\mathbf{u}, \mathbf{v} \in V$, consider the function $p(t) = \|t\mathbf{u} + \mathbf{v}\|^2$ of a real variable t. (a) Why is $p(t) \ge 0$ for all real t? (b) If $p(t) = 0$ has a real root, why are $\mathbf{u}$ and $\mathbf{v}$ linearly dependent? (c) Show that $p(t) = \|\mathbf{u}\|^2 t^2 + 2\langle \mathbf{u}, \mathbf{v}\rangle t + \|\mathbf{v}\|^2$ is a polynomial of degree 2 with real coefficients. (d) If $\mathbf{u}$ and $\mathbf{v}$ are linearly independent, why does $p(t) = 0$ have no real roots? (e) Use the quadratic formula to deduce the Cauchy–Schwarz inequality for a real inner product space.

P.5.5 Modify the argument in the preceding problem to prove the Cauchy–Schwarz inequality in a complex inner product space $\mathcal{V}$. Redefine $p(t) = \|t\mathbf{u} + e^{i\theta}\mathbf{v}\|^2$, in which θ and t are real, and $e^{-i\theta}\langle\mathbf{u}, \mathbf{v}\rangle = |\langle\mathbf{u}, \mathbf{v}\rangle|$. (a) Explain why such a choice of θ is possible and why $p(t) = \|\mathbf{u}\|^2 t^2 + 2|\langle\mathbf{u}, \mathbf{v}\rangle|t + \|\mathbf{v}\|^2$ is a polynomial of degree 2 with real coefficients. (b) If $\mathbf{u}$ and $\mathbf{v}$ are linearly independent, why does $p(t) = 0$ have no real roots? (c) Use the quadratic formula to deduce the Cauchy–Schwarz inequality for a complex inner product space.

P.5.6 Let x and y be nonnegative real numbers. Use the fact that $(a - b)^2 \geq 0$ for real a, b to show that

$$\sqrt{xy} \leq \frac{x + y}{2}, \tag{5.6.1}$$

with equality if and only if $x = y$. This inequality is the *arithmetic–geometric mean inequality*. The left side of (5.6.1) is the *geometric mean* of x and y; the right side is their *arithmetic mean*.

P.5.7 Let $\mathcal{V}$ be an $\mathbb{F}$-inner product space and let $\mathbf{u}, \mathbf{v} \in \mathcal{V}$. (a) Expand the inequality $0 \leq \|\mathbf{x} - \mathbf{y}\|^2$ and choose suitable $\mathbf{x}, \mathbf{y} \in \mathcal{V}$ to obtain

$$|\langle\mathbf{u}, \mathbf{v}\rangle| \leq \frac{\lambda^2}{2}\|\mathbf{u}\|^2 + \frac{1}{2\lambda^2}\|\mathbf{v}\|^2 \tag{5.6.2}$$

for all $\lambda > 0$. Be sure your proof covers both cases $\mathbb{F} = \mathbb{R}$ and $\mathbb{F} = \mathbb{C}$. (b) Use (5.6.2) to prove the arithmetic–geometric mean inequality (5.6.1). (c) Use (5.6.2) to prove the Cauchy–Schwarz inequality.

P.5.8 Let $x, y \geq 0$, and consider $\mathbf{u} = [\sqrt{x} \ \sqrt{y}]^\mathsf{T}$ and $\mathbf{v} = [\sqrt{y} \ \sqrt{x}]^\mathsf{T}$ in $\mathbb{R}^2$. Use the Cauchy–Schwarz inequality to prove the arithmetic–geometric mean inequality (5.6.1).

P.5.9 Let $\mathcal{V}$ be an $\mathbb{F}$-inner product space and let $\mathbf{u}, \mathbf{v} \in \mathcal{V}$. Show that $2|\langle\mathbf{u}, \mathbf{v}\rangle| \leq \|\mathbf{u}\|^2 + \|\mathbf{v}\|^2$. When is this inequality an equality?

P.5.10 Let $\mathcal{V}$ be an $\mathbb{F}$-inner product space and let $\mathbf{u}, \mathbf{v} \in \mathcal{V}$. Show that $\|\mathbf{u} + \mathbf{v}\| \, \|\mathbf{u} - \mathbf{v}\| \leq \|\mathbf{u}\|^2 + \|\mathbf{v}\|^2$. When is this inequality an equality?

P.5.11 Let $\mathcal{V}$ be an $\mathbb{F}$-inner product space and let $\mathbf{u}, \mathbf{v} \in \mathcal{V}$. Prove that $\mathbf{u}$ and $\mathbf{v}$ are orthogonal if and only if $\|\mathbf{v}\| \leq \|c\mathbf{u} + \mathbf{v}\|$ for all $c \in \mathbb{F}$. Be sure your proof covers both cases $\mathbb{F} = \mathbb{R}$ and $\mathbb{F} = \mathbb{C}$. Draw a diagram illustrating what this means if $\mathcal{V} = \mathbb{R}^2$.

P.5.12 (a) Let $f, g \in C_{\mathbb{R}}[0, 1]$ (see Example 5.3.8) be the functions depicted in Figure 5.9. Does there exist a $c \in \mathbb{R}$ such that $\|f + cg\| < \|f\|$? (b) Consider the functions $f(x) = x(1 - x)$ and $g(x) = \sin(2\pi x)$ in $C_{\mathbb{R}}[0, 1]$. Does there exist a $c \in \mathbb{R}$ such that $\|f + cg\| < \|f\|$?

P.5.13 Let $\mathcal{V}$ be a complex inner product space and let $\mathbf{u}, \mathbf{v} \in \mathcal{V}$. Show that

$$\langle\mathbf{u}, \mathbf{v}\rangle = \frac{1}{2\pi} \int_{-\pi}^{\pi} e^{i\theta} \|\mathbf{u} + e^{i\theta}\mathbf{v}\|^2 \, d\theta.$$

P.5.14 Let $\mathcal{V}$ be the real inner product space $C_{\mathbb{R}}[0, 1]$, let $f \in \mathcal{V}$, and suppose that $f(t) > 0$ for all $t \in [0, 1]$. Use the Cauchy–Schwarz inequality to deduce that

$$\frac{1}{\int_0^1 f(t)\,dt} \leq \int_0^1 \frac{1}{f(t)}\,dt.$$

P.5.15 Let $\mathcal{V}$ be the real inner product space $C_{\mathbb{R}}[0, 1]$ and let $a \in [0, 1]$. Show that there is no $f \in \mathcal{V}$ such that $f(t) \geq 0$ for all $t \in [0, 1]$, $\int_0^1 f(t)\,dt = 1$, $\int_0^1 tf(t)\,dt = a$, and $\int_0^1 t^2 f(t)\,dt = a^2$.

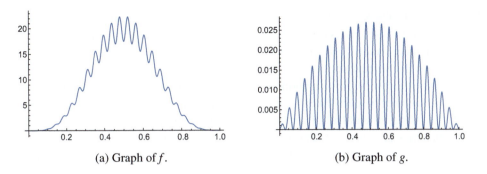

(a) Graph of f.   (b) Graph of g.

Figure 5.9 Graphs for P.5.12.

P.5.16 Let $\langle \cdot, \cdot \rangle$ be an inner product on $\mathbb{F}^n$ and suppose that $A \in M_n(\mathbb{F})$ is invertible. Define $\langle \mathbf{u}, \mathbf{v} \rangle_A = \langle A\mathbf{u}, A\mathbf{v} \rangle$. Show that $\langle \cdot, \cdot \rangle_A$ is an inner product on $\mathbb{F}^n$.

P.5.17 Let $\mathbf{x} \in \mathbb{C}^n$ and $A \in M_n$. Let $\| \cdot \|_F$ denote the Frobenius norm. (a) Show that $\|A\mathbf{x}\|_2 \le \|A\|_F \|\mathbf{x}\|_2$ and $\|AB\|_F \le \|A\|_F \|B\|_F$; in particular, $\|A^2\|_F \le \|A\|_F^2$. (b) If $A^2 = A \ne 0$, show that $\|A\|_F \ge 1$ and give an example to show that equality is possible.

P.5.18 Let $\mathbf{u}$, $\mathbf{v}$, and $\mathbf{w}$ be vectors in an inner product space. (a) If $\mathbf{u} \perp \mathbf{v}$ and $\mathbf{v} \perp \mathbf{w}$, is $\mathbf{u} \perp \mathbf{w}$? (b) If $\mathbf{u} \perp \mathbf{v}$ and $\mathbf{u} \perp \mathbf{w}$, is $\mathbf{u} \perp (\mathbf{v} + \mathbf{w})$?

P.5.19 (a) Let $\mathbf{u}$ and $\mathbf{v}$ be vectors in a real inner product space $\mathcal{V}$. If $\|\mathbf{u} + \mathbf{v}\|^2 = \|\mathbf{u}\|^2 + \|\mathbf{v}\|^2$, show that $\mathbf{u} \perp \mathbf{v}$. (b) What can you say if $\mathcal{V}$ is a complex inner product space?

P.5.20 Let $\mathbf{u}_1, \mathbf{u}_2, \ldots, \mathbf{u}_k$ be vectors in an inner product space, and let $c_1, c_2, \ldots, c_k$ be scalars. Suppose that $\mathbf{u}_i \perp \mathbf{u}_j$ for all $i \ne j$. Show that

$$\|c_1\mathbf{u}_1 + c_2\mathbf{u}_2 + \cdots + c_k\mathbf{u}_k\|^2 = |c_1|^2 \|\mathbf{u}_1\|^2 + |c_2|^2 \|\mathbf{u}_2\|^2 + \cdots + |c_k|^2 \|\mathbf{u}_k\|^2.$$

P.5.21 Show that the triangle inequality $a + b \ge c$ in Corollary 5.1.3 is a strict inequality if the angle between the sides of length a and b is less than π. Sketch what the case of equality looks like.

P.5.22 Let $\mathcal{V}$ be an inner product space and let $\mathbf{u}, \mathbf{v}, \mathbf{w} \in \mathcal{V}$. Show that

$$\|\mathbf{u} + \mathbf{v} + \mathbf{w}\|^2 + \|\mathbf{u}\|^2 + \|\mathbf{v}\|^2 + \|\mathbf{w}\|^2 = \|\mathbf{u} + \mathbf{v}\|^2 + \|\mathbf{u} + \mathbf{w}\|^2 + \|\mathbf{v} + \mathbf{w}\|^2.$$

P.5.23 Let $\mathcal{V}$ be an inner product space and let $\mathbf{u}, \mathbf{v} \in \mathcal{V}$. Show that $\big| \|\mathbf{u}\| - \|\mathbf{v}\| \big| \le \|\mathbf{u} - \mathbf{v}\|$.

P.5.24 Let $\mathcal{V} = \mathcal{P}_{n-1}$. If $p(z) = \sum_{k=0}^{n-1} p_k z^k$ and $q(z) = \sum_{k=0}^{n-1} q_k z^k$, define

$$\langle p, q \rangle = \sum_{i,j=0}^{n-1} \frac{p_i \overline{q_j}}{i + j + 1}.$$

(a) Show that $\langle \cdot, \cdot \rangle : \mathcal{V} \times \mathcal{V} \to \mathbb{C}$ is an inner product. *Hint*: Example 5.3.4; compute $\langle z^i, z^j \rangle$. (b) Deduce that $A = [(i + j - 1)^{-1}] \in M_n$ has the property that $\langle A\mathbf{x}, \mathbf{x} \rangle \ge 0$ for all $\mathbf{x} \in \mathbb{C}^n$.

P.5.25 Let $\mathcal{V}$ be a normed vector space with norm $\| \cdot \|$ that is derived from an inner product $\langle \cdot, \cdot \rangle_1$ on $\mathcal{V} \times \mathcal{V}$. Could there be a different inner product $\langle \cdot, \cdot \rangle_2$ on $\mathcal{V} \times \mathcal{V}$ such that $\|\mathbf{u}\| = \langle \mathbf{u}, \mathbf{u} \rangle_2^{1/2}$ for all $\mathbf{u} \in \mathcal{V}$?

P.5.26 Consider the function $\| \cdot \| : \mathbb{R}^2 \to [0, \infty)$ defined by $\|\mathbf{u}\| = |u_1|$, in which $\mathbf{u} = [u_1 \ u_2]^\mathsf{T}$. Show that it satisfies three of the axioms for a norm in Definition 5.5.1. Show that it does not satisfy the remaining axiom.

P.5.27 Let $\mathcal{V}$ be the real vector space $\mathbb{R}^2$ and let $\mathbf{u} = [u_1 \ u_2]^\mathsf{T} \in \mathcal{V}$. Show that the function $\|\mathbf{u}\| = 2|u_1| + 5|u_2|$ is a norm on $\mathcal{V}$. Is it derived from an inner product? If so, what is that inner product? Sketch the unit ball for this norm.

P.5.28 Let $\mathcal{V}$ be the real vector space $\mathbb{R}^2$ and let $\mathbf{u} = [u_1 \ u_2]^\mathsf{T} \in \mathcal{V}$. Show that the function $\|\mathbf{u}\| = (2u_1^2 + 5u_2^2)^{1/2}$ is a norm on $\mathcal{V}$. Is it derived from an inner product? If so, what is that inner product? Sketch the unit ball for this norm.

P.5.29 Let $\mathcal{B}$ be the unit ball of a normed vector space. If $\mathbf{u}, \mathbf{v} \in \mathcal{B}$ and $0 \le t \le 1$, show that $t\mathbf{u} + (1 - t)\mathbf{v} \in \mathcal{B}$. This shows that $\mathcal{B}$ is a *convex* set.

5.7 Notes

A norm is derived from an inner product if and only if it satisfies the parallelogram identity (Theorem 5.4.9.e); see [HJ13, 5.1.P12].

5.8 Some Important Concepts

- Axioms for an inner product.

- Orthogonality.

- Parallelogram identity and derived norms.

- Cauchy–Schwarz inequality.

- Triangle inequality.

- Polarization identity and derived norms.

- Axioms for a norm.

- Normalization and the unit ball of a normed vector space.

- The ℓ_1, ℓ_2, and ℓ_∞ norms and their unit balls.

6 Orthonormal Vectors

In this chapter, we explore the role of orthonormal (orthogonal and normalized) vectors in an $\mathbb{F}$-inner-product space V ($\mathbb{F} = \mathbb{R}$ or $\mathbb{C}$). Matrix representations of linear transformations with respect to orthonormal bases are of particular importance. They are associated with the notion of an adjoint transformation. We give a brief introduction to Fourier series that highlights the orthogonality properties of sine and cosine functions. In the final section of the chapter, we discuss orthogonal polynomials and the remarkable numerical integration rules associated with them.

6.1 Orthonormal Sequences

Definition 6.1.1 A sequence of vectors $\mathbf{u}_1, \mathbf{u}_2, \ldots$ (finite or infinite) in an inner product space is *orthonormal* if

$$\langle \mathbf{u}_i, \mathbf{u}_j \rangle = \delta_{ij} \text{ for all } i, j. \tag{6.1.2}$$

A finite orthonormal sequence of vectors is an *orthonormal list*.

If $\mathbf{u}_1, \mathbf{u}_2, \ldots$ is an orthonormal sequence and $1 \leq i_1 < i_2 < \cdots$, then the subsequence $\mathbf{u}_{i_1}, \mathbf{u}_{i_2}, \ldots$ is also an orthonormal sequence. It is convenient to say that "$\mathbf{u}_1, \mathbf{u}_2, \ldots$ are orthonormal" rather than "the sequence of vectors $\mathbf{u}_1, \mathbf{u}_2, \ldots$ is orthonormal."

Any sequence $\mathbf{v}_1, \mathbf{v}_2, \ldots$ of nonzero pairwise orthogonal vectors can be turned into an orthonormal sequence by normalizing each $\mathbf{v}_i$; see (5.4.11).

Example 6.1.3 In $\mathbb{R}^2$ with the standard inner product, the vectors $\mathbf{u}_1 = \frac{1}{\sqrt{2}} \begin{bmatrix} 1 \\ 1 \end{bmatrix}$ and $\mathbf{u}_2 = \frac{1}{\sqrt{2}} \begin{bmatrix} -1 \\ 1 \end{bmatrix}$ are orthonormal.

Example 6.1.4 The infinite sequence of vectors

$$\mathbf{u}_1 = (1, 0, 0, 0, \ldots), \quad \mathbf{u}_2 = (0, 1, 0, 0, \ldots), \quad \mathbf{u}_3 = (0, 0, 1, 0, \ldots), \ldots$$

is orthonormal in the inner product space of finitely nonzero sequences; see Example 5.3.12.

Example 6.1.5 We claim that the polynomials $f_1(x) = 1, f_2(x) = 2x - 1$, and $f_3(x) = 6x^2 - 6x + 1$ are pairwise orthogonal vectors in $C[0, 1]$, with the L^2 inner product

$$\langle f, g \rangle = \int_0^1 f(x)\overline{g(x)} \, dx. \tag{6.1.6}$$

To verify this, compute

$$\langle f_1, f_2 \rangle = \int_0^1 (1)\overline{(2x-1)}\, dx = \int_0^1 (2x-1)\, dx = x^2 - x \Big|_0^1 = 0,$$

$$\langle f_1, f_3 \rangle = \int_0^1 (1)\overline{(6x^2 - 6x + 1)}\, dx = \int_0^1 (6x^2 - 6x + 1)\, dx$$

$$= 2x^3 - 3x^2 + x \Big|_0^1 = 2 - 3 + 1 = 0, \quad \text{and}$$

$$\langle f_2, f_3 \rangle = \int_0^1 (2x-1)\overline{(6x^2 - 6x + 1)}\, dx = \int_0^1 (12x^3 - 18x^2 + 8x - 1)\, dx$$

$$= 3x^4 - 6x^3 + 4x^2 - x \Big|_0^1 = 3 - 6 + 4 - 1 = 0.$$

A further computation shows that

$$\|f_1\|^2 = \int_0^1 dx = 1,$$

$$\|f_2\|^2 = \int_0^1 (2x-1)^2\, dx = \frac{1}{3}, \quad \text{and}$$

$$\|f_3\|^2 = \int_0^1 (6x^2 - 6x + 1)^2\, dx = \frac{1}{5}.$$

Normalize the vectors f_i to obtain the orthonormal list

$$u_1 = 1, \qquad u_2 = \sqrt{3}(2x-1), \qquad u_3 = \sqrt{5}(6x^2 - 6x + 1). \tag{6.1.7}$$

The following theorem generalizes Theorem 5.4.9.d to an identity that involves n orthonormal vectors.

Theorem 6.1.8 *If* $\mathbf{u}_1, \mathbf{u}_2, \ldots, \mathbf{u}_n$ *is an orthonormal list and* $a_1, a_2, \ldots, a_n \in \mathbb{F}$, *then*

$$\left\| \sum_{i=1}^n a_i \mathbf{u}_i \right\|^2 = \sum_{i=1}^n |a_i|^2. \tag{6.1.9}$$

Proof Compute

$$\left\| \sum_{i=1}^n a_i \mathbf{u}_i \right\|^2 = \left\langle \sum_{i=1}^n a_i \mathbf{u}_i, \sum_{j=1}^n a_j \mathbf{u}_j \right\rangle = \sum_{i=1}^n a_i \left\langle \mathbf{u}_i, \sum_{j=1}^n a_j \mathbf{u}_j \right\rangle$$

$$= \sum_{i=1}^n a_i \sum_{j=1}^n \overline{a_j} \langle \mathbf{u}_i, \mathbf{u}_j \rangle = \sum_{i=1}^n a_i \sum_{j=1}^n \overline{a_j} \delta_{ij}$$

$$= \sum_{i=1}^n a_i \overline{a_i} = \sum_{i=1}^n |a_i|^2. \qquad \square$$

A linearly independent list of vectors need not be orthonormal; its vectors need not be normalized or orthogonal. However, an orthonormal list is linearly independent.

Theorem 6.1.10 *If* $\mathbf{u}_1, \mathbf{u}_2, \ldots, \mathbf{u}_n$ *is an orthonormal list, then* $\mathbf{u}_1, \mathbf{u}_2, \ldots, \mathbf{u}_n$ *are linearly independent.*

Proof If $a_1, a_2, \ldots, a_n \in \mathbb{F}$ and $a_1 \mathbf{u}_1 + a_2 \mathbf{u}_2 + \cdots + a_n \mathbf{u}_n = \mathbf{0}$, then (6.1.9) ensures that $|a_1|^2 + |a_2|^2 + \cdots + |a_n|^2 = \|\mathbf{0}\|^2 = 0$, so $a_1 = a_2 = \cdots = a_n = 0$. $\qquad \square$

6.2 Orthonormal Bases

Definition 6.2.1 An *orthonormal basis* for a finite-dimensional inner product space is a basis that is an orthonormal list.

Since an orthonormal list in an inner product space V is linearly independent (Theorem 6.1.10), it is a basis for its span.

Example 6.2.2 In the $\mathbb{F}$-inner product space $M_{m \times n}(\mathbb{F})$ with the Frobenius inner product, the basis comprised of the matrices E_{pq} defined in Example 2.2.3 is orthonormal.

Example 6.2.3 We claim that

$$\mathbf{u}_1 = \frac{1}{2}\begin{bmatrix} 1 \\ 1 \\ 1 \\ 1 \end{bmatrix}, \quad \mathbf{u}_2 = \frac{1}{2}\begin{bmatrix} 1 \\ i \\ -1 \\ -i \end{bmatrix}, \quad \mathbf{u}_3 = \frac{1}{2}\begin{bmatrix} 1 \\ -1 \\ 1 \\ -1 \end{bmatrix}, \quad \mathbf{u}_4 = \frac{1}{2}\begin{bmatrix} 1 \\ -i \\ -1 \\ i \end{bmatrix} \tag{6.2.4}$$

are an orthonormal basis for $\mathbb{C}^4$. Since $\langle \mathbf{u}_i, \mathbf{u}_j \rangle = \delta_{ij}$, the list $\beta = \mathbf{u}_1, \mathbf{u}_2, \mathbf{u}_3, \mathbf{u}_4$ is orthonormal; Theorem 6.1.10 ensures that β is linearly independent. Since it is a maximal linearly independent list in $\mathbb{C}^4$, Corollary 2.2.9 ensures that β is a basis. The vectors (6.2.4) are the columns of the 4×4 Fourier matrix; see (7.1.19).

The principle invoked in the preceding example is important. If V is an n-dimensional inner product space, then a list of any n orthonormal vectors is an orthonormal basis for V. An orthonormal basis is desirable because determining coordinate vectors is straightforward.

Theorem 6.2.5 *Let V be a finite-dimensional inner product space, let $\beta = \mathbf{u}_1, \mathbf{u}_2, \ldots, \mathbf{u}_n$ be an orthonormal basis for V, and let $\mathbf{v} \in V$. Then*

$$\mathbf{v} = \sum_{i=1}^{n} \langle \mathbf{v}, \mathbf{u}_i \rangle \mathbf{u}_i, \tag{6.2.6}$$

$$\|\mathbf{v}\|^2 = \sum_{i=1}^{n} |\langle \mathbf{v}, \mathbf{u}_i \rangle|^2, \tag{6.2.7}$$

and

$$[\mathbf{v}]_\beta = \begin{bmatrix} \langle \mathbf{v}, \mathbf{u}_1 \rangle \\ \langle \mathbf{v}, \mathbf{u}_2 \rangle \\ \vdots \\ \langle \mathbf{v}, \mathbf{u}_n \rangle \end{bmatrix}. \tag{6.2.8}$$

Proof Since β is a basis for V, there exist scalars $a_i \in \mathbb{F}$ such that $\mathbf{v} = \sum_{i=1}^{n} a_i \mathbf{u}_i$. The inner product of both sides with $\mathbf{u}_j$ is

$$\langle \mathbf{v}, \mathbf{u}_j \rangle = \left\langle \sum_{i=1}^{n} a_i \mathbf{u}_i, \mathbf{u}_j \right\rangle = \sum_{i=1}^{n} a_i \langle \mathbf{u}_i, \mathbf{u}_j \rangle = \sum_{i=1}^{n} a_i \delta_{ij} = a_j,$$

which implies (6.2.6). The identity (6.2.7) follows from (6.2.6) and Theorem 6.1.8. The β-coordinate vector (6.2.8) restates the information in (6.2.6). $\qquad \square$

Example 6.2.9 The vectors

$$\mathbf{u}_1 = \frac{1}{3}\begin{bmatrix} 1 \\ 2 \\ 2 \end{bmatrix}, \qquad \mathbf{u}_2 = \frac{1}{3}\begin{bmatrix} -2 \\ 2 \\ -1 \end{bmatrix}, \qquad \text{and} \qquad \mathbf{u}_3 = \frac{1}{3}\begin{bmatrix} -2 \\ -1 \\ 2 \end{bmatrix}$$

are an orthonormal basis for $\mathbb{R}^3$. How can we express $\mathbf{v} = [1 \ 2 \ 3]^\mathsf{T}$ as a linear combination of these basis vectors? The preceding theorem tells us that $\mathbf{v} = \sum_{i=1}^{3} \langle \mathbf{v}, \mathbf{u}_i \rangle \mathbf{u}_i$, so we compute

$$\langle \mathbf{v}, \mathbf{u}_1 \rangle = \tfrac{1}{3}(1 + 4 + 6) = \frac{11}{3},$$

$$\langle \mathbf{v}, \mathbf{u}_2 \rangle = \tfrac{1}{3}(-2 + 4 - 3) = -\frac{1}{3},$$

$$\langle \mathbf{v}, \mathbf{u}_3 \rangle = \tfrac{1}{3}(-2 - 2 + 6) = \frac{2}{3},$$

and find that $\mathbf{v} = \frac{11}{3}\mathbf{u}_1 - \frac{1}{3}\mathbf{u}_2 + \frac{2}{3}\mathbf{u}_3$.

In the vector space $V = \mathbb{F}^n$ with the standard inner product $\langle \mathbf{v}, \mathbf{u} \rangle = \mathbf{u}^*\mathbf{v}$, the identity (6.2.8) becomes an intriguing matrix equation. If the columns of $U = [\mathbf{u}_1 \ \mathbf{u}_2 \ \dots \ \mathbf{u}_n] \in \mathsf{M}_n$ are an orthonormal basis β of $\mathbb{F}^n$, it says that

$$[\mathbf{v}]_\beta = U^*\mathbf{v} = \begin{bmatrix} \mathbf{u}_1^*\mathbf{v} \\ \mathbf{u}_2^*\mathbf{v} \\ \vdots \\ \mathbf{u}_n^*\mathbf{v} \end{bmatrix}.$$

However, Theorem 2.4.6 tells us that $[\mathbf{v}]_\beta = U^{-1}\mathbf{v}$, which suggests that U has the remarkable property that $U^{-1} = U^*$. We have much more to say about this in Chapter 7.

6.3 The Gram–Schmidt Process

If we are given a linearly independent list in an inner product space, there are several ways to construct an orthonormal list with the same span. Before describing one of them, let us consider an example.

Example 6.3.1 Consider the set $\mathcal{U}$ in $\mathbb{R}^3$ defined by

$$x + 2y + 3z = 0. \tag{6.3.2}$$

If $A = [1 \ 2 \ 3] \in \mathsf{M}_{1\times3}$ and $\mathbf{x} = [x \ y \ z]^\mathsf{T}$, then (6.3.2) is equivalent to $A\mathbf{x} = \mathbf{0}$, so $\mathcal{U} = \text{null } A$. Consequently, $\mathcal{U}$ is a subspace of $\mathbb{R}^3$. Since $\text{rank } A = 1$, the rank-nullity theorem ensures that $\dim \mathcal{U} = 2$.

The linearly independent vectors $\mathbf{v}_1 = [3 \ 0 \ -1]^\mathsf{T}$ and $\mathbf{v}_2 = [-2 \ 1 \ 0]^\mathsf{T}$ are in null A, so they are a basis for $\mathcal{U}$. Since $\|\mathbf{v}_1\|_2 = \sqrt{10}$, $\|\mathbf{v}_2\|_2 = \sqrt{5}$, and $\langle \mathbf{v}_1, \mathbf{v}_2 \rangle = -6$, our basis vectors are not normalized and they are not orthogonal, but we can use them to construct an orthonormal basis for $\mathcal{U}$.

First normalize $\mathbf{v}_1$ to obtain a unit vector $\mathbf{u}_1$:

$$\mathbf{u}_1 = \frac{1}{\sqrt{10}}\mathbf{v}_1 = \frac{1}{\sqrt{10}}\begin{bmatrix} 3 \\ 0 \\ -1 \end{bmatrix}. \tag{6.3.3}$$

In our derivation of (5.4.13), we discovered how to use $\mathbf{v}_2$ to construct a vector that is orthogonal to $\mathbf{u}_1$: find the projection of $\mathbf{v}_2$ onto $\mathbf{u}_1$ and subtract it from $\mathbf{v}_2$. This vector,

$$\mathbf{x}_2 = \mathbf{v}_2 - \underbrace{\langle \mathbf{v}_2, \mathbf{u}_1 \rangle \mathbf{u}_1}_{\text{Projection of } \mathbf{v}_2 \text{ onto } \mathbf{u}_1}, \tag{6.3.4}$$

belongs to $\mathcal{U}$ because it is a linear combination of vectors in $\mathcal{U}$. Using (6.3.3), we compute

$$\mathbf{x}_2 = \mathbf{v}_2 - \frac{1}{10}\langle \mathbf{v}_2, \mathbf{v}_1 \rangle \mathbf{v}_1 = \mathbf{v}_2 + \frac{3}{5}\mathbf{v}_1 = \begin{bmatrix} -2 \\ 1 \\ 0 \end{bmatrix} + \frac{3}{5}\begin{bmatrix} 3 \\ 0 \\ -1 \end{bmatrix} = \frac{1}{5}\begin{bmatrix} -1 \\ 5 \\ -3 \end{bmatrix}.$$

We have $\|\mathbf{x}_2\|_2^2 = \frac{35}{25}$, so the unit vector $\mathbf{u}_2 = \mathbf{x}_2/\|\mathbf{x}_2\|_2$ is

$$\mathbf{u}_2 = \frac{1}{\sqrt{35}}\begin{bmatrix} -1 \\ 5 \\ -3 \end{bmatrix}.$$

By construction, $\|\mathbf{u}_2\|_2 = 1$ and $\mathbf{u}_2$ is orthogonal to $\mathbf{u}_1$. As a check, we compute $\langle \mathbf{u}_1, \mathbf{u}_2 \rangle = \frac{1}{\sqrt{350}}(-3 + 0 + 3) = 0$.

The Gram–Schmidt process is a systematic implementation of the ideas employed in the preceding example. It starts with a linearly independent list of vectors and produces an orthonormal list with the same span. Details of the process are explained in the proof of the following theorem.

Theorem 6.3.5 (Gram–Schmidt Process) *Let V be an inner product space, and suppose that $\mathbf{v}_1, \mathbf{v}_2, \ldots, \mathbf{v}_n \in V$ are linearly independent. There is an orthonormal list $\mathbf{u}_1, \mathbf{u}_2, \ldots, \mathbf{u}_n$ such that*

$$\mathrm{span}\{\mathbf{v}_1, \mathbf{v}_2, \ldots, \mathbf{v}_k\} = \mathrm{span}\{\mathbf{u}_1, \mathbf{u}_2, \ldots, \mathbf{u}_k\} \quad \text{for } k = 1, 2, \ldots, n. \tag{6.3.6}$$

Proof We proceed by induction on n. In the base case $n = 1$, $\mathbf{u}_1 = \mathbf{v}_1/\|\mathbf{v}_1\|$ is a unit vector and $\mathrm{span}\{\mathbf{v}_1\} = \mathrm{span}\{\mathbf{u}_1\}$.

For the induction step, let $2 \leq m \leq n$. Suppose that, given $m - 1$ linearly independent vectors $\mathbf{v}_1, \mathbf{v}_2, \ldots, \mathbf{v}_{m-1}$, there are orthonormal vectors $\mathbf{u}_1, \mathbf{u}_2, \ldots, \mathbf{u}_{m-1}$ such that

$$\mathrm{span}\{\mathbf{v}_1, \mathbf{v}_2, \ldots, \mathbf{v}_k\} = \mathrm{span}\{\mathbf{u}_1, \mathbf{u}_2, \ldots, \mathbf{u}_k\} \quad \text{for } k = 1, 2, \ldots, m - 1. \tag{6.3.7}$$

Since $\mathbf{v}_1, \mathbf{v}_2, \ldots, \mathbf{v}_m$ are linearly independent,

$$\mathbf{v}_m \notin \mathrm{span}\{\mathbf{v}_1, \mathbf{v}_2, \ldots, \mathbf{v}_{m-1}\} = \mathrm{span}\{\mathbf{u}_1, \mathbf{u}_2, \ldots, \mathbf{u}_{m-1}\},$$

so $\mathbf{x}_m = \mathbf{v}_m - \sum_{i=1}^{m-1} \langle \mathbf{v}_m, \mathbf{u}_i \rangle \mathbf{u}_i \neq \mathbf{0}$ and we may define $\mathbf{u}_m = \mathbf{x}_m/\|\mathbf{x}_m\|$.

We claim that $\mathbf{x}_m$ (and hence also $\mathbf{u}_m$) is orthogonal to $\mathbf{u}_1, \mathbf{u}_2, \ldots, \mathbf{u}_{m-1}$. Indeed, if $1 \leq j \leq m - 1$, then

$$\langle \mathbf{u}_j, \mathbf{x}_m \rangle = \left\langle \mathbf{u}_j, \mathbf{v}_m - \sum_{i=1}^{m-1} \langle \mathbf{v}_m, \mathbf{u}_i \rangle \mathbf{u}_i \right\rangle = \langle \mathbf{u}_j, \mathbf{v}_m \rangle - \left\langle \mathbf{u}_j, \sum_{i=1}^{m-1} \langle \mathbf{v}_m, \mathbf{u}_i \rangle \mathbf{u}_i \right\rangle$$

$$= \langle \mathbf{u}_j, \mathbf{v}_m \rangle - \sum_{i=1}^{m-1} \overline{\langle \mathbf{v}_m, \mathbf{u}_i \rangle} \langle \mathbf{u}_j, \mathbf{u}_i \rangle = \langle \mathbf{u}_j, \mathbf{v}_m \rangle - \sum_{i=1}^{m-1} \langle \mathbf{u}_i, \mathbf{v}_m \rangle \delta_{ij}$$

$$= \langle \mathbf{u}_j, \mathbf{v}_m \rangle - \langle \mathbf{u}_j, \mathbf{v}_m \rangle = 0.$$

Since $\mathbf{u}_m$ is a linear combination of $\mathbf{v}_1, \mathbf{v}_2, \ldots, \mathbf{v}_m$, the induction hypothesis (6.3.6) ensures that

$$\text{span}\{\mathbf{u}_1, \mathbf{u}_2, \ldots, \mathbf{u}_m\} \subseteq \text{span}\{\mathbf{v}_1, \mathbf{v}_2, \ldots, \mathbf{v}_m\}. \tag{6.3.8}$$

The vectors $\mathbf{u}_1, \mathbf{u}_2, \ldots, \mathbf{u}_m$ are orthonormal and hence they are linearly independent. Thus, $\dim \text{span}\{\mathbf{u}_1, \mathbf{u}_2, \ldots, \mathbf{u}_m\} = m$. The containment (6.3.8) is an equality since both spans are m-dimensional vector spaces; see Theorem 2.2.11. $\qquad\square$

The Gram–Schmidt process takes a linearly independent list $\mathbf{v}_1, \mathbf{v}_2, \ldots, \mathbf{v}_n$ and constructs orthonormal vectors $\mathbf{u}_1, \mathbf{u}_2, \ldots, \mathbf{u}_n$ using the following algorithm. Start by setting $\mathbf{x}_1 = \mathbf{v}_1$ and then normalize it to obtain $\mathbf{u}_1 = \frac{\mathbf{x}_1}{\|\mathbf{x}_1\|}$. Then, for each $k = 2, 3, \ldots, n$, compute

$$\mathbf{x}_k = \mathbf{v}_k - \langle \mathbf{v}_k, \mathbf{u}_1 \rangle \mathbf{u}_1 - \cdots - \langle \mathbf{v}_k, \mathbf{u}_{k-1} \rangle \mathbf{u}_{k-1} \tag{6.3.9}$$

and normalize it: $\mathbf{u}_k = \frac{\mathbf{x}_k}{\|\mathbf{x}_k\|}$.

What does the Gram–Schmidt process do to an orthogonal list of nonzero vectors?

Lemma 6.3.10 *Let V be an inner product space and suppose that $\mathbf{v}_1, \mathbf{v}_2, \ldots, \mathbf{v}_n \in V$ are nonzero and pairwise orthogonal. The Gram–Schmidt process constructs the orthonormal list*

$$\mathbf{u}_1 = \frac{\mathbf{v}_1}{\|\mathbf{v}_1\|}, \qquad \mathbf{u}_2 = \frac{\mathbf{v}_2}{\|\mathbf{v}_2\|}, \ldots, \qquad \mathbf{u}_n = \frac{\mathbf{v}_n}{\|\mathbf{v}_n\|}.$$

If $\mathbf{v}_1, \mathbf{v}_2, \ldots, \mathbf{v}_n$ are orthonormal, then each $\mathbf{u}_i = \mathbf{v}_i$.

Proof It suffices to show that the vectors (6.3.9) constructed by the Gram–Schmidt process are $\mathbf{x}_i = \mathbf{v}_i$. We proceed by induction. In the base case $n = 1$, we have $\mathbf{x}_1 = \mathbf{v}_1$. For the induction step, suppose that $\mathbf{x}_i = \mathbf{v}_i$ whenever $1 \le i \le m < n$. Then

$$\mathbf{x}_{m+1} = \mathbf{v}_{m+1} - \sum_{i=1}^{m} \langle \mathbf{v}_{m+1}, \mathbf{u}_i \rangle \mathbf{u}_i = \mathbf{v}_{m+1} - \sum_{i=1}^{m} \left\langle \mathbf{v}_{m+1}, \frac{\mathbf{v}_i}{\|\mathbf{v}_i\|} \right\rangle \frac{\mathbf{v}_i}{\|\mathbf{v}_i\|}$$

$$= \mathbf{v}_{m+1} - \sum_{i=1}^{m} \langle \mathbf{v}_{m+1}, \mathbf{v}_i \rangle \frac{\mathbf{v}_i}{\|\mathbf{v}_i\|^2} = \mathbf{v}_{m+1}. \qquad\square$$

Example 6.3.11 The polynomials $1, x, x^2, x^3, \ldots, x^n$ are linearly independent in $C[0, 1]$ for each $n = 1, 2, \ldots$ (see Example 1.6.12), but they are not orthonormal with respect to the L^2 inner product (6.1.6). For example, $\|x\| = \frac{1}{\sqrt{3}}$ and $\langle x, x^2 \rangle = \frac{1}{4}$. To construct an orthonormal list with the same span as $1, x, x^2, x^3, \ldots, x^n$, we label $v_1 = 1$, $v_2 = x$, $v_3 = x^2, \ldots$, apply the Gram–Schmidt process, and obtain the orthonormal sequence of polynomials

$$u_1 = 1, \quad u_2 = \sqrt{3}(2x - 1), \quad u_3 = \sqrt{5}(6x^2 - 6x + 1),$$
$$u_4 = \sqrt{7}(20x^3 - 30x^2 + 12x - 1), \ldots \tag{6.3.12}$$

This is how the orthogonal polynomials in Example 6.1.5 were constructed.

An important consequence of the Gram–Schmidt process is the following.

Corollary 6.3.13 *Every finite-dimensional inner product space has an orthonormal basis.*

Proof Start with any basis $\mathbf{v}_1, \mathbf{v}_2, \ldots, \mathbf{v}_n$ and apply the Gram–Schmidt process to obtain an orthonormal list $\mathbf{u}_1, \mathbf{u}_2, \ldots, \mathbf{u}_n$, which is linearly independent and has the same span as $\mathbf{v}_1, \mathbf{v}_2, \ldots, \mathbf{v}_n$. $\qquad\square$

Corollary 6.3.14 *Every orthonormal list in a finite-dimensional inner product space can be extended to an orthonormal basis.*

Proof Given an orthonormal list $\mathbf{v}_1, \mathbf{v}_2, \ldots, \mathbf{v}_r$, extend it to a basis $\mathbf{v}_1, \mathbf{v}_2, \ldots, \mathbf{v}_n$. Apply the Gram–Schmidt process to this basis and obtain an orthonormal basis $\mathbf{u}_1, \mathbf{u}_2, \ldots, \mathbf{u}_n$. Since $\mathbf{v}_1, \mathbf{v}_2, \ldots, \mathbf{v}_r$ are already orthonormal, Lemma 6.3.10 ensures that the Gram–Schmidt process leaves them unchanged, that is, $\mathbf{u}_i = \mathbf{v}_i$ for $i = 1, 2, \ldots, r$. □

6.4 The Riesz Representation Theorem

Since a finite-dimensional $\mathbb{F}$-inner product space $\mathcal{V}$ has an orthonormal basis, we can use (6.2.6) to obtain a remarkable representation for any linear transformation from $\mathcal{V}$ to $\mathbb{F}$.

Definition 6.4.1 Let $\mathcal{V}$ be an $\mathbb{F}$-vector space. A *linear functional* is a linear transformation $\phi : \mathcal{V} \to \mathbb{F}$.

Example 6.4.2 Let $\mathcal{V}$ be an $\mathbb{F}$-inner product space. If $\mathbf{w} \in \mathcal{V}$, then $\phi(\mathbf{v}) = \langle \mathbf{v}, \mathbf{w} \rangle$ defines a linear functional on $\mathcal{V}$.

Example 6.4.3 Let $\mathcal{V} = C[0, 1]$. Then $\phi(f) = f(\frac{1}{2})$ defines a linear functional on $\mathcal{V}$.

Theorem 6.4.4 (Riesz Representation Theorem) *Let $\mathcal{V}$ be a finite-dimensional $\mathbb{F}$-inner product space and let $\phi : \mathcal{V} \to \mathbb{F}$ be a linear functional.*

(a) *There is a unique $\mathbf{w} \in \mathcal{V}$ such that*

$$\phi(\mathbf{v}) = \langle \mathbf{v}, \mathbf{w} \rangle \quad \text{for all } \mathbf{v} \in \mathcal{V}. \tag{6.4.5}$$

(b) *Let $\mathbf{u}_1, \mathbf{u}_2, \ldots, \mathbf{u}_n$ be an orthonormal basis for $\mathcal{V}$. The vector $\mathbf{w}$ in (a) is*

$$\mathbf{w} = \overline{\phi(\mathbf{u}_1)}\mathbf{u}_1 + \overline{\phi(\mathbf{u}_2)}\mathbf{u}_2 + \cdots + \overline{\phi(\mathbf{u}_n)}\mathbf{u}_n. \tag{6.4.6}$$

Proof For any $\mathbf{v} \in \mathcal{V}$ and any orthonormal basis $\mathbf{u}_1, \mathbf{u}_2, \ldots, \mathbf{u}_n$ for $\mathcal{V}$, use (6.2.6) to compute

$$\phi(\mathbf{v}) = \phi\left(\sum_{i=1}^{n} \langle \mathbf{v}, \mathbf{u}_i \rangle \mathbf{u}_i\right) = \sum_{i=1}^{n} \phi\big(\langle \mathbf{v}, \mathbf{u}_i \rangle \mathbf{u}_i\big) = \sum_{i=1}^{n} \langle \mathbf{v}, \mathbf{u}_i \rangle \phi(\mathbf{u}_i)$$

$$= \sum_{i=1}^{n} \langle \mathbf{v}, \overline{\phi(\mathbf{u}_i)}\mathbf{u}_i \rangle = \left\langle \mathbf{v}, \sum_{i=1}^{n} \overline{\phi(\mathbf{u}_i)}\mathbf{u}_i \right\rangle.$$

Therefore, $\mathbf{w}$ in (6.4.6) satisfies (6.4.5). If $\mathbf{y} \in \mathcal{V}$ and $\phi(\mathbf{v}) = \langle \mathbf{v}, \mathbf{y} \rangle$ for all $\mathbf{v} \in \mathcal{V}$, then $\langle \mathbf{v}, \mathbf{y} \rangle = \langle \mathbf{v}, \mathbf{w} \rangle$ for all $\mathbf{v} \in \mathcal{V}$. Corollary 5.3.15 ensures that $\mathbf{y} = \mathbf{w}$. □

Definition 6.4.7 The vector $\mathbf{w}$ in (6.4.6) is the *Riesz vector* for the linear functional ϕ.

The formula (6.4.6) may give the impression that the Riesz vector for ϕ depends on the choice of orthonormal basis $\mathbf{u}_1, \mathbf{u}_2, \ldots, \mathbf{u}_n$, but it does not.

Example 6.4.8 Let $\mathbf{v} = [v_1\ v_2\ v_3]^T \in \mathbb{R}^3$. Consider the linear functional $\phi(\mathbf{v}) = v_1$. If we use the standard inner product and orthonormal basis, the Riesz vector (6.4.6) is

$$\mathbf{w} = \phi(\mathbf{e}_1)\mathbf{e}_1 + \phi(\mathbf{e}_2)\mathbf{e}_2 + \phi(\mathbf{e}_3)\mathbf{e}_3 = 1\mathbf{e}_1 + 0\mathbf{e}_2 + 0\mathbf{e}_3 = \mathbf{e}_1.$$

If we use the orthonormal basis in Example 6.2.9, the Riesz vector is

$$\mathbf{w} = \phi(\mathbf{u}_1)\mathbf{u}_1 + \phi(\mathbf{u}_2)\mathbf{u}_2 + \phi(\mathbf{u}_3)\mathbf{u}_3 = \frac{1}{3}\mathbf{u}_1 - \frac{2}{3}\mathbf{u}_2 - \frac{2}{3}\mathbf{u}_3$$

$$= \frac{1}{9}\begin{bmatrix} 1 \\ 2 \\ 2 \end{bmatrix} - \frac{2}{9}\begin{bmatrix} -2 \\ 2 \\ -1 \end{bmatrix} - \frac{2}{9}\begin{bmatrix} -2 \\ -1 \\ 2 \end{bmatrix} = \frac{1}{9}\begin{bmatrix} 9 \\ 0 \\ 0 \end{bmatrix} = \mathbf{e}_1.$$

Example 6.4.9 What is the Riesz vector for the linear functional $A \mapsto \operatorname{tr} A$ on M_n with the Frobenius inner product? The matrices E_{pq} defined in Example 2.2.3 are an orthonormal basis for M_n, so the Riesz vector (6.4.6) is

$$\sum_{p,q=1}^{n} \overline{(\operatorname{tr} E_{pq})} E_{pq} = \sum_{p,q=1}^{n} \delta_{pq} E_{pq} = \sum_{p=1}^{n} E_{pp} = I_n,$$

that is, $\operatorname{tr} A = \operatorname{tr} I^* A = \langle A, I \rangle_{\mathrm{F}}$ for each $A \in \mathsf{M}_n$.

Example 6.4.10 What is the Riesz vector for the linear functional $\phi(p) = p(\frac{1}{2})$ on $\mathcal{P}_2$ with the L^2 inner product (6.1.6)? The polynomials u_1, u_2, and u_3 in (6.3.12) are an orthonormal basis for $\mathcal{P}_2$, and we have

$$\phi(u_1) = 1, \qquad \phi(u_2) = u_2(\tfrac{1}{2}) = 0, \qquad \text{and} \qquad \phi(u_3) = u_3(\tfrac{1}{2}) = -\frac{\sqrt{5}}{2}.$$

The Riesz vector (6.4.6) is

$$w(x) = \overline{\phi(u_1)}u_1 + \overline{\phi(u_2)}u_2 + \overline{\phi(u_3)}u_3 = 1 - \frac{5}{2}(6x^2 - 6x + 1) = -15x^2 + 15x - \frac{3}{2}.$$

A computation reveals that

$$\int_0^1 1 w(x)\,dx = 1, \qquad \int_0^1 x w(x)\,dx = \frac{1}{2}, \qquad \text{and} \qquad \int_0^1 x^2 w(x)\,dx = \frac{1}{4}.$$

If $p(x) = ax^2 + bx + c \in \mathcal{P}_2$, then $\phi(p) = p(\frac{1}{2}) = \frac{a}{4} + \frac{b}{2} + c$ and

$$\langle p, w \rangle = \int_0^1 p(x)w(x)\,dx = a \int_0^1 x^2 w(x)\,dx + b \int_0^1 x w(x)\,dx + c \int_0^1 1 w(x)\,dx$$

$$= \frac{a}{4} + \frac{b}{2} + c = p(\tfrac{1}{2}).$$

6.5 Orthonormal Bases and Linear Transformations

Let $\mathcal{V}$ and $\mathcal{W}$ be finite-dimensional vector spaces over the same field $\mathbb{F}$ and let $T \in \mathcal{L}(\mathcal{V}, \mathcal{W})$. The following theorem provides a way to compute the β-γ matrix representation of T with respect to a basis $\beta = \mathbf{v}_1, \mathbf{v}_2, \ldots, \mathbf{v}_n$ of $\mathcal{V}$ and an orthonormal basis $\gamma = \mathbf{w}_1, \mathbf{w}_2, \ldots, \mathbf{w}_m$ of $\mathcal{W}$.

If $\mathbf{v} = a_1 \mathbf{v}_1 + a_2 \mathbf{v}_2 + \cdots + a_n \mathbf{v}_n$ is the unique representation of $\mathbf{v} \in \mathcal{V}$ as a linear combination of the vectors in β, then the β-coordinate vector of $\mathbf{v}$ is $[\mathbf{v}]_\beta = [a_1 \ a_2 \ \cdots \ a_n]^{\mathsf{T}}$. There is a coordinate vector for $\mathbf{v}$ associated with any basis. The coordinate vector for $\mathbf{v}$ associated with an orthonormal basis has the special form (6.2.8). What does this special form tell us about the matrix representation of a linear transformation?

The matrix that represents $T \in \mathcal{L}(\mathcal{V}, \mathcal{W})$ with respect to the bases β and γ is

$$_\gamma[T]_\beta = \left[[T\mathbf{v}_1]_\gamma \ [T\mathbf{v}_2]_\gamma \ \cdots \ [T\mathbf{v}_n]_\gamma \right] \in \mathsf{M}_{m \times n}(\mathbb{F}). \tag{6.5.1}$$

It is a consequence of linearity that

$$[T\mathbf{v}]_\gamma = {}_\gamma[T]_\beta[\mathbf{v}]_\beta \tag{6.5.2}$$

for each $\mathbf{v} \in \mathcal{V}$; see (2.4.22).

Theorem 6.5.3 *Let $\mathcal{V}$ and $\mathcal{W}$ be finite-dimensional vector spaces over the same field $\mathbb{F}$ and let $\langle \cdot, \cdot \rangle$ be an inner product on $\mathcal{W}$. Let $\beta = \mathbf{v}_1, \mathbf{v}_2, \ldots, \mathbf{v}_n$ be a basis for $\mathcal{V}$ and let $\gamma = \mathbf{w}_1, \mathbf{w}_2, \ldots, \mathbf{w}_m$ be an orthonormal basis for $\mathcal{W}$.*

(a) *If $T \in \mathcal{L}(\mathcal{V}, \mathcal{W})$, then*

$${}_\gamma[T]_\beta = [\langle T\mathbf{v}_j, \mathbf{w}_i \rangle] = \begin{bmatrix} \langle T\mathbf{v}_1, \mathbf{w}_1 \rangle & \langle T\mathbf{v}_2, \mathbf{w}_1 \rangle & \cdots & \langle T\mathbf{v}_n, \mathbf{w}_1 \rangle \\ \langle T\mathbf{v}_1, \mathbf{w}_2 \rangle & \langle T\mathbf{v}_2, \mathbf{w}_2 \rangle & \cdots & \langle T\mathbf{v}_n, \mathbf{w}_2 \rangle \\ \vdots & \vdots & \ddots & \vdots \\ \langle T\mathbf{v}_1, \mathbf{w}_m \rangle & \langle T\mathbf{v}_2, \mathbf{w}_m \rangle & \cdots & \langle T\mathbf{v}_n, \mathbf{w}_m \rangle \end{bmatrix} \in \mathsf{M}_{m \times n}(\mathbb{F}). \tag{6.5.4}$$

(b) *Let $A \in \mathsf{M}_{m \times n}(\mathbb{F})$. There is a unique $T \in \mathcal{L}(\mathcal{V}, \mathcal{W})$ such that ${}_\gamma[T]_\beta = A$.*

Proof (a) It suffices to show that each column of ${}_\gamma[T]_\beta$ has the asserted form. Since γ is an orthonormal basis for $\mathcal{W}$, Theorem 6.2.5 ensures that

$$T\mathbf{v}_j = \langle T\mathbf{v}_j, \mathbf{w}_1 \rangle \mathbf{w}_1 + \langle T\mathbf{v}_j, \mathbf{w}_2 \rangle \mathbf{w}_2 + \cdots + \langle T\mathbf{v}_j, \mathbf{w}_m \rangle \mathbf{w}_m,$$

and hence

$$[T\mathbf{v}_j]_\gamma = \begin{bmatrix} \langle T\mathbf{v}_j, \mathbf{w}_1 \rangle \\ \langle T\mathbf{v}_j, \mathbf{w}_2 \rangle \\ \vdots \\ \langle T\mathbf{v}_j, \mathbf{w}_m \rangle \end{bmatrix}.$$

(b) Let $A = [\mathbf{a}_1 \ \mathbf{a}_2 \ \cdots \ \mathbf{a}_n]$. Because a vector in $\mathcal{W}$ is uniquely determined by its γ-coordinate vector, we may define a function $T : \mathcal{V} \to \mathcal{W}$ by $[T\mathbf{v}]_\gamma = A[\mathbf{v}]_\beta$. Then

$$[T(c\mathbf{v} + \mathbf{u})]_\gamma = A[c\mathbf{v} + \mathbf{u}]_\beta = A(c[\mathbf{v}]_\beta + [\mathbf{u}]_\beta)$$
$$= cA[\mathbf{v}]_\beta + A[\mathbf{u}]_\beta = c[T\mathbf{v}]_\gamma + [T\mathbf{u}]_\gamma,$$

so T is a linear transformation. Since

$$[T\mathbf{v}_i]_\gamma = A[\mathbf{v}_i]_\beta = A\mathbf{e}_i = \mathbf{a}_i \qquad \text{for } i = 1, 2, \ldots, n,$$

we have ${}_\gamma[T]_\beta = A$. If $S \in \mathcal{L}(\mathcal{V}, \mathcal{W})$ and ${}_\gamma[S]_\beta = A$, then

$$[S\mathbf{v}_i]_\gamma = {}_\gamma[S]_\beta[\mathbf{v}_i]_\beta = A[\mathbf{v}_i]_\beta = [T\mathbf{v}_i]_\gamma$$

and hence $S\mathbf{v}_i = T\mathbf{v}_i$ for each $i = 1, 2, \ldots, n$. Thus, S and T are identical linear transformations because they agree on a basis. $\square$

6.6 Adjoints of Linear Transformations and Matrices

In this section, $\mathcal{V}$ and $\mathcal{W}$ are inner product spaces over the same field $\mathbb{F}$. We denote the respective inner products on $\mathcal{V}$ and $\mathcal{W}$ by $\langle \cdot, \cdot \rangle_\mathcal{V}$ and $\langle \cdot, \cdot \rangle_\mathcal{W}$.

Definition 6.6.1 A function $f : \mathcal{W} \to \mathcal{V}$ is an *adjoint* of a linear transformation $T : \mathcal{V} \to \mathcal{W}$ if $\langle T\mathbf{v}, \mathbf{w} \rangle_{\mathcal{W}} = \langle \mathbf{v}, f(\mathbf{w}) \rangle_{\mathcal{V}}$ for all $\mathbf{v} \in \mathcal{V}$ and all $\mathbf{w} \in \mathcal{W}$.

Adjoints arise in differential equations, integral equations, and functional analysis. Our first observation is that if an adjoint exists, then it is unique.

Lemma 6.6.2 *Let f and g be functions from $\mathcal{W}$ to $\mathcal{V}$. If $\langle \mathbf{v}, f(\mathbf{w}) \rangle_{\mathcal{V}} = \langle \mathbf{v}, g(\mathbf{w}) \rangle_{\mathcal{V}}$ for all $\mathbf{v} \in \mathcal{V}$ and all $\mathbf{w} \in \mathcal{W}$, then $f = g$. In particular, if $T \in \mathcal{L}(\mathcal{V}, \mathcal{W})$, then it has at most one adjoint.*

Proof Corollary 5.3.15 ensures that $f(\mathbf{w}) = g(\mathbf{w})$ for all $\mathbf{w} \in \mathcal{W}$. □

Since a linear transformation T has at most one adjoint, we refer to it as the adjoint of T and denote it by T^*. Our second observation is that if a linear transformation has an adjoint, then that adjoint is a linear transformation.

Theorem 6.6.3 *If $T \in \mathcal{L}(\mathcal{V}, \mathcal{W})$ has an adjoint $T^* : \mathcal{W} \to \mathcal{V}$, then T^* is a linear transformation.*

Proof Let $\mathbf{u}, \mathbf{w} \in \mathcal{W}$ and $c \in \mathbb{F}$. Then for all $\mathbf{v} \in \mathcal{V}$,

$$\langle \mathbf{v}, T^*(c\mathbf{u} + \mathbf{w}) \rangle = \langle T\mathbf{v}, c\mathbf{u} + \mathbf{w} \rangle = \overline{c}\langle T\mathbf{v}, \mathbf{u} \rangle + \langle T\mathbf{v}, \mathbf{w} \rangle$$

$$= \overline{c}\langle \mathbf{v}, T^*(\mathbf{u}) \rangle + \langle \mathbf{v}, T^*(\mathbf{w}) \rangle = \langle \mathbf{v}, cT^*(\mathbf{u}) \rangle + \langle \mathbf{v}, T^*(\mathbf{w}) \rangle$$

$$= \langle \mathbf{v}, cT^*(\mathbf{u}) + T^*(\mathbf{w}) \rangle.$$

Corollary 5.3.15 ensures that $T^*(c\mathbf{u} + \mathbf{w}) = cT^*(\mathbf{u}) + T^*(\mathbf{w})$. □

Example 6.6.4 Let $\mathcal{V} = \mathbb{F}^n$ and $\mathcal{W} = \mathbb{F}^m$, both with the standard inner product. Let $A \in \mathsf{M}_{m \times n}(\mathbb{F})$ and consider the linear transformation $T_A : \mathcal{V} \to \mathcal{W}$ induced by A; see Definition 2.4.13. For all $\mathbf{x} \in \mathcal{V}$ and $\mathbf{y} \in \mathcal{W}$,

$$\langle A\mathbf{x}, \mathbf{y} \rangle = \mathbf{y}^*(A\mathbf{x}) = (A^*\mathbf{y})^*\mathbf{x} = \langle \mathbf{x}, A^*\mathbf{y} \rangle; \tag{6.6.5}$$

see Example 5.3.3. We conclude that T_{A^*} is the adjoint of T_A. The terms *adjoint* and *conjugate transpose* of a matrix are synonyms.

Example 6.6.6 If $\mathbb{F} = \mathbb{C}$, $m = n = 1$, and $A = [a]$, the linear transformation $T_A : \mathbb{C} \to \mathbb{C}$ is $T_A(z) = az$. Its adjoint is $T_{A^*}(z) = \overline{a}z$.

Examples of matrices $A \in \mathsf{M}_n$ whose definitions involve adjoints are: normal matrices ($A^*A = AA^*$; see Chapter 14), Hermitian matrices ($A = A^*$; see Definition 6.6.9), and unitary matrices ($A^* = A^{-1}$; see Chapter 7).

Example 6.6.7 Let $\mathcal{V} = \mathcal{W} = \mathcal{P}$ be the inner product space of all polynomials, endowed with the L^2 inner product (6.1.6) on $[0, 1]$, and let $p \in \mathcal{P}$ be given. Consider $T \in \mathcal{L}(\mathcal{P})$ defined by $Tf = pf$ for each $f \in \mathcal{P}$. Then for all $g \in \mathcal{P}$,

$$\langle Tf, g \rangle = \int_0^1 p(t)f(t)\overline{g(t)}\,dt = \int_0^1 f(t)\overline{\overline{p(t)}g(t)}\,dt = \langle f, \overline{p}g \rangle. \tag{6.6.8}$$

Define the function $\Phi : \mathcal{P} \to \mathcal{P}$ by $\Phi g = \overline{p}g$. Then (6.6.8) says that $\langle Tf, g \rangle = \langle f, \Phi g \rangle$ for all $g \in \mathcal{P}$. We conclude that T has an adjoint, and $T^* = \Phi$. If p is a real polynomial, then $\overline{p} = p$ and $T^* = T$, and vice versa.

Definition 6.6.9 Suppose that T^* is the adjoint of $T \in \mathcal{L}(\mathcal{V})$. If $T^* = T$, then T is *self-adjoint*. A square matrix A is *Hermitian* if $A^* = A$. The terms Hermitian and selfadjoint are used interchangeably for linear transformations and matrices.

The linear transformation $f \mapsto pf$ in Example 6.6.7 is selfadjoint if and only if p is a real polynomial. In Example 6.6.4, the linear transformation $\mathbf{x} \mapsto A\mathbf{x}$ is selfadjoint if and only if $A = A^*$, that is, if and only if A is square and Hermitian. It makes no sense to speak of a selfadjoint operator $T \in \mathcal{L}(\mathcal{V}, \mathcal{W})$ unless $\mathcal{V} = \mathcal{W}$; nor does it make sense to speak of a Hermitian matrix $A \in \mathsf{M}_{m \times n}$ unless $m = n$.

Adjoints are unique when they exist, but existence cannot be taken for granted. The following example shows that some linear transformations do not have an adjoint.

Example 6.6.10 Let $\mathcal{P}$ be as in Example 6.6.7. Define $T \in \mathcal{L}(\mathcal{P})$ by $Tf = f'$, that is, the action of T on polynomials is differentiation. Suppose that T has an adjoint T^*, let g be the constant polynomial $g(t) = 1$, and let $T^*g = h$. Then for each $f \in \mathcal{P}$, the fundamental theorem of calculus ensures that

$$\langle Tf, g \rangle = \int_0^1 f'(t)\, dt = f(t)\Big|_0^1 = f(1) - f(0)$$

and therefore

$$f(1) - f(0) = \langle Tf, g \rangle = \langle f, T^*g \rangle = \langle f, h \rangle = \int_0^1 f(t)\overline{h(t)}\, dt.$$

Now let $f(t) = t^2(t-1)^2 h(t)$, so $f(0) = f(1) = 0$ and

$$0 = \int_0^1 f(t)\overline{h(t)}\, dt = \int_0^1 t^2(t-1)^2 |h(t)|^2\, dt = \|t(t-1)h(t)\|^2.$$

We conclude that $t(t-1)h(t)$ is the zero polynomial (see P.B.8), which implies that h is the zero polynomial. It follows that $p(1) - p(0) = \langle Tp, g \rangle = \langle p, T^*g \rangle = \langle p, h \rangle = 0$ for every $p \in \mathcal{P}$, which is false for $p(t) = t$. We conclude that T does not have an adjoint.

For an example of a linear functional that does not have an adjoint, see P.6.14. It is not an accident that our examples of linear transformations without adjoints involve infinite-dimensional inner product spaces: every linear transformation between finite-dimensional inner product spaces has an adjoint.

Theorem 6.6.11 *Let $\mathcal{V}$ and $\mathcal{W}$ be finite-dimensional inner product spaces over the same field $\mathbb{F}$ and let $T \in \mathcal{L}(\mathcal{V}, \mathcal{W})$. Let $\beta = \mathbf{v}_1, \mathbf{v}_2, \ldots, \mathbf{v}_n$ be an orthonormal basis for $\mathcal{V}$, let $\gamma = \mathbf{w}_1, \mathbf{w}_2, \ldots, \mathbf{w}_m$ be an orthonormal basis for $\mathcal{W}$, and let $_\gamma[T]_\beta = [\langle T\mathbf{v}_j, \mathbf{w}_i \rangle]$ be the β-γ matrix representation of T, as in (6.5.4).*

(a) *T has an adjoint.*

(b) *The β-γ matrix representation of T^* is $_\gamma[T]_\beta^*$, that is,*

$$_\beta[T^*]_\gamma = {}_\gamma[T]_\beta^*. \tag{6.6.12}$$

Proof (a) If $\mathbf{w} \in \mathcal{W}$, then $\phi_{\mathbf{w}}(\mathbf{v}) = \langle T\mathbf{v}, \mathbf{w} \rangle$ defines a linear functional on $\mathcal{V}$. Theorem 6.4.4 ensures that there is a unique vector $S(\mathbf{w}) \in \mathcal{V}$ such that $\langle T\mathbf{v}, \mathbf{w} \rangle = \langle \mathbf{v}, S(\mathbf{w}) \rangle$ for all $\mathbf{v} \in \mathcal{V}$. This construction defines a function $S : \mathcal{W} \to \mathcal{V}$ that is an adjoint of T, according to Definition 6.6.1. Theorem 6.6.3 ensures that $S \in \mathcal{L}(\mathcal{W}, \mathcal{V})$ and $S = T^*$.

(b) Theorem 6.5.3 and the definition of S tell us that

$$_\beta[T^*]_\gamma = _\beta[S]_\gamma = [\langle S\mathbf{w}_j, \mathbf{v}_i\rangle] = [\overline{\langle \mathbf{v}_i, S\mathbf{w}_j\rangle}] = [\overline{\langle T\mathbf{v}_i, \mathbf{w}_j\rangle}] = _\gamma[T]_\beta^*. \qquad \square$$

The identity (6.6.12) ensures that the β-γ matrix representation of the adjoint of a linear transformation T is the conjugate transpose of the β-γ matrix representation of T, if both bases involved are orthonormal. The restriction to orthonormal bases is essential; see P.6.22.

The notation "$*$" is used in two different ways in (6.6.12). In the expression $_\beta[T^*]_\gamma$ it indicates the adjoint of a linear transformation; in the expression $_\gamma[T]_\beta^*$ it indicates the conjugate transpose of a matrix. This abuse of notation is forgivable, as it reminds us that the conjugate transpose of an $m \times n$ matrix A represents the adjoint of the linear transformation $T_A : \mathbb{F}^m \to \mathbb{F}^n$.

The conjugate transpose operation on matrices shares with the inverse operation the property that it is *product reversing*: $(AB)^* = B^*A^*$ and $(AB)^{-1} = B^{-1}A^{-1}$. Do not let this common property lead you to confuse these two operations, which coincide only for unitary matrices; see Chapter 7. Other basic properties of the conjugate transpose are listed in Appendix C.2.

Example 6.6.13 Let V be a finite-dimensional $\mathbb{F}$-inner product space and let $\phi : V \to \mathbb{F}$ be a linear functional. The preceding theorem ensures that ϕ has an adjoint $\phi^* \in \mathcal{L}(\mathbb{F}, V)$. What is it? Theorem 6.4.4 says that there is a $\mathbf{w} \in V$ such that $\phi(\mathbf{v}) = \langle \mathbf{v}, \mathbf{w}\rangle_V$ for all $\mathbf{v} \in V$. Use the definition of the adjoint to compute

$$\langle \mathbf{v}, \phi^*(c)\rangle_V = \langle \phi(\mathbf{v}), c\rangle_{\mathbb{F}} = \overline{c}\phi(\mathbf{v}) = \overline{c}\langle \mathbf{v}, \mathbf{w}\rangle_V = \langle \mathbf{v}, c\mathbf{w}\rangle_V,$$

which is valid for all $\mathbf{v} \in V$. We conclude that $\phi^*(c) = c\mathbf{w}$ for all $c \in \mathbb{F}$, in which $\mathbf{w}$ is the Riesz vector for ϕ.

Example 6.6.14 Let $V = \mathsf{M}_n$, with the Frobenius inner product, and let $A \in \mathsf{M}_n$. What is the adjoint of the linear operator defined by $T(X) = AX$? Compute

$$\langle T(X), Y\rangle_{\mathrm{F}} = \langle AX, Y\rangle_{\mathrm{F}} = \operatorname{tr} Y^*AX = \operatorname{tr}((A^*Y)^*X) = \langle X, A^*Y\rangle_{\mathrm{F}}.$$

We conclude that $T^*(Y) = A^*Y$. For the linear transformation $S(X) = XA$, we have

$$\langle S(X), Y\rangle_{\mathrm{F}} = \langle XA, Y\rangle_{\mathrm{F}} = \operatorname{tr} Y^*XA = \operatorname{tr} AY^*X = \operatorname{tr}((YA^*)^*X) = \langle X, YA^*\rangle_{\mathrm{F}}$$

and hence $S^*(Y) = YA^*$.

6.7 Parseval's Identity and Bessel's Inequality

Parseval's identity is a generalization of (6.2.7), and it plays a role in the theory of Fourier series. It says that an inner product of two vectors in a finite-dimensional inner product space equals the standard inner product of their coordinate vectors with respect to an orthonormal basis.

Theorem 6.7.1 (Parseval's Identity) *Let $\beta = \mathbf{u}_1, \mathbf{u}_2, \ldots, \mathbf{u}_n$ be an orthonormal basis for an inner product space V with inner product $\langle \cdot, \cdot \rangle$. For all $\mathbf{v}, \mathbf{w} \in V$,*

$$\langle \mathbf{v}, \mathbf{w}\rangle = \sum_{i=1}^n \langle \mathbf{v}, \mathbf{u}_i\rangle \overline{\langle \mathbf{w}, \mathbf{u}_i\rangle} = \big\langle [\mathbf{v}]_\beta, [\mathbf{w}]_\beta\big\rangle_{\mathbb{F}^n}. \tag{6.7.2}$$

Proof Theorem 6.2.5 ensures that $\mathbf{v} = \sum_{i=1}^{n} \langle \mathbf{v}, \mathbf{u}_i \rangle \mathbf{u}_i$ and $\mathbf{w} = \sum_{j=1}^{n} \langle \mathbf{w}, \mathbf{u}_j \rangle \mathbf{u}_j$. Compute

$$\langle \mathbf{v}, \mathbf{w} \rangle = \left\langle \sum_{i=1}^{n} \langle \mathbf{v}, \mathbf{u}_i \rangle \mathbf{u}_i, \sum_{j=1}^{n} \langle \mathbf{w}, \mathbf{u}_j \rangle \mathbf{u}_j \right\rangle = \sum_{i,j=1}^{n} \langle \mathbf{v}, \mathbf{u}_i \rangle \overline{\langle \mathbf{w}, \mathbf{u}_j \rangle} \langle \mathbf{u}_i, \mathbf{u}_j \rangle$$

$$= \sum_{i,j=1}^{n} \langle \mathbf{v}, \mathbf{u}_i \rangle \overline{\langle \mathbf{w}, \mathbf{u}_j \rangle} \delta_{ij} = \sum_{i=1}^{n} \langle \mathbf{v}, \mathbf{u}_i \rangle \overline{\langle \mathbf{w}, \mathbf{u}_i \rangle} = \langle [\mathbf{v}]_\beta, [\mathbf{w}]_\beta \rangle_{\mathbb{F}^n}. \qquad \square$$

Corollary 6.7.3 *Let $\beta = \mathbf{u}_1, \mathbf{u}_2, \ldots, \mathbf{u}_n$ be an orthonormal basis for an inner product space $\mathcal{V}$ and let $T \in \mathcal{L}(\mathcal{V})$. For all $\mathbf{v}, \mathbf{w} \in \mathcal{V}$,*

$$\langle T\mathbf{v}, \mathbf{w} \rangle_{\mathcal{V}} = [\mathbf{w}]^*_\beta {}_\beta [T]_\beta [\mathbf{v}]_\beta = \langle {}_\beta [T]_\beta [\mathbf{v}]_\beta, [\mathbf{w}]_\beta \rangle_{\mathbb{F}^n}.$$

Proof The assertion follows from (6.7.2) and (6.5.2). $\qquad \square$

Bessel's inequality is another generalization of (6.2.7). It is valid for all inner product spaces, and it has special significance for infinite-dimensional spaces.

Theorem 6.7.4 (Bessel's Inequality) *Let $\mathbf{u}_1, \mathbf{u}_2, \ldots, \mathbf{u}_n$ be an orthonormal list in an inner product space $\mathcal{V}$. Then for every $\mathbf{v} \in \mathcal{V}$,*

$$\sum_{i=1}^{n} |\langle \mathbf{v}, \mathbf{u}_i \rangle|^2 \leq \|\mathbf{v}\|^2. \qquad (6.7.5)$$

Proof Use Theorem 6.1.8 and obtain (6.7.5) as follows:

$$0 \leq \left\| \mathbf{v} - \sum_{i=1}^{n} \langle \mathbf{v}, \mathbf{u}_i \rangle \mathbf{u}_i \right\|^2$$

$$= \|\mathbf{v}\|^2 - \left\langle \mathbf{v}, \sum_{i=1}^{n} \langle \mathbf{v}, \mathbf{u}_i \rangle \mathbf{u}_i \right\rangle - \left\langle \sum_{i=1}^{n} \langle \mathbf{v}, \mathbf{u}_i \rangle \mathbf{u}_i, \mathbf{v} \right\rangle + \left\| \sum_{i=1}^{n} \langle \mathbf{v}, \mathbf{u}_i \rangle \mathbf{u}_i \right\|^2$$

$$= \|\mathbf{v}\|^2 - 2\operatorname{Re} \left\langle \mathbf{v}, \sum_{i=1}^{n} \langle \mathbf{v}, \mathbf{u}_i \rangle \mathbf{u}_i \right\rangle + \sum_{i=1}^{n} |\langle \mathbf{v}, \mathbf{u}_i \rangle|^2$$

$$= \|\mathbf{v}\|^2 - 2\operatorname{Re} \sum_{i=1}^{n} \langle \mathbf{v}, \langle \mathbf{v}, \mathbf{u}_i \rangle \mathbf{u}_i \rangle + \sum_{i=1}^{n} |\langle \mathbf{v}, \mathbf{u}_i \rangle|^2$$

$$= \|\mathbf{v}\|^2 - 2\sum_{i=1}^{n} |\langle \mathbf{v}, \mathbf{u}_i \rangle|^2 + \sum_{i=1}^{n} |\langle \mathbf{v}, \mathbf{u}_i \rangle|^2$$

$$= \|\mathbf{v}\|^2 - \sum_{i=1}^{n} |\langle \mathbf{v}, \mathbf{u}_i \rangle|^2. \qquad \square$$

If $\mathcal{V}$ is infinite dimensional and $\mathbf{u}_1, \mathbf{u}_2, \ldots$ is an infinite orthonormal sequence in $\mathcal{V}$, then (6.7.5) (and the fact that a bounded monotone sequence of real numbers converges) implies that the infinite series $\sum_{i=1}^{\infty} |\langle \mathbf{v}, \mathbf{u}_i \rangle|^2$ is convergent for each $\mathbf{v} \in \mathcal{V}$. A weaker (but easier to prove) consequence of (6.7.5) is the following corollary.

Corollary 6.7.6 *Let* $\mathbf{u}_1, \mathbf{u}_2, \ldots$ *be an infinite orthonormal sequence in an inner product space* $\mathcal{V}$. *Then* $\lim_{n\to\infty} \langle \mathbf{v}, \mathbf{u}_n \rangle = 0$ *for each* $\mathbf{v} \in \mathcal{V}$.

Proof Let $\mathbf{v} \in \mathcal{V}$ and $\varepsilon > 0$ be given. It suffices to show that there are only finitely many indices $i = 1, 2, \ldots$ such that $|\langle \mathbf{v}, \mathbf{u}_i \rangle| \geq \varepsilon$. Let $N(n, \varepsilon, \mathbf{v})$ denote the number of indices $i \in \{1, 2, \ldots, n\}$ such that $|\langle \mathbf{v}, \mathbf{u}_i \rangle| \geq \varepsilon$. Then

$$\|\mathbf{v}\|^2 \geq \sum_{i=1}^{n} |\langle \mathbf{v}, \mathbf{u}_i \rangle|^2 \geq N(n, \varepsilon, \mathbf{v})\varepsilon^2,$$

so $N(n, \varepsilon, \mathbf{v}) \leq \frac{\|\mathbf{v}\|^2}{\varepsilon^2}$ for all $n = 1, 2, \ldots$. $\qquad\square$

6.8 Fourier Series

In this section, we use inner product spaces to discuss Fourier series, which attempt to represent (or approximate) periodic functions as linear combinations of sines and cosines.

Definition 6.8.1 A function $f : \mathbb{R} \to \mathbb{C}$ is *periodic* if there is a nonzero $\tau \in \mathbb{R}$ such that $f(x) = f(x + \tau)$ for all $x \in \mathbb{R}$; τ is a *period* of f.

If f is periodic with period τ and $n \in \mathbb{N}$, then

$$f(x) = f(x + \tau) = f(x + \tau + \tau) = f(x + 2\tau) = \cdots = f(x + n\tau)$$

and

$$f(x - n\tau) = f(x - n\tau + \tau) = f(x - (n-1)\tau) = \cdots = f(x - 2\tau) = f(x - \tau) = f(x).$$

Thus, $n\tau$ is a period for f for each nonzero $n \in \mathbb{Z}$.

Example 6.8.2 Since $\sin(x + 2\pi n) = \sin x$ for all $x \in \mathbb{R}$ and $n \in \mathbb{Z}$, the function $\sin x$ is periodic and $\tau = 2\pi n$ is a period for each nonzero $n \in \mathbb{Z}$.

Example 6.8.3 If $n \in \mathbb{Z}$ and $n \neq 0$, then $\sin nx = \sin(nx + 2\pi) = \sin(n(x + 2\pi/n))$. Therefore, $\sin nx$ is periodic with period $\tau = \frac{2\pi}{n}$. It is also periodic with period $-\frac{2\pi}{n}, \pm\frac{4\pi}{n}, \pm\frac{6\pi}{n}, \ldots$.

Periodic functions are ubiquitous in nature: anything that vibrates (a musical instrument), trembles (an earthquake), oscillates (a pendulum), or is connected to electromagnetism involves periodic functions.

The trigonometric functions $\sin nx$ and $\cos nx$ are familiar examples of periodic functions. They have period 2π, so their behavior everywhere is determined by their behavior on any real interval of length 2π. It is convenient to study them on the interval $[-\pi, \pi]$ and rescale the usual L^2 inner product on $[-\pi, \pi]$ so that $\sin nx$ and $\cos nx$ have unit norm; see Example 5.3.10. In the proof of the following lemma, we make repeated use of the facts $\sin n\pi = \sin(-n\pi)$ and $\cos n\pi = \cos(-n\pi)$ for $n \in \mathbb{Z}$.

Lemma 6.8.4 *With respect to the inner product*

$$\langle f, g \rangle = \frac{1}{\pi} \int_{-\pi}^{\pi} f(x)g(x)\, dx \qquad (6.8.5)$$

on $C_{\mathbb{R}}[-\pi, \pi]$, *the following functions are orthonormal:*

$$\frac{1}{\sqrt{2}}, \cos x, \cos 2x, \cos 3x, \ldots, \sin x, \sin 2x, \sin 3x, \ldots. \qquad (6.8.6)$$

Proof We begin by verifying that $\langle \frac{1}{\sqrt{2}}, \frac{1}{\sqrt{2}} \rangle = \frac{1}{\pi} \int_{-\pi}^{\pi} (\frac{1}{\sqrt{2}})^2 \, dx = 1$. If $n \neq 0$, then

$$\left\langle \frac{1}{\sqrt{2}}, \sin nx \right\rangle = \frac{1}{\pi\sqrt{2}} \int_{-\pi}^{\pi} \sin nx \, dx = -\left. \frac{1}{\pi n\sqrt{2}} \cos nx \right|_{-\pi}^{\pi} = 0.$$

A similar calculation shows that $\langle \frac{1}{\sqrt{2}}, \cos nx \rangle = 0$ for $n = 1, 2, \ldots$. The remaining assertions are that, if m and n are positive integers, then

$$\langle \sin mx, \cos nx \rangle = \frac{1}{\pi} \int_{-\pi}^{\pi} \sin mx \cos nx \, dx = 0, \tag{6.8.7}$$

$$\langle \sin mx, \sin nx \rangle = \frac{1}{\pi} \int_{-\pi}^{\pi} \sin mx \sin nx \, dx = \delta_{mn}, \quad \text{and} \tag{6.8.8}$$

$$\langle \cos mx, \cos nx \rangle = \frac{1}{\pi} \int_{-\pi}^{\pi} \cos mx \cos nx \, dx = \delta_{mn}. \tag{6.8.9}$$

(a) Equation (6.8.7) follows from the fact that $\sin mx \cos nx$ is an odd function and the interval $[-\pi, \pi]$ is symmetric with respect to 0.

(b) To verify (6.8.8), suppose that $m \neq n$ and compute

$$\begin{aligned}
\langle \sin mx, \sin nx \rangle &= \frac{1}{\pi} \int_{-\pi}^{\pi} \sin mx \sin nx \, dx \\
&= \frac{1}{\pi} \int_{-\pi}^{\pi} \frac{1}{2} [\cos(m-n)x - \cos(m+n)x] \, dx \\
&= \left. \frac{\sin(m-n)x}{2\pi(m-n)} - \frac{\sin(m+n)x}{2\pi(m+n)} \right|_{-\pi}^{\pi} = 0.
\end{aligned}$$

If $m = n$, then

$$\begin{aligned}
\langle \sin mx, \sin nx \rangle &= \frac{1}{\pi} \int_{-\pi}^{\pi} \sin^2 nx \, dx = \frac{1}{\pi} \int_{-\pi}^{\pi} \left(\frac{1 - \cos 2nx}{2} \right) dx \\
&= \frac{1}{2\pi} \int_{-\pi}^{\pi} dx - \frac{1}{2\pi} \int_{-\pi}^{\pi} \cos 2nx \, dx \\
&= 1 - \left. \frac{1}{4\pi n} \sin 2nx \right|_{-\pi}^{\pi} = 1.
\end{aligned}$$

(c) Equation (6.8.9) can be verified in a similar fashion. $\square$

The equations (6.8.7), (6.8.8), and (6.8.9) are the *orthonormality relations* for the sine and cosine functions. The fact that n in the expressions $\cos nx$ and $\sin nx$ is an integer is important; it ensures that these functions have period 2π.

It is convenient to label the orthonormal vectors (6.8.6) as follows:

$$u_0 = \frac{1}{\sqrt{2}}, \qquad u_n = \cos nx, \qquad u_{-n} = \sin nx, \quad \text{for } n = 1, 2, \ldots.$$

For $N = 1, 2, \ldots$, the $2N + 1$ vectors $u_{-N}, u_{-N+1}, \ldots, u_N$ form an orthonormal basis for the subspace $V_N = \operatorname{span}\{u_{-N}, u_{-N+1}, \ldots, u_N\}$. Consider

$$f = \frac{a_0}{\sqrt{2}} + \sum_{n=1}^{N} (a_n \cos nx + a_{-n} \sin nx) = \sum_{n=-N}^{N} a_n u_n.$$

Then $f \in C_{\mathbb{R}}[-\pi, \pi]$ and (6.2.6) ensures that the coefficients are

$$a_{\pm n} = \langle f, u_{\pm n} \rangle \quad \text{for } n = 0, 1, 2, \ldots. \tag{6.8.10}$$

These inner products are integrals, namely,

$$a_{-n} = \langle f, u_{-n} \rangle = \frac{1}{\pi} \int_{-\pi}^{\pi} f(x) \sin nx\, dx, \tag{6.8.11}$$

$$a_0 = \langle f, u_0 \rangle = \frac{1}{\pi} \int_{-\pi}^{\pi} \frac{f(x)}{\sqrt{2}}\, dx, \tag{6.8.12}$$

and

$$a_n = \langle f, u_n \rangle = \frac{1}{\pi} \int_{-\pi}^{\pi} f(x) \cos nx\, dx \tag{6.8.13}$$

for $1 \leq n \leq N$. These integrals are defined not only for $f \in V_N$, but also for any $f \in C_{\mathbb{R}}[-\pi, \pi]$. We might ask (as Fourier did) how the function

$$f_N = \sum_{n=-N}^{N} \langle f, u_n \rangle u_n \quad \text{for } f \in C_{\mathbb{R}}[-\pi, \pi] \tag{6.8.14}$$

is related to f. We know that $f_N = f$ if $f \in V_N$, so is f_N some sort of approximation to f if $f \notin V$? For an answer, see Example 8.4.5. What happens as $N \to \infty$?

Definition 6.8.15 Let $f : [-\pi, \pi] \to \mathbb{R}$ and suppose that the integrals (6.8.11), (6.8.12), (6.8.13) exist and are finite. The *Fourier series* associated with f is the infinite series

$$\frac{a_0}{\sqrt{2}} + \sum_{n=1}^{\infty} (a_n \cos nx + a_{-n} \sin nx). \tag{6.8.16}$$

If this series converges, it is a periodic function with period 2π. Does it converge to f? What does it mean for a series of functions to "converge"? What happens if f is not continuous but the integrals (6.8.10) are defined? Fourier was not able to answer these questions in his lifetime, and mathematicians have been working on them ever since. Their attempts to find answers have led to fundamental discoveries about sets, functions, measurement, integration, and convergence. The following definitions permit us to state a useful sufficient condition for the convergence of the series (6.8.16).

Definition 6.8.17 Let $-\infty < a < b < +\infty$ and let $f : (a, b) \to \mathbb{R}$. If $x \in [a, b)$, define

$$f(x^+) = \lim_{\substack{\varepsilon \to 0 \\ \varepsilon > 0}} f(x + \varepsilon),$$

and if $x \in (a, b]$, define

$$f(x^-) = \lim_{\substack{\varepsilon \to 0 \\ \varepsilon > 0}} f(x - \varepsilon),$$

provided that the respective one-sided limits exist and are finite. Then f is *piecewise* C^1 on $[a, b]$ if there are finitely many points $x_1, x_2, \ldots, x_n \in [a, b]$ such that

(a) $a = x_1 < x_2 < \cdots < x_n = b$;

(b) the derivative f' exists and is continuous on each open interval $(x_1, x_2), (x_2, x_3), \ldots, (x_{n-1}, x_n)$;

(c) finite one-sided limits $f(x_i^+)$ and $f'(x_i^+)$ exist for each $i = 1, 2, \ldots, n - 1$; and

(d) finite one-sided limits $f(x_i^-)$ and $f'(x_i^-)$ exist for each $i = 2, 3, \ldots, n$.

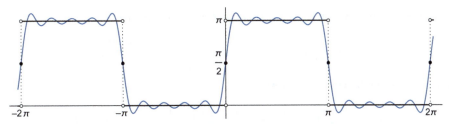

Figure 6.1 The graph of the function f from Example 6.8.19 and its Fourier approximation $\frac{\pi}{2} + 2\sin x + \frac{2}{3}\sin 3x + \frac{2}{5}\sin 5x + \frac{2}{7}\sin 7x + \frac{2}{9}\sin 9x$.

The differentiability of f on an interval (x_i, x_{i+1}) implies that f is continuous there. Notice that the values of f at the points $x_1, x_2, \ldots, x_n$ play no role in the definition. The following theorem provides answers to many questions about Fourier series; for a proof see [Bha05, 2.3.10].

Theorem 6.8.18 *Let* $f : [-\pi, \pi] \to \mathbb{R}$ *be piecewise* C^1 *on* $[-\pi, \pi]$. *Define* $f((-\pi)^-) = f(\pi^-)$ *and* $f(\pi^+) = f((-\pi)^+)$. *Then the Fourier series* (6.8.16) *converges to* $\frac{1}{2}(f(x^+) + f(x^-))$ *for each* $x \in [-\pi, \pi]$.

If f is piecewise C^1 on $[-\pi, \pi]$, if $f(-\pi) = f(\pi)$, and if f is continuous at a point $x \in [-\pi, \pi]$, then $f(x^-) = f(x^+) = f(x)$, so the Fourier series (6.8.16) converges to $f(x)$.

Example 6.8.19 Let $f : \mathbb{R} \to \mathbb{R}$ be periodic with period 2π and defined on $[-\pi, \pi]$ by

$$f(x) = \begin{cases} 0 & \text{if } -\pi < x < 0, \\ \pi & \text{if } 0 < x < \pi, \\ \frac{\pi}{2} & \text{if } x = 0 \text{ or } x = \pm\pi; \end{cases}$$

see Figure 6.1. Use (6.8.11), (6.8.12), and (6.8.13) to compute

$$a_0 = \left\langle f, \frac{1}{\sqrt{2}} \right\rangle = \frac{1}{\pi} \int_{-\pi}^{\pi} \frac{f(x)\,dx}{\sqrt{2}} = \frac{1}{\pi} \int_0^{\pi} \frac{\pi}{\sqrt{2}}\,dx = \frac{\pi}{\sqrt{2}},$$

$$a_n = \langle f, \cos nx \rangle = \frac{1}{\pi} \int_{-\pi}^{\pi} f(x) \cos nx\,dx = \frac{1}{\pi} \int_0^{\pi} \pi \cos nx\,dx = \int_0^{\pi} \cos nx = 0,$$

and

$$a_{-n} = \langle f, \sin nx \rangle = \frac{1}{\pi} \int_0^{\pi} \pi \sin nx\,dx = \frac{1}{n}(1 - \cos n\pi) = \begin{cases} 0 & \text{if } n \text{ is even,} \\ \frac{2}{n} & \text{if } n \text{ is odd,} \end{cases}$$

for $n = 1, 2, 3, \ldots$.

Let $x_1 = -\pi$, $x_2 = 0$, and $x_3 = \pi$. Then $f' = 0$ on (x_1, x_2) and (x_2, x_3),

$$f(x_1^+) = f(x_2^-) = f(x_3^+) = 0, \qquad f(x_1^-) = f(x_2^+) = f(x_3^-) = \pi,$$

and

$$f'(x_1^+) = f'(x_2^-) = f'(x_2^+) = f'(x_3^-) = 0.$$

Moreover, $f(x) = \frac{1}{2}(f(x^-) + f(x^+))$ for all $x \in [-\pi, \pi]$.

Since f is piecewise C^1 on $[-\pi, \pi]$, Theorem 6.8.18 ensures that

$$f(x) = \frac{a_0}{\sqrt{2}} + \sum_{n=1}^{\infty} (a_n \cos nx + a_{-n} \sin nx) = \frac{\pi}{2} + \sum_{n=1}^{\infty} \frac{2 \sin(2n-1)x}{2n-1}$$

for all $x \in [-\pi, \pi]$. Since $f(\frac{\pi}{2}) = \pi$, we have

$$\pi = \frac{\pi}{2} + 2 \sum_{n=1}^{\infty} \frac{(-1)^{n+1}}{(2n-1)} = \frac{\pi}{2} + 2 \left(1 - \frac{1}{3} + \frac{1}{5} - \frac{1}{7} + \cdots \right),$$

which implies Leibniz's 1674 discovery that

$$1 - \frac{1}{3} + \frac{1}{5} - \frac{1}{7} + \cdots = \frac{\pi}{4}.$$

6.9 Orthogonal Polynomial Bases and Gaussian Quadrature

In this section, we consider a finite real interval $[a, b]$ $(-\infty < a < b < +\infty)$ and the real inner product space of real polynomials with the L^2 inner product

$$\langle p, q \rangle = \int_a^b p(x)q(x)\, dx. \tag{6.9.1}$$

Definition 6.9.2 The *orthogonal polynomials* associated with $[a, b]$ and the inner product (6.9.1) are the unique sequence of real polynomials $p_0, p_1, p_2, \ldots$ obtained by applying the Gram–Schmidt process (Theorem 6.3.5) to the linearly independent polynomials $1, x, x^2, \ldots$.

Each orthogonal polynomial p_n has degree n, and Theorem 6.3.5 ensures that $\mathrm{span}\{1, x, \ldots, x^n\} = \mathrm{span}\{p_0, p_1, \ldots, p_n\}$ for each $n = 0, 1, 2, \ldots$.

Example 6.9.3 The lowest-degree orthogonal polynomial on $[a, b]$ is $p_0 = 1/\|1\| = 1/\sqrt{b-a}$. The next one is obtained by normalizing

$$v_1 = x - \langle x, p_0 \rangle p_0 = x - \left(\int_a^b \frac{x}{\sqrt{b-a}}\, dx \right) \frac{1}{\sqrt{b-a}} = x - \frac{1}{2}(a+b).$$

We have

$$\|v_1\|^2 = \int_a^b \left(x - \frac{1}{2}(a+b) \right)^2 dx = \int_{-\frac{b-a}{2}}^{\frac{b-a}{2}} u^2\, du = \frac{1}{12}(b-a)^3$$

and hence

$$p_1 = \frac{v_1}{\|v_1\|} = \frac{2\sqrt{3}}{(b-a)^{3/2}} \left(x - \frac{1}{2}(a+b) \right).$$

If $a = 0$ and $b = 1$, we obtain the polynomials $p_0 = 1$ and $p_1 = \sqrt{3}(2x-1)$ in Example 6.3.11.

Since $\langle p_k, p_n \rangle = 0$ for each $k = 0, 1, \ldots, n-1$, it follows that $\langle q, p_n \rangle = 0$ for every real polynomial q of degree less than n. In particular, $\langle 1, p_n \rangle = 0$ if $n \geq 1$, that is,

$$\int_a^b p_n(x)\, dx = 0$$

for each $n = 1, 2, \ldots$. If no zero of p_n in (a, b) has odd multiplicity (see Appendix B.3), then p_n does not change sign in $[a, b]$ and hence its integral is zero only if $p_n(x) = 0$ for all $x \in [a, b]$.

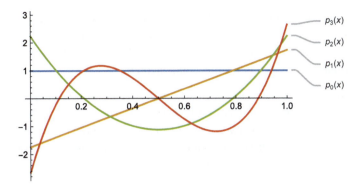

Figure 6.2 The polynomials of p_0, p_1, p_2, p_3 from Example 6.9.4 on $[0, 1]$.

This is not possible for a polynomial of positive degree (see Appendix B.4), so it follows that p_n has at least one zero in (a, b) with odd multiplicity. Since p_n is a polynomial of degree $n \geq 1$, it has no more than n zeros.

Example 6.9.4 In Example 6.3.11 we found the first four orthonormal polynomials if $a = 0$ and $b = 1$. Their zeros (rounded to four decimal places) are shown in the following table:

Polynomial	Zeros (rounded)	Zeros (exact)
$p_0(x) = 1$	none	none
$p_1(x) = \sqrt{3}(2x - 1)$	0.5000	$\frac{1}{2}$
$p_2(x) = \sqrt{5}(6x^2 - 6x + 1)$	0.2113, 0.7887	$\frac{1}{6}(3 \pm \sqrt{3})$
$p_3(x) = \sqrt{7}(20x^3 - 30x^2 + 12x - 1)$	0.1127, 0.5000, 0.8873	$\frac{1}{2}, \frac{1}{10}(5 \pm \sqrt{15})$

All the zeros are in $(0, 1)$, and each has multiplicity one; see Figure 6.2.

Theorem 6.9.5 *Let $p_0, p_1, p_2, \ldots$ be the orthogonal polynomials associated with $[a, b]$ and the inner product (6.9.1). If $n \geq 1$, then every zero of p_n has multiplicity one and is in (a, b).*

Proof Suppose that there are exactly k distinct points in (a, b) that are zeros of p_n with odd multiplicity, and denote them by $x_1, x_2, \ldots, x_k$. We know that $1 \leq k \leq n$. Suppose that $k < n$ and let $q(x) = (x - x_1)(x - x_2) \cdots (x - x_k)$. Every zero of $p_n q$ in (a, b) has even multiplicity, so $p_n q$ does not change sign in $[a, b]$. It follows that

$$\langle p_n, q \rangle = \int_a^b p_n(x) q(x) \, dx \neq 0.$$

However, q is a polynomial of degree less than n, so $\langle p_n, q \rangle = 0$. This contradiction shows that $k = n$. Therefore, every zero of p_n is in (a, b) and these zeros are distinct, so each has multiplicity one. $\square$

In Section 2.7, we discussed n-point rules (2.7.16), which provide numerical approximations to a definite integral $\int_a^b f(x) \, dx$. The basic idea of an n-point rule is to interpolate f by a polynomial f_n of degree $n - 1$ that is a linear combination of Lagrange basis functions at n distinct points in $[a, b]$, and then integrate f_n; see (2.7.14) and (2.7.16). If f is a polynomial of degree $n - 1$ or less, then $f_n = f$ and consequently an n-point rule gives the exact value of the definite integral of a polynomial of degree $n - 1$ or less, for any choice of the interpolation points. The trapezoidal rule ($n = 2$) and Simpson's rule ($n = 3$) are n-point rules in which the interpolation points are equally spaced. We now describe a special choice of n interpolation points for which the corresponding n-point rule gives the exact value of the integral for every polynomial of degree $2n - 1$ or less. We make essential use of the facts about the zeros of orthogonal polynomials revealed in the preceding theorem.

Theorem 6.9.6 (Gaussian Quadrature) *Let $p_0, p_1, p_2, \ldots$ be the orthogonal polynomials associated with $[a, b]$ and the inner product (6.9.1). Let $n \geq 1$, let $x_1, x_2, \ldots, x_n$ be the zeros of p_n, let f be a continuous real-valued function on $[a, b]$, and let $f_n = f(x_1)\ell_1 + f(x_2)\ell_2 + \cdots + f(x_n)\ell_n$, in which $\ell_1, \ell_2, \ldots, \ell_n$ is the Lagrange basis for $\mathcal{P}_{n-1}$ associated with the nodes $x_1, x_2, \ldots, x_n \in (a, b)$. If f is a polynomial of degree $2n - 1$ or less, then*

$$\int_a^b f(x)\,dx = \int_a^b f_n(x)\,dx = \sum_{k=1}^n f(x_k) \underbrace{\int_a^b \ell_k(x)\,dx}_{w_k} = \sum_{k=1}^n w_k f(x_k). \qquad (6.9.7)$$

Moreover, the weights w_k are positive and $w_1 + w_2 + \cdots + w_n = b - a$.

Proof If f is a real polynomial of degree $2n - 1$ or less, then there are unique real polynomials $q, r \in \mathcal{P}_{n-1}$ such that $f = q p_n + r$; see Appendix B.2. We have $\langle q, p_n \rangle = 0$ and

$$f(x_k) = q(x_k)p_n(x_k) + r(x_k) = q(x_k)0 + r(x_k) = r(x_k)$$

for each $k = 1, 2, \ldots, n$. Therefore,

$$\int_a^b f(x)\,dx = \int_a^b (q(x)p_n(x) + r(x))\,dx = \langle q, p_n \rangle + \int_a^b r(x)\,dx = \int_a^b r(x)\,dx$$

$$= \sum_{k=1}^n r(x_k) \int_a^b \ell_k(x)\,dx = \sum_{k=1}^n f(x_k)w_k,$$

in which we have used the fact that any n-point rule gives the exact value for $\int_a^b r(x)\,dx$. Each Lagrange basis polynomial $\ell_k(x)$ has degree $n - 1$, so $\ell_k^2(x) \in \mathcal{P}_{2n-2}$ and (6.9.7) ensures that

$$0 < \int_a^b \ell_k^2(x)\,dx = \sum_{i=1}^n \ell_k^2(x_i)w_i = \sum_{i=1}^n \delta_{ki}w_i = w_k.$$

Finally, $b - a = \int_0^1 1\,dx = \sum_{i=1}^n 1 w_i = \sum_{i=1}^n w_i$. $\qquad \square$

The approximation

$$\int_a^b f(x)\,dx \approx w_1 f(x_1) + w_2 f(x_2) + \cdots + w_n f(x_n) \qquad (6.9.8)$$

is an *n-point Gaussian rule*. The positivity of the Gaussian weights w_k for any number of nodes helps n-point Gaussian rules to be more suitable for computation than n-point Newton–Cotes rules when n is large. Some Newton–Cotes weights can be negative if $n \geq 7$.

Example 6.9.9 The zeros of the orthogonal polynomial $p_3(x)$ on $[0, 1]$ with the inner product (6.9.1) are

$$x_1 = \frac{1}{10}(5 - \sqrt{15}), \qquad x_2 = \frac{1}{2}, \quad \text{and} \quad x_3 = \frac{1}{10}(5 + \sqrt{15}).$$

The Lagrange basis polynomials for these nodes are

$$\ell_1(x) = \frac{10}{3}x^2 - \frac{1}{3}(\sqrt{15} + 10)x + \frac{1}{6}(5 + \sqrt{15}),$$

$$\ell_2(x) = -\frac{20}{3}x^2 + \frac{20}{3}x - \frac{2}{3}, \quad \text{and}$$

$$\ell_3(x) = \frac{10}{3}x^2 + \frac{1}{3}(\sqrt{15} - 10)x + \frac{1}{6}(5 - \sqrt{15}),$$

and the 3-point Gaussian rule weights are

$$w_1 = \int_0^1 \ell_1(x)\,dx = \frac{5}{18}, \quad w_2 = \int_0^1 \ell_2(x)\,dx = \frac{4}{9}, \quad \text{and} \quad w_3 = \int_0^1 \ell_3(x)\,dx = \frac{5}{18}.$$

Therefore, the 3-point Gaussian rule (6.9.8) on $[0, 1]$ is

$$\int_0^1 f(x)\,dx \approx \frac{5}{18}f(x_1) + \frac{4}{9}f(x_2) + \frac{5}{18}f(x_3). \tag{6.9.10}$$

Theorem 6.9.6 ensures that this rule gives the exact value of the integral whenever f is a polynomial of degree five or less.

Example 6.9.11 For the equally spaced nodes $x_1 = 0$, $x_2 = \frac{1}{2}$, and $x_3 = 1$ in $[0, 1]$, the Lagrange basis functions are

$$\ell_1(x) = 2x^2 - 3x + 1, \quad \ell_2(x) = -4x^2 + 4x, \quad \text{and} \quad \ell_3(x) = 2x^2 - x.$$

The weights for a 3-point Newton–Cotes rule on $[0, 1]$ are

$$w_1 = \int_0^1 \ell_1(x)\,dx = \frac{1}{6}, \quad w_2 = \int_0^1 \ell_2(x)\,dx = \frac{2}{3}, \quad \text{and} \quad w_3 = \int_0^1 \ell_3(x)\,dx = \frac{1}{6}.$$

Therefore, the 3-point rule with equally spaced nodes on $[0, 1]$ is

$$\int_0^1 f(x)\,dx \approx \frac{1}{6}f(0) + \frac{2}{3}f(\tfrac{1}{2}) + \frac{1}{6}f(1), \tag{6.9.12}$$

which is Simpson's rule.

Example 6.9.13 Let $f(x) = 128x^4 - 256x^3 + 160x^2 - 32x + 1$. The exact value of $\int_0^1 f(x)\,dx$ is $-1/15$, which is the value produced by the 3-point Gaussian rule (6.9.10). However, Simpson's 3-point rule (6.9.12) gives the value 1. The polynomial (2.7.14) that interpolates f at the three Gaussian nodes is $g(x) = (-64x^2 + 64x - 11)/5$; the polynomial (2.7.14) that interpolates f at the three Simpson nodes is $h(x) = 1$; see Figure 6.3.

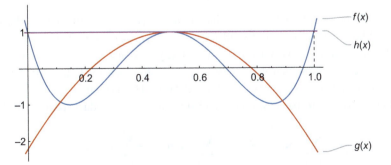

Figure 6.3 The function f in Example 6.9.13 and two 3-point interpolants.

6.10 Problems

P.6.1 Show that

$$
\mathbf{u}_1 = \frac{1}{\sqrt{2}} \begin{bmatrix} 1 \\ 1 \\ 0 \end{bmatrix}, \qquad
\mathbf{u}_2 = \frac{1}{\sqrt{3}} \begin{bmatrix} 1 \\ -1 \\ 1 \end{bmatrix}, \qquad
\mathbf{u}_3 = \frac{1}{\sqrt{6}} \begin{bmatrix} -1 \\ 1 \\ 2 \end{bmatrix},
$$

form an orthonormal basis for $\mathbb{R}^3$ and use Theorem 6.2.5 to find scalars a_1, a_2, a_3 such that $a_1\mathbf{u}_1 + a_2\mathbf{u}_2 + a_3\mathbf{u}_3 = [2\ 1\ 3]^\mathsf{T}$. Do not solve a 3×3 system of equations!

P.6.2 Let β be the orthonormal basis for M_n comprised of the matrices E_{pq} defined in Example 2.2.3. (a) Show that $E_{pq}E_{rs} = \delta_{qr}E_{ps}$ for all $p, q, r, s \in \{1, 2, \ldots, n\}$. (b) Use (a) to show that $\operatorname{tr} E_{pq}E_{rs} = \operatorname{tr} E_{rs}E_{pq}$ for all $p, q, r, s \in \{1, 2, \ldots, n\}$. (c) Use (b) to show that $\operatorname{tr} AB = \operatorname{tr} BA$ for all $A, B \in \mathsf{M}_n$.

P.6.3 Let $\mathcal{V} = \mathbb{R}^3$ and $\mathcal{W} = \mathbb{R}^2$, each with the standard inner product. Let $\beta = \mathbf{v}_1, \mathbf{v}_2, \mathbf{v}_3$, in which $\mathbf{v}_1 = [2\ 3\ 5]^\mathsf{T}$, $\mathbf{v}_2 = [7\ 11\ 13]^\mathsf{T}$, and $\mathbf{v}_3 = [17\ 19\ 23]^\mathsf{T}$. Let $\gamma = \mathbf{u}_1, \mathbf{u}_2$, in which $\mathbf{u}_1 = [\frac{1}{\sqrt{2}}\ \frac{1}{\sqrt{2}}]^\mathsf{T}$ and $\mathbf{u}_2 = [-\frac{1}{\sqrt{2}}\ \frac{1}{\sqrt{2}}]^\mathsf{T}$. Let $T \in \mathcal{L}(\mathcal{V}, \mathcal{W})$ be such that $T\mathbf{v}_1 = \begin{bmatrix} 2 \\ 3 \end{bmatrix}$, $T\mathbf{v}_2 = \begin{bmatrix} 7 \\ 11 \end{bmatrix}$, and $T\mathbf{v}_3 = \begin{bmatrix} 17 \\ 19 \end{bmatrix}$. Show that

$$
_\gamma[T]_\beta = \begin{bmatrix} \frac{5}{\sqrt{2}} & 9\sqrt{2} & 18\sqrt{2} \\ \frac{1}{\sqrt{2}} & 2\sqrt{2} & \sqrt{2} \end{bmatrix}.
$$

P.6.4 What does the Gram–Schmidt process do to a linearly dependent list of nonzero vectors $\mathbf{v}_1, \mathbf{v}_2, \ldots, \mathbf{v}_n$ in an inner product space $\mathcal{V}$? Suppose that $q \in \{2, 3, \ldots, n\}$, $\mathbf{v}_1, \mathbf{v}_2, \ldots, \mathbf{v}_{q-1}$ are linearly independent, and $\mathbf{v}_1, \mathbf{v}_2, \ldots, \mathbf{v}_q$ are linearly dependent. (a) Show that the orthogonal vectors $\mathbf{x}_k$ in (6.3.9) can be computed for $k = 2, 3, \ldots, q - 1$, but $\mathbf{x}_q = \mathbf{0}$, so the Gram–Schmidt algorithm cannot proceed. Why not? (b) Describe how to compute the coefficients in a linear combination $c_1\mathbf{v}_1 + c_2\mathbf{v}_2 + \cdots + c_{q-1}\mathbf{v}_{q-1} + \mathbf{v}_q = \mathbf{0}$. Why are these coefficients unique?

P.6.5 Let $A = [\mathbf{a}_1\ \mathbf{a}_2\ \ldots\ \mathbf{a}_n] \in \mathsf{M}_n(\mathbb{F})$ be invertible. Let $\mathcal{V} = \mathbb{F}^n$ with the inner product $\langle \mathbf{u}, \mathbf{v} \rangle_{A^{-1}} = \langle A^{-1}\mathbf{u}, A^{-1}\mathbf{v} \rangle$; see P.5.16. Let $A_i(\mathbf{u}) = [\mathbf{a}_1\ \ldots\ \mathbf{a}_{i-1}\ \mathbf{u}\ \mathbf{a}_{i+1}\ \ldots\ \mathbf{a}_n]$ denote the matrix obtained by replacing the ith column of A with $\mathbf{u} \in \mathbb{F}^n$. Let $\phi_i : \mathcal{V} \to \mathbb{F}$ be the linear functional

$$
\phi_i(\mathbf{u}) = \frac{\det A_i(\mathbf{u})}{\det A} \quad \text{for } i = 1, 2, \ldots, n;
$$

see Theorem 3.1.25 (a) Show that $\mathbf{a}_1, \mathbf{a}_2, \ldots, \mathbf{a}_n$ is an orthonormal basis for $\mathcal{V}$. (b) Show that $\phi_i(\mathbf{a}_j) = \delta_{ij}$ for all $i, j = 1, 2, \ldots, n$. (c) For each $i = 1, 2, \ldots, n$, show that $\mathbf{a}_i$ is the Riesz vector for ϕ_i. (d) Let $\mathbf{x}, \mathbf{y} \in \mathcal{V}$ and $\mathbf{x} = [x_1\ x_2\ \ldots\ x_n]^\mathsf{T}$. Suppose that $A\mathbf{x} = \mathbf{y}$. Show that $\phi_i(\mathbf{y}) = \langle \mathbf{y}, \mathbf{a}_i \rangle_{A^{-1}} = x_i$ for each $i = 1, 2, \ldots, n$. Hint: (6.2.6). (e) Conclude that $\mathbf{x} = [\phi_1(\mathbf{y})\ \phi_2(\mathbf{y})\ \ldots\ \phi_n(\mathbf{y})]^\mathsf{T}$. This is Cramer's rule.

P.6.6 Let $\mathbf{u}_1, \mathbf{u}_2, \ldots, \mathbf{u}_n$ be an orthonormal list in an $\mathbb{F}$-inner product space $\mathcal{V}$. (a) Prove that

$$
\left\| \mathbf{v} - \sum_{i=1}^{n} a_i\mathbf{u}_i \right\|^2 = \|\mathbf{v}\|^2 - \sum_{i=1}^{n} |\langle \mathbf{v}, \mathbf{u}_i \rangle|^2 + \sum_{i=1}^{n} |\langle \mathbf{v}, \mathbf{u}_i \rangle - a_i|^2 \tag{6.10.1}
$$

for all $a_1, a_2, \ldots, a_n \in \mathbb{F}$ and all $\mathbf{v} \in \mathcal{V}$. (b) Let $\mathbf{v} \in \mathcal{V}$ be given and let $\mathcal{W} = \text{span}\{\mathbf{u}_1, \mathbf{u}_2, \ldots, \mathbf{u}_n\}$. Prove that there is a unique $\mathbf{x} \in \mathcal{W}$ such that $\|\mathbf{v} - \mathbf{x}\| \leq \|\mathbf{v} - \mathbf{w}\|$ for all $\mathbf{w} \in \mathcal{W}$. Why is $\mathbf{x} = \sum_{i=1}^{n} \langle \mathbf{v}, \mathbf{u}_i \rangle \mathbf{u}_i$? This vector $\mathbf{x}$ is the orthogonal projection of $\mathbf{v}$ onto the subspace $\mathcal{W}$; see Section 8.3. (c) Deduce Bessel's inequality (6.7.5) from (6.10.1).

P.6.7 Let $\mathbf{u}_1, \mathbf{u}_2, \ldots, \mathbf{u}_n$ be an orthonormal list in an $\mathbb{F}$-inner product space $\mathcal{V}$, let $\mathcal{U} = \text{span}\{\mathbf{u}_1, \mathbf{u}_2, \ldots, \mathbf{u}_n\}$, and let $\mathbf{v} \in \mathcal{V}$. Provide details for the following approach to Bessel's inequality (6.7.5): (a) If $\mathbf{v} \in \mathcal{U}$, why is $\|\mathbf{v}\|^2 = \sum_{i=1}^{n} |\langle \mathbf{v}, \mathbf{u}_i \rangle|^2$? (b) Suppose that $\mathbf{v} \notin \mathcal{U}$, let $\mathcal{W} = \text{span}\{\mathbf{u}_1, \mathbf{u}_2, \ldots, \mathbf{u}_n, \mathbf{v}\}$, and apply the Gram–Schmidt process to the linearly independent list $\mathbf{u}_1, \mathbf{u}_2, \ldots, \mathbf{u}_n, \mathbf{v}$. Why do you obtain an orthonormal basis of the form $\mathbf{u}_1, \mathbf{u}_2, \ldots, \mathbf{u}_n, \mathbf{u}_{n+1}$ for $\mathcal{W}$? (c) Why is $\|\mathbf{v}\|^2 = \sum_{i=1}^{n+1} |\langle \mathbf{v}, \mathbf{u}_i \rangle|^2 \geq \sum_{i=1}^{n} |\langle \mathbf{v}, \mathbf{u}_i \rangle|^2$?

P.6.8 Deduce the Cauchy–Schwarz inequality from Bessel's inequality.

P.6.9 Let $A, B \in \mathsf{M}_n$. Define $T \in \mathcal{L}(\mathsf{M}_n)$ by $T(X) = AXB$. Find the adjoint of T with respect to the Frobenius inner product.

P.6.10 Show that the functions $1, e^{\pm ix}, e^{\pm 2ix}, e^{\pm 3ix}, \ldots$ are orthonormal in the inner product space $C[-\pi, \pi]$ endowed with the inner product $\langle f, g \rangle = \frac{1}{2\pi} \int_{-\pi}^{\pi} f(x)\overline{g(x)} \, dx$.

P.6.11 Let $A \in \mathsf{M}_n$ be invertible. Show that $(A^{-1})^* = (A^*)^{-1}$. *Hint*: Compute $(A^{-1}A)^*$.

P.6.12 Let $\mathcal{V} = C[0, 1]$ with the L^2 inner product (6.1.6), and let T be the *Volterra operator* on $\mathcal{V}$ defined by

$$(Tf)(t) = \int_0^t f(s) \, ds.$$

Show that the adjoint of T is the linear operator on $\mathcal{V}$ defined by

$$(T^*g)(s) = \int_s^1 g(t) \, dt.$$

P.6.13 Let $K(s, t)$ be a continuous complex-valued function on $[0, 1] \times [0, 1]$ and let $\mathcal{V} = C[0, 1]$ with the L^2 inner product (6.1.6). Define $(Tf)(t) = \int_0^1 K(s, t)f(s) \, ds$. Why does T have an adjoint and what is it? What would you have to assume to make T selfadjoint?

P.6.14 Let $\mathcal{P}$ be the complex inner product space of polynomials of all degrees, with the L^2 inner product (6.1.6). Let $\phi : \mathcal{P} \to \mathbb{C}$ be the linear functional defined by $\phi(p) = p(0)$. Suppose that ϕ has an adjoint, let $p \in \mathcal{P}$, and let $g = \phi^*(1)$. (a) Explain why $p(0) = \langle p, \phi^*(1) \rangle = \int_0^1 p(t)\overline{g(t)} \, dt$. (b) Let $p(t) = t^2 g(t)$ and explain why $0 = \int_0^1 t^2 |g(t)|^2 \, dt = \|tg(t)\|^2$. Why must g be the zero polynomial? (c) Explain why ϕ does not have an adjoint. *Hint*: Let $p = 1$.

P.6.15 Let $\mathcal{V}$ be a finite-dimensional inner product space over $\mathbb{F}$, let $\phi \in \mathcal{L}(\mathcal{V}, \mathbb{F})$ be a linear functional, and let $\mathbf{w}$ be the Riesz vector (6.4.6) for ϕ. (a) Show that $\phi(\mathbf{w}) = \|\mathbf{w}\|^2$. (b) Show that $\max\{|\phi(\mathbf{v})| : \|\mathbf{v}\| = 1\} = \|\mathbf{w}\|$.

P.6.16 Verify the computations in Example 6.4.10.

P.6.17 What is the Riesz vector for the linear functional $\phi(p) = \int_0^1 p(x) \, dx$ on $\mathcal{P}_3$ with the inner product (6.1.6)?

P.6.18 Let $\mathcal{V} = \mathcal{P}_2$ be the real inner product space of real polynomials of degree at most two, with the L^2 inner product (6.1.6). Let $\phi \in \mathcal{L}(\mathcal{V}, \mathbb{R})$ be the linear functional defined by $\phi(p) = p(0)$. (a) Use the orthonormal basis $\beta = u_1, u_2, u_3$ in (6.1.7) to show that the Riesz vector for ϕ is $w(x) = 30x^2 - 36x + 9$. (b) If $p(x) = ax^2 + bx + c \in \mathcal{V}$, verify that $\phi(p) = \langle p, w \rangle$.

P.6.19 Let $V = P_2$ as in the preceding problem. Let $\psi \in \mathcal{L}(V, \mathbb{R})$ be the linear functional defined by $\psi(p) = \int_0^1 \sqrt{x} p(x)\, dx$. (a) Show that the Riesz vector for ψ is $w(x) = -\frac{4}{7}x^2 + \frac{48}{35}x + \frac{6}{35}$. (b) If $p(x) = ax^2 + bx + c \in V$, verify that $\psi(p) = \langle p, w \rangle$.

P.6.20 Does there exist a function $g : \mathbb{R} \to \mathbb{R}$ such that

$$f(-47) - 3f'(0) + 5f''(\pi) = \int_{-2}^{2} f(x)g(x)\, dx$$

for all functions $f : \mathbb{R} \to \mathbb{R}$ of the form $f(x) = a_1 e^x + a_2 x e^x \cos x + a_3 x^2 e^x \sin x$?

P.6.21 Let ϕ be a linear functional on the inner product space M_n (with the Frobenius inner product) and suppose that $\phi(C) = 0$ for every commutator $C \in \mathsf{M}_n$. (a) Show that $\phi(AB) = \phi(BA)$ for all $A, B \in \mathsf{M}_n$. (b) If Y is the Riesz vector for ϕ, show that Y commutes with every matrix in M_n. (c) Deduce that Y is a scalar matrix and ϕ is a scalar multiple of the trace functional. (d) If $\phi(I) = 1$, show that $\phi(A) = \frac{1}{n} \operatorname{tr} A$ for all $A \in \mathsf{M}_n$.

P.6.22 Let V be a two-dimensional inner product space with orthonormal basis $\beta = \mathbf{u}_1, \mathbf{u}_2$. Define $T \in \mathcal{L}(V)$ by $T\mathbf{u}_1 = 2\mathbf{u}_1 + \mathbf{u}_2$ and $T\mathbf{u}_2 = \mathbf{u}_1 - \mathbf{u}_2$. (a) Use Definition 6.6.1 to show that T is selfadjoint. (b) Compute the matrix representation $_\beta[T]_\beta$ of T with respect to the orthonormal basis β. Is it Hermitian? (c) Define $\mathbf{v}_1 = \mathbf{u}_1$ and $\mathbf{v}_2 = \frac{1}{2}\mathbf{u}_1 + \frac{1}{2\sqrt{3}}\mathbf{u}_2$. Show that $\gamma = \mathbf{v}_1, \mathbf{v}_2$ is a basis for V. (d) Compute the matrix representation $_\gamma[T]_\gamma$ of T with respect to the basis γ. Is it Hermitian? (e) Let $V = P_2$ (see P.6.18 and P.6.19) with the L^2 inner product (6.1.6) and let $\mathbf{u}_1 = 1$ and $\mathbf{u}_2 = \sqrt{3}(2x - 1)$ as in (6.1.7). What are $\mathbf{v}_1$ and $\mathbf{v}_2$? What are the actions of T and T^* on a polynomial $ax + b \in V$?

P.6.23 Let $f \in C_{\mathbb{R}}[-\pi, \pi]$. (a) For each $n \in \mathbb{N}$, show that $\lim_{n \to \infty} \int_{-\pi}^{\pi} f(x) \sin nx\, dx = 0$. *Remark*: This is an instance of the *Riemann–Lebesgue lemma*. (b) What happens if $\sin nx$ is replaced by $\cos nx$? *Hint*: Corollary 6.7.6.

P.6.24 Let f be a differentiable real-valued function on $[-\pi, \pi]$. Suppose that $f(-\pi) = f(\pi)$ and f' is continuous. (a) For each $n \in \mathbb{N}$, show that $\lim_{n \to \infty} (n \int_{-\pi}^{\pi} f(x) \sin nx\, dx) = 0$. (b) What can you say if f is twice differentiable on $[-\pi, \pi]$, $f(-\pi) = f(\pi)$, $f'(-\pi) = f'(\pi)$, and f'' is continuous? (c) What happens if $\sin nx$ is replaced by $\cos nx$? *Hint*: Integrate by parts.

P.6.25 Let $f : [-\pi, \pi] \to \mathbb{R}$ be piecewise C^1 on $[-\pi, \pi]$ and let (6.8.16) be the Fourier series associated with f. (a) If f is even, show that $a_{-n} = 0$ for all $n = 1, 2, \ldots$. (b) If f is odd, show that $a_n = 0$ for all $n = 0, 1, 2, \ldots$.

P.6.26 Let $f(x) = x^2$ for $x \in [-\pi, \pi]$. (a) Use Theorem 6.8.18 to show that

$$f(x) = \frac{\pi^2}{3} + 4 \sum_{n=1}^{\infty} \frac{(-1)^n}{n^2} \cos nx$$

for all $x \in [-\pi, \pi]$. (b) Deduce that $\sum_{n=1}^{\infty} 1/n^2 = \pi^2/6$, which was discovered by Euler in 1735.

P.6.27 Let $f : \mathbb{R} \to \mathbb{R}$ be periodic with period 2π and suppose that $f(x) = |x|$ for $x \in [-\pi, \pi]$. (a) Show that f is piecewise C^1 on $[-\pi, \pi]$ (what are the points x_i?) and use Theorem 6.8.18 to show that

$$f(x) = \frac{\pi}{2} - \frac{4}{\pi} \sum_{n=1}^{\infty} \frac{\cos(2n - 1)x}{(2n - 1)^2}$$

for all $x \in [-\pi, \pi]$. (b) Use (a) to deduce that $1 + \dfrac{1}{3^2} + \dfrac{1}{5^2} + \dfrac{1}{7^2} + \cdots = \dfrac{\pi^2}{8}$.

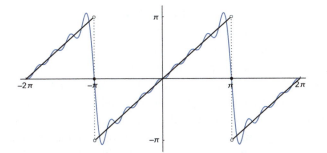

Figure 6.4 The graph of the function f in P.6.28 and its Fourier approximation $2 \sin x - \sin 2x +$ $\frac{2}{3} \sin 3x - \frac{1}{2} \sin 4x + \frac{2}{5} \sin 5x - \frac{1}{3} \sin 6x + \frac{2}{7} \sin 7x - \frac{1}{4} \sin 8x$.

P.6.28 Let $f : \mathbb{R} \to \mathbb{R}$ be periodic with period 2π, suppose that $f(x) = x$ for $x \in (-\pi, \pi)$, and let $f(\pm \pi) = 0$; see Figure 6.4. (a) Show that f is piecewise C^1 on $[-\pi, \pi]$ (what are the points x_i?) and use Theorem 6.8.18 to show that

$$f(x) = 2 \sum_{n=1}^{\infty} \frac{(-1)^{n+1} \sin nx}{n}$$

for all $x \in [-\pi, \pi]$. (b) Use (a) to deduce that $1 - \dfrac{1}{3} + \dfrac{1}{5} - \dfrac{1}{7} + \cdots = \dfrac{\pi}{4}$.

P.6.29 Let $\mathcal{V}$ be an n-dimensional $\mathbb{F}$-inner product space. If $\mathbf{v}_1, \mathbf{v}_2, \ldots, \mathbf{v}_n \in \mathcal{V}$ are linearly independent, show that there exists an inner product $\langle \cdot, \cdot \rangle$ on $\mathcal{V}$ with respect to which $\mathbf{v}_1, \mathbf{v}_2, \ldots, \mathbf{v}_n$ is an orthonormal basis for $\mathcal{V}$.

P.6.30 Let $\mathbf{v}_1, \mathbf{v}_2, \ldots, \mathbf{v}_n$ be linearly independent vectors in an inner product space $\mathcal{V}$. The *modified Gram–Schmidt process* produces an orthonormal basis for $\text{span}\{\mathbf{v}_1, \mathbf{v}_2, \ldots, \mathbf{v}_n\}$ via the following algorithm. For $k = 1, 2, \ldots, n$, first replace $\mathbf{v}_k$ by $\mathbf{v}_k / \|\mathbf{v}_k\|$ and then for $j = k + 1, k + 2, \ldots, n$ replace $\mathbf{v}_j$ by $\mathbf{v}_j - \langle \mathbf{v}_j, \mathbf{v}_k \rangle \mathbf{v}_k$. At the conclusion of the algorithm, the vectors $\mathbf{v}_1, \mathbf{v}_2, \ldots, \mathbf{v}_n$ are orthonormal. (a) Let $n = 3$ and follow the steps in the modified Gram–Schmidt process. (b) Do the same for the classical Gram–Schmidt process described in the proof of Theorem 6.3.5. (c) Explain why the former involves the same computations as the latter, but in a different order.

P.6.31 Let $p_0, p_1, p_2, \ldots$ be orthogonal polynomials as in Definition 6.9.2. In Example 6.9.3, we found that the only zero of $p_1(x)$ is at $x = (a + b)/2$. Let $n \geq 2$. In the following three parts, prove by contradiction that p_n has no zeros that are not real, no real zeros that have multiplicity greater than one, and no real zeros outside the interval (a, b). (a) If $c \in \mathbb{C}$ is not real and $p_n(c) = 0$, show that $q(x) = p_n(x)/|x - c|^2 \in \mathcal{P}_{n-2}$, $\langle q, p_n \rangle = 0$, and $\langle q, p_n \rangle = \langle 1, |\frac{p_n}{x-c}|^2 \rangle > 0$. (b) If c is a real zero of p_n with multiplicity greater than one, show that $q(x) = p_n(x)/(x - c)^2 \in \mathcal{P}_{n-2}$, $\langle q, p_n \rangle = 0$, and $\langle q, p_n \rangle = \langle 1, (\frac{p_n}{x-c})^2 \rangle > 0$. (c) If c is a real zero of p_n and $c \notin (a, b)$, show that $q(x) = p_n(x)/(x - c) \in \mathcal{P}_{n-1}$, $\langle q, p_n \rangle = 0$, and $\langle q, p_n \rangle = \langle 1, p_n^2/(x - c) \rangle \neq 0$.

P.6.32 Derive the 2-point Gaussian rule for $[0, 1]$ and the inner product (6.9.1) as follows: (a) Show that the zeros of the orthogonal polynomial $p_2(x)$ are $x_1 = (3 - \sqrt{3})/6$ and $x_2 = (3 + \sqrt{3})/6$. (b) Show that the Lagrange basis polynomials for the nodes x_1, x_2 are $\ell_1(x) = -\sqrt{3}(x - x_2)$ and $\ell_2(x) = \sqrt{3}(x - x_1)$. (c) Show that the weights for the 2-point Gaussian rule are $w_1 = w_2 = 1/2$. (d) Calculate $\int_0^1 x^3 \, dx$ and $\frac{1}{2}x_1^3 + \frac{1}{2}x_2^3$ to show that the 2-point Gaussian rule gives the exact value for $f(x) = x^3$. (e) What about $f(x) = x$ and $f(x) = 1$?

6.11 Notes

For an exposition of the fundamentals of Fourier series, see [Bha05].

The Gram–Schmidt process is a powerful theoretical tool, but it is not the algorithm of choice for orthogonalizing large lists of vectors. Algorithms based on the QR and singular value decompositions (see Chapters 7 and 16) are more reliable in practice.

The two algorithms in P.6.30 produce the same lists of orthonormal vectors in exact arithmetic, but in floating point computations the modified Gram–Schmidt algorithm can sometimes produce better results than the classical algorithm. However, there are some problems for which the modified Gram–Schmidt algorithm and floating point arithmetic fail to produce a final set of vectors that are nearly orthonormal.

There is a rich theory of orthogonal polynomials in which the inner product (6.9.1) can incorporate a positive weight function and the interval need not be finite. For example, the orthogonal polynomials associated with the inner product $\langle p, q \rangle = \int_{-1}^{1} p(x)q(x)(1 - x^2)^{-1/2}\, dx$ are the (normalized) Chebyshev polynomials; see [Dav63, Chapter 10]. The first four are $p_0(x) = \frac{1}{\sqrt{\pi}}, p_1(x) = \sqrt{\frac{2}{\pi}}x, p_2(x) = \sqrt{\frac{2}{\pi}}(2x^2 - 1)$, and $p_3(x) = \sqrt{\frac{2}{\pi}}(4x^3 - 3x)$. The zeros of $p_n(x)$ were employed in Figure 2.5; they are $x_k = \cos(\frac{2k-1}{2n}\pi)$ for $k = 1, 2, \ldots, n$.

6.12 Some Important Concepts

- Orthonormal vectors and linear independence.

- Orthogonalization of a linearly independent list via the Gram–Schmidt process.

- Linear functionals and the Riesz representation theorem.

- Matrix representation of a linear transformation with respect to orthonormal bases.

- Adjoint of a linear transformation or matrix.

- Parseval's identity, Bessel's inequality, and the Cauchy–Schwarz inequality.

- Fourier series.

- Orthogonal polynomials and n-point Gaussian rules.

Unitary Matrices

Unitary matrices play important roles in theory and computation. The adjoint of a unitary matrix is its inverse, so unitary matrices are easy to invert. They preserve lengths and angles, and have remarkable stability properties in many numerical algorithms. In this chapter, we explore the properties of unitary matrices and present several special cases. We derive an explicit formula for a unitary matrix whose first column is given. We give a constructive proof of the QR factorization and show that every square complex matrix is unitarily similar to an upper Hessenberg matrix.

7.1 Unitary Matrices

Definition 7.1.1 A square complex matrix U is *unitary* if $U^*U = I$. A real unitary matrix Q is *real orthogonal*; it satisfies $Q^{\mathsf{T}}Q = I$.

A 1×1 unitary matrix is a complex number u such that $|u|^2 = \bar{u}u = 1$. Unitary matrices may be thought of as matrix analogs of complex numbers with modulus 1. Since $U^*U = I$ if and only if U is invertible and U^* is its inverse, it is easy to invert a unitary matrix.

Example 7.1.2 The matrices $\operatorname{diag}(e^{i\theta_1}, e^{i\theta_2}, \dots, e^{i\theta_n})$ are unitary for any $\theta_1, \theta_2, \dots, \theta_n \in \mathbb{R}$. In particular, the complex rotation matrix $e^{i\theta}I \in \mathsf{M}_n$ is unitary for any $\theta \in \mathbb{R}$.

Example 7.1.3 The *Hadamard gate*

$$ H = \begin{bmatrix} \frac{1}{\sqrt{2}} & \frac{1}{\sqrt{2}} \\ \frac{1}{\sqrt{2}} & -\frac{1}{\sqrt{2}} \end{bmatrix} $$

is a real orthogonal matrix that arises in quantum information theory.

Example 7.1.4 Three 2×2 unitary matrices occur in the Pauli equation in quantum mechanics. The Pauli equation is a nonrelativistic version of the Schrödinger equation for spin-$\frac{1}{2}$ particles in an external electromagnetic field. It involves the three *Pauli spin matrices*

$$ \sigma_x = \begin{bmatrix} 0 & 1 \\ 1 & 0 \end{bmatrix}, \quad \sigma_y = \begin{bmatrix} 0 & -i \\ i & 0 \end{bmatrix}, \quad \text{and} \quad \sigma_z = \begin{bmatrix} 1 & 0 \\ 0 & -1 \end{bmatrix}. $$

These matrices also arise in quantum information theory, where they are known, respectively, as the *Pauli-X gate*, *Pauli-Y gate*, and *Pauli-Z gate*.

Example 7.1.5 The *Householder matrix*

$$ U_{\mathbf{v}} = I - 2\mathbf{v}\mathbf{v}^* \in \mathsf{M}_n(\mathbb{F}) \qquad \text{for } \mathbf{v} \in \mathbb{F}^n \text{ such that } \|\mathbf{v}\|_2 = 1 \qquad (7.1.6) $$

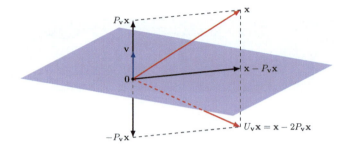

Figure 7.1 The Householder transformation $\mathbf{x} \mapsto U_{\mathbf{v}}\mathbf{x}$ on $\mathbb{R}^3$ reflects $\mathbf{x}$ across a plane that is orthogonal to the unit vector $\mathbf{v}$.

is Hermitian, unitary, and involutory:

$$U_{\mathbf{v}}U_{\mathbf{v}}^* = U_{\mathbf{v}}^2 = (I - 2\mathbf{vv}^*)(I - 2\mathbf{vv}^*) = I - 4\mathbf{vv}^* + 4\mathbf{vv}^*\mathbf{vv}^* = I - 4\mathbf{vv}^* + 4\mathbf{vv}^* = I.$$

The associated Householder transformation $T_{U_{\mathbf{v}}} \in \mathcal{L}(\mathbb{F}^n)$ is

$$\mathbf{x} \mapsto U_{\mathbf{v}}\mathbf{x} = \mathbf{x} - 2\mathbf{vv}^*\mathbf{x} = \mathbf{x} - 2\langle \mathbf{x}, \mathbf{v}\rangle\mathbf{v} = \mathbf{x} - 2P_{\mathbf{v}}\mathbf{x}, \tag{7.1.7}$$

in which

$$P_{\mathbf{v}}\mathbf{x} = \mathbf{vv}^*\mathbf{x} = \langle \mathbf{x}, \mathbf{v}\rangle\mathbf{v} \tag{7.1.8}$$

is the projection of $\mathbf{x}$ onto the unit vector $\mathbf{v}$; see (5.4.12). The action of the Householder transformation $T_{U_{\mathbf{v}}}$ on $\mathbb{R}^3$ is illustrated in Figure 7.1. Householder matrices are employed by many numerical linear algebra algorithms; see P.7.30. A Householder matrix can be used to construct a unitary matrix whose first column is a given unit vector; see P.7.31.

Products and direct sums of unitary matrices are unitary.

Theorem 7.1.9 *Let $U, V \in \mathsf{M}_n$ and $W \in \mathsf{M}_m$.*

(a) *If U and V are unitary, then UV is unitary.*

(b) *$V \oplus W$ is unitary if and only if V and W are unitary.*

(c) *If U is unitary, then $|\det U| = 1$.*

Proof (a) Use Definition 7.1.1 to compute $(UV)^*(UV) = V^*U^*UV = V^*IV = V^*V = I$.

(b) Compute $(V \oplus W)^*(V \oplus W) = (V^* \oplus W^*)(V \oplus W) = V^*V \oplus W^*W$. If V and W are unitary, then $V^*V \oplus W^*W = I_n \oplus I_m = I_{n+m}$, so $V \oplus W$ is unitary. Conversely, if $V \oplus W$ is unitary, then $(V \oplus W)^*(V \oplus W) = I_{n+m} = V^*V \oplus W^*W$, so $V^*V = I_n$ and $W^*W = I_m$.

(c) Use Definition 7.1.1 and the product rule for the determinant to compute

$$1 = \det I = \det(U^*U) = (\det U^*)(\det U) = \overline{(\det U)}(\det U) = |\det U|^2. \qquad \square$$

Example 7.1.10 The $n \times n$ *reversal matrix*

$$K_n = \begin{bmatrix} & & & 1 \\ & & \reflectbox{$\ddots$} & \\ & 1 & & \\ 1 & & & \end{bmatrix} \tag{7.1.11}$$

is a real orthogonal, symmetric involution: $K_n^\mathsf{T} K_n = K_n^2 = I$. Its action on the standard basis for $\mathbb{F}^n$ is $K_n \mathbf{e}_j = \mathbf{e}_{n-j+1}$ for $j = 1, 2, \ldots, n$.

Example 7.1.12 For any $\theta_1, \theta_2, \ldots, \theta_n \in \mathbb{R}$,

$$\begin{bmatrix} & & & e^{i\theta_n} \\ & & \cdot^{\cdot^{\cdot}} & \\ & e^{i\theta_2} & & \\ e^{i\theta_1} & & & \end{bmatrix} = K_n \operatorname{diag}(e^{i\theta_1}, e^{i\theta_2}, \ldots, e^{i\theta_n})$$

is unitary because it is the product of two unitary matrices.

The defining identity $U^*U = I$ for a unitary matrix in $\mathsf{M}_n(\mathbb{F})$ has a useful geometric interpretation: In the $\mathbb{F}$-inner product space $\mathbb{F}^n$ with inner product $\langle \mathbf{x}, \mathbf{y} \rangle = \mathbf{y}^*\mathbf{x}$, the columns of U are orthonormal. In the following theorem, we learn that the rows of U are also orthonormal.

Theorem 7.1.13 *Let $U \in \mathsf{M}_n(\mathbb{F})$. The following are equivalent:*

(a) *U is unitary, that is, $U^*U = I$.*

(b) *The columns of U are orthonormal.*

(c) *$UU^* = I$.*

(d) *U^* is unitary.*

(e) *The rows of U are orthonormal.*

(f) *U is invertible and $U^{-1} = U^*$.*

(g) *$\|\mathbf{x}\|_2 = \|U\mathbf{x}\|_2$ for all $\mathbf{x} \in \mathbb{F}^n$.*

(h) *$\langle U\mathbf{x}, U\mathbf{y} \rangle = \langle \mathbf{x}, \mathbf{y} \rangle$ for all $\mathbf{x}, \mathbf{y} \in \mathbb{F}^n$.*

Proof (a) $\Leftrightarrow$ (b) Let $U = [\mathbf{u}_1 \ \mathbf{u}_2 \ \ldots \ \mathbf{u}_n]$. Then $U^*U = [\mathbf{u}_i^*\mathbf{u}_j] = [\langle \mathbf{u}_j, \mathbf{u}_i \rangle]$, so $U^*U = I$ if and only if $\langle \mathbf{u}_j, \mathbf{u}_i \rangle = \delta_{ij}$ for all $i, j = 1, 2, \ldots, n$.

(a) $\Leftrightarrow$ (c) U^* is a left inverse of U if and only if it is a right inverse; see Theorem 2.2.13.

(c) $\Leftrightarrow$ (d) Write (c) as $(U^*)^*(U^*) = I$. The asserted equivalence follows from Definition 7.1.1.

(d) $\Leftrightarrow$ (e) The rows of U are the conjugates of the columns of U^*, so the equivalence of (d) and (e) follows from the equivalence of (a) and (b).

(f) $\Leftrightarrow$ (a) If $U^*U = I$, then U^* is a left inverse, and hence an inverse, of U; see Theorem 2.2.13. If $U^* = U^{-1}$, then $I = U^{-1}U = U^*U$.

(a) $\Rightarrow$ (g) $\|U\mathbf{x}\|_2^2 = \langle U\mathbf{x}, U\mathbf{x} \rangle = \langle \mathbf{x}, U^*U\mathbf{x} \rangle = \langle \mathbf{x}, \mathbf{x} \rangle = \|\mathbf{x}\|_2^2$.

(g) $\Rightarrow$ (h) If $\mathbb{F} = \mathbb{R}$ or $\mathbb{C}$, then $\|U\mathbf{x} \pm U\mathbf{y}\|_2 = \|U(\mathbf{x} \pm \mathbf{y})\|_2 = \|\mathbf{x} \pm \mathbf{y}\|_2$. If $\mathbb{F} = \mathbb{C}$, then $\|U\mathbf{x} \pm iU\mathbf{y}\|_2 = \|U(\mathbf{x} \pm i\mathbf{y})\|_2 = \|\mathbf{x} \pm i\mathbf{y}\|_2$. In both cases, the polarization identities (5.4.24) and (5.4.25) ensure that $\langle U\mathbf{x}, U\mathbf{y} \rangle = \langle \mathbf{x}, \mathbf{y} \rangle$.

(h) $\Rightarrow$ (a) Since $\langle \mathbf{x}, \mathbf{y} \rangle = \langle U\mathbf{x}, U\mathbf{y} \rangle = \langle U^*U\mathbf{x}, \mathbf{y} \rangle$, it follows that

$$0 = \langle U^*U\mathbf{x}, \mathbf{y} \rangle - \langle \mathbf{x}, \mathbf{y} \rangle = \langle U^*U\mathbf{x} - \mathbf{x}, \mathbf{y} \rangle = \langle (U^*U - I)\mathbf{x}, \mathbf{y} \rangle \quad \text{for all } \mathbf{x}, \mathbf{y} \in \mathbb{F}^n.$$

Let $\mathbf{y} = (U^*U - I)\mathbf{x}$ and conclude that

$$\|(U^*U - I)\mathbf{x}\|_2^2 = \langle (U^*U - I)\mathbf{x}, (U^*U - I)\mathbf{x} \rangle = 0 \quad \text{for all } \mathbf{x} \in \mathbb{F}^n.$$

Therefore, $U^*U - I = 0$, so U is unitary. $\qquad \square$

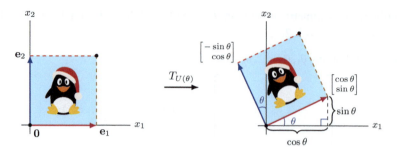

Figure 7.2 The linear transformation $T_{U(\theta)} : \mathbb{R}^2 \to \mathbb{R}^2$ induced by the matrix $U(\theta)$ defined by (7.1.16) is a rotation around the origin through an angle of θ.

Example 7.1.14 The 3×3 matrices

$$V = [\mathbf{v}_1 \; \mathbf{v}_2 \; \mathbf{v}_3] = \frac{1}{3} \begin{bmatrix} 1 & -2 & -2 \\ 2 & 2 & -1 \\ 2 & -1 & 2 \end{bmatrix} \quad \text{and} \quad W = [\mathbf{w}_1 \; \mathbf{w}_2 \; \mathbf{w}_3] = \frac{1}{2} \begin{bmatrix} -1 & \sqrt{2} & 1 \\ 1 & \sqrt{2} & -1 \\ \sqrt{2} & 0 & \sqrt{2} \end{bmatrix}$$

are real and have orthonormal columns, so V and W are real orthogonal.

Example 7.1.15 The *plane rotation* matrices (see Figure 7.2)

$$U(\theta) = \begin{bmatrix} \cos\theta & -\sin\theta \\ \sin\theta & \cos\theta \end{bmatrix} \qquad \text{for } \theta \in \mathbb{R}, \tag{7.1.16}$$

are real orthogonal:

$$
\begin{aligned}
U(\theta)^\mathsf{T} U(\theta) &= \begin{bmatrix} \cos\theta & \sin\theta \\ -\sin\theta & \cos\theta \end{bmatrix} \begin{bmatrix} \cos\theta & -\sin\theta \\ \sin\theta & \cos\theta \end{bmatrix} \\
&= \begin{bmatrix} \cos^2\theta + \sin^2\theta & -\cos\theta\sin\theta + \sin\theta\cos\theta \\ -\sin\theta\cos\theta + \cos\theta\sin\theta & \cos^2\theta + \sin^2\theta \end{bmatrix} = \begin{bmatrix} 1 & 0 \\ 0 & 1 \end{bmatrix}.
\end{aligned}
$$

Computations with plane rotations can be used to prove many trigonometric identities; see P.7.14.

Example 7.1.17 The $n \times n$ *Fourier matrix* is the symmetric matrix defined by

$$F_n = \frac{1}{\sqrt{n}} \big[\omega^{(j-1)(k-1)} \big]_{j,k=1}^n, \qquad \text{in which } \omega = e^{2\pi i/n}. \tag{7.1.18}$$

For example,

$$F_2 = \frac{1}{\sqrt{2}} \begin{bmatrix} \omega^0 & \omega^0 \\ \omega^0 & \omega^1 \end{bmatrix} = \frac{1}{\sqrt{2}} \begin{bmatrix} 1 & 1 \\ 1 & -1 \end{bmatrix}, \qquad \text{in which } \omega = e^{\pi i} = -1,$$

$$F_3 = \frac{1}{\sqrt{3}} \begin{bmatrix} \omega^0 & \omega^0 & \omega^0 \\ \omega^0 & \omega^1 & \omega^2 \\ \omega^0 & \omega^2 & \omega^4 \end{bmatrix} = \frac{1}{\sqrt{3}} \begin{bmatrix} 1 & 1 & 1 \\ 1 & e^{2\pi i/3} & e^{-2\pi i/3} \\ 1 & e^{-2\pi i/3} & e^{2\pi i/3} \end{bmatrix}, \qquad \text{in which } \omega = e^{2\pi i/3},$$

and

$$F_4 = \frac{1}{2} \begin{bmatrix} \omega^0 & \omega^0 & \omega^0 & \omega^0 \\ \omega^0 & \omega^1 & \omega^2 & \omega^3 \\ \omega^0 & \omega^2 & \omega^4 & \omega^6 \\ \omega^0 & \omega^3 & \omega^6 & \omega^9 \end{bmatrix} = \frac{1}{2} \begin{bmatrix} 1 & 1 & 1 & 1 \\ 1 & i & -1 & -i \\ 1 & -1 & 1 & -1 \\ 1 & -i & -1 & i \end{bmatrix}, \qquad \text{in which } \omega = e^{\pi i/2} = i. \tag{7.1.19}$$

The identity

$$\sum_{k=1}^{n} z^{k-1} = \begin{cases} \dfrac{1-z^n}{1-z} & \text{if } z \neq 1, \\ n & \text{if } z = 1, \end{cases}$$

for finite geometric series is in P.D.5. Since $\omega^{\ell} = (e^{2\pi i/n})^{\ell} = 1$ if and only if ℓ is an integer multiple of n,

$$\sum_{k=1}^{n} \omega^{(k-1)\ell} = \sum_{k=1}^{n} (\omega^{\ell})^{k-1} = \begin{cases} 0 & \text{if } \ell \neq pn, \\ n & \text{if } \ell = pn, \end{cases} \quad \text{in which } \omega = e^{2\pi i/n} \text{ and } p = 0, \pm 1, \pm 2, \ldots .$$

Partition $F_n = [\mathbf{f}_1 \ \mathbf{f}_2 \ \ldots \ \mathbf{f}_n]$ according to its columns, in which the jth column of F_n is

$$\mathbf{f}_j = \frac{1}{\sqrt{n}} \begin{bmatrix} 1 & \omega^{j-1} & \omega^{2(j-1)} & \omega^{3(j-1)} & \ldots & \omega^{(n-1)(j-1)} \end{bmatrix}^{\mathsf{T}}, \quad \text{for } j = 1, 2, \ldots, n.$$

For $1 \leq p, q \leq n$, the (p, q) entry of $F_n^* F_n$ is

$$\mathbf{f}_p^* \mathbf{f}_q = \frac{1}{n} \sum_{k=1}^{n} \overline{\omega^{(p-1)(k-1)}} \omega^{(q-1)(k-1)} = \frac{1}{n} \sum_{k=1}^{n} \omega^{-(p-1)(k-1)} \omega^{(q-1)(k-1)}$$

$$= \frac{1}{n} \sum_{k=1}^{n} (\omega^{1-p})^{k-1} (\omega^{q-1})^{k-1} = \frac{1}{n} \sum_{k=1}^{n} (\omega^{q-p})^{k-1} = \begin{cases} 0 & \text{if } p \neq q, \\ 1 & \text{if } p = q. \end{cases}$$

This calculation shows that $F_n^* F_n = I$, so F_n is unitary. Since F_n is symmetric, we have $F_n^{-1} = F_n^* = \overline{F_n}$. In particular,

$$F_2^{-1} = \frac{1}{\sqrt{2}} \begin{bmatrix} 1 & 1 \\ 1 & -1 \end{bmatrix} \quad \text{and} \quad F_4^{-1} = \frac{1}{2} \begin{bmatrix} 1 & 1 & 1 & 1 \\ 1 & -i & -1 & i \\ 1 & -1 & 1 & -1 \\ 1 & i & -1 & -i \end{bmatrix}.$$

Fourier matrices play a key role in our discussion of circulant matrices; see Example 7.2.11 and Section 14.5.

An orthonormal list of vectors in a finite-dimensional inner product space can be extended to an orthonormal basis; see Corollary 6.3.14. The following theorem is a block-matrix version of this principle.

Theorem 7.1.20 *If* $1 \leq n \leq m$, $X \in \mathsf{M}_{m \times n}(\mathbb{F})$, *and* $X^* X = I_n$, *then there is an* $X' \in \mathsf{M}_{m \times (m-n)}(\mathbb{F})$ *such that* $U = [X \ X'] \in \mathsf{M}_m(\mathbb{F})$ *is unitary.*

Proof Partition $X = [\mathbf{x}_1 \ \mathbf{x}_2 \ \ldots \ \mathbf{x}_n]$ according to its columns. Since $I_n = X^* X = [\mathbf{x}_i^* \mathbf{x}_j]$, the vectors $\mathbf{x}_1, \mathbf{x}_2, \ldots, \mathbf{x}_n \in \mathbb{F}^m$ are orthonormal. Corollary 6.3.14 says that there are $\mathbf{x}_{n+1}, \mathbf{x}_{n+2}, \ldots, \mathbf{x}_m \in \mathbb{F}^m$ such that $\mathbf{x}_1, \mathbf{x}_2, \ldots, \mathbf{x}_n, \mathbf{x}_{n+1}, \ldots, \mathbf{x}_m$ is an orthonormal basis for $\mathbb{F}^m$. Let $X' = [\mathbf{x}_{n+1} \ \mathbf{x}_{n+2} \ \ldots \ \mathbf{x}_m]$. Theorem 7.1.13 ensures that $U = [X \ X'] \in \mathsf{M}_m(\mathbb{F})$ is unitary. $\square$

The case $n = 1$ of the preceding theorem ensures that there is a unitary matrix whose first column is any given unit vector. The following example provides an explicit formula for one such matrix.

Example 7.1.21 (A unitary $n \times n$ matrix whose first column is a given unit vector) If $n = 1$ and $v \in \mathbb{F}$ has $|v| = 1$, then $U = [v]$ is a unitary matrix whose first column is v. Suppose that $n \geq 2$ and let $\mathbf{v} = [v_i] \in \mathbb{F}^n$ be a given unit vector. Let $x = |v_1| = e^{-i\theta} v_1$, in which $\theta \in [0, 2\pi)$. Partition

$$e^{-i\theta} \mathbf{v} = \begin{bmatrix} x \\ \mathbf{w} \end{bmatrix}, \qquad \text{in which } \mathbf{w} \in \mathbb{F}^{n-1} \text{ and } x \geq 0,$$

and let

$$V = \begin{bmatrix} x & \mathbf{w}^* \\ \mathbf{w} & -I_{n-1} + \frac{1}{1+x}\mathbf{w}\mathbf{w}^* \end{bmatrix} \in \mathsf{M}_n(\mathbb{F}).$$

Then V is Hermitian, $1 = \|e^{-i\theta}\mathbf{v}\|_2^2 = [x \ \mathbf{w}^*]\begin{bmatrix} x \\ \mathbf{w} \end{bmatrix} = x^2 + \mathbf{w}^*\mathbf{w}$. Therefore, $\mathbf{w}^*\mathbf{w} = 1 - x^2$ and

$$V^*V = \begin{bmatrix} x & \mathbf{w}^* \\ \mathbf{w} & -I_{n-1} + \frac{1}{1+x}\mathbf{w}\mathbf{w}^* \end{bmatrix} \begin{bmatrix} x & \mathbf{w}^* \\ \mathbf{w} & -I_{n-1} + \frac{1}{1+x}\mathbf{w}\mathbf{w}^* \end{bmatrix} = \begin{bmatrix} a & \mathbf{z}^* \\ \mathbf{z} & C \end{bmatrix},$$

in which

$$a = x^2 + \mathbf{w}^*\mathbf{w} = x^2 + (1 - x^2) = 1,$$

$$\mathbf{z} = x\mathbf{w} - \mathbf{w} + \frac{1}{1+x}\mathbf{w}\mathbf{w}^*\mathbf{w} = \left(x - 1 + \frac{1 - x^2}{1 + x}\right)\mathbf{w} = \mathbf{0}, \qquad \text{and}$$

$$C = \mathbf{w}\mathbf{w}^* + I_{n-1} - \frac{2}{1+x}\mathbf{w}\mathbf{w}^* + \frac{1}{(1+x)^2}\mathbf{w}\mathbf{w}^*\mathbf{w}\mathbf{w}^*$$

$$= I_{n-1} + \left(1 - \frac{2}{1+x} + \frac{1 - x^2}{(1+x)^2}\right)\mathbf{w}\mathbf{w}^* = I_{n-1}.$$

Therefore, V is unitary, as is $U = e^{i\theta}V$. The first column of U is the given unit vector $\mathbf{v} = e^{i\theta}\begin{bmatrix} x \\ \mathbf{w} \end{bmatrix}$. Notice that $\mathbf{v} = U\mathbf{e}_1$ and $\mathbf{e}_1 = U^*\mathbf{v}$. Finally, observe that if $\mathbf{v}$ is a real vector, then $e^{i\theta} = \pm 1$ and the unitary matrices U and V are real orthogonal.

Example 7.1.22 Using the construction in the preceding example, we can find a real unitary matrix whose first column is $[\frac{1}{3} \ \frac{2}{3} \ \frac{2}{3}]^\mathsf{T}$ as follows:

$$\theta = 0, \qquad x = \frac{1}{3}, \qquad \mathbf{w} = \begin{bmatrix} \frac{2}{3} \\ \frac{2}{3} \end{bmatrix}, \qquad \text{and} \qquad \frac{1}{1+x} = \frac{3}{4}.$$

Then

$$V = \begin{bmatrix} x & \mathbf{w}^\mathsf{T} \\ \mathbf{w} & -I_2 + \frac{3}{4}\mathbf{w}\mathbf{w}^\mathsf{T} \end{bmatrix} = \begin{bmatrix} \frac{1}{3} & \frac{2}{3} & \frac{2}{3} \\ \frac{2}{3} & -\frac{2}{3} & \frac{1}{3} \\ \frac{2}{3} & \frac{1}{3} & -\frac{2}{3} \end{bmatrix} = \frac{1}{3}\begin{bmatrix} 1 & 2 & 2 \\ 2 & -2 & 1 \\ 2 & 1 & -2 \end{bmatrix}.$$

This unitary matrix is different from the matrix V in Example 7.1.14, which has the same first column.

Theorem 7.1.23 *Let* $\mathbf{x}, \mathbf{y} \in \mathbb{F}^n$. *There is a unitary* $U \in \mathsf{M}_n(\mathbb{F})$ *such that* $U\mathbf{x} = \mathbf{y}$ *if and only if* $\|\mathbf{x}\|_2 = \|\mathbf{y}\|_2$.

Proof ($\Rightarrow$) This is Theorem 7.1.13.g.
($\Leftarrow$) Let $c = \|\mathbf{x}\|_2 = \|\mathbf{y}\|_2$. If $c = 0$, then $\mathbf{x} = \mathbf{y} = \mathbf{0}$ and $U\mathbf{x} = \mathbf{y}$ for any unitary $U \in \mathsf{M}_n(\mathbb{F})$. If $c \neq 0$, use the construction in the preceding example to find unitary $V, W \in \mathsf{M}_n(\mathbb{F})$

such that $c^{-1}\mathbf{x}$ is the first column of V and $c^{-1}\mathbf{y}$ is the first column of W. Then $V\mathbf{e}_1 = c^{-1}\mathbf{x}$, $V^*\mathbf{x} = c\mathbf{e}_1$, and $W\mathbf{e}_1 = c^{-1}\mathbf{y}$. Consequently, $U = WV^*$ is a unitary matrix and $U\mathbf{x} = WV^*\mathbf{x} = W(c\mathbf{e}_1) = cc^{-1}\mathbf{y} = \mathbf{y}$. □

7.2 Change of Orthonormal Basis and Unitary Similarity

The change-of-basis matrix from one basis to another is invertible, and every invertible matrix arises in this way (Theorem 2.5.7). If both bases are orthonormal, then the change-of-basis matrix is unitary, and every unitary matrix arises in this way.

Theorem 7.2.1 *Let n be a positive integer, let $\mathcal{V}$ be an n-dimensional $\mathbb{F}$-inner product space, and let $\beta = \mathbf{v}_1, \mathbf{v}_2, \dots, \mathbf{v}_n$ be an orthonormal basis for $\mathcal{V}$.*

(a) *Let $\gamma = \mathbf{w}_1, \mathbf{w}_2, \dots, \mathbf{w}_n$ be an orthonormal basis for $\mathcal{V}$. The change-of-basis matrix $U = {}_\gamma[I]_\beta \in M_n(\mathbb{F})$ is unitary, $U = [\langle \mathbf{v}_j, \mathbf{w}_i \rangle]$, and ${}_\beta[I]_\gamma = U^*$.*

(b) *If $U \in M_n(\mathbb{F})$ is unitary, then there is an orthonormal basis γ for $\mathcal{V}$ such that $U = {}_\gamma[I]_\beta$ and $U^* = {}_\beta[I]_\gamma$.*

Proof (a) The identity (2.5.1) ensures that the jth column of $U = {}_\gamma[I]_\beta$ is $[\mathbf{v}_j]_\gamma$, which (6.2.8) identifies as

$$[\mathbf{v}_j]_\gamma = \begin{bmatrix} \langle \mathbf{v}_j, \mathbf{w}_1 \rangle \\ \langle \mathbf{v}_j, \mathbf{w}_2 \rangle \\ \vdots \\ \langle \mathbf{v}_j, \mathbf{w}_n \rangle \end{bmatrix}.$$

Therefore, $U = [\langle \mathbf{v}_j, \mathbf{w}_i \rangle]$. Parseval's identity (6.7.2) lets us compute the inner product of the jth and kth columns of U as

$$[\mathbf{v}_j]_\gamma^* [\mathbf{v}_k]_\gamma = \sum_{i=1}^n \overline{\langle \mathbf{v}_j, \mathbf{w}_i \rangle} \langle \mathbf{v}_k, \mathbf{w}_i \rangle = \langle \mathbf{v}_k, \mathbf{v}_j \rangle = \delta_{jk},$$

which proves that U is unitary. Theorem 2.5.7.a ensures that ${}_\beta[I]_\gamma = U^{-1}$, which is U^*.

(b) Let $U^* = [\mathbf{u}_1 \ \mathbf{u}_2 \ \dots \ \mathbf{u}_n]$ and define $\mathbf{w}_1, \mathbf{w}_2, \dots, \mathbf{w}_n \in \mathcal{V}$ by $[\mathbf{w}_j]_\beta = \mathbf{u}_j$. Parseval's identity (6.7.2) ensures that $\langle \mathbf{w}_i, \mathbf{w}_j \rangle = [\mathbf{w}_j]_\beta^* [\mathbf{w}_i]_\beta = \mathbf{u}_j^* \mathbf{u}_i = \delta_{ij}$ since U^* is unitary. Therefore, $\gamma = \mathbf{w}_1, \mathbf{w}_2, \dots, \mathbf{w}_n$ is orthonormal and hence it is linearly independent. It is a basis for $\mathcal{V}$ since $\dim \mathcal{V} = n$. Finally, $U^* = [[\mathbf{w}_1]_\beta \ [\mathbf{w}_2]_\beta \ \dots \ [\mathbf{w}_n]_\beta] = {}_\beta[I]_\gamma$ and $U = ({}_\beta[I]_\gamma)^{-1} = {}_\gamma[I]_\beta$. □

Example 7.2.2 In Example 7.1.14, let $\beta = \mathbf{v}_1, \mathbf{v}_2, \mathbf{v}_3$ and $\gamma = \mathbf{w}_1, \mathbf{w}_2, \mathbf{w}_3$. Then β and γ are orthonormal bases of $\mathbb{R}^3$ and

$$_\gamma[I]_\beta = [\langle \mathbf{w}_i^* \mathbf{v}_j \rangle] = W^* V = \frac{1}{6} \begin{bmatrix} 2\sqrt{2}+1 & 4-\sqrt{2} & 2\sqrt{2}+1 \\ 3\sqrt{2} & 0 & -3\sqrt{2} \\ 2\sqrt{2}-1 & -4-\sqrt{2} & 2\sqrt{2}-1 \end{bmatrix}$$

is the change-of-basis matrix from β to γ. It is a product of two unitary matrices, so it is unitary.

Now we can relate the matrix representations of a linear transformation with respect to two orthonormal bases.

Corollary 7.2.3 *Let $n \geq 1$, let $\mathcal{V}$ be an n-dimensional $\mathbb{F}$-inner product space, and let $T \in \mathcal{L}(\mathcal{V})$.*

(a) *Let β and γ be orthonormal bases for $\mathcal{V}$. Then $U = {}_\gamma[I]_\beta$ is unitary and*

$$_\gamma[T]_\gamma = U_\beta[T]_\beta U^*. \tag{7.2.4}$$

(b) *Let $U \in \mathsf{M}_n(\mathbb{F})$ be unitary and let β be an orthonormal basis for $\mathcal{V}$. Then there is an orthonormal basis γ for $\mathcal{V}$ such that $_\gamma[T]_\gamma = U_\beta[T]_\beta U^*$.*

Proof (a) and (b) follow from the preceding theorem and Theorem 2.6.1. $\square$

The identity (7.2.4) motivates the following definition, which is a specialization of Definition 2.6.4.

Definition 7.2.5 $A, B \in \mathsf{M}_n$ are *unitarily similar* if there is a unitary $U \in \mathsf{M}_n$ such that $A = UBU^*$; they are *real orthogonally similar* if there is a real orthogonal $Q \in \mathsf{M}_n$ such that $A = QBQ^\mathsf{T}$.

Example 7.2.6 Consider $A = \begin{bmatrix} 1 & 1 \\ 1 & 1 \end{bmatrix}$ and $U = \frac{1}{\sqrt{2}} \begin{bmatrix} 1 & 1 \\ 1 & -1 \end{bmatrix}$. Then U is real orthogonal and A is real orthogonally similar to $UAU^\mathsf{T} = \frac{1}{2} \begin{bmatrix} 1 & 1 \\ 1 & -1 \end{bmatrix} \begin{bmatrix} 1 & 1 \\ 1 & 1 \end{bmatrix} \begin{bmatrix} 1 & 1 \\ 1 & -1 \end{bmatrix} = \begin{bmatrix} 2 & 0 \\ 0 & 0 \end{bmatrix}$.

Example 7.2.7 Consider $A = \begin{bmatrix} i & 1 \\ 1 & -i \end{bmatrix}$ and $U = \frac{1}{\sqrt{2}} \begin{bmatrix} 1 & i \\ i & 1 \end{bmatrix}$. Then U is unitary and A is unitarily similar to $UAU^* = \frac{1}{2} \begin{bmatrix} 1 & i \\ i & 1 \end{bmatrix} \begin{bmatrix} i & 1 \\ 1 & -i \end{bmatrix} \begin{bmatrix} 1 & -i \\ -i & 1 \end{bmatrix} = \begin{bmatrix} 0 & 2 \\ 0 & 0 \end{bmatrix}$.

Two matrices $A, B \in \mathsf{M}_n$ are similar if and only if they are matrix representations for the same linear operator. Corollary 7.2.3 shows that both bases can be chosen to be orthonormal if and only if A and B are unitarily similar. In Chapter 12, we present a simple necessary and sufficient condition for A and B to be similar. However, there is no analogous condition known for unitary similarity. Unitarily similar matrices are similar, but similar matrices need not be unitarily similar. The necessary condition for unitary similarity in the following theorem involves the Frobenius norm; for other necessary conditions see P.7.23.

Theorem 7.2.8 *If $A, B \in \mathsf{M}_n$ are unitarily similar, then $\|A\|_\mathrm{F}^2 = \|B\|_\mathrm{F}^2$; that is,*

$$\operatorname{tr} A^*A = \operatorname{tr} B^*B. \tag{7.2.9}$$

Proof Suppose that $A = UBU^*$, in which $U \in \mathsf{M}_n$ is unitary. Then $U(B^*B)U^*$ is similar to B^*B, so Theorem 2.6.8 ensures that

$$\|A\|_\mathrm{F}^2 = \operatorname{tr} A^*A = \operatorname{tr} UB^*U^*UBU^* = \operatorname{tr} UB^*BU^* = \operatorname{tr} B^*B = \|B\|_\mathrm{F}^2. \qquad \square$$

Example 7.2.10 Let $A = \begin{bmatrix} 0 & 1 \\ 0 & 0 \end{bmatrix}$, $B = \begin{bmatrix} 0 & 2 \\ 0 & 0 \end{bmatrix}$, and $S = \begin{bmatrix} 1 & 0 \\ 0 & 2 \end{bmatrix}$. Then

$$SBS^{-1} = \begin{bmatrix} 1 & 0 \\ 0 & 2 \end{bmatrix} \begin{bmatrix} 0 & 2 \\ 0 & 0 \end{bmatrix} \begin{bmatrix} 1 & 0 \\ 0 & \frac{1}{2} \end{bmatrix} = \begin{bmatrix} 1 & 0 \\ 0 & 2 \end{bmatrix} \begin{bmatrix} 0 & 1 \\ 0 & 0 \end{bmatrix} = \begin{bmatrix} 0 & 1 \\ 0 & 0 \end{bmatrix} = A,$$

so A and B are similar. However, $\|A\|_\mathrm{F}^2 = 1$ and $\|B\|_\mathrm{F}^2 = 4$, so the preceding theorem ensures that A and B are not unitarily similar.

Like similarity, unitary similarity is an equivalence relation; see Definition 2.6.11.

Example 7.2.11 A complex matrix whose rows exhibit the cyclic pattern

$$A = \begin{bmatrix} a & b & c & d \\ d & a & b & c \\ c & d & a & b \\ b & c & d & a \end{bmatrix} \tag{7.2.12}$$

is a *circulant matrix*. It is remarkable that a unitary similarity of A, via a Fourier matrix (7.1.19), achieves a diagonal form:

$$F_4^* A F_4 = \frac{1}{2} \begin{bmatrix} 1 & 1 & 1 & 1 \\ 1 & -i & -1 & i \\ 1 & -1 & 1 & -1 \\ 1 & i & -1 & -i \end{bmatrix} \begin{bmatrix} a & b & c & d \\ d & a & b & c \\ c & d & a & b \\ b & c & d & a \end{bmatrix} F_4$$

$$= \frac{1}{4} \begin{bmatrix} a+b+c+d & a+b+c+d & a+b+c+d & a+b+c+d \\ a+ib-c-id & -ia+b+ic-d & -a-ib+c+id & ia-b-ic+d \\ a-b+c-d & -a+b-c+d & a-b+c-d & -a+b-c+d \\ a-ib-c+id & ia+b-ic-d & -a+ib+c-id & -ia-b+ic+d \end{bmatrix}$$

$$\begin{bmatrix} 1 & 1 & 1 & 1 \\ 1 & i & -1 & -i \\ 1 & -1 & 1 & -1 \\ 1 & -i & -1 & i \end{bmatrix}$$

$$= \begin{bmatrix} a+b+c+d & 0 & 0 & 0 \\ 0 & a+ib-c-id & 0 & 0 \\ 0 & 0 & a-b+c-d & 0 \\ 0 & 0 & 0 & a-ib-c+id \end{bmatrix}. \tag{7.2.13}$$

The linear transformation whose matrix representation with respect to the standard orthonormal basis is (7.2.12) is represented as a diagonal matrix with respect to the orthonormal basis whose elements are the columns of the Fourier matrix F_4. Furthermore, let $p(z) = dz^3 + cz^2 + bz + a$ and let $\omega = e^{2\pi i/4} = i$. The diagonal entries of (7.2.13) are $p(\omega^0) = a + b + c + d$, $p(\omega) = a + ib - c - id$, $p(\omega^2) = a - b + c - d$, and $p(\omega^3) = a - ib - c + id$. This is not an accident; see Section 14.5.

7.3 Permutation Matrices

Definition 7.3.1 A square matrix A is a *permutation matrix* if exactly one entry in each row and in each column is 1; all other entries are 0.

Multiplying a matrix on the left by a permutation matrix permutes rows:

$$\begin{bmatrix} 0 & 1 & 0 \\ 0 & 0 & 1 \\ 1 & 0 & 0 \end{bmatrix} \begin{bmatrix} 1 & 2 & 3 \\ 4 & 5 & 6 \\ 7 & 8 & 9 \end{bmatrix} = \begin{bmatrix} 4 & 5 & 6 \\ 7 & 8 & 9 \\ 1 & 2 & 3 \end{bmatrix}.$$

Multiplying a matrix on the right by a permutation matrix permutes columns:

$$\begin{bmatrix} 1 & 2 & 3 \\ 4 & 5 & 6 \\ 7 & 8 & 9 \end{bmatrix} \begin{bmatrix} 0 & 1 & 0 \\ 0 & 0 & 1 \\ 1 & 0 & 0 \end{bmatrix} = \begin{bmatrix} 3 & 1 & 2 \\ 6 & 4 & 5 \\ 9 & 7 & 8 \end{bmatrix}.$$

The columns of an $n \times n$ permutation matrix are a permutation of the standard basis vectors in $\mathbb{R}^n$. If $\sigma : \{1, 2, \ldots, n\} \to \{1, 2, \ldots, n\}$ is a function, then

$$P = [\mathbf{e}_{\sigma(1)} \ \mathbf{e}_{\sigma(2)} \ \cdots \ \mathbf{e}_{\sigma(n)}] \in \mathsf{M}_n \tag{7.3.2}$$

is a permutation matrix if and only if σ is a permutation of the list $1, 2, \ldots, n$.

The transpose of a permutation matrix is a permutation matrix. If P is the permutation matrix (7.3.2), then

$$P^{\mathsf{T}} P = [\mathbf{e}_{\sigma(i)}^{\mathsf{T}} \mathbf{e}_{\sigma(j)}] = [\delta_{ij}] = I_n.$$

This says that P^{T} is the inverse of P, so permutation matrices are real orthogonal.

Definition 7.3.3 Square matrices A, B are *permutation similar* if there is a permutation matrix P such that $A = PBP^{\mathsf{T}}$.

A permutation similarity rearranges columns and rows:

$$\begin{bmatrix} 0 & 1 & 0 \\ 0 & 0 & 1 \\ 1 & 0 & 0 \end{bmatrix} \begin{bmatrix} 1 & 2 & 3 \\ 4 & 5 & 6 \\ 7 & 8 & 9 \end{bmatrix} \begin{bmatrix} 0 & 0 & 1 \\ 1 & 0 & 0 \\ 0 & 1 & 0 \end{bmatrix} = \begin{bmatrix} 5 & 6 & 4 \\ 8 & 9 & 7 \\ 2 & 3 & 1 \end{bmatrix}.$$

Entries that start on the diagonal remain on the diagonal, but their positions within the diagonal may be rearranged. Entries that start in a row together remain together in a row, but the row may be moved and the entries in the row may be rearranged. Entries that start in a column together remain together in a column, but the column may be moved and the entries in the column may be rearranged.

Permutation similarities can permute the diagonal entries of a matrix to achieve a particular pattern. For example, we might want equal diagonal entries grouped together. If a matrix has real diagonal entries, we might want to rearrange them in increasing order. If $\Lambda = \mathrm{diag}(\lambda_1, \lambda_2, \ldots, \lambda_n)$, inspection of the permutation similarity

$$\begin{aligned} P^{\mathsf{T}} \Lambda P &= [\mathbf{e}_{\sigma(1)} \ \mathbf{e}_{\sigma(2)} \ \cdots \ \mathbf{e}_{\sigma(n)}]^{\mathsf{T}} \mathrm{diag}(\lambda_1, \lambda_2, \ldots, \lambda_n)[\mathbf{e}_{\sigma(1)} \ \mathbf{e}_{\sigma(2)} \ \cdots \ \mathbf{e}_{\sigma(n)}] \\ &= [\lambda_{\sigma(1)} \mathbf{e}_{\sigma(1)} \ \lambda_{\sigma(2)} \mathbf{e}_{\sigma(2)} \ \cdots \ \lambda_{\sigma(n)} \mathbf{e}_{\sigma(n)}]^{\mathsf{T}} [\mathbf{e}_{\sigma(1)} \ \mathbf{e}_{\sigma(2)} \ \cdots \ \mathbf{e}_{\sigma(n)}] \\ &= [\lambda_{\sigma(i)} \mathbf{e}_{\sigma(i)}^{\mathsf{T}} \mathbf{e}_{\sigma(j)}] = [\lambda_{\sigma(i)} \delta_{ij}] \\ &= \mathrm{diag}(\lambda_{\sigma(1)}, \lambda_{\sigma(2)}, \ldots, \lambda_{\sigma(n)}) \end{aligned} \tag{7.3.4}$$

reveals how to arrange the standard basis vectors in the columns of P to achieve a particular pattern for the rearranged diagonal entries of a square matrix whose diagonal is Λ.

Example 7.3.5 A permutation similarity via the permutation matrix $K_n = [\mathbf{e}_n \ \mathbf{e}_{n-1} \ \cdots \ \mathbf{e}_1]$ in (7.1.11) reverses the orders of both rows and columns, so it reverses the order of diagonal entries. If $\Lambda = \mathrm{diag}(\lambda_1, \lambda_2, \ldots, \lambda_n)$, then $K_n^{\mathsf{T}} \Lambda K_n = K_n \Lambda K_n = \mathrm{diag}(\lambda_n, \lambda_{n-1}, \ldots, \lambda_1)$.

Permutation matrices can be used to rearrange diagonal blocks of a matrix.

Example 7.3.6 Let $A = B \oplus C$, in which $B \in \mathsf{M}_p$ and $C \in \mathsf{M}_q$. Let

$$P = \begin{bmatrix} 0_{p \times q} & I_p \\ I_q & 0_{q \times p} \end{bmatrix} \in \mathsf{M}_{p+q}.$$

Then P is a permutation matrix and

$$
\begin{aligned}
P^\mathsf{T} A P &= \begin{bmatrix} 0_{q \times p} & I_q \\ I_p & 0_{p \times q} \end{bmatrix} \begin{bmatrix} B & 0 \\ 0 & C \end{bmatrix} \begin{bmatrix} 0_{p \times q} & I_p \\ I_q & 0_{q \times p} \end{bmatrix} \\
&= \begin{bmatrix} 0_{q \times p} & C \\ B & 0_{p \times q} \end{bmatrix} \begin{bmatrix} 0_{p \times q} & I_p \\ I_q & 0_{q \times p} \end{bmatrix} = \begin{bmatrix} C & 0 \\ 0 & B \end{bmatrix}.
\end{aligned}
$$

Another useful permutation similarity involves a 2×2 block matrix of diagonal matrices.

Example 7.3.7 The matrices

$$
A = \begin{bmatrix} a & 0 & c & 0 \\ 0 & b & 0 & d \\ e & 0 & g & 0 \\ 0 & f & 0 & h \end{bmatrix} \quad \text{and} \quad B = \begin{bmatrix} a & c & 0 & 0 \\ e & g & 0 & 0 \\ 0 & 0 & b & d \\ 0 & 0 & f & h \end{bmatrix}
$$

are permutation similar via $P = [\mathbf{e}_1 \ \mathbf{e}_3 \ \mathbf{e}_2 \ \mathbf{e}_4]$, that is, $B = PAP^\mathsf{T}$. In general, if $\Lambda = \operatorname{diag}(\lambda_1, \lambda_2, \ldots, \lambda_n)$, $M = \operatorname{diag}(\mu_1, \mu_2, \ldots, \mu_n)$, $N = \operatorname{diag}(\nu_1, \nu_2, \ldots, \nu_n)$, and $T = \operatorname{diag}(\tau_1, \tau_2, \ldots, \tau_n)$, then

$$
\begin{bmatrix} \Lambda & M \\ N & T \end{bmatrix} \quad \text{and} \quad \begin{bmatrix} \lambda_1 & \mu_1 \\ \nu_1 & \tau_1 \end{bmatrix} \oplus \begin{bmatrix} \lambda_2 & \mu_2 \\ \nu_2 & \tau_2 \end{bmatrix} \oplus \cdots \oplus \begin{bmatrix} \lambda_n & \mu_n \\ \nu_n & \tau_n \end{bmatrix} \tag{7.3.8}
$$

are permutation similar.

Example 7.3.9 Let $n \geq 2$ and let $A, I \in \mathsf{M}_{n-1}$. Then $P = \begin{bmatrix} \mathbf{0}^\mathsf{T} & 1 \\ I & \mathbf{0} \end{bmatrix}$ is a permutation matrix (see Example 7.3.6) and a computation reveals that $P \begin{bmatrix} A & \mathbf{x} \\ \mathbf{y}^\mathsf{T} & c \end{bmatrix} P^\mathsf{T} = \begin{bmatrix} c & \mathbf{y}^\mathsf{T} \\ \mathbf{x} & A \end{bmatrix}$. This permutation similarity and (3.3.17) show that

$$
\det \begin{bmatrix} A & \mathbf{x} \\ \mathbf{y}^\mathsf{T} & c \end{bmatrix} = \det \begin{bmatrix} c & \mathbf{y}^\mathsf{T} \\ \mathbf{x} & A \end{bmatrix} = c \det A - \mathbf{y}^\mathsf{T}(\operatorname{adj} A)\mathbf{x}.
$$

7.4 The *QR* Factorization

The *QR factorization* of an $m \times n$ matrix A (also called the *QR decomposition*) presents it as a product of an $m \times n$ matrix Q with orthonormal columns and a square upper triangular matrix R with nonnegative diagonal entries; this presentation requires that $m \geq n$. If A is square, the factor Q is unitary. If A is real, both factors Q and R may be chosen to be real.

The *QR* factorization is an important tool for numerical solution of least-squares problems, for transforming a basis into an orthonormal basis, and for numerical computation of eigenvalues.

Example 7.4.1 A computation verifies that $A = \begin{bmatrix} 3 & 1 \\ 4 & 2 \end{bmatrix} = \begin{bmatrix} \frac{3}{5} & -\frac{4}{5} \\ \frac{4}{5} & \frac{3}{5} \end{bmatrix} \begin{bmatrix} 5 & \frac{11}{5} \\ 0 & \frac{2}{5} \end{bmatrix}$ is a *QR* factorization of A.

The algorithm described in the following theorem uses unitary matrices with prescribed first columns to annihilate the lower entries of a sequence of vectors, thereby transforming a given matrix into an upper triangular matrix.

Theorem 7.4.2 (*QR* Factorization) *Let $A \in \mathsf{M}_{m \times n}(\mathbb{F})$ and suppose that $m \geq n$.*

(a) *There is a unitary $V \in \mathsf{M}_m(\mathbb{F})$ and an upper triangular $R \in \mathsf{M}_n(\mathbb{F})$ with real nonnegative diagonal entries such that*

$$A = V \begin{bmatrix} R \\ 0 \end{bmatrix} \tag{7.4.3}$$

if $m > n$; if $m = n$, then $A = VR$. If $V = [Q \ Q']$, in which $Q \in \mathsf{M}_{m \times n}(\mathbb{F})$ contains the first n columns of V, then Q has orthonormal columns and

$$A = QR. \tag{7.4.4}$$

(b) *If $\operatorname{rank} A = n$, then the factors Q and R in (7.4.4) are unique, R has positive diagonal entries, and the columns of Q are an orthonormal basis for $\operatorname{col} A$.*

Proof (a) Suppose that $m > n$. Let $\mathbf{a}_1$ be the first column of A. If $\mathbf{a}_1 = \mathbf{0}$, let $U_1 = I_m$. Then $U_1^* A$ has a zero first column. If $\mathbf{a}_1 \neq \mathbf{0}$, let $U_1 \in \mathsf{M}_m(\mathbb{F})$ be a unitary matrix whose first column is $\mathbf{a}_1 / \|\mathbf{a}_1\|_2$. The algorithms in Example 7.1.21 and P.7.31 show how to construct such a matrix. In either case, $U_1^* \mathbf{a}_1 = \|\mathbf{a}_1\|_2 \mathbf{e}_1$. Then

$$U_1^* A = \begin{bmatrix} r_{11} & \star \\ \mathbf{0} & A' \end{bmatrix}, \qquad \text{in which } A' = [\mathbf{a}_1' \ \mathbf{a}_2' \ \dots \ \mathbf{a}_{n-1}'] \in \mathsf{M}_{(m-1) \times (n-1)}(\mathbb{F}) \tag{7.4.5}$$

and $r_{11} = \|\mathbf{a}_1\|_2 \geq 0$. If $\mathbf{a}_1' = \mathbf{0}$, let $V = I_{n-1}$. If $\mathbf{a}_1' \neq \mathbf{0}$, let $V \in \mathsf{M}_{m-1}(\mathbb{F})$ be a unitary matrix whose first column is $\mathbf{a}_1' / \|\mathbf{a}_1'\|_2$. In either case, $V^* \mathbf{a}_1' = \|\mathbf{a}_1'\|_2 \mathbf{e}_1 \in \mathbb{F}^{m-1}$. Let $r_{22} = \|\mathbf{a}_1'\|_2$ and let $U_2 = I_1 \oplus V$. Then $r_{22} \geq 0$ and

$$U_2^* U_1^* A = \begin{bmatrix} r_{11} & \star & \star \\ 0 & r_{22} & \star \\ \mathbf{0} & \mathbf{0} & A'' \end{bmatrix}, \qquad \text{in which } A'' \in \mathsf{M}_{(m-2) \times (n-2)}(\mathbb{F}). \tag{7.4.6}$$

The direct-sum structure of U_2 ensures that the reduction achieved in (7.4.6) affects only the lower-right block A' in (7.4.5). After n reduction steps we obtain $U_n^* \cdots U_2^* U_1^* A = \begin{bmatrix} R \\ 0 \end{bmatrix}$, in which

$$R = \begin{bmatrix} r_{11} & \star & \star & \star \\ & r_{22} & \star & \star \\ & & \ddots & \star \\ 0 & & & r_{nn} \end{bmatrix} \in \mathsf{M}_n(\mathbb{F})$$

is upper triangular and has diagonal entries that are nonnegative (each is the Euclidean length of some vector). Let $U = U_1 U_2 \cdots U_n$ and partition $U = [Q \ Q']$, in which $Q \in \mathsf{M}_{m \times n}$. The block Q has orthonormal columns since it comprises the first n columns of the unitary matrix U. Then

$$U^* A = \begin{bmatrix} R \\ 0 \end{bmatrix} \qquad \text{and} \qquad A = U \begin{bmatrix} R \\ 0 \end{bmatrix} = [Q \ Q'] \begin{bmatrix} R \\ 0 \end{bmatrix} = QR.$$

If $m = n$, then the zero block below R in (7.4.3) is missing.

(b) The factor Q in (7.4.4) has linearly independent columns, so it has full column rank. If $\operatorname{rank} A = n$, then (4.2.5) ensures that $n = \operatorname{rank} A = \operatorname{rank} QR \leq \min\{n, \operatorname{rank} R\} = \operatorname{rank} R \leq n$. Therefore, the upper triangular matrix R is invertible and hence its diagonal entries are positive. Suppose that $A = Q_1 R_1 = Q_2 R_2$, in which Q_1 and Q_2 have orthonormal columns, and

R_1 and R_2 are upper triangular and have positive diagonal entries. Then $A^* = R_1^* Q_1^* = R_2^* Q_2^*$, so $A^*A = R_1^* Q_1^* Q_1 R_1 = R_1^* R_1$ and $A^*A = R_2^* Q_2^* Q_2 R_2 = R_2^* R_2$. This means that $R_1^* R_1 = R_2^* R_2$, and hence

$$R_1 R_2^{-1} = R_1^{-*} R_2^* = (R_2 R_1^{-1})^*. \tag{7.4.7}$$

We know that R_1^{-1} and R_2^{-1} are upper triangular and have positive diagonal entries (Theorem 3.2.9). Thus, $R_1 R_2^{-1}$ and $R_2 R_1^{-1}$ (products of upper triangular matrices with positive diagonal entries) are upper triangular. The identity (7.4.7) says that an upper triangular matrix equals a lower triangular matrix. Therefore, $R_1 R_2^{-1} = R_1^{-*} R_2^* = D$ is a diagonal matrix with positive diagonal entries. Then

$$D = R_1^{-*} R_2^* = (DR_2)^{-*} R_2^* = D^{-1} R_2^{-*} R_2^* = D^{-1},$$

so $D^2 = I$. Since D is diagonal and has positive diagonal entries, we conclude that $D = I$. Therefore, $R_1 = R_2$ and $Q_1 = Q_2$. Finally, the identities $A = QR$ and $Q = AR^{-1}$ ensure that $\mathrm{col}\, A \subseteq \mathrm{col}\, Q$ and $\mathrm{col}\, Q \subseteq \mathrm{col}\, A$, so $\mathrm{col}\, A = \mathrm{col}\, Q$. □

An important, if subtle, part of the statement of the preceding theorem is that if A is real, then the matrices U, R, and Q in (7.4.3) and (7.4.4) are real. The final sentence in Example 7.1.21 is the key to understanding why. If A is real, then every step in the constructive proof of Theorem 7.4.2 can be performed with real vectors and matrices.

Definition 7.4.8 The factorization (7.4.3) is the *wide QR factorization* of A; (7.4.4) is the *(narrow) QR factorization* of A.

Example 7.4.9 The factorizations

$$A = \begin{bmatrix} 1 & -5 \\ -2 & 4 \\ 2 & 2 \end{bmatrix} = \begin{bmatrix} \frac{1}{3} & -\frac{2}{3} & \frac{2}{3} \\ -\frac{2}{3} & \frac{1}{3} & \frac{2}{3} \\ \frac{2}{3} & \frac{2}{3} & \frac{1}{3} \end{bmatrix} \underbrace{\begin{bmatrix} 3 & -3 \\ 0 & 6 \\ 0 & 0 \end{bmatrix}}_{U} = U \begin{bmatrix} R \\ 0 \end{bmatrix} \tag{7.4.10}$$

and

$$A = \begin{bmatrix} 1 & -5 \\ -2 & 4 \\ 2 & 2 \end{bmatrix} = \underbrace{\begin{bmatrix} \frac{1}{3} & -\frac{2}{3} \\ -\frac{2}{3} & \frac{1}{3} \\ \frac{2}{3} & \frac{2}{3} \end{bmatrix}}_{Q} \begin{bmatrix} 3 & -3 \\ 0 & 6 \end{bmatrix} = QR \tag{7.4.11}$$

are examples of wide and narrow QR factorizations, respectively.

Example 7.4.12 The display (7.4.13) illustrates the algorithm in the preceding theorem with a 4×3 matrix A. The symbol $\star$ indicates an entry that is not necessarily zero; $\bigstar$ indicates an entry that has just been changed and is not necessarily zero; $\tilde{r}_{ii}$ indicates a nonnegative diagonal entry that has just been created; $\tilde{0}$ indicates a zero entry that has just been created.

$$\begin{bmatrix} \star & \star & \star \\ \star & \star & \star \\ \star & \star & \star \\ \star & \star & \star \end{bmatrix} \xrightarrow{U_1^*} \begin{bmatrix} \tilde{r}_{11} & \bigstar & \bigstar \\ \tilde{0} & \bigstar & \bigstar \\ \tilde{0} & \bigstar & \bigstar \\ \tilde{0} & \bigstar & \bigstar \end{bmatrix} \xrightarrow{U_2^*} \begin{bmatrix} r_{11} & \star & \star \\ 0 & \tilde{r}_{22} & \bigstar \\ 0 & \tilde{0} & \bigstar \\ 0 & \tilde{0} & \bigstar \end{bmatrix} \xrightarrow{U_3^*} \begin{bmatrix} r_{11} & \star & \star \\ 0 & r_{22} & \star \\ 0 & 0 & \tilde{r}_{33} \\ 0 & 0 & \tilde{0} \end{bmatrix} \tag{7.4.13}$$
$$\quad\ A \qquad\qquad\quad U_1^* A \qquad\qquad\quad U_2^* U_1^* A \qquad\qquad U_3^* U_2^* U_1^* A$$

The 4×4 unitary matrices U_1, $U_2 = I_1 \oplus U'$, and $U_3 = I_2 \oplus U''$ annihilate the lower entries of successive columns of A without disturbing previously created lower column zeros. The transformation $A \mapsto U_3^* U_2^* U_1^* A$ creates an upper triangular matrix with nonnegative diagonal entries that are positive if A has full column rank. In the QR decomposition of A, Q consists of the first three columns of the 4×4 unitary matrix $U_1 U_2 U_3$, and R is the upper 3×3 block of the 4×3 matrix $U_3^* U_2^* U_1^* A$.

Example 7.4.14 Let us apply the algorithm in Theorem 7.4.2 to find the unique QR factorization of the invertible matrix $A = \begin{bmatrix} 3 & 1 \\ 4 & 2 \end{bmatrix}$. Use the construction in Example 7.1.21 to find a unitary matrix whose first column is $\frac{1}{5}\begin{bmatrix} 3 \\ 4 \end{bmatrix}$. We have $x = \frac{3}{5}$, $\mathbf{w} = [\frac{4}{5}]$, $(1 + x)^{-1} = \frac{5}{8}$, $\mathbf{w}\mathbf{w}^* = [\frac{16}{25}]$, and

$$U_1 = \begin{bmatrix} x & \mathbf{w}^* \\ \mathbf{w} & -I_1 + \frac{1}{1+x}\mathbf{w}\mathbf{w}^* \end{bmatrix} = \begin{bmatrix} \frac{3}{5} & \frac{4}{5} \\ \frac{4}{5} & -1 + \frac{5}{8} \cdot \frac{16}{25} \end{bmatrix} = \frac{1}{5}\begin{bmatrix} 3 & 4 \\ 4 & -3 \end{bmatrix}.$$

Then

$$U_1^{\mathsf{T}} A = \frac{1}{5}\begin{bmatrix} 3 & 4 \\ 4 & -3 \end{bmatrix}\begin{bmatrix} 3 & 1 \\ 4 & 2 \end{bmatrix} = \begin{bmatrix} 5 & \frac{11}{5} \\ 0 & -\frac{2}{5} \end{bmatrix}.$$

Finally, $U_2 = I_1 \oplus [-1]$ and

$$U_2^{\mathsf{T}} U_1^{\mathsf{T}} A = \begin{bmatrix} 1 & 0 \\ 0 & -1 \end{bmatrix}\begin{bmatrix} 5 & \frac{11}{5} \\ 0 & -\frac{2}{5} \end{bmatrix} = \begin{bmatrix} 5 & \frac{11}{5} \\ 0 & \frac{2}{5} \end{bmatrix} = R.$$

The QR factorization

$$A = (U_1 U_2)R = \underbrace{\begin{bmatrix} \frac{3}{5} & -\frac{4}{5} \\ \frac{4}{5} & \frac{3}{5} \end{bmatrix}}_{Q} \underbrace{\begin{bmatrix} 5 & \frac{11}{5} \\ 0 & \frac{2}{5} \end{bmatrix}}_{R}$$

is unique since rank $A = 2$.

Let $A \in \mathsf{M}_n$, let $A = QR$ be a QR factorization, and let $\mathbf{x}, \mathbf{y} \in \mathbb{C}^n$. If $A\mathbf{x} = \mathbf{y}$, then $QR\mathbf{x} = \mathbf{y}$ and hence $R\mathbf{x} = Q^* QR\mathbf{x} = Q^* A\mathbf{x} = Q^* \mathbf{y}$, in which Q is unitary. Conversely, if $R\mathbf{x} = Q^* \mathbf{y}$, then $A\mathbf{x} = QR\mathbf{x} = QQ^*\mathbf{y} = \mathbf{y}$. This shows that the linear systems $A\mathbf{x} = \mathbf{y}$ and $R\mathbf{x} = Q^*\mathbf{y}$ are *equivalent*: they have the same solutions. If A (and hence R) is invertible, then $R\mathbf{x} = Q^*\mathbf{y}$ can be solved using backward substitution.

Example 7.4.15 Let A be the 2×2 matrix in Example 7.4.14 and let $\mathbf{y} = [1 \ -1]^{\mathsf{T}}$. The linear systems $A\mathbf{x} = \mathbf{y}$ and $R\mathbf{x} = Q^*\mathbf{y}$ are equivalent. We have

$$R\mathbf{x} = \frac{1}{5}\begin{bmatrix} 25 & 11 \\ 0 & 2 \end{bmatrix}\begin{bmatrix} x_1 \\ x_2 \end{bmatrix} = \frac{1}{5}\begin{bmatrix} 25x_1 + 11x_2 \\ 2x_2 \end{bmatrix}$$

and

$$Q^*\mathbf{y} = \frac{1}{5}\begin{bmatrix} 3 & 4 \\ -4 & 3 \end{bmatrix}\begin{bmatrix} 1 \\ -1 \end{bmatrix} = \frac{1}{5}\begin{bmatrix} -1 \\ -7 \end{bmatrix}.$$

Then $2x_2 = -7$, so $x_2 = -\frac{7}{2}$. Consequently, $25x_1 + 11x_2 = 25x_1 - \frac{77}{2} = -1$, so $x_1 = \frac{3}{2}$.

Caution: If $A \in \mathsf{M}_{m \times n}$, $m > n$, and $A = QR$ is a QR factorization, the linear systems $A\mathbf{x} = \mathbf{y}$ and $R\mathbf{x} = Q^*\mathbf{y}$ are not equivalent; see P.7.41. The issue here is that $Q^*Q = I_n$ but $QQ^* \neq I_m$.

For a discussion of the role of QR factorizations in numerical computation of a least-squares solution to a linear system, see Section 8.5. Their role in orthogonalizing a list of linearly independent vectors is revealed in the following corollary.

Corollary 7.4.16 *Let $A \in \mathsf{M}_{m \times n}(\mathbb{F})$. Suppose that $m \geq n$ and rank $A = n$. Let $A = QR$, in which $Q \in \mathsf{M}_{m \times n}(\mathbb{F})$ has orthonormal columns and $R \in \mathsf{M}_n(\mathbb{F})$ is upper triangular and has positive diagonal entries. For each $k = 1, 2, \ldots, n$, let A_k and Q_k denote the submatrices of A and Q, respectively, comprising their first k columns. Then*

$$\operatorname{col} A_k = \operatorname{col} Q_k \quad \text{for } k = 1, 2, \ldots, n,$$

that is, the columns of each submatrix Q_k are an orthonormal basis for $\operatorname{col} A_k$.

Proof The preceding theorem ensures that a factorization of the stated form exists. Let R_k denote the leading $k \times k$ principal submatrix of R; it has positive diagonal entries, so it is invertible. Then

$$[A_k \; \star] = A = [Q_k \; \star] \begin{bmatrix} R_k & \star \\ 0 & \star \end{bmatrix} = [Q_k R_k \; \star],$$

so $A_k = Q_k R_k$ and $\operatorname{col} A_k \subseteq \operatorname{col} Q_k$ for each $k = 1, 2, \ldots, n$. Since $Q_k = A_k R_k^{-1}$, it follows that $\operatorname{col} Q_k \subseteq \operatorname{col} A_k$ for each $k = 1, 2, \ldots, n$. $\square$

7.5 Upper Hessenberg Matrices

Reduction of a matrix to some standard form that contains many zero entries is a common procedure in numerical linear algebra algorithms. One of those standard forms is described in the following definition.

Definition 7.5.1 A matrix $A = [a_{ij}] \in \mathsf{M}_{m \times n}$ is *upper Hessenberg* if $a_{ij} = 0$ whenever $i > j + 1$.

Example 7.5.2 5×6 and 6×5 upper Hessenberg matrices have the following patterns of zero entries

$$\begin{bmatrix} \star & \star & \star & \star & \star & \star \\ \star & \star & \star & \star & \star & \star \\ 0 & \star & \star & \star & \star & \star \\ 0 & 0 & \star & \star & \star & \star \\ 0 & 0 & 0 & \star & \star & \star \end{bmatrix} \quad \text{and} \quad \begin{bmatrix} \star & \star & \star & \star & \star \\ \star & \star & \star & \star & \star \\ 0 & \star & \star & \star & \star \\ 0 & 0 & \star & \star & \star \\ 0 & 0 & 0 & \star & \star \\ 0 & 0 & 0 & 0 & \star \end{bmatrix},$$

respectively. All entries below the first subdiagonal are zero.

Every square matrix is unitarily similar to an upper Hessenberg matrix, and the unitary similarity can be constructed from a sequence of unitary matrices with prescribed first columns. If the matrix is real, all the unitary matrices can be chosen to be real orthogonal. Just as in our construction of the QR factorization, the basic idea is unitary annihilation of the lower entries of a sequence of vectors. However, those vectors must now be chosen in a different way to ensure that previous annihilations are not disturbed by subsequent unitary similarities.

Theorem 7.5.3 *If $n \geq 2$ and $A = [a_{ij}] \in M_n(\mathbb{F})$, there is a unitary $U \in M_n(\mathbb{F})$ such that $U^*AU = [b_{ij}]$ is upper Hessenberg and each subdiagonal entry $b_{2,1}, b_{3,2}, \ldots, b_{n,n-1}$ is real and nonnegative.*

Proof If $n = 2$ and $a_{21} = e^{i\theta}|a_{21}|$, let $U = \mathrm{diag}(1, e^{i\theta})$. Then $U^*AU = \begin{bmatrix} a_{11} & a_{12}e^{i\theta} \\ |a_{21}| & a_{22} \end{bmatrix}$. Now suppose that $n \geq 3$. We are not concerned with the entries in the first row of A. Let $A = \begin{bmatrix} \star \\ A' \end{bmatrix}$ and $A' \in M_{(n-1) \times n}$. Let $\mathbf{a}_1' \in \mathbb{F}^{n-1}$ be the first column of A'. If $\mathbf{a}_1' = \mathbf{0}$, let $V_1 = I_{n-1}$. If $\mathbf{a}_1' \neq \mathbf{0}$, let $V_1 \in M_{n-1}(\mathbb{F})$ be a unitary matrix whose first column is $\mathbf{a}_1'/\|\mathbf{a}_1'\|_2$. Example 7.1.21 provides one way to construct such a matrix. Then $V_1\mathbf{e}_1 = \mathbf{a}_1'/\|\mathbf{a}_1'\|_2$ and hence $V_1^*\mathbf{a}_1' = \|\mathbf{a}_1'\|_2\mathbf{e}_1 \in \mathbb{F}^{n-1}$. Let $U_1 = I_1 \oplus V_1$ and compute

$$U_1^*A = \begin{bmatrix} \star & \star \\ \|\mathbf{a}_1'\|_2 & \star \\ \mathbf{0} & \star \end{bmatrix} \quad \text{and} \quad U_1^*AU_1 = \begin{bmatrix} \star & \star \\ \|\mathbf{a}_1'\|_2 & \star \\ \mathbf{0} & A'' \end{bmatrix}, \quad \text{in which } A'' \in M_{(n-2) \times (n-1)}.$$

Our construction ensures that right multiplication by $U_1 = I_1 \oplus V_1$ does not disturb the first column of U_1^*A.

Let $\mathbf{a}_1''$ be the first column of A''. If $\mathbf{a}_1'' = \mathbf{0}$, let $V_2 = I_{n-2}$. If $\mathbf{a}_1'' \neq \mathbf{0}$, let $V_2 \in M_{n-2}(\mathbb{F})$ be a unitary matrix whose first column is $\mathbf{a}_1''/\|\mathbf{a}_1''\|_2$. Let $U_2 = I_2 \oplus V_2$ and form $U_2^*(U_1^*AU_1)U_2$. This unitary similarity does not disturb any entries in the first column, puts a nonnegative entry in the $(3,2)$ position, and puts zeros in the entries below it. After $n-2$ steps, this algorithm produces an upper Hessenberg matrix with nonnegative entries in the first subdiagonal except possibly for the entry in position $(n, n-1)$. If necessary, a final unitary similarity by $I_{n-1} \oplus [e^{i\theta}]$ (for a suitable real θ) makes the entry in position $(n, n-1)$ real and nonnegative. $\qquad\square$

We illustrate the algorithm described in the preceding theorem with a 4×4 matrix A and the same symbol conventions that we used in the illustration (7.4.13).

$$\begin{bmatrix} \star & \star & \star & \star \\ \star & \star & \star & \star \\ \star & \star & \star & \star \\ \star & \star & \star & \star \end{bmatrix} \rightarrow \begin{bmatrix} \star & \star & \star & \star \\ \star & \star & \star & \star \\ \tilde{0} & \star & \star & \star \\ \tilde{0} & \star & \star & \star \end{bmatrix} \rightarrow \begin{bmatrix} \star & \star & \star & \star \\ \star & \star & \star & \star \\ 0 & \star & \star & \star \\ 0 & \tilde{0} & \star & \star \end{bmatrix} \rightarrow \begin{bmatrix} \star & \star & \star & \star \\ \star & \star & \star & \star \\ 0 & \star & \star & \star \\ 0 & 0 & \star & \star \end{bmatrix}$$
$$\qquad A \qquad\qquad\qquad U_1^*AU_1 \qquad\qquad U_2^*U_1^*AU_1U_2 \qquad U_3^*U_2^*U_1^*AU_1U_2U_3$$

The final unitary similarity uses a matrix of the form $U_3 = I_3 \oplus [e^{i\theta}]$ to make the $(4,3)$ entry real and nonnegative, if necessary.

Corollary 7.5.4 *Let $A = [a_{ij}] \in M_n(\mathbb{F})$ be Hermitian. There is a unitary $U \in M_n(\mathbb{F})$ such that $U^*AU = [b_{ij}]$ is real, symmetric, and tridiagonal. Moreover, $b_{i,i+1} = b_{i+1,i} \geq 0$ for all $i = 1, 2, \ldots, n-1$.*

Proof The preceding theorem ensures that there is a unitary $U \in M_n(\mathbb{F})$ such that $B = U^*AU$ is upper Hessenberg. Notice that $B^* = (U^*AU)^* = U^*A^*U = U^*AU = B$, so $B = [b_{ij}]$ is Hermitian. Since $b_{ij} = \overline{b_{ji}}$ for all i, j and $b_{ij} = 0$ for all $i > j + 1$, it follows that B is tridiagonal. That is, all entries above its first superdiagonal and below its first subdiagonal are zero. The subdiagonal entries are real and nonnegative, so the superdiagonal entries have the same properties. The diagonal entries of any Hermitian matrix are real. $\qquad\square$

7.6 Problems

P.7.1 Show that the Hadamard gate matrix in Example 7.1.3 is unitary.

P.7.2 Show that the Pauli spin matrices in Example 7.1.4 are unitary.

P.7.3 Use the definition to show that the matrix in Example 7.1.12 is unitary.

P.7.4 Verify the factorization in Example 7.4.1.

P.7.5 Suppose that $U \in M_n(\mathbb{F})$ is unitary. If there is a nonzero $\mathbf{x} \in \mathbb{F}^n$ and a $\lambda \in \mathbb{F}$ such that $U\mathbf{x} = \lambda\mathbf{x}$, show that $|\lambda| = 1$.

P.7.6 Let $U, V \in M_n$ be unitary. Must $U + V$ be unitary?

P.7.7 Let K_n be the reversal matrix in (7.1.11). If $n = 2m$ or $n = 2m + 1$, show in two ways that $\det K_n = (-1)^m$. (a) Use mathematical induction and a Laplace expansion. (b) Show that K_n is permutation similar to the direct sum of some copies of K_2 and K_1.

P.7.8 Let $A = [a_{ij}] \in M_n$ and suppose that $a_{ij} = 0$ if $i + j > n + 1$. (a) Give a numerical example of a 4×4 matrix that satisfies this condition. (b) If $n = 2m$ or $n = 2m + 1$, show that $\det A = (-1)^m a_{n1} a_{n-1,2} \cdots a_{1n}$. (c) Calculate the determinant of your matrix in (a).

P.7.9 Let $n \geq 2$ and define the *cyclic permutation matrix* $S_n = \begin{bmatrix} 0 & I_{n-1} \\ 1 & 0^T \end{bmatrix}$. (a) Write down S_5 in full detail. (b) Show that S_n is a permutation matrix. (c) If $\Lambda = \text{diag}(1, 2, \ldots, n)$, what is $S_n^T \Lambda S_n$? (d) What is the permutation σ in (7.3.4) that corresponds to S_n?

P.7.10 Show that $U = \begin{bmatrix} a & b \\ c & d \end{bmatrix} \in M_2$ is unitary if and only if $U = \begin{bmatrix} a & -e^{i\phi}\bar{c} \\ c & e^{i\phi}\bar{a} \end{bmatrix}$, in which $\phi \in \mathbb{R}$ and $|a|^2 + |c|^2 = 1$. Deduce that $|a| = |d|$ and $|b| = |c|$.

P.7.11 Show that $Q \in M_2(\mathbb{R})$ is real orthogonal if and only if there is a $\theta \in \mathbb{R}$ such that either $Q = \begin{bmatrix} \cos\theta & -\sin\theta \\ \sin\theta & \cos\theta \end{bmatrix}$ or $Q = \begin{bmatrix} \cos\theta & \sin\theta \\ \sin\theta & -\cos\theta \end{bmatrix}$.

P.7.12 Show that $U \in M_2$ is unitary if and only if $|a| \leq 1$ and there are diagonal unitary matrices $X, Y \in M_2$ and $\theta \in \mathbb{R}$ such that

$$U = X \begin{bmatrix} |a| & -\sqrt{1 - |a|^2} \\ \sqrt{1 - |a|^2} & |a| \end{bmatrix} Y = X \begin{bmatrix} \cos\theta & -\sin\theta \\ \sin\theta & \cos\theta \end{bmatrix} Y. \qquad (7.6.1)$$

P.7.13 Suppose that $U = [u_{ij}] \in M_2$ is unitary and one of its entries is zero. What can you say about the remaining three entries?

P.7.14 How are the plane rotation matrices $U(\theta)$ and $U(\phi)$ (see (7.1.16)) related to $U(\theta + \phi)$? Use these three matrices to prove the addition formula $\cos(\theta + \phi) = \cos\theta \cos\phi - \sin\theta \sin\phi$ and its analog for the sine function.

P.7.15 Let $\mathbf{u}_1, \mathbf{u}_2, \ldots, \mathbf{u}_n$ be an orthonormal basis for $\mathbb{C}^n$ with respect to the standard inner product, let $c_1, c_2, \ldots, c_n$ be complex scalars with modulus 1, and let $P_{\mathbf{u}_i}$ be the projection onto the unit vector $\mathbf{u}_i$; see (7.1.8). Show that $P_{\mathbf{u}_1} + P_{\mathbf{u}_2} + \cdots + P_{\mathbf{u}_n} = I$ and that $c_1 P_{\mathbf{u}_1} + c_2 P_{\mathbf{u}_2} + \cdots + c_n P_{\mathbf{u}_n}$ is unitary.

P.7.16 Let $\mathbf{a}_1, \mathbf{a}_2, \ldots, \mathbf{a}_n$ and $\mathbf{b}_1, \mathbf{b}_2, \ldots, \mathbf{b}_n$ be orthonormal bases of $\mathbb{F}^n$. Describe how to construct a unitary matrix U such that $U\mathbf{b}_k = \mathbf{a}_k$ for each $k = 1, 2, \ldots, n$.

P.7.17 What are all possible entries of a real diagonal $n \times n$ unitary matrix? How many such matrices are there?

P.7.18 If $U \in M_n$ is unitary, show that U^*, U^T, and $\overline{U}$ are unitary. What is the inverse of $\overline{U}$? What is the inverse of U^T?

P.7.19 If $n \geq 2$, let $U \in M_n$ be unitary and partitioned as $U = \begin{bmatrix} u & \mathbf{x}^* \\ \mathbf{y} & V \end{bmatrix}$, in which $\mathbf{x}, \mathbf{y} \in \mathbb{C}^{n-1}$. Show that $V^*V = I_{n-1} - \mathbf{x}\mathbf{x}^*$, $V^*\mathbf{y} = -u\mathbf{x}$, and $V\mathbf{x} = -\bar{u}\mathbf{y}$. Verify these identities for one of the real orthogonal matrices in Example 7.1.14.

P.7.20 If $n \geq 2$, let $U \in M_n$ be unitary and partitioned as in the preceding problem. Show that $u = (\det U)(\overline{\det V})$. Verify this identity for one of the matrices in Example 7.1.14 and for the Fourier matrix F_3 in Example 7.1.17. *Hint*: Apply the adjugate formula (C.4.4) to $A = U^*$.

P.7.21 If $n \geq 2$, let $U \in M_n$ be unitary and partitioned as $U = \begin{bmatrix} A & B \\ C & D \end{bmatrix}$, in which A and D are square. (a) Show that $|\det A| = |\det D|$ and verify this identity for two choices of partitions for one of the real orthogonal matrices in Example 7.1.14. (b) If B is square and A is invertible, show that $|\det B| = |\det C|$. *Hint*: Compute UU^* and U^*U; use the Sylvester determinant identity.

P.7.22 Consider the real 3×3 circulant matrix

$$A = \begin{bmatrix} a & b & c \\ c & a & b \\ b & c & a \end{bmatrix} \in M_3(\mathbb{R})$$

and the 3×3 Fourier matrix (7.1.18). (a) Compute the entries of $B = F_3 A F_3^*$ and conclude that A is unitarily similar to a diagonal matrix. (b) Show that the diagonal entries of B are $a + b + c$ and $a - (b+c)/2 \pm i\sqrt{3}(b-c)/2$. (c) Compute $\|A\|_F^2$ and $\|B\|_F^2$ and comment on what you find.

P.7.23 Suppose that $A, B \in M_n$ are unitarily similar. (a) Show that $\operatorname{tr} A^2 A^* = \operatorname{tr} B^2 B^*$ and $\operatorname{tr}(A^2(A^*)^2 AA^*) = \operatorname{tr}(B^2(B^*)^2 BB^*)$. (b) State and prove a necessary condition for unitary similarity that is different from these two and (7.2.9).

P.7.24 Review the discussion of the identity (3.1.19). Suppose that each of the matrices $A, B \in M_{m \times n}$ has orthonormal columns. Show that $\operatorname{col} A = \operatorname{col} B$ if and only if there is a unitary $U \in M_n$ such that $A = BU$. See P.3.29.

P.7.25 If $U \in M_n$ is unitary, compute $\|U\|_F$, the Frobenius norm of U; see Example 5.4.5.

P.7.26 Let $\beta = I_2, \sigma_x, \sigma_y, \sigma_z$ (the identity matrix and the three Pauli spin matrices). (a) Show that β is an orthogonal basis for the real vector space of 2×2 complex Hermitian matrices with the Frobenius inner product. (b) Show that β is an orthogonal basis for the complex vector space $M_2(\mathbb{C})$ with the Frobenius inner product. See P.15.61 for a generalization to M_n.

P.7.27 Show that unitary similarity is an equivalence relation on $M_n(\mathbb{C})$ and real orthogonal similarity is an equivalence relation on $M_n(\mathbb{R})$. Is real orthogonal similarity an equivalence relation on $M_n(\mathbb{C})$?

P.7.28 If $U \in M_n$ is upper triangular and unitary, what can you say about its entries?

P.7.29 If the construction in Example 7.1.21 is performed using the first column of the matrix W in Example 7.1.14, verify that the unitary matrix produced is

$$\frac{1}{2} \begin{bmatrix} -1 & 1 & \sqrt{2} \\ 1 & 5/3 & -\sqrt{2}/3 \\ \sqrt{2} & -\sqrt{2}/3 & 4/3 \end{bmatrix}.$$

P.7.30 Let $\mathbf{v} \in \mathbb{F}^n$ be a unit vector, let $U_\mathbf{v}$ be the Householder matrix (7.1.6), and let $X = [\mathbf{x}_1 \ \mathbf{x}_2 \ \dots \ \mathbf{x}_n] \in M_n(\mathbb{F})$. Show that $U_\mathbf{v}X = X - 2[\langle \mathbf{x}_1, \mathbf{v}\rangle\mathbf{v} \ \langle \mathbf{x}_2, \mathbf{v}\rangle\mathbf{v} \ \dots \ \langle \mathbf{x}_n, \mathbf{v}\rangle\mathbf{v}]$, which requires the computation of only n inner products. In general, straightforward multiplication of two $n \times n$ matrices involves the computation of n^2 inner products. Why?

P.7.31 Let $n \geq 2$, let $\mathbf{v} \in \mathbb{F}^n$ be a unit vector, and choose $\theta \in [0, 2\pi)$ so that the first entry of $\mathbf{z} = e^{-i\theta}\mathbf{v}$ is nonnegative. Show that $\mathbf{e}_1 + \mathbf{z} \neq \mathbf{0}$, let $\mathbf{u} = (\mathbf{e}_1 + \mathbf{z})/\|\mathbf{e}_1 + \mathbf{z}\|_2$, and compute the Householder matrix $H_\mathbf{u} = I - 2\mathbf{u}\mathbf{u}^*$; see Example 7.1.5. Show that $H_\mathbf{u}\mathbf{e}_1 = -\mathbf{z}$ and conclude that $-e^{-i\theta}H_\mathbf{u}$ is a unitary matrix whose first column is $\mathbf{v}$.

P.7.32 Use Householder matrices and the algorithm in the preceding problem to obtain the QR factorization of the matrix in Example 7.4.14.

P.7.33 Let $A \in \mathsf{M}_n$ and let $A = QR$ be a QR factorization. Let $A = [\mathbf{a}_1 \ \mathbf{a}_2 \ \ldots \ \mathbf{a}_n]$, $Q = [\mathbf{q}_1 \ \mathbf{q}_2 \ \ldots \ \mathbf{q}_n]$, and $R = [\mathbf{r}_1 \ \mathbf{r}_2 \ \ldots \ \mathbf{r}_n] = [r_{ij}]$. (a) Explain why $|\det A| = \det R = r_{11}r_{22}\cdots r_{nn}$. (b) Show that $\|\mathbf{a}_i\|_2 = \|\mathbf{r}_i\|_2 \geq r_{ii}$ for each $i = 1, 2, \ldots, n$, with equality for some i if and only if $\mathbf{a}_i = r_{ii}\mathbf{q}_i$. (c) Conclude that

$$|\det A| \leq \|\mathbf{a}_1\|_2\|\mathbf{a}_2\|_2 \cdots \|\mathbf{a}_n\|_2 \tag{7.6.2}$$

with equality if and only if either A has a zero column or A has orthogonal columns (that is, $A^*A = \operatorname{diag}(\|\mathbf{a}_1\|_2^2, \|\mathbf{a}_2\|_2^2, \ldots, \|\mathbf{a}_n\|_2^2)$). This is *Hadamard's inequality*.

P.7.34 Let $A = [\mathbf{a}_1 \ \mathbf{a}_2 \ \ldots \ \mathbf{a}_n] \in \mathsf{M}_n$ and let $A = QR$ be a QR factorization in which $R = [r_{ij}]$. Suppose that $1 < k \leq n$. (a) If $\mathbf{a}_k \in \operatorname{span}\{\mathbf{a}_1, \mathbf{a}_2, \ldots, \mathbf{a}_{k-1}\}$, show that $r_{kk} = 0$. $r_{kk} = 0$. (b) Give an example in which $r_{kk} = 0$ but $\mathbf{a}_k \notin \operatorname{span}\{\mathbf{a}_1, \mathbf{a}_2, \ldots, \mathbf{a}_{k-1}\}$. (c) If $r_{11}r_{22}\cdots r_{k-1,k-1} > 0$ and $r_{kk} = 0$, prove that $\mathbf{a}_k \in \operatorname{span}\{\mathbf{a}_1, \mathbf{a}_{k-1}\}$.

P.7.35 Suppose that $A \in \mathsf{M}_{m \times n}(\mathbb{F})$ and rank $A = n$. If $A = QR$ is a QR factorization, then the columns of Q are an orthonormal basis for col A. Explain why this orthonormal basis is identical to the orthonormal basis produced by applying the Gram–Schmidt process to the list of columns of A.

P.7.36 If $A \in \mathsf{M}_n(\mathbb{F})$ has orthonormal rows and the upper triangular factor in the wide QR factorization of $B \in \mathsf{M}_n(\mathbb{F})$ is I, show that $|\operatorname{tr} ABAB| \leq n$.

P.7.37 Let $A = [a_{ij}] \in \mathsf{M}_n$ be Hermitian and tridiagonal. Find a diagonal unitary $D \in \mathsf{M}_n$ such that the tridiagonal matrix $B = D^*AD = [b_{ij}]$ has $b_{ii} = a_{ii}$ for all $i = 1, 2, \ldots, n$ and $b_{i,i+1} = b_{i+1,i} = |a_{i,i+1}|$ for all $i = 1, 2, \ldots, n - 1$. Compare this result with Corollary 7.5.4.

P.7.38 Let $A \in \mathsf{M}_n$ and let $A_0 = A = Q_0R_0$ be a QR factorization. Define $A_1 = R_0Q_0$ and let $A_1 = Q_1R_1$ be a QR factorization. Define $A_2 = R_1Q_1$. Continue this construction, so at the kth step $A_k = Q_kR_k$ is a QR factorization and $A_{k+1} = R_kQ_k$. For each $k = 1, 2, \ldots$, show that A_k is unitarily similar to A. This construction is at the heart of the QR *algorithm* for computing eigenvalues; see Chapter 9. In practice, the algorithm in Theorem 7.5.3 is often employed to reduce A to upper Hessenberg form before the QR algorithm is started.

P.7.39 Let $\mathbf{u} \in \mathbb{C}^n$ be a unit vector and let c be a scalar with modulus 1. Show that $P_\mathbf{u} = P_{c\mathbf{u}}$ and $P_{\bar{\mathbf{u}}} = P_\mathbf{u}^\mathsf{T}$.

P.7.40 Let $\mathbf{a} \in \mathbb{F}^m$ be a unit vector and let $A = [\mathbf{a}] \in \mathsf{M}_{m \times 1}(\mathbb{F})$. Let

$$A = V \begin{bmatrix} R \\ \mathbf{0} \end{bmatrix} \tag{7.6.3}$$

be a wide QR factorization of A. Explain why $R = [1]$ and $V \in \mathsf{M}_m(\mathbb{F})$ is a unitary matrix whose first column is $\mathbf{a}$.

P.7.41 Let $A = QR$ be the 3×2 matrix and QR factorization in (7.4.11), and let $\mathbf{y} = [2 \ 2 \ 1]^\mathsf{T}$. If $\mathbf{x}$ is a solution of the linear system $A\mathbf{x} = \mathbf{y}$, then $R\mathbf{x} = Q^*\mathbf{y}$. (a) Show that $Q^*\mathbf{y} = \mathbf{0}$

and the solution of $Rx = Q^*y$ is $x = 0$. (b) Is $x = 0$ a solution of $Ax = y$? Explain. (c) Are the linear systems $Ax = QRx = y$ and $Rx = Q^*y$ equivalent? Why?

P.7.42 Let $a_1, a_2, \ldots, a_n \in \mathbb{F}^m$ be an orthonormal list of vectors and let $A = [a_1 \ a_2 \ \ldots \ a_n] \in M_{m \times n}(\mathbb{F})$. Let (7.6.3) be a wide QR factorization of A. Show that $R = I_n$ and $V \in M_m(\mathbb{F})$ is a unitary matrix whose first n columns are the columns of A. Describe how one might use this result to create an algorithm that extends a given orthonormal list to an orthonormal basis.

P.7.43 Let $A \in M_{m \times n}(\mathbb{R})$ with $m \geq n$ and let $A = QR$, as in (7.4.4). Let $R = [r_{ij}] \in M_n$, $Q = [q_1 \ q_2 \ \ldots \ q_n]$, and $A = [a_1 \ a_2 \ \ldots \ a_n]$. Show that $r_{ij} = \langle a_i, q_j \rangle$ for all $i, j = 1, 2, \ldots, n$. Why is $\text{span}\{a_1, a_2, \ldots, a_j\} \subseteq \text{span}\{q_1, q_2, \ldots, q_j\}$ for each $j = 1, 2, \ldots, n$? Discuss the case of equality.

P.7.44 Prove that a square matrix is similar to an upper triangular matrix if and only if it is unitarily similar to some upper triangular matrix. *Hint*: If $A = SBS^{-1}$, consider the QR factorization of S.

P.7.45 Consider the 4×4 Fourier matrix defined in (7.1.19). (a) Verify that

$$F_4^2 = \begin{bmatrix} 1 & 0 & 0 & 0 \\ 0 & 0 & 0 & 1 \\ 0 & 0 & 1 & 0 \\ 0 & 1 & 0 & 0 \end{bmatrix} = \begin{bmatrix} 1 & 0 \\ 0 & K_3 \end{bmatrix};$$

see (7.1.11). (b) Deduce that $F_4^4 = I$.

P.7.46 Consider the Fourier matrix F_n defined in (7.1.18). (a) Show that

$$F_n^2 = \frac{1}{n} \left[\sum_{k=1}^n \left(\omega^{i+j-2} \right)^{k-1} \right]_{i,j=1}^n = \begin{bmatrix} 1 & 0 \\ 0 & K_{n-1} \end{bmatrix};$$

see (7.1.11). (b) Deduce that $F_n^4 = I$.

P.7.47 Verify the following identity, which connects the Fourier matrices F_4 and F_2:

$$F_4 = \frac{1}{\sqrt{2}} \begin{bmatrix} I & D_2 \\ I & -D_2 \end{bmatrix} \begin{bmatrix} F_2 & 0 \\ 0 & F_2 \end{bmatrix} P_4, \tag{7.6.4}$$

in which $D_2 = \text{diag}(1, \omega)$, $\omega = e^{\pi i/2} = i$, and $P_4 = [e_1 \ e_3 \ e_2 \ e_4]$ is the permutation matrix such that $P_4[x_1 \ x_2 \ x_3 \ x_4]^\mathsf{T} = [x_1 \ x_3 \ x_2 \ x_4]^\mathsf{T}$.

P.7.48 Let $A \in M_n$. If B is permutation similar to A, show that each entry of A is an entry of B and it occurs in the same number of distinct positions in both matrices.

P.7.49 Let $A \in M_n$. Each entry of A is an entry of A^T and it occurs in the same number of distinct positions in both matrices. Are A and A^T always permutation similar? Why?

P.7.50 Let $A \in M_n$ and suppose that $A \neq 0$. (a) If $\text{rank } A = r$, show that A has an $r \times r$ submatrix that is invertible. *Hint*: Why is there a permutation matrix P such that $PA = \begin{bmatrix} A_1 \\ A_2 \end{bmatrix}$ and $A_1 \in M_{r \times n}$ has linearly independent rows? Why is there a permutation matrix Q such that $A_1 Q = [A_{11} \ A_{22}]$ and $A_{11} \in M_r$ has linearly independent columns? (b) If $k \geq 1$ and A has a $k \times k$ invertible submatrix, show that $\text{rank } A \geq k$. *Hint*: Why are at least k rows of A linearly independent? (c) Since $A \neq 0$ the set $\mathscr{S} = \{k : A \text{ has an invertible } k \times k \text{ submatrix}\}$ is nonempty. Show that

$$\text{rank } A = \max\{k : B \text{ is a } k \times k \text{ submatrix of } A \text{ and } \det B \neq 0\} \tag{7.6.5}$$

and $\mathcal{S} = \{1, 2, \ldots, \text{rank}\, A\}$. (d) If $E \in \mathsf{M}_n$ is the all-ones matrix, use (c) to show that rank $E = 1$. (e) If $n \geq 3$ and rank $A \leq n-2$, use (c) to show that adj $A = 0$; see (C.4.3). (f) If $n \geq 2$ and rank $A = n - 1$, use (c) to show that rank adj $A \geq 1$. Use (C.4.3) to show that the nullity of adj A is at least $n - 1$ and conclude that rank adj $A = 1$.

P.7.51 If $U \in \mathsf{M}_m$ and $V \in \mathsf{M}_n$ are unitary, show that $U \otimes V \in \mathsf{M}_{mn}$ is unitary.

7.7 Notes

If the Gram–Schmidt process is used to orthogonalize a linearly independent list of vectors with floating point arithmetic on a computer, roundoff errors can cause the list of vectors obtained to be far from orthogonal. Despite the result in P.7.35, numerical orthogonalization is typically much more satisfactory via a QR factorization obtained with a sequence of unitary transformations; see [GVL13].

The identity (7.6.4) is a special case of

$$F_{2n} = \frac{1}{\sqrt{2}} \begin{bmatrix} I & D_n \\ I & -D_n \end{bmatrix} \begin{bmatrix} F_n & 0 \\ 0 & F_n \end{bmatrix} P_{2n}, \tag{7.7.1}$$

in which $D_n = \text{diag}(1, \omega, \omega^2, \ldots, \omega^{n-1})$, $\omega = e^{i\pi/n}$, and P_{2n} is the permutation matrix such that

$$P_{2n}[x_1 \ x_2 \ \ldots \ x_{2n-1} \ x_{2n}]^{\mathsf{T}} = [x_1 \ x_3 \ x_5 \ \ldots \ x_{2n-1} \ x_2 \ x_4 \ x_6 \ \ldots \ x_{2n}]^{\mathsf{T}},$$

which places the odd-indexed entries first, followed by the even-indexed entries. The identity (7.7.1) suggests a recursive scheme to calculate the matrix-vector product $F_{2^m}\mathbf{x}$, in which $\mathbf{x} \in \mathbb{C}^{2^n}$. This scheme is at the heart of the celebrated fast Fourier transform (FFT) algorithm.

The factorization (7.6.1) of a 2×2 unitary matrix is a special case of the CS decomposition of an $n \times n$ unitary matrix that is presented as a 2×2 block matrix; see [HJ13, Section 2.7].

There are necessary and sufficient conditions for unitary similarity that involve the ideas in P.7.23, but they are of practical value only for matrices of small size; see [HJ13, Theorem 2.2.8].

7.8 Some Important Concepts

- The adjoint (conjugate transpose) of a unitary matrix is its inverse.

- Characterizations of unitary matrices.

- Unitary similarity and permutation similarity.

- Fourier matrices, Householder matrices, and permutation matrices are unitary.

- Each unit vector is the first column of some unitary matrix.

- If $\mathbf{x}, \mathbf{y} \in \mathbb{F}^n$ and $\|\mathbf{x}\|_2 = \|\mathbf{y}\|_2$, then there is a unitary $U \in \mathsf{M}_n(\mathbb{F})$ such that $U\mathbf{x} = \mathbf{y}$.

- QR factorizations and orthogonalization of a linearly independent list of vectors.

- Each square matrix is unitarily similar to an upper Hessenberg matrix.

- Each Hermitian matrix is unitarily similar to a real symmetric tridiagonal matrix with nonnegative off-diagonal entries.

Orthogonal Complements and Orthogonal Projections

Many problems in applied mathematics involve finding a minimum-norm solution or a best approximation, subject to certain constraints. Orthogonal subspaces arise frequently in solving such problems. Among the topics we discuss in this chapter are the minimum-norm solution to a consistent linear system, a least-squares solution to an inconsistent linear system, and orthogonal projections.

8.1 Orthogonal Complements

Orthogonal projections are simple operators that are building blocks for a variety of other operators (see Section 14.9). They are also a fundamental tool in real-world optimization problems and in many applications. To define orthogonal projections, we require the following notion.

Definition 8.1.1 Let $\mathcal{U}$ be a subset of an inner product space $\mathcal{V}$. If $\mathcal{U}$ is nonempty, then

$$\mathcal{U}^\perp = \{\mathbf{v} \in \mathcal{V} : \langle \mathbf{u}, \mathbf{v} \rangle = 0 \text{ for all } \mathbf{u} \in \mathcal{U}\}.$$

If $\mathcal{U} = \varnothing$, we define $\mathcal{U}^\perp = \mathcal{V}$. The set $\mathcal{U}^\perp$ (read as "$\mathcal{U}$-perp") is the *orthogonal complement* of $\mathcal{U}$ in $\mathcal{V}$.

Figure 8.1 and the following examples illustrate several basic properties of orthogonal complements.

Example 8.1.2 Let $\mathcal{V} = \mathbb{R}^3$, $\mathbf{u} = [1\ 2\ 3]^\mathsf{T}$, and $\mathcal{U} = \{\mathbf{u}\}$. Then $\mathbf{x} = [x_1\ x_2\ x_3]^\mathsf{T} \in \mathcal{U}^\perp$ if and only if $x_1 + 2x_2 + 3x_3 = \langle \mathbf{u}, \mathbf{x} \rangle = 0$. Thus,

$$\mathcal{U}^\perp = \{[x_1\ x_2\ x_3]^\mathsf{T} : x_1 + 2x_2 + 3x_3 = 0\} \tag{8.1.3}$$

is the plane in $\mathbb{R}^3$ that passes through $\mathbf{0} = [0\ 0\ 0]^\mathsf{T}$ and has $\mathbf{u}$ as a normal vector. Observe that $\mathcal{U}^\perp$ is a subspace of $\mathbb{R}^3$ even though $\mathcal{U}$ is merely a subset of $\mathbb{R}^3$; see Theorem 8.1.5.

Example 8.1.4 Consider $\mathcal{V} = \mathsf{M}_n$ with the Frobenius inner product. If $\mathcal{U}$ is the subspace of all upper triangular matrices, then $\mathcal{U}^\perp$ is the subspace of all strictly lower triangular matrices; see P.8.10.

The following theorem lists some important properties of orthogonal complements.

Theorem 8.1.5 *Let $\mathcal{U}, \mathcal{W}$ be subsets of an inner product space $\mathcal{V}$.*

(a) *$\mathcal{U}^\perp$ is a subspace of $\mathcal{V}$. In particular, $\mathbf{0} \in \mathcal{U}^\perp$.*

(b) *If $\mathcal{U} \subseteq \mathcal{W}$, then $\mathcal{W}^\perp \subseteq \mathcal{U}^\perp$.*

(c) *$\mathcal{U}^\perp = (\operatorname{span} \mathcal{U})^\perp$.*

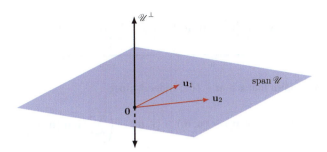

Figure 8.1 The orthogonal complement of the set $\mathscr{U} = \{\mathbf{u}_1, \mathbf{u}_2\}$ is the subspace $\mathscr{U}^\perp$.

(d) *If $\mathscr{U} \cap \mathscr{U}^\perp \neq \varnothing$, then $\mathscr{U} \cap \mathscr{U}^\perp = \{\mathbf{0}\}$.*

(e) *If $\mathbf{0} \in \mathscr{U}$, then $\mathscr{U} \cap \mathscr{U}^\perp = \{\mathbf{0}\}$.*

(f) $\{\mathbf{0}\}^\perp = \mathcal{V}$.

(g) $\mathcal{V}^\perp = \{\mathbf{0}\}$.

(h) $\mathscr{U} \subseteq \operatorname{span} \mathscr{U} \subseteq (\mathscr{U}^\perp)^\perp$.

Proof (a) If $\mathscr{U} = \varnothing$, then $\mathscr{U}^\perp = \mathcal{V}$ is a subspace. If $\mathscr{U} \neq \varnothing$, if $\mathbf{u}, \mathbf{v} \in \mathscr{U}^\perp$, and if a is a scalar, then $\langle \mathbf{w}, \mathbf{u} + a\mathbf{v} \rangle = \langle \mathbf{w}, \mathbf{u} \rangle + \overline{a}\langle \mathbf{w}, \mathbf{v} \rangle = 0$ for all $\mathbf{w} \in \mathscr{U}$. Thus, $\mathbf{u} + a\mathbf{v} \in \mathscr{U}^\perp$, so $\mathscr{U}^\perp$ is a subspace.

(b) If $\mathscr{U} = \varnothing$, then $\mathcal{W}^\perp \subseteq \mathcal{V} = \varnothing^\perp$. If $\mathscr{U} \neq \varnothing$ and $\mathbf{v} \in \mathcal{W}^\perp$, then $\langle \mathbf{u}, \mathbf{v} \rangle = 0$ for all $\mathbf{u} \in \mathcal{U}$ since $\mathcal{U} \subseteq \mathcal{W}$. Thus, $\mathbf{v} \in \mathcal{U}^\perp$ and we conclude that $\mathcal{W}^\perp \subseteq \mathcal{U}^\perp$.

(c) If $\mathscr{U} = \varnothing$, then $\operatorname{span} \mathscr{U} = \{\mathbf{0}\}$ (Definition 1.4.7) and $\mathscr{U}^\perp = \mathcal{V} = \{\mathbf{0}\}^\perp$. Now suppose that $\mathscr{U} \neq \varnothing$. Since $\mathscr{U} \subseteq \operatorname{span} \mathscr{U}$, (b) ensures that $(\operatorname{span} \mathscr{U})^\perp \subseteq \mathscr{U}^\perp$. If $\mathbf{v} \in \mathscr{U}^\perp$, then $\langle \sum_{i=1}^r c_i \mathbf{u}_i, \mathbf{v} \rangle = \sum_{i=1}^r c_i \langle \mathbf{u}_i, \mathbf{v} \rangle = 0$ for any scalars $c_1, c_2, \ldots, c_r$ and any $\mathbf{u}_1, \mathbf{u}_2, \ldots, \mathbf{u}_r \in \mathcal{U}$. Thus, $\mathscr{U}^\perp \subseteq (\operatorname{span} \mathscr{U})^\perp$. We conclude that $\mathscr{U}^\perp = (\operatorname{span} \mathscr{U})^\perp$.

(d) If $\mathbf{u} \in \mathscr{U} \cap \mathscr{U}^\perp$, then $\mathbf{u}$ is orthogonal to itself. Therefore, $\|\mathbf{u}\|^2 = \langle \mathbf{u}, \mathbf{u} \rangle = 0$ and $\mathbf{u} = \mathbf{0}$.

(e) Let $\mathbf{0} \in \mathscr{U}$. Since $\mathbf{0} \in \mathscr{U}^\perp$ by (a), $\mathscr{U} \cap \mathscr{U}^\perp \neq \varnothing$ and hence $\mathscr{U} \cap \mathscr{U}^\perp = \{\mathbf{0}\}$ by (d).

(f) Since $\langle \mathbf{v}, \mathbf{0} \rangle = 0$ for all $\mathbf{v} \in \mathcal{V}$, we see that $\mathcal{V} \subseteq \{\mathbf{0}\}^\perp \subseteq \mathcal{V}$, so $\mathcal{V} = \{\mathbf{0}\}^\perp$.

(g) Since $\mathbf{0} \in \mathcal{V}^\perp$ by (a), $\mathcal{V}^\perp = \mathcal{V} \cap \mathcal{V}^\perp = \{\mathbf{0}\}$ by (d).

(h) If $\mathscr{U} = \varnothing$, then $\mathscr{U} \subseteq \operatorname{span} \mathscr{U} = \{\mathbf{0}\} = \mathcal{V}^\perp = (\mathscr{U}^\perp)^\perp$. Now suppose that $\mathscr{U} \neq \varnothing$. Since every vector in $\operatorname{span} \mathscr{U}$ is orthogonal to every vector in $(\operatorname{span} \mathscr{U})^\perp$, (c) ensures that $\mathscr{U} \subseteq \operatorname{span} \mathscr{U} \subseteq ((\operatorname{span} \mathscr{U})^\perp)^\perp = (\mathscr{U}^\perp)^\perp$. $\qquad\square$

The principal assertion of the following theorem is that an inner product space is the direct sum of any finite-dimensional subspace and its orthogonal complement.

Theorem 8.1.6 *Let $\mathcal{V}$ be an inner product space. If $\mathcal{U} = \mathcal{V}$ or $\mathcal{U}$ is a finite-dimensional subspace of $\mathcal{V}$, then $\mathcal{V} = \mathcal{U} \oplus \mathcal{U}^\perp$. In either case, for each $\mathbf{v} \in \mathcal{V}$ there is a unique $\mathbf{u} \in \mathcal{U}$ such that $\mathbf{v} - \mathbf{u} \in \mathcal{U}^\perp$.*

Proof Parts (f) and (g) of Theorem 8.1.5 ensure that $\mathcal{V} = \mathcal{U} \oplus \mathcal{U}^\perp$ if $\mathcal{U} = \{\mathbf{0}\}$ or $\mathcal{U} = \mathcal{V}$. Now suppose that $\mathcal{U} \neq \{\mathbf{0}\}$, $\mathcal{U} \neq \mathcal{V}$, and that $\mathcal{U}$ is finite dimensional. Let $\mathbf{u}_1, \mathbf{u}_2, \ldots, \mathbf{u}_r$ be an orthonormal basis for $\mathcal{U}$. For any $\mathbf{v}$ in $\mathcal{V}$, write

$$\mathbf{v} = \underbrace{\left(\sum_{i=1}^{r} \langle \mathbf{v}, \mathbf{u}_i \rangle \mathbf{u}_i \right)}_{\mathbf{u}} + \underbrace{\left(\mathbf{v} - \sum_{i=1}^{r} \langle \mathbf{v}, \mathbf{u}_i \rangle \mathbf{u}_i \right)}_{\mathbf{v} - \mathbf{u}}, \tag{8.1.7}$$

in which $\mathbf{u} \in \mathcal{U}$. We claim that $\mathbf{v} - \mathbf{u} \in \mathcal{U}^\perp$. Indeed, since

$$\langle \mathbf{v} - \mathbf{u}, \mathbf{u}_j \rangle = \langle \mathbf{v}, \mathbf{u}_j \rangle - \langle \mathbf{u}, \mathbf{u}_j \rangle = \langle \mathbf{v}, \mathbf{u}_j \rangle - \left\langle \sum_{i=1}^{r} \langle \mathbf{v}, \mathbf{u}_i \rangle \mathbf{u}_i, \mathbf{u}_j \right\rangle$$

$$= \langle \mathbf{v}, \mathbf{u}_j \rangle - \sum_{i=1}^{r} \langle \mathbf{v}, \mathbf{u}_i \rangle \langle \mathbf{u}_i, \mathbf{u}_j \rangle = \langle \mathbf{v}, \mathbf{u}_j \rangle - \langle \mathbf{v}, \mathbf{u}_j \rangle = 0,$$

Theorem 8.1.5.c ensures that $\mathbf{v} - \mathbf{u} \in \{\mathbf{u}_1, \mathbf{u}_2, \ldots, \mathbf{u}_r\}^\perp = \left(\operatorname{span} \{\mathbf{u}_1, \mathbf{u}_2, \ldots, \mathbf{u}_r\} \right)^\perp = \mathcal{U}^\perp$. Thus, $\mathcal{V} = \mathcal{U} + \mathcal{U}^\perp$. Since $\mathcal{U}$ is a subspace, $\mathbf{0} \in \mathcal{U}$ and hence Theorem 8.1.5.e ensures that $\mathcal{U} \cap \mathcal{U}^\perp = \{\mathbf{0}\}$. We conclude that $\mathcal{V} = \mathcal{U} \oplus \mathcal{U}^\perp$. The uniqueness of $\mathbf{u}$ follows from Theorem 1.5.9. $\square$

Theorem 8.1.8 *If $\mathcal{U}$ is a finite-dimensional subspace of an inner product space, then $(\mathcal{U}^\perp)^\perp = \mathcal{U}$.*

Proof Theorem 8.1.5.h asserts that $\mathcal{U} \subseteq (\mathcal{U}^\perp)^\perp$. Let $\mathbf{v} \in (\mathcal{U}^\perp)^\perp$ and use Theorem 8.1.6 to write $\mathbf{v} = \mathbf{u} + (\mathbf{v} - \mathbf{u})$, in which $\mathbf{u} \in \mathcal{U}$ and $(\mathbf{v} - \mathbf{u}) \in \mathcal{U}^\perp$. Since

$$0 = \langle \mathbf{v}, \mathbf{v} - \mathbf{u} \rangle = \langle \mathbf{u} + (\mathbf{v} - \mathbf{u}), \mathbf{v} - \mathbf{u} \rangle = \langle \mathbf{u}, \mathbf{v} - \mathbf{u} \rangle + \|\mathbf{v} - \mathbf{u}\|^2 = \|\mathbf{v} - \mathbf{u}\|^2,$$

it follows that $\mathbf{v} = \mathbf{u}$. Thus, $\mathbf{v} \in \mathcal{U}$ and $(\mathcal{U}^\perp)^\perp \subseteq \mathcal{U}$. $\square$

Corollary 8.1.9 *Let $\mathscr{U}$ be a subset of an inner product space. If $\operatorname{span} \mathscr{U}$ is finite dimensional, then $(\mathscr{U}^\perp)^\perp = \operatorname{span} \mathscr{U}$.*

Proof Since $\operatorname{span} \mathscr{U}$ is a finite-dimensional subspace, Theorems 8.1.8 and 8.1.5.c tell us that $\operatorname{span} \mathscr{U} = ((\operatorname{span} \mathscr{U})^\perp)^\perp = (\mathscr{U}^\perp)^\perp$. $\square$

Example 8.1.10 Let $\mathscr{U}$ be the set defined in Example 8.1.2. Why is $(\mathscr{U}^\perp)^\perp = \operatorname{span} \mathscr{U}$? Since $\mathscr{U}^\perp$ is the null space of the rank-1 matrix $[1 \ 2 \ 3]$, the rank-nullity theorem ensures that $\dim \mathscr{U}^\perp = 2$. Therefore, any two linearly independent vectors in $\mathscr{U}^\perp$, such as $\mathbf{v}_1 = [3 \ 0 \ {-}1]^\mathsf{T}$ and $\mathbf{v}_2 = [2 \ {-}1 \ 0]^\mathsf{T}$, form a basis for $\mathscr{U}^\perp$. Thus $(\mathscr{U}^\perp)^\perp = (\operatorname{span} \{\mathbf{v}_1, \mathbf{v}_2\})^\perp = \{\mathbf{v}_1, \mathbf{v}_2\}^\perp$. However, $\{\mathbf{v}_1, \mathbf{v}_2\}^\perp$ is the null space of the 2×3 matrix $[\mathbf{v}_1 \ \mathbf{v}_2]^\mathsf{T}$, which row reduction confirms is $\operatorname{span} \mathscr{U}$.

If $\mathcal{V}$ is infinite dimensional and $\mathscr{U} \subseteq \mathcal{V}$, then $\operatorname{span} \mathscr{U}$ can be a proper subset of $(\mathscr{U}^\perp)^\perp$; see P.8.28.

8.2 The Minimum-Norm Solution of a Consistent Linear System

Definition 8.2.1 Let $A \in \mathsf{M}_{m \times n}(\mathbb{F})$ and suppose that $\mathbf{y} \in \operatorname{col} A$. A solution $\mathbf{s}$ of the consistent linear system $A\mathbf{x} = \mathbf{y}$ is a *minimum-norm solution* if $\|\mathbf{s}\|_2 \leq \|\mathbf{u}\|_2$ whenever $A\mathbf{u} = \mathbf{y}$.

There are many reasons why we might be interested in finding minimum-norm solutions to a consistent system of equations that has infinitely many solutions. For example, if the entries of its solution vectors represent economic quantities, perhaps total expenditures are minimized with a minimum-norm solution. Fortunately, every consistent linear system has a unique minimum-norm solution. To prove this, we begin with the following lemma.

Lemma 8.2.2 *If $\mathcal{V}, \mathcal{W}$ are finite-dimensional inner product spaces and $T \in \mathcal{L}(\mathcal{V}, \mathcal{W})$, then*

$$\ker T = (\operatorname{ran} T^*)^\perp \quad and \quad \operatorname{ran} T = (\ker T^*)^\perp. \qquad (8.2.3)$$

Proof The equivalences

$$
\begin{aligned}
\mathbf{v} \in \ker T \quad &\Longleftrightarrow \quad T\mathbf{v} = \mathbf{0} \\
&\Longleftrightarrow \quad \langle T\mathbf{v}, \mathbf{w} \rangle_\mathcal{W} = 0 \text{ for all } \mathbf{w} \in \mathcal{W} \\
&\Longleftrightarrow \quad \langle \mathbf{v}, T^*\mathbf{w} \rangle_\mathcal{V} = 0 \text{ for all } \mathbf{w} \in \mathcal{W} \\
&\Longleftrightarrow \quad \mathbf{v} \in (\operatorname{ran} T^*)^\perp
\end{aligned}
$$

tell us that $\ker T = (\operatorname{ran} T^*)^\perp$. To see that $\operatorname{ran} T = (\ker T^*)^\perp$, replace T with T^* and use Theorem 8.1.8. $\qquad \square$

If $\mathcal{V} = \mathbb{F}^n$ and $\mathcal{W} = \mathbb{F}^m$, and if $T_A \in \mathcal{L}(\mathcal{V}, \mathcal{W})$ is the linear operator induced by $A \in \mathsf{M}_{m \times n}(\mathbb{F})$ (see Definition 2.4.13), the preceding lemma ensures that

$$\operatorname{null} A = (\operatorname{col} A^*)^\perp \quad \text{and} \quad \operatorname{col} A = (\operatorname{null} A^*)^\perp. \qquad (8.2.4)$$

If we combine these identities with Theorem 8.1.6 and Corollary 8.1.9, we obtain the following presentation of $\mathbb{F}^n$ as a direct sum of orthogonal subspaces:

$$\mathbb{F}^n = \operatorname{col} A^* \oplus \underbrace{\operatorname{null} A}_{(\operatorname{col} A^*)^\perp}. \qquad (8.2.5)$$

We can now prove our assertion about minimum-norm solutions; see Figure 8.2.

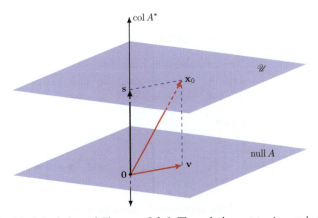

Figure 8.2 Graphical depiction of Theorem 8.2.6. The solution set to $A\mathbf{x} = \mathbf{y}$ is $\mathcal{U}$, which is a copy of null A that has been translated by $\mathbf{x}_0$ (a solution to $A\mathbf{x} = \mathbf{y}$). The set $\mathcal{U}$ is not a subspace of $\mathbb{F}^n$ unless $\mathbf{y} = \mathbf{0}$, in which case $\mathcal{U} = \operatorname{null} A$. The minimum-norm solution $\mathbf{s}$ to $A\mathbf{x} = \mathbf{y}$ is the unique vector in $\mathcal{U}$ that belongs to $\operatorname{col} A^* = (\operatorname{null} A)^\perp$.

Theorem 8.2.6 *Suppose that* $\mathbf{y} \in \mathbb{F}^m$, $A \in \mathsf{M}_{m \times n}(\mathbb{F})$, *and*

$$A\mathbf{x} = \mathbf{y} \tag{8.2.7}$$

is consistent.

(a) *Equation (8.2.7) has a unique minimum-norm solution* $\mathbf{s} \in \mathbb{F}^n$. *Moreover,* $\mathbf{s} \in \operatorname{col} A^*$.

(b) $\mathbf{s}$ *is the only solution to (8.2.7) that lies in* $\operatorname{col} A^*$.

(c) $AA^*\mathbf{u} = \mathbf{y}$ *is consistent. If* $\mathbf{u}_0 \in \mathbb{F}^m$ *and* $AA^*\mathbf{u}_0 = \mathbf{y}$, *then* $\mathbf{s} = A^*\mathbf{u}_0$.

Proof (a) Let $\mathbf{x}_0$ be a solution (any solution will do) to (8.2.7) and use (8.2.5) to write $\mathbf{x}_0 = \mathbf{s} + \mathbf{v}$, in which $\mathbf{s} \in \operatorname{col} A^*$ and $\mathbf{v} \in \operatorname{null} A$. It follows that $\mathbf{y} = A\mathbf{x}_0 = A(\mathbf{s} + \mathbf{v}) = A\mathbf{s} + A\mathbf{v} = A\mathbf{s}$, so $\mathbf{s}$ is a solution to (8.2.7). Every solution $\mathbf{x}$ to (8.2.7) has the form $\mathbf{x} = \mathbf{s} + \mathbf{v}$ for some $\mathbf{v} \in \operatorname{null} A$; see Appendix C.3. The Pythagorean theorem ensures that $\|\mathbf{x}\|_2^2 = \|\mathbf{s} + \mathbf{v}\|_2^2 = \|\mathbf{s}\|_2^2 + \|\mathbf{v}\|_2^2 \geq \|\mathbf{s}\|_2^2$, so $\mathbf{s}$ is a minimum-norm solution to (8.2.7).

If $\mathbf{s}'$ is any minimum-norm solution to (8.2.7), then $\mathbf{s}' = \mathbf{s} + \mathbf{v}$, in which $\mathbf{v} \in \operatorname{null} A$ and $\|\mathbf{s}\|_2 = \|\mathbf{s}'\|_2$. The Pythagorean theorem ensures that $\|\mathbf{s}\|_2^2 = \|\mathbf{s}'\|_2^2 = \|\mathbf{s} + \mathbf{v}\|_2^2 = \|\mathbf{s}\|_2^2 + \|\mathbf{v}\|_2^2$. Consequently, $\mathbf{v} = \mathbf{0}$ and $\mathbf{s} = \mathbf{s}'$.

(b) Let $\mathbf{s}'$ be any vector in $\operatorname{col} A^*$. Then $\mathbf{s} - \mathbf{s}' \in \operatorname{col} A^*$ since $\operatorname{col} A^*$ is a subspace of $\mathbb{F}^n$. If $\mathbf{s}'$ is a solution to (8.2.7), then $A\mathbf{s} = A\mathbf{s}' = \mathbf{y}$ and $A(\mathbf{s} - \mathbf{s}') = \mathbf{0}$. Thus, $\mathbf{s} - \mathbf{s}' \in \operatorname{col} A^* \cap \operatorname{null} A = \{\mathbf{0}\}$. We conclude that $\mathbf{s} = \mathbf{s}'$ and hence $\mathbf{s}$ is the only solution to (8.2.7) in $\operatorname{col} A^*$.

(c) Because the minimum-norm solution $\mathbf{s}$ to (8.2.7) is in $\operatorname{col} A^*$, there is a $\mathbf{w}$ such that $\mathbf{s} = A^*\mathbf{w}$. Since $AA^*\mathbf{w} = A\mathbf{s} = \mathbf{y}$, the linear system $AA^*\mathbf{u} = \mathbf{y}$ is consistent. If $\mathbf{u}_0$ is any solution to $AA^*\mathbf{u} = \mathbf{y}$, then $A^*\mathbf{u}_0$ is a solution to (8.2.7) that lies in $\operatorname{col} A^*$. It follows from (b) that $A^*\mathbf{u}_0 = \mathbf{s}$. $\qquad\square$

Theorem 8.2.6 provides a recipe to find the minimum-norm solution of the consistent linear system (8.2.7). First find a $\mathbf{u}_0 \in \mathbb{F}^n$ such that $AA^*\mathbf{u}_0 = \mathbf{y}$; existence of a solution is guaranteed. Then $\mathbf{s} = A^*\mathbf{u}_0$ is the minimum-norm solution to (8.2.7). If A and $\mathbf{y}$ are real, then $\mathbf{s}$ is real. If A or $\mathbf{y}$ is not real, then $\mathbf{s}$ need not be real. An alternative approach to finding the minimum-norm solution is in Theorem 17.4.13.

Example 8.2.8 Consider the real linear system

$$\begin{array}{rrrrrrl} x_1 & + & 2x_2 & + & 3x_3 & = & 3, \\ 4x_1 & + & 5x_2 & + & 6x_3 & = & 3, \\ 7x_1 & + & 8x_2 & + & 9x_3 & = & 3, \end{array} \tag{8.2.9}$$

which can be written as $A\mathbf{x} = \mathbf{y}$ with

$$A = \begin{bmatrix} 1 & 2 & 3 \\ 4 & 5 & 6 \\ 7 & 8 & 9 \end{bmatrix}, \quad \mathbf{x} = \begin{bmatrix} x_1 \\ x_2 \\ x_3 \end{bmatrix}, \quad \text{and} \quad \mathbf{y} = \begin{bmatrix} 3 \\ 3 \\ 3 \end{bmatrix}.$$

The reduced row echelon form of the augmented matrix $[A \ \mathbf{y}]$ is

$$\begin{bmatrix} 1 & 0 & -1 & -3 \\ 0 & 1 & 2 & 3 \\ 0 & 0 & 0 & 0 \end{bmatrix},$$

from which we obtain the general solution

$$\mathbf{x}(t) = \begin{bmatrix} x_1 \\ x_2 \\ x_3 \end{bmatrix} = \begin{bmatrix} -3 \\ 3 \\ 0 \end{bmatrix} + t \begin{bmatrix} 1 \\ -2 \\ 1 \end{bmatrix} \qquad \text{for } t \in \mathbb{R}.$$

It is tempting to think that the solution with $t = 0$ might be the minimum-norm solution. We have $\mathbf{x}(0) = [-3 \ 3 \ 0]^{\mathsf{T}}$ and $\|\mathbf{x}\|_2 = 3\sqrt{2} \approx 4.24264$. However, we can do better.

To find the minimum-norm solution to (8.2.9), we begin by finding a solution to the (necessarily consistent) linear system $(AA^*)\mathbf{u} = \mathbf{y}$. We have

$$AA^* = \begin{bmatrix} 14 & 32 & 50 \\ 32 & 77 & 122 \\ 50 & 122 & 194 \end{bmatrix}.$$

The linear system

$$\begin{array}{rcrcrcl} 14u_1 & + & 32u_2 & + & 50u_3 & = & 3, \\ 32u_1 & + & 77u_2 & + & 122u_3 & = & 3, \\ 50u_1 & + & 122u_2 & + & 194u_3 & = & 3. \end{array} \qquad (8.2.10)$$

has infinitely many solutions, one of which is $\mathbf{u}_0 = [0 \ 4 \ -\frac{5}{2}]^{\mathsf{T}}$. The minimum-norm solution to (8.2.9) is

$$\mathbf{s} = A^* \mathbf{u}_0 = \begin{bmatrix} 1 & 4 & 7 \\ 2 & 5 & 8 \\ 3 & 6 & 9 \end{bmatrix} \begin{bmatrix} 0 \\ 4 \\ -\frac{5}{2} \end{bmatrix} = \begin{bmatrix} -\frac{3}{2} \\ 0 \\ \frac{3}{2} \end{bmatrix}.$$

Note that $\|\mathbf{s}\|_2 = \frac{3\sqrt{2}}{2} \approx 2.12132$. The vector $\mathbf{u}_1 = [2 \ 0 \ -\frac{1}{2}]^{\mathsf{T}}$ is also a solution to (8.2.10), but $A^* \mathbf{u}_1 = \mathbf{s}$, as predicted by Theorem 8.2.6.

8.3 Orthogonal Projections

Throughout this section, $\mathcal{U}$ is a subspace of an inner product space $\mathcal{V}$ with inner product $\langle \cdot, \cdot \rangle$. Theorem 8.1.6 ensures that if $\mathcal{U}$ is finite dimensional, then $\mathcal{V} = \mathcal{U} \oplus \mathcal{U}^\perp$. In this event, each $\mathbf{v} \in \mathcal{V}$ can be written uniquely as

$$\mathbf{v} = \mathbf{u} + \mathbf{w}, \qquad (8.3.1)$$

in which $\mathbf{u} \in \mathcal{U}$ and $\mathbf{w} \in \mathcal{U}^\perp$ (Theorem 8.1.6). The vector $\mathbf{u} = P_{\mathcal{U}}\mathbf{v}$ is the *orthogonal projection* of $\mathbf{v}$ onto $\mathcal{U}$; see Figure 8.3. If $\mathcal{U} = \{\mathbf{0}\}$, then the only choice for the vector $\mathbf{u}$ in (8.3.1) is the zero vector. Consequently, $P_{\{0\}} = 0$ (the zero transformation).

Theorem 8.3.2 *Let $\mathcal{U}$ be a nonzero finite-dimensional subspace of an inner product space $\mathcal{V}$ and suppose that $\mathbf{u}_1, \mathbf{u}_2, \ldots, \mathbf{u}_r$ is an orthonormal basis for $\mathcal{U}$. Then*

$$P_{\mathcal{U}}\mathbf{v} = \sum_{i=1}^{r} \langle \mathbf{v}, \mathbf{u}_i \rangle \mathbf{u}_i \qquad (8.3.3)$$

for every $\mathbf{v} \in \mathcal{V}$.

Proof This follows from (8.1.7) and the uniqueness of the decomposition (8.3.1). $\qquad \square$

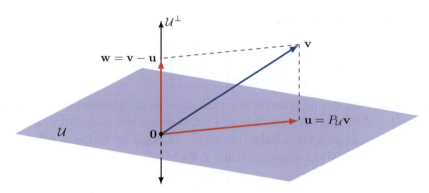

Figure 8.3 An illustration of the decomposition (8.3.1).

The uniqueness of the representation (8.3.1) ensures that $P_{\mathcal{U}}\mathbf{v}$ does not depend on the choice of orthonormal basis for $\mathcal{U}$, something that is not evident from (8.3.3). Furthermore, (8.3.3) tells us that the map $\mathbf{v} \mapsto P_{\mathcal{U}}\mathbf{v}$ is a linear transformation since each $\langle \mathbf{v}, \mathbf{u}_i \rangle$ is linear in $\mathbf{v}$.

Definition 8.3.4 The linear operator $P_{\mathcal{U}} \in \mathcal{L}(\mathcal{V})$ defined by (8.3.3) is the *orthogonal projection* from $\mathcal{V}$ onto $\mathcal{U}$.

Example 8.3.5 Suppose that $\mathbf{u}_1, \mathbf{u}_2, \ldots, \mathbf{u}_r$ is an orthonormal basis for a subspace $\mathcal{U}$ of $\mathbb{F}^n$ and let $U = [\mathbf{u}_1 \ \mathbf{u}_2 \ \ldots \ \mathbf{u}_r] \in \mathsf{M}_{n \times r}(\mathbb{F})$. For any $\mathbf{v} \in \mathbb{F}^n$,

$$P_{\mathcal{U}}\mathbf{v} = \sum_{i=1}^{r} \langle \mathbf{v}, \mathbf{u}_i \rangle \mathbf{u}_i = \sum_{i=1}^{r} \mathbf{u}_i(\mathbf{u}_i^*\mathbf{v}) = \sum_{i=1}^{r} (\mathbf{u}_i\mathbf{u}_i^*)\mathbf{v} = \left(\sum_{i=1}^{r} \mathbf{u}_i\mathbf{u}_i^* \right)\mathbf{v} = UU^*\mathbf{v}, \qquad (8.3.6)$$

in which we invoke (3.1.17) for the final equality. For example, if $A \in \mathsf{M}_{m \times n}(\mathbb{F})$ has full column rank and $A = QR$ is a QR factorization, then $\operatorname{col} A = \operatorname{col} Q$ (Theorem 7.4.2.b). Therefore,

$$P_{\operatorname{col} A} = QQ^* \qquad (8.3.7)$$

is the orthogonal projection onto the column space of A.

Let $\beta = \mathbf{e}_1, \mathbf{e}_2, \ldots, \mathbf{e}_n$ be the standard basis for $\mathbb{F}^n$. Theorem 6.5.3 tells us that $_\beta[P_{\mathcal{U}}]_\beta = [\langle P_{\mathcal{U}}\mathbf{e}_j, \mathbf{e}_i \rangle]$. Use (8.3.6) to compute $[\langle P_{\mathcal{U}}\mathbf{e}_j, \mathbf{e}_i \rangle] = [\langle UU^*\mathbf{e}_j, \mathbf{e}_i \rangle] = [\mathbf{e}_i^* UU^* \mathbf{e}_j] = UU^*$. Thus, UU^* is the matrix representation of the orthogonal projection $P_{\mathcal{U}}$ with respect to the standard basis.

Example 8.3.8 If $\mathbf{u} \neq \mathbf{0}$, then $\mathbf{u}/\|\mathbf{u}\|_2$ is an orthonormal basis for $\mathcal{U} = \operatorname{span}\{\mathbf{u}\} \subseteq \mathbb{F}^n$. The preceding example includes as a special case the formula (5.4.12) for projecting one vector in $\mathbb{F}^n$ onto the span of the other. Indeed, for any $\mathbf{v} \in \mathbb{F}^n$,

$$P_{\mathcal{U}}\mathbf{v} = \frac{\mathbf{u}\mathbf{u}^*}{\|\mathbf{u}\|_2^2}\mathbf{v} = \frac{\mathbf{u}(\mathbf{u}^*\mathbf{v})}{\|\mathbf{u}\|_2^2} = \frac{\langle \mathbf{v}, \mathbf{u} \rangle \mathbf{u}}{\|\mathbf{u}\|_2^2}.$$

If an orthonormal basis for a subspace $\mathcal{U}$ of $\mathbb{F}^n$ is not readily available, one can apply the QR factorization or another orthogonalization algorithm to a basis for $\mathcal{U}$ to obtain the matrix representation for $P_{\mathcal{U}}$. Even though this orthogonalization can be avoided in some applications (see Sections 8.4 and 8.5), in actual numerical work it is often advisable to do it to enhance the stability of computations.

The following theorem returns to an issue that was addressed differently in P.7.24.

Theorem 8.3.9 *Let* $X, Y \in \mathsf{M}_{n \times r}$ *have orthonormal columns. Then* $\operatorname{col} X = \operatorname{col} Y$ *if and only if there is a unitary* $U \in \mathsf{M}_r$ *such that* $X = YU$.

Proof If $\operatorname{col} X = \operatorname{col} Y$, Example 8.3.5 shows that $XX^* = YY^*$, since both matrices represent the projection onto the same subspace of $\mathbb{C}^n$. Let $U = Y^*X$ and compute $X = XX^*X = YY^*X = YU$. Then $I_r = X^*X = (YU)^*(YU) = U^*Y^*YU = U^*U$, so U is unitary. Conversely, if $X = YU$ for some unitary $U \in \mathsf{M}_r$, then $\operatorname{col} X \subseteq \operatorname{col} Y$. Since U is unitary, we also have $Y = XU^*$, which implies that $\operatorname{col} Y \subseteq \operatorname{col} X$. $\square$

Orthogonal projections permit us to associate an operator (an algebraic object) with each finite-dimensional subspace (a geometric object) of an inner product space. Many statements about subspaces and their relationships can be translated into statements about the corresponding orthogonal projections. Moreover, orthogonal projections form the basic building blocks from which many other operators are built (see Section 14.9). Some of their properties are listed in the following two theorems.

Theorem 8.3.10 *Let* $\mathcal{U}$ *be a finite-dimensional subspace of an inner product space* $\mathcal{V}$.

(a) $P_{\{0\}} = 0$ *and* $P_\mathcal{V} = I$.

(b) $\operatorname{ran} P_\mathcal{U} = \mathcal{U}$.

(c) $\ker P_\mathcal{U} = \mathcal{U}^\perp$.

(d) $\mathbf{v} - P_\mathcal{U}\mathbf{v} \in \mathcal{U}^\perp$ *for all* $\mathbf{v} \in \mathcal{V}$.

(e) $\|P_\mathcal{U}\mathbf{v}\| \leq \|\mathbf{v}\|$ *for all* $\mathbf{v} \in \mathcal{V}$ *with equality if and only if* $\mathbf{v} \in \mathcal{U}$.

Proof Assertion (a) follows from Theorem 8.1.6. The assertions (b), (c), and (d) follow from the fact that if we represent $\mathbf{v} \in \mathcal{V}$ as in (8.3.1), then $P_\mathcal{U}\mathbf{v} = \mathbf{u}$. Using the fact that $\mathbf{u}$ and $\mathbf{w}$ in (8.3.1) are orthogonal, (e) follows from the Pythagorean theorem and the calculation $\|P_\mathcal{U}\mathbf{v}\|^2 = \|\mathbf{u}\|^2 \leq \|\mathbf{u}\|^2 + \|\mathbf{w}\|^2 = \|\mathbf{u}+\mathbf{w}\|^2 = \|\mathbf{v}\|^2$. Finally, $\|P_\mathcal{U}\mathbf{v}\| = \|\mathbf{v}\|$ if and only if $\mathbf{w} = \mathbf{0}$, which occurs if and only if $\mathbf{v} \in \mathcal{U}$. $\square$

Theorem 8.3.11 *Let* $\mathcal{V}$ *be a finite-dimensional inner product space and let* $\mathcal{U}$ *be a subspace of* $\mathcal{V}$.

(a) $P_{\mathcal{U}^\perp} = I - P_\mathcal{U}$.

(b) $P_\mathcal{U} P_{\mathcal{U}^\perp} = P_{\mathcal{U}^\perp} P_\mathcal{U} = 0$.

(c) $P_\mathcal{U}^2 = P_\mathcal{U}$.

(d) $P_\mathcal{U} = P_\mathcal{U}^*$.

Proof Let $\mathbf{v} \in \mathcal{V}$ and write $\mathbf{v} = \mathbf{u} + \mathbf{w}$, in which $\mathbf{u} \in \mathcal{U}$ and $\mathbf{w} \in \mathcal{U}^\perp$.

(a) Compute $P_{\mathcal{U}^\perp}\mathbf{v} = \mathbf{w} = \mathbf{v} - \mathbf{u} = \mathbf{v} - P_\mathcal{U}\mathbf{v} = (I - P_\mathcal{U})\mathbf{v}$. Consequently, $P_{\mathcal{U}^\perp} = I - P_\mathcal{U}$.

(b) Since $\mathbf{w} \in \mathcal{U}^\perp$, we have $P_\mathcal{U} P_{\mathcal{U}^\perp}\mathbf{v} = P_\mathcal{U}\mathbf{w} = \mathbf{0}$. Similarly, $P_{\mathcal{U}^\perp} P_\mathcal{U}\mathbf{v} = P_{\mathcal{U}^\perp}\mathbf{u} = \mathbf{0}$ since $\mathbf{u} \in \mathcal{U}$. Thus, $P_\mathcal{U} P_{\mathcal{U}^\perp} = P_{\mathcal{U}^\perp} P_\mathcal{U} = 0$.

(c) From (a) and (b) we see that $0 = P_\mathcal{U} P_{\mathcal{U}^\perp} = P_\mathcal{U}(I - P_\mathcal{U}) = P_\mathcal{U} - P_\mathcal{U}^2$, so $P_\mathcal{U}^2 = P_\mathcal{U}$.

(d) Let $\mathbf{v}_1, \mathbf{v}_2 \in \mathcal{V}$ and write $\mathbf{v}_1 = \mathbf{u}_1 + \mathbf{w}_1$ and $\mathbf{v}_2 = \mathbf{u}_2 + \mathbf{w}_2$, in which $\mathbf{u}_1, \mathbf{u}_2 \in \mathcal{U}$ and $\mathbf{w}_1, \mathbf{w}_2 \in \mathcal{U}^\perp$. Then $\langle P_{\mathcal{U}}\mathbf{v}_1, \mathbf{v}_2 \rangle = \langle \mathbf{u}_1, \mathbf{u}_2 + \mathbf{w}_2 \rangle = \langle \mathbf{u}_1, \mathbf{u}_2 \rangle = \langle \mathbf{u}_1 + \mathbf{w}_1, \mathbf{u}_2 \rangle = \langle \mathbf{v}_1, P_{\mathcal{U}}\mathbf{v}_2 \rangle$, so $P_{\mathcal{U}} = P_{\mathcal{U}}^*$. □

Example 8.3.12 Let $n > r \geq 1$, suppose that $\mathcal{V}$ is an n-dimensional inner product space, let $\mathcal{U}$ be a subspace of $\mathcal{V}$, let $\mathbf{u}_1, \mathbf{u}_2, \ldots, \mathbf{u}_r \in \mathcal{V}$ be an orthonormal basis for $\mathcal{U}$, and let $\mathbf{u}_{r+1}, \mathbf{u}_{r+2}, \ldots, \mathbf{u}_n$ be an orthonormal basis for $\mathcal{U}^\perp$. Let $\beta = \mathbf{u}_1, \mathbf{u}_2, \ldots, \mathbf{u}_r, \mathbf{u}_{r+1}, \mathbf{u}_{r+2}, \ldots, \mathbf{u}_n$. The β-β matrix representation of $P_{\mathcal{U}}$ is

$$_\beta[P_{\mathcal{U}}]_\beta = [\langle P_{\mathcal{U}}\mathbf{u}_j, \mathbf{u}_i \rangle] = \begin{bmatrix} I_r & 0 \\ 0 & 0_{n-r} \end{bmatrix}$$

and the β-β matrix representation of $P_{\mathcal{U}^\perp}$ is

$$_\beta[P_{\mathcal{U}^\perp}]_\beta = {}_\beta[I - P_{\mathcal{U}}]_\beta = I - {}_\beta[P_{\mathcal{U}}]_\beta = \begin{bmatrix} 0_r & 0 \\ 0 & I_{n-r} \end{bmatrix}.$$

Example 8.3.13 Let $\mathcal{U}$ be a nonzero subspace of $\mathbb{F}^n$ and let $\mathbf{u}_1, \mathbf{u}_2, \ldots, \mathbf{u}_r$ be an orthonormal basis for $\mathcal{U}$. Let $U = [\mathbf{u}_1 \ \mathbf{u}_2 \ \ldots \ \mathbf{u}_r] \in \mathsf{M}_{n \times r}(\mathbb{F})$. Then $P_{\mathcal{U}} = UU^*$ (see Example 8.3.5), so $I - UU^* = P_{\mathcal{U}^\perp}$ is the projection onto the orthogonal complement of $\mathcal{U}$.

Theorem 8.3.14 *Let $\mathcal{V}$ be a finite-dimensional inner product space and let $P \in \mathcal{L}(\mathcal{V})$. Then P is the orthogonal projection onto* $\operatorname{ran} P$ *if and only if P is selfadjoint and idempotent.*

Proof In light of Theorems 8.3.10.b, 8.3.11.c, and 8.3.11.d, it suffices to prove the reverse implication. Let $P \in \mathcal{L}(\mathcal{V})$ and let $\mathcal{U} = \operatorname{ran} P$. For any $\mathbf{v} \in \mathcal{V}$,

$$\mathbf{v} = P\mathbf{v} + (\mathbf{v} - P\mathbf{v}), \tag{8.3.15}$$

in which $P\mathbf{v} \in \operatorname{ran} P$. Now assume that $P^2 = P$ and compute $P(\mathbf{v} - P\mathbf{v}) = P\mathbf{v} - P^2\mathbf{v} = P\mathbf{v} - P\mathbf{v} = \mathbf{0}$. This shows that $\mathbf{v} - P\mathbf{v} \in \ker P$, which is $(\operatorname{ran} P^*)^\perp$ by (8.2.3). If we now assume that $P = P^*$, then $\mathbf{v} - P\mathbf{v} \in (\operatorname{ran} P)^\perp$. If P is idempotent and selfadjoint, then (8.3.15) expresses $\mathbf{v}$ as the sum of something in $\mathcal{U}$, namely $P\mathbf{v}$, and something in $\mathcal{U}^\perp$, namely $\mathbf{v} - P\mathbf{v}$. It follows from the definition of an orthogonal projection that $P\mathbf{v} = P_{\mathcal{U}}\mathbf{v}$ for all $\mathbf{v} \in \mathcal{V}$, so $P = P_{\mathcal{U}}$. □

Our proof of the preceding theorem identifies the role of each of the two key assumptions. The idempotence of P ensures that (8.3.15) provides a way to express $\mathcal{V}$ as a direct sum, namely $\mathcal{V} = \operatorname{ran} P \oplus \ker P = \operatorname{ran} P \oplus (\operatorname{ran} P^*)^\perp$. However, this direct sum need not have orthogonal direct summands. Orthogonality is ensured by the selfadjointness of P, for then we have $\mathcal{V} = \operatorname{ran} P \oplus (\operatorname{ran} P)^\perp$.

Example 8.3.16 Let $\mathcal{V} = \mathbb{R}^2$, let

$$A = \begin{bmatrix} -1 & 1 \\ -2 & 2 \end{bmatrix}, \qquad \text{and} \qquad B = \frac{1}{2}\begin{bmatrix} 1 & 1 \\ 1 & 1 \end{bmatrix}.$$

Let $P = T_A : \mathcal{V} \to \mathcal{V}$ and $Q = T_B : \mathcal{V} \to \mathcal{V}$. Since $A^2 = A$ and $B^2 = B$, the operators P and Q are idempotent. We have $\ker P = \operatorname{span}\{[1 \ 1]^\mathsf{T}\}$ and $\operatorname{ran} P = \operatorname{span}\{[1 \ 2]^\mathsf{T}\}$. The vectors $[1 \ 1]^\mathsf{T}$ and $[1 \ 2]^\mathsf{T}$ are linearly independent, so $\operatorname{ran} P \oplus \ker P = \mathbb{R}^2$, but the direct summands are not orthogonal; see Figure 8.4a and P.8.4. We also have $\ker Q = \operatorname{span}\{[1 \ -1]^\mathsf{T}\}$ and $\operatorname{ran} Q = \operatorname{span}\{[1 \ 1]^\mathsf{T}\}$. The vectors $[1 \ -1]^\mathsf{T}$ and $[1 \ 1]^\mathsf{T}$ are orthogonal, so $\operatorname{ran} Q \oplus \ker Q = \mathbb{R}^2$ and the direct summands are orthogonal; see Figure 8.4b. Since B is Hermitian and idempotent, it is an orthogonal projection; A is idempotent, but not Hermitian, so it is not an orthogonal projection.

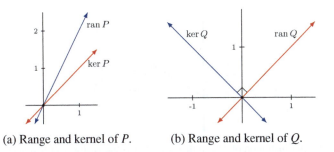

(a) Range and kernel of P.   (b) Range and kernel of Q.

Figure 8.4 Ranges and kernels of the idempotent operators in Example 8.3.16.

8.4 Best Approximation

Many practical applications of orthogonal projections stem from the following theorem. It states that the best approximation to a given vector $\mathbf{v}$ by a vector lying in a finite-dimensional subspace $\mathcal{U}$ is the orthogonal projection of $\mathbf{v}$ onto $\mathcal{U}$.

Theorem 8.4.1 (Best Approximation Theorem) *Let $\mathcal{U}$ be a finite-dimensional subspace of an inner product space $\mathcal{V}$ and let $P_{\mathcal{U}}$ be the orthogonal projection onto $\mathcal{U}$. Then*

$$\|\mathbf{v} - P_{\mathcal{U}}\mathbf{v}\| \leq \|\mathbf{v} - \mathbf{u}\|$$

for all $\mathbf{v} \in \mathcal{V}$ and $\mathbf{u} \in \mathcal{U}$, with equality if and only if $\mathbf{u} = P_{\mathcal{U}}\mathbf{v}$; see Figure 8.5.

Proof For each $\mathbf{v} \in \mathcal{V}$, we have the orthogonal decomposition

$$\mathbf{v} = \underbrace{P_{\mathcal{U}}\mathbf{v}}_{\in \mathcal{U}} + \underbrace{(\mathbf{v} - P_{\mathcal{U}}\mathbf{v})}_{\in \mathcal{U}^{\perp}}.$$

The Pythagorean theorem asserts that

$$\|\mathbf{v} - P_{\mathcal{U}}\mathbf{v}\|^2 \leq \|\mathbf{v} - P_{\mathcal{U}}\mathbf{v}\|^2 + \|P_{\mathcal{U}}\mathbf{v} - \mathbf{u}\|^2 \qquad (8.4.2)$$
$$= \|(\mathbf{v} - P_{\mathcal{U}}\mathbf{v}) + (P_{\mathcal{U}}\mathbf{v} - \mathbf{u})\|^2$$
$$= \|\mathbf{v} - \mathbf{u}\|^2$$

since $P_{\mathcal{U}}\mathbf{v} - \mathbf{u}$ belongs to $\mathcal{U}$, with equality in (8.4.2) if and only if $\mathbf{u} = P_{\mathcal{U}}\mathbf{v}$. $\square$

The preceding theorem can be interpreted as a statement about distances from a subspace. Let

$$d(\mathbf{v}, \mathcal{U}) = \min_{\mathbf{u} \in \mathcal{U}} \|\mathbf{v} - \mathbf{u}\| \qquad (8.4.3)$$

denote the distance between $\mathbf{v}$ and the subspace $\mathcal{U}$. The best approximation theorem tells us that $d(\mathbf{v}, \mathcal{U}) = \|\mathbf{v} - P_{\mathcal{U}}\mathbf{v}\|$.

Example 8.4.4 Let $A \in \mathsf{M}_{m \times n}(\mathbb{F})$ have rank $A = r \geq 1$. Suppose that $m < n$ and let $\mathcal{U} = \operatorname{null} A$. If $\mathbf{v} \in \mathbb{F}^n$, then $\mathbf{v} = P_{\mathcal{U}}\mathbf{v} + P_{\mathcal{U}^{\perp}}\mathbf{v}$ and the distance from $\mathbf{v}$ to $\mathcal{U}$ is $\|\mathbf{v} - P_{\mathcal{U}}\mathbf{v}\|_2 = \|P_{\mathcal{U}^{\perp}}\mathbf{v}\|_2$. The identity (8.2.4) tells us that $\mathcal{U}^{\perp} = \operatorname{col} A^*$. If $\beta = \mathbf{u}_1, \mathbf{u}_2, \ldots, \mathbf{u}_r$ is an orthonormal basis for $\operatorname{col} A^*$, then (8.3.3) ensures that $d(\mathbf{v}, \mathcal{U})^2 = \sum_{i=1}^{r} |\mathbf{u}_i^*\mathbf{v}|^2$. We can obtain β by

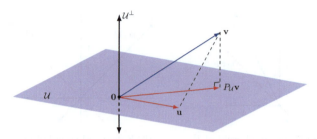

Figure 8.5 An illustration of Theorem 8.4.1. $P_{\mathcal{U}}\mathbf{v}$ is the vector in $\mathcal{U}$ that is closest to $\mathbf{v}$.

using the Gram–Schmidt process or another orthogonalization algorithm to orthogonalize the columns of A^*. For example, if A has full row rank and $A^* = QR$ is a QR factorization, then we may take β to be the columns of Q. In this case, $d(\mathbf{v}, \mathcal{U}) = \|Q^*\mathbf{v}\|_2$. See P.8.7 for a geometric application of this example.

Example 8.4.5 Let $\mathcal{V} = C_{\mathbb{R}}[-\pi, \pi]$ with the inner product (6.8.5), and let

$$\mathcal{U} = \operatorname{span}\left\{\frac{1}{\sqrt{2}}, \cos x, \cos 2x, \ldots, \cos Nx, \sin x, \sin 2x, \ldots, \sin Nx\right\}.$$

Theorems 8.3.2 and 8.4.1 ensure that the function (6.8.14) (a *finite Fourier series*) is the function in $\mathcal{U}$ that is closest to f.

If $\mathbf{u}_1, \mathbf{u}_2, \ldots, \mathbf{u}_r$ is an orthonormal basis for a subspace $\mathcal{U}$ of $\mathcal{V}$, then (8.3.3) gives the orthogonal projection onto $\mathcal{U}$. If we only have a basis for $\mathcal{U}$, or perhaps just a spanning list, we can use an orthogonalization algorithm to obtain an orthonormal basis, or we can proceed as follows:

Theorem 8.4.6 (Normal Equations) *Let $\mathcal{U}$ be a finite-dimensional subspace of an inner product space $\mathcal{V}$ and suppose that $\operatorname{span}\{\mathbf{u}_1, \mathbf{u}_2, \ldots, \mathbf{u}_n\} = \mathcal{U}$. The projection of $\mathbf{v} \in \mathcal{V}$ onto $\mathcal{U}$ is*

$$P_{\mathcal{U}}\mathbf{v} = \sum_{j=1}^{n} c_j \mathbf{u}_j, \tag{8.4.7}$$

in which $[c_1 \ c_2 \ \ldots \ c_n]^{\mathsf{T}}$ *is a solution of the* normal equations

$$\begin{bmatrix} \langle\mathbf{u}_1,\mathbf{u}_1\rangle & \langle\mathbf{u}_2,\mathbf{u}_1\rangle & \cdots & \langle\mathbf{u}_n,\mathbf{u}_1\rangle \\ \langle\mathbf{u}_1,\mathbf{u}_2\rangle & \langle\mathbf{u}_2,\mathbf{u}_2\rangle & \cdots & \langle\mathbf{u}_n,\mathbf{u}_2\rangle \\ \vdots & \vdots & \ddots & \vdots \\ \langle\mathbf{u}_1,\mathbf{u}_n\rangle & \langle\mathbf{u}_2,\mathbf{u}_n\rangle & \cdots & \langle\mathbf{u}_n,\mathbf{u}_n\rangle \end{bmatrix} \begin{bmatrix} c_1 \\ c_2 \\ \vdots \\ c_n \end{bmatrix} = \begin{bmatrix} \langle\mathbf{v},\mathbf{u}_1\rangle \\ \langle\mathbf{v},\mathbf{u}_2\rangle \\ \vdots \\ \langle\mathbf{v},\mathbf{u}_n\rangle \end{bmatrix}. \tag{8.4.8}$$

The system (8.4.8) is consistent. If $\mathbf{u}_1, \mathbf{u}_2, \ldots, \mathbf{u}_n$ are linearly independent, then (8.4.8) has a unique solution.

Proof Since $P_{\mathcal{U}}\mathbf{v} \in \mathcal{U}$ and $\operatorname{span}\{\mathbf{u}_1, \mathbf{u}_2, \ldots, \mathbf{u}_n\} = \mathcal{U}$, there exist scalars $c_1, c_2, \ldots, c_n$ such that

$$P_{\mathcal{U}}\mathbf{v} = \sum_{j=1}^{n} c_j \mathbf{u}_j. \tag{8.4.9}$$

Then $\langle\mathbf{v}, \mathbf{u}_i\rangle = \langle\mathbf{v}, P_{\mathcal{U}}\mathbf{u}_i\rangle = \langle P_{\mathcal{U}}\mathbf{v}, \mathbf{u}_i\rangle = \langle\sum_{j=1}^{n} c_j\mathbf{u}_j, \mathbf{u}_i\rangle = \sum_{j=1}^{n} c_j\langle\mathbf{u}_j, \mathbf{u}_i\rangle$ for $i = 1, 2, \ldots, n$. These equations are the system (8.4.8), which is consistent since $c_1, c_2, \ldots, c_n$ are already known to exist by (8.4.9). If the list $\mathbf{u}_1, \mathbf{u}_2, \ldots, \mathbf{u}_n$ is linearly independent, it is a basis for $\mathcal{U}$. In this case, (8.4.8) has a unique solution since each vector in $\mathcal{U}$ can be written as a linear combination of basis vectors in exactly one way. $\qquad\square$

Example 8.4.10 Let $\mathbf{v} = [1\ 1\ 1]^T$ and let $\mathcal{U} = \{[x_1\ x_2\ x_3]^T \in \mathbb{R}^3 : x_1 + 2x_2 + 3x_3 = 0\}$. Then $\mathcal{U} = \operatorname{span}\{\mathbf{u}_1, \mathbf{u}_2\}$, in which $\mathbf{u}_1 = [3\ 0\ -1]^T$ and $\mathbf{u}_2 = [2\ -1\ 0]^T$ (see Examples 8.1.2 and 8.1.10). To find the projection of $\mathbf{v}$ onto $\mathcal{U}$, solve the normal equations (8.4.8), which are

$$\begin{bmatrix} 10 & 6 \\ 6 & 5 \end{bmatrix} \begin{bmatrix} c_1 \\ c_2 \end{bmatrix} = \begin{bmatrix} 2 \\ 1 \end{bmatrix}.$$

We obtain $c_1 = \frac{2}{7}$ and $c_2 = -\frac{1}{7}$, so (8.4.7) ensures that $P_{\mathcal{U}}\mathbf{v} = \frac{2}{7}\mathbf{u}_1 - \frac{1}{7}\mathbf{u}_2 = [\frac{4}{7}\ \frac{1}{7}\ -\frac{2}{7}]^T$.

Example 8.4.11 Let $\mathcal{U}$, $\mathbf{v}$, $\mathbf{u}_1$, and $\mathbf{u}_2$ be as in the preceding example. Let $\mathbf{u}_3 = \mathbf{u}_1 + \mathbf{u}_2$, and consider the list $\mathbf{u}_1, \mathbf{u}_2, \mathbf{u}_3$, which spans $\mathcal{U}$ but is not a basis. The normal equations (8.4.8) are

$$\begin{bmatrix} 10 & 6 & 16 \\ 6 & 5 & 11 \\ 16 & 11 & 27 \end{bmatrix} \begin{bmatrix} c_1 \\ c_2 \\ c_3 \end{bmatrix} = \begin{bmatrix} 2 \\ 1 \\ 3 \end{bmatrix}.$$

A solution is $c_1 = \frac{5}{21}$, $c_2 = -\frac{4}{21}$, $c_3 = \frac{1}{21}$. Thus,

$$P_{\mathcal{U}}\mathbf{v} = \frac{5}{21}\mathbf{u}_1 - \frac{4}{21}\mathbf{u}_2 + \frac{1}{21}\mathbf{u}_3 = [\frac{4}{7}\ \frac{1}{7}\ -\frac{2}{7}]^T.$$

Another solution is $c_1 = \frac{1}{3}$, $c_2 = -\frac{2}{21}$, $c_3 = -\frac{1}{21}$, which also gives

$$P_{\mathcal{U}}\mathbf{v} = \frac{1}{3}\mathbf{u}_1 - \frac{2}{21}\mathbf{u}_2 - \frac{1}{21}\mathbf{u}_3 = [\frac{4}{7}\ \frac{1}{7}\ -\frac{2}{7}]^T.$$

The projection of $\mathbf{v}$ onto $\mathcal{U}$ is unique, even if the normal equations do not have a unique solution.

Example 8.4.12 Consider the function $f(x) = \sin x$ on the interval $[-\pi, \pi]$. How can it be approximated by real polynomials of degree 5 or less? The Taylor polynomial $x - \frac{x^3}{6} + \frac{x^5}{120}$ is one possibility. However, Taylor polynomials approximate well only near the center of the Taylor expansion; see Figure 8.6a. We can do better by considering f to be an element of the inner product space $V = C_{\mathbb{R}}[-\pi, \pi]$ with the inner product (5.3.9) and derived norm (5.4.7).

Let $\mathcal{U} = \operatorname{span}\{1, x, x^2, x^3, x^4, x^5\}$ denote the subspace of V consisting of all polynomials of degree at most five. We want to find the unique polynomial $p(x) \in \mathcal{U}$ such that

$$\|\sin x - p(x)\| = \min_{q \in \mathcal{U}} \|\sin x - q(x)\| = \min_{q \in \mathcal{U}} \sqrt{\int_{-\pi}^{\pi} |\sin x - q(x)|^2\, dx}.$$

Theorem 8.4.1 tells us that $p(x) = P_{\mathcal{U}}(\sin x)$. We can use the normal equations (Theorem 8.4.6) to compute $p(x)$. Let $\mathbf{u}_i = x^{i-1}$ for $i = 1, 2, \ldots, 6$, so that

$$\langle \mathbf{u}_j, \mathbf{u}_i \rangle = \int_{-\pi}^{\pi} x^{i+j-2}\, dx = \frac{x^{i+j-1}}{i+j-1}\Big|_{-\pi}^{\pi} = \begin{cases} 0 & \text{if } i+j \text{ is odd,} \\ \dfrac{2\pi^{i+j-1}}{i+j-1} & \text{if } i+j \text{ is even.} \end{cases}$$

We have $\langle \sin x, 1 \rangle = \int_{-\pi}^{\pi} (\sin x)(1)\, dx = -\cos x\big|_{-\pi}^{\pi} = 0$, which we could have deduced without a computation by observing that the function $1 \cdot \sin x$ is odd. This observation tells us

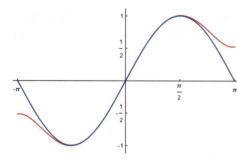
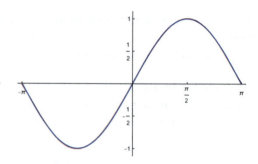

(a) The graphs of $\sin x$ (blue) and its Taylor approximation $x - \frac{x^3}{6} + \frac{x^5}{120}$ (red). The quality of the approximation deteriorates rapidly away from $x = 0$.

(b) The graphs of $\sin x$ (blue) and $p(x) = P_{\mathcal{U}}(\sin x)$ (red) are nearly indistinguishable on $[-\pi, \pi]$.

Figure 8.6 Approximations of $\sin x$ by a polynomial of degree 5: Taylor series vs. least squares.

that $\langle \sin x, x^2 \rangle = \langle \sin x, x^4 \rangle = 0$ as well, since $x^2 \sin x$ and $x^4 \sin x$ are odd functions. Some computations reveal that

$$\langle \sin x, x \rangle = 2\pi, \quad \langle \sin x, x^3 \rangle = 2\pi(\pi^2 - 6), \quad \text{and} \quad \langle \sin x, x^5 \rangle = 2\pi(120 - 20\pi^2 + \pi^4).$$

We now have all the entries of the 6×6 coefficient matrix and the right side of the linear system (8.4.8). Its solution is

$$P_{\mathcal{U}}(\sin x) = \frac{105(1485 - 153\pi^2 + \pi^4)}{8\pi^6}x - \frac{315(1155 - 125\pi^2 + \pi^4)}{4\pi^8}x^3$$
$$+ \frac{693(945 - 105\pi^2 + \pi^4)}{8\pi^{10}}x^5 \tag{8.4.13}$$
$$\approx 0.987862x - 0.155271x^3 + 0.00564312x^5.$$

Figure 8.6b shows that (8.4.13) is an excellent approximation to $\sin x$ over the interval $[-\pi, \pi]$. The matrix that appears in the normal equations (8.4.8) is an important one.

Definition 8.4.14 If $\mathbf{u}_1, \mathbf{u}_2, \ldots, \mathbf{u}_n$ are vectors in an inner product space, then

$$G(\mathbf{u}_1, \mathbf{u}_2, \ldots, \mathbf{u}_n) = \begin{bmatrix} \langle \mathbf{u}_1, \mathbf{u}_1 \rangle & \langle \mathbf{u}_2, \mathbf{u}_1 \rangle & \cdots & \langle \mathbf{u}_n, \mathbf{u}_1 \rangle \\ \langle \mathbf{u}_1, \mathbf{u}_2 \rangle & \langle \mathbf{u}_2, \mathbf{u}_2 \rangle & \cdots & \langle \mathbf{u}_n, \mathbf{u}_2 \rangle \\ \vdots & \vdots & \ddots & \vdots \\ \langle \mathbf{u}_1, \mathbf{u}_n \rangle & \langle \mathbf{u}_2, \mathbf{u}_n \rangle & \cdots & \langle \mathbf{u}_n, \mathbf{u}_n \rangle \end{bmatrix}$$

is the *Gram matrix* (or *Grammian*) of $\mathbf{u}_1, \mathbf{u}_2, \ldots, \mathbf{u}_n$. The *Gram determinant* of the list $\mathbf{u}_1, \mathbf{u}_2, \ldots, \mathbf{u}_n$ is $g(\mathbf{u}_1, \mathbf{u}_2, \ldots, \mathbf{u}_n) = \det G(\mathbf{u}_1, \mathbf{u}_2, \ldots, \mathbf{u}_n)$.

A Gram matrix is Hermitian ($G = G^*$) and positive semidefinite (see Chapter 15). Gram matrices are closely related to covariance matrices from statistics. If the vectors $\mathbf{u}_1, \mathbf{u}_2, \ldots, \mathbf{u}_n$ are centered random variables, then $G(\mathbf{u}_1, \mathbf{u}_2, \ldots, \mathbf{u}_n)$ is the corresponding covariance matrix.

8.5 The Least-Squares Solution of an Inconsistent Linear System

Suppose that $Ax = y$ is an inconsistent $m \times n$ system of linear equations, that is, $\mathbf{y} - A\mathbf{x} \neq \mathbf{0}$ for all $\mathbf{x}$ in $\mathbb{F}^n$. We want to identify an $\mathbf{x}_0 \in \mathbb{F}^n$ such that $\|\mathbf{y} - A\mathbf{x}_0\|_2$ is as small as possible. If we find such a vector, we can regard it as a "best approximate solution" to the inconsistent system $Ax = y$; see Figure 8.7.

Theorem 8.5.1 *If $A \in \mathsf{M}_{m \times n}(\mathbb{F})$ and $\mathbf{y} \in \mathbb{F}^m$, then $\mathbf{x}_0 \in \mathbb{F}^n$ satisfies*

$$\min_{\mathbf{x} \in \mathbb{F}^n} \|\mathbf{y} - A\mathbf{x}\|_2 = \|\mathbf{y} - A\mathbf{x}_0\|_2 \tag{8.5.2}$$

if and only if

$$A^* A\mathbf{x}_0 = A^* \mathbf{y}. \tag{8.5.3}$$

The system (8.5.3) is always consistent; it has a unique solution if rank $A = n$.

Proof Partition $A = [\mathbf{a}_1 \ \mathbf{a}_2 \ \ldots \ \mathbf{a}_n]$ according to its columns and let $\mathcal{U} = \operatorname{col} A = \operatorname{span}\{\mathbf{a}_1, \mathbf{a}_2, \ldots, \mathbf{a}_n\}$. We are asked to find an $A\mathbf{x}_0$ in $\mathcal{U}$ that is closest to $\mathbf{y}$ with respect to the Euclidean norm. Theorem 8.4.1 ensures that $P_{\mathcal{U}}\mathbf{y}$ is the unique such vector, and Theorem 8.4.6 tells us how to compute it. First find a solution $\mathbf{x}_0 = [c_1 \ c_2 \ \ldots \ c_n]^{\mathsf{T}}$ of the normal equations

$$\begin{bmatrix} \mathbf{a}_1^* \mathbf{a}_1 & \mathbf{a}_1^* \mathbf{a}_2 & \cdots & \mathbf{a}_1^* \mathbf{a}_n \\ \mathbf{a}_2^* \mathbf{a}_1 & \mathbf{a}_2^* \mathbf{a}_2 & \cdots & \mathbf{a}_2^* \mathbf{a}_n \\ \vdots & \vdots & \ddots & \vdots \\ \mathbf{a}_n^* \mathbf{a}_1 & \mathbf{a}_n^* \mathbf{a}_2 & \cdots & \mathbf{a}_n^* \mathbf{a}_n \end{bmatrix} \begin{bmatrix} c_1 \\ c_2 \\ \vdots \\ c_n \end{bmatrix} = \begin{bmatrix} \mathbf{a}_1^* \mathbf{y} \\ \mathbf{a}_2^* \mathbf{y} \\ \vdots \\ \mathbf{a}_n^* \mathbf{y} \end{bmatrix},$$

that is, $A^* A\mathbf{x}_0 = A^* \mathbf{y}$. Then

$$P_{\mathcal{U}}\mathbf{y} = \sum_{i=1}^n c_i \mathbf{a}_i = A\mathbf{x}_0. \tag{8.5.4}$$

Theorem 8.4.6 ensures that the normal equations (8.5.3) are consistent. They have a unique solution if the columns of A are linearly independent, that is, if rank $A = n$. $\square$

If the null space of A is nontrivial, there can be many vectors $\mathbf{x}_0$ that satisfy (8.5.3). However, $P_{\mathcal{U}}\mathbf{y} = A\mathbf{x}_0$ is the same for all of them.

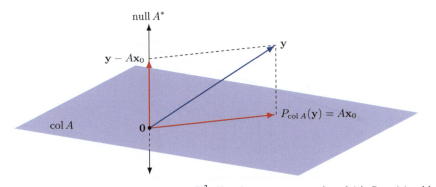

Figure 8.7 Illustration of Theorem 8.5.1 in $\mathbb{R}^3$. The closest vector to $\mathbf{y}$ in col A is $P_{\operatorname{col} A}(\mathbf{y})$, which is of the form $A\mathbf{x}_0$ for some $\mathbf{x}_0 \in \mathbb{R}^3$.

If A has full column rank and $A = QR$ is a QR factorization, then the square upper triangular factor R is invertible; see Theorem 7.4.2.b. Consequently, $A^*A = R^*Q^*QR = R^*R$ is invertible and (8.5.3) has the unique solution $\mathbf{x}_0 = (A^*A)^{-1}A^*\mathbf{y}$. In this case, $P_{\mathcal{U}}\mathbf{y} = A\mathbf{x}_0 = A(A^*A)^{-1}A^*\mathbf{y}$, that is, the orthogonal projection onto the column space of A is

$$P_{\text{col}\,A} = A(A^*A)^{-1}A^* = (QR)(R^{-1}R^{-*})(R^*Q^*) = QQ^*. \tag{8.5.5}$$

Suppose that we are given some real data points $(x_1, y_1), (x_2, y_2), \ldots, (x_n, y_n)$ in the plane and we want to model them as a (perhaps approximate) linear relationship. If all the points lie on a vertical line $x = c$, there is nothing to do. If the points do not lie on a vertical line, consider the system of linear equations

$$
\begin{array}{rcl}
y_1 & = & ax_1 + b, \\
y_2 & = & ax_2 + b, \\
\vdots & \vdots & \vdots \\
y_n & = & ax_n + b,
\end{array}
\tag{8.5.6}
$$

which we write as $A\mathbf{x} = \mathbf{y}$, with

$$
A = \begin{bmatrix} x_1 & 1 \\ x_2 & 1 \\ \vdots & \vdots \\ x_n & 1 \end{bmatrix}, \qquad \mathbf{x} = \begin{bmatrix} a \\ b \end{bmatrix}, \quad \text{and} \quad \mathbf{y} = \begin{bmatrix} y_1 \\ y_2 \\ \vdots \\ y_n \end{bmatrix}.
$$

Because the data points do not lie on a vertical line, the columns of A are linearly independent and hence rank $A = 2$. The system $A\mathbf{x} = \mathbf{y}$ involves n equations in two unknowns, so if $n > 2$ it might not have a solution. However, Theorem 8.5.1 ensures that any solution of the normal equations $A^*A\mathbf{x} = A^*\mathbf{y}$ minimizes

$$\|\mathbf{y} - A\mathbf{x}\|_2 = \left(\sum_{i=1}^{n} \big(y_i - (ax_i + b)\big)^2 \right)^{1/2}; \tag{8.5.7}$$

see Figure 8.8. The normal equations have a unique solution because rank $A = 2$.

Example 8.5.8 Find a least-squares line $y = ax + b$ to model the data

$$(0, 1), \quad (1, 1), \quad (2, 3), \quad (3, 3), \quad (4, 4).$$

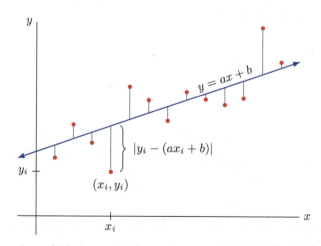

Figure 8.8 $|y_i - (ax_i + b)|$ is the vertical distance between the data point (x_i, y_i) and the graph of the line $y = ax + b$.

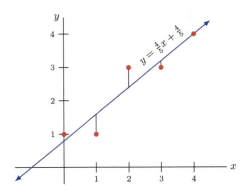

Figure 8.9 The least-squares line corresponding to the data in Example 8.5.8.

According to the preceding recipe we have

$$
A = \begin{bmatrix} 0 & 1 \\ 1 & 1 \\ 2 & 1 \\ 3 & 1 \\ 4 & 1 \end{bmatrix}, \qquad \mathbf{x} = \begin{bmatrix} a \\ b \end{bmatrix}, \qquad \mathbf{y} = \begin{bmatrix} 1 \\ 1 \\ 3 \\ 3 \\ 4 \end{bmatrix},
$$

and we solve

$$
\underbrace{\begin{bmatrix} 30 & 10 \\ 10 & 5 \end{bmatrix}}_{A^*A} \begin{bmatrix} a \\ b \end{bmatrix} = \underbrace{\begin{bmatrix} 32 \\ 12 \end{bmatrix}}_{A^*\mathbf{y}}
$$

for $a = b = \frac{4}{5}$. Therefore, the least-squares line is $y = \frac{4}{5}x + \frac{4}{5}$; see Figure 8.9.

In numerical work, there are hazards associated with forming and then solving directly the normal equations (8.5.3) or (8.4.8) to obtain a solution to the least-squares problem (8.5.2); see Section 17.5. A safer route is via the QR factorization of A; see Theorem 7.4.2. If $m \geq n$ and $A \in \mathsf{M}_{m \times n}(\mathbb{F})$ has full column rank, then $A = QR$, in which $Q \in \mathsf{M}_{m \times n}(\mathbb{F})$ has orthonormal columns and $R \in \mathsf{M}_n(\mathbb{F})$ is upper triangular and has positive diagonal entries. Then $A^*A = R^*Q^*QR = R^*I_nR = R^*R$, so the normal equations (8.5.3) are

$$
R^*R\mathbf{u} = R^*Q^*\mathbf{y}. \tag{8.5.9}
$$

Since R is invertible, the system (8.5.9) has the same solutions as the upper triangular system

$$
R\mathbf{u} = Q^*\mathbf{y}, \tag{8.5.10}
$$

which can be solved by backward substitution.

The formulation $P_{\mathrm{col}\,A} = A(A^*A)^{-1}A^*$ should also be avoided in practical computations. Again, a safer route is via the QR factorization of A. If $A \in \mathsf{M}_{m \times n}(\mathbb{F})$, $\operatorname{rank} A = n$, and $A = QR$ is a QR factorization, then (8.5.5) shows that $P_{\mathrm{col}\,A} = QQ^*$. With a QR factorization, the orthogonal projection onto the column space of A can be computed without a matrix inversion.

8.6 Invariant Subspaces

If $V = \mathbb{F}^n$, then Theorem 8.3.14 says that $P \in M_n(\mathbb{F})$ is the matrix representation of an orthogonal projection with respect to an orthonormal basis for $\mathbb{F}^n$ if and only if P is Hermitian and idempotent. We therefore make the following definition.

Definition 8.6.1 A square matrix is an *orthogonal projection* if it is Hermitian and idempotent.

If an orthogonal projection $P \in M_n(\mathbb{F})$ is identified with the linear operator $T_P : \mathbb{F}^n \to \mathbb{F}^n$ induced by it, then P is the orthogonal projection onto col P.

Suppose that $\mathcal{U}$ is an r-dimensional subspace of $\mathbb{F}^n$ with $1 \leq r \leq n - 1$. Let $\mathbf{u}_1, \mathbf{u}_2, \ldots, \mathbf{u}_r$ be an orthonormal basis for $\mathcal{U}$ and let $U_1 = [\mathbf{u}_1 \ \mathbf{u}_2 \ \cdots \ \mathbf{u}_r] \in M_{n \times r}$. Then Example 8.3.5 tells us that $P_1 = U_1 U_1^*$ is the orthogonal projection onto $\mathcal{U}$. Similarly, if $\mathbf{u}_{r+1}, \mathbf{u}_{r+2}, \ldots, \mathbf{u}_n$ is an orthonormal basis for $\mathcal{U}^\perp$ and $U_2 = [\mathbf{u}_{r+1} \ \mathbf{u}_{r+2} \ \cdots \ \mathbf{u}_n] \in M_{n \times (n-r)}$, then $P_2 = U_2 U_2^*$ is the orthogonal projection onto $\mathcal{U}^\perp$. The matrix $U = [U_1 \ U_2]$ is unitary since its columns form an orthonormal basis for $\mathbb{F}^n$.

It is instructive to use block matrices to rederive the algebraic properties of orthogonal projections from Theorem 8.3.11. For example,

$$P_1 + P_2 = U_1 U_1^* + U_2 U_2^* = [U_1 \ U_2] \begin{bmatrix} U_1^* \\ U_2^* \end{bmatrix} = UU^* = I_n,$$

which is Theorem 8.3.11.a. Now observe that the block-matrix identity

$$\begin{bmatrix} I_r & 0 \\ 0 & I_{n-r} \end{bmatrix} = U^* U = \begin{bmatrix} U_1^* \\ U_2^* \end{bmatrix} [U_1 \ U_2] = \begin{bmatrix} U_1^* U_1 & U_1^* U_2 \\ U_2^* U_1 & U_2^* U_2 \end{bmatrix}$$

implies that

$$U_1^* U_1 = I_r, \qquad U_2^* U_2 = I_{n-r}, \qquad U_1^* U_2 = 0, \qquad \text{and} \qquad U_2^* U_1 = 0. \tag{8.6.2}$$

Consequently,

$$P_1 P_2 = (U_1 U_1^*)(U_2 U_2^*) = U_1 (U_1^* U_2) U_2^* = 0$$

and

$$P_2 P_1 = (U_2 U_2^*)(U_1 U_1^*) = U_2 (U_2^* U_1) U_1^* = 0,$$

which implies Theorem 8.3.11.b. Moreover,

$$P_1^2 = (U_1 U_1^*)(U_1 U_1^*) = U_1 (U_1^* U_1) U_1^* = U_1 I_r U_1^* = U_1 U_1^* = P_1,$$

which is Theorem 8.3.11.c. The matrices $P_1 = U_1 U_1^*$ and $P_2 = U_2 U_2^*$ are Hermitian, which is Theorem 8.3.11.d.

Another useful representation for an orthogonal projection follows from the observation

$$U \begin{bmatrix} I_r & 0 \\ 0 & 0 \end{bmatrix} U^* = [U_1 \ U_2] \begin{bmatrix} I_r & 0 \\ 0 & 0 \end{bmatrix} \begin{bmatrix} U_1^* \\ U_2^* \end{bmatrix} = U_1 U_1^* = P_1. \tag{8.6.3}$$

This is important enough to formalize as a theorem.

Theorem 8.6.4 *Let $P \in M_n$ be an orthogonal projection and let $r = $ rank P. If $1 \leq r \leq n - 1$, then P is unitarily similar to $I_r \oplus 0_{n-r}$.*

Example 8.6.5 Consider the plane $\mathcal{U}$ in $\mathbb{R}^3$ determined by the equation $-2x_1 + 2x_2 - x_3 = 0$. Let $\mathbf{u}_1 = \frac{1}{3}[1 \ 2 \ 2]^\mathsf{T}$, $\mathbf{u}_2 = \frac{1}{3}[-2 \ -1 \ 2]^\mathsf{T}$, and $\mathbf{u}_3 = \frac{1}{3}[-2 \ 2 \ -1]^\mathsf{T}$. Then $\mathbf{u}_1, \mathbf{u}_2$ is an orthonormal basis for $\mathcal{U}$ and $\mathbf{u}_3$ is an orthonormal basis for $\mathcal{U}^\perp$. We have

$$U = \frac{1}{3}\begin{bmatrix} 1 & -2 & -2 \\ 2 & -1 & 2 \\ 2 & 2 & -1 \end{bmatrix}, \quad U_1 = \frac{1}{3}\begin{bmatrix} 1 & -2 \\ 2 & -1 \\ 2 & 2 \end{bmatrix}, \quad \text{and} \quad U_2 = \frac{1}{3}\begin{bmatrix} -2 \\ 2 \\ -1 \end{bmatrix}.$$

Consequently,

$$P_1 = U_1 U_1^* = \frac{1}{9}\begin{bmatrix} 5 & 4 & -2 \\ 4 & 5 & 2 \\ -2 & 2 & 8 \end{bmatrix} \quad \text{and} \quad P_2 = U_2 U_2^* = \frac{1}{9}\begin{bmatrix} 4 & -4 & 2 \\ -4 & 4 & -2 \\ 2 & -2 & 1 \end{bmatrix}.$$

Definition 8.6.6 Let $A \in \mathsf{M}_n(\mathbb{F})$, let $\mathcal{U}$ be a subspace of $\mathbb{F}^n$, and let $A\mathcal{U} = \{A\mathbf{x} : \mathbf{x} \in \mathcal{U}\}$ (see Example 1.3.12). If $A\mathcal{U} \subseteq \mathcal{U}$, then $\mathcal{U}$ is *invariant under* A; alternatively, we say that $\mathcal{U}$ is A-*invariant*.

The subspaces $\{\mathbf{0}\}$ and $\mathbb{F}^n$ are A-invariant for any $A \in \mathsf{M}_n(\mathbb{F})$.

Theorem 8.6.7 *Let $A \in \mathsf{M}_n(\mathbb{F})$, let $\mathcal{U}$ be an r-dimensional subspace of $\mathbb{F}^n$ such that $1 \leq r \leq n - 1$. Let $U = [U_1 \ U_2] \in \mathsf{M}_n(\mathbb{F})$ be unitary and such that the columns of $U_1 \in \mathsf{M}_{n \times r}(\mathbb{F})$ and $U_2 \in \mathsf{M}_{n \times (n-r)}(\mathbb{F})$ are orthonormal bases for $\mathcal{U}$ and $\mathcal{U}^\perp$, respectively. Let $P = U_1 U_1^*$ denote the orthogonal projection onto $\mathcal{U}$. Then the following are equivalent:*

(a) $\mathcal{U}$ *is A-invariant.*

(b) $U^*AU = \begin{bmatrix} B & X \\ 0 & C \end{bmatrix}$, *in which $B \in \mathsf{M}_r(\mathbb{F})$ and $C \in \mathsf{M}_{n-r}(\mathbb{F})$.*

(c) $PAP = AP$.

Proof (a) $\Leftrightarrow$ (b) Since

$$U^*AU = \begin{bmatrix} U_1^* \\ U_2^* \end{bmatrix} A[U_1 \ U_2] = \begin{bmatrix} U_1^* \\ U_2^* \end{bmatrix} [AU_1 \ AU_2] = \begin{bmatrix} U_1^*AU_1 & U_1^*AU_2 \\ U_2^*AU_1 & U_2^*AU_2 \end{bmatrix},$$

it suffices to prove that $U_2^*AU_1 = 0$ if and only if $A\mathcal{U} \subseteq \mathcal{U}$. Let $U_1 = [\mathbf{u}_1 \ \mathbf{u}_2 \ \ldots \ \mathbf{u}_r] \in \mathsf{M}_{n \times r}(\mathbb{F})$. Using (8.2.4), we have null $U_2^* = (\mathrm{col} \ U_2)^\perp = (\mathcal{U}^\perp)^\perp = \mathcal{U}$, so

$$\begin{aligned} U_2^*AU_1 = 0 &\iff U_2^*[A\mathbf{u}_1 \ A\mathbf{u}_2 \ \ldots \ A\mathbf{u}_r] = 0 \\ &\iff [U_2^*(A\mathbf{u}_1) \ U_2^*(A\mathbf{u}_2) \ \ldots \ U_2^*(A\mathbf{u}_r)] = 0 \\ &\iff \mathrm{span}\{A\mathbf{u}_1, A\mathbf{u}_2, \ldots, A\mathbf{u}_r\} \subseteq \mathrm{null} \ U_2^* \\ &\iff A\mathcal{U} \subseteq \mathcal{U}. \end{aligned}$$

(b) $\Leftrightarrow$ (c) The representation (8.6.3) ensures that $U^*PU = I_r \oplus 0_{n-r}$. Partition

$$U^*AU = \begin{bmatrix} B & X \\ Y & C \end{bmatrix}$$

conformally with $I_r \oplus I_{n-r}$. It suffices to prove that $PAP = AP$ if and only if $Y = 0$, which follows from the block-matrix computation

$$PAP = AP \iff (U^*PU)(U^*AU)(U^*PU) = (U^*AU)(U^*PU)$$

$$\iff \begin{bmatrix} I_r & 0 \\ 0 & 0_{n-r} \end{bmatrix} \begin{bmatrix} B & X \\ Y & C \end{bmatrix} \begin{bmatrix} I_r & 0 \\ 0 & 0_{n-r} \end{bmatrix} = \begin{bmatrix} B & X \\ Y & C \end{bmatrix} \begin{bmatrix} I_r & 0 \\ 0 & 0_{n-r} \end{bmatrix}$$

$$\iff \begin{bmatrix} B & 0 \\ 0 & 0_{n-r} \end{bmatrix} = \begin{bmatrix} B & 0 \\ Y & 0_{n-r} \end{bmatrix}$$

$$\iff Y = 0. \qquad \square$$

Corollary 8.6.8 *Maintain the notation of Theorem 8.6.7. The following are equivalent:*

(a) $\mathcal{U}$ *is invariant under A and A^*.*

(b) $U^*AU = B \oplus C$, *in which $B \in \mathsf{M}_r(\mathbb{F})$ and $C \in \mathsf{M}_{n-r}(\mathbb{F})$.*

(c) $PA = AP$.

Proof (a) $\Leftrightarrow$ (b) Theorem 8.6.7 ensures that $\mathcal{U}$ is invariant under A and A^* if and only if

$$U^*AU = \begin{bmatrix} B & X \\ 0 & C \end{bmatrix} \quad \text{and} \quad U^*A^*U = \begin{bmatrix} B' & X' \\ 0 & C' \end{bmatrix}, \qquad (8.6.9)$$

in which $B, B' \in \mathsf{M}_r(\mathbb{F})$, $C, C' \in \mathsf{M}_{n-r}(\mathbb{F})$, and $X, X' \in \mathsf{M}_{r \times (n-r)}(\mathbb{F})$. Since the block matrices in (8.6.9) must be adjoints of each other, $B' = B^*$, $C' = C^*$, and $X = X' = 0$. Thus, $\mathcal{U}$ is invariant under A and A^* if and only if $U^*AU = B \oplus C$.

(a) $\Rightarrow$ (c) Since $\mathcal{U}$ is invariant under A and A^*, Theorem 8.6.7 ensures that $PAP = AP$ and $PA^*P = A^*P$. Therefore, $PA = (A^*P)^* = (PA^*P)^* = PAP = AP$.

(c) $\Rightarrow$ (a) Suppose that $PA = AP$. Then $PAP = AP^2 = AP$ and Theorem 8.6.7 ensures that $\mathcal{U}$ is A-invariant. Moreover, $A^*P = PA^*$ so that $PA^*P = A^*P^2 = A^*P$. Hence $\mathcal{U}$ is A^*-invariant (Theorem 8.6.7 again). $\qquad \square$

8.7 Problems

P.8.1 Consider $A_1 = \begin{bmatrix} 1 & 0 \\ 0 & 0 \end{bmatrix}$, $A_2 = \begin{bmatrix} \frac{1}{2} & \frac{1}{2} \\ \frac{1}{2} & \frac{1}{2} \end{bmatrix}$, and $A_3 = \begin{bmatrix} 1 & 1 \\ 0 & 0 \end{bmatrix}$. (a) Verify that A_1, A_2, A_3 are idempotent. Draw a diagram illustrating $\text{null } A_i$ and $\text{col } A_i$ for $i = 1, 2, 3$. (b) Which, if any, of A_1, A_2, A_3 are orthogonal projections? (c) Verify that (8.3.6) produces the orthogonal projections found in (b).

P.8.2 If $A \in \mathsf{M}_n$, show that

$$M = \begin{bmatrix} A & A \\ I - A & I - A \end{bmatrix} \qquad (8.7.1)$$

is idempotent. When is M an orthogonal projection?

P.8.3 If $A \in \mathsf{M}_n$, show that

$$M = \begin{bmatrix} I & A \\ 0 & 0 \end{bmatrix} \qquad (8.7.2)$$

is idempotent. When is M an orthogonal projection?

P.8.4 Verify that the matrix A in Example 8.3.16 is idempotent, not of the form (8.7.1) or (8.7.2), and not an orthogonal projection.

P.8.5 Suppose that $A \in \mathsf{M}_{m \times n}$ has full row rank and let $A^* = QR$ be a QR factorization. If $\mathbf{y} \in \operatorname{col} A$, show that the minimum-norm solution of $A\mathbf{x} = \mathbf{y}$ is $\mathbf{s} = QR^{-*}\mathbf{y}$.

P.8.6 Let $\mathbf{u} = [a \ b \ c]^\mathsf{T} \in \mathbb{R}^3$ be a unit vector and let $\mathcal{U}$ denote the plane in $\mathbb{R}^3$ defined by $ax_1 + bx_2 + cx_3 = 0$. Find the 3×3 matrix P that represents, with respect to the standard basis for $\mathbb{R}^3$, the orthogonal projection from $\mathbb{R}^3$ onto $\mathcal{U}$. Verify that $P\mathbf{u} = \mathbf{0}$ and that P is Hermitian and idempotent.

P.8.7 Let $A = [a \ b \ c] \in \mathsf{M}_{1 \times 3}(\mathbb{R})$ be nonzero, let $\mathcal{P}$ denote the plane in $\mathbb{R}^3$ determined by $ax_1 + bx_2 + cx_3 + d = 0$, let $\mathbf{x}_0 \in \mathcal{P}$, and let $\mathbf{v} = [v_1 \ v_2 \ v_3]^\mathsf{T} \in \mathbb{R}^3$. Review Example 8.4.4 and show the following: (a) The distance from $\mathbf{v}$ to $\mathcal{P}$ is the projection of $\mathbf{v} - \mathbf{x}_0$ onto the subspace $\operatorname{col} A^\mathsf{T}$. (b) $(a^2 + b^2 + c^2)^{-1/2}[a \ b \ c]^\mathsf{T}$ is an orthonormal basis for $\operatorname{col} A^\mathsf{T}$. (c) The distance from $\mathbf{v}$ to $\mathcal{P}$ is $(a^2 + b^2 + c^2)^{-1/2}|av_1 + bv_2 + cv_3 + d|$.

P.8.8 Find the minimum-norm solution to the system (8.2.9) from Example 8.2.8 by minimizing $\|\mathbf{x}(t)\|_2$ using calculus. Does your answer agree with the answer obtained from Theorem 8.2.6?

P.8.9 Let
$$A = \begin{bmatrix} 1 & 2 & 1 \\ 2 & 4 & 2 \\ 0 & 1 & 0 \end{bmatrix} \quad \text{and} \quad \mathbf{y} = \begin{bmatrix} 1 \\ 2 \\ 1 \end{bmatrix}.$$
Show that $A\mathbf{x} = \mathbf{y}$ is consistent and show that $\mathbf{s} = [-\frac{1}{2} \ 1 \ -\frac{1}{2}]^\mathsf{T}$ is its minimum-norm solution.

P.8.10 Let $\mathcal{V} = \mathsf{M}_n(\mathbb{R})$ with the Frobenius inner product. (a) Let $\mathcal{U}_+$ denote the subspace of all symmetric matrices in $\mathcal{V}$ and let $\mathcal{U}_-$ denote the subspace of all skew-symmetric matrices in $\mathcal{V}$. Show that $A = \frac{1}{2}(A + A^\mathsf{T}) + \frac{1}{2}(A - A^\mathsf{T})$ for any $A \in \mathcal{V}$ and deduce that $\mathcal{V} = \mathcal{U}_+ \oplus \mathcal{U}_-$. (b) Show that $\mathcal{U}_- = \mathcal{U}_+^\perp$ and $\mathcal{U}_+ = \mathcal{U}_-^\perp$. (c) Show that $P_{\mathcal{U}_+}A = \frac{1}{2}(A + A^\mathsf{T})$ and $P_{\mathcal{U}_-}A = \frac{1}{2}(A - A^\mathsf{T})$ for all $A \in \mathsf{M}_n(\mathbb{R})$.

P.8.11 Let $\mathcal{V} = \mathsf{M}_n(\mathbb{R})$ with the Frobenius inner product. Let $\mathcal{U}_1$ denote the subspace of upper triangular matrices in $\mathcal{V}$ and let $\mathcal{U}_2$ denote the subspace of all strictly lower triangular matrices in $\mathcal{V}$. Show that $\mathcal{V} = \mathcal{U}_1 \oplus \mathcal{U}_2$ and $\mathcal{U}_2 = \mathcal{U}_1^\perp$.

P.8.12 Let $\mathcal{V} = \mathsf{M}_n(\mathbb{R})$ with the Frobenius inner product. Partition each $M \in \mathsf{M}_n(\mathbb{R})$ as $M = \begin{bmatrix} A & B \\ C & D \end{bmatrix}$, in which $A \in \mathsf{M}_{p \times q}$, and let $P(M) = \begin{bmatrix} A & 0 \\ 0 & 0 \end{bmatrix}$. Show that $P \in \mathcal{L}(\mathsf{M}_n(\mathbb{R}))$ is an orthogonal projection.

P.8.13 Let $A \in \mathsf{M}_n$ be an orthogonal projection. Show that the operator $T \in \mathcal{L}(\mathbb{F}^n)$ defined by $T(\mathbf{x}) = A\mathbf{x}$ is the orthogonal projection onto $\operatorname{ran} T$.

P.8.14 Let $P \in \mathsf{M}_n$ be an orthogonal projection. Use (8.6.3) to show that $\operatorname{tr} P = \dim \operatorname{col} P$.

P.8.15 Let $\mathcal{V}$ be a finite-dimensional inner product space and let $P, Q \in \mathcal{L}(\mathcal{V})$ be orthogonal projections that commute. Show that PQ is the orthogonal projection onto $\operatorname{ran} P \cap \operatorname{ran} Q$. *Hint*: First show that PQ is selfadjoint and idempotent.

P.8.16 Let $\mathcal{V}$ be a finite-dimensional inner product space and let $P, Q \in \mathcal{L}(\mathcal{V})$ be orthogonal projections. Show that the following are equivalent: (a) $\operatorname{ran} P \perp \operatorname{ran} Q$. (b) $\operatorname{ran} Q \subseteq \ker P$. (c) $PQ = 0$. (d) $PQ + QP = 0$. (e) $P + Q$ is an orthogonal projection. *Hint*: (d) implies that $-PQP = QP^2 = QP$ is selfadjoint.

P.8.17 Let $\mathcal{V}$ be a finite-dimensional inner product space and let $P, Q \in \mathcal{L}(\mathcal{V})$ be orthogonal projections. Show that $(\operatorname{ran} P \cap \operatorname{ran} Q)^\perp = \ker P + \ker Q$.

P.8.18 Let $\mathcal{V}$ be a finite-dimensional inner product space, let $P \in \mathcal{L}(\mathcal{V})$, and suppose that $P^2 = P$. (a) Show that $\mathcal{V} = \operatorname{ran} P \oplus \ker P$. (b) Show that $\operatorname{ran} P \perp \ker P$ if and only if $P = P^*$. *Hint*: See the proof of Theorem 8.3.11.d.

P.8.19 Let V be a finite-dimensional inner product space and let $P \in \mathcal{L}(V)$ be idempotent. Show that $\|Pv\| \leq \|v\|$ for every $v \in V$ if and only if P is an orthogonal projection. *Hint*: Use the preceding problem and P.5.11.

P.8.20 (a) Let V be an inner product space and let $u_1, u_2, \ldots, u_n \in V$ be linearly independent. Show that the Gram matrix $G(u_1, u_2, \ldots, u_n)$ is invertible. *Hint*: If $x = [x_i] \in \mathbb{F}^n$, then $x^* G(u_1, u_2, \ldots, u_n)x = \| \sum_{i=1}^{n} x_i u_i \|^2$. (b) Suppose that $A \in M_{m \times n}$ and $\operatorname{rank} A = n$. Deduce from (a) that $A^* A$ is invertible.

P.8.21 Let $\mathcal{U}$ be a finite-dimensional subspace of an inner product space V, let $v \in V$, and define $d(v, \mathcal{U})$ as in (8.4.3). Let $u_1, u_2, \ldots, u_n$ be a basis for $\mathcal{U}$, and suppose that $P_{\mathcal{U}} v = \sum_{i=1}^{n} c_i u_i$. (a) Show that $d(v, \mathcal{U})^2 = \|v\|^2 - \sum_{i=1}^{n} c_i \langle u_i, v \rangle$ for all $v \in V$. *Hint*: $v - P_{\mathcal{U}} v \in \mathcal{U}^{\perp}$. (b) Combine (a) and the normal equations (8.4.8) to obtain an $(n+1) \times (n+1)$ linear system for the unknowns $c_1, c_2, \ldots, c_n, d(v, \mathcal{U})^2$. Use Cramer's rule to obtain

$$d(v, \mathcal{U})^2 = \frac{g(v, u_1, u_2, \ldots, u_n)}{g(u_1, u_2, \ldots, u_n)},$$

which expresses $d(v, \mathcal{U})^2$ as the quotient of two Gram determinants. (c) Let $u, v \in V$ and suppose that $u \neq 0$. Show that $d(v, \operatorname{span}\{u\})^2 = g(v, u)/g(u)$ and conclude that $g(v, u) \geq 0$. Deduce that $|\langle u, v \rangle| \leq \|u\| \|v\|$, which is the Cauchy–Schwarz inequality.

P.8.22 In this problem, we approach the theory of least-squares approximation from a different perspective. Let V, W be finite-dimensional inner product spaces with $\dim V \leq \dim W$, let $T \in \mathcal{L}(V, W)$, and use (8.2.3) and Corollary 4.1.7. (a) Show that $\ker T = \ker T^* T$. (b) Prove that if $\dim \operatorname{ran} T = \dim V$, then $T^* T$ is invertible. (c) Prove that if $\dim \operatorname{ran} T = \dim V$, then $P = T(T^* T)^{-1} T^* \in \mathcal{L}(W)$ is the orthogonal projection onto $\operatorname{ran} T$. (d) If $T \in \mathcal{L}(V, W)$ and $\dim \operatorname{ran} T = \dim V$, prove that for each $y \in V$ there exists a unique vector $x \in V$ such that $\|Tx - y\|$ is minimized. Show that this vector x satisfies $T^* Tx = T^* y$.

P.8.23 Compute the least-squares line for the data $(-2, -3), (-1, -1), (0, 1), (1, 1), (2, 3)$.

P.8.24 Find the polynomial $y = ax^2 + bx + c$ that minimizes $\sum_{i=1}^{5} (y_i - (ax_i^2 + bx_i + c))^2$ for the data $(-2, 3), (-1, 1), (0, 1), (1, 2), (2, 4)$.

P.8.25 In linear regression, one is given real data points $(x_1, y_1), (x_2, y_2), \ldots, (x_m, y_m)$ and must find parameters a and b such that $\sum_{i=1}^{m} (y_i - ax_i - b)^2$ is minimized. This is the same problem as in Example 8.5.8. Derive explicit formulas for a and b that involve the quantities

$$S_x = \frac{1}{m} \sum_{i=1}^{m} x_i, \qquad\qquad S_y = \frac{1}{m} \sum_{i=1}^{m} y_i,$$

$$S_{x^2} = \frac{1}{m} \sum_{i=1}^{m} x_i^2, \qquad\qquad S_{xy} = \frac{1}{m} \sum_{i=1}^{m} x_i y_i.$$

P.8.26 Let $A \in M_{m \times n}$ and suppose that $\operatorname{rank} A = n$. Let $P = A(A^* A)^{-1} A^*$. (a) Show that P is Hermitian and idempotent. (b) Show that $\operatorname{col} P = \operatorname{col} A$, and conclude that P is the orthogonal projection onto $\operatorname{col} A$.

P.8.27 Let $a_1, a_2, \ldots, a_n \in \mathbb{F}^m$ be linearly independent. Let $A = [a_1 \ a_2 \ \ldots \ a_n] \in M_{m \times n}(\mathbb{F})$ and let $A = QR$ be a QR factorization, so $R = [r_{ij}] \in M_n(\mathbb{F})$ is upper triangular. (a) For each $k = 2, 3, \ldots, n$, show that r_{kk} is the distance from a_k to $\operatorname{span}\{a_1, a_2, \ldots, a_{k-1}\}$.

(b) Discuss how the *QR* factorization can be used to compute the quantity on the right side of (8.4.3).

P.8.28 Let $\mathcal{V} = C_{\mathbb{R}}[0, 1]$ and let $\mathcal{P}$ be the subspace of $\mathcal{V}$ consisting of all real polynomials. (a) Let $f \in \mathcal{V}$. The *Weierstrass approximation theorem* says that for any given $\varepsilon > 0$ there is a polynomial $p_\varepsilon \in \mathcal{P}$ such that $|f(t) - p_\varepsilon(t)| \leq \varepsilon$ for all $t \in [0, 1]$. If $f \in \mathcal{P}^\perp$, show that the L^2 norm of f satisfies the inequality $\|f\| \leq \varepsilon$ for every $\varepsilon > 0$; see (5.4.7). *Hint*: Consider $\|p_\varepsilon - f\|^2$. (b) Show that $\mathcal{P}^\perp = \{0\}$ and conclude that $\mathcal{P} \neq (\mathcal{P}^\perp)^\perp$. This does not contradict Corollary 8.1.9 because span $\mathcal{P}$ is not finite dimensional.

P.8.29 Let $A \in \mathsf{M}_{m \times n}(\mathbb{F})$ and $\mathbf{y} \in \mathbb{F}^m$. Use (8.2.4) to prove the *Fredholm alternative*: There is some $\mathbf{x}_0 \in \mathbb{F}^n$ such that $A\mathbf{x}_0 = \mathbf{y}$ if and only if $\mathbf{z}^*\mathbf{y} = 0$ for every $\mathbf{z} \in \mathbb{F}^m$ such that $A^*\mathbf{z} = \mathbf{0}$.

P.8.30 Let $A, B \in \mathsf{M}_{m \times n}(\mathbb{F})$. Use (8.2.4) and P.3.26 to show that null $A = $ null B if and only if there is an invertible $S \in \mathsf{M}_m(\mathbb{F})$ such that $A = SB$. Notice that the proof of Theorem 4.4.1 makes no use of orthogonality.

8.8 Notes

For a proof of the Weierstrass approximation theorem, cited in P.8.28, see [Dav63, Theorem 6.1.1] or [Dur12, Chapter 6].

8.9 Some Important Concepts

- Orthogonal complements of sets and subspaces.

- Minimum-norm solution of a consistent linear system.

- How to use an orthonormal basis for a subspace to construct an orthogonal projection.

- Best approximation of a given vector by a vector in a given subspace.

- Least-squares solution of an inconsistent linear system.

- Orthogonal-projection matrices are Hermitian and idempotent.

- Invariant subspaces and block triangular matrices.

9 Eigenvalues, Eigenvectors, and Geometric Multiplicity

In the next four chapters, we develop tools to show (in Chapter 12) that each square complex matrix is similar to an essentially unique direct sum of special bidiagonal matrices (the Jordan canonical form). The first step is to show that each square complex matrix has a one-dimensional invariant subspace and explore some consequences of that fact.

9.1 Eigenvalue-Eigenvector Pairs

Definition 9.1.1 Let $A \in M_n$, let $\lambda \in \mathbb{C}$, and let $\mathbf{x}$ be a nonzero vector. Then $(\lambda, \mathbf{x})$ is an *eigenpair* of A if

$$A\mathbf{x} = \lambda\mathbf{x} \quad \text{and} \quad \mathbf{x} \neq \mathbf{0}. \tag{9.1.2}$$

If $(\lambda, \mathbf{x})$ is an eigenpair of A, then λ is an *eigenvalue* of A and $\mathbf{x}$ is an *eigenvector* of A.

We cannot emphasize strongly enough that an eigenvector must be nonzero.

Do not confuse the eigenpair equation $A\mathbf{x} = \lambda\mathbf{x}$ with the linear system $A\mathbf{x} = \mathbf{b}$. In the linear system, $\mathbf{b}$ is known, but in the eigenpair equation, $\mathbf{x}$ and λ are unknown.

Many different vectors can be eigenvectors of A associated with a given eigenvalue. For example, if $(\lambda, \mathbf{x})$ is an eigenpair of A and if $c \neq 0$, then $c\mathbf{x} \neq \mathbf{0}$ and $A(c\mathbf{x}) = cA\mathbf{x} = c\lambda\mathbf{x} = \lambda(c\mathbf{x})$, so $(\lambda, c\mathbf{x})$ is also an eigenpair of A. However, only one scalar can be an eigenvalue of A associated with a given eigenvector. If $(\lambda, \mathbf{x})$ and $(\mu, \mathbf{x})$ are eigenpairs of A, then $\lambda\mathbf{x} = A\mathbf{x} = \mu\mathbf{x}$, so $(\lambda - \mu)\mathbf{x} = \mathbf{0}$. Since $\mathbf{x} \neq \mathbf{0}$, it follows that $\lambda = \mu$.

Eigenvalues and eigenvectors are important tools that help us understand the behavior of matrices by resolving them into simple components that we can exploit for algorithms, data analysis, approximation, data compression, and other purposes.

Before exploring the theory and applications of eigenvectors, we first consider several examples.

Example 9.1.3 Since $I\mathbf{x} = 1\mathbf{x}$ for all $\mathbf{x}$, it follows that $(1, \mathbf{x})$ is an eigenpair of I for any nonzero $\mathbf{x}$. Moreover, $I\mathbf{x} = \lambda\mathbf{x}$ for $\mathbf{x} \neq \mathbf{0}$ implies that $\mathbf{x} = \lambda\mathbf{x}$ and $\lambda = 1$, so 1 is the only eigenvalue of I.

Example 9.1.4 Since

$$A = \begin{bmatrix} -3 & 0 \\ 0 & 2 \end{bmatrix} = [-3\mathbf{e}_1 \; 2\mathbf{e}_2] \tag{9.1.5}$$

satisfies $A\mathbf{e}_1 = -3\mathbf{e}_1$ and $A\mathbf{e}_2 = 2\mathbf{e}_2$, we see that $(-3, \mathbf{e}_1)$ and $(2, \mathbf{e}_2)$ are eigenpairs for A. For real matrices, it is often instructive to visualize eigenvectors and eigenvalues by studying the associated linear transformation; see Figure 9.1.

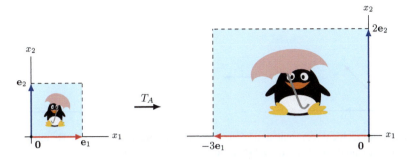

Figure 9.1 A graphical representation of the linear transformation $T_A : \mathbb{R}^2 \to \mathbb{R}^2$ induced by the matrix from Example 9.1.4; $(-3, \mathbf{e}_1)$ and $(2, \mathbf{e}_2)$ are eigenpairs.

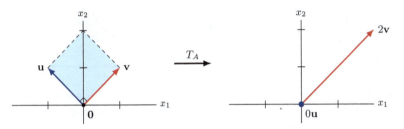

Figure 9.2 A graphical representation of the linear transformation $T_A : \mathbb{R}^2 \to \mathbb{R}^2$ induced by the matrix from Example 9.1.6; $(0, \mathbf{u})$ and $(2, \mathbf{v})$ are eigenpairs.

Example 9.1.6 Consider $A = \begin{bmatrix} 1 & 1 \\ 1 & 1 \end{bmatrix}$ and observe that, unlike the case in Example 9.1.4, $\mathbf{e}_1$ and $\mathbf{e}_2$ are not eigenvectors of A. Let $\mathbf{x} = [x_1 \ x_2]^\mathsf{T}$ be nonzero and examine (9.1.2):

$$x_1 + x_2 = \lambda x_1,$$
$$x_1 + x_2 = \lambda x_2. \tag{9.1.7}$$

These equations tell us that $\lambda x_1 = \lambda x_2$, that is, $\lambda(x_1 - x_2) = 0$. There are two cases. If $\lambda = 0$, then $x_1 + x_2 = 0$ is the only constraint imposed by (9.1.7); hence any nonzero vector whose entries satisfy this equation, such as $\mathbf{u} = [-1 \ 1]^\mathsf{T}$, is an eigenvector corresponding to the eigenvalue 0. If $x_1 - x_2 = 0$, then x_1 and x_2 are equal and nonzero since $\mathbf{x} \neq \mathbf{0}$. Return to (9.1.7) and see that $2x_1 = \lambda x_1$, from which it follows that $\lambda = 2$. Any nonzero $\mathbf{x}$ whose entries satisfy $x_1 = x_2$, such as $\mathbf{v} = [1 \ 1]^\mathsf{T}$, is an eigenvector corresponding to the eigenvalue 2; see Figure 9.2. This example shows that, although an eigenvector may never be a zero vector, an eigenvalue may be a zero scalar.

Example 9.1.8 Consider $A = \begin{bmatrix} 1 & 1 \\ 0 & 2 \end{bmatrix}$. Let $\mathbf{x} = \begin{bmatrix} x_1 \\ x_2 \end{bmatrix}$ be nonzero and examine (9.1.2):

$$x_1 + x_2 = \lambda x_1,$$
$$2x_2 = \lambda x_2. \tag{9.1.9}$$

If $x_2 = 0$, then $x_1 \neq 0$ since $\mathbf{x} \neq \mathbf{0}$. Thus, the system reduces to $x_1 = \lambda x_1$, so 1 is an eigenvalue of A and every eigenvector corresponding to this eigenvalue is a nonzero multiple of

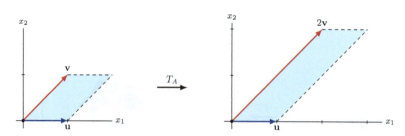

Figure 9.3 A graphical representation of the linear transformation $T_A : \mathbb{R}^2 \to \mathbb{R}^2$ induced by the matrix from Example 9.1.8; $(1, \mathbf{u})$ and $(2, \mathbf{v})$ are eigenpairs.

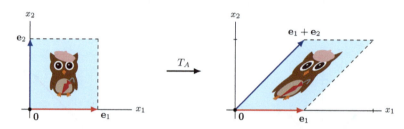

Figure 9.4 The linear transformation $T_A : \mathbb{R}^2 \to \mathbb{R}^2$ induced by the matrix A from Example 9.1.10 is a *shear* in the x_1-direction. The only eigenvalue of A is 1 and the corresponding eigenvectors are the nonzero multiples of $\mathbf{e}_1$.

$\mathbf{u} = [1 \ 0]^\mathsf{T}$. If $x_2 \neq 0$, then the second equation in (9.1.9) tells us that $\lambda = 2$. The first equation in (9.1.9) reveals that $x_1 = x_2$, that is, every eigenvector corresponding to the eigenvalue 2 is a nonzero multiple of $\mathbf{v} = [1 \ 1]^\mathsf{T}$; see Figure 9.3.

Example 9.1.10 Consider $A = \begin{bmatrix} 1 & 1 \\ 0 & 1 \end{bmatrix}$. Let $\mathbf{x} = \begin{bmatrix} x_1 \\ x_2 \end{bmatrix}$ be nonzero and examine (9.1.2):

$$x_1 + x_2 = \lambda x_1,$$
$$x_2 = \lambda x_2. \tag{9.1.11}$$

If $x_2 \neq 0$, then the second equation in (9.1.11) tells us that $\lambda = 1$. Substitute this into the first equation and obtain $x_1 + x_2 = x_1$. We conclude that $x_2 = 0$, which is a contradiction. Therefore, $x_2 = 0$, from which we deduce that $x_1 \neq 0$ since $\mathbf{x} \neq \mathbf{0}$. The first equation in (9.1.11) ensures that $x_1 = \lambda x_1$, and hence $\lambda = 1$ is the only eigenvalue of A. The corresponding eigenvectors are the nonzero multiples $[x_1 \ 0]^\mathsf{T}$ of $\mathbf{e}_1$; see Figure 9.4.

Example 9.1.12 Consider $A = \begin{bmatrix} 1 & i \\ -i & 1 \end{bmatrix}$. Let $\mathbf{x} = \begin{bmatrix} x_1 \\ x_2 \end{bmatrix}$ be nonzero and examine (9.1.2):

$$x_1 + ix_2 = \lambda x_1,$$
$$-ix_1 + x_2 = \lambda x_2.$$

Multiply the first equation by i and add the result to the second to get $0 = \lambda(x_2 + ix_1)$. There are two cases: $\lambda = 0$, or $\lambda \neq 0$ and $x_2 + ix_1 = 0$. In the first case, the original system reduces to the single equation $x_2 = ix_1$, so $(0, [1 \ i]^\mathsf{T})$ is an eigenpair of A. In the second case, $x_2 = -ix_1$,

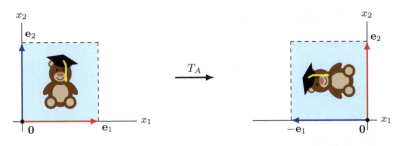

Figure 9.5 The linear transformation $T_A : \mathbb{R}^2 \to \mathbb{R}^2$ induced by the matrix from Example 9.1.13 is a rotation around the origin through an angle of $\frac{\pi}{2}$. A has no real eigenvalues, although it does have the non-real eigenpairs $(i, [1 \; -i]^\mathsf{T})$ and $(-i, [1 \; i]^\mathsf{T})$.

to which $[1 \; -i]^\mathsf{T}$ is a solution. Substitute this into the original system to find that $\lambda = 2$. Thus, $(2, [1 \; -i]^\mathsf{T})$ is an eigenpair of A.

Example 9.1.13 Consider $A = \begin{bmatrix} 0 & -1 \\ 1 & 0 \end{bmatrix}$. Let $\mathbf{x} = \begin{bmatrix} x_1 \\ x_2 \end{bmatrix}$ be nonzero and examine (9.1.2):

$$-x_2 = \lambda x_1, \tag{9.1.14}$$

$$x_1 = \lambda x_2. \tag{9.1.15}$$

If $x_1 \neq 0$, substitute the first equation into the second and obtain $x_1 = -\lambda^2 x_1$, from which we conclude that any eigenvalue λ of A satisfies $\lambda^2 = -1$. If $x_2 \neq 0$, then substitute (9.1.15) into (9.1.14) to obtain the same conclusion. It follows that A has no real eigenvalues; see Figure 9.5. The equation $\lambda^2 = -1$ has non-real solutions $\lambda_\pm = \pm i$. Substitute $\lambda = i$ into (9.1.15) and (9.1.14). Conclude that $-x_2 = ix_1$ and $x_1 = ix_2$. These equations are multiples of each other; multiply the first equation by $-i$ to obtain the second. A nonzero solution to these equations is $[1 \; -i]^\mathsf{T}$, so $(i, [1 \; -i]^\mathsf{T})$ is an eigenpair of A. Similarly, $(-i, [1 \; i]^\mathsf{T})$ is an eigenpair of A. These two eigenpairs are complex conjugates; see P.9.9 for an explanation of this phenomenon.

The preceding example illustrates that eigenvalues and eigenvectors of a real matrix need not be real. That is one reason why complex numbers play a central role in linear algebra. Example 9.1.13 also suggests that the eigenvalues of a square matrix might be obtained by finding the zeros of an associated polynomial. We investigate this possibility in the following chapter.

The following variations on Definition 9.1.1 are useful in our exposition.

Theorem 9.1.16 *Let $A \in \mathsf{M}_n$ and let $\lambda \in \mathbb{C}$. The following statements are equivalent:*

(a) *λ is an eigenvalue of A.*

(b) *$A\mathbf{x} = \lambda\mathbf{x}$ for some nonzero $\mathbf{x} \in \mathbb{C}^n$.*

(c) *$(A - \lambda I)\mathbf{x} = \mathbf{0}$ has a nontrivial solution, that is,* nullity$(A - \lambda I) > 0$.

(d) rank$(A - \lambda I) < n$.

(e) *$A - \lambda I$ is not invertible.*

(f) *$A^\mathsf{T} - \lambda I$ is not invertible.*

(g) *λ is an eigenvalue of A^T.*

Proof (a) $\Leftrightarrow$ (b) This is Definition 9.1.1.

(b) $\Leftrightarrow$ (c) These are restatements of each other.

(c) $\Leftrightarrow$ (d) This follows from the rank-nullity theorem.

(d) $\Leftrightarrow$ (e) See Theorem 2.3.13.

(e) $\Leftrightarrow$ (f) See Theorem 2.3.1.

(f) $\Leftrightarrow$ (g) Apply the equivalence of (a) and (e) to A^{T}. $\qquad\qquad\square$

Corollary 9.1.17 *Let $A \in \mathsf{M}_n$. Then A is invertible if and only if 0 is not an eigenvalue of A.*

Proof This follows from the equivalence of (a) and (e) in the preceding theorem. $\quad\square$

9.2 Every Square Complex Matrix Has an Eigenvalue

The title of this section makes our aim clear. We want to prove that every square matrix has an eigenvalue. The relationship among matrices, eigenvalues, and polynomials is crucial to this endeavor. For a polynomial $p(z) = c_k z^k + c_{k-1} z^{k-1} + \cdots + c_1 z + c_0$ and $A \in \mathsf{M}_n$, Appendix B.5 defines

$$p(A) = c_k A^k + c_{k-1} A^{k-1} + \cdots + c_1 A + c_0 I \in \mathsf{M}_n. \tag{9.2.1}$$

Example 9.2.2 The matrix $A = \begin{bmatrix} 0 & -1 \\ 1 & 0 \end{bmatrix}$ from Example 9.1.13 satisfies $A^2 = -I$. The linear transformation induced by A is a rotation of $\mathbb{R}^2$ about the origin through an angle of $\frac{\pi}{2}$; see Figure 9.5. Thus, A^2 induces a rotation through an angle of π, which is represented by the matrix $-I$. If $p(z) = z^2 + 1$, then $p(A) = A^2 + I = 0$.

The preceding example motivates the following definition.

Definition 9.2.3 If $A \in \mathsf{M}_n$ and p is a polynomial such that $p(A) = 0$, then p *annihilates* A (p is an *annihilating polynomial for* A).

The following lemma shows that each $A \in \mathsf{M}_n$ is annihilated by some nonconstant polynomial; it relies on the fact that M_n is a vector space of dimension n^2 (Example 2.2.3).

Lemma 9.2.4 *Let $A \in \mathsf{M}_n(\mathbb{F})$. There is a nonconstant polynomial p of degree at most n^2 that annihilates A; if A is real, p may be chosen to have real coefficients.*

Proof The $\mathbb{F}$-vector space $\mathsf{M}_n(\mathbb{F})$ has dimension n^2, so the $n^2 + 1$ matrices $I, A, A^2, \ldots, A^{n^2}$ are linearly dependent. Consequently, there are $c_0, c_1, c_2, \ldots, c_{n^2} \in \mathbb{F}$, not all zero, such that

$$c_0 I + c_1 A + c_2 A^2 + \cdots + c_{n^2} A^{n^2} = 0. \tag{9.2.5}$$

If $c_1 = c_2 = \cdots = c_{n^2} = 0$, then $c_0 I = 0$ and $c_0 = 0$, which is a contradiction. Let $r = \max\{k : 1 \le k \le n^2 \text{ and } c_k \ne 0\}$. Then $c_r \ne 0$ and $c_0 I + c_1 A + c_2 A^2 + \cdots + c_r A^r = 0$, so the nonconstant polynomial $p(z) = c_r z^r + c_{r-1} z^{r-1} + \cdots + c_1 z + c_0$ has degree at most n^2 and annihilates A. Its coefficients belong to $\mathbb{F}$. $\qquad\square$

For $A \in \mathsf{M}_2$, there is an explicit polynomial p such that $p(A) = 0$; it is a special case of a principle that is explained in Section 11.2.

Example 9.2.6 Let $A = \begin{bmatrix} a & b \\ c & d \end{bmatrix}$. Lemma 9.2.4 says that I, A, A^2, A^3, A^4 are linearly dependent. Then

$$A^2 = \begin{bmatrix} a & b \\ c & d \end{bmatrix} \begin{bmatrix} a & b \\ c & d \end{bmatrix} = \begin{bmatrix} a^2 + bc & ab + bd \\ ac + cd & bc + d^2 \end{bmatrix} = \begin{bmatrix} a^2 + bc & b(a+d) \\ c(a+d) & bc + d^2 \end{bmatrix},$$

so

$$A^2 - (a+d)A = \begin{bmatrix} a^2 + bc & b(a+d) \\ c(a+d) & bc + d^2 \end{bmatrix} - \begin{bmatrix} a(a+d) & b(a+d) \\ c(a+d) & d(a+d) \end{bmatrix} = \begin{bmatrix} bc - ad & 0 \\ 0 & bc - ad \end{bmatrix},$$

from which it follows that

$$A^2 - (a+d)A + (ad - bc)I = 0. \tag{9.2.7}$$

Thus, $p(z) = z^2 - (\operatorname{tr} A)z + \det A$ annihilates A.

Our interest in polynomials that annihilate a given matrix stems from the following fact.

Lemma 9.2.8 *Let $A \in \mathsf{M}_n$ and let p be a nonconstant polynomial such that $p(A) = 0$. Then some root of $p(z) = 0$ is an eigenvalue of A.*

Proof Suppose that $k \geq 1$, $c_k \neq 0$, and

$$p(z) = c_k z^k + c_{k-1} z^{k-1} + \cdots + c_1 z + c_0 \tag{9.2.9}$$

annihilates A. Factor p as

$$p(z) = c_k(z - \lambda_1)(z - \lambda_2) \cdots (z - \lambda_k). \tag{9.2.10}$$

Equality of the representations (9.2.9) and (9.2.10), together with the fact that powers of A commute, ensure that we can rewrite $p(A) = 0$ in the factored form

$$0 = p(A) = c_k(A - \lambda_1 I)(A - \lambda_2 I) \cdots (A - \lambda_k I). \tag{9.2.11}$$

Since $p(A) = 0$ is not invertible, at least one of the factors $A - \lambda_i I$ is not invertible. Theorem 9.1.16 ensures that λ_i is an eigenvalue of A. $\qquad\square$

Example 9.2.12 Let $A = \begin{bmatrix} 1 & 1 \\ 1 & 1 \end{bmatrix}$. In Example 9.1.6 we found that 0 and 2 are the eigenvalues of A. The identity (9.2.7) tells us A is annihilated by $p(z) = z^2 - 2z = z(z - 2)$. In this case, the roots of $p(z) = 0$ are the eigenvalues of A. However, not every zero of a polynomial that annihilates A need be an eigenvalue of A. For example, $q(z) = z^3 - 3z^2 + 2z = z(z - 1)(z - 2) = (z - 1)p(z)$ satisfies

$$q(A) = A^3 - 3A^2 + 2A = (A - I)p(A) = (A - I)0 = 0,$$

but $q(1) = 0$ and 1 is not an eigenvalue of A.

Example 9.2.13 Example 9.2.6 ensures that $p(z) = z^2 - (a + d)z + (ad - bc)$ annihilates $A = \begin{bmatrix} a & b \\ c & d \end{bmatrix}$. The roots of $p(z) = 0$ are

$$\frac{1}{2}\left(a + d \pm \sqrt{(a+d)^2 - 4(ad - bc)}\right) = \frac{1}{2}\left(a + d \pm \sqrt{(a - d)^2 + 4bc}\right).$$

Let $s = (a - d)^2 + 4bc$ and let

$$\lambda_+ = \frac{1}{2}(a + d + \sqrt{s}) \quad \text{and} \quad \lambda_- = \frac{1}{2}(a + d - \sqrt{s}). \tag{9.2.14}$$

Lemma 9.2.8 ensures that at least one of λ_+ and λ_- is an eigenvalue of A. If $s=0$, then $\lambda_+=\lambda_-=\frac{1}{2}(a+d)$ is an eigenvalue. If $s\neq 0$ and λ_+ is an eigenvalue of A, but λ_- is not, then Theorem 9.1.16 ensures that $A-\lambda_-I$ is invertible. From $0=p(A)=(A-\lambda_-I)(A-\lambda_+I)$, we conclude that $0=(A-\lambda_-I)^{-1}p(A)=A-\lambda_+I$. Consequently, $A=\lambda_+I$, so $b=c=0$, $a=d$, and $s=0$, which is a contradiction. If $s\neq 0$ and λ_- is an eigenvalue of A, but λ_+ is not, then similar reasoning leads to the conclusion that $A-\lambda_-I=0$, which is another contradiction. Thus, both λ_+ and λ_- are eigenvalues of A.

If $c=0$, then $s=(a-d)^2$, $\lambda_+=a$, and $\lambda_-=d$. Thus, if A is upper triangular, its diagonal entries are eigenvalues of A. If $b=0$, then A is lower triangular and its diagonal entries are its eigenvalues.

The following theorem is the main result of this section.

Theorem 9.2.15 *Every square complex matrix has an eigenvalue.*

Proof Let $A\in M_n$. Lemma 9.2.4 ensures that there is a nonconstant polynomial p such that $p(A)=0$. Among the roots of $p(z)=0$, at least one is an eigenvalue of A (Lemma 9.2.8). □

Although every square matrix has at least one eigenvalue, there is much to learn. For example, (9.2.14) is a formula for eigenvalues of a 2×2 matrix, but are these the only eigenvalues? How many eigenvalues can an $n\times n$ matrix have?

9.3 How Many Eigenvalues Are There?

Definition 9.3.1 The set of eigenvalues of $A\in M_n$ is the *spectrum* of A, denoted by $\operatorname{spec}A$.

The spectrum of a matrix is a set, so it is characterized by its distinct elements. For example, the sets $\{1,2\}$ and $\{1,1,2\}$ are identical.

Theorem 9.2.15 says that the spectrum of a square matrix is nonempty. How many elements are in $\operatorname{spec}A$? The following lemma is a first step toward an answer to this question.

Lemma 9.3.2 *Let $(\lambda,\mathbf{x})$ be an eigenpair of $A\in M_n$ and let p be a polynomial. Then $p(A)\mathbf{x}=p(\lambda)\mathbf{x}$, that is, $(p(\lambda),\mathbf{x})$ is an eigenpair of $p(A)$.*

Proof Let $(\lambda,\mathbf{x})$ be an eigenpair of $A\in M_n$. We use induction to show that $A^j\mathbf{x}=\lambda^j\mathbf{x}$ for all $j\geq 0$. For the base case $j=0$, observe that $A^0\mathbf{x}=I\mathbf{x}=\mathbf{x}=1\mathbf{x}=\lambda^0\mathbf{x}$. For the inductive step, suppose that $A^j\mathbf{x}=\lambda^j\mathbf{x}$ for some $j\geq 0$. Then

$$A^{j+1}\mathbf{x}=A(A^j\mathbf{x})=A(\lambda^j\mathbf{x})=\lambda^j(A\mathbf{x})=\lambda^j(\lambda\mathbf{x})=\lambda^{j+1}\mathbf{x},$$

and the induction is complete. If p is given by (9.2.9), then

$$\begin{aligned}
p(A)\mathbf{x}&=(c_kA^k+c_{k-1}A^{k-1}+\cdots+c_1A+c_0I)\mathbf{x}\\
&=c_kA^k\mathbf{x}+c_{k-1}A^{k-1}\mathbf{x}+\cdots+c_1A\mathbf{x}+c_0I\mathbf{x}\\
&=c_k\lambda^k\mathbf{x}+c_{k-1}\lambda^{k-1}\mathbf{x}+\cdots+c_1\lambda\mathbf{x}+c_0\mathbf{x}\\
&=(c_k\lambda^k+c_{k-1}\lambda^{k-1}+\cdots+c_1\lambda+c_0)\mathbf{x}\\
&=p(\lambda)\mathbf{x}.
\end{aligned}$$

□

One application of the preceding lemma is to provide a complement to Lemma 9.2.8. If $A \in M_n$ and p is a polynomial that annihilates A, then not only is some root of $p(z) = 0$ an eigenvalue, every eigenvalue of A is a root of $p(z) = 0$. Here is a formal statement of this observation.

Theorem 9.3.3 *Let $A \in M_n$ and let p be a nonconstant polynomial that annihilates A.*

(a) *Every eigenvalue of A is a root of $p(z) = 0$.*

(b) *A has finitely many eigenvalues.*

Proof (a) Let $(\lambda, \mathbf{x})$ be an eigenpair of A. Then $\mathbf{0} = 0_n \mathbf{x} = p(A)\mathbf{x} = p(\lambda)\mathbf{x}$, so $p(\lambda) = 0$.

(b) Lemma 9.2.4 ensures that there is a nonconstant polynomial of degree at most n^2 that annihilates A. Such a polynomial has at most n^2 zeros, so A has at most n^2 eigenvalues.   □

Example 9.3.4 Consider the matrix A in Example 9.2.13. We showed that both roots of $p(z) = z^2 - (\operatorname{tr} A)z + \det A = 0$ are eigenvalues of A. Part (a) of the preceding theorem shows that they are its only eigenvalues.

The next application of Lemma 9.3.2 leads to a sharp bound for the number of different eigenvalues of a matrix.

Theorem 9.3.5 *Let $(\lambda_1, \mathbf{x}_1), (\lambda_2, \mathbf{x}_2), \ldots, (\lambda_d, \mathbf{x}_d)$ be eigenpairs of $A \in M_n$, in which $\lambda_1, \lambda_2, \ldots, \lambda_d$ are distinct. Then $\mathbf{x}_1, \mathbf{x}_2, \ldots, \mathbf{x}_d$ are linearly independent.*

Proof The Lagrange interpolation theorem (Theorem 2.7.4) ensures that there are polynomials $p_1, p_2, \ldots, p_d$ such that $p_i(\lambda_j) = \delta_{ij}$ for each $i = 1, 2, \ldots, d$. Suppose that $c_1, c_2, \ldots, c_d \in \mathbb{F}$ and $c_1 \mathbf{x}_1 + c_2 \mathbf{x}_2 + \cdots + c_d \mathbf{x}_d = \mathbf{0}$. Lemma 9.3.2 ensures that $p_i(A)\mathbf{x}_j = p_i(\lambda_j)\mathbf{x}_j = \delta_{ij}\mathbf{x}_j$, so for each $i = 1, 2, \ldots, d$,

$$
\begin{aligned}
\mathbf{0} = p_i(A)\mathbf{0} &= p_i(A)(c_1 \mathbf{x}_1 + c_2 \mathbf{x}_2 + \cdots + c_d \mathbf{x}_d) \\
&= c_1 p_i(A)\mathbf{x}_1 + c_2 p_i(A)\mathbf{x}_2 + \cdots + c_d p_i(A)\mathbf{x}_d \\
&= c_1 p_i(\lambda_1)\mathbf{x}_1 + c_2 p_i(\lambda_2)\mathbf{x}_2 + \cdots + c_d p_i(\lambda_d)\mathbf{x}_d \\
&= c_i \mathbf{x}_i.
\end{aligned}
$$

Therefore, each $c_i = 0$ and hence $\mathbf{x}_1, \mathbf{x}_2, \ldots, \mathbf{x}_d$ are linearly independent.   □

Corollary 9.3.6 *Each $A \in M_n$ has at most n distinct eigenvalues. If A has n distinct eigenvalues, then $\mathbb{C}^n$ has a basis consisting of eigenvectors of A.*

Proof The preceding theorem says that if $(\lambda_1, \mathbf{x}_1), (\lambda_2, \mathbf{x}_2), \ldots, (\lambda_d, \mathbf{x}_d)$ are eigenpairs of A and $\lambda_1, \lambda_2, \ldots, \lambda_d \in \mathbb{F}$ are distinct, then $\mathbf{x}_1, \mathbf{x}_2, \ldots, \mathbf{x}_d$ are linearly independent. Consequently, $d \leq n$. If $d = n$, then $\mathbf{x}_1, \mathbf{x}_2, \ldots, \mathbf{x}_n$ is a maximal linearly independent list, so it is a basis; see Corollary 2.2.9.   □

Definition 9.3.7 If $A \in M_n$ has n distinct eigenvalues, then A *has distinct eigenvalues.*

Example 9.3.8 The formula (9.2.14) ensures that

$$
A = \begin{bmatrix} 9 & 8 \\ 2 & -6 \end{bmatrix} \tag{9.3.9}
$$

has eigenvalues $\lambda_+ = 10$ and $\lambda_- = -7$. Corollary 9.3.6 says that these are the only eigenvalues of A, so spec $A = \{10, -7\}$ and A has distinct eigenvalues. To find corresponding eigenvectors, solve the homogeneous systems $(A - 10I)\mathbf{x} = \mathbf{0}$ and $(A + 7I)\mathbf{x} = \mathbf{0}$. This results in the eigenpairs $(10, [8\ 1]^T)$ and $(-7, [-1\ 2]^T)$ for A and the basis $\beta = [8\ 1]^T, [-1\ 2]^T$ for $\mathbb{C}^2$. Now use β to construct the (necessarily invertible) matrix $S = \begin{bmatrix} 8 & -1 \\ 1 & 2 \end{bmatrix}$. With respect to the standard basis for $\mathbb{C}^2$, the matrix representation of $T_A : \mathbb{C}^2 \to \mathbb{C}^2$ is A; see Definition 2.4.13. Its representation with respect to β (see Theorem 2.6.1.a) is

$$S^{-1}AS = \frac{1}{17}\begin{bmatrix} 2 & 1 \\ -1 & 8 \end{bmatrix}\begin{bmatrix} 9 & 8 \\ 2 & -6 \end{bmatrix}\begin{bmatrix} 8 & -1 \\ 1 & 2 \end{bmatrix} = \frac{1}{17}\begin{bmatrix} 2 & 1 \\ -1 & 8 \end{bmatrix}\begin{bmatrix} 80 & 7 \\ 10 & -14 \end{bmatrix}$$

$$= \frac{1}{17}\begin{bmatrix} 170 & 0 \\ 0 & -119 \end{bmatrix} = \begin{bmatrix} 10 & 0 \\ 0 & -7 \end{bmatrix}.$$

Thus, the existence of a basis of eigenvectors of A implies that A is similar to a diagonal matrix, in which the diagonal entries are the eigenvalues of A. An equivalent statement is that the β-β matrix representation of the linear operator T_A is a diagonal matrix. We have more to say about this in Section 10.4.

Example 9.3.10 The matrix

$$U(\theta) = \begin{bmatrix} \cos\theta & -\sin\theta \\ \sin\theta & \cos\theta \end{bmatrix}, \qquad \text{in which } 0 < \theta \le \frac{\pi}{2},$$

induces a counterclockwise rotation of $\mathbb{R}^2$ through θ radians around the origin (see Figure 9.6). The formula (9.2.14) ensures that

$$\lambda_\pm = \frac{2\cos\theta \pm \sqrt{-4\sin^2\theta}}{2} = \cos\theta \pm i\sin\theta = e^{\pm i\theta}$$

are distinct eigenvalues of $U(\theta)$. Corollary 9.3.6 permits us to conclude that spec $U(\theta) = \{e^{i\theta}, e^{-i\theta}\}$. To find eigenvectors corresponding to $\lambda_+ = e^{i\theta}$, compute nonzero solutions of the homogeneous system $(U(\theta) - \lambda_+ I)\mathbf{x} = \mathbf{0}$:

$$-i(\sin\theta)x_1 - (\sin\theta)x_2 = 0,$$
$$(\sin\theta)x_1 - i(\sin\theta)x_2 = 0,$$

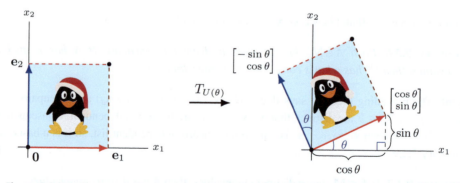

Figure 9.6 The linear transformation $T_{U(\theta)} : \mathbb{R}^2 \to \mathbb{R}^2$ induced by $U(\theta)$ from Example 9.3.10 is a counterclockwise rotation around the origin through θ radians, in which $0 < \theta \le \frac{\pi}{2}$. The matrix $U(\theta)$ has the distinct non-real eigenvalues $\lambda_\pm = e^{\pm i\theta}$ and eigenpairs $(e^{i\theta}, [1\ -i]^T)$ and $(e^{-i\theta}, [1\ i]^T)$.

in which $\mathbf{x} = [x_1 \ x_2]^\mathsf{T}$. Since i times the first equation equals the second equation and $\sin \theta \neq 0$, this system is equivalent to the single equation $x_1 = ix_2$. Thus $(e^{i\theta}, [1 \ {-i}]^\mathsf{T})$ is an eigenpair of $U(\theta)$. A similar computation (or P.9.9) shows that $(e^{-i\theta}, [1 \ i]^\mathsf{T})$ is an eigenpair of $U(\theta)$. Observe that $[1 \ i]^\mathsf{T}, [1 \ {-i}]^\mathsf{T}$ is a basis for $\mathbb{C}^2$. It is noteworthy that the eigenvalues of $U(\theta)$ depend on $\theta \in (0, \frac{\pi}{2}]$, but the associated eigenvectors do not. Each of the vectors $[1 \ \pm i]^\mathsf{T}$ is an eigenvector of all the matrices $U(\theta)$. Corollary 9.5.4 provides an explanation for this phenomenon.

Example 9.3.11 If $n \geq 2$, then I_n does not have distinct eigenvalues. Nevertheless, $\mathbb{C}^n$ has a basis consisting of eigenvectors of I_n. Any basis for $\mathbb{C}^n$ comprises eigenvectors of I_n since any nonzero vector in $\mathbb{C}^n$ is an eigenvector of I_n associated with the eigenvalue $\lambda = 1$.

Definition 9.3.12 Let $A \in \mathsf{M}_n$ and let $\lambda \in \mathbb{C}$. Then $\mathcal{E}_\lambda(A) = \text{null}(A - \lambda I)$ is the *eigenspace* of A *associated with* λ.

The term eigenspace suggests that $\mathcal{E}_\lambda(A)$ is a subspace of $\mathbb{C}^n$, which it is; the null space of any $n \times n$ matrix is a subspace of $\mathbb{C}^n$ (Example 1.3.4). If $\lambda \notin \text{spec} \, A$, then $A - \lambda I$ is invertible (Theorem 9.1.16) and hence $\mathcal{E}_\lambda(A) = \{\mathbf{0}\}$ is the zero subspace. If $\lambda \in \text{spec} \, A$, then $\mathcal{E}_\lambda(A)$ consists of the zero vector and all eigenvectors of A associated with λ (Theorem 9.1.16). Thus, $\mathcal{E}_\lambda(A) \neq \{\mathbf{0}\}$ if and only if $\lambda \in \text{spec} \, A$.

If $\lambda \in \text{spec} \, A$, every nonzero vector in $\mathcal{E}_\lambda(A)$ is an eigenvector of A associated with λ. Consequently, any basis for $\mathcal{E}_\lambda(A)$ is a linearly independent list of eigenvectors of A associated with λ.

Definition 9.3.13 Let $A \in \mathsf{M}_n$. The *geometric multiplicity* of λ as an eigenvalue of A is the dimension of the subspace $\mathcal{E}_\lambda(A)$.

If $\lambda \notin \text{spec} \, A$, its geometric multiplicity is 0. If $A \in \mathsf{M}_n$, the geometric multiplicity of $\lambda \in \text{spec} \, A$ is between 1 and n because $\mathcal{E}_\lambda(A)$ is a nonzero subspace of $\mathbb{C}^n$.

The rank-nullity theorem says that $\text{nullity}(A - \lambda I) + \text{rank}(A - \lambda I) = n$, that is, $\dim \mathcal{E}_\lambda(A) + \text{rank}(A - \lambda I) = n$. Therefore, the geometric multiplicity of λ as an eigenvalue of $A \in \mathsf{M}_n$ is

$$\dim \mathcal{E}_\lambda(A) = n - \text{rank}(A - \lambda I). \tag{9.3.14}$$

Example 9.3.15 Let $A = \begin{bmatrix} 1 & 1 \\ 0 & 1 \end{bmatrix}$. Examples 9.1.10 and 9.2.13 show that $\text{spec} \, A = \{1\}$ and $\mathcal{E}_1(A) = \text{span}\{[1 \ 0]^\mathsf{T}\}$. Thus, the geometric multiplicity of 1 as an eigenvalue of A is 1. This agrees with (9.3.14) since $\text{rank}(A - I) = 1$.

Example 9.3.16 Consider the complex symmetric matrix $A = \begin{bmatrix} 1 & i \\ i & -1 \end{bmatrix}$. Let $\mathbf{x} = \begin{bmatrix} x_1 \\ x_2 \end{bmatrix}$ be nonzero and suppose that $A\mathbf{x} = \lambda \mathbf{x}$. Multiply the second equation in

$$x_1 + ix_2 = \lambda x_1, \tag{9.3.17}$$

$$ix_1 - x_2 = \lambda x_2, \tag{9.3.18}$$

by i and add it to the first. The result is that $0 = \lambda(x_1 + ix_2)$. Thus, $\lambda = 0$ or $x_1 + ix_2 = 0$. If $\lambda = 0$, then (9.3.17) reduces to $x_1 + ix_2 = 0$. If $x_1 + ix_2 = 0$, then (9.3.17) and (9.3.18) reduce to $0 = \lambda x_1$ and $0 = -i\lambda x_2$. Then $\lambda = 0$ since $\mathbf{x} \neq \mathbf{0}$. Thus, $\text{spec} \, A = \{0\}$ and $\mathcal{E}_0(A) = \text{span}\{[1 \ i]^\mathsf{T}\}$. The geometric multiplicity of 0 as an eigenvalue of A is one. As a check, observe that $\text{rank} \, A = 1$ (its second column is i times its first column), so (9.3.14) says that the eigenvalue $\lambda = 0$ has geometric multiplicity 1.

Theorem 9.1.16 says that a square matrix and its transpose have the same eigenvalues. The following lemma refines this observation.

Lemma 9.3.19 *Let $A \in \mathsf{M}_n$ and let $\lambda \in \operatorname{spec} A$. Then $\dim \mathcal{E}_\lambda(A) = \dim \mathcal{E}_\lambda(A^\mathsf{T})$.*

Proof Compute

$$
\begin{aligned}
\dim \mathcal{E}_\lambda(A) &= \operatorname{nullity}(A - \lambda I) = n - \operatorname{rank}(A - \lambda I) \\
&= n - \operatorname{rank}(A - \lambda I)^\mathsf{T} = n - \operatorname{rank}(A^\mathsf{T} - \lambda I) \\
&= \operatorname{nullity}(A^\mathsf{T} - \lambda I) = \dim \mathcal{E}_\lambda(A^\mathsf{T}).
\end{aligned}
$$
$\qquad\square$

9.4 The Eigenvalues Are in Gershgorin Disks

Examples 9.1.3 and 9.1.4 illustrate the fact that the diagonal entries of a diagonal matrix are eigenvalues whose associated eigenvectors are the standard basis vectors. If we modify a diagonal matrix by inserting some nonzero off-diagonal entries, can we give quantitative bounds for how much each diagonal entry can differ from an eigenvalue? Gershgorin's disk theorem provides an answer to this question.

Definition 9.4.1 Let $n \geq 2$ and let $A = [a_{ij}] \in \mathsf{M}_n$. For $k = 1, 2, \ldots, n$, define the *kth absolute row sum* $R_k(A) = \sum_{j=1}^n |a_{kj}|$, and the *$k$th deleted absolute row sum* $R_k'(A) = \sum_{j \neq k} |a_{kj}|$, the *$k$th Gershgorin disk* $\mathcal{G}_k(A) = \{z \in \mathbb{C} : |z - a_{kk}| \leq R_k'(A)\}$, and the *Gershgorin region* $\mathcal{G}(A) = \bigcup_{k=1}^n \mathcal{G}_k(A)$.

Theorem 9.4.2 (Gershgorin Disk Theorem) *If $n \geq 2$ and $A = [a_{ij}] \in \mathsf{M}_n$, then*

$$
\operatorname{spec} A \subseteq \mathcal{G}(A) = \bigcup_{k=1}^n \mathcal{G}_k(A). \tag{9.4.3}
$$

Proof Let $(\lambda, \mathbf{x})$ be an eigenpair of $A = [a_{ij}] \in \mathsf{M}_n$ and let $\mathbf{x} = [x_i]$. Since $\lambda \mathbf{x} = A\mathbf{x}$, we have

$$
\lambda x_i = \sum_{j=1}^n a_{ij} x_j = a_{ii} x_i + \sum_{j \neq i} a_{ij} x_j \qquad \text{for } i = 1, 2, \ldots, n,
$$

which we express as

$$
(\lambda - a_{ii}) x_i = \sum_{j \neq i} a_{ij} x_j \qquad \text{for } i = 1, 2, \ldots, n. \tag{9.4.4}
$$

Let $k \in \{1, 2, \ldots, n\}$ be any index such that $|x_k| = \max_{1 \leq i \leq n} |x_i|$. Since $\mathbf{x} \neq \mathbf{0}$, it follows that $|x_k| > 0$ and $\frac{|x_j|}{|x_k|} \leq 1$ for all $i = 1, 2, \ldots, n$. Set $i = k$ in (9.4.4) and divide by x_k to obtain

$$
\lambda - a_{kk} = \sum_{j \neq k} a_{kj} \frac{x_j}{x_k}.
$$

The triangle inequality ensures that

$$
|\lambda - a_{kk}| = \left| \sum_{j \neq k} a_{kj} \frac{x_j}{x_k} \right| \leq \sum_{j \neq k} |a_{kj}| \left| \frac{x_j}{x_k} \right| \leq \sum_{j \neq k} |a_{kj}| = R_k'(A)
$$

and hence $\lambda \in \mathcal{G}_k(A) \subseteq \mathcal{G}(A)$. $\qquad\square$

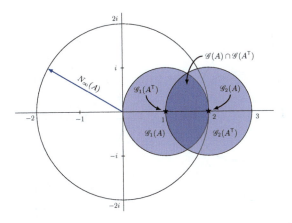

Figure 9.7 $\mathscr{G}(A)$ and $\mathscr{G}(A^\mathsf{T})$ for the matrix (9.4.9). The lens-shaped region is $\mathscr{G}(A) \cap \mathscr{G}(A^\mathsf{T})$. The eigenvalues (denoted by ★) are 1 and 2.

Each Gershgorin disk $\mathscr{G}_k(A)$ is contained in the disk centered at the origin whose radius $R_k(A)$ is the kth absolute row sum of A. Therefore, the Gershgorin region $\mathscr{G}(A)$ (and hence each eigenvalue of A) is contained in the disk centered at the origin whose radius

$$N_\infty(A) = \max_{1 \le k \le n} R_k(A) = \max \left\{ \sum_{j=1}^{n} |a_{kj}| : 1 \le k \le n \right\} \tag{9.4.5}$$

is the maximum absolute row sum of A; see Figure 9.7.

If we apply the preceding theorem to A^T, we obtain Gershgorin disks centered at the diagonal entries of A whose radii are the deleted absolute row sums of A^T.

Corollary 9.4.6 *If $A \in \mathsf{M}_n$, then*

$$\operatorname{spec} A \subseteq \mathscr{G}(A) \cap \mathscr{G}(A^\mathsf{T}) \subseteq \left\{ z \in \mathbb{C} : |z| \le \min \left\{ N_\infty(A), N_\infty(A^\mathsf{T}) \right\} \right\}. \tag{9.4.7}$$

Proof Theorem 9.4.2 says that $\operatorname{spec} A \subseteq \mathscr{G}(A)$, and Theorem 9.1.16 ensures that $\operatorname{spec} A = \operatorname{spec} A^\mathsf{T} \subseteq \mathscr{G}(A^\mathsf{T})$. Now apply (9.4.5) to A and A^T. □

Example 9.4.8 The matrix

$$A = \begin{bmatrix} 1 & 1 \\ 0 & 2 \end{bmatrix} \tag{9.4.9}$$

has $\operatorname{spec} A = \{1, 2\}$, $R_1'(A) = 1$, $R_2'(A) = 0$, $R_1'(A^\mathsf{T}) = 0$, and $R_2'(A^\mathsf{T}) = 1$. Consequently,

$$\mathscr{G}_1(A) = \{z \in \mathbb{C} : |z - 1| \le 1\}, \qquad \mathscr{G}_1(A^\mathsf{T}) = \{1\},$$
$$\mathscr{G}_2(A) = \{2\}, \qquad\qquad\qquad \mathscr{G}_2(A^\mathsf{T}) = \{z \in \mathbb{C} : |z - 2| \le 1\}.$$

The region $\mathscr{G}(A) \cap \mathscr{G}(A^\mathsf{T})$ contains $\operatorname{spec} A$ and is much smaller than either $\mathscr{G}(A)$ or $\mathscr{G}(A^\mathsf{T})$. Also,

$$R_1(A) = R_2(A) = 2, \quad R_1(A^\mathsf{T}) = 1, \quad \text{and} \quad R_2(A^\mathsf{T}) = 3,$$

so that $N_\infty(A) = 2$ and $N_\infty(A^\mathsf{T}) = 3$. Since $\min\{N_\infty(A), N_\infty(A^\mathsf{T})\} = 2$, $\operatorname{spec} A$ is contained in the disk $\{z \in \mathbb{C} : |z| \le 2\}$; see Figure 9.7.

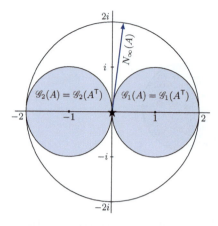

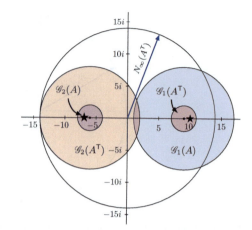

(a) $\mathscr{G}(A)$ for the matrix (9.4.11); 0 is the only eigenvalue.

(b) $\mathscr{G}(A)$ and $\mathscr{G}(A^{\mathsf{T}})$ for the matrix (9.4.13). The eigenvalues are 10 and -7.

Figure 9.8 Gershgorin regions for the matrices (9.4.11) and (9.4.13).

Example 9.4.10 The matrix

$$A = \begin{bmatrix} 1 & i \\ i & -1 \end{bmatrix} \tag{9.4.11}$$

in Example 9.3.16 has $\operatorname{spec} A = \{0\}$, $R_1'(A) = R_2'(A) = 1$ and $R_1'(A^{\mathsf{T}}) = R_2'(A^{\mathsf{T}}) = 1$, so $\mathscr{G}_1(A) = \mathscr{G}_1(A^{\mathsf{T}}) = \{z \in \mathbb{C} : |z - 1| \leq 1\}$ and $\mathscr{G}_2(A) = \mathscr{G}_2(A^{\mathsf{T}}) = \{z \in \mathbb{C} : |z + 1| \leq 1\}$. Also, $R_1(A) = R_2(A) = 2$ and $R_1(A^{\mathsf{T}}) = R_2(A^{\mathsf{T}}) = 2$, so $N_\infty(A) = N_\infty(A^{\mathsf{T}}) = 2$. See Figure 9.8a, which illustrates that an eigenvalue of a matrix can lie on the boundary of its Gershgorin region.

Example 9.4.12 The matrix

$$A = \begin{bmatrix} 9 & 8 \\ 2 & -6 \end{bmatrix} \tag{9.4.13}$$

in Example 9.3.8 has $\operatorname{spec} A = \{10, -7\}$, $R_1'(A) = 8$, $R_2'(A) = 2$, $R_1'(A^{\mathsf{T}}) = 2$, and $R_2'(A^{\mathsf{T}}) = 8$. Then

$$\mathscr{G}_1(A) = \{z \in \mathbb{C} : |z - 9| \leq 8\}, \qquad \mathscr{G}_1(A^{\mathsf{T}}) = \{z \in \mathbb{C} : |z - 9| \leq 2\},$$
$$\mathscr{G}_2(A) = \{z \in \mathbb{C} : |z + 6| \leq 2\}, \qquad \mathscr{G}_2(A^{\mathsf{T}}) = \{z \in \mathbb{C} : |z + 6| \leq 8\},$$

$N_\infty(A) = 17$, and $N_\infty(A^{\mathsf{T}}) = 14$. Thus, $\operatorname{spec} A \subseteq \{z \in \mathbb{C} : |z| \leq 14\}$ and, even better,

$$\operatorname{spec} A \subseteq \mathscr{G}(A) \cap \mathscr{G}(A^{\mathsf{T}}) = \mathscr{G}_2(A) \cup \mathscr{G}_1(A^{\mathsf{T}}) \cup \left(\mathscr{G}_1(A) \cap \mathscr{G}_2(A^{\mathsf{T}})\right);$$

see Figure 9.8b. The Gershgorin disks $\mathscr{G}_1(A)$ and $\mathscr{G}_2(A)$ do not contain 0, so 0 is not an eigenvalue of A. Corollary 9.1.17 ensures that A is invertible.

Definition 9.4.14 $A = [a_{ij}] \in \mathsf{M}_n$ is *diagonally dominant* if $|a_{kk}| \geq R_k'(A)$ for each $k = 1, 2, \ldots, n$; A is *strictly diagonally dominant* if $|a_{kk}| > R_k'(A)$ for each $k = 1, 2, \ldots, n$.

Corollary 9.4.15 *Let $A \in M_n$.*

(a) *If A is strictly diagonally dominant, then it is invertible.*

(b) *Suppose that A has real eigenvalues and real nonnegative diagonal entries. If A is diagonally dominant, then all of its eigenvalues are nonnegative. If A is strictly diagonally dominant, then all of its eigenvalues are positive.*

Proof (a) The strict diagonal dominance of A ensures that $0 \notin \mathscr{G}_k(A)$ for each $k = 1, 2, \ldots, n$. Theorem 9.4.2 tells us that 0 is not an eigenvalue of A and Corollary 9.1.17 says that A is invertible.

(b) The hypotheses ensure that each $\mathscr{G}_k(A)$ is a disk in the right half plane. Theorem 9.4.2 ensures that the eigenvalues of A are in the right half plane. Since the eigenvalues of A are all real, they must be nonnegative. If A is strictly diagonally dominant, then it is invertible, so 0 is not an eigenvalue. $\square$

Example 9.4.16 We can determine whether

$$A = \begin{bmatrix} 20 & 2 & -2 & 2 \\ -1 & 15 & 2 & 0 \\ 3 & -3 & 10 & 2 \\ -3 & 0 & 0 & 5 \end{bmatrix} \tag{9.4.17}$$

is invertible without row reduction or determinants. Since $a_{11} = 20$, $a_{22} = 15$, $a_{33} = 10$, $a_{44} = 5$, and $R'_1(A) = 6$, $R'_2(A) = 3$, $R'_3(A) = 8$, $R'_4(A) = 3$, each eigenvalue of A is contained in one of the Gershgorin disks

$$\mathscr{G}_1(A) = \{z : |z - 20| \le 6\},$$
$$\mathscr{G}_2(A) = \{z : |z - 15| \le 3\},$$
$$\mathscr{G}_3(A) = \{z : |z - 10| \le 8\},$$
$$\mathscr{G}_4(A) = \{z : |z - 5| \le 3\};$$

see Figure 9.9a. Since 0 is not contained in any of the Gershgorin disks, it is not an eigenvalue of A. Thus, A is invertible.

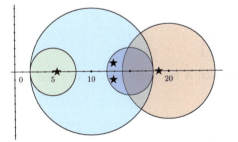
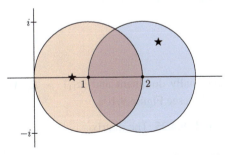

(a) Gershgorin disks for the matrix (9.4.17). To two decimal places, its eigenvalues are 18.72, $12.86 + 1.05i$, $12.86 - 1.05i$, and 5.56.

(b) Gershgorin disks for the matrix (9.4.19). To two decimal places, its eigenvalues are 0.70 and $2.30 + 0.62i$.

Figure 9.9 Gershgorin regions for the matrices (9.4.17) and (9.4.19).

Example 9.4.18 The matrices in (9.4.13) and (9.4.17) are strictly diagonally dominant and invertible. The matrix (9.4.11) is diagonally dominant and not invertible. The matrix

$$A = \begin{bmatrix} 1 & i \\ 1 & 2 \end{bmatrix} \tag{9.4.19}$$

is diagonally dominant and $0 \in \mathcal{G}(A)$. Nevertheless, the following theorem ensures that A is invertible.

Theorem 9.4.20 *Let $n \geq 2$ and let $A = [a_{ij}] \in \mathsf{M}_n$ be diagonally dominant. If $a_{ij} \neq 0$ for all $i, j \in \{1, 2, \ldots, n\}$ and $|a_{ii}| > R_i'(A)$ for at least one $i \in \{1, 2, \ldots, n\}$, then A is invertible.*

Proof If $0 \in \operatorname{spec} A$, then there is a nonzero $\mathbf{x} = [x_i] \in \mathbb{C}^n$ such that $A\mathbf{x} = \mathbf{0}$, that is,

$$-a_{ii}x_i = \sum_{j \neq i} a_{ij}x_j$$

for each $i = 1, 2, \ldots, n$. Let $k \in \{1, 2, \ldots, n\}$ be any index such that $|x_k| = \max_{1 \leq i \leq n} |x_i|$. Then

$$|a_{kk}||x_k| = |-a_{kk}x_k| = \left| \sum_{j \neq k} a_{kj}x_j \right| \leq \sum_{j \neq k} |a_{kj}||x_j| \leq \sum_{j \neq k} |a_{kj}||x_k| = R_k'(A)|x_k|. \tag{9.4.21}$$

Since $|x_k| \neq 0$, it follows that $|a_{kk}| \leq R_k'(A)$. However, A is diagonally dominant, so $|a_{ii}| \geq R_i'(A)$ for every $i = 1, 2, \ldots, n$. Thus,

$$|a_{kk}| = R_k'(A) \tag{9.4.22}$$

for any k such that $|x_k| = \max_{1 \leq i \leq n} |x_i|$. It follows that the inequality in (9.4.21) is an equality. Therefore, each inequality $|a_{kj}||x_j| \leq |a_{kj}||x_k|$ is an equality, that is,

$$|a_{kj}||x_j| = |a_{kj}||x_k| \qquad \text{for } j = 1, 2, \ldots, n.$$

Because A has no zero entries, it follows that $|x_j| = |x_k|$ for each $j = 1, 2, \ldots, n$. Therefore, every entry of $\mathbf{x}$ has maximum modulus and (9.4.22) ensures that $|a_{jj}| = R_j'(A)$ for all $j = 1, 2, \ldots, n$. Since $|a_{ii}| > R_i'(A)$ for some $i \in \{1, 2, \ldots, n\}$, we have a contradiction. It follows that $0 \notin \operatorname{spec} A$ and Theorem 9.1.16 ensures that A is invertible. $\square$

Example 9.4.23 The matrix

$$A = \begin{bmatrix} 3 & 1 & 1 \\ 0 & 1 & 1 \\ 0 & 1 & 1 \end{bmatrix} \tag{9.4.24}$$

is diagonally dominant and $|a_{11}| > R_1'(A)$. However, A has some zero entries and is not invertible; see Figure 9.10a.

Example 9.4.25 The matrix

$$A = \begin{bmatrix} 3 & 1 & 1 \\ 1 & 2 & 1 \\ 1 & 1 & 2 \end{bmatrix} \tag{9.4.26}$$

has no zero entry. It is diagonally dominant and $|a_{11}| > R_1'(A)$. Even though $0 \in \mathcal{G}(A)$, Theorem 9.4.20 ensures that A is invertible; see Figure 9.10b.

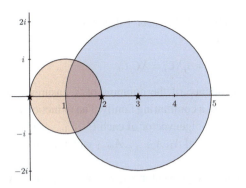

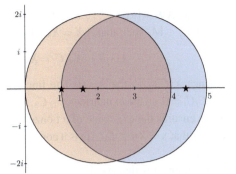

(a) $\mathscr{G}(A)$ for the matrix (9.4.24). A is diagonally dominant and $|a_{11}| > R_1'(A)$. Its eigenvalues are 0, 2, and 3.

(b) $\mathscr{G}(A)$ for the matrix (9.4.26). A is diagonally dominant and $|a_{11}| > R_1'(A)$. To two decimal places, its eigenvalues are 1, 1.59, and 4.41.

Figure 9.10 Gershgorin regions for the matrices (9.4.24) and (9.4.26).

9.5 Eigenvectors and Commuting Matrices

Commuting matrices often share properties of interest. The following generalization of Theorem 9.2.15 says that the matrices in a commuting family share a common eigenvector.

Theorem 9.5.1 *Let $k \geq 2$, let $A_1, A_2, \ldots, A_k \in M_n$, let λ be an eigenvalue of A_1, and suppose that $A_iA_j = A_jA_i$ for all $i, j \in \{1, 2, \ldots, k\}$.*

(a) *Some nonzero $\mathbf{x}$ in $\mathcal{E}_\lambda(A_1)$ is an eigenvector of each $A_2, A_3, \ldots, A_k$.*

(b) *If each A_i is real and has only real eigenvalues, then some real nonzero $\mathbf{x} \in \mathcal{E}_\lambda(A_1)$ is an eigenvector of each $A_2, A_3, \ldots, A_k$.*

Proof (a) Proceed by induction. Let $k = 2$ and suppose that the geometric multiplicity of λ as an eigenvalue of A_1 is r. Let $\mathbf{x}_1, \mathbf{x}_2, \ldots, \mathbf{x}_r$ be a basis for $\mathcal{E}_\lambda(A_1)$ and let $X = [\mathbf{x}_1 \ \mathbf{x}_2 \ \ldots \ \mathbf{x}_r] \in M_{n \times r}$. For each $j = 1, 2, \ldots, r$,

$$A_1(A_2\mathbf{x}_j) = A_1A_2\mathbf{x}_j = A_2A_1\mathbf{x}_j = A_2\lambda\mathbf{x}_j = \lambda(A_2\mathbf{x}_j),$$

so $A_2\mathbf{x}_j \in \mathcal{E}_\lambda(A_1)$. Consequently, $A_2\mathbf{x}_j$ is a linear combination of the columns of X for each $j = 1, 2, \ldots, r$. Therefore, (3.1.19) ensures that there is a $C_2 \in M_r$ such that $A_2X = XC_2$. Invoke Theorem 9.2.15 and let $(\mu, \mathbf{u})$ be an eigenpair of C_2. Observe that $X\mathbf{u} \neq \mathbf{0}$ because the columns of X are linearly independent and $\mathbf{u} \neq \mathbf{0}$. Thus,

$$A_2(X\mathbf{u}) = A_2X\mathbf{u} = XC_2\mathbf{u} = X\mu\mathbf{u} = \mu(X\mathbf{u}).$$

Finally, $A_1(X\mathbf{u}) = \lambda(X\mathbf{u})$ because $X\mathbf{u} \in \mathcal{E}_\lambda(A_1)$. Then $\mathbf{x} = X\mathbf{u}$ is a common eigenvector of A_1 and A_2 with the asserted properties.

Suppose that for some $m \geq 2$, the theorem has been proved for commuting families of m or fewer matrices. Let $A_1, A_2, \ldots, A_{m+1} \in M_n$, suppose that $A_iA_j = A_jA_i$ for all $i, j \in \{1, 2, \ldots, m + 1\}$, and let λ be an eigenvalue of A_1. Let $\mathbf{x}_1, \mathbf{x}_2, \ldots, \mathbf{x}_r$ be a basis for

$\mathcal{E}_\lambda(A_1)$ and let $X = [\mathbf{x}_1 \ \mathbf{x}_2 \ \ldots \ \mathbf{x}_r] \in \mathsf{M}_{n \times r}$. For each $j = 2, 3, \ldots, m + 1$, we have shown that there is a $C_j \in \mathsf{M}_r$ such that $A_j X = X C_j$. Consequently,

$$A_i A_j X = A_i X C_j = X C_i C_j = A_j A_i X = A_j X C_i = X C_j C_i$$

and hence $X(C_i C_j - C_j C_i) = 0$. The linear independence of the columns of X ensures that $C_i C_j - C_j C_i = 0$, so the m matrices $C_2, C_3, \ldots, C_{m+1}$ are a commuting family. The induction hypothesis ensures the existence of a nonzero $\mathbf{u}$ that is an eigenvector of each $C_2, C_3, \ldots, C_{m+1}$. It follows that $\mathbf{x} = X\mathbf{u} \in \mathcal{E}_\lambda(A_1)$ is a common eigenvector of $A_1, A_2, \ldots, A_{m+1}$ with the required properties.

(b) Part (a) ensures that $A_1, A_2, \ldots, A_k$ have a common eigenvector $\mathbf{x} \in \mathbb{C}^n$. Write $\mathbf{x} = \mathbf{u} + i\mathbf{v}$ with $\mathbf{u}, \mathbf{v} \in \mathbb{R}^n$; at least one of $\mathbf{u}$ or $\mathbf{v}$ is nonzero. For each $j = 1, 2, \ldots, k$, there is a $\lambda_j \in \mathbb{R}$ such that

$$\lambda_j \mathbf{u} + i\lambda_j \mathbf{v} = \lambda_j \mathbf{x} = A_j \mathbf{x} = A_j \mathbf{u} + iA_j \mathbf{v}.$$

Therefore, $A_j \mathbf{u} = \lambda_j \mathbf{u}$ and $A_j \mathbf{v} = \lambda_j \mathbf{v}$ for all $j = 1, 2, \ldots, k$. If $\mathbf{u} \neq \mathbf{0}$, it is a common real eigenvector for $A_1, A_2, \ldots, A_k$; if $\mathbf{u} = \mathbf{0}$, then $\mathbf{v}$ is a common real eigenvector. $\square$

The preceding theorem concerns finite families of commuting matrices, but sometimes we encounter infinite families of commuting matrices.

Example 9.5.2 Consider the matrices $U(\theta)$ in Example 9.3.10, one for each of the infinitely many values of θ in the interval $(0, \pi/2]$. A computation reveals that

$$U(\theta)U(\phi) = \begin{bmatrix} \cos\theta\cos\phi - \sin\theta\sin\phi & -(\sin\theta\cos\phi + \cos\theta\sin\phi) \\ \sin\theta\cos\phi + \cos\theta\sin\phi & \cos\theta\cos\phi - \sin\theta\sin\phi \end{bmatrix} = U(\phi)U(\theta),$$

which reflects the geometric fact that any two rotations about the origin in $\mathbb{R}^2$ commute. Thus, $\mathcal{F} = \{U(\theta) : 0 < \theta \leq \pi/2\}$ is an infinite commuting family. Moreover, each of the vectors $[1 \ \pm i]^\mathsf{T}$ is an eigenvector of every $U(\theta) \in \mathcal{F}$ even though a key hypothesis of Theorem 9.5.1 is not satisfied. This is not an accident.

The key to extending Theorem 9.5.1 to infinite families is in the following lemma, which relies on the finite dimensionality of M_n.

Lemma 9.5.3 *Let $\mathcal{F} \subseteq \mathsf{M}_n$ be a nonempty set of matrices.*

(a) *There are finitely many matrices in $\mathcal{F}$ whose span contains $\mathcal{F}$.*

(b) *Let $A_1, A_2, \ldots, A_k \in \mathsf{M}_n$. If a nonzero $\mathbf{x} \in \mathbb{C}^n$ is an eigenvector of each $A_1, A_2, \ldots, A_k$, then it is an eigenvector of each matrix in $\mathrm{span}\{A_1, A_2, \ldots, A_k\}$.*

Proof (a) If $\mathcal{F}$ is a finite set, there is nothing to prove. If $\mathcal{F}$ has infinitely many elements, then observe that $\mathcal{F} \subseteq \mathrm{span}\,\mathcal{F} \subseteq \mathsf{M}_n$. Theorem 2.2.11 ensures that there are at most n^2 elements of $\mathcal{F}$ in any basis for $\mathrm{span}\,\mathcal{F}$.

(b) If $A_i \mathbf{x} = \lambda_i \mathbf{x}$ for each $i = 1, 2, \ldots, k$, then

$$\begin{aligned} (c_1 A_1 + c_2 A_2 + \cdots + c_k A_k)\mathbf{x} &= c_1 A_1 \mathbf{x} + c_2 A_2 \mathbf{x} + \cdots + c_k A_k \mathbf{x} \\ &= c_1 \lambda_1 \mathbf{x} + c_2 \lambda_2 \mathbf{x} + \cdots + c_k \lambda_k \mathbf{x} \\ &= (c_1 \lambda_1 + c_2 \lambda_2 + \cdots + c_k \lambda_k)\mathbf{x}, \end{aligned}$$

so $\mathbf{x}$ is an eigenvector of every matrix in $\mathrm{span}\{A_1, A_2, \ldots, A_k\}$. $\square$

Corollary 9.5.4 *Let $\mathscr{F} \subseteq \mathsf{M}_n$ be a nonempty family of commuting matrices, let $A \in \mathscr{F}$ be given, and let λ be an eigenvalue of A.*

(a) *Some nonzero vector in $\mathcal{E}_\lambda(A)$ is an eigenvector of every matrix in $\mathscr{F}$.*

(b) *If every matrix in $\mathscr{F}$ is real and has only real eigenvalues, then some real vector in $\mathcal{E}_\lambda(A)$ is an eigenvector of every matrix in $\mathscr{F}$.*

Proof Invoke Lemma 9.5.3.a to obtain finitely many matrices $A_2, A_3, \ldots, A_k \in \mathscr{F}$ whose span contains $\mathscr{F}$. Let $A_1 = A$. Then

$$\mathscr{F} \subseteq \mathrm{span}\{A_2, A_3, \ldots, A_k\} \subseteq \mathrm{span}\{A, A_2, \ldots, A_k\}$$

and $A_i A_j = A_j A_i$ for all $i, j = 1, 2, \ldots, k$. Theorem 9.5.1 ensures that there is a nonzero vector $\mathbf{x}$ (real under the hypotheses of (b)) such that $A\mathbf{x} = \lambda \mathbf{x}$ and $\mathbf{x}$ is an eigenvector of each $A_2, A_3, \ldots, A_k$. Finally, Lemma 9.5.3.b ensures that $\mathbf{x}$ is an eigenvector of each matrix in $\mathrm{span}\{A, A_2, \ldots, A_k\}$, which includes every matrix in $\mathscr{F}$. $\qquad\square$

In both Theorem 9.5.1 and Corollary 9.5.4, the initial choices of a matrix A and its eigenvalue λ are unrestricted. Once we have made these choices, all we know about the common eigenvector is that it is in $\mathcal{E}_\lambda(A)$. However, if $\dim \mathcal{E}_\lambda(A) = 1$, then it is unique up to a nonzero scalar factor.

Example 9.5.5 Among the matrices $U(\theta)$ defined in Example 9.3.10, $A_{\frac{\pi}{2}} = \begin{bmatrix} 0 & -1 \\ 1 & 0 \end{bmatrix}$ has eigenvalues $\lambda_\pm = \pm i$ and corresponding eigenvectors $\mathbf{x}_\pm = [1 \ \mp i]^\mathsf{T}$. Because the eigenspaces of $A_{\pi/2}$ are one dimensional, common eigenvectors for all the matrices $U(\theta)$ can only be nonzero scalar multiples of $\mathbf{x}_\pm$.

Example 9.5.6 The matrices

$$A_1 = \begin{bmatrix} 1 & 0 & 0 \\ 0 & 1 & 0 \\ 0 & 0 & 2 \end{bmatrix} \quad \text{and} \quad A_2 = \begin{bmatrix} 3 & 1 & 0 \\ 0 & 3 & 0 \\ 0 & 0 & 1 \end{bmatrix}$$

commute and $\mathcal{E}_1(A_1) = \mathrm{span}\{\mathbf{e}_1, \mathbf{e}_2\}$. However, not every nonzero vector in $\mathcal{E}_1(A_1)$ is an eigenvector of A_2. For example, $\mathbf{e}_2 \in \mathcal{E}_1(A_1)$ but $A_2 \mathbf{e}_2 = [1 \ 3 \ 0]^\mathsf{T}$ is not a scalar multiple of $\mathbf{e}_2$. The only vectors in $\mathcal{E}_1(A_1)$ that are eigenvectors of A_2 are the nonzero vectors in $\mathrm{span}\{\mathbf{e}_1\}$.

9.6 Real Similarity of Real Matrices

Let

$$A = \begin{bmatrix} 1 & 1 \\ 0 & 1 \end{bmatrix}, \qquad B = \begin{bmatrix} 1 & 2 \\ 0 & 1 \end{bmatrix}, \tag{9.6.1}$$

$$S = \begin{bmatrix} 1 & 1+i \\ 0 & 2 \end{bmatrix} \quad \text{and} \quad S^{-1} = \begin{bmatrix} 1 & -\frac{1}{2}(1+i) \\ 0 & \frac{1}{2} \end{bmatrix}. \tag{9.6.2}$$

Then $A = SBS^{-1}$, so the real matrices A and B are similar over $\mathbb{C}$; see Definition 2.6.4. Are they similar over $\mathbb{R}$?

Theorem 9.6.3 *If two real matrices are similar, then they are similar via a real matrix.*

Proof Let $A, B \in M_n(\mathbb{R})$. Let $S \in M_n$ be invertible and such that $A = SBS^{-1}$. For any $\theta \in (-\pi, \pi]$, let $S_\theta = e^{i\theta}S$ and compute $A = (e^{i\theta}S)B(e^{-i\theta}S^{-1}) = S_\theta BS_\theta^{-1}$. Then $AS_\theta = S_\theta B$. The complex conjugate of this identity is $A\overline{S_\theta} = \overline{S_\theta}B$ since A and B are real. Let $R_\theta = S_\theta + \overline{S_\theta} = 2 \operatorname{Re} S_\theta$, which is real for all $\theta \in (-\pi, \pi]$. Then $AR_\theta = A(S_\theta + \overline{S_\theta}) = (S_\theta + \overline{S_\theta})B = R_\theta B$. The computation

$$R_\theta = S_\theta(I + S_\theta^{-1}\overline{S_\theta}) = S_\theta(I + e^{-2i\theta}S^{-1}\overline{S}) = e^{-2i\theta}S_\theta(e^{2i\theta}I + S^{-1}\overline{S})$$

shows that R_θ is invertible if $-e^{-2i\theta}$ is not an eigenvalue of $S^{-1}\overline{S}$; see Theorem 9.1.16. Since $S^{-1}\overline{S}$ has at most n distinct eigenvalues (Corollary 9.3.6), there is a $\phi \in (-\pi, \pi]$ such that $-e^{-2i\phi}$ is not an eigenvalue of $S^{-1}\overline{S}$. Then $AR_\phi = R_\phi B$, R_ϕ is real and invertible, and $A = R_\phi B R_\phi^{-1}$. □

Example 9.6.4 With the matrices in (9.6.1) and (9.6.2), we have

$$S^{-1}\overline{S} = \begin{bmatrix} 1 & -2i \\ 0 & 1 \end{bmatrix} \quad \text{and} \quad \operatorname{spec} S^{-1}\overline{S} = \{1\}.$$

A computation and Euler's formula shows that

$$R_\theta = S_\theta + \overline{S_\theta} = \begin{bmatrix} 2\cos\theta & 2\cos\theta - 2\sin\theta \\ 0 & 4\cos\theta \end{bmatrix}$$

is invertible for every $\theta \in (-\pi, \pi]$ except $\pm\pi/2$. With the choice $\theta = 0$ we have

$$R_0 = \begin{bmatrix} 2 & 2 \\ 0 & 4 \end{bmatrix} \quad \text{and} \quad R_0^{-1} = \frac{1}{4}\begin{bmatrix} 2 & -1 \\ 0 & 1 \end{bmatrix}.$$

A computation confirms that $A = R_0 B R_0^{-1}$.

9.7 Problems

P.9.1 Let $A \in M_8$ and suppose that $A^5 + 2A + I = 0$. Is A invertible?

P.9.2 Find the eigenvalues of $\begin{bmatrix} 1 & 2 \\ 3 & 4 \end{bmatrix}$, $\begin{bmatrix} 1 & 3 \\ 2 & 4 \end{bmatrix}$, $\begin{bmatrix} 3 & 4 \\ 1 & 2 \end{bmatrix}$, $\begin{bmatrix} 2 & 1 \\ 4 & 3 \end{bmatrix}$, and $\begin{bmatrix} 4 & 3 \\ 2 & 1 \end{bmatrix}$. Discuss any patterns that you observe in your answers.

P.9.3 Let $A = \begin{bmatrix} 1 & 2 \\ 0 & 4 \end{bmatrix}$ and $B = \begin{bmatrix} 2 & 3 \\ 0 & 1 \end{bmatrix}$. (a) Show that 1 is an eigenvalue of A and 2 is an eigenvalue of B. (b) Show that $1 + 2 = 3$ is an eigenvalue of $A + B$. (c) Show that $1 \cdot 2 = 2$ is an eigenvalue of AB. (d) Do A and B commute?

P.9.4 Is

$$A = \begin{bmatrix} 5 & 1 & 2 & 1 & 0 \\ 1 & 7 & 1 & 0 & 1 \\ 1 & 1 & 12 & 1 & 1 \\ 1 & 0 & 1 & 13 & 1 \\ 0 & 1 & 2 & 1 & 14 \end{bmatrix}$$

invertible? Why? Do not row reduce A or compute $\det A$.

P.9.5 (a) Let $p(z) = z^2 - tz + d$. Use the division algorithm (Appendix B.2) to show that $z^3 = p(z)(z + t) + (t^2 - d)z - td$. (b) Let $A \in M_2$, let $t = \operatorname{tr} A$, and let $d = \det A$. Use (9.2.7) to show that A^3 can be computed without any matrix multiplications via the formula $A^3 = (t^2 - d)A - td I_2$. (c) Verify this formula for the matrices (9.3.9) and (9.4.11).

P.9.6 Using the notation in the preceding problem, show that $A^4 = (t^3 - 2td)A + d(d - t^2)I_2$ and verify this formula for the matrices (9.4.11) and (9.4.13).

P.9.7 The commutativity of powers of A is essential for the identity (9.2.11). Consider the two presentations $z^2 - w^2 = (z + w)(z - w)$ of a two-variable polynomial and the matrices $A = \begin{bmatrix} 0 & 1 \\ 0 & 0 \end{bmatrix}$ and $B = \begin{bmatrix} 0 & 0 \\ 1 & 0 \end{bmatrix}$. Compute $A^2 - B^2$ and $(A + B)(A - B)$, and show that they are unequal. Discuss.

P.9.8 Let $A \in M_n$ be invertible and let $(\lambda, \mathbf{x})$ be an eigenpair of A. Explain why $\lambda \neq 0$ and show that $(\lambda^{-1}, \mathbf{x})$ is an eigenpair of A^{-1}.

P.9.9 Suppose that $A \in M_n(\mathbb{R})$ has a non-real eigenvalue λ. (a) Show that no eigenvector $\mathbf{x}$ associated with λ can be real. (b) If $(\lambda, \mathbf{x})$ is an eigenpair of A, then so is $(\bar{\lambda}, \bar{\mathbf{x}})$.

P.9.10 Suppose that $A \in M_n(\mathbb{R})$ has a real eigenvalue λ and an associated non-real eigenvector $\mathbf{x}$. Write $\mathbf{x} = \mathbf{u} + i\mathbf{v}$, in which $\mathbf{u}$ and $\mathbf{v}$ are real vectors. Show that at least one of $(\lambda, \mathbf{u})$ or $(\lambda, \mathbf{v})$ is a real eigenpair of A.

P.9.11 Suppose that $A \in M_n(\mathbb{R})$ and $k \geq 1$. Show that λ is an eigenvalue of A with geometric multiplicity k if and only if $\bar{\lambda}$ is an eigenvalue of A with geometric multiplicity k.

P.9.12 Let $A = [a_{ij}] \in M_n$, let $\mathbf{e} = [1 \ 1 \ \dots \ 1]^\mathsf{T} \in \mathbb{C}^n$, and let γ be a scalar. Show that $\sum_{j=1}^n a_{ij} = \gamma$ for each $i = 1, 2, \dots, n$ (all row sums equal to γ) if and only if $(\gamma, \mathbf{e})$ is an eigenpair of A. What can you say about the column sums of A if $(\gamma, \mathbf{e})$ is an eigenpair of A^T?

P.9.13 Let $n \geq 2$, let $c \in \mathbb{C}$, and let $A_c \in M_n$, whose off-diagonal entries are all 1, and whose diagonal entries are c. (a) Sketch the Gershgorin region $\mathscr{G}(A_c)$. (b) Why is A_c invertible if $|c| > n - 1$? (c) Show that A_{1-n} is not invertible.

P.9.14 Let $A \in M_n$ be diagonally dominant and have real nonnegative diagonal entries. Let $\lambda \in \operatorname{spec} A$. (a) Show that $\operatorname{Re} \lambda \geq 0$. What can you say if $\operatorname{Re} \lambda = 0$? (b) If A has no zero entries and $a_{kk} > R'_k(A)$ for some $k \in \{1, 2, \dots, n\}$, show that $\operatorname{Re} \lambda > 0$.

P.9.15 Let $E = \mathbf{e}\mathbf{e}^\mathsf{T} \in M_n$ be the all-ones matrix. (a) Why is $(n, \mathbf{e})$ an eigenpair of E? (b) If $\mathbf{v}$ is nonzero and orthogonal to $\mathbf{e}$, why is $(0, \mathbf{v})$ an eigenpair of E? (c) Let $p(z) = z^2 - nz$ and show that $p(E) = 0$. (d) Use Theorem 9.3.3 to explain why 0 and n are the only eigenvalues of E. What are their respective geometric multiplicities?

P.9.16 Consider $A = [a_{ij}] \in M_n$, in which $a_{ii} = a$ for each $i = 1, 2, \dots, n$ and $a_{ij} = b \neq 0$ if $i \neq j$. Find the eigenvalues of A and their geometric multiplicities. *Hint*: Use the preceding problem.

P.9.17 Let A be the matrix in (9.4.11). (a) If $(\lambda, \mathbf{x})$ is an eigenpair of A, simplify $A(A\mathbf{x})$ and show that $(\lambda^2, \mathbf{x})$ is an eigenpair of A^2. (b) Compute A^2, deduce that $\lambda^2 = 0$, and explain why $\operatorname{spec} A = \{0\}$.

P.9.18 Suppose that each row and each column of $A \in M_{n+1}$ contains all of the entries $1, 2, 2^2, \dots, 2^n$ in some order. Prove that A is invertible.

P.9.19 Let $A = [a_{ij}] \in M_n$ and let $\lambda \in \mathbb{C}$. Suppose that $n \geq 2$, $a_{ij} \neq 0$ for $i \neq j$, $A - \lambda I$ is diagonally dominant, and $|a_{kk} - \lambda| > \sum_{j \neq k} |a_{kj}|$ for some $k \in \{1, 2, \dots, n\}$. Prove that $\lambda \notin \operatorname{spec} A$.

P.9.20 Let $A \in M_m$, $B \in M_n$, and $C = A \oplus B \in M_{m+n}$. Use eigenpairs (do not use Theorem 10.2.11) to show the following. (a) If $(\lambda, \mathbf{u})$ is an eigenpair of A, show that λ is an eigenvalue of C. What is a corresponding eigenvector? (b) If $(\lambda, \mathbf{v})$ is an eigenpair of B, show that λ is an eigenvalue of C. What is a corresponding eigenvector? (c) If

$(\lambda, \mathbf{w})$ is an eigenpair of C, show that λ is an eigenvalue of A or B (perhaps both). What is a corresponding eigenvector? (d) Deduce that $\operatorname{spec} C = \operatorname{spec} A \cup \operatorname{spec} B$.

P.9.21 Let $A \in \mathsf{M}_n$, suppose that A is idempotent, and let $(\lambda, \mathbf{x})$ be an eigenpair of A. Show that $\lambda = 1$ or $\lambda = 0$.

P.9.22 Let $A = \begin{bmatrix} 1 & 1 \\ 0 & 1 \end{bmatrix}$ and $B = \begin{bmatrix} 1 & 0 \\ 1 & 1 \end{bmatrix}$, and let $\lambda \in \operatorname{spec} A$. (a) Show that there is no eigenvalue μ of B such that $\lambda + \mu$ is an eigenvalue of $A + B$. (b) Show that there is no eigenvalue μ of B such that $\lambda\mu$ is an eigenvalue of AB. (c) Do A and B commute?

P.9.23 Let $A, B \in \mathsf{M}_n$ and let $\lambda \in \operatorname{spec} A$. If A commutes with B, prove the following: (a) There is a $\mu \in \operatorname{spec} B$ such that $\lambda + \mu \in \operatorname{spec}(A + B)$. (b) There is a $\mu \in \operatorname{spec} B$ such that $\lambda\mu \in \operatorname{spec} AB$.

P.9.24 Using the notation of Example 9.3.10, show that $U(\theta)U(\phi) = U(\theta + \phi)$. Deduce that $U(\theta)$ commutes with $U(\phi)$.

P.9.25 Let $A_1, A_2, \ldots, A_k \in \mathsf{M}_n$ and suppose that $A_i A_j = A_j A_i$ for all $i, j \in \{1, 2, \ldots, k\}$. Show that $\operatorname{span}\{A_1, A_2, \ldots, A_k\}$ is a commuting family of matrices.

P.9.26 Let A_1 be the matrix in Example 9.1.13 and let $A_2 = I \in \mathsf{M}_2$. (a) Show that $\operatorname{span}\{A_1, A_2\} = \{Z_{a,b} : a, b \in \mathbb{C}\}$, in which $Z_{a,b} = \begin{bmatrix} a & -b \\ b & a \end{bmatrix}$, and explain why $\operatorname{span}\{A_1, A_2\}$ is a commuting family. (b) Show that $\operatorname{spec} Z_{a,b} = \{a \pm ib\}$ with associated eigenvectors $\mathbf{x}_\pm = [1 \mp i]^\mathsf{T}$. (c) What is $Z_{\cos\theta, \sin\theta}$?

P.9.27 Consider the Fourier matrix F_n defined in (7.1.18). Show that $\operatorname{spec} F_n \subseteq \{\pm 1, \pm i\}$. *Hint*: Theorem 9.3.3 and P.7.46.

P.9.28 Let $n \geq 2$, let $m \geq n \geq 1$, and let $X = [x_{ij}] = [\mathbf{x}_1 \ \mathbf{x}_2 \ \cdots \ \mathbf{x}_n] \in \mathsf{M}_{m \times n}$ have full column rank. Show that there is an invertible $B \in \mathsf{M}_n$ and distinct indices $i_1, i_2, \ldots, i_n \in \{1, 2, \ldots, m\}$ such that the columns of

$$XB = Y = [\mathbf{y}_1 \ \mathbf{y}_2 \ \cdots \ \mathbf{y}_n] \tag{9.7.1}$$

have the following properties: For each $j = 1, 2, \ldots, n$, (a) Every entry of $\mathbf{y}_j$ has modulus at most 1; (b) $\mathbf{y}_j$ has an entry with modulus 1 in position i_j; and (c) If $j \geq 2$, then $\mathbf{y}_j$ has the entry 0 in position i_k for each $k < j$. *Hint*: Let i_1 be the index of a largest-modulus entry of $\mathbf{x}_1$. Let $\mathbf{y}_1 = \mathbf{x}_1 / |x_{i_1 1}|$ and add suitable scalar multiples of $\mathbf{y}_1$ to $\mathbf{x}_2, \mathbf{x}_3, \ldots, \mathbf{x}_n$ to obtain columns that have a 0 entry in position i_1. Repeat, starting with the second column.

P.9.29 Let $A \in \mathsf{M}_n$ and suppose that $\lambda \in \operatorname{spec} A$ has geometric multiplicity at least $m \geq 1$. Adopt the notation of Theorem 9.4.2. (a) Show that there are distinct indices $i_1, i_2, \ldots, i_m \in \{1, 2, \ldots, n\}$ such that $\lambda \in \mathscr{G}_{i_j}$ for each $j = 1, 2, \ldots, m$. (b) Show that λ is contained in the union of any $n - m + 1$ of the disks $\mathscr{G}_k$. (c) What do (a) and (b) say if $m = 1$? If $m = 2$? If $m = n$? *Hint*: Let the columns of $X \in \mathsf{M}_{n \times m}$ be a basis for $\mathcal{E}_\lambda(A)$ and construct the matrix Y in (9.7.1). Use a column of Y as the eigenvector in the proof of Theorem 9.4.2.

P.9.30 The *Volterra operator* is the linear operator $T : C[0, 1] \to C[0, 1]$ defined by $(Tf)(t) = \int_0^t f(s)\, ds$; see P.6.12. The pair (λ, f) is an *eigenpair* of T if $\lambda \in \mathbb{C}$, $f \in C[0, 1]$ is not the zero function, and $(Tf)(t) = \lambda f(t)$ for all $t \in [0, 1]$. (a) Show that 0 is not an eigenvalue of T. (b) Find the eigenpairs of the Volterra operator. Compare your results with the situation for eigenvalues of a matrix. *Hint*: Consider the equation $Tf = \lambda f$ and use the fundamental theorem of calculus.

P.9.31 Suppose that $(\lambda, \mathbf{x})$ and $(\mu, \mathbf{y})$ are eigenpairs of $A \in \mathsf{M}_n$ and $B \in \mathsf{M}_m$, respectively. Show that (a) $(\lambda\mu, \mathbf{x} \otimes \mathbf{y})$ is an eigenpair of $A \otimes B \in \mathsf{M}_{nm}$ and (b) $(\lambda + \mu, \mathbf{x} \otimes \mathbf{y})$ is an eigenpair of $(A \otimes I_m) + (I_n \otimes B) \in \mathsf{M}_{nm}$.

9.8 Notes

The term *spectrum* (in the context of eigenvalues) was introduced by D. Hilbert in his 1912 book on integral equations. In the mid-1920s, W. Heisenberg, M. Born, and P. Jordan discovered the matrix-mechanics formulation of quantum mechanics, after which W. Pauli identified the wavelengths of the spectral lines of hydrogen (Balmer series) with the eigenvalues of the energy matrix.

If $A \in \mathsf{M}_n$, it can happen that $\mathscr{G}(A) = \mathscr{S}_1 \cup \mathscr{S}_2$ is the union of two disjoint sets, in which $\mathscr{S}_1$ is the union of n_1 Gershgorin disks, and $\mathscr{S}_2$ is the union of n_2 Gershgorin disks, with $n_1 + n_2 = n$. In this case, A has n_1 eigenvalues in $\mathscr{S}_1$ and n_2 eigenvalues in $\mathscr{S}_2$. For example, in Figure 9.8b, $\mathscr{G}(A) = \mathscr{S}_1(A) \cup \mathscr{S}_2(A)$, in which $n = 2$ and $\mathscr{S}_1(A)$ is disjoint from $\mathscr{S}_2(A)$. There must be one eigenvalue of A in $\mathscr{S}_1(A)$ and one in $\mathscr{S}_2(A)$. Disjointness of $\mathscr{S}_1$ and $\mathscr{S}_2$ is an essential hypothesis. For example, $A = \begin{bmatrix} 0 & 1 \\ 2 & 0 \end{bmatrix}$ has $\mathscr{G}_1(A) \subset \mathscr{G}_2(A)$ and spec $A \subset \mathscr{G}_2(A)$, but no eigenvalue of A is in $\mathscr{G}_1(A)$. For details about this and other results associated with Gershgorin's theorem, see [HJ13, Chapter 6].

9.9 Some Important Concepts

- Eigenpair of a matrix.

- Eigenvectors must be nonzero.

- Characterizations of eigenvalues.

- Every square complex matrix has an eigenvalue.

- Eigenvectors corresponding to distinct eigenvalues are linearly independent.

- An $n \times n$ complex matrix has at most n distinct eigenvalues.

- Geometric multiplicity of an eigenvalue.

- The Gershgorin region of a matrix contains all its eigenvalues.

- A strictly diagonally dominant matrix is invertible.

- A diagonally dominant matrix with no zero entries is invertible if the dominance is strict in some row.

- Commuting matrices share a common eigenvector.

The Characteristic Polynomial and Algebraic Multiplicity

In this chapter, we identify the eigenvalues of a square complex matrix as the zeros of its characteristic polynomial. We show that an $n \times n$ complex matrix is diagonalizable (similar to a diagonal matrix) if and only if it has n linearly independent eigenvectors. If A is a diagonalizable matrix and if f is a complex-valued function on the spectrum of A, we discuss a way to define $f(A)$ that has many desirable properties.

10.1 The Characteristic Polynomial

A systematic method to determine all the eigenvalues of $A = [a_{ij}] \in M_n$ is based on the observation that the conditions in Theorem 9.1.16 are equivalent to $\det(\lambda I - A) = 0$. This suggests that a careful study of the function $p_A(z) = \det(zI - A)$ could be fruitful, since the roots of $p_A(z) = 0$ are the eigenvalues of A.

If $n = 1$, then $p_A(z) = \det(zI - A) = \det[z - a_{11}] = z - a_{11}$, so $p_A(z)$ is a monic polynomial in z of degree 1; see Appendix B.1.

If $n = 2$, use a Laplace expansion (C.4.1) along the first column to compute

$$
\begin{aligned}
p_A(z) = \det(zI - A) &= \det\begin{bmatrix} z - a_{11} & -a_{12} \\ -a_{21} & z - a_{22} \end{bmatrix} \\
&= (z - a_{11})\det[z - a_{22}] + a_{21}\det[-a_{12}] \\
&= (z - a_{11})(z - a_{22}) - a_{21}a_{12} \\
&= z^2 - (a_{11} + a_{22})z + (a_{11}a_{22} - a_{21}a_{12}) \\
&= z^2 - (\operatorname{tr} A)z + \det A,
\end{aligned}
\tag{10.1.1}
$$

so $p_A(z)$ is a monic polynomial in z of degree 2 whose coefficients are polynomials in the entries of A.

If $n = 3$, a Laplace expansion along the first column gives

$$
\begin{aligned}
p_A(z) = \det(zI - A) &= \det\begin{bmatrix} z - a_{11} & -a_{12} & -a_{13} \\ -a_{21} & z - a_{22} & -a_{23} \\ -a_{31} & -a_{32} & z - a_{33} \end{bmatrix} \\
&= (z - a_{11})\det\begin{bmatrix} z - a_{22} & -a_{23} \\ -a_{32} & z - a_{33} \end{bmatrix} + a_{21}\det\begin{bmatrix} -a_{12} & -a_{13} \\ -a_{32} & z - a_{33} \end{bmatrix} \\
&\quad - a_{31}\det\begin{bmatrix} -a_{12} & -a_{13} \\ z - a_{22} & -a_{23} \end{bmatrix}
\end{aligned}
$$

$$= (z - a_{11})(z - a_{22})(z - a_{33}) - (z - a_{11})a_{32}a_{23} - (z - a_{33})a_{21}a_{12} - a_{21}a_{32}a_{13}$$
$$\quad - a_{12}a_{23}a_{31} - (z - a_{22})a_{31}a_{13}$$
$$= z^3 + c_2 z^2 + c_1 z + c_0, \tag{10.1.2}$$

in which

$$c_2 = -(a_{11} + a_{22} + a_{33}) = -\operatorname{tr} A,$$
$$c_1 = a_{11}a_{22} - a_{12}a_{21} + a_{11}a_{33} - a_{13}a_{31} + a_{22}a_{33} - a_{23}a_{32}, \quad \text{and}$$
$$c_0 = -\det A.$$

The terms in (10.1.2) of degrees three and two in z are obtained solely from the summand

$$(z - a_{11})(z - a_{22})(z - a_{33}) = z^3 - (a_{11} + a_{22} + a_{33})z^2 + \cdots.$$

The other summands contribute only to the terms of degree one or less.

Theorem 10.1.3 *Let $A \in \mathsf{M}_n$. The function $p_A(z) = \det(zI - A)$ is a monic polynomial*

$$p_A(z) = z^n + c_{n-1}z^{n-1} + \cdots + c_1 z + c_0 \tag{10.1.4}$$

in z of degree n, in which each coefficient is a polynomial function of the entries of A, $c_{n-1} = -\operatorname{tr} A$, and $c_0 = (-1)^n \det A$. Moreover, $p_A(\lambda) = 0$ if and only if λ is an eigenvalue of A.

Proof We have proved the assertion for $n = 1, 2, 3$. Proceed by induction on n and suppose that we have proved it for some $n \geq 3$. Let $A \in \mathsf{M}_{n+1}$ and let A_{ij} and $C_{ij}(z)$, respectively, denote the $n \times n$ submatrices of A and $zI - A$, respectively, obtained by deleting their ith row and jth column. Observe that $C_{11}(z) = zI - A_{11}$, so $\det C_{11}(z) = p_{A_{11}}(z)$. The Laplace expansion of $\det(zI - A)$ along the first column is

$$p_A(z) = \det(zI - A) = (z - a_{11})\det C_{11}(z) + \sum_{i=2}^{n+1} (-1)^{i+1} a_{i1} \det C_{i1}(z)$$

$$= (z - a_{11})p_{A_{11}}(z) + \underbrace{\sum_{i=2}^{n+1} (-1)^{i+1} a_{i1} \det C_{i1}(z)}_{g(z)}.$$

For $i = 2, 3, \ldots, n+1$, the $n \times n$ submatrices $C_{i1}(z)$ do not contain $z - a_{11}$ and $z - a_{ii}$ as entries. Since $\det C_{i1}(z)$ is a sum of signed products of entries from n distinct matrix positions, it is a polynomial in z of degree at most $n - 1$. The induction hypothesis is that

$$p_{A_{11}}(z) = z^n - (\operatorname{tr} A_{11})z^{n-1} + f(z),$$

in which f is a polynomial in z of degree at most $n - 2$. Therefore,

$$p_A(z) = (z - a_{11})p_{A_{11}}(z) + g(z)$$
$$= (z - a_{11})\big(z^n - (\operatorname{tr} A_{11})z^{n-1} + f(z)\big) + g(z)$$
$$= z^{n+1} - (a_{11} + \operatorname{tr} A_{11})z^n + a_{11}(\operatorname{tr} A_{11})z^{n-1} + (z - a_{11})f(z) + g(z)$$
$$= z^{n+1} - (\operatorname{tr} A)z^n + h(z),$$

in which g and h are polynomials in z of degree at most $n - 1$. The presentation (10.1.4) ensures that

$$c_0 = p_A(0) = \det(0I - A) = \det(-A) = (-1)^n \det A.$$

The final assertion is a restatement of the equivalence of the condition $\det(\lambda I - A) = 0$ and the conditions in Theorem 9.1.16. □

Definition 10.1.5 The polynomial (10.1.4) is the *characteristic polynomial* of A.

Example 10.1.6 In (10.1.1) we found that the characteristic polynomial of $A = \begin{bmatrix} a & b \\ c & d \end{bmatrix}$ is $p_A(z) = z^2 - (a+d)z + (ad - bc) = z^2 - (\operatorname{tr} A)z + \det A$. This is the same polynomial that appears in Example 9.2.13. We showed that the two roots of $p_A(z) = 0$ (see (9.2.14)) are eigenvalues of A. Theorem 10.1.3 ensures that these two roots are the only eigenvalues of A.

Example 10.1.7 Suppose that $A = [a_{ij}] \in M_n$ is upper triangular. Its determinant is the product of its main diagonal entries: $\det A = a_{11}a_{22} \cdots a_{nn}$. The matrix $zI - A$ is also upper triangular, so its determinant is

$$p_A(z) = \det(zI - A) = (z - a_{11})(z - a_{22}) \cdots (z - a_{nn}).$$

Thus, λ is an eigenvalue of an upper triangular matrix if and only if it is a main diagonal entry. Lower triangular and diagonal matrices have the same property. For matrices without these special structures, a main diagonal entry need not be an eigenvalue.

Example 10.1.8 The characteristic polynomial of

$$A = \begin{bmatrix} 1 & 1 & 1 \\ 1 & 1 & 0 \\ 1 & 0 & 0 \end{bmatrix}$$

is $p_A(z) = z^3 - 2z^2 - z + 1$. The roots of $p_A(z) = 0$ are approximately 2.25, 0.55, and −0.80. The diagonal entries of A are not eigenvalues, and A is neither upper nor lower triangular.

10.2 Algebraic Multiplicity

The characteristic polynomial of $A \in M_n$ has degree n and is monic. We factor it as

$$p_A(z) = (z - \lambda_1)(z - \lambda_2) \cdots (z - \lambda_n), \tag{10.2.1}$$

in which $\lambda_1, \lambda_2, \ldots, \lambda_n$ is a list of the roots of $p_A(z) = 0$. Some roots may be repeated in this list, and it is useful to indicate the distinct roots and their repetitions. Suppose that there are d distinct roots $\mu_1, \mu_2, \ldots, \mu_d$. That is,

- $\mu_i \in \{\lambda_1, \lambda_2, \ldots, \lambda_n\}$ for each $i = 1, 2, \ldots, d$;
- $\lambda_i \in \{\mu_1, \mu_2, \ldots, \mu_d\}$ for each $i = 1, 2, \ldots, n$; and
- $\mu_i \neq \mu_j$ for all $i, j \in \{1, 2, \ldots, d\}$ such that $i \neq j$.

If μ_i appears n_i times in the list $\lambda_1, \lambda_2, \ldots, \lambda_n$, then

$$p_A(z) = (z - \mu_1)^{n_1}(z - \mu_2)^{n_2} \cdots (z - \mu_d)^{n_d}. \tag{10.2.2}$$

Example 10.2.3 If $A \in M_8$ and $1, 2, 1, 2, 3, 2, 3, 4$ is a list of the roots of $p_A(z) = 0$, then $1, 2, 3, 4$ is a list of the four distinct roots, repeated $2, 3, 2, 1$ times, respectively. We have

$$p_A(z) = (z - 1)(z - 2)(z - 1)(z - 2)(z - 3)(z - 2)(z - 3)(z - 4)$$
$$= (z - 1)^2(z - 2)^3(z - 3)^2(z - 4).$$

In the notation of (10.2.2), we have $\mu_1 = 1$, $\mu_2 = 2$, $\mu_3 = 3$, $\mu_4 = 4$, $n_1 = 2$, $n_2 = 3$, $n_3 = 2$, and $n_4 = 1$.

Definition 10.2.4 The scalars $\mu_1, \mu_2, \ldots, \mu_d$ in (10.2.2) are the *distinct eigenvalues of A*. The exponent n_i in (10.2.2) is the number of times that each μ_i appears in the list $\lambda_1, \lambda_2, \ldots, \lambda_n$. We say that n_i is the *algebraic multiplicity* of the eigenvalue μ_i. If $n_i = 1$, then μ_i is a *simple eigenvalue*. The scalars $\lambda_1, \lambda_2, \ldots, \lambda_n$ in (10.2.1) are the eigenvalues of A *including multiplicities*, so $\operatorname{spec} A = \{\lambda_1, \lambda_2, \ldots, \lambda_n\} = \{\mu_1, \mu_2, \ldots, \mu_d\}$.

Definition 10.2.5 If we refer to the *multiplicity* of an eigenvalue without qualification (algebraic or geometric), we mean the algebraic multiplicity. Two matrices have the *same spectrum* if their distinct eigenvalues are the same; they have the *same eigenvalues* if their distinct eigenvalues are the same with the same algebraic multiplicities.

Example 10.2.6 The matrices $\begin{bmatrix} 0 & 1 \\ 0 & 0 \end{bmatrix}$ and 0_2 have the same eigenvalues. The matrices 0_2 and 0_3 have the same spectrum, namely $\{0\}$, but they do not have the same eigenvalues because the multiplicities of 0 are different.

The highest-order term in (10.2.1) is z^n; the highest-order term in (10.2.2) is $z^{n_1 + n_2 + \cdots + n_d}$. These two terms must be equal, so

$$n = n_1 + n_2 + \cdots + n_d, \tag{10.2.7}$$

that is, the sum of the multiplicities of the eigenvalues of $A \in \mathsf{M}_n$ is n.

Example 10.2.8 The characteristic polynomial of $\begin{bmatrix} 1 & 1 \\ 0 & 1 \end{bmatrix}$ is $p_A(z) = (z - 1)^2$, so $\lambda = 1$ is its only eigenvalue. It has algebraic multiplicity 2, but we saw in Example 9.1.10 that its geometric multiplicity is 1. Thus, the geometric multiplicity of an eigenvalue can be less than its algebraic multiplicity.

How are eigenvalues of $A^\mathsf{T}, \overline{A}$, and A^* related to the eigenvalues of A? The relationship is a consequence of the identities $\det A^\mathsf{T} = \det A$ and $\det \overline{A} = \overline{\det A}$.

Theorem 10.2.9 *Let $\lambda_1, \lambda_2, \ldots, \lambda_n$ be the eigenvalues of $A \in \mathsf{M}_n$.*

(a) *The eigenvalues of A^T are $\lambda_1, \lambda_2, \ldots, \lambda_n$.*

(b) *The eigenvalues of $\overline{A}$ and A^* are $\overline{\lambda_1}, \overline{\lambda_2}, \ldots, \overline{\lambda_n}$.*

(c) *If A has real entries and λ is a non-real eigenvalue of A with multiplicity k, then $\overline{\lambda}$ is an eigenvalue of A with multiplicity k.*

Proof (a) Compute $p_{A^\mathsf{T}}(z) = \det(zI - A^\mathsf{T}) = \det(zI - A)^\mathsf{T} = \det(zI - A) = p_A(z)$, so A and A^T have the same characteristic polynomial. Therefore, they have the same eigenvalues with the same multiplicities.

(b) Observe that $p_{\overline{A}}(z) = \det(zI - \overline{A}) = \det(\overline{\overline{z}I - A}) = \overline{\det(\overline{z}I - A)} = \overline{p_A(\overline{z})}$. Use the presentation (10.2.1) to compute

$$p_{\overline{A}}(z) = \overline{p_A(\overline{z})} = \overline{(\overline{z} - \lambda_1)(\overline{z} - \lambda_2) \cdots (\overline{z} - \lambda_n)} = (z - \overline{\lambda_1})(z - \overline{\lambda_2}) \cdots (z - \overline{\lambda_n}),$$

which shows that the eigenvalues of $\overline{A}$ are $\overline{\lambda_1}, \overline{\lambda_2}, \ldots, \overline{\lambda_n}$. The assertion in (a) ensures that the eigenvalues of $A^* = \overline{A}^\mathsf{T}$ are the same as those of $\overline{A}$.

(c) If $A \in \mathsf{M}_n(\mathbb{R})$, then $A = \overline{A}$ and it follows from (b) that the elements of the two lists $\lambda_1, \lambda_2, \ldots, \lambda_n$ and $\overline{\lambda_1}, \overline{\lambda_2}, \ldots, \overline{\lambda_n}$ are the same, though one list might appear as a reordered presentation of the other. Thus, any non-real eigenvalues of A occur in complex-conjugate pairs with equal multiplicities.   □

Two useful identities between matrix entries and eigenvalues can be deduced from a comparison of the presentations (10.1.4) and (10.2.1) of the characteristic polynomial.

Theorem 10.2.10 *Let* $\lambda_1, \lambda_2, \ldots, \lambda_n$ *be the eigenvalues of* $A \in \mathsf{M}_n$. *Then* $\operatorname{tr} A = \lambda_1 + \lambda_2 + \cdots + \lambda_n$ *and* $\det A = \lambda_1 \lambda_2 \cdots \lambda_n$.

Proof Expand (10.2.1) to obtain

$$p_A(z) = z^n - (\lambda_1 + \lambda_2 + \cdots + \lambda_n)z^{n-1} + \cdots + (-1)^n \lambda_1 \lambda_2 \cdots \lambda_n$$

and invoke Theorem 10.1.3 to write $p_A(z) = z^n - (\operatorname{tr} A)z^{n-1} + \cdots + (-1)^n \det A$.   □

Block upper triangular matrices arise frequently, so it is important to understand how their eigenvalues are related to the eigenvalues of their diagonal blocks; a special case is discussed in Example 10.1.7.

Theorem 10.2.11 *Let*

$$A = \begin{bmatrix} B & C \\ 0 & D \end{bmatrix}, \tag{10.2.12}$$

in which B and D are square. Then $\operatorname{spec} A = \operatorname{spec} B \cup \operatorname{spec} D$. *Moreover,* $p_A(z) = p_B(z)p_D(z)$, *so the eigenvalues of A are the eigenvalues of B together with the eigenvalues of D, including multiplicities in each case.*

Proof Use Theorem 3.3.4 and compute

$$p_A(z) = \det \begin{bmatrix} zI - B & -C \\ 0 & zI - D \end{bmatrix} = \det(zI - B) \det(zI - D) = p_B(z)p_D(z).$$

The roots of $p_A(z) = 0$ are the roots of $p_B(z) = 0$ together with the roots of $p_D(z) = 0$, including their respective multiplicities. Each distinct root of $p_A(z) = 0$ is a root of at least one of $p_B(z) = 0$ or $p_D(z) = 0$.   □

Example 10.2.13 Consider the block matrix A in (10.2.12), in which $B = \begin{bmatrix} 1 & 1 \\ 1 & 1 \end{bmatrix}$ and $D = \begin{bmatrix} 1 & i \\ i & -1 \end{bmatrix}$ are the matrices in Examples 9.1.6 and 9.3.16, respectively. The eigenvalues of B are 2 and 0; the only eigenvalue of D is 0 (with multiplicity 2). Thus, the eigenvalues of A are 2 (with multiplicity 1) and 0 (with multiplicity 3), regardless of what C is.

10.3 Similarity and Eigenvalue Multiplicities

Similar matrices represent the same linear transformation with respect to different bases; see Corollary 2.6.5. Therefore, similar matrices share many properties. For example, they have the same characteristic polynomials, eigenvalues, and eigenvalue multiplicities.

Theorem 10.3.1 *Let* $A, S \in \mathsf{M}_n$, *let S be invertible, and let* $B = SAS^{-1}$. *Then A and B have the same characteristic polynomial and the same eigenvalues, with the same multiplicities, both algebraic and geometric.*

Proof Compute

$$
\begin{aligned}
p_B(z) &= \det(zI - B) = \det(zI - SAS^{-1})\\
&= \det(zSS^{-1} - SAS^{-1}) = \det(S(zI - A)S^{-1})\\
&= (\det S)\det(zI - A)(\det S^{-1}) = (\det S)\det(zI - A)(\det S)^{-1}\\
&= \det(zI - A) = p_A(z).
\end{aligned}
$$

Since A and B have the same characteristic polynomials, they have the same eigenvalues with the same algebraic multiplicities.

Theorems 2.6.9 and 4.2.1 ensure that $A - \lambda I$ and $B - \lambda I$ are similar and have the same rank. The identity (9.3.14) implies that

$$
\dim \mathcal{E}_\lambda(A) = n - \operatorname{rank}(A - \lambda I) = n - \operatorname{rank}(B - \lambda I) = \dim \mathcal{E}_\lambda(B),
$$

which is the asserted equality of geometric multiplicities. $\qquad\square$

The preceding result suggests a strategy that is often effective in proving theorems about eigenvalues: transform by similarity to a new matrix for which the theorem is easier to prove. We employ this strategy to prove an inequality between the algebraic and geometric multiplicities of an eigenvalue, of which Example 10.2.8 is a special case.

Theorem 10.3.2 *The geometric multiplicity of an eigenvalue is less than or equal to its algebraic multiplicity.*

Proof Let λ be an eigenvalue of $A \in \mathsf{M}_n$ and suppose that $\dim \mathcal{E}_\lambda(A) = k$. Let $\mathbf{x}_1, \mathbf{x}_2, \ldots, \mathbf{x}_k$ be a basis for $\mathcal{E}_\lambda(A)$, let $X = [\mathbf{x}_1 \ \mathbf{x}_2 \ \ldots \ \mathbf{x}_k] \in \mathsf{M}_{n \times k}$, and compute

$$
AX = [A\mathbf{x}_1 \ A\mathbf{x}_2 \ \ldots \ A\mathbf{x}_k] = [\lambda\mathbf{x}_1 \ \lambda\mathbf{x}_2 \ \ldots \ \lambda\mathbf{x}_k] = \lambda X.
$$

Let $S = [X \ X'] \in \mathsf{M}_n$ be invertible (see Theorem 3.1.22). Then

$$
[S^{-1}X \ \ S^{-1}X'] = S^{-1}S = I_n = \begin{bmatrix} I_k & 0\\ 0 & I_{n-k} \end{bmatrix},
$$

so $S^{-1}X = \begin{bmatrix} I_k\\ 0 \end{bmatrix}$. Compute

$$
S^{-1}AS = S^{-1}[AX \ AX'] = S^{-1}[\lambda X \ AX'] = [\lambda S^{-1}X \ \ S^{-1}AX'] = \begin{bmatrix} \lambda I_k & \star\\ 0 & C \end{bmatrix}, \tag{10.3.3}
$$

in which $C \in \mathsf{M}_{n-k}$. Since similar matrices have the same characteristic polynomial, Theorems 10.2.11 and 3.3.4 ensure that $p_A(z) = p_{S^{-1}AS}(z) = p_{\lambda I_k}(z)p_C(z) = (z - \lambda)^k p_C(z)$. Consequently, λ is a root of $p_A(z) = 0$ with multiplicity at least k. $\qquad\square$

10.4 Diagonalization and Eigenvalue Multiplicities

What is special about a matrix if every eigenvalue has equal geometric and algebraic multiplicities?

Lemma 10.4.1 *Let $A \in \mathsf{M}_n$ and suppose that for each $\lambda \in \operatorname{spec} A$, the algebraic and geometric multiplicities of λ are equal. Then $\mathbb{C}^n$ has a basis consisting of eigenvectors of A.*

Proof Let $\mu_1, \mu_2, \ldots, \mu_d$ be the distinct eigenvalues of A and let $n_1, n_2, \ldots, n_d$ be their respective algebraic multiplicities. Then $n_1 + n_2 + \cdots + n_d = n$, and the assumption of equal algebraic and geometric multiplicities ensures that each eigenspace $\mathcal{E}_{\mu_i}(A)$ has dimension n_i. For each $i = 1, 2, \ldots, d$, let the columns of $X_i \in \mathsf{M}_{n \times n_i}$ be a basis for $\mathcal{E}_{\mu_i}(A)$. Since $AX_i = \mu_i X_i$, we have $AX_i \mathbf{y}_i = \mu_i X_i \mathbf{y}_i$ for any $\mathbf{y}_i \in \mathbb{C}^{n_i}$. Moreover, X_i has full column rank, so $X_i \mathbf{y}_i = \mathbf{0}$ only if $\mathbf{y}_i = \mathbf{0}$; that is, $(\mu_i, X_i \mathbf{y}_i)$ is an eigenpair of A whenever $\mathbf{y}_i \neq \mathbf{0}$.

Let $X = [X_1 \ X_2 \ \ldots \ X_d] \in \mathsf{M}_n$. Every column of X is an eigenvector of A. We claim that $\operatorname{rank} X = n$, which ensures that the columns of X are a basis for $\mathbb{C}^n$. Let $\mathbf{y} \in \mathbb{C}^n$ and suppose that $X\mathbf{y} = \mathbf{0}$. We must show that $\mathbf{y} = \mathbf{0}$. Partition $\mathbf{y} = [\mathbf{y}_1^\mathsf{T} \ \mathbf{y}_2^\mathsf{T} \ \ldots \ \mathbf{y}_d^\mathsf{T}]^\mathsf{T}$ conformally with X. Then $X\mathbf{y}_i \in \mathcal{E}_{\mu_i}(A)$ for each $i = 1, 2, \ldots, d$ and

$$\mathbf{0} = X\mathbf{y} = X_1 \mathbf{y}_1 + X_2 \mathbf{y}_2 + \cdots + X_d \mathbf{y}_d. \tag{10.4.2}$$

If some $X_i \mathbf{y}_i$ is nonzero, then (10.4.2) expresses $\mathbf{0}$ as a nontrivial linear combination of eigenvectors of A corresponding to distinct eigenvalues. This contradicts Theorem 9.3.5, so each $X_i \mathbf{y}_i = \mathbf{0}$, and hence each $\mathbf{y}_i = \mathbf{0}$. Thus, $\mathbf{y} = \mathbf{0}$. $\qquad\qquad\qquad \square$

Definition 10.4.3 An eigenvalue is *semisimple* if its geometric and algebraic multiplicities are equal.

The preceding lemma says that if every eigenvalue of A is semisimple, then $\mathbb{C}^n$ has a basis consisting of eigenvectors of A. This is a generalization of Corollary 9.3.6 since a simple eigenvalue is semisimple.

We can now characterize the matrices $A \in \mathsf{M}_n$ such that $\mathbb{C}^n$ has a basis consisting of eigenvectors of A: they are similar to diagonal matrices.

Theorem 10.4.4 *Let $A \in \mathsf{M}_n$. Then $\mathbb{C}^n$ has a basis consisting of eigenvectors of A if and only if there are $S, \Lambda \in \mathsf{M}_n$ such that S is invertible, Λ is diagonal, and*

$$A = S\Lambda S^{-1}. \tag{10.4.5}$$

Proof Suppose that $\mathbf{s}_1, \mathbf{s}_2, \ldots, \mathbf{s}_n$ is a basis for $\mathbb{C}^n$ and $A\mathbf{s}_j = \lambda_j \mathbf{s}_j$ for each $j = 1, 2, \ldots, n$. Let $\Lambda = \operatorname{diag}(\lambda_1, \lambda_2, \ldots, \lambda_n)$ and $S = [\mathbf{s}_1 \ \mathbf{s}_2 \ \ldots \ \mathbf{s}_n]$. Then S is invertible and

$$AS = [A\mathbf{s}_1 \ A\mathbf{s}_2 \ \ldots \ A\mathbf{s}_n] = [\lambda_1 \mathbf{s}_1 \ \lambda_2 \mathbf{s}_2 \ \ldots \ \lambda_n \mathbf{s}_n] = [\mathbf{s}_1 \ \mathbf{s}_2 \ \ldots \ \mathbf{s}_n] \begin{bmatrix} \lambda_1 & & 0 \\ & \ddots & \\ 0 & & \lambda_n \end{bmatrix} = S\Lambda.$$

Thus, $A = S\Lambda S^{-1}$. Conversely, if $S = [\mathbf{s}_1 \ \mathbf{s}_2 \ \ldots \ \mathbf{s}_n]$ is invertible, $\Lambda = \operatorname{diag}(\lambda_1, \lambda_2, \ldots, \lambda_n)$, and $A = S\Lambda S^{-1}$, then the columns of S are a basis for $\mathbb{C}^n$ and

$$[A\mathbf{s}_1 \ A\mathbf{s}_2 \ \ldots \ A\mathbf{s}_n] = AS = S\Lambda = [\lambda_1 \mathbf{s}_1 \ \lambda_2 \mathbf{s}_2 \ \ldots \ \lambda_n \mathbf{s}_n].$$

Thus, $A\mathbf{s}_j = \lambda_j \mathbf{s}_j$ for each $j = 1, 2, \ldots, n$. Since S has no zero columns, $\mathbf{s}_j$ is an eigenvector of A for each $j = 1, 2, \ldots, n$, so $\mathbb{C}^n$ has a basis of eigenvectors of A. $\qquad \square$

Definition 10.4.6 A square matrix is *diagonalizable* if it is similar to a diagonal matrix, that is, if it can be factored as in (10.4.5).

Example 10.4.7 A diagonal matrix $\Lambda \in \mathsf{M}_n$ is diagonalizable since $\Lambda = I\Lambda I^{-1}$ is diagonal.

Example 10.4.8 $A = \begin{bmatrix} 1 & 1 \\ 1 & 1 \end{bmatrix}$ is diagonalizable since

$$A = \underbrace{\begin{bmatrix} 1 & 1 \\ 1 & -1 \end{bmatrix}}_{S} \underbrace{\begin{bmatrix} 2 & 0 \\ 0 & 0 \end{bmatrix}}_{\Lambda} \underbrace{\begin{bmatrix} \frac{1}{2} & \frac{1}{2} \\ \frac{1}{2} & -\frac{1}{2} \end{bmatrix}}_{S^{-1}}$$

and Λ is diagonal.

Corollary 10.4.9 *If $A \in \mathsf{M}_n$ has distinct eigenvalues, then it is diagonalizable.*

Proof If A has distinct eigenvalues, Corollary 9.3.6 ensures that $\mathbb{C}^n$ has a basis of eigenvectors of A. Theorem 10.4.4 says that A is similar to a diagonal matrix. $\quad\square$

Not every matrix is diagonalizable (Example 10.4.11). One way to produce examples of diagonalizable matrices is via the following theorem, which provides a converse to Lemma 10.4.1. In the proof, we use the facts that (a) similar matrices have the same rank, and hence the same nullity, and (b) the nullity of a diagonal matrix is the number of zeros that appear along its diagonal.

Theorem 10.4.10 *Let $A \in \mathsf{M}_n$. Then A is diagonalizable if and only if each $\lambda \in \mathrm{spec}\, A$ is semisimple.*

Proof If each eigenvalue of A is semisimple, Lemma 10.4.1 and Theorem 10.4.4 ensure that A is diagonalizable. Conversely, suppose that $A = S\Lambda S^{-1}$, in which S is invertible and $\Lambda = \mathrm{diag}(\lambda_1, \lambda_2, \ldots, \lambda_n)$. The geometric multiplicity of λ is

$$
\begin{aligned}
\dim \mathcal{E}_\lambda(A) &= \mathrm{nullity}(A - \lambda I) \\
&= \mathrm{nullity}(S\Lambda S^{-1} - \lambda SIS^{-1}) \\
&= \mathrm{nullity}(S(\Lambda - \lambda I)S^{-1}) \\
&= \mathrm{nullity}(\Lambda - \lambda I) \qquad\qquad\qquad \text{(Theorem 4.2.1)} \\
&= \mathrm{nullity}\, \mathrm{diag}(\lambda_1 - \lambda, \lambda_2 - \lambda, \ldots, \lambda_n - \lambda) \\
&= \text{the number of eigenvalues of } A \text{ that are equal to } \lambda \\
&= \text{algebraic multiplicity of } \lambda. \qquad\qquad\qquad\qquad\qquad\square
\end{aligned}
$$

Example 10.4.11 The matrix A in Example 9.1.10 is not diagonalizable. Its eigenvalue $\lambda = 1$ has geometric multiplicity 1 and algebraic multiplicity 2. Alternatively, one could argue as follows: If A were diagonalizable and $A = S\Lambda S^{-1}$, in which Λ is diagonal, then since the algebraic multiplicities of $\lambda = 1$ for the similar matrices A and Λ are the same, we would have $\Lambda = I$ and $A = S\Lambda S^{-1} = SIS^{-1} = SS^{-1} = I$, which is not the case.

Corollary 10.4.12 *Suppose that $A \in \mathsf{M}_n$ is diagonalizable, let $\lambda_1, \lambda_2, \ldots, \lambda_n$ be a list of its eigenvalues in any given order, and let $D = \mathrm{diag}(\lambda_1, \lambda_2, \ldots, \lambda_n)$. Then there is an invertible $R \in \mathsf{M}_n$ such that $A = RDR^{-1}$.*

Proof The hypothesis is that $A = S\Lambda S^{-1}$, in which S is invertible and Λ is a diagonal matrix. Moreover, there is a permutation matrix P such that $\Lambda = PDP^{-1}$; see (7.3.4). Then $A = S\Lambda S^{-1} = SPDP^{-1}S^{-1} = (SP)D(SP)^{-1}$. Let $R = SP$. $\quad\square$

Suppose that $A \in \mathsf{M}_n$ is diagonalizable. If A has only one eigenvalue λ, then there is an invertible $S \in \mathsf{M}_n$ such that $A = S(\lambda I)S^{-1} = \lambda SS^{-1} = \lambda I$, so A is a scalar matrix. If A has d distinct eigenvalues $\mu_1, \mu_2, \ldots, \mu_d$ with respective multiplicities $n_1, n_2, \ldots, n_d$, then Corollary 10.4.12 ensures that there is an invertible $S \in \mathsf{M}_n$ such that $A = S\Lambda S^{-1}$ and $\Lambda = \mu_1 I_{n_1} \oplus \mu_2 I_{n_2} \oplus \cdots \oplus \mu_d I_{n_d}$. Thus, we may group equal eigenvalues together as diagonal entries of Λ.

Theorem 10.4.13 *Let* $A = A_1 \oplus A_2 \oplus \cdots \oplus A_k \in \mathsf{M}_n$, *in which each* $A_i \in \mathsf{M}_{n_i}$. *Then* A *is diagonalizable if and only if each direct summand* A_i *is diagonalizable.*

Proof If each A_i is diagonalizable, then there are invertible matrices $R_i \in \mathsf{M}_{n_i}$ and diagonal matrices $\Lambda_i \in \mathsf{M}_{n_i}$ such that $A_i = R_i \Lambda_i R_i^{-1}$. Let $R = R_1 \oplus R_2 \oplus \cdots \oplus R_k$ and $\Lambda = \Lambda_1 \oplus \Lambda_2 \oplus \cdots \oplus \Lambda_k$. Then $A = R\Lambda R^{-1}$.

Conversely, if A is diagonalizable, let $A = S\Lambda S^{-1}$ as in (10.4.5) and partition

$$S = \begin{bmatrix} S_1 \\ S_2 \\ \vdots \\ S_k \end{bmatrix} \in \mathsf{M}_n, \qquad \text{in which } S_i \in \mathsf{M}_{n_i \times n} \text{ for each } i = 1, 2, \ldots, k.$$

Then S has linearly independent rows (because it is invertible), so each S_i has linearly independent rows, that is, $\operatorname{rank} S_i = n_i$ for each $i = 1, 2, \ldots, k$. Compute

$$\begin{bmatrix} A_1 S_1 \\ A_2 S_2 \\ \vdots \\ A_k S_k \end{bmatrix} = AS = S\Lambda = \begin{bmatrix} S_1 \Lambda \\ S_2 \Lambda \\ \vdots \\ S_k \Lambda \end{bmatrix}.$$

Each identity $A_i S_i = S_i \Lambda$ says that every nonzero column of S_i is an eigenvector of A_i. Since $\operatorname{rank} S_i = n_i$, each S_i has n_i linearly independent columns. Thus, each $A_i \in \mathsf{M}_{n_i}$ has n_i linearly independent eigenvectors, so $\mathbb{C}^{n_i}$ has a basis of eigenvectors of A_i. Theorem 10.4.4 ensures that A_i is diagonalizable. $\square$

Similar matrices have the same rank, and the rank of a diagonal matrix is the number of nonzero entries on its diagonal. These observations are synthesized in the following theorem.

Theorem 10.4.14 *If* $A \in \mathsf{M}_n$ *is diagonalizable, then the number of nonzero eigenvalues of* A *equals its rank.*

Proof According to the rank-nullity theorem, it suffices to show that nullity A (the geometric multiplicity of the eigenvalue $\lambda = 0$ of A) equals the number of zero eigenvalues of A (the algebraic multiplicity of $\lambda = 0$). But every eigenvalue (zero or not) of a diagonalizable matrix has equal geometric and algebraic multiplicities; see Theorem 10.4.10. $\square$

We have learned that commuting matrices share a common eigenvector; see Corollary 9.5.4. If those commuting matrices are diagonalizable, we can make a stronger statement: they share a common basis of eigenvectors.

Definition 10.4.15 A nonempty set of matrices $\mathscr{F} \subseteq \mathsf{M}_n$ is *simultaneously diagonalizable* if there is an invertible $S \in \mathsf{M}_n$ such that $S^{-1}AS$ is diagonal for each $A \in \mathscr{F}$.

The simultaneous diagonalizability of A and B means that there is a basis for $\mathbb{C}^n$ with respect to which each basis vector is an eigenvector of A and B.

Lemma 10.4.16 *Let* $A, B, S, X, Y \in \mathsf{M}_n$. *Let* S *be invertible and suppose that* $A = SXS^{-1}$ *and* $B = SYS^{-1}$. *Then* $AB = BA$ *if and only if* $XY = YX$.

Proof $AB = (SXS^{-1})(SYS^{-1}) = SXYS^{-1}$ and $BA = (SYS^{-1})(SXS^{-1}) = SYXS^{-1}$. Thus $AB = BA$ if and only if $SXYS^{-1} = SYXS^{-1}$, which is equivalent to $XY = YX$. □

Theorem 10.4.17 *Let* $\mathscr{F} \subseteq \mathsf{M}_n$ *be a nonempty set of diagonalizable matrices. Then* $\mathscr{F}$ *is simultaneously diagonalizable if and only if* $AB = BA$ *for all* $A, B \in \mathscr{F}$.

Proof Suppose that $S \in \mathsf{M}_n$ is invertible and $S^{-1}AS$ is diagonal for every $A \in \mathscr{F}$. Then $S^{-1}AS$ commutes with $S^{-1}BS$ for every $A, B \in \mathscr{F}$ since both matrices are diagonal. The preceding lemma ensures that $AB = BA$ for every $A, B \in \mathscr{F}$.

Conversely, suppose that $AB = BA$ for all $A, B \in \mathscr{F}$. Proceed by induction on n. If $n = 1$, there is nothing to prove. Suppose that $n \geq 2$ and any nonempty commuting set of diagonalizable matrices of size at most $n - 1$ is simultaneously diagonalizable. If every matrix in $\mathscr{F}$ is a scalar matrix, there is nothing to prove. If not, let $\mu_1, \mu_2, \ldots, \mu_d$ be the distinct eigenvalues of some $A \in \mathscr{F}$ with $d \geq 2$ and respective multiplicities $n_1, n_2, \ldots, n_d$. Let $S \in \mathsf{M}_n$ be invertible and such that $A = S\Lambda S^{-1}$ with $\Lambda = \mu_1 I_{n_1} \oplus \mu_2 I_{n_2} \oplus \cdots \oplus \mu_d I_{n_d}$. Since A commutes with every $B \in \mathscr{F}$, Lemma 10.4.16 ensures that Λ commutes with $S^{-1}BS$. Lemma 3.2.17.b tells us that $SBS^{-1} = B_1 \oplus B_2 \oplus \cdots \oplus B_d$ is block diagonal and conformal with Λ. For each $i = 1, 2, \ldots, d$, let $\mathscr{F}_i = \{B_i : B \in \mathscr{F}\} \subseteq \mathsf{M}_{n_i}$. The matrices in $\mathscr{F}_i$ commute, and Theorem 10.4.13 ensures that they are diagonalizable. The induction hypothesis is that they are simultaneously diagonalizable, so for each $i = 1, 2, \ldots, d$ there is an invertible $R_i \in \mathsf{M}_{n_i}$ such that $R_i B_i R_i^{-1}$ is diagonal for all $B_i \in \mathscr{F}_i$. Let $R = R_1 \oplus R_2 \oplus \cdots \oplus R_d$. Then $(RS)B(RS)^{-1}$ is diagonal for all $B \in \mathscr{F}$. □

10.5 The Functional Calculus for Diagonalizable Matrices

The next theorem states the *polynomial functional calculus* for diagonalizable matrices.

Theorem 10.5.1 *Let* $A \in \mathsf{M}_n$ *be diagonalizable and write* $A = S\Lambda S^{-1}$, *in which* $S \in \mathsf{M}_n$ *is invertible and* $\Lambda = \mathrm{diag}(\lambda_1, \lambda_2, \ldots, \lambda_n)$. *If* p *is a polynomial, then*

$$p(A) = S\,\mathrm{diag}\big(p(\lambda_1), p(\lambda_2), \ldots, p(\lambda_n)\big)S^{-1}. \tag{10.5.2}$$

Proof Theorem 2.6.9 ensures that $p(A) = Sp(\Lambda)S^{-1}$ and

$$p(\Lambda) = \mathrm{diag}(p(\lambda_1), p(\lambda_2), \ldots, p(\lambda_n)). \qquad □$$

Example 10.5.3 The eigenvalues of $A = \begin{bmatrix} 1 & 2 \\ 2 & 4 \end{bmatrix}$ are 0 and 5, with corresponding eigenvectors $\begin{bmatrix} -2 \\ 1 \end{bmatrix}$ and $\begin{bmatrix} 1 \\ 2 \end{bmatrix}$. Let $\Lambda = \mathrm{diag}(0, 5)$ and $S = \begin{bmatrix} -2 & 1 \\ 1 & 2 \end{bmatrix}$. Then $A = S\Lambda S^{-1}$, so for each $k = 2, 3, \ldots,$

$$A^k = S\Lambda^k S^{-1} = \frac{1}{5}\begin{bmatrix} -2 & 1 \\ 1 & 2 \end{bmatrix}\begin{bmatrix} 0 & 0 \\ 0 & 5^k \end{bmatrix}\begin{bmatrix} -2 & 1 \\ 1 & 2 \end{bmatrix} = \frac{1}{5}\begin{bmatrix} 5^k & 2 \cdot 5^k \\ 2 \cdot 5^k & 4 \cdot 5^k \end{bmatrix} = 5^{k-1}A.$$

Example 10.5.4 Consider the following real symmetric matrix and a few of its powers:

$$A = \begin{bmatrix} 1 & 1 \\ 1 & 0 \end{bmatrix}, \quad A^2 = \begin{bmatrix} 2 & 1 \\ 1 & 1 \end{bmatrix}, \quad A^3 = \begin{bmatrix} 3 & 2 \\ 2 & 1 \end{bmatrix}, \quad A^4 = \begin{bmatrix} 5 & 3 \\ 3 & 2 \end{bmatrix}, \quad \text{and} \quad A^{-1} = \begin{bmatrix} 0 & 1 \\ 1 & -1 \end{bmatrix}.$$

Let f_k denote the $(1, 1)$ entry of A^{k-1} for $k = 0, 1, 2, \ldots$. We have $f_0 = 0, f_1 = 1, f_2 = 1$, and

$$A^{k-1} = \begin{bmatrix} f_k & f_{k-1} \\ f_{k-1} & f_{k-2} \end{bmatrix} \tag{10.5.5}$$

for $k = 2, 3, 4, 5$. Since

$$\begin{bmatrix} f_{k+1} & f_k \\ f_k & f_{k-1} \end{bmatrix} = A^k = AA^{k-1} = \begin{bmatrix} 1 & 1 \\ 1 & 0 \end{bmatrix} \begin{bmatrix} f_k & f_{k-1} \\ f_{k-1} & f_{k-2} \end{bmatrix} = \begin{bmatrix} f_k + f_{k-1} & f_{k-1} + f_{k-2} \\ f_k & f_{k-1} \end{bmatrix},$$

we see that $f_0 = 0, f_1 = 1, f_2 = 1$, and $f_{k+1} = f_k + f_{k-1}$ for $k = 2, 3, \ldots$, which is the recurrence relation that defines the *Fibonacci numbers*

$$0, \ 1, \ 1, \ 2, \ 3, \ 5, \ 8, \ 13, \ 21, \ 34, \ 55, \ldots.$$

The characteristic polynomial of A is $p_A(z) = z^2 - z - 1$, so the eigenvalues of A are $\lambda_\pm = \frac{1}{2}(1 \pm \sqrt{5})$. Corollary 10.4.9 ensures that A is diagonalizable. The eigenpair equation $A\mathbf{x} = \lambda\mathbf{x}$ is

$$\begin{bmatrix} 1 & 1 \\ 1 & 0 \end{bmatrix} \begin{bmatrix} x_1 \\ x_2 \end{bmatrix} = \begin{bmatrix} x_1 + x_2 \\ x_1 \end{bmatrix} = \begin{bmatrix} \lambda_\pm x_1 \\ \lambda_\pm x_2 \end{bmatrix}.$$

It has solutions $\mathbf{x}_\pm = [\lambda_\pm \ 1]^\mathsf{T}$, so (10.4.5) tells us that $A = S\Lambda S^{-1}$, in which

$$S = \begin{bmatrix} \lambda_+ & \lambda_- \\ 1 & 1 \end{bmatrix}, \quad \Lambda = \begin{bmatrix} \lambda_+ & 0 \\ 0 & \lambda_- \end{bmatrix}, \quad \text{and} \quad S^{-1} = \frac{1}{\sqrt{5}} \begin{bmatrix} 1 & -\lambda_- \\ -1 & \lambda_+ \end{bmatrix}.$$

Theorem 10.5.1 shows how to express powers of A as a function of powers of Λ:

$$\begin{bmatrix} f_k & \star \\ \star & \star \end{bmatrix} = A^{k-1} = S\Lambda^{k-1}S^{-1}$$

$$= \frac{1}{\sqrt{5}} \begin{bmatrix} \lambda_+ & \lambda_- \\ 1 & 1 \end{bmatrix} \begin{bmatrix} \lambda_+^{k-1} & 0 \\ 0 & \lambda_-^{k-1} \end{bmatrix} \begin{bmatrix} 1 & -\lambda_- \\ -1 & \lambda_+ \end{bmatrix}$$

$$= \frac{1}{\sqrt{5}} \begin{bmatrix} \lambda_+ & \lambda_- \\ \star & \star \end{bmatrix} \begin{bmatrix} \lambda_+^{k-1} & \star \\ -\lambda_-^{k-1} & \star \end{bmatrix}$$

$$= \begin{bmatrix} \frac{1}{\sqrt{5}}(\lambda_+^k - \lambda_-^k) & \star \\ \star & \star \end{bmatrix}.$$

This identity reveals *Binet's formula* for the Fibonacci numbers:

$$f_k = \frac{1}{\sqrt{5}} \left(\left(\frac{1 + \sqrt{5}}{2} \right)^k - \left(\frac{1 - \sqrt{5}}{2} \right)^k \right) \qquad \text{for } k = 0, 1, 2, \ldots. \tag{10.5.6}$$

The number $\phi = \frac{1}{2}(1 + \sqrt{5}) = 1.618\ldots$ is the *golden ratio*.

For diagonalizable matrices it is possible to develop a broader functional calculus, in which expressions such as $\sin A$, $\cos A$, and e^A are unambiguously defined.

Let $A \in M_n$ be diagonalizable and write $A = S\Lambda S^{-1}$, in which $S \in M_n$ is invertible and $\Lambda = \mathrm{diag}(\lambda_1, \lambda_2, \ldots, \lambda_n)$. Let f be a complex-valued function on $\mathrm{spec}\, A$ and define

$$f(\Lambda) = \mathrm{diag}\big(f(\lambda_1), f(\lambda_2), \ldots, f(\lambda_n)\big).$$

The Lagrange interpolation theorem (Theorem 2.7.4) provides an algorithm to construct a polynomial p such that $f(\lambda_i) = p(\lambda_i)$ for each $i = 1, 2, \ldots, n$. Therefore, $f(\Lambda) = p(\Lambda)$. Moreover, if M is a diagonal matrix that is obtained by permuting the diagonal entries of Λ, then $f(M) = p(M)$.

Suppose that A is diagonalized in two ways: $A = S\Lambda S^{-1} = RMR^{-1}$, in which $R, S \in \mathsf{M}_n$ are invertible and $\Lambda, M \in \mathsf{M}_n$ are diagonal. Then Theorem 2.6.9 ensures that

$$S\Lambda S^{-1} = RMR^{-1} \implies (R^{-1}S)\Lambda = M(R^{-1}S) \implies (R^{-1}S)p(\Lambda) = p(M)(R^{-1}S)$$
$$\implies (R^{-1}S)f(\Lambda) = f(M)(R^{-1}S) \implies Sf(\Lambda)S^{-1} = Rf(M)R^{-1}.$$

Thus, we may define

$$f(A) = Sf(\Lambda)S^{-1}. \tag{10.5.7}$$

We obtain the same matrix no matter what diagonalization of A is chosen. Since $f(A)$ is a polynomial in A, Theorem 2.6.9 tells us that $f(A)$ commutes with any matrix that commutes with A. Theorem 10.5.1 ensures that if f is a polynomial, then the definitions of $f(A)$ in (10.5.7) and (9.2.1) are not in conflict.

Example 10.5.8 Suppose that $A \in \mathsf{M}_n$ is diagonalizable and has real eigenvalues. Consider the functions $\sin t$ and $\cos t$ on $\operatorname{spec} A$. Let $A = S\Lambda S^{-1}$, in which $\Lambda = \operatorname{diag}(\lambda_1, \lambda_2, \ldots, \lambda_n)$ is real. Since $\cos^2 \lambda_i + \sin^2 \lambda_i = 1$ for each i, the definition (10.5.7) permits us to compute

$$\cos^2 A + \sin^2 A = S(\cos^2 \Lambda)S^{-1} + S(\sin^2 \Lambda)S^{-1} = S(\cos^2 \Lambda + \sin^2 \Lambda)S^{-1}$$
$$= S\operatorname{diag}(\cos^2 \lambda_1 + \sin^2 \lambda_1, \ldots, \cos^2 \lambda_n + \sin^2 \lambda_n)S^{-1}$$
$$= SIS^{-1} = SS^{-1} = I.$$

Example 10.5.9 Let $A \in \mathsf{M}_n$ be diagonalizable and suppose that $A = S\Lambda S^{-1}$, in which $\Lambda = \operatorname{diag}(\lambda_1, \lambda_2, \ldots, \lambda_n)$. Consider the function $f(z) = e^z$ on $\operatorname{spec} A$. Using the definition (10.5.7), we compute $e^A = Se^\Lambda S^{-1} = S\operatorname{diag}(e^{\lambda_1}, e^{\lambda_2}, \ldots, e^{\lambda_n})S^{-1}$, so Theorem 10.2.10 ensures that

$$\det e^A = (\det S)(e^{\lambda_1}e^{\lambda_2}\cdots e^{\lambda_n})(\det S^{-1}) = e^{\lambda_1 + \lambda_2 + \cdots + \lambda_n} = e^{\operatorname{tr} A}.$$

For an important application of the functional calculus, in which the function is $f(t) = \sqrt{t}$ on $[0, \infty)$, see Section 15.3.

10.6 Commutants

Let $A \in \mathsf{M}_n$. Since $p(A)$ is a polynomial in A, it commutes with A. This observation and Theorem 10.4.17 motivate the following definition.

Definition 10.6.1 Let $\mathscr{F}$ be a nonempty subset of M_n. The *commutant* of $\mathscr{F}$ is the set $\mathscr{F}'$ of all matrices that commute with every element of $\mathscr{F}$, that is, $\mathscr{F}' = \{X \in \mathsf{M}_n : AX = XA \text{ for all } A \in \mathscr{F}\}$.

The commutant of any nonempty subset $\mathscr{F}$ of M_n is a subspace of M_n. Indeed, if $X, Y \in \mathscr{F}'$ and $c \in \mathbb{C}$, then $(X + cY)A = XA + cYA = AX + cAY = A(X + cY)$ for all $A \in \mathscr{F}$ (see also P.10.34).

Theorem 10.6.2 *Let $A \in \mathsf{M}_n$ be diagonalizable and suppose that the distinct eigenvalues of A are $\mu_1, \mu_2, \ldots, \mu_d$ with respective multiplicities $n_1, n_2, \ldots, n_d$. Write $A = S\Lambda S^{-1}$, in which $S \in \mathsf{M}_n$ is invertible and $\Lambda = \mu_1 I_{n_1} \oplus \mu_2 I_{n_2} \oplus \cdots \oplus \mu_d I_{n_d}$.*

(a) $B \in \{A\}'$ if and only if $B = SXS^{-1}$, in which $X = X_{11} \oplus \cdots \oplus X_{dd}$ and each $X_{ii} \in M_{n_i}$.

(b) $\dim\{A\}' = n_1^2 + n_2^2 + \cdots + n_d^2$.

Proof Lemma 10.4.16 ensures that A and B commute if and only if Λ and $S^{-1}BS$ commute. It follows from Lemma 3.2.17 that $S^{-1}BS = X_{11} \oplus X_{22} \oplus \cdots \oplus X_{dd}$, in which $X_{ii} \in M_{n_i}$ (all $X_{ii} \in M_{n_i}$ are possible) for $i = 1, 2, \ldots, d$. Since $\dim M_{n_i} = n_i^2$, the claim in (b) follows. $\quad\square$

Corollary 10.6.3 *Suppose that $A \in M_n$ has distinct eigenvalues.*

(a) $\{A\}' = \{p(A) : p \text{ is a polynomial}\}$.

(b) $\dim\{A\}' = n$.

Proof Since A has distinct eigenvalues, it is diagonalizable (Corollary 10.4.9). Write $A = S\Lambda S^{-1}$, in which $S \in M_n$ is invertible and $\Lambda = \mathrm{diag}(\lambda_1, \lambda_2, \ldots, \lambda_n)$ has distinct diagonal entries.

(a) Theorem 10.6.2 ensures that B commutes with A if and only if $B = SXS^{-1}$, in which $X = \mathrm{diag}(\xi_1, \xi_2, \ldots, \xi_n)$ for some $\xi_1, \xi_2, \ldots, \xi_n \in \mathbb{C}$. The Lagrange interpolation theorem (Theorem 2.7.4) provides a polynomial p such that $p(\lambda_i) = \xi_i$ for $i = 1, 2, \ldots, n$. Thus, $p(A) = Sp(\Lambda)S^{-1} = SXS^{-1} = B$. Conversely, $p(A) \in \{A\}'$ for any polynomial p.

(b) This follows from Theorem 10.6.2.b. $\quad\square$

Example 10.6.4 In Corollary 10.6.3, the hypothesis that A has distinct eigenvalues cannot be omitted. For example, $\{I_n\}' = M_n$, which has dimension n^2. Any matrix that is not a scalar multiple of I_n belongs to $\{I_n\}'$ yet is not a polynomial in I_n.

10.7 The Eigenvalues of *AB* and *BA*

The identity

$$\begin{bmatrix} I_k & 0 \\ X & I_{n-k} \end{bmatrix}^{-1} = \begin{bmatrix} I_k & 0 \\ -X & I_{n-k} \end{bmatrix}, \qquad \text{in which } X \in M_{(n-k) \times k}, \qquad (10.7.1)$$

is a useful companion to (3.2.6). We can use it to clarify the relationship between the eigenvalues of AB and BA. These two products need not be matrices of the same size, and even if they are, they need not be equal. Nevertheless, their nonzero eigenvalues are the same.

Theorem 10.7.2 *Let $A \in M_{m \times n}$, $B \in M_{n \times m}$, and $n \geq m$.*

(a) *The nonzero eigenvalues of $AB \in M_m$ and $BA \in M_n$ are the same, with the same algebraic multiplicities.*

(b) *If 0 is an eigenvalue of AB with algebraic multiplicity $k \geq 0$, then 0 is an eigenvalue of BA with algebraic multiplicity $k + n - m$.*

(c) *If $m = n$, then the eigenvalues of AB and BA are the same, with the same algebraic multiplicities.*

Proof Let $X = \begin{bmatrix} AB & A \\ 0 & 0_n \end{bmatrix}$ and $Y = \begin{bmatrix} 0_m & A \\ 0 & BA \end{bmatrix}$ and consider the following similarity transformation applied to X:

$$\begin{bmatrix} I_m & 0 \\ B & I_n \end{bmatrix} \begin{bmatrix} AB & A \\ 0 & 0_n \end{bmatrix} \begin{bmatrix} I_m & 0 \\ B & I_n \end{bmatrix}^{-1} = \begin{bmatrix} I_m & 0 \\ B & I_n \end{bmatrix} \begin{bmatrix} AB & A \\ 0 & 0_n \end{bmatrix} \begin{bmatrix} I_m & 0 \\ -B & I_n \end{bmatrix}$$

$$= \begin{bmatrix} AB & A \\ BAB & BA \end{bmatrix} \begin{bmatrix} I_m & 0 \\ -B & I_n \end{bmatrix} = \begin{bmatrix} AB - AB & A \\ BAB - BAB & BA \end{bmatrix}$$

$$= \begin{bmatrix} 0_m & A \\ 0 & BA \end{bmatrix} = Y.$$

Because X and Y are similar, $p_X(z) = p_Y(z)$. Since $p_X(z) = p_{AB}(z)p_{0_n}(z) = z^n p_{AB}(z)$ and $p_Y(z) = p_{0_m}(z)p_{BA}(z) = z^m p_{BA}(z)$, it follows that $p_{BA}(z) = z^{n-m} p_{AB}(z)$. Thus, if $\lambda_1, \lambda_2, \ldots, \lambda_m$ are the roots of $p_{AB}(z) = 0$ (the eigenvalues of AB), then the n roots of $p_{BA}(z) = 0$ (the eigenvalues of BA) are $\lambda_1, \lambda_2, \ldots, \lambda_m, 0, 0, \ldots, 0$ ($n - m$ zeros). The nonzero eigenvalues in the two lists are identical, with the same multiplicities. However, if k is the multiplicity of 0 as an eigenvalue of AB, its multiplicity as an eigenvalue of BA is $k + n - m$. $\qquad\square$

Example 10.7.3 Let $A = \begin{bmatrix} 1 & 0 \\ 0 & 0 \end{bmatrix}$ and $B = \begin{bmatrix} 0 & 1 \\ 0 & 0 \end{bmatrix}$. Then $AB = \begin{bmatrix} 0 & 1 \\ 0 & 0 \end{bmatrix}$ and $BA = \begin{bmatrix} 0 & 0 \\ 0 & 0 \end{bmatrix}$. Although zero is an eigenvalue of both AB and BA with algebraic multiplicity 2, it has different geometric multiplicities (1 and 2, respectively). This does not happen for nonzero eigenvalues; see Theorem 13.5.1.

Example 10.7.4 Consider $\mathbf{e} = [1 \ 1 \ \ldots \ 1]^T \in \mathbb{R}^n$. Every entry of the $n \times n$ matrix $\mathbf{e}\mathbf{e}^T$ is 1. What are its eigenvalues? Theorem 10.7.2 tells us that they are the eigenvalue of the 1×1 matrix $\mathbf{e}^T\mathbf{e} = [n]$ (namely, n), together with $n - 1$ zeros.

Example 10.7.5 Let $\mathbf{r} = [1 \ 2 \ \ldots \ n]^T \in \mathbb{R}^n$ and let

$$A = \mathbf{r}\mathbf{e}^T = \begin{bmatrix} 1 & 1 & \cdots & 1 \\ 2 & 2 & \cdots & 2 \\ \vdots & \vdots & & \vdots \\ n & n & \cdots & n \end{bmatrix} \in \mathsf{M}_n. \tag{10.7.6}$$

Theorem 10.7.2 tells us that the eigenvalues of A are $n - 1$ zeros together with $\mathbf{e}^T\mathbf{r} = 1 + 2 + \cdots + n = n(n+1)/2$.

Example 10.7.7 Let A be the matrix (10.7.6) and observe that

$$A + A^T = [i+j] = \begin{bmatrix} 2 & 3 & 4 & \cdots & n+1 \\ 3 & 4 & 5 & \cdots & n+2 \\ 4 & 5 & 6 & \cdots & n+3 \\ \vdots & \vdots & \vdots & \ddots & \vdots \\ n+1 & n+2 & n+3 & \cdots & 2n \end{bmatrix} = \mathbf{r}\mathbf{e}^T + \mathbf{e}\mathbf{r}^T = [\mathbf{r} \ \mathbf{e}]\begin{bmatrix} \mathbf{e}^T \\ \mathbf{r}^T \end{bmatrix}. \tag{10.7.8}$$

Theorem 10.7.2 tells us that the eigenvalues of $A + A^T$ are the two eigenvalues of

$$\begin{bmatrix} \mathbf{e}^T \\ \mathbf{r}^T \end{bmatrix}[\mathbf{r} \ \mathbf{e}] = \begin{bmatrix} \mathbf{e}^T\mathbf{r} & \mathbf{e}^T\mathbf{e} \\ \mathbf{r}^T\mathbf{r} & \mathbf{r}^T\mathbf{e} \end{bmatrix} \tag{10.7.9}$$

together with $n - 2$ zeros; see P.10.24.

The three preceding examples suggest a strategy for finding eigenvalues. If $A \in M_n$, rank $A \leq r$, and $A = XY^\mathsf{T}$ for some $X, Y \in M_{n \times r}$, then the n eigenvalues of A are the r eigenvalues of $Y^\mathsf{T}X \in M_r$, together with $n - r$ zeros. If r is much smaller than n, one might prefer to compute the eigenvalues of $Y^\mathsf{T}X$ instead of the eigenvalues of A. For example, $A = XY^\mathsf{T}$ could be a full-rank factorization.

10.8 Problems

Remember: The "multiplicity" of an eigenvalue, with no modifier, means algebraic multiplicity. Two matrices have the same spectrum if their distinct eigenvalues are the same; they have the same eigenvalues if their distinct eigenvalues are the same with the same algebraic multiplicities.

P.10.1 Find the characteristic polynomials, eigenvalues, and eigenspaces of (a) $\begin{bmatrix} 3 & -2 \\ 1 & 0 \end{bmatrix}$ and (b) $\begin{bmatrix} 2 & -1 \\ 1 & 0 \end{bmatrix}$.

P.10.2 Find the characteristic polynomials, eigenvalues, and eigenspaces of (a) $\begin{bmatrix} 2 & 2i \\ i & -1 \end{bmatrix}$ and (b) $\begin{bmatrix} 2 & 1 & 1 \\ -4 & -3 & 0 \\ -2 & -2 & 1 \end{bmatrix}$.

P.10.3 Show that $\begin{bmatrix} 1 & 0 \\ 0 & 1 \end{bmatrix}$ and $\begin{bmatrix} 1 & 1 \\ 0 & 1 \end{bmatrix}$ have the same rank, trace, determinant, characteristic polynomial, and eigenvalues, but are not similar.

P.10.4 Show that $\begin{bmatrix} 1 & 10^{20} \\ 0 & 2 \end{bmatrix}$ and $\begin{bmatrix} 3 & -2 \\ 1 & 0 \end{bmatrix}$ are similar.

P.10.5 (a) Provide details for the assertions in Example 10.5.3. (b) Compute A^3 with two matrix multiplications and verify that it is $25A$.

P.10.6 Find an $A \in M_3$ that has eigenvalues 0, 1, and -1 and associated eigenvectors $[0 \ 1 \ -1]^\mathsf{T}$, $[1 \ -1 \ 1]^\mathsf{T}$, and $[0 \ 1 \ 1]^\mathsf{T}$, respectively.

P.10.7 Suppose that $A \in M_5$ has eigenvalues $-4, -1, 0, 1, 4$. Is there a $B \in M_5$ such that $B^2 = A$? Justify your answer.

P.10.8 Let $\lambda_1, \lambda_2, \ldots, \lambda_n$ be the scalars in (10.2.1) and let $\mu_1, \mu_2, \ldots, \mu_d$ be the scalars in (10.2.2). Explain why $\{\lambda_1, \lambda_2, \ldots, \lambda_n\} = \{\mu_1, \mu_2, \ldots, \mu_d\}$.

P.10.9 Let $A, B \in M_n$. (a) If spec $A =$ spec B, do A and B have the same characteristic polynomials? Why? (b) If A and B have the same characteristic polynomials, is spec $A =$ spec B? Why?

P.10.10 Review the statement of Theorem 10.4.14. (a) If $A \in M_n$ and rank A is the number of nonzero eigenvalues of A, is A diagonalizable? Why? (b) What is the rank of $\begin{bmatrix} 0 & 1 \\ 0 & 0 \end{bmatrix}$? How many nonzero eigenvalues does it have? Is it diagonalizable?

P.10.11 Let $n \geq 2$. (a) Give an example of a nonzero nilpotent matrix in M_n. (b) Show that the eigenvalues of an $n \times n$ nilpotent matrix are $0, 0, \ldots, 0$. (c) What is the characteristic polynomial of a nilpotent matrix?

P.10.12 If $A \in M_n$ is diagonalizable, show that A is nilpotent if and only if $A = 0$.

P.10.13 If $A \in M_n$ is similar to cA for every nonzero $c \in \mathbb{C}$, show that spec $A = 0$.

P.10.14 Let $\lambda_1, \lambda_2, \ldots, \lambda_n$ be the eigenvalues of $A \in M_n$ and let $c \in \mathbb{C}$. Show that $p_{A+cI}(z) = p_A(z - c)$ and deduce that the eigenvalues of $A + cI$ are $\lambda_1 + c, \lambda_2 + c, \ldots, \lambda_n + c$.

P.10.15 The eigenvalues of $A \in M_2$ are $\lambda_\pm = \frac{1}{2}(\operatorname{tr} A \pm \sqrt{s})$, in which $s = (\operatorname{tr} A)^2 - 4 \det A$ is the *discriminant* of A. If A has real entries, show that its eigenvalues are real if and only if its discriminant is nonnegative.

P.10.16 Verify that the asserted values of $\lambda_\pm$, $\mathbf{x}_\pm$, S, Λ, and S^{-1} in Example 10.5.4 are correct, and show that the kth Fibonacci number is $f_k = \frac{1}{\sqrt{5}}\left(\phi^k + (-1)^{k+1}\phi^{-k}\right)$ for $k = 0, 1, 2, \ldots$. *Remark*: Since $\phi \approx 1.6180$, $\phi^{-k} \to 0$ as $k \to \infty$ and hence $\frac{1}{\sqrt{5}}\phi^k$ is a good approximation to f_k for large k. For example, $f_{10} = 55$, $\frac{1}{\sqrt{5}}\phi^{10} \approx 55.004$, and $f_{11} = 89$, $\frac{1}{\sqrt{5}}\phi^{11} \approx 88.998$.

P.10.17 Let $A, B \in M_n$ and let $C = \begin{bmatrix} 0 & A \\ B & 0 \end{bmatrix} \in M_{2n}$. (a) Use the identities in P.3.8 to show that $p_C(z) = \det(zI_{2n} - C) = p_{AB}(z^2) = p_{BA}(z^2)$. (b) Deduce from this identity that AB and BA have the same eigenvalues. This is the square case of Theorem 10.7.2. (c) If λ is an eigenvalue of C with multiplicity k, why is $-\lambda$ also an eigenvalue of C with multiplicity k? If zero is an eigenvalue of C, why must it have even multiplicity? (d) Let $\pm \lambda_1, \pm \lambda_2, \ldots, \pm \lambda_n$ be the $2n$ eigenvalues of C. Explain why the n eigenvalues of AB are $\lambda_1^2, \lambda_2^2, \ldots, \lambda_n^2$. (e) Let $\mu_1, \mu_2, \ldots, \mu_n$ be the eigenvalues of AB. Show that the eigenvalues of C are $\pm\sqrt{\mu_1}, \pm\sqrt{\mu_2}, \ldots, \pm\sqrt{\mu_n}$. (f) Show that $\det C = (-1)^n (\det A)(\det B)$.

P.10.18 Let $A \in M_n$. Let $X = \operatorname{Re} A$ and $Y = \operatorname{Im} A$, so $X, Y \in M_n(\mathbb{R})$ and $A = X + iY$. Let $V = \frac{1}{\sqrt{2}}\begin{bmatrix} -iI_n & -iI_n \\ I_n & -I_n \end{bmatrix} \in M_{2n}$ and $C = \begin{bmatrix} X & Y \\ Y & -X \end{bmatrix} \in M_{2n}(\mathbb{R})$. (a) Show that V is unitary and $V^* C V = \begin{bmatrix} 0 & \overline{A} \\ A & 0 \end{bmatrix}$. (b) Show that the characteristic polynomials of C, $A\overline{A}$, and $\overline{A}A$ satisfy $p_C(z) = p_{A\overline{A}}(z^2) = p_{\overline{A}A}(z^2)$. (c) If λ is an eigenvalue of C with multiplicity k, why are $-\lambda$, $\overline{\lambda}$, and $-\overline{\lambda}$ also eigenvalues of C, each with multiplicity k? If zero is an eigenvalue of C, why must it have even multiplicity? (d) If $\pm \lambda_1, \pm \lambda_2, \ldots, \pm \lambda_n$ are the $2n$ eigenvalues of C, explain why the n eigenvalues of $A\overline{A}$ are $\lambda_1^2, \lambda_2^2, \ldots, \lambda_n^2$. Why do the non-real eigenvalues of $A\overline{A}$ occur in conjugate pairs?

P.10.19 For $A \in M_n$, let $X = \operatorname{Re} A$ and $Y = \operatorname{Im} A$, and define $C(A) = \begin{bmatrix} X & -Y \\ Y & X \end{bmatrix}$, which is a matrix of complex type; see P.3.15 and P.3.16. Let $U = \frac{1}{\sqrt{2}}\begin{bmatrix} I_n & iI_n \\ iI_n & I_n \end{bmatrix}$. (a) Show that U is unitary and $U^* C(A) U = \overline{A} \oplus A$. (b) If $\lambda_1, \lambda_2, \ldots, \lambda_n$ are the eigenvalues of A, show that $\lambda_1, \lambda_2, \ldots, \lambda_n, \overline{\lambda}_1, \overline{\lambda}_2, \ldots, \overline{\lambda}_n$ are the eigenvalues of $C(A)$. (c) Why is $\det C(A) \geq 0$? (d) Show that the characteristic polynomials of $C(A)$, A, and $\overline{A}$ satisfy $p_C(z) = p_A(z)p_{\overline{A}}(z)$. (e) What can you say if $n = 1$? (f) Show that $C(A+B) = C(A) + C(B)$ and $C(A)C(B) = C(AB)$. (g) Show that $C(I_n) = I_{2n}$. (h) If A is invertible, explain why the real and imaginary parts of A^{-1} are, respectively, the $(1, 1)$ and $(2, 1)$ blocks of the 2×2 block matrix $C(A)^{-1}$.

P.10.20 Let $A, B \in M_n$ and let $Q = \frac{1}{\sqrt{2}}\begin{bmatrix} I_n & I_n \\ I_n & -I_n \end{bmatrix} \in M_{2n}$. Consider $C = \begin{bmatrix} A & B \\ B & A \end{bmatrix} \in M_{2n}$, which is a block centrosymmetric matrix; see P.3.17 and P.3.18. (a) Show that Q is real orthogonal and $Q^\mathsf{T} C Q = (A + B) \oplus (A - B)$. (b) Show that every eigenvalue of $A + B$ and every eigenvalue of $A - B$ is an eigenvalue of C. What about the converse? (c) Show that $\det C = \det(A^2 - AB + BA - B^2)$. If A and B commute, compare this identity with the identities in P.3.8.

P.10.21 Let $\lambda_1, \lambda_2, \ldots, \lambda_n$ be the eigenvalues of $A \in M_n$. Use the preceding problem to determine the eigenvalues of $C = \begin{bmatrix} A & A \\ A & A \end{bmatrix} \in M_{2n}$. What does P.9.31 say about the eigenvalues of C? Discuss.

P.10.22 Let $A, B \in M_n$ and suppose that either A or B is invertible. Show that AB is similar to BA, and conclude that these two products have the same eigenvalues.

P.10.23 (a) Let $A = \begin{bmatrix} 0 & 1 \\ 0 & 0 \end{bmatrix}$ and $B = \begin{bmatrix} 0 & 0 \\ 1 & 0 \end{bmatrix}$. Is AB similar to BA? Do these two products have the same eigenvalues? (b) Answer the same questions for $A = \begin{bmatrix} 0 & 1 \\ 0 & 0 \end{bmatrix}$ and $B = \begin{bmatrix} 1 & 0 \\ 0 & 0 \end{bmatrix}$.

P.10.24 Verify that the matrix (10.7.9) equals $\begin{bmatrix} \frac{1}{2}n(n+1) & n \\ \frac{1}{6}n(n+1)(2n+1) & \frac{1}{2}n(n+1) \end{bmatrix}$ and that its eigenvalues are $n(n+1)\left(\frac{1}{2} \pm \sqrt{\frac{2n+1}{6(n+1)}} \right)$. What are the eigenvalues of the matrix (10.7.8)?

P.10.25 Use the vectors $\mathbf{e}$ and $\mathbf{r}$ in Example 10.7.7. (a) Verify that

$$A = [i - j] = \begin{bmatrix} 0 & -1 & -2 & \cdots & -n+1 \\ 1 & 0 & -1 & \cdots & -n+2 \\ 2 & 1 & 0 & \cdots & -n+3 \\ \vdots & \vdots & \vdots & \ddots & \vdots \\ n-1 & n-2 & n-3 & \cdots & 0 \end{bmatrix} = \mathbf{r}\mathbf{e}^{\mathsf{T}} - \mathbf{e}\mathbf{r}^{\mathsf{T}} = ZY^{\mathsf{T}},$$

in which $Y = [\mathbf{e} \; \mathbf{r}]$ and $Z = [\mathbf{r} \; -\mathbf{e}]$. (b) Show that the eigenvalues of A are the two eigenvalues of

$$Y^{\mathsf{T}}Z = \begin{bmatrix} \mathbf{e}^{\mathsf{T}}\mathbf{r} & -\mathbf{e}^{\mathsf{T}}\mathbf{e} \\ \mathbf{r}^{\mathsf{T}}\mathbf{r} & -\mathbf{r}^{\mathsf{T}}\mathbf{e} \end{bmatrix} = \begin{bmatrix} \frac{1}{2}n(n+1) & -n \\ \frac{1}{6}n(n+1)(2n+1) & -\frac{1}{2}n(n+1) \end{bmatrix},$$

together with $n - 2$ zeros. (c) Show that the discriminant of $Y^{\mathsf{T}}Z$ is negative (see P.10.15) and explain what this implies about the eigenvalues. (d) Show that the eigenvalues of $Y^{\mathsf{T}}Z$ are $\pm i \frac{n}{2} \sqrt{\frac{n^2-1}{3}}$.

P.10.26 Let $n \geq 3$ and consider

$$A = \begin{bmatrix} a & a & a & \cdots & a \\ a & b & b & \cdots & b \\ \vdots & \vdots & \vdots & \ddots & \vdots \\ a & b & b & \cdots & b \\ a & a & a & \cdots & a \end{bmatrix} \in M_n, \quad B = \begin{bmatrix} 1 & 0 \\ 0 & 1 \\ \vdots & \vdots \\ 0 & 1 \\ 1 & 0 \end{bmatrix} \in M_{n,2}, \quad \text{and} \quad C^{\mathsf{T}} = \begin{bmatrix} a & a \\ a & b \\ \vdots & \vdots \\ a & b \\ a & b \end{bmatrix} \in M_{n,2}.$$

(a) Show that $A = BC$. (b) Show that the eigenvalues of A are $n - 2$ zeros and the eigenvalues of $\begin{bmatrix} 2a & (n-2)a \\ a+b & (n-2)b \end{bmatrix}$. (c) If a and b are real, show that all the eigenvalues of A are real.

P.10.27 Let $n \geq 2$ and let $A = [(i-1)n + j] \in M_n$. If $n = 3$, then

$$A = \begin{bmatrix} 1 & 2 & 3 \\ 4 & 5 & 6 \\ 7 & 8 & 9 \end{bmatrix}.$$

(a) What is A if $n = 4$? (b) Let $\mathbf{v} = [0 \; 1 \; 2 \; \ldots \; n-1]^{\mathsf{T}}$ and $\mathbf{r} = [1 \; 2 \; 3 \; \ldots \; n]^{\mathsf{T}}$. Let $X = [\mathbf{v} \; \mathbf{e}]^{\mathsf{T}}$ and $Y = [n\mathbf{e} \; \mathbf{r}]^{\mathsf{T}}$. Show that $A = XY^{\mathsf{T}}$ and rank $A = 2$. (c) Show that the eigenvalues of A are $n-2$ zeros and the eigenvalues of $\begin{bmatrix} n\mathbf{e}^T\mathbf{v} & n^2 \\ \mathbf{r}^T\mathbf{v} & \mathbf{e}^T\mathbf{r} \end{bmatrix}$. (d) Why are all the eigenvalues of A real?

P.10.28 Use Theorems 10.2.10 and 10.7.2 to prove the Sylvester determinant identity (3.5.5). *Hint:* What are the eigenvalues of $I+AB$ and $I+BA$?

P.10.29 Let $n \geq 2$. Let λ and μ be eigenvalues of $A \in M_n$. Let $(\lambda, \mathbf{x})$ be an eigenpair of A and let $(\overline{\mu}, \mathbf{y})$ be an eigenpair of A^*. (a) Show that $\mathbf{y}^*A = \mu\mathbf{y}^*$. (b) If $\lambda \neq \mu$, prove that $\mathbf{y}^*\mathbf{x} = 0$. This is the *principle of biorthogonality*.

P.10.30 Suppose that $A, B \in M_n$ commute and A has distinct eigenvalues. Use Theorem 9.5.1 to show that B is diagonalizable. Moreover, show that there is an invertible $S \in M_n$ such that $S^{-1}AS$ and $S^{-1}BS$ are diagonal matrices.

P.10.31 Use Lemma 3.2.17 to give another proof of the assertion in the preceding problem.

P.10.32 If $A \in M_n(\mathbb{R})$ and n is odd, show that A has at least one real eigenvalue in two ways: (a) Use the intermediate value theorem from calculus; and (b) use the fact that non-real eigenvalues occur in complex-conjugate pairs.

P.10.33 Let $f(z) = z^n + c_{n-1}z^{n-1} + \cdots + c_1z + c_0$ and let

$$C_f = \begin{bmatrix} 0 & 0 & \cdots & 0 & -c_0 \\ 1 & 0 & \cdots & 0 & -c_1 \\ 0 & 1 & \cdots & 0 & -c_2 \\ \vdots & \vdots & \ddots & \vdots & \vdots \\ 0 & 0 & \cdots & 1 & -c_{n-1} \end{bmatrix}.$$

(a) Show that $p_{C_f} = f$ as follows: Add z times row n of $zI - C_f$ to row $n-1$; then add z times row $n-1$ to row $n-2$. (b) Show that $p_{C_f} = f$ by using induction and (3.3.14). (c) Theorems about location of eigenvalues of a matrix can be used to say something about zeros of a polynomial f. Show that every zero of f is in the disk with radius $\max\{|c_0|, 1+|c_1|, \ldots, 1+|c_{n-1}|\}$ centered at the origin. (d) Show that every zero of f is in the disk with radius $\max\{1, |c_0|+|c_1|+\cdots+|c_{n-1}|\}$ centered at the origin.

P.10.34 Let $\mathcal{F}$ be a nonempty subset of M_n and let $\mathcal{F}'$ be its commutant. If $A, B \in \mathcal{F}'$, show that $AB \in \mathcal{F}'$.

P.10.35 Let $A \in M_n$ be diagonalizable. Show that e^A is invertible and e^{-A} is its inverse.

P.10.36 We say that $B \in M_n$ is a *square root* of $A \in M_n$ if $B^2 = A$. (a) Show that $\begin{bmatrix} 1 & 1 \\ 0 & 1 \end{bmatrix}$ is a square root of $\begin{bmatrix} 1 & 2 \\ 0 & 1 \end{bmatrix}$. (b) Show that $\begin{bmatrix} 0 & 1 \\ 0 & 0 \end{bmatrix}$ does not have a square root. (c) Show that each of the three matrices in (a) and (b) is nondiagonalizable.

P.10.37 Show that every diagonalizable matrix in M_n has a square root in M_n.

P.10.38 Let $A \in M_n$ and assume that $\operatorname{tr} A = 0$. Incorporate the following ideas into a proof that A is a commutator of matrices in M_n (Shoda's theorem; Theorem 4.5.7). (a) Use induction and Lemma 4.5.4 to show that it suffices to consider a block matrix of the form $A = \begin{bmatrix} 0 & \mathbf{x}^* \\ \mathbf{y} & BC-CB \end{bmatrix}$ and $B, C \in M_{n-1}$. (b) Consider $B + \lambda I$ and show that B may be assumed to be invertible. (c) Let $X = \begin{bmatrix} 0 & 0 \\ 0 & B \end{bmatrix}$ and $Y = \begin{bmatrix} 0 & \mathbf{u}^* \\ \mathbf{v} & C \end{bmatrix}$. Show that there are $\mathbf{u}, \mathbf{v} \in \mathbb{C}^{n-1}$ such that $A = XY - YX$.

P.10.39 Let $A \in M_2$ and let f be a complex-valued function on $\operatorname{spec} A = \{\lambda, \mu\}$. (a) If $\lambda \neq \mu$ and $f(A)$ is defined by (10.5.7), show that

$$f(A) = \frac{f(\lambda) - f(\mu)}{\lambda - \mu} A + \frac{\lambda f(\mu) - \mu f(\lambda)}{\lambda - \mu} I. \tag{10.8.1}$$

(b) If $\lambda = \mu$ and A is diagonalizable, why is $f(A) = f(\lambda)I$?

P.10.40 The coefficients c_k of the characteristic polynomial (10.1.4) of $A \in \mathsf{M}_n$ are related to its principal submatrices in the following way: For each $k = 1, 2, \ldots, n$, the sum of the determinants of the $k \times k$ principal submatrices of A is $(-1)^k c_{n-k}$. Inspect the calculations at the beginning of Section 10.1 and verify this assertion for $n = 1, 2, 3$. See [HJ13, (1.2.13)] for the general case.

10.9 Notes

None of the schemes discussed in this chapter should be programmed to compute eigenpairs of real-world data matrices. Different strategies have been found to be better for numerical computations. Improving algorithms for numerical calculation of eigenvalues has been a top-priority research goal for numerical analysts since the mid-twentieth century [GVL13, Chapters 7 and 8].

Suppose that $(\lambda, \mathbf{x})$ is an eigenpair of $A \in \mathsf{M}_n$. Theorem 10.2.9.b ensures that there is a nonzero $\mathbf{y} \in \mathbb{C}^n$ such that $(\overline{\lambda}, \mathbf{y})$ is an eigenpair of A^*, that is, $A^*\mathbf{y} = \overline{\lambda}\mathbf{y}$ or $\mathbf{y}^*A = \lambda\mathbf{y}^*$. The identities $A\mathbf{x} = \lambda\mathbf{x}$ and $\mathbf{y}^*A = \lambda\mathbf{y}^*$ motivate us to call $\mathbf{y}$ a *left eigenvector* of A; $\mathbf{x}$ is a *right eigenvector*. See Theorem 20.2.3 for a result about left and right eigenvectors of a positive matrix.

Example 10.6.4 reveals only part of the story about Corollary 10.6.3.a. The commutant of A is the set of all polynomials in A if and only if every eigenvalue of A has geometric multiplicity 1; see [HJ13, Theorem 3.2.4.2].

It is known that $A\overline{A}$ is similar to the square of a real matrix, which implies results in P.10.18 and a little more: real negative eigenvalues of $A\overline{A}$ have even multiplicity; see [HJ13, Corollaries 4.4.13 and 4.6.16].

P.10.33 suggests two bounds on the zeros of a given polynomial. For many others, see [HJ13, 5.6.P27–35].

If the matrix A in P.10.39 is not diagonalizable and f is differentiable, it turns out that

$$f(A) = f'(\lambda)A + (f(\lambda) - \lambda f'(\lambda))I. \tag{10.9.1}$$

For an explanation of this remarkable formula, see [HJ94, P.11, Section 6.1].

10.10 Some Important Concepts

- Characteristic polynomial of a matrix.
- Algebraic multiplicity of an eigenvalue.
- Trace, determinant, and eigenvalues.
- Diagonalizable matrices and eigenvalue multiplicities.
- Simultaneous diagonalization of commuting matrices.
- Functional calculus for diagonalizable matrices.
- Eigenvalues of AB and BA.

Unitary Triangularization and Block Diagonalization

Many facts about matrices can be revealed (or questions about them answered) by performing a suitable transformation that puts them into a special form. Such a form typically contains many zero entries in strategic locations. For example, Theorem 10.4.4 says that some matrices are similar to diagonal matrices, and Theorem 7.5.3 says that every square matrix is unitarily similar to an upper Hessenberg matrix. In this chapter, we show that every square complex matrix is unitarily similar to an upper triangular matrix. This is a powerful result with a host of important consequences.

11.1 Schur's Triangularization Theorem

Theorem 11.1.1 (Schur Triangularization) *Let the eigenvalues of $A \in M_n$ be arranged in any given order $\lambda_1, \lambda_2, \ldots, \lambda_n$ (including multiplicities), and let $(\lambda_1, \mathbf{x})$ be an eigenpair of A, in which $\mathbf{x}$ is a unit vector.*

(a) *There is a unitary $U \in M_n$ with first column $\mathbf{x}$ such that $A = UTU^*$, in which $T = [t_{ij}]$ is upper triangular and has diagonal entries $t_{ii} = \lambda_i$ for $i = 1, 2, \ldots, n$.*

(b) *If A is real, $\lambda_1, \lambda_2, \ldots, \lambda_n$ are real, and $\mathbf{x}$ is real, then there is a real orthogonal $Q \in M_n(\mathbb{R})$ with first column $\mathbf{x}$ such that $A = QTQ^\mathsf{T}$, in which $T = [t_{ij}]$ is real upper triangular and has diagonal entries $t_{ii} = \lambda_i$ for $i = 1, 2, \ldots, n$.*

Proof (a) We proceed by induction on n. In the base case $n = 1$, there is nothing to prove. For our induction hypothesis, assume that $n \geq 2$ and every matrix in M_{n-1} can be factored as asserted. Suppose that $A \in M_n$ has eigenvalues $\lambda_1, \lambda_2, \ldots, \lambda_n$ and let $\mathbf{x}$ be a unit eigenvector of A associated with the eigenvalue λ_1. Example 7.1.21 shows how to construct a unitary $V = [\mathbf{x} \ V_2] \in M_n$ whose first column is $\mathbf{x}$. Since the columns of V are orthonormal, $V_2^*\mathbf{x} = \mathbf{0}$. Then $AV = [A\mathbf{x} \ AV_2] = [\lambda_1\mathbf{x} \ AV_2]$ and

$$V^*AV = \begin{bmatrix} \mathbf{x}^* \\ V_2^* \end{bmatrix} [\lambda_1 \mathbf{x} \ AV_2] = \begin{bmatrix} \lambda_1 \mathbf{x}^*\mathbf{x} & \mathbf{x}^*AV_2 \\ \lambda_1 V_2^*\mathbf{x} & V_2^*AV_2 \end{bmatrix} = \begin{bmatrix} \lambda_1 & \star \\ \mathbf{0} & A' \end{bmatrix}. \tag{11.1.2}$$

Because A and V^*AV are (unitarily) similar, they have the same eigenvalues. The 1×1 block $[\lambda_1]$ is one of the two diagonal blocks of (11.1.2); its other diagonal block $A' \in M_{n-1}$ has eigenvalues $\lambda_2, \lambda_3, \ldots, \lambda_n$ (see Theorem 10.2.11).

The induction hypothesis ensures that there is a unitary $W \in M_{n-1}$ such that $W^*A'W = T'$, in which $T' = [\tau_{ij}] \in M_{n-1}$ is upper triangular and has diagonal entries $\tau_{ii} = \lambda_{i+1}$ for $i = 1, 2, \ldots, n-1$. Then

$$U = V \begin{bmatrix} 1 & \mathbf{0}^\mathsf{T} \\ \mathbf{0} & W \end{bmatrix}$$

is unitary (it is a product of unitaries), has the same first column as V, and satisfies

$$U^*AU = \begin{bmatrix} 1 & \mathbf{0}^\mathsf{T} \\ \mathbf{0} & W \end{bmatrix}^* V^*AV \begin{bmatrix} 1 & \mathbf{0}^\mathsf{T} \\ \mathbf{0} & W \end{bmatrix} = \begin{bmatrix} 1 & \mathbf{0}^\mathsf{T} \\ \mathbf{0} & W^* \end{bmatrix} \begin{bmatrix} \lambda_1 & \star \\ \mathbf{0} & A' \end{bmatrix} \begin{bmatrix} 1 & \mathbf{0}^\mathsf{T} \\ \mathbf{0} & W \end{bmatrix}$$

$$= \begin{bmatrix} \lambda_1 & \star \\ \mathbf{0} & W^*A'W \end{bmatrix} = \begin{bmatrix} \lambda_1 & \star \\ \mathbf{0} & T' \end{bmatrix},$$

which has the asserted form. This completes the induction.

(b) Since A is real and λ is a real eigenvalue, there is a real unit eigenvector $\mathbf{x}$ associated with λ; see P.9.10. Example 7.1.21 shows how to construct a real unitary matrix with $\mathbf{x}$ as its first column. Now proceed as in (a). $\qquad\square$

As a first application of Theorem 11.1.1, we have a transparent demonstration (independent of the characteristic polynomial; see Theorem 10.2.10) that the trace and determinant of a matrix are the sum and product, respectively, of its eigenvalues.

Corollary 11.1.3 *Let* $\lambda_1, \lambda_2, \ldots, \lambda_n$ *be the eigenvalues of* $A \in \mathsf{M}_n$. *Then* $\operatorname{tr} A = \lambda_1 + \lambda_2 + \cdots + \lambda_n$ *and* $\det A = \lambda_1 \lambda_2 \cdots \lambda_n$.

Proof Let $A = UTU^*$, as in the preceding theorem. Then (C.2.9) ensures that

$$\operatorname{tr} A = \operatorname{tr} UTU^* = \operatorname{tr} U(TU^*) = \operatorname{tr}(TU^*)U = \operatorname{tr} T(U^*U) = \operatorname{tr} T = \lambda_1 + \lambda_2 + \cdots + \lambda_n$$

and

$$\det A = \det(UTU^*) = (\det U)(\det T)(\det U^{-1})$$
$$= (\det U)(\det T)(\det U)^{-1} = \det T = \lambda_1 \lambda_2 \cdots \lambda_n. \qquad\square$$

Lemma 9.3.2 says that if λ is an eigenvalue of $A \in \mathsf{M}_n$ and p is a polynomial, then $p(\lambda)$ is an eigenvalue of $p(A)$. What is its algebraic multiplicity? Schur's triangularization theorem permits us to answer that question. The key observation is that if $T = [t_{ij}] \in \mathsf{M}_n$ is upper triangular, then the diagonal entries of $p(T)$ are $p(t_{11}), p(t_{22}), \ldots, p(t_{nn})$; see P.C.20. These are all the eigenvalues of $p(T)$, including multiplicities.

Corollary 11.1.4 *Let* $A \in \mathsf{M}_n$ *and let* p *be a polynomial. If* $\lambda_1, \lambda_2, \ldots, \lambda_n$ *are the eigenvalues of* A, *then* $p(\lambda_1), p(\lambda_2), \ldots, p(\lambda_n)$ *are the eigenvalues of* $p(A)$ *(including multiplicities in both cases).*

Proof Let $A = UTU^*$, as in Theorem 11.1.1. Since $U^* = U^{-1}$, Theorem 2.6.9 ensures that $p(A) = p(UTU^*) = Up(T)U^*$, so the eigenvalues of $p(A)$ are the same as the eigenvalues of $p(T)$, which are $p(\lambda_1), p(\lambda_2), \ldots, p(\lambda_n)$. $\qquad\square$

Corollary 11.1.5 *Let* $A \in \mathsf{M}_n$ *and let* p *be a polynomial. If* $p(\lambda) \neq 0$ *for every* $\lambda \in \operatorname{spec} A$, *then* $p(A)$ *is invertible.*

Proof The preceding corollary tells us that $0 \notin \operatorname{spec} p(A)$. Theorem 9.1.16 ensures that $p(A)$ is invertible. $\qquad\square$

If $(\lambda, \mathbf{x})$ is an eigenpair of an invertible $A \in \mathsf{M}_n$, then $A\mathbf{x} = \lambda\mathbf{x}$ implies that $A^{-1}\mathbf{x} = \lambda^{-1}\mathbf{x}$. We conclude that λ^{-1} is an eigenvalue of A^{-1}, but what is its algebraic multiplicity? Schur's triangularization theorem permits us to answer this question, too.

Corollary 11.1.6 *Let $A \in M_n$ be invertible. If $\lambda_1, \lambda_2, \ldots, \lambda_n$ are the eigenvalues of A, then $\lambda_1^{-1}, \lambda_2^{-1}, \ldots, \lambda_n^{-1}$ are the eigenvalues of A^{-1} (including multiplicities in both cases).*

Proof Let $A = UTU^*$, as in Theorem 11.1.1. Then $A^{-1} = UT^{-1}U^*$. Theorem 3.2.9 ensures that T^{-1} is upper triangular and its diagonal entries are $\lambda_1^{-1}, \lambda_2^{-1}, \ldots, \lambda_n^{-1}$. These are the eigenvalues of A^{-1}.   □

11.2 The Cayley–Hamilton Theorem

Our development of the theory of eigenvalues and eigenvectors relies on the existence of annihilating polynomials. We now use Schur's theorem to construct an annihilating polynomial for $A \in M_n$ that has degree n; see (9.2.1) and Definition 9.2.3.

Theorem 11.2.1 (Cayley–Hamilton) *Let*

$$p_A(z) = \det(zI - A) = z^n + c_{n-1}z^{n-1} + \cdots + c_1 z + c_0 \tag{11.2.2}$$

be the characteristic polynomial of $A \in M_n$. Then $p_A(A) = 0_n$.

Proof Let $\lambda_1, \lambda_2, \ldots, \lambda_n$ be the eigenvalues of A and write

$$p_A(z) = (z - \lambda_1)(z - \lambda_2) \cdots (z - \lambda_n), \tag{11.2.3}$$

as in (10.2.1). Schur's triangularization theorem says that there is a unitary $U \in M_n$ and an upper triangular $T = [t_{ij}] \in M_n$ such that $A = UTU^*$ and $t_{ii} = \lambda_i$ for $i = 1, 2, \ldots, n$. Since Theorem 2.6.9 ensures that $p_A(A) = p_A(UTU^*) = Up_A(T)U^*$, it suffices to show that

$$p_A(T) = (T - \lambda_1 I)(T - \lambda_2 I) \cdots (T - \lambda_n I) = 0. \tag{11.2.4}$$

Our strategy is to show that for each $j = 1, 2, \ldots, n$,

$$P_j = (T - \lambda_1 I)(T - \lambda_2 I) \cdots (T - \lambda_j I)$$

is a block matrix of the form $P_j = [0_{n \times j} \;\; \star]$. If we can do this, then (11.2.4) is proved since $p_A(T) = P_n = 0_n$.

We proceed by induction on j. In the base case $j = 1$, the upper triangularity of T and the presence of λ_1 in its $(1, 1)$ entry ensure that $P_1 = (T - \lambda_1 I) = [\mathbf{0} \;\; \star]$ has a zero first column. The induction hypothesis is that for some $j \in \{1, 2, \ldots, n - 1\}$,

$$P_j = \begin{bmatrix} 0_{j \times j} & \star \\ 0_{(n-j) \times j} & \star \end{bmatrix}.$$

The upper triangular matrix $T - \lambda_{j+1} I$ has zero entries in and below its $(j+1, j+1)$ diagonal position, so

$$P_{j+1} = P_j(T - \lambda_{j+1} I) = \begin{bmatrix} 0_{j \times j} & \star \\ 0_{(n-j) \times j} & \star \end{bmatrix} \begin{bmatrix} \star & \star & \star \\ 0_{(n-j) \times j} & \mathbf{0} & \star \end{bmatrix}$$

$$= \begin{bmatrix} 0_{j \times j} & \mathbf{0} & \star \\ 0_{(n-j) \times j} & \mathbf{0} & \star \end{bmatrix} = [0_{n \times (j+1)} \;\; \star].$$

This completes the induction.   □

Example 11.2.5 If $A \in \mathsf{M}_n$, then p_A is a monic polynomial of degree n that annihilates A. But a monic polynomial of lesser degree might also annihilate A. For example, the characteristic polynomial of

$$A = \begin{bmatrix} 0 & 1 & 0 \\ 0 & 0 & 0 \\ 0 & 0 & 0 \end{bmatrix}$$

is $p_A(z) = z^3$, but z^2 also annihilates A.

Example 11.2.6 Consider

$$A = \begin{bmatrix} 1 & 2 \\ 3 & 4 \end{bmatrix} \tag{11.2.7}$$

and its characteristic polynomial $p_A(z) = z^2 - 5z - 2$. Then

$$A^2 - 5A - 2I = \begin{bmatrix} 7 & 10 \\ 15 & 22 \end{bmatrix} - \begin{bmatrix} 5 & 10 \\ 15 & 20 \end{bmatrix} - \begin{bmatrix} 2 & 0 \\ 0 & 2 \end{bmatrix} = \begin{bmatrix} 0 & 0 \\ 0 & 0 \end{bmatrix}.$$

Rewriting $p_A(A) = 0$ as $A^2 = 5A + 2I$ suggests other identities, for example

$$A^3 = A(A^2) = A(5A + 2I) = 5A^2 + 2A = 5(5A + 2I) + 2A = 27A + 10I$$

and

$$A^4 = (A^2)^2 = (5A + 2I)^2 = 25A^2 + 20A + 4I = 25(5A + 2I) + 20A + 4I = 145A + 54I.$$

Our first corollary of the Cayley–Hamilton theorem provides a systematic way to derive and understand identities like those in the preceding example. It relies on the division algorithm; see Appendix B.2. The degree of a polynomial p is denoted by $\deg p$; see Appendix B.1.

Corollary 11.2.8 *Let $A \in \mathsf{M}_n$ and let f be a polynomial with $\deg f \geq n$. Write $f = p_A q + r$, in which q and r are polynomials and $\deg r < n$. Then $f(A) = r(A)$.*

Proof The degree of p_A is n, so the division algorithm ensures that there are (unique) polynomials q and r such that $f = p_A q + r$ and $\deg r < n$. Then $f(A) = p_A(A)q(A) + r(A) = 0q(A) + r(A) = r(A)$. □

Example 11.2.9 Let A be the matrix (11.2.7). If $f(z) = z^3$, then $f = p_A q + r$ with $q(z) = z + 5$ and $r(z) = 27z + 10$. Consequently, $A^3 = 27A + 10I$. If $f(z) = z^4$, then $f = p_A q + r$ with $q(z) = z^2 + 5z + 27$ and $r(z) = 145z + 54$, so $A^4 = 145A + 54I$. Both of these computations agree with the identities in Example 11.2.6.

Corollary 11.2.10 *Let $A \in \mathsf{M}_n$. Then $\operatorname{span}\{I, A, A^2, ...\} = \operatorname{span}\{I, A, A^2, \ldots, A^{n-1}\}$, which has dimension at most n.*

Proof For each integer $k \geq 0$, the preceding corollary ensures that there is a polynomial r_k with $\deg r_k \leq n - 1$ such that $A^{n+k} = r_k(A)$. Thus, $A^{n+k} \in \operatorname{span}\{I, A, A^2, \ldots, A^{n-1}\}$. □

If A is invertible, then the Cayley–Hamilton theorem permits us to express A^{-1} as a polynomial in A that is closely related to its characteristic polynomial.

Corollary 11.2.11 *If $A \in \mathsf{M}_n$ is invertible and (11.2.2) is its characteristic polynomial, then*

$$A^{-1} = -c_0^{-1}(A^{n-1} + c_{n-1}A^{n-2} + \cdots + c_2 A + c_1 I). \tag{11.2.12}$$

Proof Theorem 10.1.3 ensures that $c_0 = (-1)^n \det A$, which is nonzero since A is invertible. Thus, we may rewrite the Cayley–Hamilton theorem

$$p_A(A) = A^n + c_{n-1}A^{n-1} + \cdots + c_2 A^2 + c_1 A + c_0 I = 0$$

as

$$I = A\left(-c_0^{-1}(A^{n-1} + c_{n-1}A^{n-2} + \cdots + c_2 A + c_1 I)\right)$$
$$= \left(-c_0^{-1}(A^{n-1} + c_{n-1}A^{n-2} + \cdots + c_2 A + c_1 I)\right)A. \qquad \square$$

Example 11.2.13 Let A be the matrix (11.2.7), for which $p_A(z) = z^2 - 5z - 2$. Then (11.2.12) ensures that $A^{-1} = \frac{1}{2}(A - 5I) = \begin{bmatrix} -2 & 1 \\ 3/2 & -1/2 \end{bmatrix}$.

11.3 The Minimal Polynomial

Example 11.2.5 shows that a square matrix might be annihilated by a monic polynomial whose degree is less than the degree of the characteristic polynomial. The following theorem establishes basic properties of an annihilating polynomial of minimum degree.

Theorem 11.3.1 *Let $A \in \mathsf{M}_n$ and let $\mu_1, \mu_2, \ldots, \mu_d$ be its distinct eigenvalues.*

(a) *There is a unique monic polynomial m_A of minimum positive degree that annihilates A.*

(b) *The degree of m_A is at most n.*

(c) *If p is a nonconstant polynomial that annihilates A, then there is a polynomial f such that $p = m_A f$. In particular, m_A divides p_A.*

(d) *There are positive integers $q_1, q_2, \ldots, q_d$ such that*

$$m_A(z) = (z - \mu_1)^{q_1}(z - \mu_2)^{q_2} \cdots (z - \mu_d)^{q_d}. \qquad (11.3.2)$$

Each q_i is at least 1 and is at most the algebraic multiplicity of μ_i. In particular, the degree of m_A is at least d.

Proof (a) Lemma 9.2.4 and Theorem 11.2.1 ensure that the set of monic nonconstant polynomials that annihilate A is nonempty. Let ℓ be the degree of a monic polynomial of minimum positive degree that annihilates A. Let g and h be monic polynomials of degree ℓ that annihilate A. The division algorithm ensures that there is a polynomial f and a polynomial r with $\deg r < \ell$ such that $g = hf + r$. Then

$$0 = g(A) = h(A)f(A) + r(A) = 0 + r(A) = r(A),$$

so r is a polynomial of degree less than ℓ that annihilates A. The definition of ℓ ensures that r has degree 0, so $r(z) = a$ for some scalar a. Then $aI = r(A) = 0$, so $a = 0$ and hence $g = hf$. But g and h are both monic polynomials of the same degree, so $f(z) = 1$ and hence $g = h$. Let $m_A = h$.

(b) Since p_A annihilates A and has degree n, it follows that $\ell \leq n$.

(c) If p is a nonconstant polynomial that annihilates A, then $\deg p \geq \deg m_A$. The division algorithm ensures that there are polynomials f and r such that $p = m_A f + r$ and the degree of

r is less than the degree of m_A. The argument in (a) shows that r is the zero polynomial and hence $p = m_A f$.

(d) Because m_A divides p_A, the factorization (11.3.2) follows from (10.2.2). Each exponent q_i can be no larger than the corresponding algebraic multiplicity n_i. Theorem 9.3.3 tells us that $z - \mu_i$ is a factor of m_A for each $i = 1, 2, \ldots, d$, so each q_i is at least 1. $\qquad\square$

Definition 11.3.3 Let $A \in M_n$. The *minimal polynomial* m_A is the unique monic polynomial of minimum positive degree that annihilates A.

The equation (11.3.2) reveals that every eigenvalue of A is a zero of m_A and that every zero of m_A is an eigenvalue of A.

Theorem 11.3.4 *If $A, B \in M_n$ are similar, then $p_A = p_B$ and $m_A = m_B$.*

Proof The first assertion is Theorem 10.3.1. If p is a polynomial, $S \in M_n$ is invertible, and $A = SBS^{-1}$, then Theorem 2.6.9 ensures that $p(A) = Sp(B)S^{-1}$. Thus, $p(A) = 0$ if and only if $p(B) = 0$. Theorem 11.3.1.c ensures that m_A divides m_B and that m_B divides m_A. Since both polynomials are monic, $m_A = m_B$. $\qquad\square$

Example 11.3.5 The previous theorem ensures that similar matrices have the same minimal and characteristic polynomials. However, two matrices with the same minimal and characteristic polynomials need not be similar. Let

$$A = \begin{bmatrix} 0 & 1 & 0 & 0 \\ 0 & 0 & 0 & 0 \\ 0 & 0 & 0 & 1 \\ 0 & 0 & 0 & 0 \end{bmatrix} \quad \text{and} \quad B = \begin{bmatrix} 0 & 1 & 0 & 0 \\ 0 & 0 & 0 & 0 \\ 0 & 0 & 0 & 0 \\ 0 & 0 & 0 & 0 \end{bmatrix}.$$

Then $p_A(z) = p_B(z) = z^4$ and $m_A(z) = m_B(z) = z^2$. However, rank $A = 2$ and rank $B = 1$, so A and B are not similar; see Theorem 2.6.9.

Example 11.3.6 Let A be the matrix in Example 11.2.5. Only three nonconstant monic polynomials divide p_A, namely z, z^2, and z^3. Since z does not annihilate A, but z^2 does, it follows that $m_A(z) = z^2$.

Example 11.3.7 Let $A \in M_5$ and suppose that $p_A(z) = (z-1)^3(z+1)^2$. Six distinct polynomials of the form $(z-1)^{q_1}(z+1)^{q_2}$ satisfy $1 \leq q_1 \leq 3$ and $1 \leq q_2 \leq 2$: one of degree 2, two of degree 3, two of degree 4, and one of degree 5. One of these six polynomials is m_A. If

$$\text{(a) } (A - I)(A + I) = 0,$$

then $m_A(z) = (z - 1)(z + 1)$. If not, then check whether

$$\text{(b) } (A - I)^2(A + I) = 0 \quad \text{or} \quad \text{(c) } (A - I)(A + I)^2 = 0. \qquad (11.3.8)$$

Theorem 11.3.1.a ensures that at most one of the identities (11.3.8) is true. If (b) is true, then $m_A(z) = (z - 1)^2(z + 1)$; if (c) is true, then $m_A(z) = (z - 1)(z + 1)^2$. If neither (b) nor (c) is true, then check whether

$$\text{(d) } (A - I)^3(A + I) = 0 \quad \text{or} \quad \text{(e) } (A - I)^2(A + I)^2 = 0. \qquad (11.3.9)$$

At most one of the identities (11.3.9) is true. If (d) is true, then $m_A(z) = (z - 1)^3(z + 1)$; if (e) is true, then $m_A(z) = (z - 1)^2(z + 1)^2$. If neither (d) nor (e) is true, then

$$\text{(f) } (A - I)^3(A + I)^2 = 0$$

and $m_A(z)=(z-1)^3(z+1)^2$. All six of the cases (a) - (f) are possible. Let $J_k(\lambda)$ be the $k \times k$ Jordan block in Definition 12.2.4. Then each of the following matrices satisfies the indicated identity, but does not satisfy any of the preceding identities:

(a) $A = I_3 \oplus (-I_2)$ and $m_A(z) = (z-1)(z+1)$;

(b) $A = J_1(1) \oplus J_2(1) \oplus (-I_2)$ and $m_A(z) = (z-1)^2(z+1)$;

(c) $A = I_3 \oplus J_2(-1)$ and $m_A(z) = (z-1)(z+1)^2$;

(d) $A = J_3(1) \oplus (-I_2)$ and $m_A(z) = (z-1)^3(z+1)$;

(e) $A = J_1(1) \oplus J_2(1) \oplus J_2(-1)$ and $m_A(z) = (z-1)^2(z+1)^2$;

(f) $A = J_3(1) \oplus J_2(-1)$ and $m_A(z) = (z-1)^3(z+1)^2$.

The trial-and-error method in the preceding example works, in principle, for matrices whose eigenvalues and algebraic multiplicities are known. However, the maximum number of trials required grows rapidly with the size of the matrix. See P.11.39 for another algorithm to compute the minimal polynomial.

Given a matrix, we can determine its minimal polynomial. Given a monic polynomial, is it the minimal polynomial of some matrix?

Definition 11.3.10 If $n \geq 2$, the *companion matrix* of the monic polynomial $f(z) = z^n + c_{n-1}z^{n-1} + \cdots + c_1 z + c_0$ is

$$C_f = \begin{bmatrix} 0 & 0 & \cdots & 0 & -c_0 \\ 1 & 0 & \cdots & 0 & -c_1 \\ 0 & 1 & \cdots & 0 & -c_2 \\ \vdots & \vdots & \ddots & \vdots & \vdots \\ 0 & 0 & \cdots & 1 & -c_{n-1} \end{bmatrix} = [\mathbf{e}_2 \ \mathbf{e}_3 \ \cdots \ \mathbf{e}_n \ -\mathbf{c}], \qquad (11.3.11)$$

in which $\mathbf{e}_1, \mathbf{e}_2, \ldots, \mathbf{e}_n$ is the standard basis for $\mathbb{C}^n$ and $\mathbf{c} = [c_0 \ c_1 \ \cdots \ c_{n-1}]^\mathsf{T}$. The companion matrix of $f(z) = z + c_0$ is $C_f = [-c_0]$.

Theorem 11.3.12 *The polynomial $f(z) = z^n + c_{n-1}z^{n-1} + \cdots + c_1 z + c_0$ is both the minimal polynomial and the characteristic polynomial of its companion matrix (11.3.11), that is, $f = p_{C_f} = m_{C_f}$.*

Proof We have $C_f \mathbf{e}_n = -\mathbf{c}$ and $C_f \mathbf{e}_{j-1} = \mathbf{e}_j$ for each $j = 2, 3, \ldots, n$. Therefore, $\mathbf{e}_j = C_f \mathbf{e}_{j-1} = C_f^2 \mathbf{e}_{j-2} = \cdots = C_f^{j-1} \mathbf{e}_1$ for each $j = 2, 3, \ldots, n$. Then

$$C_f^n \mathbf{e}_1 = C_f C_f^{n-1} \mathbf{e}_1 = C_f \mathbf{e}_n = -\mathbf{c}$$
$$= -c_0 \mathbf{e}_1 - c_1 \mathbf{e}_2 - c_2 \mathbf{e}_3 - \cdots - c_{n-1} \mathbf{e}_n$$
$$= -c_0 I \mathbf{e}_1 - c_1 C_f \mathbf{e}_1 - c_2 C_f^2 \mathbf{e}_1 + \cdots - c_{n-1} C_f^{n-1} \mathbf{e}_1,$$

which shows that $f(C_f)\mathbf{e}_1 = C_f^n \mathbf{e}_1 + c_{n-1} C_f^{n-1} \mathbf{e}_1 + \cdots + c_1 C_f \mathbf{e}_1 + c_0 I \mathbf{e}_1 = \mathbf{0}$. For each $j = 2, 3, \ldots, n$, we have $f(C_f)\mathbf{e}_j = f(C_f)C_f^{j-1} \mathbf{e}_1 = C_f^{j-1} f(C_f)\mathbf{e}_1 = \mathbf{0}$. Thus, $f(C_f) = 0$. If $g(z) = z^m + b_{m-1} z^{m-1} + \cdots + b_1 z + b_0$ and $m < n$, then

$$g(C_f)\mathbf{e}_1 = C_f^m \mathbf{e}_1 + b_{m-1} C_f^{m-1} \mathbf{e}_1 + \cdots + b_1 C_f \mathbf{e}_1 + b_0 \mathbf{e}_1$$
$$= \mathbf{e}_{m+1} + b_{m-1} \mathbf{e}_m + \cdots + b_1 \mathbf{e}_2 + b_0 \mathbf{e}_1.$$

The linear independence of $\mathbf{e}_1, \mathbf{e}_2, \ldots, \mathbf{e}_{m+1}$ ensures that $g(C_f)\mathbf{e}_1 \neq \mathbf{0}$, so g cannot annihilate C_f. We conclude that $f = m_{C_f}$, which has degree n. Since p_{C_f} is a monic polynomial that annihilates C_f (Theorem 11.2.1) and has the same degree as m_{C_f}, it must be m_{C_f}. $\qquad\square$

The final result in this section identifies the minimal polynomial of a diagonalizable matrix and provides a criterion for a matrix to be nondiagonalizable.

Theorem 11.3.13 *Let* $\mu_1, \mu_2, \ldots, \mu_d$ *be the distinct eigenvalues of* $A \in \mathsf{M}_n$ *and let*

$$p(z) = (z - \mu_1)(z - \mu_2) \cdots (z - \mu_d). \qquad (11.3.14)$$

If A *is diagonalizable, then* $p(A) = 0$ *and* p *is the minimal polynomial of* A.

Proof Let $n_1, n_2, \ldots, n_d$ be the respective multiplicities of $\mu_1, \mu_2, \ldots, \mu_d$. There is an invertible $S \in \mathsf{M}_n$ such that $A = S\Lambda S^{-1}$ and $\Lambda = \mu_1 I_{n_1} \oplus \mu_2 I_{n_2} \oplus \cdots \oplus \mu_d I_{n_d}$. Theorem 10.5.1 tells us that

$$p(A) = Sp(\Lambda)S^{-1} = S\big(p(\mu_1)I_{n_1} \oplus p(\mu_2)I_{n_2} \oplus \cdots \oplus p(\mu_d)I_{n_d}\big)S^{-1}. \qquad (11.3.15)$$

Since $p(\mu_j) = 0$ for each $j = 1, 2, \ldots, d$, each direct summand in (11.3.15) is a zero matrix. Therefore, $p(A) = 0$. Theorem 11.3.1.d says that no monic polynomial of positive degree less than d annihilates A. Since p is a monic polynomial of degree d that annihilates A, it is the minimal polynomial of A. $\qquad\square$

In the next section, we learn that the converse of Theorem 11.3.13 is true. For now, we know that if p is the polynomial (11.3.14) and if $p(A) \neq 0$, then A is not diagonalizable.

Example 11.3.16 The characteristic polynomial of $A = \begin{bmatrix} 3 & i \\ i & 1 \end{bmatrix}$ is $p_A(z) = z^2 - 4z + 4 = (z - 2)^2$. Since $p(z) = z - 2$ does not annihilate A, we conclude that A is not diagonalizable.

11.4 Linear Matrix Equations and Block Diagonalization

If a matrix has d distinct eigenvalues, it is similar to an upper triangular matrix that is block diagonal with d diagonal blocks. To prove this, we use a theorem that employs the Cayley–Hamilton theorem in a clever way.

Theorem 11.4.1 (Sylvester's Theorem on Linear Matrix Equations) *Let* $A \in \mathsf{M}_m$ *and* $B \in \mathsf{M}_n$, *and suppose that* $\operatorname{spec} A \cap \operatorname{spec} B = \varnothing$. *For each* $C \in \mathsf{M}_{m \times n}$,

$$AX - XB = C \qquad (11.4.2)$$

has a unique solution $X \in \mathsf{M}_{m \times n}$. *In particular, the only solution to* $AX - XB = 0$ *is* $X = 0$.

Proof Define the linear operator $T : \mathsf{M}_{m \times n} \to \mathsf{M}_{m \times n}$ by $T(X) = AX - XB$. We claim that T is onto and one to one. Since $\mathsf{M}_{m \times n}$ is finite dimensional, Corollary 4.1.7 ensures that T is onto if and only if it is one to one. Thus, it suffices to show that $T(X) = 0$ implies $X = 0$.

If $AX = XB$, then Theorems 2.6.9 and 11.2.1 ensure that $p_B(A)X = Xp_B(B) = X0 = 0$. The zeros of p_B are the eigenvalues of B, so the hypotheses imply that $p_B(\lambda) \neq 0$ for all $\lambda \in \operatorname{spec} A$. Corollary 11.1.5 tells us that $p_B(A)$ is invertible, so $p_B(A)X = 0$ implies that $X = 0$. $\qquad\square$

Our theorem on block diagonalization is the following.

Theorem 11.4.3 *Let $\mu_1, \mu_2, \ldots, \mu_d$ be the distinct eigenvalues of $A \in \mathsf{M}_n$, in any given order and with respective algebraic multiplicities $n_1, n_2, \ldots, n_d$. Then A is unitarily similar to a block upper triangular matrix*

$$
T = \begin{bmatrix} T_{11} & T_{12} & \cdots & T_{1d} \\ 0 & T_{22} & \cdots & T_{2d} \\ \vdots & & \ddots & \vdots \\ 0 & 0 & \cdots & T_{dd} \end{bmatrix}, \qquad \textit{in which } T_{ii} \in \mathsf{M}_{n_i} \textit{ and } i = 1, 2, \ldots, d. \qquad (11.4.4)
$$

Each diagonal block T_{ii} in (11.4.4) is upper triangular and all of its diagonal entries are μ_i. Moreover, A is similar to

$$
T_{11} \oplus T_{22} \oplus \cdots \oplus T_{dd}, \qquad (11.4.5)
$$

which is the direct sum of the diagonal blocks in (11.4.4).

Proof Schur's triangularization theorem ensures that A is unitarily similar to an upper triangular matrix T with the stated properties, so we must prove that T is similar to the block diagonal matrix (11.4.5). We proceed by induction on d. In the base case $d = 1$, there is nothing to prove. For the inductive step, assume that $d \geq 2$ and the asserted block diagonalization has been established for matrices with at most $d - 1$ distinct eigenvalues.

Partition (11.4.4) as

$$
T = \begin{bmatrix} T_{11} & C \\ 0 & T' \end{bmatrix}, \qquad \textit{in which } C = [T_{12}\ T_{13}\ \ldots\ T_{1d}] \in \mathsf{M}_{n_1 \times (n - n_1)}.
$$

Then $\operatorname{spec} T_{11} = \{\mu_1\}$ and $\operatorname{spec} T' = \{\mu_2, \mu_3, \ldots, \mu_d\}$, so $\operatorname{spec} T_{11} \cap \operatorname{spec} T' = \varnothing$. Theorem 11.4.1 ensures that there is an $X \in \mathsf{M}_{n_1 \times (n - n_1)}$ such that $T_{11} X - X T' = C$. Theorem 3.2.10 says that T is similar to

$$
\begin{bmatrix} T_{11} & -T_{11}X + XT' + C \\ 0 & T' \end{bmatrix} = \begin{bmatrix} T_{11} & 0 \\ 0 & T' \end{bmatrix}.
$$

The induction hypothesis ensures that there is an invertible matrix $S \in \mathsf{M}_{n - n_1}$ such that $S^{-1} T' S = T_{22} \oplus T_{33} \oplus \cdots \oplus T_{dd}$. Thus, T is similar to

$$
\begin{bmatrix} I & 0 \\ 0 & S \end{bmatrix}^{-1} \begin{bmatrix} T_{11} & 0 \\ 0 & T' \end{bmatrix} \begin{bmatrix} I & 0 \\ 0 & S \end{bmatrix} = \begin{bmatrix} T_{11} & 0 \\ 0 & S^{-1} T' S \end{bmatrix} = T_{11} \oplus T_{22} \oplus \cdots \oplus T_{dd}. \qquad \square
$$

Definition 11.4.6 $A \in \mathsf{M}_n$ is *unispectral* if $\operatorname{spec} A = \{\lambda\}$ for some scalar λ.

Theorem 11.4.3 says something of great importance. A square matrix with d distinct eigenvalues is similar (but not necessarily unitarily similar) to a direct sum of d unispectral matrices whose spectra are pairwise disjoint. These direct summands need not be unique, but one choice of them can be computed using only unitary similarities; they are the diagonal blocks of the block upper triangular matrix (11.4.4).

Lemma 11.4.7 *A square matrix is unispectral and diagonalizable if and only if it is a scalar matrix.*

Proof If $A \in \mathsf{M}_n$ is diagonalizable, then there is an invertible $S \in \mathsf{M}_n$ and a diagonal $\Lambda \in \mathsf{M}_n$ such that $A = S \Lambda S^{-1}$. If A is also unispectral, then $\Lambda = \lambda I$ for some scalar λ, so $A = S \Lambda S^{-1} = S(\lambda I) S^{-1} = \lambda S S^{-1} = \lambda I$. Conversely, if $A = \lambda I$ for some scalar λ, then it is diagonal and $\operatorname{spec} A = \{\lambda\}$. $\square$

Theorem 11.4.8 *Suppose that $A \in M_n$ is unitarily similar to a block upper triangular matrix (11.4.4), in which the diagonal blocks T_{ii} are unispectral and have pairwise disjoint spectra. Then A is diagonalizable if and only if each T_{ii} is a scalar matrix.*

Proof Theorem 11.4.3 says that A is similar to $T_{11} \oplus T_{22} \oplus \cdots \oplus T_{dd}$, in which each direct summand is unispectral. If each T_{ii} is a scalar matrix, then A is similar to a diagonal matrix. Conversely, if A is diagonalizable, then Theorem 10.4.13 tells us that each T_{ii} is diagonalizable. But each T_{ii} is unispectral, so the preceding lemma ensures that it is a scalar matrix. $\square$

The preceding theorem provides a criterion for diagonalizability of a given square complex matrix A. Via a sequence of unitary similarities (for example, use the algorithm in Schur's triangularization theorem), reduce A to upper triangular form, in which equal eigenvalues are grouped together. Examine the unispectral diagonal blocks with pairwise disjoint spectra. They are all diagonal if and only if A is diagonalizable.

We can use the same ideas in a different way to formulate a criterion for diagonalizability that involves the minimal polynomial rather than unitary similarities.

Theorem 11.4.9 *Let $\mu_1, \mu_2, \ldots, \mu_d$ be the distinct eigenvalues of $A \in M_n$ and let*

$$p(z) = (z - \mu_1)(z - \mu_2) \cdots (z - \mu_d). \tag{11.4.10}$$

Then A is diagonalizable if and only if $p(A) = 0$, that is, if and only if p is the minimal polynomial of A.

Proof If A is diagonalizable, then Theorem 11.3.13 tells us that $p(A) = 0$. Conversely, suppose that $p(A) = 0$. If $d = 1$, then $p(z) = z - \mu_1$ and $p(A) = A - \mu_1 I = 0$, so $A = \mu_1 I$. If $d > 1$, let

$$p_i(z) = \frac{p(z)}{z - \mu_i} \qquad \text{for } i = 1, 2, \ldots, d.$$

Thus, p_i is the polynomial of degree $d - 1$ obtained from p by omitting the factor $z - \mu_i$. Then $p(z) = (z - \mu_i)p_i(z)$ for each $i = 1, 2, \ldots, d$, and

$$p_i(\mu_i) = \prod_{j \neq i} (\mu_j - \mu_i) \neq 0 \qquad \text{for } i = 1, 2, \ldots, d.$$

Theorem 11.4.3 ensures that A is similar to $T_{11} \oplus T_{22} \oplus \cdots \oplus T_{dd}$, in which spec $T_{ii} = \{\mu_i\}$ for each $i = 1, 2, \ldots, d$. Therefore, $p(A)$ is similar to $p(T_{11}) \oplus p(T_{22}) \oplus \cdots \oplus p(T_{dd})$; see Theorem 2.6.9. Since $p(A) = 0$, we must have $p(T_{ii}) = 0$ for each $i = 1, 2, \ldots, d$. Because $p_i(\mu_i) \neq 0$, Corollary 11.1.5 ensures that each $p_i(T_{ii})$ is invertible. But the invertibility of $p_i(T_{ii})$ and the identity $0 = p(T_{ii}) = (T_{ii} - \mu_i I)p_i(T_{ii})$ imply that $T_{ii} - \mu_i I = 0$, that is, $T_{ii} = \mu_i I$. Since A is similar to a direct sum of scalar matrices, it is diagonalizable. $\square$

Corollary 11.4.11 *Let $A \in M_n$ and let f be a polynomial, each of whose zeros has multiplicity 1. If $f(A) = 0$, then A is diagonalizable.*

Proof Since m_A divides f, the zeros of m_A also have multiplicity 1. Therefore, the exponents in (11.3.2) all equal 1 and m_A is the polynomial in (11.4.10). $\square$

Example 11.4.12 If $A \in M_n$ is idempotent, then $A^2 = A$. Hence $f(A) = 0$, in which $f(z) = z^2 - z = z(z - 1)$. The preceding corollary ensures that A is diagonalizable. Theorems 11.4.3 and 11.4.8 tell us that A is unitarily similar to a 2×2 block upper triangular matrix whose diagonal blocks are scalar matrices. That is, an idempotent matrix in M_n with rank r is unitarily similar to $\begin{bmatrix} I_r & X \\ 0 & 0_{n-r} \end{bmatrix}$, in which $X \in M_{r \times (n-r)}$. Moreover, $r=0$ (that is, $A=0$) and $r=n$ (that is, $A=I$) are possible. Theorem 11.4.3 ensures that A is similar to $I_r \oplus 0_{n-r}$; see (11.4.5).

Since each square complex matrix is similar to a direct sum of unispectral matrices, it is natural to ask whether a single unispectral matrix is similar to a direct sum of simpler unispectral matrices in some useful way. We answer that question in the following chapter.

The direct sum in (11.4.5) motivates the following observation.

Theorem 11.4.13 *Let $A = A_1 \oplus A_2 \oplus \cdots \oplus A_d$, in which* $\operatorname{spec} A_i = \{\mu_i\}$ *for each $i = 1, 2, \ldots, d$ and $\mu_i \neq \mu_j$ for all $i \neq j$. Then $m_A = m_{A_1} m_{A_2} \cdots m_{A_d}$.*

Proof Let $f = m_{A_1} m_{A_2} \cdots m_{A_d}$. Then $f(A_i) = 0$ for each $i = 1, 2, \ldots, d$, and hence

$$f(A) = f(A_1) \oplus f(A_2) \oplus \cdots \oplus f(A_d) = 0.$$

Moreover, there are positive integers q_i such that $m_{A_i} = (z - \mu_i)^{q_i}$ for each $i = 1, 2, \ldots, d$. Since m_A divides f,

$$m_A(z) = (z - \mu_1)^{r_1} (z - \mu_2)^{r_2} \cdots (z - \mu_d)^{r_d},$$

in which $1 \leq r_i \leq q_i$ for each $i = 1, 2, \ldots, d$. Define $h_i(z) = \frac{m_A(z)}{(z - \mu_i)^{r_i}}$ for $i = 1, 2, \ldots, d$. Then $h_i(\mu_i) \neq 0$, so $h_i(A_i)$ is invertible; see Corollary 11.1.5. Since

$$0 = m_A(A_i) = (A_i - \mu_i I)^{r_i} h_i(A_i) \qquad \text{for } i = 1, 2, \ldots, d,$$

it follows that each $(A_i - \mu_i I)^{r_i} = 0$. The definition of the minimal polynomial ensures that each $r_i = q_i$, so $f = m_A$. $\qquad \square$

A final consequence of Sylvester's theorem is a generalization of Lemma 3.2.17.

Corollary 11.4.14 *Let $A = A_{11} \oplus A_{22} \oplus \cdots \oplus A_{dd}$, in which each $A_{ii} \in M_{n_i}$ and $n_1 + n_2 + \cdots + n_d = n$. Partition $B = [B_{ij}] \in M_n$ conformally with A. Suppose that $\operatorname{spec} A_{ii} \cap \operatorname{spec} A_{jj} = \varnothing$ for all $i \neq j$. If $AB = BA$, then $B_{ij} = 0$ for all $i \neq j$.*

Proof Equate the (i, j) blocks of both sides of the identity $AB = BA$ and obtain $A_{ii} B_{ij} - B_{ij} A_{jj} = 0$ for all $i \neq j$. This is a linear matrix equation of the form (11.4.2) with $C = 0$. Theorem 11.4.1 ensures that $B_{ij} = 0$ for all $i \neq j$. $\qquad \square$

11.5 Commuting Matrices and Triangularization

The algorithm described in the proof of Schur's triangularization theorem can be performed simultaneously on two or more matrices, provided that they have a common eigenvector. Commuting matrices have common eigenvectors (Corollary 9.5.4). These two observations are the basis for the following theorem, which gives a sufficient condition for a family of matrices to be simultaneously unitarily upper triangularizable.

Theorem 11.5.1 *Let $n \geq 2$ and let $\mathscr{F} \subseteq M_n$ be a nonempty set of commuting matrices.*

(a) *There is a unitary $U \in M_n$ such that U^*AU is upper triangular for all $A \in \mathscr{F}$.*

(b) *If every matrix in $\mathscr{F}$ is real and has only real eigenvalues, then there is a real orthogonal $Q \in M_n(\mathbb{R})$ such that $Q^\mathsf{T} A Q$ is upper triangular for all $A \in \mathscr{F}$.*

Proof (a) Corollary 9.5.4 ensures the existence of a unit vector $\mathbf{x}$ that is an eigenvector of every matrix in $\mathscr{F}$. Let $V \in \mathsf{M}_n$ be any unitary matrix whose first column is $\mathbf{x}$; see Example 7.1.21. If $A, B \in \mathscr{F}$, then $A\mathbf{x} = \lambda\mathbf{x}$ and $B\mathbf{x} = \mu\mathbf{x}$ for some scalars λ and μ. Just as in the proof of Schur's triangularization theorem,

$$V^*AV = \begin{bmatrix} \lambda & \star \\ 0 & A' \end{bmatrix} \quad \text{and} \quad V^*BV = \begin{bmatrix} \mu & \star \\ 0 & B' \end{bmatrix},$$

so V achieves a simultaneous reduction of every matrix in $\mathscr{F}$. Now compute

$$\begin{bmatrix} \lambda\mu & \star \\ 0 & A'B' \end{bmatrix} = V^*ABV = V^*BAV = \begin{bmatrix} \lambda\mu & \star \\ 0 & B'A' \end{bmatrix},$$

which tells us that $A'B' = B'A'$ for all $A, B \in \mathscr{F}$. An induction argument completes the proof. (b) If every matrix in $\mathscr{F}$ is real and has only real eigenvalues, Corollary 9.5.4 ensures the existence of a real unit vector $\mathbf{x}$ that is an eigenvector of every matrix in $\mathscr{F}$. Example 7.1.21 shows how to construct a real unitary matrix V whose first column is $\mathbf{x}$. The rest of the argument is the same as in (a); all the matrices involved are real. □

Commuting matrices have the pleasant property that the eigenvalues of their sums and products are, respectively, sums and products of their eigenvalues in some order.

Corollary 11.5.2 *Let* $A, B \in \mathsf{M}_n$, *and suppose that* $AB = BA$. *There is some ordering* $\lambda_1, \lambda_2, \ldots, \lambda_n$ *of the eigenvalues of* A *and some ordering* $\mu_1, \mu_2, \ldots, \mu_n$ *of the eigenvalues of* B *such that* $\lambda_1 + \mu_1, \lambda_2 + \mu_2, \ldots, \lambda_n + \mu_n$ *are the eigenvalues of* $A + B$ *and* $\lambda_1\mu_1, \lambda_2\mu_2, \ldots, \lambda_n\mu_n$ *are the eigenvalues of* AB.

Proof The preceding theorem says that there is a unitary U such that $U^*AU = T = [t_{ij}]$ and $U^*BU = R = [r_{ij}]$ are upper triangular. The entries $t_{11}, t_{22}, \ldots, t_{nn}$ are the eigenvalues of A in some order and the entries $r_{11}, r_{22}, \ldots, r_{nn}$ are the eigenvalues of B in some order. Then

$$A + B = U(T + R)U^* = U[t_{ij} + r_{ij}]U^*,$$

so $A + B$ is similar to the upper triangular matrix $T + R$. Its diagonal entries are $t_{ii} + r_{ii}$ for $i = 1, 2, \ldots, n$; they are the eigenvalues of $A + B$. Also, $AB = U(TR)U^*$, so AB is similar to the upper triangular matrix TR. Its diagonal entries are $t_{ii}r_{ii}$ for $i = 1, 2, \ldots, n$; they are the eigenvalues of AB. □

Example 11.5.3 If A and B do not commute, then the eigenvalues of $A + B$ need not be sums of eigenvalues of A and B. For example, all the eigenvalues of the noncommuting matrices $A = \begin{bmatrix} 0 & 1 \\ 0 & 0 \end{bmatrix}$ and $B = \begin{bmatrix} 0 & 0 \\ 1 & 0 \end{bmatrix}$ are zero. The eigenvalues of $A + B$ are ± 1, neither of which is a sum of eigenvalues of A and B. The eigenvalue 1 of AB is not a product of eigenvalues of A and B.

Example 11.5.4 If A and B do not commute, the eigenvalues of $A + B$ might nevertheless be sums of eigenvalues of A and B. Consider the noncommuting matrices $A = \begin{bmatrix} 1 & 1 \\ 0 & 2 \end{bmatrix}$ and $B = \begin{bmatrix} 3 & 2 \\ 0 & 6 \end{bmatrix}$. The eigenvalues of $A + B$ are $1 + 3 = 4$ and $2 + 6 = 8$. The eigenvalues of AB are $1 \cdot 3 = 3$ and $2 \cdot 6 = 12$.

11.6 Eigenvalue Adjustments and the Google Matrix

In some applications, we would like to adjust an eigenvalue of a matrix by adding a suitable rank-1 matrix. The following theorem describes one way to do this.

Theorem 11.6.1 (Brauer) *Let* $(\lambda, \mathbf{x})$ *be an eigenpair of* $A \in M_n$ *and let* $\lambda, \lambda_2, \ldots, \lambda_n$ *be its eigenvalues. For any* $\mathbf{y} \in \mathbb{F}^n$, *the eigenvalues of* $A + \mathbf{x}\mathbf{y}^*$ *are* $\lambda + \mathbf{y}^*\mathbf{x}, \lambda_2, \ldots, \lambda_n$ *and* $(\lambda + \mathbf{y}^*\mathbf{x}, \mathbf{x})$ *is an eigenpair of* $A + \mathbf{x}\mathbf{y}^*$.

Proof Let $\mathbf{u} = \mathbf{x}/\|\mathbf{x}\|_2$. Then $(\lambda, \mathbf{u})$ is an eigenpair of A, in which $\mathbf{u}$ is a unit vector. Schur's triangularization theorem says that there is a unitary $U = [\mathbf{u} \ U_2] \in M_n$ whose first column is $\mathbf{u}$ and is such that

$$U^*AU = T = \begin{bmatrix} \lambda & \star \\ \mathbf{0} & T' \end{bmatrix}$$

is upper triangular. The eigenvalues (diagonal entries) of T' are $\lambda_2, \lambda_3, \ldots, \lambda_n$. Compute

$$U^*(\mathbf{x}\mathbf{y}^*)U = (U^*\mathbf{x})(\mathbf{y}^*U) = \begin{bmatrix} \mathbf{u}^*\mathbf{x} \\ U_2^*\mathbf{x} \end{bmatrix} [\mathbf{y}^*\mathbf{u} \ \ \mathbf{y}^*U_2]$$

$$= \begin{bmatrix} \|\mathbf{x}\|_2 \\ \mathbf{0} \end{bmatrix} [\mathbf{y}^*\mathbf{u} \ \ \mathbf{y}^*U_2] = \begin{bmatrix} \mathbf{y}^*\mathbf{x} & \star \\ \mathbf{0} & \mathbf{0} \end{bmatrix}.$$

Therefore,

$$U^*(A + \mathbf{x}\mathbf{y}^*)U = \begin{bmatrix} \lambda & \star \\ \mathbf{0} & T' \end{bmatrix} + \begin{bmatrix} \mathbf{y}^*\mathbf{x} & \star \\ \mathbf{0} & \mathbf{0} \end{bmatrix} = \begin{bmatrix} \lambda + \mathbf{y}^*\mathbf{x} & \star \\ \mathbf{0} & T' \end{bmatrix},$$

which has eigenvalues $\lambda + \mathbf{y}^*\mathbf{x}, \lambda_2, \ldots, \lambda_n$. These are also the eigenvalues of $A + \mathbf{x}\mathbf{y}^*$. Since $(A + \mathbf{x}\mathbf{y}^*)\mathbf{x} = A\mathbf{x} + \mathbf{x}\mathbf{y}^*\mathbf{x} = (\lambda + \mathbf{y}^*\mathbf{x})\mathbf{x}$, it follows that $(\lambda + \mathbf{y}^*\mathbf{x}, \mathbf{x})$ is an eigenpair of $A + \mathbf{x}\mathbf{y}^*$. $\square$

A notable application of Brauer's theorem involves a famous matrix that played a key role in early methods to rank websites.

Corollary 11.6.2 *Let* $(\lambda, \mathbf{x})$ *be an eigenpair of* $A \in M_n$ *and let* $\lambda, \lambda_2, \ldots, \lambda_n$ *be the eigenvalues of* A. *Let* $\mathbf{y} \in \mathbb{C}^n$ *be such that* $\mathbf{y}^*\mathbf{x} = 1$ *and let* $\tau \in \mathbb{C}$. *Then the eigenvalues of* $A_\tau = \tau A + (1 - \tau)\lambda \mathbf{x}\mathbf{y}^*$ *are* $\lambda, \tau\lambda_2, \ldots, \tau\lambda_n$.

Proof The eigenvalues of τA are $\tau\lambda, \tau\lambda_2, \ldots, \tau\lambda_n$. The preceding theorem says that the eigenvalues of $A_\tau = \tau A + \mathbf{x}((1 - \overline{\tau})\overline{\lambda}\mathbf{y})^*$ are $\tau\lambda + ((1 - \overline{\tau})\overline{\lambda}\mathbf{y})^*\mathbf{x} = \tau\lambda + (1 - \tau)\lambda = \lambda$, together with $\tau\lambda_2, \tau\lambda_3, \ldots, \tau\lambda_n$. $\square$

Example 11.6.3 Let $A = [a_{ij}] \in M_n(\mathbb{R})$ have nonnegative entries and suppose that all of its row sums are equal to 1. Then $(1, \mathbf{e})$ is an eigenpair of A. Let $1, \lambda_2, \ldots, \lambda_n$ be the eigenvalues of A. Corollary 9.4.6 ensures that $|\lambda_i| \leq 1$ for each $i = 2, 3, \ldots, n$. Some of the eigenvalues $\lambda_2, \lambda_3, \ldots, \lambda_n$ can have modulus 1. For example,

$$\begin{bmatrix} 0 & 0 & 1 \\ 1 & 0 & 0 \\ 0 & 1 & 0 \end{bmatrix} \quad\quad (11.6.4)$$

has nonnegative entries and row sums equal to 1; its eigenvalues are 1 and $e^{\pm 2\pi i/3}$, all of which have modulus 1. For computational reasons (see Section 19.6), we might wish to adjust our matrix A so that $\lambda = 1$ is the only eigenvalue of maximum modulus, and we would like the adjusted matrix to continue to have nonnegative entries and row sums equal to 1. Let $\mathbf{x} = \mathbf{e}$, let $\mathbf{y} = \mathbf{e}/n$, let $E = \mathbf{e}\mathbf{e}^\mathsf{T}$, and suppose that $0 < \tau < 1$. Then $\mathbf{y}^\mathsf{T}\mathbf{x} = 1$, so the preceding corollary says that the eigenvalues of

$$A_\tau = \tau A + (1 - \tau)\mathbf{x}\mathbf{y}^\mathsf{T} = \tau A + \frac{1 - \tau}{n} E \tag{11.6.5}$$

are $1, \tau\lambda_2, \ldots, \tau\lambda_n$. Thus, $A_\tau \in M_n(\mathbb{R})$ has positive entries and row sums equal to 1. Its only maximum-modulus eigenvalue is $\lambda = 1$ (with algebraic multiplicity 1); its other $n - 1$ eigenvalues have modulus at most $\tau < 1$.

A matrix of the form $A_\tau \in M_n(\mathbb{R})$ in (11.6.5) is often called a *Google matrix* because of its association with the problem of ranking web pages. The page rankings are determined by the entries of the vector $\mathbf{z}$ such that $A_\tau^\mathsf{T}\mathbf{z} = \mathbf{z}$ and $\mathbf{e}^\mathsf{T}\mathbf{z} = 1$; see Theorem 20.2.3. Folklore has it that the founders of Google used the value $\tau = 0.85$.

11.7 Problems

In Problems 1–5, let $A = \begin{bmatrix} 5 & -3 \\ 1 & 1 \end{bmatrix}$ and let λ_1, λ_2 be its eigenvalues.

P.11.1 Compute the characteristic polynomial of A and use it to find λ_1 and λ_2.

P.11.2 Find a real orthogonal matrix U such that $T_1 = U^\mathsf{T}AU$ is upper triangular and has λ_1 as its $(1, 1)$ entry. What matrix T_1 do you obtain?

P.11.3 Find a real orthogonal matrix V such that $T_2 = V^\mathsf{T}AV$ is upper triangular and has λ_2 as its $(1, 1)$ entry. What matrix T_2 do you obtain?

P.11.4 Find an invertible S such that $S^{-1}AS$ is diagonal.

P.11.5 Can you find a real orthogonal matrix W such that $W^\mathsf{T}AW$ is diagonal? Why?

P.11.6 Let $B = \begin{bmatrix} 3 & -1 \\ -1 & 3 \end{bmatrix}$. Find a real orthogonal matrix Q such that $T = Q^\mathsf{T}BQ$ is upper triangular. What matrix T do you obtain? How might you have predicted the form of T?

P.11.7 Let $A \in M_3$ and suppose that $\operatorname{spec} A = \{1\}$. Show that A is invertible and express A^{-1} as a linear combination of I, A, and A^2.

P.11.8 Let $A \in M_n$. Prove in two ways that A is nilpotent if and only if $\operatorname{spec} A = \{0\}$.
(a) Use Theorem 11.1.1 and consider powers of a strictly upper triangular matrix.
(b) Use Theorems 11.2.1 and 9.3.3.

P.11.9 Let $A \in M_n$. Show that the following statements are equivalent:

(a) A is nilpotent.

(b) A is unitarily similar to a strictly upper triangular matrix.

(c) A is similar to a strictly upper triangular matrix.

P.11.10 Suppose that an upper triangular matrix $T \in M_n$ has ν nonzero diagonal entries. Show that rank $T \geq \nu$ and give an example for which rank $T > \nu$.

P.11.11 Suppose that $A \in M_n$ has ν nonzero eigenvalues. Explain why rank $A \geq \nu$ and give an example for which rank $A > \nu$.

P.11.12 Let $A \in M_2$. (a) Show that $\det A = \frac{1}{2}((\operatorname{tr}A)^2 - \operatorname{tr}A^2)$. (b) Show that A is unitarily similar to an upper triangular matrix with a real nonnegative entry in position $(1, 1)$. *Hint*: Consider a diagonal unitary similarity.

P.11.13 Let $A, B \in M_2$. (a) Show that A and B have the same eigenvalues if and only if $\operatorname{tr} A = \operatorname{tr} B$ and $\operatorname{tr} A^2 = \operatorname{tr} B^2$. (b) Show that A and B are unitarily similar if and only if $\operatorname{tr} A = \operatorname{tr} B$, $\operatorname{tr} A^2 = \operatorname{tr} B^2$, and $\operatorname{tr} A^*A = \operatorname{tr} B^*B$.

P.11.14 Let $A = [a_{ij}] \in M_n$ and write $A = UTU^*$, in which U is unitary, $T = [t_{ij}]$ is upper triangular, and $|t_{11}| \geq |t_{22}| \geq \cdots \geq |t_{nn}|$. Suppose that A has exactly $k \geq 1$ nonzero eigenvalues $\lambda_1, \lambda_2, \ldots, \lambda_k$ (including multiplicities). Explain why

$$|\operatorname{tr} A|^2 = \left| \sum_{i=1}^{k} \lambda_i \right|^2 \leq k \sum_{i=1}^{k} |\lambda_i|^2 = k \sum_{i=1}^{k} |t_{ii}|^2 \leq k \sum_{i,j=1}^{n} |t_{ij}|^2 = k \sum_{i,j=1}^{n} |a_{ij}|^2 = k\|A\|_F^2$$

and deduce that $\operatorname{rank} A \geq k \geq |\operatorname{tr} A|^2/(\operatorname{tr} A^*A)$, with equality if and only if $T = cI_k \oplus 0_{n-k}$ for some nonzero scalar c.

P.11.15 Let $\lambda_1, \lambda_2, \ldots, \lambda_n$ be the eigenvalues of $A \in M_n$. (a) Show that $\operatorname{tr} A^k = \lambda_1^k + \lambda_2^k + \cdots + \lambda_n^k$ for each $k = 1, 2, \ldots$. (b) Use the presentation (11.2.3) of the characteristic polynomial (11.2.2) to show that the coefficient of its z^{n-2} term is $c_{n-2} = \sum_{i<j} \lambda_i \lambda_j$. (c) Show that $c_{n-2} = \frac{1}{2}((\operatorname{tr} A)^2 - \operatorname{tr} A^2)$. (d) If $A \in M_3$, show that $\det A = \frac{1}{6}(2\operatorname{tr} A^3 - 3(\operatorname{tr} A)(\operatorname{tr} A^2) + (\operatorname{tr} A)^3)$. *Hint*: Compute $\operatorname{tr} p_A(A)$.

P.11.16 If $A \in M_n$ is invertible and has characteristic polynomial (11.2.2), show that the characteristic polynomial of A^{-1} is $z^n p_A(z^{-1})/c_0$.

P.11.17 Let $A \in M_n$ be invertible. Show that $\operatorname{span}\{I, A, A^{-1}, A^2, A^{-2}, \ldots\} = \operatorname{span}\{I, A, A^2, \ldots, A^{n-1}\}$, which has dimension at most n.

P.11.18 Let $A \in M_m$ and $B \in M_n$. If $\operatorname{spec} A \cap \operatorname{spec} B \neq \varnothing$, show that there is a nonzero $X \in M_{m \times n}$ such that $AX - XB = 0$. *Hint*: You may find it helpful to investigate $X = \mathbf{xy}^*$, in which $(\lambda, \mathbf{x})$ is an eigenpair of A and $(\bar{\lambda}, \mathbf{y})$ is an eigenpair of B^*.

P.11.19 Show that the Google matrix A_τ in (11.6.5) has nonnegative real entries and row sums equal to 1. What is A_τ for the matrix in (11.6.4)? What are its eigenvalues if $\tau = 0.85$? What are their moduli?

P.11.20 If you wanted to use the method in Theorem 11.6.1 to adjust an eigenvalue λ of A, you would have to choose $\mathbf{y}$ somehow. Why should you avoid choosing it to be an eigenvector of A^* associated with an eigenvalue of A^* different from $\bar{\lambda}$? Why is it always safe to take $\mathbf{y}$ to be a nonzero scalar multiple of $\mathbf{x}$? *Hint*: P.10.29.

P.11.21 Let $r \in \{1, 2, \ldots, n\}$ and let $\lambda_1, \lambda_2, \ldots, \lambda_n$ be the eigenvalues of $A \in M_n$. Suppose that $\mathbf{x}_1, \mathbf{x}_2, \ldots, \mathbf{x}_r \in \mathbb{C}^n$ are linearly independent and $A\mathbf{x}_i = \lambda_i \mathbf{x}_i$ for each $i = 1, 2, \ldots, r$. Let $X = [\mathbf{x}_1 \ \mathbf{x}_2 \ \cdots \ \mathbf{x}_r]$, $\Lambda = \operatorname{diag}(\lambda_1, \lambda_2, \ldots, \lambda_r)$, and $Y \in M_{n \times r}$. (a) Prove that the n eigenvalues of $A + XY^*$ are $\lambda_{r+1}, \lambda_{r+2}, \ldots, \lambda_n$ together with the r eigenvalues of $\Lambda + Y^*X$. *Hint*: Consider a similarity via $S = [X \ X']$. (b) Deduce Theorem 11.6.1 (Brauer's Theorem) from (a).

P.11.22 Let $A, B^\mathsf{T} \in M_{m \times n}$. Prove that AB is nilpotent if and only if BA is nilpotent in two ways. (a) Use Theorem 10.7.2. (b) Use the identity $(BA)^{m+1} = B(AB)^m A$.

P.11.23 Suppose that $A, B \in M_n$ commute and $\operatorname{spec} A \cap \operatorname{spec}(-B) = \varnothing$. Explain why $A + B$ is invertible.

P.11.24 Let $A = \begin{bmatrix} 1 & 0 \\ 2 & 1 \end{bmatrix}$ and $B = \begin{bmatrix} 1 & 2 \\ 0 & 1 \end{bmatrix}$. Show that $\operatorname{spec} A \cap \operatorname{spec}(-B) = \varnothing$, but $A + B$ is not invertible. Does this contradict the preceding problem?

P.11.25 Let $A = [a_{ij}] \in M_n$ be upper triangular and invertible. Use Corollary 11.2.11 to show that A^{-1} is upper triangular. What are its diagonal entries?

P.11.26 Let $A, B \in M_n$ and suppose that there is an invertible $S \in M_n$ such that $S^{-1}AS$ and $S^{-1}BS$ are upper triangular. Show that there is a unitary $U \in M_n$ such that U^*AU and U^*BU are upper triangular. *Hint*: Consider the QR factorization of S.

P.11.27 Explain what is wrong with the following false "proof" of the Cayley–Hamilton theorem: $p_A(z) = \det(zI - A)$, so $p_A(A) = \det(AI - A) = \det(A - A) = \det 0 = 0$.

P.11.28 If $A \in M_n$, spec $A = \{-1, 1\}$, and $p(z) = (z-1)^{n-1}(z+1)^{n-1}$, show that $p(A) = 0$.

P.11.29 Let $A, B^\mathsf{T} \in M_{m \times n}$. Use Corollary 11.1.3 and Theorem 10.7.2 to explain why $\operatorname{tr} AB = \operatorname{tr} BA$. Can you explain this identity in an elementary way that does not involve eigenvalues?

P.11.30 Let $A \in M_{10}$, let $f(z) = z^4 + 11z^3 - 7z^2 + 5z + 3$, and suppose that $f(A) = 0$. Prove that A is invertible and find a polynomial g of degree 3 or less such that $A^{-1} = g(A)$.

P.11.31 Let $A, B \in M_n$ and suppose that $AB = BA$. Define spec $A + $ spec $B = \{\lambda + \mu \;:\; \lambda \in $ spec A and $\mu \in$ spec $B\}$ and (spec A)(spec B) $= \{\lambda\mu \;:\; \lambda \in$ spec A and $\mu \in$ spec $B\}$. (a) Show that spec$(A + B) \subseteq$ spec $A + $ spec B. (b) Give an example of diagonal matrices A, B such that spec$(A + B) \neq$ spec $A + $ spec B. (c) What can you say about (spec A)(spec B)?

P.11.32 Let $A = A_1 \oplus A_2 \oplus \cdots \oplus A_d$ and $B = B_1 \oplus B_2 \oplus \cdots \oplus B_d$ be $n \times n$ matrices that are conformally partitioned and block diagonal. Suppose that spec $A_i \cap$ spec $B_j = \varnothing$ for all $i \neq j$. If $C \in M_n$ and $AC = CB$, show that $C = C_1 \oplus C_2 \oplus \cdots \oplus C_d$ is block diagonal and conformal with A and B.

P.11.33 Suppose that $A, B \in M_n$ commute and let $\mu_1, \mu_2, \ldots, \mu_d$ be the distinct eigenvalues of A. Prove that there is an invertible $S \in M_n$ such that $S^{-1}AS = T_1 \oplus T_2 \oplus \cdots \oplus T_d$ and $S^{-1}BS = B_1 \oplus B_2 \oplus \cdots \oplus B_d$ are conformally partitioned and block diagonal, each T_j is upper triangular, and spec $T_j = \{\mu_j\}$ for each $j = 1, 2, \ldots, d$.

P.11.34 Let $A \in M_n$ be diagonalizable. Prove that rank $A = $ rank A^k for $k = 1, 2, \ldots$.

P.11.35 Let $C = \begin{bmatrix} 0 & I_n \\ I_n & 0 \end{bmatrix}$. Use both P.11.48.f and P.10.17.e to evaluate $\det C$.

P.11.36 Verify the claim at the end of Example 11.3.7 about the six matrices listed (a)–(f).

P.11.37 If $A \in M_5$ is diagonalizable and $p_A(z) = (z-2)^3(z-3)^2$, show that $m_A(z) = (z-2)(z-3)$.

P.11.38 If $A \in M_n$ is an involution $(A^2 = I)$, show that it is unitarily similar to $\begin{bmatrix} I_k & X \\ 0 & -I_{n-k} \end{bmatrix}$, in which $X \in M_{k \times (n-k)}$. *Hint*: Study Example 11.4.12.

P.11.39 It is possible to determine the minimal polynomial of a matrix without knowing its eigenvalues or characteristic polynomial. Let $A \in M_n$, let m_A be its minimal polynomial, and suppose that m_A has degree ℓ. Let $\mathbf{v}_1 = \operatorname{vec} I$, $\mathbf{v}_2 = \operatorname{vec} A$, $\mathbf{v}_3 = \operatorname{vec} A^2, \ldots, \mathbf{v}_{n+1} = \operatorname{vec} A^n$. (a) Show that $\mathbf{v}_1, \mathbf{v}_2, \ldots, \mathbf{v}_{n+1}$ is a linearly dependent list of vectors in $\mathbb{C}^{n^2}$. (b) Show that

$$\min\{k : \mathbf{v}_1, \mathbf{v}_2, \ldots, \mathbf{v}_{k+1} \text{ is linearly dependent and } 1 \leq k \leq n\} = \ell$$

and let $c_1, c_2, \ldots, c_{k+1}$ be scalars, not all zero, such that $c_1\mathbf{v}_1 + c_2\mathbf{v}_2 + \cdots + c_{\ell+1}\mathbf{v}_{\ell+1} = \mathbf{0}$. (c) Why is $c_{\ell+1} \neq 0$ and what are the coefficients of m_A? (d) Formulate an algorithm, based on (b) and an orthogonalization process, to compute m_A; see P.6.4.

P.11.40 Use the algorithm in the preceding problem to compute the minimal polynomials of $\begin{bmatrix} 0 & 1 \\ 0 & 1 \end{bmatrix}$, $\begin{bmatrix} 1 & 1 \\ 0 & 1 \end{bmatrix}$, and $\begin{bmatrix} 1 & 0 \\ 0 & 1 \end{bmatrix}$.

P.11.41 Let $A \in M_n$ and $B \in M_m$. (a) Show that the characteristic polynomial of $A \oplus B$ is $p_A p_B$. (b) Show that the minimal polynomial of $A \oplus B$ is the least common multiple of m_A and m_B (the monic polynomial of minimum degree that is divisible by both m_A and m_B). (c) If $A \oplus B$ is diagonalizable, use (b) to prove that A and B are diagonalizable.

P.11.42 If $A \in M_n$, show that $\dim \text{span}\{I, A, A^2, \ldots\} = \deg m_A$.

P.11.43 Let $p(z) = z^2 + 4$. (a) Is there an $A \in M_3(\mathbb{R})$ such that $m_A(z) = p(z)$? (b) Is there an $A \in M_2(\mathbb{R})$ with $m_A(z) = p(z)$? (c) Is there an $A \in M_3(\mathbb{C})$ such that $m_A(z) = p(z)$? In each case, provide a proof or a counterexample.

P.11.44 Let $A \in M_n$. Use Theorem 11.4.3 to show that there is a diagonalizable $B \in M_n$ and a nilpotent $C \in M_n$ such that $A = B + C$ and $BC = CB$. *Hint:* $T_{ii} = \lambda_i I + (T_{ii} - \lambda_i I)$.

P.11.45 Let f be a monic polynomial of degree n, let $\lambda \in \mathbb{C}$, and let $\mathbf{x}_\lambda = [1 \ \lambda \ \lambda^2 \ \ldots \ \lambda^{n-1}]^T$. (a) Show that $\mathbf{x}_\lambda$ is an eigenvector of C_f^T if and only if $\lambda \in \text{spec} \, C_f$. (b) Show that $\mathbf{y} \in \mathbb{C}^n$ is an eigenvector of C_f^T if and only if $\mathbf{y}$ is a nonzero scalar multiple of $\mathbf{x}_\lambda$ for some $\lambda \in \text{spec} \, C_f$. (c) Deduce from (b) that every eigenvalue of C_f has geometric multiplicity 1.

P.11.46 Let f and g be monic polynomials of the same degree. Prove that C_f commutes with C_g if and only if $f = g$. *Hint:* What is column $n - 1$ of $C_f C_g$?

P.11.47 Use the companion matrix to show that the fundamental theorem of algebra is equivalent to the statement "every square complex matrix has an eigenvalue."

P.11.48 This problem relies on problems in Chapters 3 and 7. Let $A \in M_m$ and $B \in M_n$ have eigenvalues $\lambda_1, \lambda_2, \ldots, \lambda_m$ and $\mu_1, \mu_2, \ldots, \mu_n$, respectively. Factor $A = UTU^*$ and $B = VT'V^*$ as in Theorem 11.1.1, in which U, V are unitary and T, T' are upper triangular. (a) Show that $(U \otimes V)^*(A \otimes B)(U \otimes V) = T \otimes T'$ is upper triangular, and its diagonal entries are the eigenvalues of $A \otimes B$. (b) Deduce that the mn eigenvalues of $A \otimes B$ are $\lambda_i \mu_j$ for all $i = 1, 2, \ldots, m$ and $j = 1, 2, \ldots, n$. (c) Show that $(U \otimes V)^*(A \otimes I_n)(U \otimes V) = T \otimes I_n$ and $(U \otimes V)^*(I_m \otimes B)(U \otimes V) = I_m \otimes T'$, which are upper triangular. (d) Deduce that the mn eigenvalues of $A \otimes I_n + I_m \otimes B$ (the *Kronecker sum* of A and B) are $\lambda_i + \mu_j$ for all $i = 1, 2, \ldots, m$ and $j = 1, 2, \ldots, n$. (e) What can you deduce by applying Corollary 11.5.2 to the commuting matrices $A \otimes I_n$ and $I_m \otimes B$? Is this as good a result as the one in (d)? Explain. (f) Deduce from (b) that $\det(A \otimes B) = (\det A)^n (\det B)^m = \det(B \otimes A)$.

P.11.49 Let A, B, X, C be as in Theorem 11.4.1. (a) Use Theorem 3.4.16 to show that $\text{vec}\,(AX - XB) = (I_n \otimes A - B^T \otimes I_m) \text{vec}\,X = \text{vec}\,C$. (b) Let $K = I_n \otimes A - B^T \otimes I_m$ and explain why the mn eigenvalues of K are $\lambda_i - \mu_j$ for all $i = 1, 2, \ldots, m$ and $j = 1, 2, \ldots, n$. *Hint:* P.11.48. (c) Deduce that K is invertible if and only if $\text{spec}\,A \cap \text{spec}\,B = \varnothing$. (d) Deduce Theorem 11.4.1 from (c). *Hint:* The linear matrix equation (11.4.2) is equivalent to the linear system $K \, \text{vec}\,X = \text{vec}\,C$.

11.8 Notes

In Section 11.5, we showed that commutativity was a sufficient condition for simultaneous triangularizability, but it is not a necessary condition. For conditions that are both necessary and sufficient, see the discussion of McCoy's theorem in [HJ13, Theorem 2.4.8.7].

The decomposition described in P.11.44 is unique. If $A = B + C = D + E$, in which B and D are diagonalizable, C and E are nilpotent, $BC = CB$, and $DE = ED$, then $B = D$ and $C = E$. For a proof, see [HJ13, 3.2.P18].

For more information about Kronecker sums and the vec operator, see [HJ94, Chapter 4].

11.9 Some Important Concepts

- Each square complex matrix is unitarily similar to an upper triangular matrix.

- Each square complex matrix is annihilated by its characteristic polynomial.

- Minimal polynomial.

- Similar matrices have the same characteristic and minimal polynomials.

- Companion matrix.

- Sylvester's theorem on linear matrix equations.

- Each square complex matrix is similar to a direct sum of unispectral matrices.

- A square complex matrix is diagonalizable if and only if every zero of its minimal polynomial has multiplicity 1.

- Commuting matrices can be simultaneously unitarily upper triangularized.

The Jordan Form: Existence and Uniqueness

In Chapter 11, we found that each square complex matrix A is similar to a direct sum of upper triangular unispectral matrices. We now show that A is similar to a direct sum of Jordan blocks (unispectral upper bidiagonal matrices with 1s in the superdiagonal) that is unique up to permutation of its direct summands.

12.1 Ranks of Powers

We begin by using Theorem 4.2.4 to establish some facts about ranks of powers.

Theorem 12.1.1 *Let $A \in \mathsf{M}_n$.*

(a) *For each $k = 1, 2, \ldots$, we have $\operatorname{rank} A^k \geq \operatorname{rank} A^{k+1}$, with equality for some k if and only if $\operatorname{col} A^k = \operatorname{col} A^{k+1}$.*

(b) *If $k \in \{0, 1, 2, \ldots\}$ and $\operatorname{rank} A^k = \operatorname{rank} A^{k+1}$, then $\operatorname{col} A^k = \operatorname{col} A^{k+p}$ and $\operatorname{rank} A^k = \operatorname{rank} A^{k+p}$ for each $p = 1, 2, \ldots$.*

(c) *If A is not invertible, then $\operatorname{rank} A \leq n - 1$ and there is a least positive integer $q \leq n$ such that $\operatorname{rank} A^q = \operatorname{rank} A^{q+1}$.*

Proof (a) Since $A^{k+1} = AA^k$, Theorem 4.2.4 ensures that $\operatorname{rank} A^{k+1} \leq \min\{\operatorname{rank} A, \operatorname{rank} A^k\} \leq \operatorname{rank} A^k$, and that $\operatorname{rank} A^k = \operatorname{rank} A^{k+1}$ if and only if

$$\operatorname{col} A^k = \operatorname{col} A^{k+1}. \tag{12.1.2}$$

(b) Suppose that $\operatorname{rank} A^k = \operatorname{rank} A^{k+1}$. Theorem 4.2.4 ensures that (12.1.2) is true. We claim that $\operatorname{col} A^k = \operatorname{col} A^{k+p}$ (and hence $\operatorname{rank} A^k = \operatorname{rank} A^{k+p}$) for $p = 1, 2, \ldots$. To prove this claim, proceed by induction on p. The base case $p = 1$ is (12.1.2). Suppose that $p \geq 1$ and

$$\operatorname{col} A^k = \operatorname{col} A^{k+p}. \tag{12.1.3}$$

Let $\mathbf{x} \in \mathbb{C}^n$ and let $\mathbf{y}, \mathbf{z} \in \mathbb{C}^n$ be such that $A^k \mathbf{x} = A^{k+p} \mathbf{y}$ (the induction hypothesis (12.1.3)) and $A^k \mathbf{y} = A^{k+1} \mathbf{z}$ (the base case (12.1.2)). Then

$$A^k \mathbf{x} = A^{k+p} \mathbf{y} = A^p (A^k \mathbf{y}) = A^p (A^{k+1} \mathbf{z}) = A^{k+p+1} \mathbf{z},$$

which shows that $\operatorname{col} A^k \subseteq \operatorname{col} A^{k+p+1}$. Since $\operatorname{col} A^{k+p+1} \subseteq \operatorname{col} A^k$, we conclude that $\operatorname{col} A^k = \operatorname{col} A^{k+p+1}$. This concludes the induction.

(c) Since A is not invertible, $\operatorname{rank} A \leq n - 1$. For each $k = 1, 2, \ldots, n$, either $\operatorname{rank} A^k = \operatorname{rank} A^{k+1}$ or $\operatorname{rank} A^k - \operatorname{rank} A^{k+1} \geq 1$. Thus, the strict inequalities in

$$\underbrace{\operatorname{rank} A}_{\leq n-1} > \underbrace{\operatorname{rank} A^2}_{\leq n-2} > \cdots > \underbrace{\operatorname{rank} A^n}_{\leq 0} > \operatorname{rank} A^{n+1}$$

cannot all be correct. We conclude that $\operatorname{rank} A^k = \operatorname{rank} A^{k+1}$ for some $k \in \{1, 2, \ldots, n\}$. Let q be the least value of k such that this occurs. $\qquad \square$

Definition 12.1.4 Let $A \in \mathsf{M}_n$. If A is invertible, define its *index* to be 0. If A is not invertible, its *index* is the least positive integer q such that $\operatorname{rank} A^q = \operatorname{rank} A^{q+1}$.

The preceding theorem ensures that the index of each $A \in \mathsf{M}_n$ is at most n. Our definition of the index of an invertible matrix is consistent with the convention that $A^0 = I$ and $\operatorname{rank} A^0 = n$.

12.2 Jordan Blocks and Jordan Matrices

How can we tell if two square matrices are similar? Similar matrices share the same characteristic and minimal polynomials (Theorem 11.3.4), but so do some matrices that are not similar (Example 11.3.5). A definitive test for similarity must involve something more than the characteristic and minimal polynomials.

A promising approach to finding a test for similarity is suggested by an important result in the preceding chapter: Each square complex matrix is similar to a direct sum of unispectral matrices with pairwise disjoint spectra (Theorem 11.4.3). This result permits us to focus on a more specific question: When are two unispectral matrices similar?

If $A, B \in \mathsf{M}_n$ and $\operatorname{spec} A = \operatorname{spec} B = \{\lambda\}$, then A is similar to B if and only if $(A - \lambda I)$ and $(B - \lambda I)$ are similar (Theorem 2.6.9). Since $\operatorname{spec}(A - \lambda I) = \operatorname{spec}(B - \lambda I) = \{0\}$, we can focus on an even more specific question: When are two nilpotent matrices similar?

Theorem 12.2.1 *Let $A \in \mathsf{M}_n$. The following are equivalent:*

(a) *A is nilpotent, that is, $A^k = 0$ for some positive integer k.*

(b) $\operatorname{spec} A = \{0\}$.

(c) $p_A(z) = z^n$.

(d) $A^n = 0$.

(e) $m_A(z) = z^q$ *for some $q \in \{1, 2, \ldots, n\}$.*

Proof (a) $\Rightarrow$ (b) Let $(\lambda, \mathbf{x})$ be an eigenpair of A. Suppose that $k \geq 1$ and $A^k = 0$. Lemma 9.3.2 ensures that $\mathbf{0} = 0_n \mathbf{x} = A^k \mathbf{x} = \lambda^k \mathbf{x}$, so $\lambda^k = 0$ and hence $\lambda = 0$.
(b) $\Rightarrow$ (c) Since 0 is the only eigenvalue of A, its algebraic multiplicity is n and (10.2.2) says that $p_A(z) = z^n$.
(c) $\Rightarrow$ (d) The Cayley–Hamilton theorem ensures that $A^n = p_A(A) = 0$.
(d) $\Rightarrow$ (e) The polynomial $p(z) = z^n$ annihilates A and m_A divides p (Theorem 11.3.1.c), so $m_A(z) = z^q$ for some $q \in \{1, 2, \ldots, n\}$.
(e) $\Rightarrow$ (a) Let $k = q$ in (a). $\qquad \square$

If $A \in \mathsf{M}_n$ is nilpotent, then Theorem 12.1.1 ensures that

$$n - 1 \geq \operatorname{rank} A \geq \operatorname{rank} A^2 \geq \cdots \geq \operatorname{rank} A^n = 0$$

and

$$\operatorname{rank} A^k = \operatorname{rank} A^{k+1} \quad \Longrightarrow \quad \operatorname{rank} A^k = \operatorname{rank} A^{k+1} = \cdots = \operatorname{rank} A^n = 0.$$

According to Definition 12.1.4, the index of A is the smallest k such that rank $A^k = 0$. The exponent in $m_A(z) = z^q$ is the smallest $r \in \{1, 2, \ldots, n\}$ such that $A^r = 0$. Since $A^k = 0$ if and only if rank $A^k = 0$, the exponent q in $m_A(z) = z^q$ is the index of A.

Example 12.2.2 A strictly upper triangular matrix is unispectral with eigenvalue 0, so it is nilpotent. Such a matrix has a lot of zero entries, but some nilpotent matrices have no zero entries. For example,

$$A = \begin{bmatrix} 1 & i \\ i & -1 \end{bmatrix} \quad \text{and} \quad B = \begin{bmatrix} 1 & 1 \\ -1 & -1 \end{bmatrix} \tag{12.2.3}$$

are nilpotent and have index two.

Definition 12.2.4 A $k \times k$ *Jordan block with eigenvalue* λ is $J_1(\lambda) = [\lambda]$ if $k = 1$; if $k \geq 2$, it is the upper bidiagonal matrix

$$J_k(\lambda) = \begin{bmatrix} \lambda & 1 & & & \\ & \lambda & 1 & & \\ & & \ddots & \ddots & \\ & & & \lambda & 1 \\ & & & & \lambda \end{bmatrix} \in \mathsf{M}_k. \tag{12.2.5}$$

Every main diagonal entry of $J_k(\lambda)$ is λ, every entry in the first superdiagonal is 1, and all other entries are 0.

Definition 12.2.6 A *Jordan matrix* J is a direct sum of Jordan blocks. If J has r direct summands, then

$$J = J_{n_1}(\lambda_1) \oplus J_{n_2}(\lambda_2) \oplus \cdots \oplus J_{n_r}(\lambda_r), \tag{12.2.7}$$

in which $n_1, n_2, \ldots, n_r$ are positive integers. If $r > 1$, the scalars $\lambda_1, \lambda_2, \ldots, \lambda_r$ need not be distinct. The number of repetitions of a Jordan block $J_k(\lambda)$ in the direct sum (12.2.7) is its *multiplicity*.

Example 12.2.8 Jordan blocks of sizes 1, 2, and 3 are

$$J_1(\lambda) = [\lambda], \quad J_2(\lambda) = \begin{bmatrix} \lambda & 1 \\ 0 & \lambda \end{bmatrix}, \quad \text{and} \quad J_3(\lambda) = \begin{bmatrix} \lambda & 1 & 0 \\ 0 & \lambda & 1 \\ 0 & 0 & \lambda \end{bmatrix}.$$

Example 12.2.9 In the 9×9 Jordan matrix $J_3(1) \oplus J_2(0) \oplus J_3(1) \oplus J_1(4)$, the Jordan block $J_3(1)$ has multiplicity 2; each of the blocks $J_2(0)$ and $J_1(4) = [4]$ has multiplicity 1.

Example 12.2.10 An $r \times r$ diagonal matrix is a Jordan matrix (12.2.7) in which $n_i = 1$ for $i = 1, 2, \ldots, r$. That is, each Jordan block is 1×1.

Jordan blocks of the form $J_k(0)$ and Jordan matrices that are direct sums of such blocks are of special importance.

Definition 12.2.11 A *nilpotent Jordan block* is a Jordan block with eigenvalue zero. We often denote the $k \times k$ nilpotent Jordan block $J_k(0)$ by J_k. A *nilpotent Jordan matrix* J is a direct sum

$$J = J_{n_1} \oplus J_{n_2} \oplus \cdots \oplus J_{n_r} \tag{12.2.12}$$

of nilpotent Jordan blocks.

We claim that each nilpotent matrix is similar to a nilpotent Jordan matrix that is unique up to permutation of its direct summands. Before we can prove this claim, we have some work to do.

Each Jordan block $J_k(\lambda) = \lambda I_k + J_k$ is the sum of a scalar matrix and a nilpotent Jordan block. If we partition J_k according to its columns, then

$$J_k = [\mathbf{0} \ \mathbf{e}_1 \ \mathbf{e}_2 \ \cdots \ \mathbf{e}_{k-1}] \in \mathsf{M}_k. \tag{12.2.13}$$

This helps us visualize the *left-shift identities*

$$J_k\mathbf{e}_1 = \mathbf{0} \quad \text{and} \quad J_k\mathbf{e}_j = \mathbf{e}_{j-1} \qquad \text{for } j = 2, 3, \ldots, k. \tag{12.2.14}$$

Example 12.2.15 If $k = 3$, then

$$J_3\mathbf{e}_2 = \begin{bmatrix} 0 & 1 & 0 \\ 0 & 0 & 1 \\ 0 & 0 & 0 \end{bmatrix}\begin{bmatrix} 0 \\ 1 \\ 0 \end{bmatrix} = \begin{bmatrix} 1 \\ 0 \\ 0 \end{bmatrix} = \mathbf{e}_1 \quad \text{and} \quad J_3\mathbf{e}_3 = \begin{bmatrix} 0 & 1 & 0 \\ 0 & 0 & 1 \\ 0 & 0 & 0 \end{bmatrix}\begin{bmatrix} 0 \\ 0 \\ 1 \end{bmatrix} = \begin{bmatrix} 0 \\ 1 \\ 0 \end{bmatrix} = \mathbf{e}_2.$$

Lemma 12.2.16 *Let $k \geq 2$. For each $p = 1, 2, \ldots, k - 1$,*

$$J_k^p = [\underbrace{\mathbf{0} \ \ldots \ \mathbf{0}}_{p} \ \mathbf{e}_1 \ \ldots \ \mathbf{e}_{k-p}], \tag{12.2.17}$$

$J_k^p = 0$ *for* $p \geq k$, *and the index of J_k is k. The geometric multiplicity of 0 as an eigenvalue of J_k is 1; its algebraic multiplicity is k. Moreover,*

$$\operatorname{rank} J_k^p = \max\{k - p, 0\} \qquad \text{for } p = 1, 2, \ldots. \tag{12.2.18}$$

Proof We proceed by induction to prove (12.2.17). The representation (12.2.13) establishes the base case $p = 1$. If $k = 2$, then $J_k^p = 0$ for $p = 2, 3, \ldots$. If $k > 2$, suppose that (12.2.17) is valid and $p \leq k - 2$. Compute

$$J_k^{p+1} = J_k J_k^p = J_k[\underbrace{\mathbf{0} \ \ldots \ \mathbf{0}}_{p} \ \mathbf{e}_1 \ \mathbf{e}_2 \ \ldots \ \mathbf{e}_{k-p}]$$

$$= [\underbrace{\mathbf{0} \ \ldots \ \mathbf{0}}_{p} \ J_k\mathbf{e}_1 \ J_k\mathbf{e}_2 \ \ldots \ J_k\mathbf{e}_{k-p}]$$

$$= [\underbrace{\mathbf{0} \ \ldots \ \mathbf{0} \ \mathbf{0}}_{p+1} \ \mathbf{e}_1 \ \ldots \ \mathbf{e}_{k-(p+1)}].$$

If $p = k$, then $J_k^k = J_k J_k^{k-1} = [\mathbf{0} \ \ldots \ \mathbf{0} \ J_k\mathbf{e}_1] = 0$. If $p > k$, then $J_k^p = J_k^k J_k^{p-k} = 0$.

The assertion (12.2.18) follows because $\mathbf{e}_1, \mathbf{e}_2, \ldots, \mathbf{e}_{k-p}$ are linearly independent. The index of J_k is k because $J_k^{k-1} = [\mathbf{0} \ \ldots \ \mathbf{0} \ \mathbf{e}_1] \neq 0$ and $J_k^k = J_k^{k+1} = \cdots = 0$. Since J_k is unispectral and has size k, the algebraic multiplicity of its eigenvalue 0 is k. The geometric multiplicity of 0 is nullity $J_k = k - \operatorname{rank} J_k = 1$. $\qquad\square$

For each $i = 1, 2, \ldots, r$, the index of the direct summand J_{n_i} in (12.2.12) is n_i. Since

$$J^p = J_{n_1}^p \oplus J_{n_2}^p \oplus \cdots \oplus J_{n_r}^p \qquad \text{for } p = 1, 2, \ldots,$$

$J^p = 0$ if and only if each $J_{n_i}^p = 0$.

(a) The index of J is $q = \max\{n_1, n_2, \ldots, n_r\}$, so $m_J(z) = z^q$.

(b) The geometric multiplicity of 0 as an eigenvalue of J is r (one for each block).

(c) The algebraic multiplicity of 0 as an eigenvalue of J is $n_1 + n_2 + \cdots + n_r$.

In computations that involve nilpotent Jordan matrices, it can be convenient to recognize them as bordered matrices.

Example 12.2.19 Here are two ways to partition J_3 as a bordered matrix:

$$\begin{bmatrix} 0 & 1 & 0 \\ 0 & 0 & 1 \\ 0 & 0 & 0 \end{bmatrix} = \begin{bmatrix} 0 & \mathbf{e}_1^{\mathsf{T}} \\ \mathbf{0} & J_2 \end{bmatrix} \quad \text{and} \quad \begin{bmatrix} 0 & 1 & 0 \\ 0 & 0 & 1 \\ 0 & 0 & 0 \end{bmatrix} = \begin{bmatrix} \mathbf{0} & I_2 \\ 0 & \mathbf{0}^{\mathsf{T}} \end{bmatrix}, \quad \text{in which } \mathbf{0}, \mathbf{e}_1 \in \mathbb{R}^2.$$

Inspection of (12.2.5) reveals analogous partitions of any nilpotent Jordan block:

$$J_{k+1} = \begin{bmatrix} 0 & \mathbf{e}_1^{\mathsf{T}} \\ \mathbf{0} & J_k \end{bmatrix} = \begin{bmatrix} \mathbf{0} & I_k \\ 0 & \mathbf{0}^{\mathsf{T}} \end{bmatrix}, \quad \text{in which } \mathbf{0}, \mathbf{e}_1 \in \mathbb{R}^k \text{ and } k = 1, 2, \dots. \tag{12.2.20}$$

12.3 Existence of a Jordan Form

Our next goal is to show that each nilpotent matrix is similar to a nilpotent Jordan matrix. We begin with three technical lemmas.

Lemma 12.3.1 $I_k - J_k^{\mathsf{T}} J_k = \mathbf{e}_1 \mathbf{e}_1^{\mathsf{T}}$ for each $k = 2, 3, \dots$.

Proof Use (12.2.20) to compute

$$J_k^{\mathsf{T}} J_k = \begin{bmatrix} \mathbf{0}^{\mathsf{T}} & 0 \\ I_{k-1} & \mathbf{0} \end{bmatrix} \begin{bmatrix} \mathbf{0} & I_{k-1} \\ 0 & \mathbf{0}^{\mathsf{T}} \end{bmatrix} = \begin{bmatrix} 0 & \mathbf{0}^{\mathsf{T}} \\ \mathbf{0} & I_{k-1} \end{bmatrix} = \operatorname{diag}(0, 1, 1, \dots, 1).$$

Then $I_k - J_k^{\mathsf{T}} J_k = I_k - \operatorname{diag}(0, 1, 1, \dots, 1) = \operatorname{diag}(1, 0, 0, \dots, 0) = \mathbf{e}_1 \mathbf{e}_1^{\mathsf{T}}$. □

Lemma 12.3.2 Let $n \geq 2$, $\mathbf{a} \in \mathbb{C}^{n-1}$, and $S, B \in \mathsf{M}_{n-1}$. Suppose that S is invertible and $SBS^{-1} = J_{n-1}$. Let $A = \begin{bmatrix} 0 & \mathbf{a}^{\mathsf{T}} \\ \mathbf{0} & B \end{bmatrix}$. Then there is a $\mathbf{z} \in \mathbb{C}^{n-1}$ and a scalar c such that

$$\begin{bmatrix} 1 & -\mathbf{z}^{\mathsf{T}} S \\ \mathbf{0} & S \end{bmatrix} \begin{bmatrix} 0 & \mathbf{a}^{\mathsf{T}} \\ \mathbf{0} & B \end{bmatrix} \begin{bmatrix} 1 & \mathbf{z}^{\mathsf{T}} \\ \mathbf{0} & S^{-1} \end{bmatrix} = \begin{bmatrix} 0 & c\mathbf{e}_1^{\mathsf{T}} \\ \mathbf{0} & J_{n-1} \end{bmatrix}.$$

If $c = 0$, then A is similar to $J_1 \oplus J_{n-1}$; otherwise, A is similar to J_n.

Proof The identity (3.2.8) ensures that $\begin{bmatrix} 1 & -\mathbf{z}^{\mathsf{T}} S \\ \mathbf{0} & S \end{bmatrix}^{-1} = \begin{bmatrix} 1 & \mathbf{z}^{\mathsf{T}} \\ \mathbf{0} & S^{-1} \end{bmatrix}$ for any $\mathbf{z} \in \mathbb{C}^{n-1}$. Let $\mathbf{z}^{\mathsf{T}} = \mathbf{a}^{\mathsf{T}} S^{-1} J_{n-1}^{\mathsf{T}}$. Then Lemma 12.3.1 shows that A is similar to

$$\begin{bmatrix} 1 & -\mathbf{z}^{\mathsf{T}} S \\ \mathbf{0} & S \end{bmatrix} \begin{bmatrix} 0 & \mathbf{a}^{\mathsf{T}} \\ \mathbf{0} & B \end{bmatrix} \begin{bmatrix} 1 & \mathbf{z}^{\mathsf{T}} \\ \mathbf{0} & S^{-1} \end{bmatrix} = \begin{bmatrix} 0 & \mathbf{a}^{\mathsf{T}} S^{-1} - \mathbf{z}^{\mathsf{T}} SBS^{-1} \\ \mathbf{0} & SBS^{-1} \end{bmatrix}$$

$$= \begin{bmatrix} 0 & \mathbf{a}^{\mathsf{T}} S^{-1} - \mathbf{a}^{\mathsf{T}} S^{-1} J_{n-1}^{\mathsf{T}} J_{n-1} \\ \mathbf{0} & J_{n-1} \end{bmatrix} = \begin{bmatrix} 0 & \mathbf{a}^{\mathsf{T}} S^{-1} (I - J_{n-1}^{\mathsf{T}} J_{n-1}) \\ \mathbf{0} & J_{n-1} \end{bmatrix}$$

$$= \begin{bmatrix} 0 & \mathbf{a}^{\mathsf{T}} S^{-1} \mathbf{e}_1 \mathbf{e}_1^{\mathsf{T}} \\ \mathbf{0} & J_{n-1} \end{bmatrix} = \begin{bmatrix} 0 & c\mathbf{e}_1^{\mathsf{T}} \\ \mathbf{0} & J_{n-1} \end{bmatrix},$$

in which $c = \mathbf{a}^{\mathsf{T}} S^{-1} \mathbf{e}_1$. If $c = 0$, then A is similar to $J_1 \oplus J_{n-1}$. If $c \neq 0$, then the bordering identity (12.2.20) and the similarity

$$\begin{bmatrix} c^{-1} & \mathbf{0} \\ \mathbf{0} & I \end{bmatrix} \begin{bmatrix} 0 & c\mathbf{e}_1^{\mathsf{T}} \\ \mathbf{0} & J_{n-1} \end{bmatrix} \begin{bmatrix} c & \mathbf{0} \\ \mathbf{0} & I \end{bmatrix} = \begin{bmatrix} 0 & \mathbf{e}_1^{\mathsf{T}} \\ \mathbf{0} & J_{n-1} \end{bmatrix} = J_n$$

show that A is similar to J_n. □

A final lemma invokes the left-shift identities (12.2.14) for a nilpotent Jordan block. We use the standard basis $\mathbf{e}_1, \mathbf{e}_2, \ldots, \mathbf{e}_{k+1}$ for $\mathbb{C}^{k+1}$ and a $\mathbf{y} \in \mathbb{C}^m$ to construct rank-1 matrices of the form $\mathbf{e}_j \mathbf{y}^{\mathsf{T}} \in \mathsf{M}_{(k+1) \times m}$; they have $\mathbf{y}^{\mathsf{T}}$ in the jth row and zeros elsewhere.

Lemma 12.3.3 *Let k and m be positive integers, let $\mathbf{y} \in \mathbb{C}^m$, let $\mathbf{e}_1 \in \mathbb{C}^{k+1}$, and let $B \in \mathsf{M}_m$ be nilpotent and have index k or less. Then*

$$A = \begin{bmatrix} J_{k+1} & \mathbf{e}_1 \mathbf{y}^{\mathsf{T}} \\ 0 & B \end{bmatrix} \quad \text{is similar to} \quad \begin{bmatrix} J_{k+1} & 0 \\ 0 & B \end{bmatrix}.$$

Proof For any $X \in \mathsf{M}_{(k+1) \times m}$, (3.2.11) ensures that A is similar to

$$\begin{bmatrix} I & X \\ 0 & I \end{bmatrix} \begin{bmatrix} J_{k+1} & \mathbf{e}_1 \mathbf{y}^{\mathsf{T}} \\ 0 & B \end{bmatrix} \begin{bmatrix} I & -X \\ 0 & I \end{bmatrix} = \begin{bmatrix} J_{k+1} & \mathbf{e}_1 \mathbf{y}^{\mathsf{T}} + XB - J_{k+1}X \\ 0 & B \end{bmatrix}.$$

Let $X = \sum_{j=1}^{k} \mathbf{e}_{j+1} \mathbf{y}^{\mathsf{T}} B^{j-1}$ and use the fact that $B^k = 0$ to compute

$$XB = \sum_{j=1}^{k} \mathbf{e}_{j+1} \mathbf{y}^{\mathsf{T}} B^j = \sum_{j=1}^{k-1} \mathbf{e}_{j+1} \mathbf{y}^{\mathsf{T}} B^j = \sum_{j=2}^{k} \mathbf{e}_j \mathbf{y}^{\mathsf{T}} B^{j-1}.$$

The left-shift identities (12.2.14) tell us that $J_{k+1} \mathbf{e}_{j+1} = \mathbf{e}_j$, so

$$J_{k+1} X = \sum_{j=1}^{k} J_{k+1} \mathbf{e}_{j+1} \mathbf{y}^{\mathsf{T}} B^{j-1} = \sum_{j=1}^{k} \mathbf{e}_j \mathbf{y}^{\mathsf{T}} B^{j-1} = \mathbf{e}_1 \mathbf{y}^{\mathsf{T}} + XB.$$

Then $\mathbf{e}_1 \mathbf{y}^{\mathsf{T}} + XB - J_{k+1}X = 0$, which shows that A is similar to $J_{k+1} \oplus B$. $\qquad \square$

Once we have proved the following theorem, it is only a short additional step to show that every square complex matrix is similar to a Jordan matrix (a direct sum of Jordan blocks with various eigenvalues).

Theorem 12.3.4 *Each nilpotent matrix is similar to a nilpotent Jordan matrix.*

Proof Let $A \in \mathsf{M}_n$ be nilpotent. Since spec $A = \{0\}$, Theorem 11.1.1 ensures that A is unitarily similar to an upper triangular matrix with all diagonal entries equal to 0. Thus, we may assume that A is strictly upper triangular. We proceed by induction on n.

If $n = 1$, then $A = J_1 = [0]$, so A is a nilpotent Jordan block.

Assume that $n \geq 2$ and that every nilpotent matrix of size $n - 1$ or less is similar to a nilpotent Jordan matrix. Partition

$$A = \begin{bmatrix} 0 & \mathbf{a}^{\mathsf{T}} \\ 0 & B \end{bmatrix},$$

in which $\mathbf{a} \in \mathbb{C}^{n-1}$ and $B \in \mathsf{M}_{n-1}$ is strictly upper triangular. Let g be the geometric multiplicity of 0 as an eigenvalue of B and let q be the index of B. The induction hypothesis ensures that there is an invertible $S \in \mathsf{M}_{n-1}$ such that $S^{-1}BS = J_{k_1} \oplus J_{k_2} \oplus \cdots \oplus J_{k_g}$. After a block permutation similarity, if necessary, we may assume that $k_1 = q$, in which case $q \geq \max\{k_2, k_3, \ldots, k_g\}$.

If $g = 1$, then B is similar to J_{n-1} and Lemma 12.3.2 ensures that A is similar to one of the nilpotent Jordan matrices $J_1 \oplus J_{n-1}$ or J_n.

If $g \geq 2$, let $J = J_{k_2} \oplus \cdots \oplus J_{k_g}$ and observe that $S^{-1}BS = J_q \oplus J$. Every direct summand in J has size q or less, so $J^q = 0$. Then A is similar to

$$\begin{bmatrix} 1 & 0^\mathsf{T} \\ 0 & S^{-1} \end{bmatrix} \begin{bmatrix} 0 & \mathbf{a}^\mathsf{T} \\ 0 & B \end{bmatrix} \begin{bmatrix} 1 & 0^\mathsf{T} \\ 0 & S \end{bmatrix} = \begin{bmatrix} 0 & \mathbf{a}^\mathsf{T}S \\ 0 & S^{-1}BS \end{bmatrix} = \begin{bmatrix} 0 & \mathbf{a}^\mathsf{T}S \\ 0 & J_q \oplus J \end{bmatrix} = \left[\begin{array}{c|c|c} 0 & \mathbf{x}^\mathsf{T} & \mathbf{y}^\mathsf{T} \\ \hline 0 & J_q & 0 \\ \hline 0 & 0 & J \end{array} \right], \quad (12.3.5)$$

in which we have partitioned $\mathbf{a}^\mathsf{T}S = [\mathbf{x}^\mathsf{T}\ \mathbf{y}^\mathsf{T}]$ with $\mathbf{x} = [x_1\ x_2\ \dots\ x_q]^\mathsf{T} \in \mathbb{C}^q$ and $\mathbf{y} \in \mathbb{C}^{n-q-1}$.

Lemma 12.3.2 tells us that there is a scalar c such that the upper-left 2×2 block matrix in (12.3.5) is similar to

$$\begin{bmatrix} 0 & c\mathbf{e}_1^\mathsf{T} \\ 0 & J_q \end{bmatrix}, \quad \text{in which } \mathbf{e}_1 \in \mathbb{C}^q$$

via a similarity matrix of the form

$$\begin{bmatrix} 1 & -\mathbf{z}^\mathsf{T} \\ 0 & I \end{bmatrix}, \quad \text{in which } \mathbf{z} \in \mathbb{C}^q.$$

This observation leads us to a similarity of the 3×3 block matrix (12.3.5):

$$\begin{bmatrix} 1 & -\mathbf{z}^\mathsf{T} & 0^\mathsf{T} \\ 0 & I & 0 \\ 0 & 0 & I \end{bmatrix} \begin{bmatrix} 0 & \mathbf{x}^\mathsf{T} & \mathbf{y}^\mathsf{T} \\ 0 & J_q & 0 \\ 0 & 0 & J \end{bmatrix} \begin{bmatrix} 1 & \mathbf{z}^\mathsf{T} & 0^\mathsf{T} \\ 0 & I & 0 \\ 0 & 0 & I \end{bmatrix} = \begin{bmatrix} 0 & c\mathbf{e}_1^\mathsf{T} & \mathbf{y}^\mathsf{T} \\ 0 & J_q & 0 \\ 0 & 0 & J \end{bmatrix}. \quad (12.3.6)$$

If $c = 0$, then (12.3.6) is block permutation similar to

$$\left[\begin{array}{c|cc} J_q & 0 & 0 \\ \hline 0^\mathsf{T} & 0 & \mathbf{y}^\mathsf{T} \\ 0 & 0 & J \end{array} \right] = J_q \oplus \begin{bmatrix} 0 & \mathbf{y}^\mathsf{T} \\ 0 & J \end{bmatrix}. \quad (12.3.7)$$

The nilpotent 2×2 block matrix in (12.3.7) has size $n - q$, so the induction hypothesis ensures that it is similar to a nilpotent Jordan matrix. Thus, the matrix (12.3.6), and therefore A itself, is similar to a nilpotent Jordan matrix.

If $c \neq 0$, use the bordering identity (12.2.20) to demonstrate that (12.3.6) is similar to

$$\begin{bmatrix} c^{-1} & 0^\mathsf{T} & 0^\mathsf{T} \\ 0 & I & 0 \\ 0 & 0 & c^{-1}I \end{bmatrix} \begin{bmatrix} 0 & c\mathbf{e}_1^\mathsf{T} & \mathbf{y}^\mathsf{T} \\ 0 & J_q & 0 \\ 0 & 0 & J \end{bmatrix} \begin{bmatrix} c & 0^\mathsf{T} & 0^\mathsf{T} \\ 0 & I & 0 \\ 0 & 0 & cI \end{bmatrix}$$

$$= \left[\begin{array}{c|c|c} 0 & \mathbf{e}_1^\mathsf{T} & \mathbf{y}^\mathsf{T} \\ \hline 0 & J_q & 0 \\ \hline 0 & 0 & J \end{array} \right], \quad \text{in which } \mathbf{e}_1 \in \mathbb{C}^q$$

$$= \begin{bmatrix} J_{q+1} & \mathbf{e}_1\mathbf{y}^\mathsf{T} \\ 0 & J \end{bmatrix}, \quad \text{in which } \mathbf{e}_1 \in \mathbb{C}^{q+1}. \quad (12.3.8)$$

Lemma 12.3.3 ensures that (12.3.8) (and hence also A) is similar to the nilpotent Jordan matrix $J_{k_1+1} \oplus J$, in which $k_1 = q$. The induction hypothesis ensures that J is similar to a nilpotent Jordan matrix and hence A is similar to a nilpotent Jordan matrix. $\qquad \square$

Definition 12.3.9 A *Jordan form* for $A \in \mathsf{M}_n$ is a Jordan matrix to which A is similar.

We have proved that each nilpotent matrix has a Jordan form; it is a direct sum of nilpotent Jordan blocks.

Example 12.3.10 What are the Jordan forms of the 2×2 nilpotent matrices A and B in (12.2.3)? There are only two possibilities: J_2 or $J_1 \oplus J_1 = 0_2$. The second is excluded since neither A nor B is the zero matrix, so J_2 must be the Jordan form for A and for B. Consequently, A and B are similar because each is similar to the same Jordan matrix.

Finally, we can show that every square matrix has a Jordan form.

Theorem 12.3.11 *Each* $A \in \mathsf{M}_n$ *is similar to a Jordan matrix.*

Proof Theorem 11.4.3 shows that A is similar to a direct sum $T_{11} \oplus T_{22} \oplus \cdots \oplus T_{dd}$ of unispectral matrices, so it suffices to show that every unispectral matrix has a Jordan form. Indeed, if each $R_i^{-1} T_{ii} R_i = J^{(i)}$, in which $J^{(i)}$ is a unispectral Jordan matrix, then A is similar to $J^{(1)} \oplus J^{(2)} \oplus \cdots \oplus J^{(d)} = R^{-1}(T_{11} \oplus T_{22} \oplus \cdots \oplus T_{dd})R$, in which $R = R_1 \oplus R_2 \oplus \cdots \oplus R_d$.

If B is unispectral and spec $B = \{\lambda\}$, then $B - \lambda I$ is nilpotent. Theorem 12.3.4 says that there is a nilpotent Jordan matrix $J = J_{k_1} \oplus J_{k_2} \oplus \cdots \oplus J_{k_g}$ and an invertible S such that $B - \lambda I = SJS^{-1}$. Therefore, $B = \lambda I + SJS^{-1} = S(\lambda I + J)S^{-1}$. The computation

$$\lambda I + J = (\lambda I_{k_1} + J_{k_1}) \oplus (\lambda I_{k_2} + J_{k_2}) \oplus \cdots \oplus (\lambda I_{k_g} + J_{k_g})$$
$$= J_{k_1}(\lambda) \oplus J_{k_2}(\lambda) \oplus \cdots \oplus J_{k_g}(\lambda) \tag{12.3.12}$$

shows that (12.3.12) is a Jordan form for B.   □

12.4 Uniqueness of a Jordan Form

In this section, we show that two Jordan matrices are similar if and only if one can be obtained from the other by permuting its direct summands.

In the preceding section, we constructed a Jordan form for a square complex matrix by combining (as a direct sum) Jordan forms of unispectral matrices with different eigenvalues. If we can show that a Jordan form for a unispectral matrix is unique up to a permutation of its direct summands, then a Jordan form for a general square matrix would have the same property. Once again, it suffices to consider nilpotent unispectral matrices. Ranks of powers of nilpotent Jordan blocks are at the heart of the matter.

Example 12.4.1 Consider

$$J = J_3 \oplus J_3 \oplus J_2 \oplus J_2 \oplus J_2 \oplus J_1 \in \mathsf{M}_{13}, \tag{12.4.2}$$

which is a direct sum of six nilpotent Jordan blocks:

> one nilpotent Jordan block of size 1,
> three nilpotent Jordan blocks of size 2,
> two nilpotent Jordan blocks of size 3.

There are

> six blocks of size 1 or greater,
> five blocks of size 2 or greater,
> two blocks of size 3 or greater.

Use Lemma 12.2.16 to compute

$$\operatorname{rank} J^0 = 13,$$
$$\operatorname{rank} J^1 = 2 + 2 + 1 + 1 + 1 = 7,$$
$$\operatorname{rank} J^2 = 1 + 1 = 2,$$
$$\operatorname{rank} J^3 = \operatorname{rank} J^4 = 0.$$

Let $w_i = \operatorname{rank} J^{i-1} - \operatorname{rank} J^i$ and compute $w_1 = 13 - 7 = 6$, $w_2 = 7 - 2 = 5$, $w_3 = 2 - 0 = 2$, and $w_4 = 0 - 0 = 0$. Notice that $w_1 = 6$ is the total number of blocks in (12.4.2), which is the geometric multiplicity of 0 as an eigenvalue of J. Moreover, each w_p is the number of Jordan blocks in J that have size p or greater. Finally, $w_3 > 0$ and $w_4 = 0$, so the largest nilpotent block in (12.4.2) is 3×3. Now compute the differences $w_1 - w_2 = 1$, $w_2 - w_3 = 3$, and $w_3 - w_4 = 2$, and observe that each difference $w_p - w_{p+1}$ is the multiplicity of J_p. This is not an accident.

The preceding example illuminates our path forward, but we pause to establish a useful fact.

Lemma 12.4.3 *Let p and k be positive integers. For each $p = 1, 2, \ldots$,*

$$\operatorname{rank} J_k^{p-1} - \operatorname{rank} J_k^p = \begin{cases} 1 & \text{if } p \le k, \\ 0 & \text{if } p > k. \end{cases} \tag{12.4.4}$$

Proof If $p \le k$, then Lemma 12.2.16 ensures that $\operatorname{rank} J_k^{p-1} - \operatorname{rank} J_k^p = (k - (p - 1)) - (k - p) = 1$, but both ranks are 0 if $p > k$. $\qquad\square$

The identity (12.4.4) leads to an algorithm to determine the number of nilpotent blocks of each size in a Jordan matrix.

Theorem 12.4.5 *Let $A \in \mathsf{M}_n$ and suppose that 0 is an eigenvalue of A with geometric multiplicity $g \ge 1$. Let*

$$J_{n_1} \oplus J_{n_2} \oplus \cdots \oplus J_{n_g} \oplus J \in \mathsf{M}_n \tag{12.4.6}$$

be a Jordan form for A, in which J is a direct sum of Jordan blocks with nonzero eigenvalues. Let

$$w_p = \operatorname{rank} A^{p-1} - \operatorname{rank} A^p \quad \text{for } p = 1, 2, \ldots. \tag{12.4.7}$$

For each $p = 1, 2, \ldots, n$, the number of nilpotent blocks in (12.4.6) that have size p or larger is w_p and the multiplicity of J_p is $w_p - w_{p+1}$.

Proof Since J is invertible, $\operatorname{rank} J = \operatorname{rank} J^p$ for all $p = 0, 1, 2, \ldots$, and

$$\begin{aligned}
\operatorname{rank} A^{p-1} - \operatorname{rank} A^p &= \left(\sum_{i=1}^{g} \operatorname{rank} J_{n_i}^{p-1} + \operatorname{rank} J^{p-1} \right) - \left(\sum_{i=1}^{g} \operatorname{rank} J_{n_i}^{p} + \operatorname{rank} J^{p} \right) \\
&= \left(\sum_{i=1}^{g} \operatorname{rank} J_{n_i}^{p-1} + \operatorname{rank} J \right) - \left(\sum_{i=1}^{g} \operatorname{rank} J_{n_i}^{p} + \operatorname{rank} J \right) \\
&= \sum_{i=1}^{g} \left(\operatorname{rank} J_{n_i}^{p-1} - \operatorname{rank} J_{n_i}^{p} \right). \tag{12.4.8}
\end{aligned}$$

The identity (12.4.4) tells us that a summand in (12.4.8) is 1 if and only if its corresponding nilpotent block has size p or larger; otherwise it is 0. Therefore, (12.4.8) counts the number of nilpotent blocks that have size p or larger. The difference $w_p - w_{p+1}$ is the number of nilpotent blocks that have size p or larger, minus the number that have size $p + 1$ or larger; this is the number of nilpotent blocks whose size is exactly p. $\qquad\square$

If q is the size of the largest nilpotent block in the Jordan matrix (12.4.6), then $J_q^{q-1} \neq 0$ and $J_q^q = 0$. Therefore, $\operatorname{rank} A^q = \operatorname{rank} A^{q+1} = \cdots = \operatorname{rank} J$. That is, q is the index of A. The

A is similar to	J_4	$J_3 \oplus J_1$	$J_2 \oplus J_2$	$J_2 \oplus J_1 \oplus J_1$	$J_1 \oplus J_1 \oplus J_1 \oplus J_1$
$\operatorname{rank} A^0 =$	4	4	4	4	4
$w_1 =$	$\}\,1$	$\}\,2$	$\}\,2$	$\}\,3$	$\}\,4$
$\operatorname{rank} A^1 =$	3 $\}\boxed{0}$	2 $\}\boxed{1}$	2 $\}\boxed{0}$	1 $\}\boxed{2}$	0 $\}\boxed{4}$
$w_2 =$	$\}\,1$	$\}\,1$	$\}\,2$	$\}\,1$	$\}\,0$
$\operatorname{rank} A^2 =$	2 $\}\boxed{0}$	1 $\}\boxed{0}$	0 $\}\boxed{2}$	0 $\}\boxed{1}$	0
$w_3 =$	$\}\,1$	$\}\,1$	$\}\,0$	$\}\,0$	
$\operatorname{rank} A^3 =$	1 $\}\boxed{0}$	0 $\}\boxed{1}$	0	0	
$w_4 =$	$\}\,1$	$\}\,0$			
$\operatorname{rank} A^4 =$	0 $\}\boxed{1}$	0			
$w_5 =$	$\}\,0$				
$\operatorname{rank} A^5 =$	0				

Figure 12.1 Weyr characteristics of 4×4 nilpotent Jordan matrices. The index of A is the largest i such that $w_i > 0$; this is the size of the largest Jordan block in the Jordan canonical form of A. Reading the boxed numbers from top to bottom reveals the multiplicities of the Jordan blocks in the Jordan canonical form of A.

ranks of powers of A decrease monotonically to a stable value, which is the sum of the sizes of the invertible Jordan blocks in a Jordan form for A. Consequently, the rank differences in (12.4.7) are eventually all zero. In fact, $w_q > 0$ and $w_{q+1} = w_{q+2} = \cdots = 0$.

Definition 12.4.9 Let $A \in \mathsf{M}_n$, suppose that $0 \in \operatorname{spec} A$, let q be the index of A, and let $w_p = \operatorname{rank} A^{p-1} - \operatorname{rank} A^p$ for each $p = 1, 2, \ldots$. The list of positive integers $w_1, w_2, \ldots, w_q$, is the *Weyr characteristic of A*. If $\lambda \in \operatorname{spec} A$, the Weyr characteristic of $A - \lambda I$ is the *Weyr characteristic of A associated with λ*.

Example 12.4.10 Suppose that $A \in \mathsf{M}_4$ is nilpotent. Figure 12.1 tabulates the Weyr characteristics of the five different 4×4 nilpotent Jordan matrices to which A can be similar. The multiplicities of the Jordan blocks (in boxes) are computed from the Weyr characteristics, as described in Theorem 12.4.5. In each column, the integers in the Weyr characteristic have the following properties:

(a) w_1 is the number of blocks in a Jordan matrix to which A is similar.

(b) $w_q > 0$ and $w_{q+1} = 0$, in which q is the index of A.

(c) $w_1 + w_2 + \cdots + w_q = 4$.

(d) $w_i \geq w_{i+1}$ for $i = 1, 2, \ldots, q$.

If $A \in \mathsf{M}_n$ and $\lambda \in \operatorname{spec} A$, the first integer in the Weyr characteristic of $A - \lambda I$ is

$$w_1 = \operatorname{rank}(A - \lambda I)^0 - \operatorname{rank}(A - \lambda I) = n - \operatorname{rank}(A - \lambda I)$$
$$= \operatorname{nullity}(A - \lambda I) = \dim \mathcal{E}_\lambda(A).$$

Therefore, w_1 is the geometric multiplicity of λ (the number of blocks with eigenvalue λ in a Jordan form for A). For each $i = 1, 2, \ldots, q$, the integer w_i is positive because is it equal to the number of blocks of size i or greater. The differences $w_i - w_{i+1}$ are nonnegative (that is, the sequence w_i is decreasing) because $w_i - w_{i+1}$ is the number of blocks of size i (a nonnegative integer). The sum

$$w_1 + w_2 + \cdots + w_q = \sum_{p=1}^{q} \left(\operatorname{rank}(A - \lambda I)^{p-1} - \operatorname{rank}(A - \lambda I)^p \right)$$

$$= \operatorname{rank}(A - \lambda I)^0 - \operatorname{rank}(A - \lambda I)^q$$

$$= n - \operatorname{rank}(A - \lambda I)^q$$

$$= \operatorname{nullity}(A - \lambda I)^q$$

is the algebraic multiplicity of λ (the sum of the sizes of the blocks with eigenvalue λ in a Jordan form for A); see P.12.19.

Similar matrices A and B have the same spectrum, and corresponding powers A^p and B^p are similar (see Theorem 2.6.9), so $\operatorname{rank} A^p = \operatorname{rank} B^p$ for all $p = 1, 2, \ldots$. Consequently, similar matrices have the same Weyr characteristic associated with corresponding eigenvalues. The following corollary provides a converse of this assertion.

Corollary 12.4.11 *Let* $J, J' \in \mathsf{M}_n$ *be Jordan matrices. Then* J *is similar to* J' *if and only if* $\operatorname{spec} J = \operatorname{spec} J'$ *and* $\operatorname{rank}(J - \lambda I)^p = \operatorname{rank}(J' - \lambda I)^p$ *for each* $\lambda \in \operatorname{spec} J$ *and each* $p = 1, 2, \ldots, n$.

Proof For each $\lambda \in \operatorname{spec} J = \operatorname{spec} J'$, the hypotheses ensure that the Weyr characteristics of J and J' associated with λ are the same. It follows that the multiplicities of the Jordan blocks $J_k(\lambda)$ in J are the same as those in J' for each $k = 1, 2, \ldots, n$. $\qquad\square$

Theorem 12.4.12 *If* $A \in \mathsf{M}_n$ *and if* J *and* J' *are Jordan matrices that are similar to* A, *then* J' *can be obtained from* J *by permuting its direct summands.*

Proof Each Jordan form of A is similar to A, so J and J' are similar. Therefore, they have the same Weyr characteristics associated with each of their eigenvalues. It follows that their block sizes and multiplicities are the same for each of their eigenvalues. Consequently, one can be obtained from the other by permuting its direct summands. $\qquad\square$

Example 12.4.13 Suppose that $A \in \mathsf{M}_4$ has eigenvalues $3, 2, 2, 2$. A Jordan form for A must contain the direct summand $J_1(3) = [3]$ as well as a 3×3 Jordan matrix whose spectrum is $\{2\}$. There are only three possibilities (up to permutation of direct summands) for the latter: $J_3(2)$, $J_1(2) \oplus J_2(2)$, or $J_1(2) \oplus J_1(2) \oplus J_1(2) = 2I_3$. Therefore, A is similar to exactly one of

$$\begin{bmatrix} 3 & 0 & 0 & 0 \\ 0 & 2 & 0 & 0 \\ 0 & 0 & 2 & 0 \\ 0 & 0 & 0 & 2 \end{bmatrix} = J_1(3) \oplus J_1(2) \oplus J_1(2) \oplus J_1(2),$$

$$\begin{bmatrix} 3 & 0 & 0 & 0 \\ 0 & 2 & 1 & 0 \\ 0 & 0 & 2 & 0 \\ 0 & 0 & 0 & 2 \end{bmatrix} = J_1(3) \oplus J_2(2) \oplus J_1(2), \text{ or}$$

$$\begin{bmatrix} 3 & 0 & 0 & 0 \\ 0 & 2 & 1 & 0 \\ 0 & 0 & 2 & 1 \\ 0 & 0 & 0 & 2 \end{bmatrix} = J_1(3) \oplus J_3(2).$$

These matrices correspond to the three possibilities: $\operatorname{rank}(A - 2I) = 1, 2,$ or 3, respectively.

Corollary 12.4.14 *Let* $A, B \in \mathsf{M}_n$ *and suppose that* $\operatorname{spec} A = \operatorname{spec} B$.

(a) *A is similar to B if and only if* $\operatorname{rank}(A - \lambda I)^p = \operatorname{rank}(B - \lambda I)^p$ *for each* $\lambda \in \operatorname{spec} A$ *and each* $p = 1, 2, \dots, n$.

(b) *A is similar to* A^{T}.

Proof (a) Theorem 12.3.11 ensures that there are Jordan matrices J_A and J_B such that A is similar to J_A and B is similar to J_B. It follows that $\operatorname{rank}(J_A - \lambda I)^p = \operatorname{rank}(J_B - \lambda I)^p$ for each $\lambda \in \operatorname{spec} A$ and each $p = 1, 2, \dots, n$. Corollary 12.4.11 tells us that J_A is similar to J_B, so the transitivity of similarity ensures that A is similar to B.

(b) We have $\operatorname{spec} A = \operatorname{spec} A^{\mathsf{T}}$ and $\operatorname{rank}(A - \lambda I)^p = \operatorname{rank}((A - \lambda I)^p)^{\mathsf{T}} = \operatorname{rank}(A^{\mathsf{T}} - \lambda I)^p$ for each $\lambda \in \operatorname{spec} A$ and each $p = 1, 2, \dots$. The assertion follows from (a). $\qquad\square$

12.5 The Jordan Canonical Form

Theorem 12.3.11 says that each square complex matrix A is similar to a Jordan matrix J. Theorem 12.4.12 says that J is unique up to permutations of its direct summands. Therefore, if we agree that permuting the direct summands of a Jordan matrix is inessential, we may speak of *the* Jordan matrix that is similar to A. It is the *Jordan canonical form* of A.

When presenting the Jordan canonical form of A, it is a useful convention to group together the Jordan blocks with the same eigenvalue, and to arrange those blocks in decreasing order of their sizes. For example, if $\mu_1, \mu_2, \dots, \mu_d$ are the distinct eigenvalues of A, we can present its Jordan canonical form as a direct sum

$$J = J(\mu_1) \oplus J(\mu_2) \oplus \cdots \oplus J(\mu_d) \tag{12.5.1}$$

of Jordan matrices with different eigenvalues. The number of Jordan blocks in a direct summand $J(\mu_i)$ in (12.5.1) equals the geometric multiplicity of μ_i, which we denote by g_i. If the index of $A - \mu_i I$ is q_i, each direct summand in (12.5.1) is a direct sum of g_i Jordan blocks with eigenvalue μ_i; the largest of these blocks has size q_i.

In principle, one can determine the Jordan canonical form of a given $A \in \mathsf{M}_n$ by doing the following for each $\lambda \in \operatorname{spec} A$:

(a) Compute $r_p = \operatorname{rank}(A - \lambda I)^p$ for $p = 0, 1, 2, \dots$. Stop when the ranks stabilize; the index of $A - \lambda I$ (this is q) is the first value of p such that $r_p = r_{p+1}$. By definition, $r_0 = n$.

(b) Compute $w_p = r_{p-1} - r_p$ for $p = 1, 2, \dots, q+1$; the list of integers $w_1, w_2, \dots, w_q$ is the Weyr characteristic of $A - \lambda I$.

(c) For each $k = 1, 2, \dots, q$ there are $w_k - w_{k+1}$ blocks of the form $J_k(\lambda)$ in the Jordan canonical form of A.

This algorithm is an elegant conceptual tool, but it is not recommended for numerical computations. The following example illustrates why computation of the Jordan canonical form of a matrix in finite-precision arithmetic can be unreliable.

Example 12.5.2 For a Jordan block $J_k(\lambda)$ with $k \geq 2$ and for any $\varepsilon > 0$, the matrix $J_k(\lambda) + \operatorname{diag}(\varepsilon, 2\varepsilon, \dots, k\varepsilon)$ has distinct eigenvalues $\lambda + \varepsilon, \lambda + 2\varepsilon, \dots, \lambda + k\varepsilon$, so Corollary 10.4.9 ensures that it is diagonalizable. Thus, its Jordan canonical form is $[\lambda + \varepsilon] \oplus [\lambda + 2\varepsilon] \oplus \cdots \oplus [\lambda + k\varepsilon]$ and all the blocks are 1×1.

Small changes in the entries of a matrix can result in major changes in its Jordan canonical form. For another example of this phenomenon, see P.12.16.

Every square complex matrix is similar to a Jordan matrix, in which every entry in the first superdiagonal of each direct summand of size two or greater is $+1$. This is a conventional choice, but it is not the only choice. In Chapter 19, we make use of the following theorem, which says that the superdiagonal entries can be any $\varepsilon \neq 0$.

Theorem 12.5.3 *Let $A \in \mathsf{M}_n$, let a nonzero $\varepsilon \in \mathbb{C}$ be given, and let (12.5.1) be the Jordan canonical form of A. Then A is similar to*

$$J_\varepsilon = J_\varepsilon(\mu_1) \oplus J_\varepsilon(\mu_2) \oplus \cdots \oplus J_\varepsilon(\mu_d), \tag{12.5.4}$$

in which each $J_\varepsilon(\mu_k)$ is obtained from $J(\mu_k)$ by replacing each of its Jordan-block summands $J_m(\lambda)$ of size $m \geq 2$ with

$$J_m(\lambda, \varepsilon) = \begin{bmatrix} \lambda & \varepsilon & & \\ & \ddots & \ddots & \\ & & \ddots & \varepsilon \\ & & & \lambda \end{bmatrix} \in \mathsf{M}_m.$$

Proof Let $D_m = \operatorname{diag}(1, \varepsilon, \varepsilon^2, \ldots, \varepsilon^{m-1})$ and observe that $J_m(\lambda)$ is similar to $D_m^{-1}J_m(\lambda)D_m = J_m(\lambda, \varepsilon)$. □

12.6 Problems

P.12.1 Find the Jordan canonical forms of

$$\begin{bmatrix} 0 & 1 & 1 & 1 \\ 0 & 0 & 1 & 1 \\ 0 & 0 & 0 & 1 \\ 0 & 0 & 0 & 0 \end{bmatrix}, \quad \begin{bmatrix} 0 & 1 & 1 & 0 \\ 0 & 0 & 0 & 0 \\ 0 & 0 & 0 & 1 \\ 0 & 0 & 0 & 0 \end{bmatrix}, \quad \begin{bmatrix} 0 & 1 & 0 & 1 \\ 0 & 0 & 0 & 0 \\ 0 & 0 & 0 & 1 \\ 0 & 0 & 0 & 0 \end{bmatrix},$$

$$\begin{bmatrix} 0 & 0 & 1 & 1 \\ 0 & 0 & 0 & 1 \\ 0 & 0 & 0 & 0 \\ 0 & 0 & 0 & 0 \end{bmatrix}, \quad \begin{bmatrix} 0 & 0 & 1 & 1 \\ 0 & 0 & 1 & 1 \\ 0 & 0 & 0 & 0 \\ 0 & 0 & 0 & 0 \end{bmatrix}, \quad \begin{bmatrix} 0 & 0 & 1 & 1 \\ 0 & 0 & 0 & 1 \\ 0 & 1 & 0 & 0 \\ 0 & 0 & 0 & 0 \end{bmatrix}.$$

Which of these matrices are similar?

P.12.2 Show that $J_{k+1} = \begin{bmatrix} \mathbf{e}_2^{\mathsf{T}} & 0 \\ J_k^2 & \mathbf{e}_{k-1} \end{bmatrix} = \begin{bmatrix} J_k & \mathbf{e}_k \\ \mathbf{0}^{\mathsf{T}} & 0 \end{bmatrix}$, in which $\mathbf{e}_2, \mathbf{e}_{k-1}, \mathbf{e}_k \in \mathbb{R}^{k-1}$.

P.12.3 Suppose that $A \in \mathsf{M}_n$ and $B \in \mathsf{M}_m$. (a) Show that $A \oplus B$ is nilpotent if and only if A and B are nilpotent. (b) If A and B are nilpotent and $m = n$, are $A + B$ and AB nilpotent?

P.12.4 If $A \in \mathsf{M}_5$, $(A - 2I)^3 = 0$, and $(A - 2I)^2 \neq 0$, what are the possible Jordan canonical forms for A?

P.12.5 Let $n \geq 2$ and let $\mathbf{x}, \mathbf{y} \in \mathbb{C}^n$ be nonzero. There are two possibilities for the Jordan canonical form of $A = I + \mathbf{x}\mathbf{y}^*$. Find them.

P.12.6 Let $A \in M_n$ and let $\lambda \in \operatorname{spec} A$. Use the Jordan canonical form of A to show that the geometric multiplicity of λ equals nullity$(A - \lambda I)$ and the algebraic multiplicity of λ equals nullity $(A - \lambda I)^n$.

P.12.7 Let $A, B \in M_n$. Prove or give a counterexample to the assertion that $ABAB = 0$ implies $BABA = 0$. Does the dimension matter?

P.12.8 (a) If $A \in M_n$ is nilpotent, show that $(I - A)^{-1} = I + A + A^2 + \cdots + A^{n-1}$. (b) Compute $(I - AB)^{-1}$, in which AB is the product in Example 13.5.5. (c) Compute $(I - J_n)^{-1}$ and $(I + J_n)^{-1}$.

P.12.9 Define the *partition function* $p : \mathbb{N} \to \mathbb{N}$ as follows: Let $p(n)$ be the number of different ways to write a positive integer n as a sum of positive integers (ignore the order of summands). For example, $p(1) = 1$, $p(2) = 2$, $p(3) = 3$, and $p(4) = 5$. It is known that $p(10) = 42$, $p(100) = 190{,}569{,}292$, and

$$p(1{,}000) = 24{,}061{,}467{,}864{,}032{,}622{,}473{,}692{,}149{,}727{,}991.$$

(a) Show that $p(5) = 7, p(6) = 11$, and $p(7) = 15$. (b) Show that there are $p(n)$ possible Jordan canonical forms for an $n \times n$ nilpotent matrix.

P.12.10 Let $A \in M_4$ and let $r_k = \operatorname{rank} A^k$ for $k = 0, 1, 2, 3, 4$. Show that $r_0 = 4, r_1 = 2, r_2 = 1, r_3 = 0$ is possible, but $r_0 = 4, r_1 = 3, r_2 = 0$ is not possible.

P.12.11 Suppose that $A \in M_n$ and $\lambda \in \operatorname{spec} A$. Show that the multiplicity of the block $J_k(\lambda)$ in the Jordan form of A is $\operatorname{rank}(A - \lambda I)^{k-1} - 2\operatorname{rank}(A - \lambda I)^k + \operatorname{rank}(A - \lambda I)^{k+1}$.

P.12.12 Let $A \in M_n$ be nilpotent. What are the possible Weyr characteristics and Jordan canonical forms for A if (a) $n = 2$? (b) $n = 3$? (c) $n = 4$? (d) $n = 5$?

P.12.13 Give examples of matrices $A, B \in M_n$ that are not similar, but that have the same characteristic polynomials and the same minimal polynomials. Why must your examples have $n \geq 4$?

P.12.14 Let p and q be positive integers and let $A \in M_{pq}$ be nilpotent. If the Weyr characteristic of A is $w_1, w_2, \ldots, w_q$ and $w_1 = w_2 = \cdots = w_q = p$, show that the Jordan canonical form of A is $J_q \oplus \cdots \oplus J_q$ (p direct summands).

P.12.15 Suppose that $B \in M_m$ and $D \in M_n$ are nilpotent, with respective indices q_1 and q_2. Let $C \in M_{m \times n}$ and let $A = \begin{bmatrix} B & C \\ 0 & D \end{bmatrix}$. (a) Show that the $(1, 2)$ block of A^k is $\sum_{j=0}^{k-1} B^j C D^{k-j-1}$. (b) Prove that A is nilpotent, with index at most $q_1 + q_2$.

P.12.16 Let $\varepsilon \geq 0$ and construct $A_\varepsilon \in M_n$ by setting the $(n, 1)$ entry of the nilpotent Jordan block $J_n(0)$ equal to ε. (a) Show that $p_{A_\varepsilon}(z) = z^n - \varepsilon$. (b) If $\varepsilon > 0$, show that A_ε is diagonalizable and the modulus of every eigenvalue is $\varepsilon^{1/n}$. If $n = 64$ and $\varepsilon = 2^{-32}$, then every eigenvalue of A_ε has modulus $2^{-1/2} > 0.7$, but every eigenvalue of A_0 is 0.

P.12.17 Suppose that $A, B \in M_6$, B has eigenvalues $1, 1, 2, 2, 3, 3$, $AB = BA$, and $A^3 = 0$. Show that there are exactly four possible Jordan canonical forms for A. What are they?

P.12.18 Let $A \in M_n$ and $\lambda \in \mathbb{C}$. Define

$$\mathcal{G}_\lambda(A) = \{\mathbf{x} \in \mathbb{C}^n : (A - \lambda I)^m \mathbf{x} = \mathbf{0} \text{ for some } m \geq 0\}.$$

Show that: (a) $\mathcal{G}_\lambda(A) \neq \{\mathbf{0}\}$ if and only if $\lambda \in \operatorname{spec} A$. (b) $\mathcal{E}_\lambda(A) \subseteq \mathcal{G}_\lambda(A)$. Give an example in which $\mathcal{E}_\lambda(A) \neq \mathcal{G}_\lambda(A)$. (c) If $\lambda \in \operatorname{spec} A$, use the Jordan canonical form to show that $\mathcal{G}_\lambda(A) = \operatorname{null}(A - \lambda I)^n$. Can you identify possibly smaller exponents that would work? Why? (d) If $\mathbf{x} \in \mathcal{G}_\lambda(A)$, then $A\mathbf{x} \in \mathcal{G}_\lambda(A)$. (e) Let $\mu_1, \mu_2, \ldots, \mu_d$ be the distinct eigenvalues of A. Use the Jordan canonical form to show that $\mathcal{G}_{\mu_i}(A) \cap \mathcal{G}_{\mu_j}(A) = \{\mathbf{0}\}$ if $i \neq j$ and $\mathcal{G}_{\mu_1}(A) \oplus \mathcal{G}_{\mu_2}(A) \oplus \cdots \oplus \mathcal{G}_{\mu_d}(A) = \mathbb{C}^n$. *Remark*: The subspace $\mathcal{G}_\lambda(A)$ is a *generalized eigenspace* of A.

P.12.19 Let $A \in \mathsf{M}_n$ and let $\lambda \in \operatorname{spec} A$. Let $w_1, w_2, \ldots, w_q$ be the Weyr characteristic of $A - \lambda I$. Show that $w_1 + w_2 + \cdots + w_q$ is the algebraic multiplicity of λ. *Hint*: If $J_k(\mu)$ is a direct summand of the Jordan canonical form of A and $\mu \neq \lambda$, why is $\operatorname{rank} J_k(\mu - \lambda)^q = k$?

P.12.20 Let f be a monic polynomial of degree n and let $\mu_1, \mu_2, \ldots, \mu_d$ be the distinct zeros of f, with respective multiplicities $n_1, n_2, \ldots, n_d$. (a) Show that the Jordan canonical form of the companion matrix C_f is $J_{n_1}(\mu_1) \oplus J_{n_2}(\mu_2) \oplus \cdots \oplus J_{n_d}(\mu_d)$, which has exactly one Jordan block corresponding to each distinct eigenvalue. (b) Show that each $\lambda \in \operatorname{spec} C_f$ has geometric multiplicity 1.

P.12.21 Let $A \in \mathsf{M}_n$. Show that the following statements are equivalent:

(a) $\deg m_A = \deg p_A$.

(b) $p_A = m_A$.

(c) A is similar to the companion matrix of p_A.

(d) Each eigenvalue of A has geometric multiplicity 1.

(e) The Jordan canonical form of A contains exactly one block corresponding to each distinct eigenvalue.

(f) The Weyr characteristic associated with each eigenvalue of A has the form $1, 1, \ldots, 1$.

We say that A is *nonderogatory* if any of these statements is true.

P.12.22 Let $A, B \in \mathsf{M}_n$. If $p_A = p_B = m_A = m_B$, show that A is similar to B.

P.12.23 If $A, B \in \mathsf{M}_n$ are nonderogatory, show that A is similar to B if and only if $p_A = p_B$.

P.12.24 Let $D = \operatorname{diag}(d_1, d_2, \ldots, d_n) \in \mathsf{M}_n$ be an invertible diagonal matrix. (a) Use Corollary 12.4.11 to show that the Jordan canonical form of $J_n D$ is J_n. (b) Find a diagonal matrix E such that $E J_n D E^{-1} = J_n$. (c) Suppose that $\varepsilon \neq 0$. What does $\lambda I + \varepsilon J_n$ look like, and why is it similar to $J_n(\lambda)$?

P.12.25 Show that $(I - J_n)$ is similar to $(I + J_n)$.

P.12.26 Let $T \in \mathcal{L}(\mathcal{P}_4)$ be the differentiation operator $T : p \to p'$ on the complex vector space of polynomials of degree at most 4, and consider the basis $\beta = 1, z, z^2, z^3, z^4$ of $\mathcal{P}_4$. (a) Show that $_\beta[T]_\beta = [\mathbf{0} \ \mathbf{e}_1 \ 2\mathbf{e}_2 \ 3\mathbf{e}_3 \ 4\mathbf{e}_4] = J_5 D$, in which $D = \operatorname{diag}(0, 1, 2, 3, 4)$. (b) Find the Jordan canonical form of $_\beta[T]_\beta$. (c) If γ is a basis for $\mathcal{P}_4$, what is the Jordan canonical form of $_\gamma[T]_\gamma$? Why?

P.12.27 Let $\beta = 1, y, z, y^2, z^2, yz$ and let $\mathcal{V} = \operatorname{span} \beta$ be the complex vector space of polynomials in two variables y and z that have degree at most 2. Let $T \in \mathcal{L}(\mathcal{V})$ be the partial differentiation operator $T : p \to \partial p / \partial y$. Find the Jordan canonical form of $_\beta[T]_\beta$.

P.12.28 If $A \in \mathsf{M}_9$ and $A^7 = A$, what can you say about the Jordan canonical form of A?

P.12.29 Let $\mu_1, \mu_2, \ldots, \mu_d$ be the distinct eigenvalues of $A \in \mathsf{M}_n$. If $\operatorname{rank}(A - \mu_1 I)^{n-2} > \operatorname{rank}(A - \mu_1 I)^{n-1}$, prove that $d \leq 2$.

P.12.30 Let $A \in \mathsf{M}_5$. If $p_A(z) = (z-2)^3(z+4)^2$ and $m_A(z) = (z-2)^2(z+4)$, what is the Jordan canonical form of A?

P.12.31 Let $\mu_1, \mu_2, \ldots, \mu_d$ be the distinct eigenvalues of $A \in M_n$, let $p_A(z) = (z - \mu_1)^{n_1}(z - \mu_2)^{n_2} \cdots (z - \mu_d)^{n_d}$ be its characteristic polynomial, and suppose that its minimal polynomial (11.3.2) has each exponent q_i equal to n_i or to $n_i - 1$. What are the possible Jordan canonical forms of A?

P.12.32 Let $n \geq 2$, let $A \in M_n$, and let k be a positive integer. (a) Show that

$$\mathbb{C}^n = \text{null } A^k \oplus \text{col } A^k \tag{12.6.1}$$

need not be true if $k = 1$. (b) Show that (12.6.1) is true if $k = n$. (c) What is the smallest k for which (12.6.1) is true?

P.12.33 If $A \in M_n$ and $B \in M_m$ and A or B is nilpotent, show that $A \otimes B$ is nilpotent.

P.12.34 Let λ be a nonzero scalar. (a) Show that the Jordan canonical form of $J_p(0) \otimes J_q(\lambda)$ is $J_p(0) \oplus \cdots \oplus J_p(0)$ (q direct summands). (b) Show that the Jordan canonical form of $J_p(\lambda) \otimes J_q(0)$ is $J_q(0) \oplus \cdots \oplus J_q(0)$ (p direct summands).

P.12.35 (a) Show that the Jordan canonical form of $J_2 \otimes J_2$ is $J_1 \oplus J_1 \oplus J_2$. (b) Show that the Jordan canonical form of $J_2 \otimes J_3$ is $J_1 \oplus J_1 \oplus J_2 \oplus J_2$. (c) What is the Jordan canonical form of $J_2 \oplus J_n$ for $n \geq 4$?

12.7 Notes

For additional information about the Jordan canonical form, see [HJ13, Chapter 3].

The Jordan canonical form was announced by Camille Jordan in an 1870 book. It is an excellent tool for problems involving powers and functions of matrices, but it is not the only way to classify matrices by similarity. In an 1885 paper, Eduard Weyr described the Weyr characteristic and used it to construct an alternative canonical form for similarity. Weyr's canonical form for $A \in M_n$ is a direct sum of blocks $W(\lambda)$, one for each $\lambda \in \text{spec } A$. Weyr's unispectral blocks are scalar matrices or block upper bidiagonal matrices that directly display the Weyr characteristic of $A - \lambda I$:

$$W(\lambda) = \begin{bmatrix} \lambda I_{w_1} & G_{w_1,w_2} & & & \\ & \lambda I_{w_2} & G_{w_2,w_3} & & \\ & & \ddots & & \\ & & & \ddots & G_{w_{q-1},w_q} \\ & & & & \lambda I_{w_q} \end{bmatrix}, \tag{12.7.1}$$

in which $q \geq 2$ and

$$G_{w_{j-1},w_j} = \begin{bmatrix} I_{w_j} \\ 0 \end{bmatrix} \in M_{w_{j-1} \times w_j} \quad \text{for } j = 2, 3, \ldots, q.$$

The Weyr and Jordan canonical forms for A are permutation similar. Weyr's canonical form is an excellent tool for problems involving commuting matrices, as illustrated in P.13.37. A matrix that commutes with a unispectral Weyr block (12.7.1) must be block upper triangular; matrices that are not block upper triangular can commute with a unispectral Jordan matrix (see P.13.37). For more information about the Weyr canonical form and a historical summary, see [HJ13, Section 3.4].

12.8 Some Important Concepts

- Index of a nilpotent matrix.

- Jordan block and Jordan matrix.

- Weyr characteristic.

- Each square complex matrix is similar to a Jordan matrix (its Jordan canonical form), which is unique up to permutation of its direct summands.

13 The Jordan Form: Applications

In this chapter, we discuss several problems in which a similarity transformation to Jordan canonical form facilitates a solution. For example, we find that A is similar to A^T; $\lim_{p\to\infty} A^p = 0$ if and only if every eigenvalue of A has modulus less than 1; and the invertible Jordan blocks of AB and BA are the same (a generalization of Theorem 10.7.2). We begin by considering coupled systems of ordinary differential equations.

13.1 Differential Equations and the Jordan Canonical Form

Let $A = [a_{ij}] \in M_n$ and consider the problem of finding a vector-valued function $\mathbf{x}(t) = [x_i(t)] \in \mathbb{C}^n$ of a real variable t such that

$$\mathbf{x}'(t) = A\mathbf{x}(t) \quad \text{and} \quad \mathbf{x}(0) \in \mathbb{C}^n \text{ is given.} \tag{13.1.1}$$

A theorem from differential equations guarantees that for each choice of the initial condition $\mathbf{x}(0)$, the *initial-value problem* (13.1.1) has a unique solution. The unknown functions $x_i(t)$ satisfy the coupled scalar equations

$$\frac{dx_i}{dt} = \sum_{j=1}^{n} a_{ij} x_j(t), \quad \text{in which } x_i(0) \text{ is given for } i = 1, 2, \dots, n.$$

If A is diagonalizable, a change of variables decouples the equations and significantly simplifies the problem. Suppose that $A = S\Lambda S^{-1}$, in which $\Lambda = \operatorname{diag}(\lambda_1 \lambda_2, \dots, \lambda_n)$ and $S = [s_{ij}]$. Let $S^{-1}\mathbf{x}(t) = \mathbf{y}(t) = [y_i(t)]$. Then

$$\mathbf{y}'(t) = (S^{-1}\mathbf{x}(t))' = S^{-1}\mathbf{x}'(t) = S^{-1}A\mathbf{x}(t) = S^{-1}AS\mathbf{y}(t) = \Lambda\mathbf{y}(t),$$

so in the new variables our problem is

$$\mathbf{y}'(t) = \Lambda\mathbf{y}(t), \quad \text{in which } \mathbf{y}(0) = S^{-1}\mathbf{x}(0) \text{ is given.} \tag{13.1.2}$$

The initial-value problem (13.1.2) is equivalent to n uncoupled scalar equations

$$\frac{dy_i}{dt} = \lambda_i y_i(t), \quad \text{in which } y_i(0) = \sum_{j=1}^{n} \sigma_{ij} x_j(0) \text{ for } i = 1, 2, \dots, n, \tag{13.1.3}$$

and $S^{-1} = [\sigma_{ij}] \in M_n$. Each equation can be solved separately: $y_i(t) = y_i(0)e^{\lambda_i t}$ for $i = 1, 2, \dots, n$. If λ_j is real, then $e^{\lambda_j t}$ is a real exponential function. If $\lambda_j = \alpha_j + i\beta_j$, in which α_j, β_j are real and $\beta_j \neq 0$, then

$$e^{\lambda_j t} = e^{\alpha_j t + i\beta_j t} = e^{\alpha_j t} e^{i\beta_j t} = e^{\alpha_j t}(\cos \beta_j t + i \sin \beta_j t) = e^{\alpha_j t} \cos \beta_j t + i e^{\alpha_j t} \sin \beta_j t \tag{13.1.4}$$

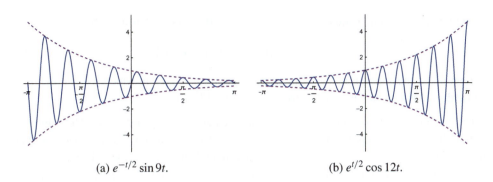

(a) $e^{-t/2}\sin 9t$.   (b) $e^{t/2}\cos 12t$.

Figure 13.1 Damped oscillating functions.

is a linear combination of damped oscillating functions $e^{\alpha_j t}\cos\beta_j t$ and $e^{\alpha_j t}\sin\beta_j t$; see Figure 13.1. Therefore, the entries of the solution $\mathbf{x}(t) = [x_i(t)] = S\mathbf{y}(t)$ to (13.1.1) are linear combinations of functions of the form (13.1.4). The parameters λ_j are the eigenvalues of A and the columns of $S = [\mathbf{s}_1\ \mathbf{s}_2\ \ldots\ \mathbf{s}_n]$ are corresponding eigenvectors of A. Thus, the solution

$$\mathbf{x}(t) = S\mathbf{y}(t) = \sum_{j=1}^{n} y_j(0)e^{\lambda_j t}\mathbf{s}_j$$

to the initial-value problem (13.1.1) is a linear combination of eigenvectors of A.

If A is not diagonalizable, let $A = SJS^{-1}$, in which J is the Jordan canonical form of A. A change of variables partially decouples the equations and yields several smaller and simpler initial-value problems, one for each block in the Jordan canonical form of A. Let $\mathbf{y}(t) = S^{-1}\mathbf{x}(t)$. Then $\mathbf{y}'(t) = (S^{-1}\mathbf{x})'(t) = S^{-1}\mathbf{x}'(t) = S^{-1}A\mathbf{x}(t) = S^{-1}AS\mathbf{y}(t) = J\mathbf{y}(t)$, so the initial-value problem is equivalent to

$$\mathbf{y}'(t) = J\mathbf{y}(t), \quad \text{in which } \mathbf{y}(0) = S^{-1}\mathbf{x}(0) \text{ is given.} \tag{13.1.5}$$

If J is a direct sum (12.2.7) of r Jordan blocks (their eigenvalues need not be distinct), the system (13.1.5) is equivalent to r smaller initial-value problems

$$\mathbf{y}'_i(t) = J_{n_i}(\lambda_i)\mathbf{y}_i(t), \quad \text{in which } \mathbf{y}_i(0) \in \mathbb{C}^{n_i} \text{ is given for } i = 1, 2, \ldots, r, \tag{13.1.6}$$

and $n_1 + n_2 + \cdots + n_r = n$. Each initial-value problem can be solved separately.

The following example illustrates how to solve a problem of the form (13.1.6).

Example 13.1.7 The system $\mathbf{y}'(t) = J_3(\lambda)\mathbf{y}(t)$ is

$$
\begin{aligned}
\frac{dy_1}{dt} &= \lambda y_1(t) + y_2(t) \\
\frac{dy_2}{dt} &= \lambda y_2(t) + y_3(t) \\
\frac{dy_3}{dt} &= \lambda y_3(t).
\end{aligned}
$$

Suppose that the initial value $\mathbf{y}(0) = [y_1(0)\ y_2(0)\ y_3(0)]^\mathsf{T}$ is given. Initial-value problems of this kind can be solved sequentially, starting at the bottom. The solution to the last equation is $y_3(t) = e^{\lambda t}y_3(0)$. The next equation is $y'_2 = \lambda y_2 + e^{\lambda t}y_3(0)$, which has the solution

$$y_2(t) = e^{\lambda t}\big(y_3(0)t + y_2(0)\big);$$

see P.13.7. The remaining equation $y'_1 = \lambda y_1 + e^{\lambda t}(y_3(0)t + y_2(0))$ has the solution

$$y_1(t) = e^{\lambda t}\left(y_3(0)\frac{t^2}{2} + y_2(0)t + y_1(0)\right).$$

The entries of the solution $\mathbf{y}(t)$ are functions of the form $e^{\lambda t}p(t)$, in which p is a polynomial whose degree is less than three (the size of the Jordan block).

The preceding example illustrates the principal features of the general case. The solution to each of the initial-value problems (13.1.6) is a vector-valued function whose entries are functions of the form $e^{\lambda_i t}p_i(t)$, in which p_i is a polynomial of degree $n_i - 1$ or less. If $\mu_1, \mu_2, \ldots, \mu_d$ are the distinct eigenvalues of A, then each entry $x_i(t)$ in the solution $\mathbf{x}(t) = [x_i(t)]$ of (13.1.1) is a linear combination of the entries of $\mathbf{y}(t)$, so for $i = 1, 2, \ldots, n$, we have

$$x_i(t) = e^{\mu_1 t}p_1(t) + e^{\mu_2 t}p_2(t) + \cdots + e^{\mu_d t}p_d(t), \tag{13.1.8}$$

in which each p_i is a polynomial with degree less than the index of $A - \mu_i I$.

If A has some non-real eigenvalues, the exponential functions $e^{\mu_i t}$ and the polynomials $p_i(t)$ in the solution (13.1.8) need not be real valued. If A is real, however, then $\operatorname{Re}\mathbf{x}(t)$ and $\operatorname{Im}\mathbf{x}(t)$ are both solutions of the differential equations in (13.1.1), with respective initial conditions $\operatorname{Re}\mathbf{x}(0)$ and $\operatorname{Im}\mathbf{x}(0)$. If the initial condition $\mathbf{x}(0)$ is real, then $\operatorname{Im}\mathbf{x}(0) = \mathbf{0}$ and the uniqueness of the solution of (13.1.1) implies that $\operatorname{Im}\mathbf{x}(t) = \mathbf{0}$ for all $t \in \mathbb{R}$. Consequently, the solution $\mathbf{x}(t) = S\mathbf{y}(t)$ must be real even if neither S nor $\mathbf{y}(t)$ is real.

Example 13.1.9 Consider the initial-value problem (13.1.1), in which $A = \begin{bmatrix} 0 & 1 \\ -1 & 0 \end{bmatrix}$ and $\mathbf{x}(0) = \begin{bmatrix} 1 \\ 2 \end{bmatrix}$. Then $A = S\Lambda S^{-1}$, in which $S = \begin{bmatrix} 1 & 1 \\ i & -i \end{bmatrix}$, $S^{-1} = \frac{1}{2}\begin{bmatrix} 1 & -i \\ 1 & i \end{bmatrix}$, and $\Lambda = \operatorname{diag}(i, -i)$. After the change of variables $\mathbf{y}(t) = S^{-1}\mathbf{x}(t)$, the decoupled equations (13.1.3) are

$$\frac{dy_1}{dt} = iy_1(t), \qquad\qquad y_1(0) = \tfrac{1}{2} - i,$$

$$\frac{dy_2}{dt} = -iy_2(t), \qquad\qquad y_2(0) = \tfrac{1}{2} + i.$$

The solution is

$$\mathbf{y}(t) = \begin{bmatrix} y_1(t) \\ y_2(y) \end{bmatrix} = \begin{bmatrix} (\tfrac{1}{2} - i)e^{it} \\ (\tfrac{1}{2} + i)e^{-it} \end{bmatrix}.$$

Thus, the solution to (13.1.1) is (using Euler's formula)

$$\mathbf{x}(t) = S\mathbf{y}(t) = \begin{bmatrix} 1 & 1 \\ i & -i \end{bmatrix}\begin{bmatrix} y_1(t) \\ y_2(t) \end{bmatrix} = \begin{bmatrix} y_1(t) + y_2(t) \\ iy_1(t) - iy_2(t) \end{bmatrix} = \begin{bmatrix} \cos t + 2\sin t \\ 2\cos t - \sin t \end{bmatrix}.$$

13.2 Convergent and Power-Bounded Matrices

In this section, we consider a sequence of matrix powers $I, A, A^2, A^3, \ldots$. What properties of A ensure that all the entries of A^p tend to zero as $p \to \infty$? What properties ensure that all the entries remain bounded as $p \to \infty$? To investigate the first question, we begin with a matrix version of the scalar geometric series.

Definition 13.2.1 Let $A_k = \left[a_{ij}^{(k)}\right] \in M_n$ for $k = 1, 2, \ldots$ and let $B = [b_{ij}] \in M_n$. If $\lim_{k \to \infty} a_{ij}^{(k)} = b_{ij}$ for each $i, j = 1, 2, \ldots, n$, then we write $\lim_{k \to \infty} A_k = B$ and say that A_k converges to B.

Two important special cases of the definition are: (a) $A_k = X^k$ is a sequence of powers of some $X \in M_n$ and $B = 0$, and (b) $A_k = \sum_{j=0}^{k} c_j X^j$ is a sequence of partial sums of a power series and $\lim_{k\to\infty} A_k$, if it exists, defines $B = \sum_{j=0}^{\infty} c_j X^j$.

Lemma 13.2.2 *Let $A \in M_n$.*

(a) *If $p \geq 1$, then $(I - A)\sum_{k=0}^{p-1} A^k = I - A^p$.*

(b) *If $p \geq 1$ and $1 \notin \mathrm{spec}\, A$, then $\sum_{k=0}^{p-1} A^k = (I - A)^{-1}(I - A^p)$.*

(c) *If $\lim_{p\to\infty} A^p = 0$, then $\sum_{k=0}^{\infty} A^k = (I - A)^{-1}$.*

Proof (a) Compute $(I - A)\sum_{k=0}^{p-1} A^k = \sum_{k=0}^{p-1} A^k - \sum_{k=1}^{p} A^k = I - A^p$.

(b) If $1 \notin \mathrm{spec}\, A$, then Theorem 9.1.16 ensures that $I - A$ is invertible, so the assertion follows from (a).

(c) If $(1, \mathbf{x})$ is an eigenpair of A, then $\mathbf{x} = A^p\mathbf{x}$ for $p \geq 1$. Thus, $\mathbf{x} = \lim_{p\to\infty} A^p\mathbf{x} = 0\mathbf{x} = \mathbf{0}$, a contradiction, so $1 \notin \mathrm{spec}\, A$. Then (b) ensures that

$$\sum_{k=0}^{\infty} A^k = \lim_{p\to\infty} \sum_{k=0}^{p-1} A^k = \lim_{p\to\infty}\left((I - A)^{-1}(I - A^p)\right) = (I - A)^{-1}(I - \lim_{p\to\infty} A^p) = (I - A)^{-1}. \quad \square$$

Definition 13.2.3 If $A \in M_n$, then A is *convergent* if $\lim_{k\to\infty} A^k = 0$.

How can we tell if a matrix is convergent? It helps to know that the property of being convergent is invariant under similarity.

Lemma 13.2.4 *If $A, B \in M_n$ are similar, then A is convergent if and only if B is convergent.*

Proof Suppose that $S = [s_{ij}] \in M_n$ is invertible, let $S^{-1} = [\sigma_{ij}]$, and suppose that $A = SBS^{-1}$. Let $A^p = \left[a_{ij}^{(p)}\right]$ and $B^p = \left[b_{ij}^{(p)}\right]$. Then $A^p = SB^pS^{-1}$, so each entry

$$a_{ij}^{(p)} = \sum_{k,\ell=1}^{n} s_{ik}b_{k\ell}^{(p)}\sigma_{\ell j} \tag{13.2.5}$$

of A^p is a fixed linear combination of the entries of B^p. If B is convergent, then $\lim_{p\to\infty} b_{k\ell}^{(p)} = 0$ for each $k, \ell = 1, 2, \ldots, n$. Therefore,

$$\lim_{p\to\infty} a_{ij}^{(p)} = \lim_{p\to\infty} \sum_{k,\ell=1}^{n} s_{ik}b_{k\ell}^{(p)}\sigma_{\ell j} = \sum_{k,\ell=1}^{n} s_{ik}\left(\lim_{p\to\infty} b_{k\ell}^{(p)}\right)\sigma_{\ell j} = 0 \quad \text{for } i, j = 1, 2, \ldots, n,$$

so A is convergent. The converse follows from interchanging A and B in this argument. $\quad \square$

Our criterion for convergence involves an eigenvalue that has the largest modulus; there could be several such eigenvalues.

Definition 13.2.6 The *spectral radius* of $A \in M_n$ is $\rho(A) = \max\{|\lambda| : \lambda \in \mathrm{spec}\, A\}$.

If $n \geq 2$, it is possible to have $A \neq 0$ and $\rho(A) = 0$. For example, $\rho(A) = 0$ for any nilpotent $A \in M_n$.

Theorem 13.2.7 *$A \in M_n$ is convergent if and only if $\rho(A) < 1$.*

Proof The preceding lemma ensures that A is convergent if and only if its Jordan canonical form is convergent. The Jordan canonical form of A is a direct sum (12.2.7) of Jordan blocks, so it is convergent if and only if each of its direct summands $J_k(\lambda)$ is convergent. If $k = 1$, then $J_1(\lambda) = [\lambda]$ is convergent if and only if $|\lambda| < 1$. Suppose that $k \geq 2$, let p be a positive integer, let $q = \min\{p, k - 1\}$, and use the binomial theorem to compute

$$J_k(\lambda)^p = (\lambda I + J_k)^p = \lambda^p I + \sum_{j=1}^{q} \binom{p}{p-j} \lambda^{p-j} J_k^j. \tag{13.2.8}$$

Each matrix J_k^j in the sum (13.2.8) has ones on its jth superdiagonal and zero entries elsewhere. Consequently, every entry in the jth superdiagonal of $J_k(\lambda)^p$ is $\binom{p}{p-j} \lambda^{p-j}$. The diagonal entries of $J_k(\lambda)^p$ are all λ^p, so $|\lambda| < 1$ is a necessary condition for $J_k(\lambda)$ to be convergent.

Suppose that $\rho(A) < 1$, so $|\lambda| < 1$ for all $\lambda \in \operatorname{spec} A$. If $\lambda = 0$, then $J_k(0)^p = 0$ for $p \geq k$, so $J_k(0)$ is convergent. If $0 < |\lambda| < 1$, then (13.2.8) ensures that $J_k(\lambda)$ is convergent if

$$\lim_{p \to \infty} \binom{p}{p-j} \lambda^{p-j} = 0 \quad \text{for } j = 1, 2, \ldots, k - 1.$$

Since

$$\left| \binom{p}{p-j} \lambda^{p-j} \right| = \left| \frac{p(p-1)(p-2)\cdots(p-j+1)\lambda^p}{j!\lambda^j} \right| \leq \frac{1}{j!|\lambda|^j} p^j |\lambda|^p,$$

it suffices to show that $\lim_{p \to \infty} p^j |\lambda|^p = 0$, that is, $\lim_{p \to \infty} \log (p^j |\lambda|^p) = -\infty$. Since $\log|\lambda| < 0$ and L'Hôpital's rule ensures that $\lim_{p \to \infty} (\log p)/p = 0$,

$$\log (p^j |\lambda|^p) = j \log p + p \log |\lambda| = p \left(j \frac{\log p}{p} + \log |\lambda| \right) \to -\infty \quad \text{as} \quad p \to \infty.$$

Thus, $J_k(\lambda)$ is convergent.   □

The following definition introduces a generalization of convergent matrices.

Definition 13.2.9 Let $A = [a_{ij}] \in \mathsf{M}_n$ and let $A^p = [a_{ij}^{(p)}]$ for $p = 0, 1, 2, \ldots$. Then A is *power bounded* if there is some $L > 0$ such that $|a_{ij}^{(p)}| \leq L$ for all $p = 1, 2, \ldots$ and all $i, j = 1, 2, \ldots, n$.

Theorem 13.2.10 $A \in \mathsf{M}_n$ *is power bounded if and only if* $\rho(A) \leq 1$ *and each* $\lambda \in \operatorname{spec} A$ *with* $|\lambda| = 1$ *is semisimple.*

Proof It follows from (13.2.5) that A is power bounded if and only if its Jordan canonical form is power bounded, so it suffices to consider power boundedness for a Jordan block $J_k(\lambda)$.

Suppose that $J_k(\lambda)$ is power bounded. The main diagonal entries of $J_k(\lambda)^p$ are λ^p, so it is necessary that $|\lambda| \leq 1$. Suppose that $|\lambda| = 1$. If $k \geq 2$, then the $(1,2)$ entry of $J_k(\lambda)^p$ is $p\lambda^{p-1}$, which is unbounded as $p \to \infty$. Since $J_k(\lambda)$ is power bounded, we conclude that $k = 1$. Therefore, every Jordan block of A with eigenvalue λ is 1×1, which means that λ is semisimple.

Conversely, suppose that λ has modulus 1 and is semisimple. Then the Jordan matrix $J(\lambda)$ in the Jordan canonical form of A is $J(\lambda) = \lambda I$, which is power bounded.   □

Example 13.2.11 The eigenvalues of $A = \begin{bmatrix} 0 & 1 \\ 1 & 0 \end{bmatrix}$ are 1 and -1, each with multiplicity one. The eigenvalues of $B = \frac{1}{4} \begin{bmatrix} 3 & 1 \\ 1 & 3 \end{bmatrix}$ are 1 and $\frac{1}{2}$, and 1 is a simple eigenvalue. The preceding theorem ensures that both A and B are power bounded. In fact,

$$A^p = \begin{cases} I_2 & \text{if } p \text{ is even,} \\ A & \text{if } p \text{ is odd,} \end{cases} \quad \text{and} \quad \lim_{p \to \infty} B^p = \frac{1}{2} \begin{bmatrix} 1 & 1 \\ 1 & 1 \end{bmatrix}.$$

13.3 The Jordan Forms of A and A^p

If $A \in \mathsf{M}_n$ is similar to a given direct sum of Jordan blocks (its Jordan canonical form) and p is a positive integer, then A^p is similar to the direct sum of the pth powers of those blocks. The nature of $J_k(\lambda)^p$ is different for $\lambda \neq 0$ and $\lambda = 0$.

Theorem 13.3.1 *Let $\lambda \in \mathbb{C}$, let k and p be positive integers, and let $q = \min\{p, k\}$.*

(a) *If $\lambda \neq 0$, the Jordan canonical form of $J_k(\lambda)^p$ is $J_k(\lambda^p)$.*

(b) *The Jordan canonical form of $J_k(0)^p$ is a direct sum of q nilpotent Jordan blocks.*

Proof If $k = 1$ or $p = 1$, there is nothing to prove, so we assume that $k \geq 2$ and $p \geq 2$. Then $J_k(\lambda)^p$ is an upper triangular matrix whose diagonal entries are λ^p. Since $J_k(\lambda)^p$ is unispectral, its Jordan canonical form is a direct sum of Jordan blocks with eigenvalue λ^p. Theorem 12.4.5 says that $w_1 = \operatorname{rank} I_k - \operatorname{rank}(J_k(\lambda^p) - \lambda^p I_k)$ is the number of these Jordan blocks.

(a) If $\lambda \neq 0$, (13.2.8) shows that the $k - 1$ entries in the first superdiagonal of $J_k(\lambda^p) - \lambda^p I_k$ are $p\lambda^{p-1}$, so $J_k(\lambda)^p$ has $w_1 = k - (k-1) = 1$ Jordan block.

(b) If $\lambda = 0$ and $p < k$, then $\lambda^{p-j} = 0$ for $j = 0, 1, \ldots, p-1$. The first $p - 1$ superdiagonals of $J_k(0)^p$ contain only zero entries. It follows from (13.2.8) that the $k - p$ entries in the pth superdiagonal of $J_k(0)^p$ are all ones, so there are $w_1 = k - (k-p) = p$ nilpotent Jordan blocks in the Jordan canonical form of $J_k(0)^p$. If $p \geq k$, then $J_k(0)^p = 0_k$ is a direct sum of k nilpotent Jordan blocks, each of which is 1×1. $\square$

Example 13.3.2 Consider $A = J_3(0)^2$. Then $\operatorname{rank} A = 1$ and $A^2 = 0$. The Weyr characteristic of A is $2, 1, 0$. Therefore, the Jordan canonical form of $J_3(0)^2$ is the direct sum of two nilpotent blocks, one of size 2 and one of size 1.

Example 13.3.3 Consider $A = J_4(0)^2$. Then $\operatorname{rank} A = 2$ and $A^2 = 0$. The Weyr characteristic of A is $2, 2, 0$. Therefore, the Jordan canonical form of $J_4(0)^2$ is the direct sum of two nilpotent blocks, each of size 2.

There is a one-to-one correspondence between the invertible Jordan blocks of A and those of A^p. They have the same sizes and multiplicities, but each nonzero eigenvalue λ of the former is replaced by λ^p in the latter. If A is not invertible, then each of its nilpotent Jordan blocks of size two or greater splits into two or more smaller nilpotent Jordan blocks of A^p. The following theorem identifies a special Jordan structure that is present in A if and only if it is present in A^p.

Theorem 13.3.4 *Let $A \in \mathsf{M}_n$ with $n \geq 2$, let $p \geq 2$ be an integer, and suppose that $1 \in \operatorname{spec} A$. Assume that the only $\lambda \in \operatorname{spec} A^p$ such that $|\lambda| = 1$ is $\lambda = 1$, and its algebraic multiplicity is 1.*

(a) *$\lambda = 1$ is the only eigenvalue of A such that $|\lambda| = 1$. Its algebraic and geometric multiplicities are 1.*

(b) *If the Jordan canonical form of A^p is $[1] \oplus K$, in which $K \in \mathsf{M}_{n-1}$ is a convergent Jordan matrix, then the Jordan canonical form of A is $[1] \oplus J$, in which $J \in \mathsf{M}_{n-1}$ is a convergent Jordan matrix.*

Proof If $1, \lambda_2, \lambda_2, \ldots, \lambda_n$ are the eigenvalues of A, then $1, \lambda_2^p, \lambda_2^p, \ldots, \lambda_n^p$ are the eigenvalues of A^p.

(a) The hypotheses ensure that $|\lambda_i^p| \neq 1$ for each $i = 2, 3, \ldots, n$. Therefore, $|\lambda_i| \neq 1$ for each $i = 2, 3, \ldots, n$, and $\lambda = 1$ is a simple eigenvalue of A.

(b) Since $\rho(K) < 1$, we have $|\lambda_i^p| < 1$ (and hence also $|\lambda_i| < 1$) for each $i = 2, 3, \ldots, n$. If $[1] \oplus J$ is the Jordan canonical form of A, it follows that $\rho(J) < 1$, so J is convergent. $\square$

13.4 Stochastic Matrices

Matrices with nonnegative entries and all column sums (or all row sums) equal to 1 play a key role in mathematical models of phenomena in genetics, economics, literary analysis, communication, finance, operations research, and biology. These matrices are all power bounded. Under some additional conditions, their powers have a remarkable limiting behavior.

Definition 13.4.1 A real matrix or vector is *nonnegative* if all of its entries are nonnegative; it is *positive* if all of its entries are positive.

Lemma 13.4.2 *Let $A \in M_n(\mathbb{R})$ and $\mathbf{x} \in \mathbb{R}^n$. If A is positive and $\mathbf{x}$ is nonnegative and nonzero, then $A\mathbf{x}$ is positive.*

Proof If $A = [a_{ij}]$, $\mathbf{x} = [x_i]$, and $x_k > 0$, then the ith entry of $A\mathbf{x}$ is $\sum_{j=1}^{n} a_{ij} x_j \geq a_{ik} x_k > 0$ for each $i = 1, 2, \ldots, n$. $\square$

Definition 13.4.3 Suppose that $A = [a_{ij}] \in M_n(\mathbb{R})$ is nonnegative.

(a) *A is column stochastic* if $\sum_{i=1}^{n} a_{ij} = 1$ for each $j = 1, 2, \ldots, n$, that is, each column sum equals 1.

(b) *A is row stochastic* if A^T is column stochastic, that is, each row sum of equals 1.

(c) *A is stochastic* if it is row stochastic or column stochastic.

(d) *A is doubly stochastic* if it is column stochastic and row stochastic.

Example 13.4.4 The matrices in Example 13.2.11 are doubly stochastic. The matrices

$$C = \begin{bmatrix} 0.25 & 0.4 \\ 0.75 & 0.6 \end{bmatrix} \quad \text{and} \quad D = \begin{bmatrix} 0.4 & 0 & 0.6 \\ 0.2 & 0.7 & 0.1 \\ 0 & 0.5 & 0.5 \end{bmatrix}$$

are column stochastic and row stochastic, respectively.

Lemma 13.4.5 *Suppose that $A = [a_{ij}] \in M_n(\mathbb{R})$ is nonnegative.*

(a) *A is row stochastic if and only if $A\mathbf{e} = \mathbf{e}$, that is, $(1, \mathbf{e})$ is an eigenpair of A.*

(b) *A is column stochastic if and only if $\mathbf{e}^\mathsf{T} A = \mathbf{e}^\mathsf{T}$, that is, $(1, \mathbf{e})$ is an eigenpair of A^T.*

(c) *If A is stochastic, then $0 \leq a_{ij} \leq 1$ for all $i, j \in \{1, 2, \ldots, n\}$.*

(d) *If A is row stochastic, then A^p is row stochastic for $p = 0, 1, 2, \ldots$.*

(e) *If A is column stochastic, then A^p is column stochastic for $p = 0, 1, 2, \ldots$.*

Proof (a) The ith entry of $A\mathbf{e}$ is the sum of the entries in the ith row of A, which is 1.

(b) The jth entry of $\mathbf{e}^\mathsf{T} A$ is the sum of the entries in the jth column of A, which is 1.

(c) For $i, j \in \{1, 2, \ldots, n\}$, we have $0 \leq a_{ij} \leq \sum_{k=1}^{n} a_{ik} = 1$ if A is row stochastic and $0 \leq a_{ij} \leq \sum_{k=1}^{n} a_{kj} = 1$ if A is column stochastic.

(d) For each $p \in \{1, 2, \ldots\}$, A^p is nonnegative and $A^p \mathbf{e} = \mathbf{e}$ (Lemma 9.3.2), so the assertion follows from (a).

(e) Apply (d) to A^{T}. $\square$

Theorem 13.4.6 *Let $A \in \mathsf{M}_n(\mathbb{R})$ be stochastic. Then $\rho(A) = 1$. Moreover, if $\lambda \in \operatorname{spec} A$ and $|\lambda| = 1$, then λ is semisimple.*

Proof Corollary 9.4.6 ensures that $\rho(A) \leq 1$. Since $1 \in \operatorname{spec} A$, we have $\rho(A) = 1$. For each $p = 1, 2, \ldots$, the preceding lemma ensures that every entry of A^p is between 0 and 1. Therefore, A is power bounded and the assertion follows from Theorem 13.2.10. $\square$

Example 13.4.7 The matrix $\begin{bmatrix} 0 & I_n \\ I_n & 0 \end{bmatrix}$ is doubly stochastic. Its eigenvalues are $\lambda = \pm 1$, each with geometric and algebraic multiplicity n. Its Jordan canonical form is $I_n \oplus (-I_n)$.

The preceding example shows that a stochastic matrix can have several eigenvalues with modulus 1. This cannot happen if every entry is positive.

Theorem 13.4.8 *Let $n \geq 2$ and let $A \in \mathsf{M}_n(\mathbb{R})$ be a positive stochastic matrix.*

(a) *If $\lambda \in \operatorname{spec} A$ and $|\lambda| = 1$, then $\lambda = 1$ and it has geometric multiplicity 1.*

(b) *$\lambda = 1$ is an eigenvalue of A with algebraic multiplicity 1.*

(c) *The Jordan canonical form of A is $\begin{bmatrix} 1 & \mathbf{0}^{\mathsf{T}} \\ \mathbf{0} & J \end{bmatrix}$, in which $J \in \mathsf{M}_{n-1}$ is a convergent Jordan matrix.*

Proof Corollary 12.4.14.b. ensures that A and A^{T} are similar, so they have the same Jordan canonical form and we may assume that A is row stochastic.

(a) Let $(\lambda, \mathbf{y})$ be an eigenpair of $A = [a_{ij}] \in \mathsf{M}_n$, in which $|\lambda| = 1$ and $\mathbf{y} = [y_i] \in \mathbb{C}^n$ is nonzero. Let $k \in \{1, 2, \ldots, n\}$ be an index such that $|y_k| = \max_{1 \leq i \leq n} |y_i|$. Since $A\mathbf{y} = \lambda\mathbf{y}$,

$$0 < |y_k| = |\lambda| |y_k| = |\lambda y_k| = \left| \sum_{j=1}^{n} a_{kj} y_j \right|$$

$$\leq \sum_{j=1}^{n} |a_{kj} y_j| = \sum_{j=1}^{n} a_{kj} |y_j| \tag{13.4.9}$$

$$\leq \sum_{j=1}^{n} a_{kj} |y_k| = |y_k|. \tag{13.4.10}$$

The inequalities (13.4.9) and (13.4.10) must be equalities. Since each $a_{kj} > 0$, equality in (13.4.10) means that $|y_j| = |y_k|$ for each $j = 1, 2, \ldots, n$. Equality in (13.4.9) is the equality case in the triangle inequality (A.3.10), so every entry of $\mathbf{y}$ is a nonnegative real multiple of y_k. Since every entry of $\mathbf{y}$ has the same positive modulus, it follows that $\mathbf{y} = y_k \mathbf{e}$. Therefore, $A\mathbf{y} = \mathbf{y}$ and $\lambda = 1$. Since $\mathbf{y} \in \operatorname{span}\{\mathbf{e}\}$, we conclude that $\mathcal{E}_1(A) = \operatorname{span}\{\mathbf{e}\}$ and $\dim \mathcal{E}_1(A) = 1$.

(b) The preceding theorem ensures that the eigenvalue $\lambda = 1$ has equal geometric and algebraic multiplicities, and (a) tells us that its geometric multiplicity is 1.

(c) Let $J_k(\lambda)$ be a Jordan block in the Jordan canonical form of A. Theorem 13.4.6 says that $|\lambda| \leq 1$. If $|\lambda| = 1$, then (a) tells us that $\lambda = 1$ and there is only one such block; (b) tells us that $k = 1$. Therefore, $J_k(1) = [1]$ and the Jordan canonical form of A is $[1] \oplus J$, in which J is a direct sum of Jordan blocks whose eigenvalues have moduli strictly less than 1. Theorem 13.2.7 ensures that J is convergent. $\qquad\qquad\qquad\qquad\qquad\qquad\qquad\qquad\qquad\qquad\qquad\square$

Theorem 13.4.11 *Let $n \geq 2$ and let $A \in \mathsf{M}_n(\mathbb{R})$ be a positive stochastic matrix.*

(a) *If A is row stochastic, there is a unique $\mathbf{x} \in \mathbb{R}^n$ such that $A^\mathsf{T}\mathbf{x} = \mathbf{x}$ and $\mathbf{e}^\mathsf{T}\mathbf{x} = 1$. Moreover, $\mathbf{x}$ is positive and $\lim_{k \to \infty} A^k = \mathbf{e}\mathbf{x}^\mathsf{T}$.*

(b) *If A is column stochastic, there is a unique $\mathbf{x} \in \mathbb{R}^n$ such that $A\mathbf{x} = \mathbf{x}$ and $\mathbf{e}^\mathsf{T}\mathbf{x} = 1$. Moreover, $\mathbf{x}$ is positive and $\lim_{k \to \infty} A^k = \mathbf{x}\mathbf{e}^\mathsf{T}$.*

Proof Since (b) follows from applying (a) to A^T, we may assume that A is row stochastic. There is an invertible $S \in \mathsf{M}_n$ such that

$$A = S \begin{bmatrix} 1 & \mathbf{0}^\mathsf{T} \\ \mathbf{0} & J \end{bmatrix} S^{-1}, \tag{13.4.12}$$

in which $\rho(J) < 1$. Partition $S = [\mathbf{y} \; Y]$, in which $Y \in \mathsf{M}_{n \times (n-1)}$. Then

$$[A\mathbf{y} \; AY] = AS = S \begin{bmatrix} 1 & \mathbf{0}^\mathsf{T} \\ \mathbf{0} & J \end{bmatrix} = [\mathbf{y} \; Y] \begin{bmatrix} 1 & \mathbf{0}^\mathsf{T} \\ \mathbf{0} & J \end{bmatrix} = [\mathbf{y} \; YJ],$$

so $A\mathbf{y} = \mathbf{y} \in \mathcal{E}_1(A)$, which has dimension one. Therefore, $\mathbf{y} = c\mathbf{e}$ for some nonzero scalar c. Since (13.4.12) is equivalent to $A = (cS) \begin{bmatrix} 1 & \mathbf{0}^\mathsf{T} \\ \mathbf{0} & J \end{bmatrix} (cS)^{-1}$, in which $cS = [c\mathbf{y} \; cY] = [\mathbf{e} \; cY]$, we may assume that $S = [\mathbf{e} \; Y]$.

Partition $S^{-\mathsf{T}} = [\mathbf{x} \; X]$, in which $X \in \mathsf{M}_{n \times (n-1)}$ and $\mathbf{x} \neq \mathbf{0}$. Then $A^\mathsf{T} = S^{-\mathsf{T}} \begin{bmatrix} 1 & \mathbf{0}^\mathsf{T} \\ \mathbf{0} & J^\mathsf{T} \end{bmatrix} S^\mathsf{T}$ and

$$[A^\mathsf{T}\mathbf{x} \; A^\mathsf{T}X] = A^\mathsf{T}S^{-\mathsf{T}} = S^{-\mathsf{T}} \begin{bmatrix} 1 & \mathbf{0}^\mathsf{T} \\ \mathbf{0} & J^\mathsf{T} \end{bmatrix} = [\mathbf{x} \; X] \begin{bmatrix} 1 & \mathbf{0}^\mathsf{T} \\ \mathbf{0} & J^\mathsf{T} \end{bmatrix} = [\mathbf{x} \; XJ^\mathsf{T}],$$

which shows that $A^\mathsf{T}\mathbf{x} = \mathbf{x} \in \mathcal{E}_1(A^\mathsf{T})$. However,

$$\begin{bmatrix} 1 & \mathbf{0}^\mathsf{T} \\ \mathbf{0} & I_{n-1} \end{bmatrix} = I_n = S^{-1}S = \begin{bmatrix} \mathbf{x}^\mathsf{T} \\ X^\mathsf{T} \end{bmatrix} [\mathbf{e} \; Y] = \begin{bmatrix} \mathbf{x}^\mathsf{T}\mathbf{e} & \star \\ \star & \star \end{bmatrix},$$

which shows that $\mathbf{x}^\mathsf{T}\mathbf{e} = \mathbf{e}^\mathsf{T}\mathbf{x} = 1$. Theorem 13.4.8.a and Lemma 9.3.19 ensure that $\dim \mathcal{E}_1(A^\mathsf{T}) = 1$. If $\mathbf{z} \in \mathcal{E}_1(A^\mathsf{T})$ and $\mathbf{z}^\mathsf{T}\mathbf{e} = 1$, then $\mathbf{z} = c\mathbf{x}$ for some nonzero scalar c, and $1 = \mathbf{z}^\mathsf{T}\mathbf{e} = c\mathbf{x}^\mathsf{T}\mathbf{e} = c$. This shows that $\mathbf{z} = \mathbf{x}$ and proves the uniqueness assertion. Since A^k is positive for each $k = 1, 2, \ldots$ and

$$A^k = S \begin{bmatrix} 1 & \mathbf{0}^\mathsf{T} \\ \mathbf{0} & J^k \end{bmatrix} S^{-1} = [\mathbf{e} \; Y] \begin{bmatrix} 1 & \mathbf{0}^\mathsf{T} \\ \mathbf{0} & J^k \end{bmatrix} \begin{bmatrix} \mathbf{x}^\mathsf{T} \\ X^\mathsf{T} \end{bmatrix} \rightarrow [\mathbf{e} \; Y] \begin{bmatrix} 1 & \mathbf{0}^\mathsf{T} \\ \mathbf{0} & 0 \end{bmatrix} \begin{bmatrix} \mathbf{x}^\mathsf{T} \\ X^\mathsf{T} \end{bmatrix} = \mathbf{e}\mathbf{x}^\mathsf{T}$$

as $k \to \infty$, it follows that $\mathbf{e}\mathbf{x}^\mathsf{T}$ (and therefore $\mathbf{x}$) is nonnegative. Because A^T is positive and $A^\mathsf{T}\mathbf{x} = \mathbf{x}$, Lemma 13.4.2 ensures that $\mathbf{x}$ is positive. $\qquad\qquad\qquad\qquad\qquad\square$

The preceding theorem ensures that a positive column stochastic matrix has a unique eigenpair $(1, \mathbf{x})$ such that $\mathbf{e}^\mathsf{T}\mathbf{x} = 1$. Moreover, $\mathbf{x}$ is positive.

Example 13.4.13 Suppose that cities C_1, C_2, and C_3 have respective initial populations n_1, n_2, and n_3. Let $\mathbf{x}_{(0)} = [n_1 \; n_2 \; n_3]^\mathsf{T}$ and $N = n_1 + n_2 + n_3$. On the first day of each month, for each

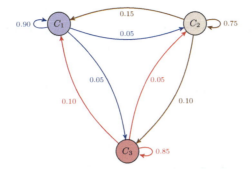

Figure 13.2 Migration pattern for Example 13.4.13.

$i \neq j$, a fixed positive fraction a_{ij} of the population of C_j migrates to C_i; for each $i = 1, 2, 3$, a positive fraction a_{ii} of the residents of C_i stays there. For example, suppose that

$$A = \begin{bmatrix} 0.9 & 0.15 & 0.10 \\ 0.05 & 0.75 & 0.05 \\ 0.05 & 0.10 & 0.85 \end{bmatrix};$$

see Figure 13.2. After the first migration, the population of C_i is the ith entry of $\mathbf{x}_{(1)} = A\mathbf{x}_{(0)}$. After the second migration, the population of C_i is the ith entry of $\mathbf{x}_{(2)} = A\mathbf{x}_{(1)} = A^2\mathbf{x}_{(0)}$. After k migrations, the population of C_i is the ith entry of $\mathbf{x}_{(k)} = A^k\mathbf{x}_{(0)}$. The unique $\mathbf{x}$ such that $A\mathbf{x} = \mathbf{x}$ and $\mathbf{e}^{\mathsf{T}}\mathbf{x} = 1$ is $\mathbf{x} = [0.54\ 0.17\ 0.29]^{\mathsf{T}}$, rounded to two decimal places. Notice that A^k is approximately $\mathbf{x}\mathbf{e}^{\mathsf{T}}$ for $k = 25$, which is consistent with Theorem 13.4.11. For example,

$$A^{24} \approx \begin{bmatrix} 0.54 & 0.54 & 0.54 \\ 0.17 & 0.17 & 0.17 \\ 0.29 & 0.29 & 0.29 \end{bmatrix} \quad \text{and} \quad A^{25} \approx \begin{bmatrix} 0.54 & 0.54 & 0.54 \\ 0.17 & 0.17 & 0.17 \\ 0.29 & 0.29 & 0.29 \end{bmatrix},$$

so the respective populations of the three cities eventually stabilizes: C_1 has a population of about $0.54N$, C_2 has a population of about $0.17N$, and C_3 has a population of about $0.29N$. These stable populations are independent of the initial distribution of population among the cities. See Figure 13.3 for illustrations of the evolution of the respective populations, for two different choices of initial populations n_1, n_2, n_3.

Theorem 13.4.11 reveals a remarkable limiting behavior for powers of stochastic matrices that are positive. The following example exhibits this behavior even though the matrix has some zero entries.

Example 13.4.14 Let

$$A = \begin{bmatrix} 0 & 0 & \frac{1}{5} \\ 1 & 0 & \frac{4}{5} \\ 0 & 1 & 0 \end{bmatrix},$$

which is column stochastic but not positive. One checks that $(\mathbf{x}, 1)$ is an eigenpair of A, in which $\mathbf{e}^{\mathsf{T}}\mathbf{x} = 1$ and $\mathbf{x} = \frac{1}{11}[1\ 5\ 5]^{\mathsf{T}} \approx [0.09091\ 0.4545\ 0.4545]^{\mathsf{T}}$. Similarly,

$$A^{25} \approx \begin{bmatrix} 0.0910 & 0.0908 & 0.0910 \\ 0.4548 & 0.4543 & 0.4547 \\ 0.4541 & 0.4548 & 0.4543 \end{bmatrix},$$

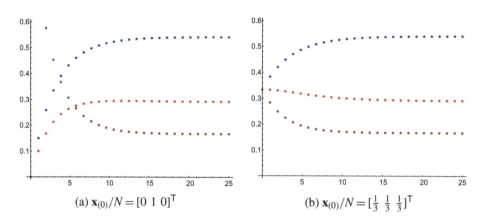

Figure 13.3 As $k \to \infty$, the distribution of populations $\mathbf{x}_{(k)}/N$ converges to the same limit for any choice of the initial distribution $\mathbf{x}_{(0)} = [n_1\ n_2\ n_3]^\mathsf{T}/N$.

so A exhibits the behavior described in Theorem 13.4.11 even though it is not positive. A calculation reveals that A^2 has four zero entries, A^3 has two, A^4 has one, and A^5 is positive.

Theorem 13.4.15 *Let $n \geq 2$, let $A \in \mathsf{M}_n(\mathbb{R})$ be stochastic, and suppose that there is a positive integer p such that A^p is positive.*

(a) *$\lambda = 1$ is the only eigenvalue of A such that $|\lambda| = 1$. Its algebraic and geometric multiplicities are 1.*

(b) *The Jordan canonical form of A is $[1] \oplus J$, in which J is a convergent Jordan matrix.*

(c) *If A is row stochastic, then $\lim_{k \to \infty} A^k = \mathbf{e}\mathbf{x}^\mathsf{T}$, in which $\mathbf{x}$ is the unique (necessarily positive) vector such that $A^\mathsf{T}\mathbf{x} = \mathbf{x}$ and $\mathbf{e}^\mathsf{T}\mathbf{x} = 1$.*

(d) *If A is column stochastic, then $\lim_{k \to \infty} A^k = \mathbf{x}\mathbf{e}^\mathsf{T}$, in which $\mathbf{x}$ is the unique (necessarily positive) vector such that $A\mathbf{x} = \mathbf{x}$ and $\mathbf{e}^\mathsf{T}\mathbf{x} = 1$.*

Proof The case $p = 1$ is dealt with in Theorems 13.4.8 and 13.4.11, so we may assume that $p \geq 2$.

(a) and (b) Since A is stochastic, Lemma 13.4.5 ensures that $1 \in \operatorname{spec} A$. If we apply Theorem 13.4.8 to A^p, we find that $\lambda = 1$ is the only eigenvalue of A^p with modulus 1; its algebraic and geometric multiplicities (as an eigenvalue of A^p) are 1. The assertions (a) and (b) follow from Theorem 13.3.4.

(c) Let $A = S([1] \oplus J)S^{-1}$, in which $S = [\mathbf{y}\ Y] \in \mathsf{M}_n$ is invertible and $S^{-\mathsf{T}} = [\mathbf{z}\ Z]$. Then $A^p = S([1] \oplus J^p)S^{-1}$ and

$$[A^p\mathbf{y}\ A^pY] = A^pS = S([1] \oplus J^p) = [\mathbf{y}\ YJ^p].$$

Thus, $A^p\mathbf{y} = \mathbf{y}$, which means that $\mathbf{y} = c\mathbf{x}$ for some scalar c, in which $\mathbf{x}$ is the unique vector such that $A^p\mathbf{x} = \mathbf{x}$ and $\mathbf{e}^\mathsf{T}\mathbf{x} = 1$. We also have $(A^p)^\mathsf{T} = S^{-\mathsf{T}}([1] + (J^p)^\mathsf{T})S^\mathsf{T}$ and

$$[(A^p)^\mathsf{T}\mathbf{z}\ (A^p)^\mathsf{T}Z] = (A^p)^\mathsf{T}S^{-\mathsf{T}} = S^{-\mathsf{T}}([1] \oplus (J^p)^\mathsf{T}) = [\mathbf{z}\ Z(J^p)^\mathsf{T}].$$

Thus, $(A^p)^\mathsf{T}\mathbf{z} = \mathbf{z}$, which means that $\mathbf{z} = d\mathbf{e}$ for some scalar d. We have $\mathbf{y}^\mathsf{T}\mathbf{z} = 1$ and $\mathbf{e}^\mathsf{T}\mathbf{x} = \mathbf{x}^\mathsf{T}\mathbf{e} = 1$, so $\mathbf{y}^\mathsf{T}\mathbf{z} = cd\mathbf{x}^\mathsf{T}\mathbf{e}$ implies that $cd = 1$. Now compute

$$A = S\begin{bmatrix} 1 & 0 \\ 0 & J \end{bmatrix} S^{-1} = [\mathbf{y}\ Y]\begin{bmatrix} 1 & 0 \\ 0 & J \end{bmatrix}\begin{bmatrix} \mathbf{z}^\mathsf{T} \\ Z^\mathsf{T} \end{bmatrix} = [c\mathbf{x}\ Y]\begin{bmatrix} 1 & 0 \\ 0 & J \end{bmatrix}\begin{bmatrix} d\mathbf{e}^\mathsf{T} \\ Z^\mathsf{T} \end{bmatrix}$$

$$= [\mathbf{x}\ Y]\begin{bmatrix} cd & 0 \\ 0 & J \end{bmatrix}\begin{bmatrix} \mathbf{e}^\mathsf{T} \\ Z^\mathsf{T} \end{bmatrix} = [\mathbf{x}\ Y]\begin{bmatrix} 1 & 0 \\ 0 & J \end{bmatrix}\begin{bmatrix} \mathbf{e}^\mathsf{T} \\ Z^\mathsf{T} \end{bmatrix}.$$

The conclusion is that we may assume that $\mathbf{y} = \mathbf{x}$ and $\mathbf{z} = \mathbf{e}$. Then

$$\lim_{k\to\infty} A^k = S\big([1] \oplus \lim_{k\to\infty} J^k\big)S^{-1} = [\mathbf{x}\ Y]\big([1] \oplus 0_{n-1}\big)[\mathbf{e}^\mathsf{T}\ Z^\mathsf{T}] = \mathbf{x}\mathbf{e}^\mathsf{T}.$$

(d) Apply (c) to A^T. □

13.5 The Invertible Jordan Blocks of *AB* and *BA*

If A and B are square matrices of the same size, then AB and BA have the same eigenvalues with the same multiplicities (Theorem 10.7.2). The following theorem reveals additional details about the nonzero eigenvalues of AB and BA.

Theorem 13.5.1 *Let* $A, B \in \mathsf{M}_n$. *If* $\operatorname{spec} AB \neq \{0\}$, *then for each nonzero* $\lambda \in \operatorname{spec} AB$ *and each* $k = 1, 2, \ldots, n$, *the Jordan canonical forms of* AB *and* BA *contain the same Jordan blocks* $J_k(\lambda)$ *with the same multiplicities.*

Proof In the proof of Theorem 10.7.2 we showed that $X = \begin{bmatrix} AB & A \\ 0 & 0 \end{bmatrix}$ and $Y = \begin{bmatrix} 0 & A \\ 0 & BA \end{bmatrix}$ are similar. Let λ be a nonzero eigenvalue of AB (and hence of BA also). Apply $f(z) = (z - \lambda)^p$ to X and Y, then use Theorem 2.6.9 to find that

$$(X - \lambda I)^p = \begin{bmatrix} (AB - \lambda I)^p & \star \\ 0 & (-\lambda)^p I \end{bmatrix} \tag{13.5.2}$$

and

$$(Y - \lambda I)^p = \begin{bmatrix} (-\lambda)^p I & \star \\ 0 & (BA - \lambda I)^p \end{bmatrix}$$

are similar. Partition (13.5.2) as $(X - \lambda I)^p = [X_1\ X_2]$, in which $X_1, X_2 \in \mathsf{M}_{2n\times n}$. Because $\lambda \neq 0$,

$$\operatorname{rank} X_2 = \operatorname{rank}\begin{bmatrix} \star \\ (-\lambda)^p I_n \end{bmatrix} = n.$$

Let $\mathbf{x}$ be a column of X_1. If $\mathbf{x} \in \operatorname{col} X_2$, then there is a $\mathbf{y} \in \mathbb{C}^n$ such that

$$\mathbf{x} = \begin{bmatrix} \star \\ 0 \end{bmatrix} = X_2\mathbf{y} = \begin{bmatrix} \star \\ (-\lambda)^p\mathbf{y} \end{bmatrix}.$$

It follows that $\mathbf{y} = \mathbf{0}$ and $\mathbf{x} = X_2\mathbf{y} = \mathbf{0}$. We conclude that $\operatorname{col} X_1 \cap \operatorname{col} X_2 = \{\mathbf{0}\}$, so

$$\operatorname{rank}(X - \lambda I)^p = \operatorname{rank} X_1 + n = \operatorname{rank}(AB - \lambda I)^p + n.$$

A similar examination of the rows of $(Y - \lambda I)^p$ shows that

$$\operatorname{rank}(Y - \lambda I)^p = \operatorname{rank}(BA - \lambda I)^p + n.$$

It follows from the similarity of $(X - \lambda I)^p$ and $(Y - \lambda I)^p$ that $\mathrm{rank}(X - \lambda I)^p = \mathrm{rank}(Y - \lambda I)^p$, so

$$\mathrm{rank}(AB - \lambda I)^p = \mathrm{rank}(BA - \lambda I)^p \quad \text{for } p = 1, 2, \ldots.$$

Therefore, the Weyr characteristics of AB and BA associated with each nonzero $\lambda \in \mathrm{spec}\, AB$ are the same, so the multiplicities of $J_k(\lambda)$ in their respective Jordan canonical forms are the same for each $k = 1, 2, \ldots$. $\qquad\square$

The preceding theorem says nothing about the nilpotent Jordan blocks of AB and BA, which can be different.

Example 13.5.3 Let $A = J_2$ and $B = \mathrm{diag}(1, 0)$. The Jordan forms of $AB = J_1 \oplus J_1$ and $BA = J_2$ do not have the same nilpotent Jordan blocks.

There is a simple definitive test for the similarity of AB and BA; it focuses only on their nilpotent Jordan blocks.

Corollary 13.5.4 *Let $A, B \in \mathsf{M}_n$. Then AB is similar to BA if and only if $\mathrm{rank}(AB)^p = \mathrm{rank}(BA)^p$ for each $p = 1, 2, \ldots, n$.*

Proof If AB is similar to BA, then $(AB)^p$ is similar to $(BA)^p$ and hence $\mathrm{rank}(AB)^p = \mathrm{rank}(BA)^p$. Conversely, if the ranks of their corresponding powers are equal, then the Weyr characteristics of AB and BA are the same. Consequently, for each $k = 1, 2, \ldots, n$, the multiplicities of each J_k in their respective Jordan canonical forms are the same. The preceding theorem ensures that the invertible Jordan blocks of AB and BA are the same, with the same multiplicities. Therefore, the Jordan canonical forms of AB and BA are the same. $\qquad\square$

Example 13.5.5 Consider

$$A = \begin{bmatrix} 0 & 0 & 0 & 0 \\ 0 & 0 & 1 & 0 \\ 0 & 0 & 0 & 1 \\ 0 & 1 & 0 & 0 \end{bmatrix} \quad \text{and} \quad B = \begin{bmatrix} 0 & 0 & 0 & 1 \\ 0 & 1 & 0 & 0 \\ 0 & 0 & 0 & 0 \\ 1 & 0 & 0 & 0 \end{bmatrix},$$

whose products are the nilpotent matrices

$$AB = \begin{bmatrix} 0 & 0 & 0 & 0 \\ 0 & 0 & 0 & 0 \\ 1 & 0 & 0 & 0 \\ 0 & 1 & 0 & 0 \end{bmatrix} \quad \text{and} \quad BA = \begin{bmatrix} 0 & 1 & 0 & 0 \\ 0 & 0 & 1 & 0 \\ 0 & 0 & 0 & 0 \\ 0 & 0 & 0 & 0 \end{bmatrix} = J_3 \oplus J_1.$$

A computation reveals that $\mathrm{rank}\, AB = \mathrm{rank}\, BA = 2$, but $\mathrm{rank}(AB)^2 = 0$ and $\mathrm{rank}(BA)^2 = 1$. The preceding corollary tells us that AB is not similar to BA. In fact, the Jordan canonical form of AB is $J_2 \oplus J_2$.

Corollary 13.5.6 *Let $A, H, K \in \mathsf{M}_n$ and suppose that H and K are Hermitian. Then HK is similar to KH and $A\overline{A}$ is similar to $\overline{A}A$.*

Proof For each $p = 1, 2, \ldots$,

$$\mathrm{rank}(HK)^p = \mathrm{rank}((HK)^p)^* = \mathrm{rank}((HK)^*)^p = \mathrm{rank}(K^*H^*)^p = \mathrm{rank}(KH)^p$$

and

$$\mathrm{rank}(A\overline{A})^p = \mathrm{rank}(\overline{A\overline{A}})^p = \mathrm{rank}(\overline{A}\,\overline{\overline{A}})^p = \mathrm{rank}(\overline{A}A)^p.$$

The preceding corollary ensures that HK is similar to KH and $A\overline{A}$ is similar to $\overline{A}A$. □

Example 13.5.7 Consider $A = \begin{bmatrix} -5 & i \\ 2i & 0 \end{bmatrix}$ and $A\overline{A} = \begin{bmatrix} 27 & 5i \\ -10i & 2 \end{bmatrix}$. A calculation reveals that $A\overline{A} = SRS^{-1}$ and $\overline{A}A = \overline{S}R\overline{S}^{-1}$, in which $R = \begin{bmatrix} 7 & 10 \\ 15 & 22 \end{bmatrix}$ and $S = \begin{bmatrix} -i & -i \\ 1 & -1 \end{bmatrix}$. This confirms that $A\overline{A}$ is similar to its complex conjugate, since it is similar to the real matrix R.

13.6 Similarity of a Matrix and Its Transpose

A square matrix A and its transpose have the same eigenvalues (Theorem 10.2.9), but this is only part of the story. With the help of the Jordan canonical form, we can show that A and A^T are similar.

Theorem 13.6.1 Let $A \in \mathsf{M}_n$.

(a) A is similar to A^T via a symmetric similarity matrix.

(b) $A = BC = DE$, in which B, C, D, and E are symmetric, and B and E are invertible.

Proof (a) The identity (12.2.5) tells us that $J_m = \begin{bmatrix} 0 & I_{m-1} \\ 0 & 0^\mathsf{T} \end{bmatrix}$ and $J_m^\mathsf{T} = \begin{bmatrix} 0^\mathsf{T} & 0 \\ I_{m-1} & 0 \end{bmatrix}$. The reversal matrix K_m defined in (7.1.11) may be written in block form as

$$K_m = \begin{bmatrix} 0 & K_{m-1} \\ 1 & 0^\mathsf{T} \end{bmatrix} = \begin{bmatrix} 0^\mathsf{T} & 1 \\ K_{m-1} & 0 \end{bmatrix}.$$

Now compute

$$K_m J_m^\mathsf{T} = \begin{bmatrix} 0 & K_{m-1} \\ 1 & 0^\mathsf{T} \end{bmatrix} \begin{bmatrix} 0^\mathsf{T} & 0 \\ I_{m-1} & 0 \end{bmatrix} = \begin{bmatrix} K_{m-1} & 0 \\ 0^\mathsf{T} & 0 \end{bmatrix}$$

and

$$J_m K_m = \begin{bmatrix} 0 & I_{m-1} \\ 0 & 0^\mathsf{T} \end{bmatrix} \begin{bmatrix} 0^\mathsf{T} & 1 \\ K_{m-1} & 0 \end{bmatrix} = \begin{bmatrix} K_{m-1} & 0 \\ 0^\mathsf{T} & 0 \end{bmatrix}. \tag{13.6.2}$$

Therefore, $K_m J_m^\mathsf{T} = J_m K_m$, so $J_m^\mathsf{T} = K_m^{-1} J_m K_m$. The computation

$$J_m(\lambda)^\mathsf{T} = (\lambda I_m + J_m)^\mathsf{T} = \lambda I_m + J_m^\mathsf{T} = \lambda I_m + K_m^{-1} J_m K_m$$
$$= K_m^{-1}(\lambda I_m + J_m) K_m = K_m^{-1} J_m(\lambda) K_m$$

shows that every Jordan block is similar to its transpose via a reversal matrix. It follows that every Jordan matrix (a direct sum of Jordan blocks) is similar to its transpose via a direct sum of reversal matrices.

Let $A \in \mathsf{M}_n$ and let $A = SJS^{-1}$, in which $S \in \mathsf{M}_n$ is invertible and J is a Jordan matrix. Let K denote a direct sum of reversal matrices such that $J^\mathsf{T} = K^{-1} J K$. Then $J = S^{-1}AS$ and

$$A^\mathsf{T} = S^{-\mathsf{T}} J^\mathsf{T} S^\mathsf{T} = S^{-\mathsf{T}} (K^{-1} J K) S^\mathsf{T} = S^{-\mathsf{T}} K^{-1} (S^{-1}AS) K S^\mathsf{T}$$
$$= (SKS^\mathsf{T})^{-1} A(SKS^\mathsf{T}).$$

Therefore, A^T is similar to A via the similarity matrix SKS^T. The matrix K is a direct sum of reversal matrices, each of which is symmetric, so K and SKS^T are symmetric.

(b) The identity (13.6.2) and the computation

$$K_m J_m = \begin{bmatrix} 0^\mathsf{T} & 1 \\ K_{m-1} & 0 \end{bmatrix} \begin{bmatrix} 0 & I_{m-1} \\ 0 & 0^\mathsf{T} \end{bmatrix} = \begin{bmatrix} 0 & 0^\mathsf{T} \\ 0 & K_{m-1} \end{bmatrix}$$

imply that JK and KJ are both symmetric. A reversal matrix is an involution, so $K^2 = I$. The computations

$$A = SJS^{-1} = SK^2 JS^{-1} = (SKS^{\mathsf{T}})(S^{-\mathsf{T}} KJS^{-1})$$

and

$$A = SJS^{-1} = SJK^2 S^{-1} = (SJKS^{\mathsf{T}})(S^{-\mathsf{T}} KS^{-1})$$

confirm the asserted factorizations.   □

13.7 Similarity of a Matrix and Its Complex Conjugate

Not every complex number equals its complex conjugate, so it is no surprise that not every square matrix A is similar to $\overline{A}$. If A is real, then A is similar to $\overline{A}$, but there are many non-real matrices with this property. For example, any matrix of the form $A\overline{A}$ is similar to its complex conjugate (Corollary 13.5.6), as is any product of two Hermitian matrices (Theorem 13.7.4).

If J is the Jordan canonical form of A, then $\overline{J}$ is the Jordan canonical form of $\overline{A}$. Therefore, A is similar to $\overline{A}$ if and only if J is similar to $\overline{J}$, which can be verified by using the test in the following theorem. The key observation is that for any Jordan block $J_k(\lambda)$, we have $\overline{J_k(\lambda)} = J_k(\overline{\lambda})$.

Theorem 13.7.1 *Let $A \in \mathsf{M}_n$. The following statements are equivalent:*

(a) *A is similar to $\overline{A}$.*

(b) *If $J_k(\lambda)$ is a direct summand of the Jordan canonical form of A, then so is $J_k(\overline{\lambda})$, with the same multiplicity.*

Proof Let $\lambda_1, \lambda_2, \ldots, \lambda_d$ be the distinct eigenvalues of A and suppose that the Jordan canonical form of A is

$$J = J(\lambda_1) \oplus J(\lambda_2) \oplus \cdots \oplus J(\lambda_d), \tag{13.7.2}$$

in which each $J(\lambda_i)$ is a Jordan matrix.

If A is similar to $\overline{A}$, then J is similar to $\overline{J}$. Moreover,

$$\overline{J} = J(\overline{\lambda_1}) \oplus J(\overline{\lambda_2}) \oplus \cdots \oplus J(\overline{\lambda_d}). \tag{13.7.3}$$

The uniqueness of the Jordan canonical form implies that the direct sum in (13.7.3) can be obtained from the direct sum in (13.7.2) by permuting its direct summands. Thus, there is a permutation σ of the list $1, 2, \ldots, d$ such that Jordan matrices $J(\lambda_j)$ and $J(\overline{\lambda_{\sigma(j)}})$ have the same Jordan canonical forms. This means that the direct summands of $J(\overline{\lambda_{\sigma(j)}})$ are obtained by permuting the direct summands of $J(\lambda_j)$. If λ_j is not real, then $\sigma(j) \neq j$. Therefore, the non-real Jordan matrices that are direct summands in (13.7.2) must occur in conjugate pairs.

Conversely, if the non-real Jordan matrices in (13.7.2) occur in conjugate pairs, then so do the non-real direct summands in (13.7.3); they can be obtained by interchanging pairs of conjugate direct summands of (13.7.2). This means that the Jordan canonical forms for A and $\overline{A}$ are the same, so A is similar to $\overline{A}$.   □

The following theorem introduces some characterizations that do not involve inspection of the Jordan canonical form.

Theorem 13.7.4 *Let $A \in \mathsf{M}_n$. The following statements are equivalent:*

(a) *A is similar to $\overline{A}$.*

(b) *A is similar to A^*.*

(c) *A is similar to A^* via a Hermitian similarity matrix.*

(d) *$A = HK = LM$, in which H, K, L, and M are Hermitian, and H and M are invertible.*

(e) *$A = HK$, in which H and K are Hermitian.*

Proof (a) $\Rightarrow$ (b) Let $S, R \in \mathsf{M}_n$ be invertible and such that $A = S\overline{A}S^{-1}$ and $A = RA^\mathsf{T}R^{-1}$ (Theorem 13.6.1). Then $\overline{A} = \overline{R}\,\overline{A}^\mathsf{T}\overline{R}^{-1}$ and

$$A = S\overline{A}S^{-1} = S\overline{R}\,\overline{A}^\mathsf{T}\overline{R}^{-1}S^{-1} = (S\overline{R})A^*(S\overline{R})^{-1}.$$

(b) $\Rightarrow$ (c) Let $S \in \mathsf{M}_n$ be invertible and such that $A = SA^*S^{-1}$. For any $\theta \in \mathbb{R}$, we have $A = (e^{i\theta}S)A^*(e^{-i\theta}S^{-1}) = S_\theta A^*S_\theta^{-1}$, in which $S_\theta = e^{i\theta}S$. Then $AS_\theta = S_\theta A^*$. The adjoint of this identity is $AS_\theta^* = S_\theta^*A^*$. Add the preceding two identities and obtain $A(S_\theta + S_\theta^*) = (S_\theta + S_\theta^*)A^*$. The matrix $H_\theta = S_\theta + S_\theta^*$ is Hermitian, and the computation

$$S_\theta^{-1}H_\theta = I + S_\theta^{-1}S_\theta^* = I + e^{-2i\theta}S^{-1}S^* = e^{-2i\theta}(e^{2i\theta}I + S^{-1}S^*)$$

tells us that H_θ is invertible for any θ such that $-e^{2i\theta} \notin \mathrm{spec}(S^{-1}S^*)$. For such a choice of θ, we have $A = H_\theta A^*H_\theta^{-1}$.

(c) $\Rightarrow$ (d) Suppose that $A = SA^*S^{-1}$, in which S is invertible and Hermitian. Then $AS = SA^*$. It follows that $(SA^*)^* = (AS)^* = S^*A^* = SA^*$, that is, SA^* is Hermitian. Then $A = (SA^*)S^{-1}$ is a product of two Hermitian factors, the second of which is invertible. Since $S^{-1}A = A^*S^{-1}$, we also have $(A^*S^{-1})^* = (S^{-1}A)^* = A^*S^{-*} = A^*S^{-1}$, so A^*S^{-1} is Hermitian and $A = S(A^*S^{-1})$ is a product of two Hermitian factors, the first of which is invertible.

(d) $\Rightarrow$ (e) There is nothing to prove.

(e) $\Rightarrow$ (a) Corollary 13.5.6 ensures that HK is similar to KH. Theorem 13.6.1 tells us that $A = KH$ is similar to $(KH)^\mathsf{T} = H^\mathsf{T}K^\mathsf{T} = \overline{H}\,\overline{K} = \overline{A}$. $\qquad\square$

13.8 Problems

P.13.1 What does Theorem 13.2.7 say if $n = 1$?

P.13.2 Let $A = [\mathbf{a}_1\ \mathbf{a}_2\ \ldots\ \mathbf{a}_n] \in \mathsf{M}_n(\mathbb{R})$ be row stochastic. Show that $\mathbf{a}_1 + \mathbf{a}_2 + \cdots + \mathbf{a}_n = \mathbf{e}$.

P.13.3 If $A, B \in \mathsf{M}_n$ are symmetric, show that AB is similar to BA.

P.13.4 Let $A, B \in \mathsf{M}_n$ and let $p \geq 2$ be an integer. (a) Show that $\rho(A^p) = \rho(A)^p$. (b) Is $\rho(AB) = \rho(A)\rho(B)$? (c) Is $\rho(AB) \leq \rho(A)\rho(B)$? (d) If A is invertible, is $\rho(A^{-1}) = \rho(A)^{-1}$?

P.13.5 Consider the 4×4 matrices A and B in Example 13.5.5. Show that $p_A(z) = m_A(z) = z^4 - z$, $p_B(z) = z^4 - z^3 - z^2 + z$, and $m_B(z) = z^3 - z$.

P.13.6 Verify that $\mathbf{x}(t) = [\cos t + 2\sin t\quad 2\cos t - \sin t]^\mathsf{T}$ satisfies the differential equations $\mathbf{x}'(t) = A\mathbf{x}(t)$ and initial condition in Example 13.1.9.

P.13.7 The initial-value problem

$$y'(t) = \lambda y(t) + f(t), \quad \text{in which } y(0) \text{ is given},\qquad (13.8.1)$$

can be solved as follows: Rewrite it as $y'(t) - \lambda y(t) = f(t)$, multiply both sides by $e^{-\lambda t}$ and use the product rule to obtain

$$\frac{d}{dt}(e^{-\lambda t}y(t)) = e^{-\lambda t}y'(t) - \lambda e^{-\lambda t}y(t) = e^{-\lambda t}f(t).$$

Integrate both sides and use the fundamental theorem of calculus:

$$e^{-\lambda t}y(t) - y(0) = \int_0^t \frac{d}{ds}(e^{-\lambda s}y(s))\,ds = \int_0^t e^{-\lambda s}f(s)\,ds.$$

Finally, multiply both sides by $e^{\lambda t}$ to obtain the solution

$$y(t) = y(0)e^{\lambda t} + e^{\lambda t}\int_0^t e^{-\lambda s}f(s)\,ds. \tag{13.8.2}$$

(a) Differentiate (13.8.2) and verify that it is the solution to (13.8.1). (b) Use (13.8.2) to derive the stated solution to the initial-value problem in Example 13.1.7.

P.13.8   Solve $\mathbf{x}'(t) = A\mathbf{x}(t)$ and $\mathbf{x}(0) = \begin{bmatrix} 4 \\ 7 \end{bmatrix}$, in which $A = \begin{bmatrix} 2 & 1 \\ 1 & 2 \end{bmatrix}$.

P.13.9   Solve $\mathbf{x}'(t) = A\mathbf{x}(t)$ and $\mathbf{x}(0) = \begin{bmatrix} 4 \\ 7 \end{bmatrix}$, in which $A = \begin{bmatrix} 2 & 1 \\ 0 & 2 \end{bmatrix}$.

P.13.10 Solve $\mathbf{x}'(t) = A\mathbf{x}(t)$ and $\mathbf{x}(0) = \begin{bmatrix} 4 \\ 7 \end{bmatrix}$, in which $A = \begin{bmatrix} 2 & -1 \\ 1 & 2 \end{bmatrix}$.

P.13.11 Solve $\mathbf{x}'(t) = A\mathbf{x}(t) + \mathbf{f}(t)$ and $\mathbf{x}(0) = \begin{bmatrix} 4 \\ 7 \end{bmatrix}$, in which $A = \begin{bmatrix} 2 & 1 \\ 1 & 2 \end{bmatrix}$ and $\mathbf{f}(t) = \begin{bmatrix} 1 \\ 2 \end{bmatrix}$.

P.13.12 Solve $\mathbf{x}'(t) = A\mathbf{x}(t) + \mathbf{f}(t)$ and $\mathbf{x}(0) = \begin{bmatrix} 4 \\ 7 \end{bmatrix}$, in which $A = \begin{bmatrix} 1 & 1 \\ 1 & 1 \end{bmatrix}$ and $\mathbf{f}(t) = \begin{bmatrix} e^t \\ t \end{bmatrix}$.

P.13.13 Solve $\mathbf{x}'(t) = A\mathbf{x}(t) + \mathbf{f}(t)$ and $\mathbf{x}(0) = \begin{bmatrix} 4 \\ 7 \end{bmatrix}$, in which $A = \begin{bmatrix} 1 & 1 \\ 0 & 0 \end{bmatrix}$ and $\mathbf{f}(t) = \begin{bmatrix} t \\ 1 \end{bmatrix}$.

P.13.14 Let $A \in \mathsf{M}_n$ be diagonalizable, use the notation in the discussion of (13.1.1), and let $E(t) = \mathrm{diag}(e^{\lambda_1 t}, e^{\lambda_2 t}, \ldots, e^{\lambda_n t})$. (a) Show that $\mathbf{x}(t) = SE(t)S^{-1}\mathbf{x}(0)$ is the solution to (13.1.1). (b) Use (10.5.7) to explain why this solution can be written as $\mathbf{x}(t) = e^{At}\mathbf{x}(0)$. (c) What does this say if $n = 1$? (d) Techniques from analysis reveal that

$$e^{At} = \sum_{k=0}^{\infty} \frac{t^k}{k!}A^k, \tag{13.8.3}$$

in which the infinite series converges entrywise for every $A \in \mathsf{M}_n$ and every $t \in \mathbb{R}$. Use (13.8.3) to show that if A and the initial values $\mathbf{x}(0)$ are real, then $\mathbf{x}(t)$ is real for all $t \in \mathbb{R}$ even if some of the eigenvalues of A (and hence also $E(t)$ and S) are not real.

P.13.15 (a) Let A and $\mathbf{x}(0)$ be the matrix and initial vector in Example 13.1.9. What are $(tA)^{4k}$, $(tA)^{4k+1}$, $(tA)^{4k+2}$, and $(tA)^{4k+3}$ for $k = 0, 1, 2, \ldots$? (b) Use the infinite series in (13.8.3) to show that $e^{At} = (\sin t)A + (\cos t)I$. This method does not require knowledge of the eigenvalues of A. (c) Compute $e^{At}\mathbf{x}(0)$ and discuss. (d) Use (10.8.1) to compute e^{At}. This method requires knowledge of the eigenvalues of A, but it does not involve infinite series.

P.13.16 Let $A \in \mathsf{M}_n$. Use (13.8.3) and Theorem 11.1.1 to prove that $\det(e^A) = e^{\mathrm{tr}\,A}$.

P.13.17 Let $A \in \mathsf{M}_n$, suppose that $\mathrm{Re}\,\lambda < 0$ for every $\lambda \in \mathrm{spec}\,A$, and let $\mathbf{x}(t)$ be the solution to (13.1.1). Use the analysis in Section 13.1 to show that $\|\mathbf{x}(t)\|_2 \to 0$ as $t \to \infty$.

P.13.18 Cecil Sagehen is either happy or sad. If he is happy one day, then he is happy the next day three times out of four. If he is sad one day, then he is sad the next day one time out of three. During the coming year, how many days do you expect Cecil to be happy?

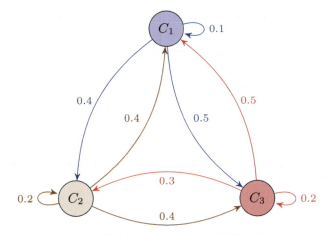

Figure 13.4 Migration pattern for P.13.20.

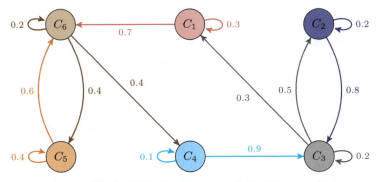

Figure 13.5 Migration pattern for P.13.21.

P.13.19 Gaul is divided into three parts, Gallia Aquitania (A), Gallia Belgica (B), and Gallia Lugdunensis (L). Each year, 5% of the residents of A move to B, and 5% move to L. Each year 15% of the residents of B move to A and 10% move to L. Finally, each year 10% of the residents of L move to A and 5% move to B. What percentage of the population should we expect to reside in each of the three regions after 50 years?

P.13.20 Each year, suppose that people migrate among cities C_1, C_2, and C_3 according to Figure 13.4. (a) What is the long-term prognosis if the initial populations are 10 million, 6 million, and 2 million, respectively? (b) What is the long-term prognosis if the initial populations are 1 million, 2 million, and 12 million, respectively?

P.13.21 Each year, suppose that people migrate among countries C_1, C_2, C_3, C_4, C_5, and C_6 according to Figure 13.5. Let $A \in M_6(\mathbb{R})$ denote the corresponding column-stochastic matrix. (a) Inspect Figure 13.5 and explain why A^k has only positive entries for some natural number k. What is the smallest such k? (b) If at least one country begins with a nonzero population, prove that the long-term relative population distribution is independent of the initial population distribution. *Hint*: Consider Theorem 13.4.15.

P.13.22 Let $A \in M_n(\mathbb{R})$ be doubly stochastic and positive. Show that $\lim_{k \to \infty} A^k = \frac{1}{n}E$, in which $E \in M_n$ is the all-ones matrix.

P.13.23 Let $A \in M_2$. (a) Show that A is unitarily similar to a unique matrix of the form $\begin{bmatrix} \lambda_1 & \alpha \\ 0 & \lambda_2 \end{bmatrix}$, in which: (i) $\alpha = (\|A\|_F^2 - |\lambda_1|^2 - |\lambda_2|^2)^{1/2} \geq 0$, (ii) $\mathrm{Re}\,\lambda_1 \geq \mathrm{Re}\,\lambda_2$, and (iii) $\mathrm{Im}\,\lambda_1 \geq \mathrm{Im}\,\lambda_2$ if $\mathrm{Re}\,\lambda_1 = \mathrm{Re}\,\lambda_2$. (b) Prove that every 2×2 complex matrix is unitarily similar to its transpose. However, see the following problem for the 3×3 case.

P.13.24 Let $A \in M_n$. (a) If $B \in M_n$ is unitarily similar to A, show that $\operatorname{tr}(A^2(A^*)^2AA^*) = \operatorname{tr}(B^2(B^*)^2BB^*)$. (b) If A is unitarily similar to A^T, show that $\operatorname{tr}(A^2(A^*)^2AA^*) = \operatorname{tr}((A^*)^2A^2A^*A)$. (c) Show that

$$A = \begin{bmatrix} 0 & 1 & 0 \\ 0 & 0 & 2 \\ 0 & 0 & 0 \end{bmatrix}$$

is not unitarily similar to A^T. Contrast this result with Theorem 13.6.1 and the preceding problem.

P.13.25 Let $\lambda \neq 0$ and define

$$S_4(\lambda) = \begin{bmatrix} \lambda^{-1} & -\lambda^{-2} & \lambda^{-3} & -\lambda^{-4} \\ 0 & \lambda^{-1} & -\lambda^{-2} & \lambda^{-3} \\ 0 & 0 & \lambda^{-1} & -\lambda^{-2} \\ 0 & 0 & 0 & \lambda^{-1} \end{bmatrix}.$$

(a) Verify that $S_4(\lambda) = J_4(\lambda)^{-1}$, that is, $J_4(\lambda)S_4(\lambda) = I_4$. The entries in the first row of $S_4(\lambda)$ are $f(\lambda)$, $f'(\lambda)$, $\frac{1}{2!}f''(\lambda)$, and $\frac{1}{3!}f'''(\lambda)$, in which $f(\lambda) = \lambda^{-1}$. (b) Let $A = J_4(\lambda) - \lambda I$. Show that $S_4(\lambda) = f(\lambda)I + f'(\lambda)A + \frac{1}{2!}f''(\lambda)A^2 + \frac{1}{3!}f'''(\lambda)A^3$, and that it is a polynomial in $J_4(\lambda)$. (c) Show that the leading 3×3 principal submatrix of $S_4(\lambda)$ is the inverse of $J_3(\lambda)$. (d) Can you exhibit the inverse of $J_5(\lambda)$?

P.13.26 Suppose that $\lambda \neq 0$ and let $\lambda^{1/2}$ be either choice of the square root. Theorem 13.3.1 ensures that $J_k(\lambda^{1/2})^2$ is similar to $J_k(\lambda)$. Let $S \in M_n$ be invertible and such that $J_k(\lambda) = SJ_k(\lambda^{1/2})^2S^{-1}$. (a) Show that $(SJ_k(\lambda^{1/2})S^{-1})^2 = J_k(\lambda)$, so $SJ_k(\lambda^{1/2})S^{-1}$ is a square root of $J_k(\lambda)$. (b) Show that every invertible matrix has a square root.

P.13.27 Let $\lambda \neq 0$ and define

$$R_4(\lambda) = \begin{bmatrix} \lambda^{1/2} & \frac{1}{2}\lambda^{-1/2} & \frac{-1}{8}\lambda^{-3/2} & \frac{1}{16}\lambda^{-5/2} \\ 0 & \lambda^{1/2} & \frac{1}{2}\lambda^{-1/2} & \frac{-1}{8}\lambda^{-3/2} \\ 0 & 0 & \lambda^{1/2} & \frac{1}{2}\lambda^{-1/2} \\ 0 & 0 & 0 & \lambda^{1/2} \end{bmatrix}.$$

(a) Verify that $R_4(\lambda)$ is a square root of $J_4(\lambda)$, that is, $R_4(\lambda)^2 = J_4(\lambda)$. The entries in the first row of $R_4(\lambda)$ are $f(\lambda), f'(\lambda), \frac{1}{2!}f''(\lambda)$, and $\frac{1}{3!}f'''(\lambda)$, in which $f(\lambda) = \lambda^{1/2}$. (b) Let $A = J_4(\lambda) - \lambda I$ and show that $R_4(\lambda) = f(\lambda)I + f'(\lambda)A + \frac{1}{2!}f''(\lambda)A^2 + \frac{1}{3!}f'''(\lambda)A^3$, and that it is a polynomial in $J_4(\lambda)$. (c) Show that the 3×3 leading principal submatrix of $R_4(\lambda)$ is a square root of $J_3(\lambda)$. (d) Can you find a square root of $J_5(\lambda)$? (e) What do you get if you compute a square root of $J_2(\lambda)$ with (10.9.1) and $f(t) = t^{1/2}$?

P.13.28 Show that J_{2k}^2 is similar to $J_k \oplus J_k$ and J_{2k+1}^2 is similar to $J_{k+1} \oplus J_k$.

P.13.29 Of the three possible Jordan canonical forms for a 3×3 nilpotent matrix, show that $J_1 \oplus J_1 \oplus J_1$ and $J_2 \oplus J_1$ have square roots, but J_3 does not.

P.13.30 Of the five possible Jordan canonical forms for a 4×4 nilpotent matrix, show that $J_3 \oplus J_1$ and J_4 do not have square roots, but the other three do.

P.13.31 Find the Jordan canonical form of J_k^3 for $k = 1, 2, \ldots$.

P.13.32 Let $A, B \in M_m$ and $C, D \in M_n$. (a) If A is similar to B and C is similar to D, show that $(A \oplus C)$ and $(B \oplus D)$ are similar. (b) Give an example to show that similarity of $(A \oplus C)$ and $(B \oplus D)$ does not imply that A is similar to B or C is similar to D. (c) If

$(A \oplus C)$ is similar to $(A \oplus D)$, show that C and D are similar. *Hint*: If $J(A)$ and $J(C)$ are the respective Jordan canonical forms of A and C, why is $J(A) \oplus J(C)$ the Jordan canonical form of both $A \oplus C$ and $A \oplus D$?

P.13.33 Let $A, B \in M_n$. Let $C = A \oplus \cdots \oplus A$ (k direct summands) and $D = B \oplus \cdots \oplus B$ (k direct summands). Show that A is similar to B if and only if C is similar to D.

P.13.34 Let $B = \begin{bmatrix} 1 & 2 \\ 3 & 4 \end{bmatrix}$. Review Example 13.5.7 and adopt its notation. Verify that $A = SB\overline{S}^{-1}$ and $R = B^2$, so $A\overline{A}$ is similar to the square of a real matrix.

P.13.35 Let $A, S, R \in M_n$. Suppose that S is invertible, R is real, and $A = SR\overline{S}^{-1}$. (a) Show that $A\overline{A}$ is similar to the square of a real matrix and deduce that $A\overline{A}$ is similar to $\overline{A}A$. (b) Show that any non-real eigenvalues of $A\overline{A}$ occur in conjugate pairs and that any negative eigenvalues of $A\overline{A}$ have even multiplicity.

P.13.36 We say that $A = [a_{ij}] \in M_n$ is a *Toeplitz matrix* if there are scalars $t_0, t_{\pm 1}, t_{\pm 2}, \ldots, t_{\pm(n-1)}$ such that $a_{ij} = t_{i-j}$. (a) Give an example of a 4×4 Toeplitz matrix. (b) Give an example of an upper triangular 4×4 Toeplitz matrix.

P.13.37 Let

$$J = \begin{bmatrix} J_2(\lambda) & 0 \\ 0 & J_2(\lambda) \end{bmatrix} \quad \text{and} \quad W = \begin{bmatrix} \lambda I_2 & I_2 \\ 0 & \lambda I_2 \end{bmatrix}, \tag{13.8.4}$$

and let $X \in M_4$. (a) Show that the Weyr characteristics of $J - \lambda I$ and $W - \lambda I$ are both $2, 2$. Deduce that J is similar to W. (b) Show that $JX = XJ$ if and only if $X = \begin{bmatrix} B & C \\ D & E \end{bmatrix}$, in which $B, C, D, E \in M_2$ are upper triangular Toeplitz matrices. (c) Show that $WX = XW$ if and only if there are $F, G \in M_2$ such that $X = \begin{bmatrix} F & G \\ 0 & F \end{bmatrix}$. (d) Show that J and W are permutation similar. (e) Show that $\dim\{J\}' = \dim\{W\}' = 8$, and exhibit bases for $\{J\}'$ and $\{W\}'$.

P.13.38 What are the possible values of $\dim\{A\}'$ if $A \in M_3$?

P.13.39 What are the possible values of $\dim\{A\}'$ if $A \in M_5$ is unispectral?

P.13.40 Let $n \geq 2$. In the proof of Lemma 12.3.1, we showed that $J_n^\mathsf{T} J_n = 0_1 \oplus I_{n-1}$, which is an orthogonal projection. (a) Without using the explicit form of $J_n J_n^\mathsf{T}$, show that it must be an orthogonal projection. (b) Use (12.2.20) to show that $J_n J_n^\mathsf{T} = I_{n-1} \oplus 0_1$. (c) Use the fact that every Jordan matrix is similar to its transpose via a reversal matrix (see the proof of Theorem 13.6.1) to deduce the identity $J_n J_n^\mathsf{T} = I_{n-1} \oplus 0_1$ from the identity $J_n^\mathsf{T} J_n = 0_1 \oplus I_{n-1}$.

P.13.41 Let $A \in M_n$ be convergent. Show that $\det(I + A) = \exp\left(\sum_{k=1}^\infty (-1)^{k-1} \frac{\operatorname{tr} A^k}{k}\right)$.

P.13.42 Let $A \in M_{m \times n}$ and $B \in M_{n \times m}$. Review Theorem 10.7.2. (a) Show that the nonzero eigenvalues of AB and BA are the same, with the same algebraic and geometric multiplicities. (b) Suppose that $m = n$. Although AB and BA have the same eigenvalues with the same algebraic multiplicities, show by example that the geometric multiplicities of a zero eigenvalue need not be the same.

P.13.43 Let K_n denote the $n \times n$ reversal matrix and let $S_n = 2^{-1/2}(I_n + K_n)$. Prove the following: (a) S_n is symmetric and unitary. (b) The matrix

$$S_n J_n(\lambda) S_n^* = \lambda I_n + \frac{1}{2}(J_n + K_n J_n K_n) + \frac{i}{2}(K_n J_n - J_n K_n)$$

is symmetric and unitarily similar to $J_n(\lambda)$. (c) Every Jordan matrix is unitarily similar to a complex symmetric matrix. (d) Every square complex matrix is similar to a complex symmetric matrix.

P.13.44 If $\lambda \neq 0$, show that $J_k(\lambda)^{-1}$ is similar to $J_k(\lambda^{-1})$.

P.13.45 If $\lambda = 0$ and p is a positive or negative integer, show that $J_k(\lambda)^p$ is similar to $J_k(\lambda^p)$. In particular, $J_k(\lambda)^2$ is similar to $J_k(\lambda^2)$.

13.9 Notes

If $A \in \mathsf{M}_n$ and p is any polynomial, then $p(A)$ commutes with A. However, a matrix that commutes with A need not be a polynomial in A. For example, $A = I_2$ commutes with J_2, which is not a polynomial in I_2 (such a matrix must be diagonal). If A is nonderogatory, however, any matrix that commutes with A must be a polynomial in A; see P.12.21 and [HJ13, Section 3.2.4].

P.13.14, P.13.25, and P.13.27 introduce ways to define $f(A)$ for $f(t) = e^t$, $f(t) = 1/t$, and $f(t) = \sqrt{t}$, respectively. They are examples of *primary matrix functions*; see [HJ94, Chapter 6].

It is known that every square complex matrix A can be factored as $A = SR\overline{S}^{-1}$, in which R is real. Consequently, the pairing of negative and non-real eigenvalues of $A\overline{A}$ discussed in P.13.35 is true for all $A \in \mathsf{M}_n$; see [HJ13, Corollaries 4.4.13 and 4.6.15].

The nilpotent Jordan blocks of AB and BA need not be the same, but they cannot differ by much. Let $m_1 \geq m_2 \geq \cdots$ and $n_1 \geq n_2 \geq \cdots$ be the decreasingly ordered sizes of the nilpotent Jordan blocks of AB and BA, respectively; if necessary, append zero sizes to one list to get lists of equal lengths. Harley Flanders showed in 1951 that $|m_i - n_i| \leq 1$ for all i. Example 13.5.3 illustrates Flanders' theorem. For a discussion and proof see [JS96].

For an approach to Theorem 13.4.11 that does not involve the Jordan canonical form and is valid for some matrices that are not stochastic, see [ĆJ15]. For example,

$$A = \frac{1}{5} \begin{bmatrix} 0 & -1 & 6 \\ 2 & -1 & 4 \\ -4 & 0 & 9 \end{bmatrix}$$

satisfies the condition $A\mathbf{e} = \mathbf{e}$, but A is not a nonnegative matrix. Nevertheless, $\lim_{k \to \infty} A^k = \mathbf{e}\mathbf{x}^\mathsf{T}$, in which $\mathbf{x} = \frac{1}{3}[-6 \ 1 \ 8]^\mathsf{T}$ and $A^\mathsf{T}\mathbf{x} = \mathbf{x}$. The eigenvalues of A are 1, $\frac{1}{5}$, and $\frac{2}{5}$. A stochastic matrix (column or row) is sometimes called a *Markov matrix*.

13.10 Some Important Concepts

- The solution $\mathbf{x}(t)$ of $\mathbf{x}'(t) = A\mathbf{x}(t)$, in which $\mathbf{x}(0)$ is given, is a combination (13.1.8) of polynomials and exponentials determined by the Jordan canonical form of A.

- Convergent matrices.

- Power-bounded matrices.

- Stochastic matrices and limits of their powers.

- Every square complex matrix is similar to its transpose.

- The Jordan canonical forms of AB and BA contain the same invertible Jordan blocks with the same multiplicities.

Normal Matrices and the Spectral Theorem

Every square complex matrix is unitarily similar to an upper triangular matrix, but which matrices are unitarily similar to a diagonal matrix? The answer is the main result of this chapter: the spectral theorem for normal matrices. Hermitian, skew-Hermitian, unitary, and circulant matrices are unitarily diagonalizable. As a consequence, they have special properties that we investigate in this and following chapters.

14.1 Normal Matrices

Definition 14.1.1 A square matrix A is *normal* if $A^*A = AA^*$.

The term "normal," as used here, is not related to the "normal equations" or to the notion of a "normal vector" to a plane. Many mathematical objects are called "normal." Normal matrices are just one instance of this usage.

Example 14.1.2 Diagonal matrices are normal. If $A = \operatorname{diag}(\lambda_1, \lambda_2, \ldots, \lambda_n)$, then $A^* = \operatorname{diag}(\overline{\lambda_1}, \overline{\lambda_2}, \ldots, \overline{\lambda_n})$ and hence $A^*A = \operatorname{diag}(|\lambda_1|^2, |\lambda_2|^2, \ldots, |\lambda_n|^2) = AA^*$.

Example 14.1.3 Unitary matrices are normal. If U is unitary, then $U^*U = I = UU^*$.

Example 14.1.4 Hermitian matrices are normal. If $A = A^*$, then $A^*A = A^2 = AA^*$.

Example 14.1.5 Real symmetric, real skew-symmetric, and real orthogonal matrices are normal.

Example 14.1.6 For any $a, b \in \mathbb{C}$,

$$A = \begin{bmatrix} a & -b \\ b & a \end{bmatrix} \tag{14.1.7}$$

is normal since

$$A^*A = \begin{bmatrix} |a|^2 + |b|^2 & 2i \operatorname{Im} b\overline{a} \\ -2i \operatorname{Im} b\overline{a} & |a|^2 + |b|^2 \end{bmatrix} = AA^*.$$

If $a, b \in \mathbb{R}$, then $A^*A = (a^2 + b^2)I = AA^*$. If $a, b \in \mathbb{R}$ and $a^2 + b^2 = 1$, then A is real orthogonal; see Example 9.3.10.

Example 14.1.8 Computations confirm that $\begin{bmatrix} 0 & 1 \\ 0 & 0 \end{bmatrix}$, $\begin{bmatrix} 1 & 1 \\ 0 & 0 \end{bmatrix}$, $\begin{bmatrix} 1 & 1 \\ 0 & 2 \end{bmatrix}$, and $\begin{bmatrix} 1 & 2 \\ 3 & 4 \end{bmatrix}$ are not normal.

Theorem 14.1.9 *Let $A \in \mathsf{M}_n$ be normal and let $\lambda \in \mathbb{C}$.*

(a) *If p is a polynomial, then $p(A)$ is normal and commutes with A.*

(b) $\|A\mathbf{x}\|_2 = \|A^*\mathbf{x}\|_2$ *for all $\mathbf{x} \in \mathbb{C}^n$.*

(c) $A\mathbf{x} = \lambda\mathbf{x}$ *if and only if* $A^*\mathbf{x} = \bar{\lambda}\mathbf{x}$.

(d) A^* *is normal.*

(e) $\operatorname{null} A = \operatorname{null} A^*$ *and* $\operatorname{col} A = \operatorname{col} A^*$.

Proof (a) We first use induction to show that $A^*A^j = A^jA^*$ for $j = 0, 1, \ldots$. The base case $j = 0$ is $A^*I = IA^*$, which is true. Suppose that $A^*A^j = A^jA^*$ for some j. Then $A^*A^{j+1} = (A^*A^j)A = (A^jA^*)A = A^j(A^*A) = A^j(AA^*) = A^{j+1}A^*$, which completes the induction. A second induction confirms that $(A^*)^iA^j = A^j(A^*)^i$ for $i, j = 0, 1, \ldots$.

Let $p(z) = c_k z^k + \cdots + c_1 z + c_0$. Then

$$(p(A))^* p(A) = \left(\sum_{i=0}^{k} \bar{c}_i (A^*)^i \right) \left(\sum_{j=0}^{k} c_j A^j \right) = \sum_{i,j=0}^{k} \bar{c}_i c_j (A^*)^i A^j$$

$$= \sum_{i,j=0}^{k} \bar{c}_i c_j A^j (A^*)^i = \left(\sum_{j=0}^{k} c_j A^j \right) \left(\sum_{i=0}^{k} \bar{c}_i (A^*)^i \right)$$

$$= p(A)(p(A))^*$$

and hence $p(A)$ is normal. The fact that $p(A)$ commutes with A is Theorem 2.6.9.

(b) $\|A\mathbf{x}\|_2^2 = \langle A\mathbf{x}, A\mathbf{x} \rangle = \langle A^*A\mathbf{x}, \mathbf{x} \rangle = \langle AA^*\mathbf{x}, \mathbf{x} \rangle = \langle A^*\mathbf{x}, A^*\mathbf{x} \rangle = \|A^*\mathbf{x}\|_2^2$.

(c) It follows from (a) and (b) that $A - \lambda I$ is normal and

$$A\mathbf{x} = \lambda\mathbf{x} \iff \|(A - \lambda I)\mathbf{x}\|_2 = 0 \iff \|(A - \lambda I)^*\mathbf{x}\|_2 = 0$$

$$\iff \|(A^* - \bar{\lambda}I)\mathbf{x}\|_2 = 0 \iff A^*\mathbf{x} = \bar{\lambda}\mathbf{x}.$$

(d) A^* is normal since $(A^*)^*(A^*) = AA^* = A^*A = A^*(A^*)^*$.

(e) The first assertion is the case $\lambda = 0$ in (c). The second assertion follows from the first and a computation: $\operatorname{col} A = \operatorname{col} A^{**} = (\operatorname{null} A^*)^{\perp} = (\operatorname{null} A)^{\perp} = \operatorname{col} A^*$. $\square$

Sums and products of normal matrices need not be normal. For example, the non-normal matrix

$$\begin{bmatrix} 0 & 1 \\ 0 & 0 \end{bmatrix} = \begin{bmatrix} 1 & 0 \\ 0 & 0 \end{bmatrix} \begin{bmatrix} 0 & 1 \\ 1 & 0 \end{bmatrix} = \begin{bmatrix} 0 & \frac{1}{2} \\ \frac{1}{2} & 0 \end{bmatrix} + \begin{bmatrix} 0 & \frac{1}{2} \\ -\frac{1}{2} & 0 \end{bmatrix}$$

is both a product of two normal matrices and a sum of two normal matrices. However, if $A, B \in \mathsf{M}_n$ are normal and commute, then AB and $A + B$ are normal; see Corollary 14.2.12.

Conditions (b) and (c) in the preceding theorem are each equivalent to normality; see Theorem 14.8.1 and P.14.9.

Theorem 14.1.10 *Direct sums of normal matrices are normal.*

Proof If $A = A_{11} \oplus A_{22} \oplus \cdots \oplus A_{kk}$ and each A_{ii} is normal, then

$$A^*A = A_{11}^*A_{11} \oplus A_{22}^*A_{22} \oplus \cdots \oplus A_{kk}^*A_{kk} = A_{11}A_{11}^* \oplus A_{22}A_{22}^* \oplus \cdots \oplus A_{kk}A_{kk}^* = AA^*,$$

so A is normal. $\square$

Lemma 14.1.11 *A matrix that is unitarily similar to a normal matrix is normal.*

Proof Let $A \in \mathsf{M}_n$ be normal, let $U \in \mathsf{M}_n$ be unitary, and let $B = UAU^*$. Then $B^* = UA^*U^*$ and $B^*B = (UA^*U^*)(UAU^*) = UA^*AU^* = UAA^*U^* = (UAU^*)(UA^*U^*) = BB^*$. $\qquad\square$

Certain patterns of zero entries cannot occur for a normal matrix.

Lemma 14.1.12 *Let $A = \begin{bmatrix} B & X \\ 0 & C \end{bmatrix}$, in which B and C are square. Then A is normal if and only if B and C are normal and $X = 0$.*

Proof If A is normal, then

$$A^*A = \begin{bmatrix} B^* & 0 \\ X^* & C^* \end{bmatrix} \begin{bmatrix} B & X \\ 0 & C \end{bmatrix} = \begin{bmatrix} B^*B & B^*X \\ X^*B & X^*X + C^*C \end{bmatrix} \tag{14.1.13}$$

equals

$$AA^* = \begin{bmatrix} B & X \\ 0 & C \end{bmatrix} \begin{bmatrix} B^* & 0 \\ X^* & C^* \end{bmatrix} = \begin{bmatrix} BB^* + XX^* & XC^* \\ CX^* & CC^* \end{bmatrix}. \tag{14.1.14}$$

Compare the $(1,1)$ blocks in (14.1.13) and (14.1.14), take traces, and obtain

$$\operatorname{tr} B^*B = \operatorname{tr}(BB^* + XX^*) = \operatorname{tr} BB^* + \operatorname{tr} XX^* = \operatorname{tr} B^*B + \operatorname{tr} X^*X,$$

which implies that $\|X\|_{\mathrm{F}}^2 = \operatorname{tr} X^*X = 0$. Thus, $X = 0$ and $A = B \oplus C$. Comparison of the $(1,1)$ and $(2,2)$ blocks of (14.1.13) and (14.1.14) tells us that B and C are normal.

If B and C are normal and $X = 0$, then Theorem 14.1.10 ensures that A is normal. $\qquad\square$

Lemma 14.1.12 can be generalized to normal block upper triangular matrices with any number of diagonal blocks.

Theorem 14.1.15 *A block upper triangular matrix is normal if and only if it is block diagonal and each of its diagonal blocks is normal. In particular, an upper triangular matrix is normal if and only if it is diagonal.*

Proof Let $A = [A_{ij}] \in \mathsf{M}_n$ be normal and $k \times k$ block upper triangular, so each $A_{ii} \in \mathsf{M}_{n_i}$, $n_1 + n_2 + \cdots + n_k = n$, and $A_{ij} = 0$ for $i > j$. Let S_k be the statement that $A_{11}, A_{22}, \ldots, A_{kk}$ are normal and $A_{ij} = 0$ for $i < j$. Proceed by induction on k. The preceding lemma shows that the base case S_2 is true. If $k > 2$ and S_{k-1} is true, write

$$A = \begin{bmatrix} A_{11} & X \\ 0 & A' \end{bmatrix},$$

in which $A' \in \mathsf{M}_{n-n_1}$ is a $(k-1) \times (k-1)$ block upper triangular matrix and $X = [A_{12}\ A_{13}\ \ldots\ A_{1k}] \in \mathsf{M}_{n_1 \times (n-n_1)}$. Lemma 14.1.12 ensures that A_{11} and A' are normal and that $X = 0$. The induction hypothesis ensures that A' is block diagonal and each of $A_{22}, A_{33}, \ldots, A_{kk}$ is normal. Therefore, $A = A_{11} \oplus A_{22} \oplus \cdots \oplus A_{kk}$, in which each direct summand is normal. This shows that S_k is true and completes the induction.

The converse is Theorem 14.1.10. $\qquad\square$

14.2 The Spectral Theorem

Definition 14.2.1 A square matrix is *unitarily diagonalizable* if it is unitarily similar to a diagonal matrix.

Normal matrices are not only diagonalizable, they are unitarily diagonalizable. The following theorem is the main result about normal matrices.

Theorem 14.2.2 (Spectral Theorem) *Let $A \in M_n$, let $\lambda_1, \lambda_2, \ldots, \lambda_n$ be its eigenvalues in any given order, and let $\Lambda = \text{diag}(\lambda_1, \lambda_2, \ldots, \lambda_n)$. The following are equivalent:*

(a) *A is normal, that is, $A^*A = AA^*$.*

(b) *A is unitarily diagonalizable: there is a unitary $U \in M_n$ such that*

$$A = U\Lambda U^*. \tag{14.2.3}$$

(c) *There is an orthonormal basis $\mathbf{u}_1, \mathbf{u}_2, \ldots, \mathbf{u}_n$ of $\mathbb{C}^n$ such that*

$$A = \lambda_1 \mathbf{u}_1 \mathbf{u}_1^* + \lambda_2 \mathbf{u}_2 \mathbf{u}_2^* + \cdots + \lambda_n \mathbf{u}_n \mathbf{u}_n^*. \tag{14.2.4}$$

(d) *$\mathbb{C}^n$ has an orthonormal basis consisting of eigenvectors of A.*

If A is real, the following are equivalent:

(a′) *A is symmetric, that is, $A = A^\mathsf{T}$.*

(b′) *A is real orthogonally diagonalizable: Λ is real and there is a real orthogonal $Q \in M_n(\mathbb{R})$ such that $A = Q\Lambda Q^\mathsf{T}$.*

(c′) *There is an orthonormal basis $\mathbf{u}_1, \mathbf{u}_2, \ldots, \mathbf{u}_n$ of $\mathbb{R}^n$ such that $A = \lambda_1 \mathbf{u}_1 \mathbf{u}_1^\mathsf{T} + \lambda_2 \mathbf{u}_2 \mathbf{u}_2^\mathsf{T} + \cdots + \lambda_n \mathbf{u}_n \mathbf{u}_n^\mathsf{T}$.*

(d′) *$\mathbb{R}^n$ has an orthonormal basis consisting of eigenvectors of A.*

Proof (a) $\Rightarrow$ (b) Suppose that A is normal. Use Theorem 11.1.1 to write $A = UTU^*$, in which U is unitary and $T = [t_{ij}]$ is upper triangular with diagonal entries $t_{ii} = \lambda_i$ for $i = 1, 2, \ldots, n$. Since T is normal (Lemma 14.1.11) and upper triangular, Theorem 14.1.15 ensures that $T = \Lambda$ is diagonal. Thus, $A = U\Lambda U^*$ is unitarily diagonalizable.

(b) $\Leftrightarrow$ (c) The list of columns of U is an orthonormal basis for $\mathbb{C}^n$, and the representation (3.1.17) of a matrix product as a sum of outer products shows that (14.2.3) and (14.2.4) are equivalent.

(c) $\Rightarrow$ (d) If A is represented as in (14.2.4), then the orthonormality of $\mathbf{u}_1, \mathbf{u}_2, \ldots, \mathbf{u}_n$ ensures that $A\mathbf{u}_k = \sum_{j=1}^{n} \lambda_j \mathbf{u}_j \mathbf{u}_j^* \mathbf{u}_k = \sum_{j=1}^{n} \lambda_j \mathbf{u}_j \delta_{jk} = \lambda_k \mathbf{u}_k$ for each $k = 1, 2, \ldots, n$.

(d) $\Rightarrow$ (a) Suppose that $\mathbf{u}_1, \mathbf{u}_2, \ldots, \mathbf{u}_n$ is an orthonormal basis for $\mathbb{C}^n$ and $\lambda_1, \lambda_2, \ldots, \lambda_n \in \mathbb{C}$ are such that $A\mathbf{u}_i = \lambda_i \mathbf{u}_i$ for each $i = 1, 2, \ldots, n$. Then $U = [\mathbf{u}_1 \ \mathbf{u}_2 \ \ldots \ \mathbf{u}_n]$ is unitary and satisfies

$$AU = A[\mathbf{u}_1 \ \mathbf{u}_2 \ \ldots \ \mathbf{u}_n] = [A\mathbf{u}_1 \ A\mathbf{u}_2 \ \ldots \ A\mathbf{u}_n] = [\lambda_1 \mathbf{u}_1 \ \lambda_2 \mathbf{u}_2 \ \ldots \ \lambda_n \mathbf{u}_n] = U\Lambda,$$

in which $\Lambda = \text{diag}(\lambda_1, \lambda_2, \ldots, \lambda_n)$. Lemma 14.1.11 and Example 14.1.2 ensure that $A = U\Lambda U^*$ is normal.

Now suppose that A is real.

(a′) $\Rightarrow$ (b′) Since A is real and symmetric, it is Hermitian ($A^* = A$). Let $(\lambda, \mathbf{x})$ be an eigenpair of A, in which $\mathbf{x} \in \mathbb{C}^n$ is a unit vector. Then $\langle A\mathbf{x}, \mathbf{x} \rangle = \langle \lambda\mathbf{x}, \mathbf{x} \rangle = \lambda \langle \mathbf{x}, \mathbf{x} \rangle = \lambda$, so the symmetry of A ensures that $\overline{\lambda} = \overline{\langle A\mathbf{x}, \mathbf{x} \rangle} = \langle \mathbf{x}, A\mathbf{x} \rangle = \langle A\mathbf{x}, \mathbf{x} \rangle = \lambda$. Therefore, Λ is real. The real case of Theorem 11.1.1 says that there is a real orthogonal matrix Q such that $Q^\mathsf{T}AQ = T = [t_{ij}]$ is upper triangular, real, and has diagonal entries $t_{ii} = \lambda_i$ for $i = 1, 2, \ldots, n$. But $Q^\mathsf{T}AQ$ is symmetric, so T is symmetric and hence $T = \Lambda$ is diagonal.

(b′) $\Leftrightarrow$ (c′) $\Rightarrow$ (d′) $\Rightarrow$ (a′) The proofs are the same as in the complex case, except that the unitary matrix is real orthogonal, $\Lambda = Q^\mathsf{T}AQ$ is real, and $A = Q\Lambda Q^\mathsf{T}$ is symmetric. $\qquad\square$

An alternative formulation of the spectral theorem represents a normal matrix as a linear combination of orthogonal projections that are pairwise orthogonal; see Theorem 14.9.6.

Example 14.2.5 The matrix A in (14.1.7) has eigenvalues $\lambda_1 = a + bi$ and $\lambda_2 = a - bi$. Let $\Lambda = \operatorname{diag}(\lambda_1, \lambda_2)$. Corresponding unit eigenvectors are $\mathbf{u}_1 = [\frac{1}{\sqrt{2}} \ -\frac{i}{\sqrt{2}}]^{\mathsf{T}}$ and $\mathbf{u}_2 = [\frac{1}{\sqrt{2}} \ \frac{i}{\sqrt{2}}]^{\mathsf{T}}$. Then $U = [\mathbf{u}_1 \ \mathbf{u}_2]$ is unitary and

$$\underbrace{\begin{bmatrix} a & -b \\ b & a \end{bmatrix}}_{A} = \underbrace{\begin{bmatrix} \frac{1}{\sqrt{2}} & \frac{1}{\sqrt{2}} \\ -\frac{i}{\sqrt{2}} & \frac{i}{\sqrt{2}} \end{bmatrix}}_{U} \underbrace{\begin{bmatrix} a+bi & 0 \\ 0 & a-bi \end{bmatrix}}_{\Lambda} \underbrace{\begin{bmatrix} \frac{1}{\sqrt{2}} & \frac{i}{\sqrt{2}} \\ \frac{1}{\sqrt{2}} & -\frac{i}{\sqrt{2}} \end{bmatrix}}_{U^*}. \tag{14.2.6}$$

Example 14.2.7 The matrix

$$A = \begin{bmatrix} 1 & 8 & -4 \\ 8 & 1 & 4 \\ -4 & 4 & 7 \end{bmatrix} \tag{14.2.8}$$

is real symmetric and hence is real orthogonally diagonalizable. Its eigenvalues are $9, 9, -9$; corresponding orthonormal eigenvectors are $\mathbf{u}_1 = [\frac{1}{3} \ \frac{2}{3} \ \frac{2}{3}]^{\mathsf{T}}$, $\mathbf{u}_2 = [-\frac{2}{3} \ -\frac{1}{3} \ \frac{2}{3}]^{\mathsf{T}}$, and $\mathbf{u}_3 = [-\frac{2}{3} \ \frac{2}{3} \ -\frac{1}{3}]^{\mathsf{T}}$. Then $U = [\mathbf{u}_1 \ \mathbf{u}_2 \ \mathbf{u}_3]$ is real orthogonal and

$$\underbrace{\begin{bmatrix} 1 & 8 & -4 \\ 8 & 1 & 4 \\ -4 & 4 & 7 \end{bmatrix}}_{A} = \underbrace{\begin{bmatrix} \frac{1}{3} & -\frac{2}{3} & -\frac{2}{3} \\ \frac{2}{3} & -\frac{1}{3} & \frac{2}{3} \\ \frac{2}{3} & \frac{2}{3} & -\frac{1}{3} \end{bmatrix}}_{U} \underbrace{\begin{bmatrix} 9 & 0 & 0 \\ 0 & 9 & 0 \\ 0 & 0 & -9 \end{bmatrix}}_{\Lambda} \underbrace{\begin{bmatrix} \frac{1}{3} & \frac{2}{3} & \frac{2}{3} \\ -\frac{2}{3} & -\frac{1}{3} & \frac{2}{3} \\ -\frac{2}{3} & \frac{2}{3} & -\frac{1}{3} \end{bmatrix}}_{U^{\mathsf{T}}}.$$

Definition 14.2.9 Let $A \in \mathsf{M}_n$ be normal. A *spectral decomposition* of A is a factorization $A = U\Lambda U^*$, in which $U \in \mathsf{M}_n$ is unitary and $\Lambda \in \mathsf{M}_n$ is diagonal.

Here is one way to compute a spectral decomposition of a normal matrix: If $A \in \mathsf{M}_n$ is normal, use the construction in the proof of Theorem 11.1.1 to factor it as $A = UTU^*$, in which U is unitary and T is upper triangular. Since A is unitarily similar to T, Lemma 14.1.11 ensures that T is normal. Theorem 14.1.15 then tells us that T is diagonal. That is, the algorithm in the proof of Theorem 11.1.1 produces a spectral decomposition of A, provided that A is normal.

An important corollary of the spectral theorem says that if A is normal, then A and A^* are simultaneously unitarily diagonalizable.

Corollary 14.2.10 Let $A \in \mathsf{M}_n$ be normal and write $A = U\Lambda U^*$, in which U is unitary and Λ is diagonal. Then $A^* = U\overline{\Lambda}U^*$.

Proof Since Λ is diagonal, $A^* = (U\Lambda U^*)^* = U\Lambda^* U^* = U\overline{\Lambda}U^*$. $\qquad\square$

The preceding corollary is a special case (in which $\mathscr{F} = \{A, A^*\}$) of the following theorem, which generalizes the spectral theorem to any set of commuting normal matrices.

Theorem 14.2.11 Let $\mathscr{F} \subseteq \mathsf{M}_n$ be nonempty.

(a) *Suppose that each matrix in $\mathscr{F}$ is normal. Then $AB = BA$ for all $A, B \in \mathscr{F}$ if and only if there is a unitary $U \in \mathsf{M}_n$ such that U^*AU is diagonal for every $A \in \mathscr{F}$.*

(b) *Suppose that each matrix in $\mathcal{F}$ is real and symmetric. Then $AB = BA$ for all $A, B \in \mathcal{F}$ if and only if there is an orthogonal $Q \in \mathsf{M}_n(\mathbb{R})$ such that $Q^\mathsf{T} AQ$ is diagonal for every $A \in \mathcal{F}$.*

Proof (a) Theorem 11.5.1 ensures that there is a unitary $U \in \mathsf{M}_n$ such that U^*AU is upper triangular for every $A \in \mathcal{F}$. However, each of these upper triangular matrices is normal (Lemma 14.1.11), so Theorem 14.1.15 ensures that each is diagonal. Conversely, any pair of simultaneously diagonalizable matrices commutes; see Theorem 10.4.17.

(b) Since all the eigenvalues of a real symmetric matrix are real, Theorem 11.5.1 ensures that there is a real orthogonal $Q \in \mathsf{M}_n(\mathbb{R})$ such that $Q^\mathsf{T} AQ$ is real and upper triangular for every $A \in \mathcal{F}$. All of these upper triangular matrices are symmetric, so they are diagonal. The converse follows as in (a). $\qquad\square$

Corollary 14.2.12 *Let $A, B \in \mathsf{M}_n$ be normal and suppose that $AB = BA$. Then AB and $A + B$ are normal.*

Proof The preceding theorem provides a unitary U and diagonal matrices Λ and M such that $A = U\Lambda U^*$ and $B = UMU^*$. Thus, $AB = (U\Lambda U^*)(UMU^*) = U(\Lambda M)U^*$ and $A + B = U\Lambda U^* + UMU^* = U(\Lambda + M)U^*$ are unitarily diagonalizable, hence normal. $\qquad\square$

Example 14.2.13 Let $A = \begin{bmatrix} 1 & 0 \\ 0 & -1 \end{bmatrix}$ and $B = \begin{bmatrix} 0 & 1 \\ 1 & 0 \end{bmatrix}$. Then A, B, AB, BA, and $A + B$ are normal, but $AB \neq BA$. Thus, commutativity is a sufficient, but not necessary, condition for the product or sum of normal matrices to be normal.

14.3 The Defect from Normality

Theorem 14.3.1 (Schur's Inequality) *Let $A \in \mathsf{M}_n$ have eigenvalues $\lambda_1, \lambda_2, \ldots, \lambda_n$. Then*

$$\sum_{i=1}^{n} |\lambda_i|^2 \leq \|A\|_\mathrm{F}^2, \tag{14.3.2}$$

with equality if and only if A is normal.

Proof Use Theorem 11.1.1 to write $A = UTU^*$, in which $T = [t_{ij}] \in \mathsf{M}_n$ is upper triangular and $t_{ii} = \lambda_i$ for $i = 1, 2, \ldots, n$. Then $A^*A = (UT^*U^*)(UTU^*) = UT^*TU^*$ and

$$\sum_{i=1}^{n} |\lambda_i|^2 \leq \sum_{i=1}^{n} |\lambda_i|^2 + \sum_{i<j} |t_{ij}|^2 = \operatorname{tr} T^*T = \operatorname{tr} UT^*TU^* = \operatorname{tr} A^*A = \|A\|_\mathrm{F}^2,$$

with equality if and only if $\sum_{i<j} |t_{ij}|^2 = 0$, that is, if and only if T is diagonal. Thus, Schur's inequality is an equality if and only if A is unitarily similar to a diagonal matrix, that is, if and only if A is normal. $\qquad\square$

Definition 14.3.3 Let $\lambda_1, \lambda_2, \ldots, \lambda_n$ be the eigenvalues of $A \in \mathsf{M}_n$. The quantity

$$\Delta(A) = \|A\|_\mathrm{F}^2 - \sum_{i=1}^{n} |\lambda_i|^2 = \operatorname{tr} A^*A - \sum_{i=1}^{n} |\lambda_i|^2 \tag{14.3.4}$$

is the *defect from normality* of A.

Schur's inequality (14.3.2) says that $\Delta(A) \geq 0$ for every $A \in M_n$, with equality if and only if A is normal.

If $A, B \in M_n$ are normal, then neither AB nor BA need be normal. However, if AB is normal, then so is BA.

Theorem 14.3.5 *Let $A, B \in M_n$ be normal. Then AB is normal if and only if BA is normal.*

Proof Since AB and BA have the same eigenvalues with the same multiplicities (Theorem 10.7.2), equation (14.3.4) ensures that $\|AB\|_F^2 - \Delta(AB) = \|BA\|_F^2 - \Delta(BA)$, which we write as $\Delta(AB) - \Delta(BA) = \|AB\|_F^2 - \|BA\|_F^2$. Now use the normality of A and B to compute

$$\|BA\|_F^2 = \operatorname{tr}((BA)^*(BA)) = \operatorname{tr} A^*B^*BA = \operatorname{tr} A^*BB^*A = \operatorname{tr} BB^*AA^*$$
$$= \operatorname{tr} BB^*A^*A = \operatorname{tr} B^*A^*AB = \operatorname{tr}((AB)^*(AB)) = \|AB\|_F^2.$$

Therefore, $\Delta(AB) = \Delta(BA)$, so $\Delta(AB) = 0$ if and only if $\Delta(BA) = 0$. □

14.4 The Fuglede–Putnam Theorem

Lemma 14.4.1 *Let $A_i \in M_{n_i}$ be normal for $i = 1, 2, \ldots, k$. Then there exists a polynomial p such that $A_i^* = p(A_i)$ for $i = 1, 2, \ldots, k$.*

Proof Theorem 2.7.4 ensures that there is a polynomial p such that $p(\lambda) = \bar{\lambda}$ for all $\lambda \in \bigcup_{i=1}^{k} \operatorname{spec} A_i$. For $i = 1, 2, \ldots, k$, write $A_i = U_i \Lambda_i U_i^*$, in which $U_i \in M_{n_i}$ is unitary and $\Lambda_i \in M_{n_i}$ is diagonal. Theorem 10.5.1 and Corollary 14.2.10 tell us that $p(A_i) = U_i p(\Lambda_i) U_i^* = U_i \overline{\Lambda_i} U_i^* = A_i^*$ for each $i = 1, 2, \ldots, k$. □

Theorem 14.4.2 (Fuglede–Putnam) *Let $A \in M_m$ and $B \in M_n$ be normal and let $X \in M_{m \times n}$. Then $AX = XB$ if and only if $A^*X = XB^*$.*

Proof Suppose that $AX = XB$. Lemma 14.4.1 provides a polynomial p such that $p(A) = A^*$ and $p(B) = B^*$. Theorem 2.6.9 ensures that $A^*X = p(A)X = Xp(B) = XB^*$. For the converse, apply the same argument with A^* and B^* in place of A and B. □

For other proofs of the Fuglede–Putnam theorem, see P.14.18 and P.14.19. Corollary 14.2.12 can also be proved using the Fuglede–Putnam theorem; see P.14.21.

Corollary 14.4.3 *Let $A \in M_{m \times n}$ and $B \in M_{n \times m}$. Then AB and BA are normal if and only if $A^*AB = BAA^*$ and $ABB^* = B^*BA$.*

Proof If AB and BA are normal, then $(AB)^*$ and $(BA)^*$ are normal (Theorem 14.1.9.d). Since $(BA)^*A^* = A^*B^*A^* = A^*(AB)^*$, the preceding theorem ensures that $BAA^* = A^*AB$. Since the roles of A and B are interchangeable, $ABB^* = B^*BA$.

Conversely, suppose that $A^*AB = BAA^*$ and $ABB^* = B^*BA$. Then

$$(AB)^*(AB) = B^*(A^*AB) = B^*(BAA^*) = (B^*BA)A^* = (ABB^*)A^* = (AB)(AB)^*,$$

that is, AB is normal. Since the roles of A and B are interchangeable, BA is also normal. □

14.5 Circulant, Fourier, and Hartley Matrices

Circulant matrices are structured matrices that arise in signal processing, finite Fourier analysis, cyclic codes for error correction, and the asymptotic analysis of Toeplitz matrices. The set of $n \times n$ circulant matrices is an example of a commuting family of normal matrices.

Definition 14.5.1 If $n \geq 2$, an $n \times n$ *circulant matrix* is a complex matrix of the form

$$\begin{bmatrix} c_0 & c_1 & c_2 & \cdots & c_{n-1} \\ c_{n-1} & c_0 & c_1 & \cdots & c_{n-2} \\ c_{n-2} & c_{n-1} & c_0 & \cdots & c_{n-3} \\ \vdots & \vdots & \vdots & \ddots & \vdots \\ c_1 & c_2 & c_3 & \cdots & c_0 \end{bmatrix}. \tag{14.5.2}$$

The entries in each row are shifted one position to the right compared to the row above, with wraparound to the left.

Linear combinations of circulant matrices are circulant matrices, so the set of $n \times n$ circulant matrices is a subspace of M_n.

If $n \geq 2$, the $n \times n$ *cyclic permutation matrix*

$$S_n = \begin{bmatrix} 0 & 1 & 0 & \cdots & 0 \\ 0 & 0 & 1 & \cdots & 0 \\ \vdots & \vdots & \vdots & \ddots & \vdots \\ 0 & 0 & 0 & \cdots & 1 \\ 1 & 0 & 0 & \cdots & 0 \end{bmatrix} \tag{14.5.3}$$

(sometimes called a *circular shift matrix*) is unitary and hence normal (see Example 14.1.3). Since $S_n[x_1 \; x_2 \; \cdots \; x_{n-1} \; x_n]^\mathsf{T} = [x_2 \; x_3 \; \cdots \; x_n \; x_1]^\mathsf{T}$, it follows that $S_n^n = I$. Thus, S_n is annihilated by the polynomial $p(z) = z^n - 1$, whose zeros are the nth roots of unity $1, \omega, \omega^2, \ldots, \omega^{n-1}$, in which $\omega = e^{2\pi i/n}$. Theorem 9.3.3 ensures that spec $S_n \subseteq \{1, \omega, \ldots, \omega^{n-1}\}$. In fact, this containment is an equality.

Let $\mathbf{x} = [x_1 \; x_2 \; \cdots \; x_n]^\mathsf{T}$ and $j \in \{1, 2, \ldots, n\}$. Then $S_n \mathbf{x} = \omega^{j-1}\mathbf{x}$ is equivalent to

$$[x_2 \; x_3 \; \cdots \; x_1]^\mathsf{T} = \omega^{j-1}[x_1 \; x_2 \; \cdots \; x_n]^\mathsf{T}. \tag{14.5.4}$$

One checks that the identity (14.5.4) is satisfied if

$$\mathbf{x} = \mathbf{f}_j = \frac{1}{\sqrt{n}}[1 \; \omega^{j-1} \; \omega^{2(j-1)} \; \cdots \; \omega^{(n-1)(j-1)}]^\mathsf{T} \qquad \text{for } j = 1, 2, \ldots, n.$$

Thus, $(\omega^{j-1}, \mathbf{f}_j)$ is an eigenpair of S_n for $j = 1, 2, \ldots, n$ and we write $S_n = F_n \Omega F_n^*$, in which $\Omega = \operatorname{diag}(1, \omega, \omega^2, \ldots, \omega^{n-1})$ and $F_n = [\mathbf{f}_1 \; \mathbf{f}_2 \; \cdots \; \mathbf{f}_n]$ is the $n \times n$ Fourier matrix (7.1.18). The Fourier matrix is unitary; see Example 7.1.17. The matrix S_n is unitary and has distinct eigenvalues.

The powers of S_n have a regular pattern. For example, S_4^0, S_4^1, S_4^2, S_4^3, and S_4^4 are

$$\begin{bmatrix} 1 & 0 & 0 & 0 \\ 0 & 1 & 0 & 0 \\ 0 & 0 & 1 & 0 \\ 0 & 0 & 0 & 1 \end{bmatrix}, \begin{bmatrix} 0 & 1 & 0 & 0 \\ 0 & 0 & 1 & 0 \\ 0 & 0 & 0 & 1 \\ 1 & 0 & 0 & 0 \end{bmatrix}, \begin{bmatrix} 0 & 0 & 1 & 0 \\ 0 & 0 & 0 & 1 \\ 1 & 0 & 0 & 0 \\ 0 & 1 & 0 & 0 \end{bmatrix}, \begin{bmatrix} 0 & 0 & 0 & 1 \\ 1 & 0 & 0 & 0 \\ 0 & 1 & 0 & 0 \\ 0 & 0 & 1 & 0 \end{bmatrix}, \begin{bmatrix} 1 & 0 & 0 & 0 \\ 0 & 1 & 0 & 0 \\ 0 & 0 & 1 & 0 \\ 0 & 0 & 0 & 1 \end{bmatrix},$$

respectively. Notice that $S_4^3 = S_4^{\mathsf{T}} = S_4^{-1}$. If

$$p(z) = c_{n-1}z^{n-1} + \cdots + c_1 z + c_0, \tag{14.5.5}$$

then $p(S_n)$ is the circulant matrix (14.5.2). Theorem 10.5.1 tells us that $p(S_n) = F_n \operatorname{diag}(p(1), p(\omega), \ldots, p(\omega^{n-1}))F_n^*$, in which $\omega = e^{2\pi i/n}$.

Corollary 10.6.3 ensures that $X \in \mathsf{M}_n$ commutes with S_n if and only if $X = q(S_n)$ for some polynomial q. Thus, a matrix commutes with S_n if and only if it is a circulant matrix. Since each $n \times n$ circulant matrix is a polynomial in S_n, Theorem 2.6.9 ensures that the product of circulant matrices is a circulant matrix and that circulant matrices of the same size commute.

We summarize the preceding discussion in the following theorem.

Theorem 14.5.6 *Let $n \geq 2$, let $C \in \mathsf{M}_n$ be the circulant matrix (14.5.2), let p be the polynomial (14.5.5), let $\omega = e^{2\pi i/n}$, and let $\Omega = \operatorname{diag}(1, \omega, \omega^2, \ldots, \omega^{n-1})$.*

(a) *The eigenvalues of C are $p(1)$, $p(\omega)$,..., $p(\omega^{n-1})$. Corresponding unit eigenvectors are $\mathbf{f}_j = \frac{1}{\sqrt{n}}[1 \ \omega^{j-1} \ \omega^{2(j-1)} \ \cdots \ \omega^{(n-1)(j-1)}]^{\mathsf{T}}$ for $j = 1, 2, \ldots, n$, that is, $C = F_n p(\Omega)F_n^*$, in which $F_n = [\mathbf{f}_1 \ \mathbf{f}_2 \ \cdots \ \mathbf{f}_n]$ is the $n \times n$ Fourier matrix (7.1.18).*

(b) *$C = p(S_n)$, in which S_n denotes the cyclic permutation matrix (14.5.3).*

(c) *The set of all $n \times n$ circulant matrices is the commutant $\{S_n\}' \subseteq \mathsf{M}_n$. It is a subspace of commuting normal matrices that are simultaneously diagonalized by the unitary matrix F_n.*

Any $n \geq 2$ complex numbers can be the first row of an $n \times n$ circulant matrix, and the preceding theorem identifies its eigenvalues. What about the converse? Can any $n \geq 2$ complex numbers be the eigenvalues of an $n \times n$ circulant matrix? If so, what is its first row?

Theorem 14.5.7 *Let $n \geq 2$, let $\lambda_1, \lambda_2, \ldots, \lambda_n \in \mathbb{C}$, let $\omega = e^{2\pi i/n}$, and let $p(z) = c_{n-1}z^{n-1} + c_{n-2}z^{n-2} + \cdots + c_1 z + c_0$ be a polynomial of degree at most $n - 1$ such that $p(\omega^{j-1}) = \lambda_j$ for all $j = 1, 2, \ldots, n$. Then $\lambda_1, \lambda_2, \ldots, \lambda_n$ are the eigenvalues of the circulant matrix with first row $[c_0 \ c_1 \ \cdots \ c_{n-1}]$.*

Proof Maintain the notation of the preceding theorem. Since the roots of unity $1, \omega, \omega^2, \ldots, \omega^{n-1}$ are distinct, Theorem 2.7.4 ensures that there is a unique polynomial p of degree at most $n - 1$ such that $p(\omega^{j-1}) = \lambda_j$ for all $j = 1, 2, \ldots, n$. The first row of the circulant matrix $p(S_n)$ is $[c_0 \ c_1 \ \cdots \ c_{n-1}]$ and since $p(S_n) = F_n p(\Omega)F_n^* = F_n \operatorname{diag}(\lambda_1, \lambda_2, \ldots, \lambda_n)F_n^*$, its eigenvalues are $\lambda_1, \lambda_2, \ldots, \lambda_n$. $\quad\square$

We have found that every circulant matrix $C \in \mathsf{M}_n$ is unitarily diagonalized by the $n \times n$ Fourier matrix, that is, $F_n^* C F_n = \Lambda$ is diagonal. If C is a complex symmetric circulant matrix, then it is symmetric normal and Corollary 14.6.9 ensures that there is some real orthogonal matrix Q such that $Q^\mathsf{T} C Q$ is diagonal. In fact, there is a single real orthogonal matrix that diagonalizes every symmetric circulant matrix (real or complex) of a given size.

Definition 14.5.8 Write the $n \times n$ Fourier matrix as $F_n = R_n + iT_n$, in which R_n and T_n are real. The $n \times n$ *Hartley matrix* is $H_n = R_n + T_n$.

Example 14.5.9 Using the Fourier matrix F_3 displayed in Example 7.1.17, we find that

$$R_3 = \frac{1}{2\sqrt{3}} \begin{bmatrix} 2 & 2 & 2 \\ 2 & -1 & -1 \\ 2 & -1 & -1 \end{bmatrix}, \qquad T_3 = \frac{1}{2} \begin{bmatrix} 0 & 0 & 0 \\ 0 & 1 & -1 \\ 0 & -1 & 1 \end{bmatrix},$$

and

$$H_3 = \frac{1}{2\sqrt{3}} \begin{bmatrix} 2 & 2 & 2 \\ 2 & \sqrt{3}-1 & -\sqrt{3}-1 \\ 2 & -\sqrt{3}-1 & \sqrt{3}-1 \end{bmatrix},$$

in which R_3, T_3, and H_3 are symmetric. A calculation reveals that $H_3^2 = I$, so H_3 is real orthogonal. Every 3×3 complex symmetric circulant matrix has the form

$$B = \begin{bmatrix} a & b & b \\ b & a & b \\ b & b & a \end{bmatrix}$$

for some $a, b \in \mathbb{C}$. A calculation reveals that

$$H_3 B H_3 = \begin{bmatrix} a+2b & 0 & 0 \\ 0 & a-b & 0 \\ 0 & 0 & a-b \end{bmatrix},$$

so every 3×3 complex symmetric circulant matrix is real orthogonally diagonalized by the 3×3 Hartley matrix. This is not an accident.

Theorem 14.5.10 *Let $n \geq 2$ and let $C \in \mathsf{M}_n(\mathbb{C})$ be a complex symmetric circulant matrix. Let F_n and H_n be the $n \times n$ Fourier and Hartley matrices, respectively, and let $\Lambda \in \mathsf{M}_n(\mathbb{C})$ be the diagonal matrix such that $C = F_n \Lambda F_n^*$. Then H_n is real symmetric orthogonal and $C = H_n \Lambda H_n^\mathsf{T}$.*

Proof We begin by showing that the Hartley matrix is real symmetric orthogonal. We have $R_n = \operatorname{Re} F_n = \operatorname{Re} F_n^\mathsf{T} = (\operatorname{Re} F_n)^\mathsf{T} = R_n^\mathsf{T}$, so R_n is real symmetric. A similar calculation using $T_n = \operatorname{Im} F_n$ shows that T_n is real symmetric. Thus, $H_n = R_n + T_n$ is real symmetric. We know that $F_n = F_n^\mathsf{T}$, $F_n F_n^* = F_n \overline{F_n} = I$ (F_n is unitary and symmetric), and $F_n^2 = \begin{bmatrix} 1 & 0 \\ 0 & K_{n-1} \end{bmatrix} \in \mathsf{M}_n(\mathbb{R})$, in which $K_{n-1} \in \mathsf{M}_{n-1}(\mathbb{R})$ is a reversal matrix (see P.7.46). Therefore,

$$I = \operatorname{Re}(F_n F_n^*) = \operatorname{Re}((R_n + iT_n)(R_n - iT_n))$$
$$= \operatorname{Re}\big((R_n^2 + T_n^2) + i(R_n T_n - T_n R_n)\big) = R_n^2 + T_n^2$$

and

$$0 = \operatorname{Im} F_n^2 = \operatorname{Im}((R_n + iT_n)(R_n + iT_n))$$
$$= \operatorname{Im}\big((R_n^2 - T_n^2) + i(R_n T_n + T_n R_n)\big) = R_n T_n + T_n R_n.$$

These identities ensure that $R_n^2 + T_n^2 = I$ and $R_n T_n + T_n R_n = 0$. Therefore,

$$H_n H_n^\mathsf{T} = H_n^2 = R_n^2 + T_n^2 + (R_n T_n + T_n R_n) = R_n^2 + T_n^2 = I,$$

which verifies that H_n is real orthogonal.

Let $P_n = F_n^2 = \begin{bmatrix} 1 & 0 \\ 0 & K_{n-1} \end{bmatrix}$, which is a symmetric permutation matrix. Then

$$I = P_n^2 = F_n^4 = F_n F_n^2 F_n = F_n P_n F_n = (F_n P_n) F_n = F_n (P_n F_n),$$

and hence $F_n^{-1} = F_n^* = F_n P_n = P_n F_n$ and $F_n = P_n F_n^* = F_n^* P_n$. Then

$$R_n + i T_n = F_n = P_n F_n^* = P_n R_n - i P_n T_n$$

and

$$R_n + i T_n = F_n = F_n^* P_n = R_n P_n - i T_n P_n.$$

It follows that $R_n P_n = P_n R_n = R_n$ and $T_n P_n = P_n T_n = -T_n$. Since A is symmetric, we have

$$F_n \Lambda F_n^* = A = A^\mathsf{T} = \overline{F_n} \Lambda F_n^\mathsf{T} = F_n^* \Lambda F_n = (F_n P_n) \Lambda (P_n F_n^*) = F_n (P_n \Lambda P_n) F_n^*,$$

so $\Lambda = P_n \Lambda P_n$. Now compute

$$R_n \Lambda T_n = R_n (P_n \Lambda P_n) T_n = (R_n P_n) \Lambda (P_n T_n) = R_n \Lambda (-T_n) = -R_n \Lambda T_n,$$

which shows that $R_n \Lambda T_n = 0$ and $T_n \Lambda R_n = (R_n \Lambda T_n)^\mathsf{T} = 0$. Finally, we have

$$\begin{aligned} A &= F_n \Lambda F_n^* = (R_n + i T_n) \Lambda (R_n - i T_n) \\ &= R_n \Lambda R_n + T_n \Lambda T_n + i(T_n \Lambda R_n - R_n \Lambda T_n) = R_n \Lambda R_n + T_n \Lambda T_n \end{aligned}$$

and

$$\begin{aligned} H_n \Lambda H_n^\mathsf{T} &= (R_n + T_n) \Lambda (R_n + T_n) \\ &= R_n \Lambda R_n + T_n \Lambda T_n + (T_n \Lambda R_n + R_n \Lambda T_n) \\ &= R_n \Lambda R_n + T_n \Lambda T_n = A. \end{aligned} \qquad \square$$

14.6 Some Special Classes of Normal Matrices

We have encountered several classes of normal matrices already: Hermitian matrices (Chapter 6), unitary matrices (Chapter 7), and orthogonal projections (Chapter 8). Each of these classes has an elegant spectral characterization.

Theorem 14.6.1 *Let $A \in \mathsf{M}_n$.*

(a) *A is unitary if and only if A is normal and $|\lambda| = 1$ for all $\lambda \in \operatorname{spec} A$.*

(b) *A is Hermitian if and only if A is normal and $\lambda \in \mathbb{R}$ for all $\lambda \in \operatorname{spec} A$.*

(c) *A is an orthogonal projection if and only if A is normal and $\lambda \in \{0, 1\}$ for all $\lambda \in \operatorname{spec} A$.*

(d) *A is skew Hermitian ($A^* = -A$) if and only if A is normal and λ is purely imaginary for all $\lambda \in \operatorname{spec} A$.*

Proof Suppose that $(\lambda, \mathbf{x})$ is an eigenpair of A and $\|\mathbf{x}\|_2 = 1$.

(a) If A is unitary, then Theorem 7.1.13.g ensures that $|\lambda| = \|\lambda\mathbf{x}\|_2 = \|A\mathbf{x}\|_2 = \|\mathbf{x}\|_2 = 1$. Conversely, if $|\lambda| = 1$ for all $\lambda \in \operatorname{spec} A$ and $A = U\Lambda U^*$, in which U is unitary and Λ is diagonal, then $\Lambda^{-1} = \overline{\Lambda} = \Lambda^*$. Thus, $A^{-1} = (U\Lambda U^*)^{-1} = U\Lambda^{-1}U^* = U\Lambda^*U^* = (U\Lambda U^*)^* = A^*$.

(b) If $A = A^*$, then $A^*A = A^2 = AA^*$ and

$$\lambda = \langle \lambda\mathbf{x}, \mathbf{x}\rangle = \langle A\mathbf{x}, \mathbf{x}\rangle = \langle \mathbf{x}, A^*\mathbf{x}\rangle = \langle \mathbf{x}, A\mathbf{x}\rangle = \langle \mathbf{x}, \lambda\mathbf{x}\rangle = \overline{\lambda}.$$

Thus, $\lambda = \overline{\lambda}$ and $\lambda \in \mathbb{R}$. Conversely, if $\lambda \in \mathbb{R}$ for all $\lambda \in \operatorname{spec} A$ and $A = U\Lambda U^*$, in which U is unitary and Λ is diagonal, then $\Lambda = \overline{\Lambda} = \Lambda^*$ and $A^* = U\Lambda^*U^* = U\Lambda U^* = A$.

(c) If A is an orthogonal projection, then A is Hermitian and idempotent, and $\lambda^2\mathbf{x} = A^2\mathbf{x} = A\mathbf{x} = \lambda\mathbf{x}$. Thus, $\lambda(\lambda - 1)\mathbf{x} = \mathbf{0}$, so $\lambda(\lambda - 1) = 0$ and hence $\lambda \in \{0, 1\}$. Conversely, if $\operatorname{spec} A \subseteq \{0, 1\}$ and $A = U\Lambda U^*$, in which U is unitary and Λ is diagonal, then $\Lambda^2 = \Lambda$. Thus, $A^2 = (U\Lambda U^*)(U\Lambda U^*) = U\Lambda^2U^* = U\Lambda U^* = A$, so A is idempotent. Part (b) shows that A is Hermitian, and Theorem 8.3.14 ensures that A is an orthogonal projection.

(d) If A is skew Hermitian, then $B = iA$ is Hermitian and has real spectrum, so $A = -iB$ is normal (Theorem 14.1.9.a) and has purely imaginary spectrum. Conversely, if $i\lambda \in \mathbb{R}$ for all $\lambda \in \operatorname{spec} A$ and $A = U\Lambda U^*$, in which U is unitary and Λ is diagonal, then $i\Lambda$ is real and $iA = U(i\Lambda)U^*$ is Hermitian. $\qquad \square$

Normal matrices whose eigenvalues are real and nonnegative are particularly important. They are the positive semidefinite matrices; see Chapter 15.

Example 14.6.2 $A = \begin{bmatrix} 1 & 1 \\ 0 & 1 \end{bmatrix}$ has $\operatorname{spec} A = \{1\}$, but it is not unitary, not Hermitian, and not an orthogonal projection. Moreover, $\operatorname{spec}(iA) = \{i\}$, but iA is not skew Hermitian. Thus, the normality hypotheses in Theorem 14.6.1 cannot be omitted.

Example 14.6.3 $A = \begin{bmatrix} 1 & 1 \\ 0 & 0 \end{bmatrix}$ has distinct eigenvalues 0 and 1, so it is diagonalizable and idempotent. But it is not Hermitian, so it is not an orthogonal projection. The hypothesis in Theorem 14.6.1.c that A is normal cannot be weakened to assume that A is diagonalizable.

Theorem 14.6.4 *If $A \in \mathsf{M}_n$, then there are unique Hermitian matrices $H, K \in \mathsf{M}_n$ such that $A = H + iK$.*

Proof We have $A = \frac{1}{2}(A + A^*) + i\frac{1}{2i}(A - A^*) = H + iK$, in which

$$H = \frac{1}{2}(A + A^*) \quad \text{and} \quad K = \frac{1}{2i}(A - A^*) \qquad (14.6.5)$$

are Hermitian. Conversely, if $A = X + iY$, in which X and Y are Hermitian, then $A^* = X - iY$. Consequently, $2X = A + A^*$ and $2iY = A - A^*$, which confirm that $X = H$ and $Y = K$. $\qquad \square$

Definition 14.6.6 The *Cartesian decomposition* of $A \in \mathsf{M}_n$ is $A = H + iK$, in which $H, K \in \mathsf{M}_n$ are Hermitian. The matrix $H = \frac{1}{2}(A + A^*)$ is the *Hermitian part* of A and $iK = \frac{1}{2}(A - A^*)$ is the *skew-Hermitian part* of A. We sometimes denote H and K by $H(A)$ and $K(A)$.

The Cartesian decomposition of A is analogous to the decomposition (A.3.8) of a complex number $z = a + bi$, in which $a = \operatorname{Re} z = \frac{1}{2}(z + \overline{z})$ and $b = \operatorname{Im} z = \frac{1}{2i}(z - \overline{z})$ are real.

Example 14.6.7 An example of a Cartesian decomposition is:

$$\underbrace{\begin{bmatrix} 0 & 1 \\ 0 & 0 \end{bmatrix}}_{A} = \underbrace{\begin{bmatrix} 0 & \frac{1}{2} \\ \frac{1}{2} & 0 \end{bmatrix}}_{H} + i \underbrace{\begin{bmatrix} 0 & -\frac{i}{2} \\ \frac{i}{2} & 0 \end{bmatrix}}_{K}.$$

Observe that A is not normal and H does not commute with K.

Theorem 14.6.8 *Let $A \in \mathsf{M}_n$ with Cartesian decomposition $A = H + iK$. Then A is normal if and only if $HK = KH$.*

Proof Observe that

$$A^*A = (H - iK)(H + iK) = H^2 + K^2 + i(HK - KH)$$

and

$$AA^* = (H + iK)(H - iK) = H^2 + K^2 - i(HK - KH).$$

Thus, $A^*A = AA^*$ if and only if $HK = KH$. $\square$

Corollary 14.6.9 *Let $A \in \mathsf{M}_n(\mathbb{C})$ be symmetric.*

(a) *A is normal if and only if there is a real orthogonal $Q \in \mathsf{M}_n(\mathbb{R})$ and a diagonal $D \in \mathsf{M}_n$ such that $A = QDQ^\mathsf{T}$.*

(b) *A is unitary if and only if there is a real orthogonal $Q \in \mathsf{M}_n(\mathbb{R})$ and a unitary diagonal $D = \mathrm{diag}(e^{i\theta_1}, e^{i\theta_2}, \ldots, e^{i\theta_n}) \in \mathsf{M}_n$ such that each $\theta_j \in [0, 2\pi)$ and $A = QDQ^\mathsf{T}$. The real numbers θ_j are uniquely determined by A.*

Proof (a) Let $A = H + iK$ be the Cartesian decomposition of A. Since A is complex symmetric,

$$H^\mathsf{T} + iK^\mathsf{T} = A^\mathsf{T} = A = H + iK,$$

so the uniqueness assertion in Theorem 14.6.4 ensures that $H^\mathsf{T} = H$. But $H = H^* = \overline{H}^\mathsf{T}$, so H is real and Hermitian, that is, it is real symmetric. The same argument shows that K is real symmetric. Now suppose that A is normal. The preceding theorem says that $HK = KH$, so Theorem 14.2.11.b ensures that H and K are simultaneously real orthogonally diagonalizable. Thus, there is a real orthogonal matrix Q and real diagonal matrices Λ and M such that $H = Q\Lambda Q^\mathsf{T}$ and $K = QMQ^\mathsf{T}$. Then $A = H + iK = Q\Lambda Q^\mathsf{T} + iQMQ^\mathsf{T} = Q(\Lambda + iM)Q^\mathsf{T} = QDQ^\mathsf{T}$, in which $D = \Lambda + iM$. Conversely, if D is diagonal and Q is real orthogonal, then QDQ^T is symmetric and normal (Lemma 14.1.11).

(b) Since a unitary matrix is normal, it follows from (a) that $A = QDQ^\mathsf{T}$. Then $Q^\mathsf{T}AQ = D$ is unitary since it is a product of unitary matrices. Since $DD^* = D\overline{D} = I$, each diagonal entry of D has modulus 1 and can be written as $e^{i\theta}$ for a unique $\theta \in [0, 2\pi)$. $\square$

The spectral decomposition of a complex symmetric unitary matrix in the preceding corollary leads to a basic result about certain matrix square roots.

Lemma 14.6.10 *Let $V \in \mathsf{M}_n(\mathbb{C})$ be unitary and symmetric. There is a polynomial p such that $S = p(V)$ is unitary and symmetric, and $S^2 = V$.*

Proof According to the preceding corollary, there is a real orthogonal matrix Q and a diagonal unitary matrix $D = \mathrm{diag}(e^{i\theta_1}, e^{i\theta_2}, \ldots, e^{i\theta_n})$ such that each $\theta_j \in [0, 2\pi)$ and $V = QDQ^{\mathsf{T}}$. Let $E = \mathrm{diag}(e^{i\theta_1/2}, e^{i\theta_2/2}, \ldots, e^{i\theta_n/2})$. Theorem 2.7.4 ensures that there is a polynomial p such that $p(e^{i\theta_j}) = e^{i\theta_j/2}$ for each $j = 1, 2, \ldots, n$, so $p(D) = E$. Then $S = QEQ^{\mathsf{T}}$ is symmetric and unitary (it is a product of unitary matrices), $S^2 = QE^2Q^{\mathsf{T}} = QDQ^{\mathsf{T}} = V$, and $S = QEQ^{\mathsf{T}} = Qp(D)Q^{\mathsf{T}} = p(QDQ^{\mathsf{T}}) = p(V)$. $\qquad\square$

We use this lemma to show that each unitary matrix is the product of a real orthogonal matrix and a complex symmetric unitary matrix.

Theorem 14.6.11 (*QS Decomposition of a Unitary Matrix*) *Let $U \in \mathsf{M}_n(\mathbb{C})$ be unitary. There are unitary $Q, S \in \mathsf{M}_n(\mathbb{C})$ and a polynomial p such that $U = QS$, Q is real orthogonal, S is symmetric, and $S = p(U^{\mathsf{T}}U)$.*

Proof Observe that $U^{\mathsf{T}}U$ is unitary and complex symmetric. The preceding lemma ensures that there is a polynomial p such that $S = p(U^{\mathsf{T}}U)$ is unitary and complex symmetric, and $S^2 = U^{\mathsf{T}}U$. Let $Q = US^* = U\overline{S}$. Then Q is a product of unitary matrices, so it is unitary and $Q^{\mathsf{T}}Q = S^*U^{\mathsf{T}}US^* = S^*S^2S^* = (S^*S)(SS^*) = I$. Therefore, $Q^{\mathsf{T}} = Q^{-1} = Q^* = \overline{Q}^{\mathsf{T}}$, so Q^{T} (and hence also Q) is real. Consequently, $Q = US^*$ is real orthogonal and $U = (US^*)S = QS$. $\qquad\square$

In Theorem 9.6.3 we learned that two similar real matrices are similar via a real matrix. The following corollary addresses unitary similarity of two real matrices.

Corollary 14.6.12 *If two real matrices are unitarily similar, then they are real orthogonally similar.*

Proof Let $A, B \in \mathsf{M}_n(\mathbb{R})$. Let $U \in \mathsf{M}_n(\mathbb{C})$ be unitary and such that $A = UBU^*$. Then

$$UBU^* = A = \overline{A} = \overline{UBU^*} = \overline{U}BU^{\mathsf{T}},$$

so $(U^{\mathsf{T}}U)B = B(U^{\mathsf{T}}U)$. The preceding theorem ensures that $U = QS$, in which Q is real orthogonal, S is unitary, and S is a polynomial in $U^{\mathsf{T}}U$. Then B commutes with S and $A = UBU^* = QSBS^*Q^{\mathsf{T}} = QBSS^*Q^{\mathsf{T}} = QBQ^{\mathsf{T}}$, which is a real orthogonal similarity. $\qquad\square$

14.7 Similarity of Normal and Other Diagonalizable Matrices

Theorem 14.7.1 *Suppose that $A, B \in \mathsf{M}_n$ have the same eigenvalues, including multiplicities.*

(a) *If A and B are diagonalizable, then they are similar.*

(b) *If A and B are normal, then they are unitarily similar.*

(c) *If A and B are real and normal, then they are real orthogonally similar.*

(d) *If A and B are diagonal, then they are permutation similar.*

Proof Let $\lambda_1, \lambda_2, \ldots, \lambda_n$ be the eigenvalues of A and B in any given order, and let $\Lambda = \mathrm{diag}(\lambda_1, \lambda_2, \ldots, \lambda_n)$.

(a) Corollary 10.4.12 ensures that there are invertible $R, S \in \mathsf{M}_n$ such that $A = R\Lambda R^{-1}$ and $B = S\Lambda S^{-1}$. Then $\Lambda = S^{-1}BS$ and $A = R\Lambda R^{-1} = R(S^{-1}BS)R^{-1} = (RS^{-1})B(RS^{-1})^{-1}$, so A is similar to B via RS^{-1}.

(b) Theorem 14.2.2 ensures that there are unitary $U, V \in \mathsf{M}_n$ such that $A = U \Lambda U^*$ and $B = V \Lambda V^*$. Then $\Lambda = V^* B V$ and $A = U \Lambda U^* = U(V^* B V)U^* = (UV^*)B(UV^*)^*$, so A is unitarily similar to B via the unitary matrix UV^*.

(c) Part (b) ensures that A and B are unitarily similar; Corollary 14.6.12 says that they are real orthogonally similar because they are real and unitarily similar.

(d) Since the main diagonal entries of A and B are their eigenvalues, the main diagonal entries of B can be obtained by permuting the main diagonal entries of A. The identity (7.3.4) shows how to construct a permutation similarity of A that effects this permutation of its diagonal entries.   □

14.8 Some Characterizations of Normality

Here are several equivalent characterizations of normality.

Theorem 14.8.1 *Let $A = [a_{ij}] \in \mathsf{M}_n$ have eigenvalues $\lambda_1, \lambda_2, \ldots, \lambda_n$. The following are equivalent:*

(a) *A is normal.*

(b) *$A = H + iK$ in which H, K are Hermitian and $HK = KH$.*

(c) *$\sum_{i,j=1}^{n} |a_{ij}|^2 = \sum_{i=1}^{n} |\lambda_i|^2$.*

(d) *There is a polynomial p such that $p(A) = A^*$.*

(e) *$\|A\mathbf{x}\|_2 = \|A^* \mathbf{x}\|_2$ for all $\mathbf{x} \in \mathbb{C}^n$.*

(f) *A commutes with $A^* A$.*

(g) *$A^* = AW$ for some unitary W.*

(h) *A commutes with a normal matrix that has distinct eigenvalues.*

Proof (a) $\Leftrightarrow$ (b) This is Theorem 14.6.8.
(a) $\Leftrightarrow$ (c) This is the case of equality in Theorem 14.3.1.
(a) $\Rightarrow$ (d) This is a special case of Lemma 14.4.1.
(d) $\Rightarrow$ (a) If $A^* = p(A)$, then $A^* A = p(A)A = Ap(A) = AA^*$, so A is normal.
(a) $\Rightarrow$ (e) This is Theorem 14.1.9.b.
(e) $\Rightarrow$ (a) Let $S = A^* A - AA^*$ and let $\mathbf{x} \in \mathbb{C}^n$. We must show that $S = 0$. Compute

$$0 = \|A\mathbf{x}\|_2^2 - \|A^* \mathbf{x}\|_2^2 = \langle A\mathbf{x}, A\mathbf{x} \rangle - \langle A^* \mathbf{x}, A^* \mathbf{x} \rangle = \langle A^* A \mathbf{x}, \mathbf{x} \rangle - \langle AA^* \mathbf{x}, \mathbf{x} \rangle = \langle S\mathbf{x}, \mathbf{x} \rangle.$$

Because $S = S^*$, it follows that $\langle S^2 \mathbf{x}, \mathbf{x} \rangle = \langle S\mathbf{x}, S\mathbf{x} \rangle = \|S\mathbf{x}\|_2^2$ and

$$0 = \langle S(\mathbf{x} + S\mathbf{x}), \mathbf{x} + S\mathbf{x} \rangle = \langle S\mathbf{x}, \mathbf{x} \rangle + \langle S\mathbf{x}, S\mathbf{x} \rangle + \langle S^2 \mathbf{x}, \mathbf{x} \rangle + \langle S(S\mathbf{x}), S\mathbf{x} \rangle$$
$$= 0 + \|S\mathbf{x}\|_2^2 + \|S\mathbf{x}\|_2^2 + 0 = 2\|S\mathbf{x}\|_2^2.$$

We conclude that $S\mathbf{x} = \mathbf{0}$ for all $\mathbf{x} \in \mathbb{C}^n$, so $S = 0$ and A is normal.
(a) $\Rightarrow$ (f) If A is normal, then $A(A^* A) = (AA^*)A = (A^* A)A$.

(f) $\Rightarrow$ (a) Suppose that A commutes with A^*A. Then we have $A^*AA = AA^*A$ and its adjoint $A^*A^*A = A^*AA^*$. Let $S = A^*A - AA^*$. We must show that $S = 0$. Since S is Hermitian,

$$\begin{aligned} S^*S = S^2 &= (A^*A)^2 - (A^*AA)A^* - A(A^*A^*A) + (AA^*)^2 \\ &= (A^*A)^2 - (AA^*A)A^* - A(A^*AA^*) + (AA^*)^2 \\ &= (A^*A)^2 - 2(AA^*)^2 + (AA^*)^2 \\ &= (A^*A)^2 - (AA^*)^2. \end{aligned}$$

Therefore, $S = 0$ because

$$\begin{aligned} \|S\|_F^2 = \mathrm{tr}(S^*S) &= \mathrm{tr}(A^*A)^2 - \mathrm{tr}(AA^*)^2 \\ &= \mathrm{tr}(A^*A)^2 - \mathrm{tr}(AA^*AA^*) = \mathrm{tr}(A^*A)^2 - \mathrm{tr}(A^*AA^*A) \\ &= \mathrm{tr}(A^*A)^2 - \mathrm{tr}(A^*A)^2 = 0. \end{aligned}$$

(g) $\Rightarrow$ (a) If $A^* = AW$ and W is unitary, then $A^*A = A^*(A^*)^* = (AW)(AW)^* = AWW^*A^* = AA^*$.

(a) $\Rightarrow$ (g) Let A be normal and write $A = U\Lambda U^*$, in which $\Lambda = \mathrm{diag}(\lambda_1, \lambda_2, \ldots, \lambda_n)$ and U is unitary. Define $X = \mathrm{diag}(\xi_1, \xi_2, \ldots, \xi_n)$, in which

$$\xi_i = \begin{cases} \overline{\lambda_i}/\lambda_i & \text{if } \lambda_i \neq 0, \\ 1 & \text{if } \lambda_i = 0, \end{cases}$$

and let $W = U\,\mathrm{diag}(\xi_1, \xi_2, \ldots, \xi_n)U^*$, which is unitary (Theorem 14.6.1.a). Then $AW = (U\Lambda U^*)(UXU^*) = U\Lambda X U^* = U\,\mathrm{diag}(\overline{\lambda_1}, \overline{\lambda_2}, \ldots, \overline{\lambda_n})U^* = (U\Lambda U^*)^* = A^*$.

(a) $\Rightarrow$ (h) If A is normal, then $A = U\Lambda U^*$, in which U is unitary and Λ is diagonal. Let $B = UNU^*$, in which $N = \mathrm{diag}(1, 2, \ldots, n)$. Then B is normal, has distinct eigenvalues, and commutes with A.

(h) $\Rightarrow$ (a) Let $B = U\Lambda U^*$, in which U is unitary and Λ is diagonal with distinct diagonal entries. If B commutes with A, then Λ commutes with U^*AU, so Lemma 3.2.17 ensures that $D = U^*AU$ is diagonal. Then $A = UDU^*$ is normal. $\qquad\square$

14.9 Spectral Resolutions

We present another version of the spectral theorem based upon orthogonal projections. This version of the spectral theorem is used in quantum computing and it generalizes to the infinite-dimensional setting.

If $P \in \mathsf{M}_n$ is an orthogonal projection ($P = P^*$ and $P^2 = P$), then $\mathbb{C}^n = \mathrm{col}\,P \oplus \mathrm{null}\,P$ is an orthogonal decomposition and the induced linear transformation $T_P : \mathbb{C}^n \to \mathbb{C}^n$ is the orthogonal projection onto $\mathrm{col}\,P$.

Definition 14.9.1 Nonzero orthogonal projections $P_1, P_2, \ldots, P_d \in \mathsf{M}_n$ are a *resolution of the identity* if $P_1 + P_2 + \cdots + P_d = I$.

If $P_1, P_2, \ldots, P_d \in \mathsf{M}_n$ are a resolution of the identity, then $\mathbf{x} = P_1\mathbf{x} + P_2\mathbf{x} + \cdots + P_d\mathbf{x}$ for all $\mathbf{x} \in \mathbb{C}^n$. Thus,

$$\mathbb{C}^n = \mathrm{col}\,P_1 + \cdots + \mathrm{col}\,P_d. \tag{14.9.2}$$

The following lemma says that (14.9.2) is a direct sum of pairwise orthogonal subspaces.

Lemma 14.9.3 *Suppose that $P_1, P_2, \ldots, P_d \in M_n$ are a resolution of the identity. Then $P_i P_j = 0$ for $i \neq j$. In particular, $\operatorname{col} P_i \perp \operatorname{col} P_j$ for $i \neq j$ and $\mathbb{C}^n = \operatorname{col} P_1 \oplus \cdots \oplus \operatorname{col} P_d$ is an orthogonal direct sum.*

Proof For $j = 1, 2, \ldots, d$,

$$0 = P_j - P_j = P_j - P_j^2 = (I - P_j)P_j = \Big(\sum_{i \neq j} P_i\Big) P_j = \sum_{i \neq j} P_i P_j = \sum_{i \neq j} P_i^2 P_j^2.$$

Therefore,

$$0 = \sum_{i \neq j} \operatorname{tr} P_i^2 P_j^2 = \sum_{i \neq j} \operatorname{tr} P_j P_i P_i P_j = \sum_{i \neq j} \operatorname{tr}\big((P_i P_j)^*(P_i P_j)\big) = \sum_{i \neq j} \|P_i P_j\|_F^2.$$

Thus, $\|P_i P_j\|_F = 0$ for all $i \neq j$, and hence $P_i P_j = 0$. If $i \neq j$, then $\operatorname{col} P_j \subseteq \operatorname{null} P_i = (\operatorname{col} P_i)^\perp$ and $\operatorname{col} P_i \subseteq \operatorname{null} P_j = (\operatorname{col} P_j)^\perp$, so $\operatorname{col} P_i \perp \operatorname{col} P_j$. $\qquad \square$

The following lemma provides a method to produce resolutions of the identity. In fact, all resolutions of the identity arise in this manner; see P.14.23.

Lemma 14.9.4 *Let $U = [U_1 \ U_2 \ \ldots \ U_d] \in M_n$ be a unitary matrix, in which each $U_i \in M_{n \times n_i}$ and $n_1 + n_2 + \cdots + n_d = n$. Then $P_i = U_i U_i^*$ is an orthogonal projection for each $i = 1, 2, \ldots, d$, and $P_1, P_2, \ldots, P_d$ is a resolution of the identity.*

Proof We have

$$I = UU^* = [U_1 \ U_2 \ \ldots \ U_d] \begin{bmatrix} U_1^* \\ U_2^* \\ \vdots \\ U_d^* \end{bmatrix} = U_1 U_1^* + U_2 U_2^* + \cdots + U_d U_d^*.$$

Example 8.3.5 ensures that each $P_i = U_i U_i^*$ is an orthogonal projection. $\qquad \square$

Example 14.9.5 Consider the unitary matrix $U = [U_1 \ U_2] \in M_3$, in which

$$U = \frac{1}{3}\begin{bmatrix} 1 & -2 & -2 \\ 2 & -1 & 2 \\ 2 & 2 & -1 \end{bmatrix}, \qquad U_1 = \frac{1}{3}\begin{bmatrix} 1 & -2 \\ 2 & -1 \\ 2 & 2 \end{bmatrix}, \qquad \text{and} \qquad U_2 = \frac{1}{3}\begin{bmatrix} -2 \\ 2 \\ -1 \end{bmatrix}.$$

Then

$$P_1 = U_1 U_1^* = \frac{1}{9}\begin{bmatrix} 5 & 4 & -2 \\ 4 & 5 & 2 \\ -2 & 2 & 8 \end{bmatrix} \qquad \text{and} \qquad P_2 = U_2 U_2^* = \frac{1}{9}\begin{bmatrix} 4 & -4 & 2 \\ -4 & 4 & -2 \\ 2 & -2 & 1 \end{bmatrix}$$

are orthogonal projections and P_1, P_2 is a resolution of the identity; P_1 is the orthogonal projection onto $\operatorname{span}\{[1 \ 2 \ 2]^T, [-2 \ -1 \ 2]^T\}$ and P_2 is the orthogonal projection onto $\operatorname{span}\{[-2 \ 2 \ -1]^T\}$.

Theorem 14.9.6 (Spectral Theorem, Version II) *Let the distinct eigenvalues of $A \in M_n$ be $\mu_1, \mu_2, \ldots, \mu_d$ with multiplicities $n_1, n_2, \ldots, n_d$. The following are equivalent:*

(a) *A is normal.*

(b) *$A = \sum_{i=1}^{d} \mu_i P_i$, in which $P_1, P_2, \ldots, P_d \in M_n$ are orthogonal projections that are a resolution of the identity.*

In (b), $\operatorname{col} P_j = \mathcal{E}_{\mu_j}(A)$.

Proof (a) $\Rightarrow$ (b) Suppose that A is normal. Theorem 14.2.2 ensures that $A = U\Lambda U^*$, in which $\Lambda = \mu_1 I_{n_1} \oplus \cdots \oplus \mu_d I_{n_d}$. Write $U = [U_1 \ \ldots \ U_d]$, in which each $U_i \in \mathsf{M}_{n\times n_i}$. Then

$$A = U\Lambda U^* = [U_1 \ \ldots \ U_d]\begin{bmatrix} \mu_1 I_{n_1} & & \\ & \ddots & \\ & & \mu_d I_{n_d} \end{bmatrix}\begin{bmatrix} U_1^* \\ \vdots \\ U_d^* \end{bmatrix}$$

$$= [\mu_1 U_1 \ \ldots \ \mu_d U_d]\begin{bmatrix} U_1^* \\ \vdots \\ U_d^* \end{bmatrix} = \mu_1 U_1 U_1^* + \cdots + \mu_d U_d U_d^*$$

$$= \mu_1 P_1 + \mu_2 P_2 + \cdots + \mu_d P_d,$$

in which $P_i = U_i U_i^*$ for $i = 1, 2, \ldots, d$. Lemma 14.9.4 ensures that $P_1, P_2, \ldots, P_d$ are a resolution of the identity.

(b) $\Rightarrow$ (a) Suppose that $A = \sum_{i=1}^d \mu_i P_i$, in which $P_1, P_2, \ldots, P_d \in \mathsf{M}_n$ are a resolution of the identity. Since $P_i P_j = 0$ for $i \neq j$ (Lemma 14.9.3) and $A^* = \sum_{j=1}^d \overline{\mu_j} P_j$, we have

$$A^*A = \sum_{i,j=1}^d \overline{\mu_j}\mu_i P_j P_i = \sum_{i=1}^d |\mu_i|^2 P_i = \sum_{i,j=1}^d \mu_i \overline{\mu_j} P_i P_j = AA^*.$$

We claim that $\mathcal{E}_{\mu_j}(A) = \mathrm{col}\, P_j$. For each $j = 1, 2, \ldots, d$ and any $\mathbf{x} \in \mathbb{C}^n$,

$$(A - \mu_j I)\mathbf{x} = \sum_{i=1}^d \mu_i P_i \mathbf{x} - \sum_{i=1}^d \mu_j P_i \mathbf{x} = \sum_{i\neq j}(\mu_i - \mu_j)P_i \mathbf{x}.$$

Since $P_i\mathbf{x} \perp P_j\mathbf{x}$ if $i \neq j$, the Pythagorean theorem ensures that

$$\|(A - \mu_j I)\mathbf{x}\|_2^2 = \left\|\sum_{i\neq j}(\mu_i - \mu_j)P_i\mathbf{x}\right\|_2^2 = \sum_{i\neq j}|\mu_i - \mu_j|^2 \|P_i\mathbf{x}\|_2^2,$$

in which $\mu_i - \mu_j \neq 0$ for $i \neq j$. Therefore, $\mathbf{x} \in \mathcal{E}_{\mu_j}(A)$ if and only if $P_i\mathbf{x} = \mathbf{0}$ for all $i \neq j$. Since $\mathbf{x} = \sum_{i=1}^d P_i\mathbf{x}$, this occurs if and only if $\mathbf{x} = P_j\mathbf{x}$. Thus, $\mathcal{E}_{\mu_j}(A) = \mathrm{col}\, P_j$. $\square$

Definition 14.9.7 The orthogonal projections $P_1, P_2, \ldots, P_d$ in the preceding theorem are *spectral projections for A*. The representation $A = \sum_{i=1}^d \mu_i P_i$ is a *spectral resolution of A*.

Theorem 14.9.6 says that a matrix is normal if and only if it is a linear combination of orthogonal projections that are a resolution of the identity. The latter condition implies that the orthogonal projections must be pairwise orthogonal (Lemma 14.9.3). If a normal matrix A has distinct eigenvalues, the representation (14.2.4) is a spectral resolution, in which each spectral projection has rank 1. Spectral resolutions are unique up to the order of the summands.

Example 14.9.8 A spectral resolution of an orthogonal projection $P \in \mathsf{M}_n$ is $P = 1P + 0(I - P)$.

Example 14.9.9 Consider the matrix A in (14.1.7) and Example 14.2.5. The orthogonal projections P_1, P_2 onto the eigenspaces $\mathcal{E}_{\lambda_1}(A) = \mathrm{span}\{[\frac{1}{\sqrt{2}} \ -\frac{i}{\sqrt{2}}]^\mathsf{T}\}$ and $\mathcal{E}_{\lambda_2}(A) = \mathrm{span}\{[\frac{1}{\sqrt{2}} \ \frac{i}{\sqrt{2}}]^\mathsf{T}\}$ are

$$P_1 = \begin{bmatrix} \frac{1}{\sqrt{2}} \\ -\frac{i}{\sqrt{2}} \end{bmatrix}[\tfrac{1}{\sqrt{2}} \ \tfrac{i}{\sqrt{2}}] = \begin{bmatrix} \frac{1}{2} & \frac{i}{2} \\ -\frac{i}{2} & \frac{1}{2} \end{bmatrix} \quad \text{and} \quad P_2 = \begin{bmatrix} \frac{1}{\sqrt{2}} \\ \frac{i}{\sqrt{2}} \end{bmatrix}[\tfrac{1}{\sqrt{2}} \ -\tfrac{i}{\sqrt{2}}] = \begin{bmatrix} \frac{1}{2} & -\frac{i}{2} \\ \frac{i}{2} & \frac{1}{2} \end{bmatrix}.$$

That $\mathcal{E}_1(A) \perp \mathcal{E}_2(A)$ and $\mathbb{C}^2 = \mathcal{E}_1(A) \oplus \mathcal{E}_2(A)$ is reflected in the equation $P_1 + P_2 = I$:

$$\begin{bmatrix} \frac{1}{2} & \frac{i}{2} \\ -\frac{i}{2} & \frac{1}{2} \end{bmatrix} + \begin{bmatrix} \frac{1}{2} & -\frac{i}{2} \\ \frac{i}{2} & \frac{1}{2} \end{bmatrix} = \begin{bmatrix} 1 & 0 \\ 0 & 1 \end{bmatrix}.$$

The spectral theorem ensures that $A = \lambda_1 P_1 + \lambda_2 P_2$:

$$\begin{bmatrix} a & -b \\ b & a \end{bmatrix} = (a + bi) \begin{bmatrix} \frac{1}{2} & \frac{i}{2} \\ -\frac{i}{2} & \frac{1}{2} \end{bmatrix} + (a - bi) \begin{bmatrix} \frac{1}{2} & -\frac{i}{2} \\ \frac{i}{2} & \frac{1}{2} \end{bmatrix}.$$

Theorem 14.9.10 *Let $A = \sum_{i=1}^d \mu_i P_i$ be a spectral resolution of a normal matrix $A \in \mathsf{M}_n$ and let f be a polynomial. Then $f(A) = \sum_{i=1}^d f(\mu_i) P_i$.*

Proof It suffices to show that $A^\ell = \sum_{i=1}^d \mu_i^\ell P_i$ for $\ell = 0, 1, \ldots$. We proceed by induction on ℓ. The base case $\ell = 0$ is $I = P_1 + \cdots + P_d$, which is the definition of a resolution of the identity. If $A^\ell = \sum_{i=1}^d \mu_i^\ell P_i$ is valid for some ℓ, then

$$A^{\ell+1} = AA^\ell = \left(\sum_{i=1}^d \mu_i P_i \right) \left(\sum_{j=1}^d \mu_j^\ell P_j \right) = \sum_{i,j=1}^d \mu_i \mu_j^\ell P_i P_j = \sum_{i=1}^d \mu_i^{\ell+1} P_i^2 = \sum_{i=1}^d \mu_i^{\ell+1} P_i,$$

since $P_i P_j = 0$ for $i \neq j$ (Lemma 14.9.3). This completes the induction. $\square$

Corollary 14.9.11 *Let $A = \sum_{i=1}^d \mu_i P_i$ be a spectral resolution of a normal matrix $A \in \mathsf{M}_n$. There are polynomials $p_1, p_2, \ldots, p_d$ such that $p_j(A) = P_j$ for each $j = 1, 2, \ldots, d$.*

Proof Theorem 2.7.4 ensures that there are unique polynomials p_j of degree at most $d - 1$ such that $p_j(\mu_i) = \delta_{ij}$ for $i, j = 1, 2, \ldots, d$. Therefore, $p_j(A) = \sum_{i=1}^d p_j(\mu_i) P_i = P_j$. $\square$

Example 14.9.12 The matrix (14.2.8) has eigenvalues $\mu_1 = 9$ and $\mu_2 = -9$, with multiplicities $n_1 = 2$ and $n_2 = 1$, respectively. The polynomials $p_1(z) = \frac{1}{18}(z+9)$ and $p_2(z) = -\frac{1}{18}(z-9)$ satisfy $p_j(\mu_i) = \delta_{ij}$ for $i, j \in \{1, 2\}$. Let $P_1 = p_1(A)$ and $P_2 = p_2(A)$. A spectral resolution of A is

$$\begin{bmatrix} 1 & 8 & -4 \\ 8 & 1 & 4 \\ -4 & 4 & 7 \end{bmatrix} = 9 \underbrace{\begin{bmatrix} \frac{5}{9} & \frac{4}{9} & -\frac{2}{9} \\ \frac{4}{9} & \frac{5}{9} & \frac{2}{9} \\ -\frac{2}{9} & \frac{2}{9} & \frac{8}{9} \end{bmatrix}}_{P_1} - 9 \underbrace{\begin{bmatrix} \frac{4}{9} & -\frac{4}{9} & \frac{2}{9} \\ -\frac{4}{9} & \frac{4}{9} & -\frac{2}{9} \\ \frac{2}{9} & -\frac{2}{9} & \frac{1}{9} \end{bmatrix}}_{P_2}.$$

If $A \in \mathsf{M}_n$ is normal, the geometric content of Theorem 14.1.9.c is that each one-dimensional invariant subspace of A is also an invariant subspace of A^*. This property of normal matrices generalizes to invariant subspaces of any dimension.

Theorem 14.9.13 *Let $A \in \mathsf{M}_n$ be normal.*

(a) *Every A-invariant subspace is A^*-invariant.*

(b) *If P is the orthogonal projection onto an A-invariant subspace, then $AP = PA$.*

Proof (a) There is nothing to prove if the A-invariant subspace is $\{\mathbf{0}\}$ or $\mathbb{C}^n$. Suppose that $\mathcal{U} \subseteq \mathbb{C}^n$ is a k-dimensional A-invariant subspace with $1 \leq k \leq n - 1$. Let $U = [U_1\ U_2] \in \mathsf{M}_n$ be a unitary matrix, in which the columns of $U_1 \in \mathsf{M}_{n \times k}$ are an orthonormal basis for $\mathcal{U}$

and the columns of $U_2 \in \mathsf{M}_{n \times (n-k)}$ are an orthonormal basis for $\mathcal{U}^\perp$. Since $\mathcal{U}$ is A-invariant, Theorem 8.6.7 ensures that

$$U^* A U = \begin{bmatrix} B & X \\ 0 & C \end{bmatrix},$$

in which $B \in \mathsf{M}_k$ and $C \in \mathsf{M}_{n-k}$. Lemma 14.1.11 ensures that $U^* A U$ is normal. Lemma 14.1.12 tells us that B and C are normal and $X = 0$. Thus, $U^* A U = B \oplus C$ and $\mathcal{U}$ is A^*-invariant by Corollary 8.6.8.

(b) This follows from (a) and Corollary 8.6.8.   □

Example 14.9.14 Let $A = \begin{bmatrix} 0 & 1 \\ 0 & 0 \end{bmatrix}$, which is not normal. The subspace $\mathrm{span}\{\mathbf{e}_1\}$ is an A-invariant subspace that is not A^*-invariant, and $\mathrm{span}\{\mathbf{e}_2\}$ is an A^*-invariant subspace that is not A-invariant.

For another proof of Theorem 14.9.13, see P.14.52. Complications arise in the infinite-dimensional setting; see P.14.51.

14.10 Problems

P.14.1 Show that the matrices in Example 14.1.8 are not normal.

P.14.2 Which of $\begin{bmatrix} 1 & 0 \\ 1 & 1 \end{bmatrix}, \begin{bmatrix} 0 & 1 \\ -1 & 0 \end{bmatrix}, \begin{bmatrix} 1 & i \\ i & 1 \end{bmatrix}, \begin{bmatrix} 1 & i \\ 1 & 1 \end{bmatrix}$, and $\begin{bmatrix} 1 & 2 \\ 1 & 1 \end{bmatrix}$, are normal?

P.14.3 If $A \in \mathsf{M}_n$ is normal and nilpotent, prove that $A = 0$.

P.14.4 If $A \in \mathsf{M}_n$ is normal and $\mathrm{spec}\, A = \{1\}$, prove that $A = I$.

P.14.5 Can the product of two nonzero non-normal matrices be normal? Why?

P.14.6 Let $A, B \in \mathsf{M}_n$ be normal. Prove that A and B are similar if and only if they are unitarily similar.

P.14.7 Let $B \in \mathsf{M}_n$ be invertible and let $A = B^{-1} B^*$. Show that A is unitary if and only if B is normal.

P.14.8 Let $A \in \mathsf{M}_n$. If there is a nonzero polynomial p such that $p(A)$ is normal, does it follow that A is normal?

P.14.9 Prove that $A \in \mathsf{M}_n$ is normal if and only if every eigenvector of A is an eigenvector of A^*. *Hint*: If $\mathbf{x}$ is a unit eigenvector of both A and A^*, then $\mathbf{x}^* A V_2 = \mathbf{0}^\mathsf{T}$ in (11.1.2).

P.14.10 (a) Give an example of a real orthogonal matrix with some non-real eigenvalues. (b) If two real orthogonal matrices are similar (perhaps via a non-real similarity) prove that they are real orthogonally similar.

P.14.11 Let $A, B \in \mathsf{M}_n$. (a) If A and B are normal, show that $AB = 0$ if and only if $BA = 0$. (b) If A is normal and B is not, does $AB = 0$ imply that $BA = 0$?

P.14.12 If $A \in \mathsf{M}_{m \times n}$ and $A^* A$ is an orthogonal projection, is AA^* an orthogonal projection?

P.14.13 Let $U, V \in \mathsf{M}_n$ be unitary. Show that $|\det(U + V)| \leq 2^n$. Find conditions for equality.

P.14.14 (a) Let $A \in \mathsf{M}_n$ be normal. Use Theorem 14.1.9.c to show that eigenvectors of A that correspond to distinct eigenvalues are orthogonal. (b) Suppose that $A \in \mathsf{M}_n$ is diagonalizable. If eigenvectors of A that correspond to distinct eigenvalues are orthogonal, show that A is normal. (c) Show that the hypothesis of diagonalizability cannot be omitted in (b).

P.14.15 (a) If $B \in \mathsf{M}_n$, show that $A = \begin{bmatrix} B & B^* \\ B^* & B \end{bmatrix}$ is normal. (b) Let $B = H + iK$ be a Cartesian decomposition (14.6.5) and let $U = \frac{1}{\sqrt{2}} \begin{bmatrix} I_n & I_n \\ -I_n & I_n \end{bmatrix}$. Show that U is unitary and $UAU^* = 2H \oplus 2iK$. (c) Show that A has n real eigenvalues and n purely imaginary eigenvalues. (d) Show that $\det A = (4i)^n (\det H)(\det K)$. (e) If B is normal and has eigenvalues $\lambda_1, \lambda_2, \ldots, \lambda_n$, show that $\det A = (4i)^n \prod_{j=1}^{n} (\operatorname{Re}\lambda_j)(\operatorname{Im}\lambda_j)$.

P.14.16 If $A \in \mathsf{M}_n$ is normal and $A^2 = A$, prove that A is an orthogonal projection. Can the hypothesis of normality be omitted?

P.14.17 (a) Show that $A \in \mathsf{M}_n$ is Hermitian if and only if $\operatorname{tr} A^2 = \operatorname{tr} A^*A$. (b) Show that Hermitian matrices $A, B \in \mathsf{M}_n$ commute if and only if $\operatorname{tr}(AB)^2 = \operatorname{tr} A^2 B^2$.

P.14.18 Let $A \in \mathsf{M}_m$ and $B \in \mathsf{M}_n$ be normal and write $A = U\Lambda U^*$ and $B = VMV^*$ in which $U \in \mathsf{M}_m$ and $V \in \mathsf{M}_n$ are unitary, $\Lambda = \operatorname{diag}(\lambda_1, \lambda_2, \ldots, \lambda_m)$ and $M = \operatorname{diag}(\mu_1, \mu_2, \ldots, \mu_n)$. Suppose that $X \in \mathsf{M}_{m \times n}$ and $AX = XB$. (a) Show that $\Lambda(U^*XV) = (U^*XV)M$. (b) Let $U^*XV = [\xi_{ij}]$ and show that $\overline{\lambda_i}\xi_{ij} = \xi_{ij}\overline{\mu_j}$ for all i, j. (c) Deduce that $A^*X = XB^*$. This provides another proof of the Fuglede–Putnam theorem.

P.14.19 Let $N \in \mathsf{M}_{m+n}$ be normal with spectral resolution $N = \sum_{i=1}^{d} \mu_i P_i$ and let $Y \in \mathsf{M}_{m+n}$. (a) If $NY = YN$, show that $P_i Y = Y P_i$ for $i = 1, 2, \ldots, d$. (b) Show that $N^*Y = YN^*$. (c) Now suppose that $A \in \mathsf{M}_m$ and $B \in \mathsf{M}_n$ are normal, $X \in \mathsf{M}_{m \times n}$, and $AX = XB$. Use the matrices $N = \begin{bmatrix} A & 0 \\ 0 & B \end{bmatrix}$ and $Y = \begin{bmatrix} 0 & X \\ 0 & 0 \end{bmatrix}$ to deduce the Fuglede–Putnam theorem.

P.14.20 In the statement of the Fuglede–Putnam theorem, do we have to assume that both A and B are normal? Consider $A = X = \begin{bmatrix} 0 & 1 \\ 0 & 0 \end{bmatrix}$ and $B = \begin{bmatrix} 1 & 0 \\ 0 & 0 \end{bmatrix}$.

P.14.21 Let $A, B \in \mathsf{M}_n$ be normal and suppose that $AB = BA$. Use the Fuglede–Putnam theorem to prove Corollary 14.2.12. *Hint*: First show that $A^*B = BA^*$ and $AB^* = B^*A$.

P.14.22 Let $A \in \mathsf{M}_n$ be normal. Show that $A\overline{A} = \overline{A}A$ if and only if $AA^\mathsf{T} = A^\mathsf{T}A$.

P.14.23 Suppose that $P_1, P_2, \ldots, P_d \in \mathsf{M}_n$ are a resolution of the identity. Prove that there exists a partitioned unitary matrix $U = [U_1 \ U_2 \ \cdots \ U_d] \in \mathsf{M}_n$ such that $P_i = U_i U_i^*$ for $i = 1, 2, \ldots, d$. Conclude that every resolution of the identity arises in the manner described in Lemma 14.9.4.

P.14.24 If $A \in \mathsf{M}_n$ is normal and has distinct eigenvalues, prove that $\{A\}'$ contains only normal matrices.

P.14.25 Find all possible values of $\dim\{A\}'$ (a) if $A \in \mathsf{M}_5$ is normal, or (b) if $A \in \mathsf{M}_6$ is normal.

P.14.26 Let $A = \begin{bmatrix} a & b \\ c & d \end{bmatrix} \in \mathsf{M}_2(\mathbb{R})$. Show that A is normal if and only if either A is symmetric or $c = -b$ and $a = d$.

P.14.27 Let $A = \begin{bmatrix} a & -b \\ b & a \end{bmatrix}$, in which $a, b \in \mathbb{C}$. Show that $\{A\}' = \mathsf{M}_2$ if $b = 0$ and $\{A\}' = \left\{ \begin{bmatrix} z & -w \\ w & z \end{bmatrix} : z, w \in \mathbb{C} \right\}$ if $b \neq 0$.

P.14.28 Consider $A = \begin{bmatrix} a & -b \\ b & a \end{bmatrix}$, in which $a, b \in \mathbb{R}$. The eigenvalues of A are $\lambda_1 = a + bi$ and $\lambda_2 = a - bi$. If p is a real polynomial, show that $\overline{p(\lambda_1)} = p(\overline{\lambda_1}) = p(\lambda_2)$ and $p(A) = \begin{bmatrix} \operatorname{Re} p(\lambda_1) & -\operatorname{Im} p(\lambda_1) \\ \operatorname{Im} p(\lambda_1) & \operatorname{Re} p(\lambda_1) \end{bmatrix}$.

P.14.29 Consider the *Volterra operator* $(Tf)(t) = \int_0^t f(s)\,ds$ on $C[0,1]$ (see P.6.12 and P.9.30). Write $T = H + iK$, in which $H = \frac{1}{2}(T + T^*)$ and $K = \frac{1}{2i}(T - T^*)$. (a) Show that $2H$ is the orthogonal projection onto the subspace of constant functions. (b) Compute the eigenvalues and eigenvectors of K. *Hint*: Use the fundamental theorem of calculus. Be sure that your prospective eigenvectors satisfy your original integral equation. (c) Verify that the eigenvalues of K are real and that the eigenvectors corresponding to distinct eigenvalues are orthogonal.

P.14.30 Let $A \in M_n$ and suppose that $-1 \notin \operatorname{spec} A$. The *Cayley transform* of A is $(I - A)$ $(I + A)^{-1}$. (a) If A is skew Hermitian, show that its Cayley transform is unitary. (b) If A is unitary, show that its Cayley transform is skew Hermitian.

P.14.31 Let $A = [a_{ij}] \in M_n$ be normal. (a) Examine the diagonal entries of $AA^* = A^*A$ and show that, for each $i = 1, 2, \ldots, n$, the Euclidean norms of the ith row and ith column of A are equal. (b) If $n = 2$, show that $|a_{21}| = |a_{12}|$. What can you say about the four matrices in Example 14.1.8? (c) If A is tridiagonal, show that $|a_{i,i+1}| = |a_{i+1,i}|$ for each $i = 1, 2, \ldots, n$.

P.14.32 Let $A \in M_n$ be symmetric, let $B = \operatorname{Re} A$, and let $C = \operatorname{Im} A$. Show that (a) B and C are real symmetric; (b) A is normal if and only if $BC = CB$.

P.14.33 The matrices $\begin{bmatrix} 1 & i \\ i & 1 \end{bmatrix}$ and $\begin{bmatrix} i & i \\ i & -1 \end{bmatrix}$ are symmetric, but only one is normal. Which one?

P.14.34 Show that the following are equivalent: (a) A is normal. (b) A^* is normal. (c) $\overline{A}$ is normal. (d) A^{T} is normal.

P.14.35 Suppose that $A \in M_n(\mathbb{R})$ is normal and stochastic. Show that A is doubly stochastic.

P.14.36 Let $A = [a_{ij}] \in M_n$ be normal. (a) Show that $\operatorname{spec} A$ lies on a line in the complex plane if and only if $A = cI + e^{i\theta}H$, in which $c \in \mathbb{C}$, $\theta \in [0, 2\pi)$, and $H \in M_n$ is Hermitian. (b) Show that $\operatorname{spec} A$ lies on a circle in the complex plane if and only if $A = cI + rU$, in which $c \in \mathbb{C}$, $r \in [0, \infty)$, and $U \in M_n$ is unitary. (c) Let $n = 2$. Show that $\operatorname{spec} A$ lies on a line and $A = cI + e^{i\theta}H$ as in (a), and deduce that $|a_{21}| = |a_{12}|$. If $a_{12} \neq 0$, show that $(a_{11} - a_{22})^2 / a_{12}a_{21}$ is real and nonnegative. (d) Let $n = 3$. Show that either $A = cI + e^{i\theta}H$ as in (a), or $A = cI + rU$ as in (b).

P.14.37 Show that the matrices A and B in Example 14.2.13 satisfy the criteria in Corollary 14.4.3 for AB and BA to be normal.

P.14.38 Suppose that $A = [a_{ij}] \in M_n$ is normal. (a) If $a_{11}, a_{22}, \ldots, a_{nn}$ are the eigenvalues of A (including multiplicities), show that A is diagonal. (b) If $n = 2$ and a_{11} is an eigenvalue of A, show that A is diagonal. (c) Give an example to show that the assertion in (b) need not be correct if A is not normal.

P.14.39 Let $A, B \in M_n$ be Hermitian. Prove that $\operatorname{rank}(AB)^k = \operatorname{rank}(BA)^k$ for all $k = 1, 2, \ldots$ and use this to show that AB is similar to BA.

P.14.40 Let $A, B \in M_n$ be normal. Let $r_1 = \operatorname{rank} A$ and $r_2 = \operatorname{rank} B$. Let $A = U\Lambda U^*$ and $B = VMV^*$, in which U and V are unitary, $\Lambda = \Lambda_1 \oplus 0_{n-r_1}$ and $M = M_1 \oplus 0_{n-r_2}$ are diagonal, and Λ_1 and M_1 are invertible. Let $W = U^*V$ and let W_{11} denote the $r_1 \times r_2$ upper-left corner of W. (a) Show that $\operatorname{rank} AB = \operatorname{rank} W_{11}$ and $\operatorname{rank} BA = \operatorname{rank} W_{11}^*$. (b) Deduce that $\operatorname{rank} AB = \operatorname{rank} BA$. See P.15.58 for a different proof.

P.14.41 Show that A and B in Example 13.5.5 are normal. Although AB and BA are not similar, observe that $\operatorname{rank} AB = \operatorname{rank} BA$, as shown in the preceding problem.

P.14.42 Let $A \in M_n$. Show that the following statements are equivalent:

(a) $\operatorname{col} A = \operatorname{col} A^*$.

(b) $\operatorname{null} A = \operatorname{null} A^*$.

(c) There is a unitary $U \in M_n$ such that $A = U(B \oplus 0_{n-r})U^*$, in which $r = \operatorname{rank} A$ and $B \in M_r$ is invertible.

A matrix that satisfies any of these conditions is an *EP matrix*. Why is every normal matrix an EP matrix?

P.14.43 Let $A, B \in M_n$ be EP matrices (see the preceding problem). Use the argument in P.14.40 to show that $\operatorname{rank} AB = \operatorname{rank} BA$.

P.14.44 Why do we assume that each $\theta_j \in [0, 2\pi)$ in the statement of Corollary 14.6.9.b? Is the corollary correct if this assumption is omitted? Discuss the example $V = I_2 = \operatorname{diag}(e^{i\theta_1}, e^{i\theta_2})$ with $\theta_1 = 0$ and $\theta_2 = 2\pi$.

P.14.45 The $n \times n$ circulant matrices are a subspace of M_n. What is its dimension?

P.14.46 Let $\lambda_1, \lambda_2, \ldots, \lambda_n \in \mathbb{R}$ and suppose that $\lambda_1 + \lambda_2 + \cdots + \lambda_n = 0$. Show that there is an $n \times n$ Hermitian matrix with zero main diagonal and eigenvalues $\lambda_1, \lambda_2, \ldots, \lambda_n$. *Hint*: Use a circulant matrix.

P.14.47 Let $C \in M_n$ be a circulant matrix whose first row is $[c_0 \ c_1 \ \ldots \ c_{n-1}]$. (a) Show that $\mathbf{e}$ is an eigenvector of C, and the corresponding eigenvalue is $\lambda = c_0 + c_1 + \cdots + c_{n-1}$. (b) If $C = F_n \Lambda F_n^*$, show that $\Lambda = \begin{bmatrix} \lambda & 0 \\ 0 & L \end{bmatrix}$, in which $L \in M_{n-1}(\mathbb{C})$ is diagonal. (c) If C is complex symmetric, show that the entries of L occur in pairs if n is odd. What can you say if n is even?

P.14.48 Let $C \in M_n$ be the circulant matrix (14.5.2). If there is an index $k \in \{0, 1, \ldots, n-1\}$ such that $|c_k| > \sum_{i \neq k} |c_i|$, prove that C is invertible. *Hint*: Theorem 14.5.6.

P.14.49 Let $A \in M_n$. (a) Suppose that $A + I$ is unitary. Sketch the circle Γ in the complex plane that contains every eigenvalue of A. (b) Suppose that $A + I$ and $A^2 + I$ are unitary. What portion of Γ cannot contain any eigenvalue of A? (c) If $A + I$, $A^2 + I$, and $A^3 + I$ are unitary, show that $A = 0$.

P.14.50 Let $A \in M_n$ be normal and let $\mathcal{U}$ be a k-dimensional A-invariant subspace with $1 \leq k \leq n - 1$. Prove the following generalization of Theorem 14.1.9.c: There is a $V \in M_{n \times k}$ and a diagonal $\Lambda \in M_k$ such that the columns of V are an orthonormal basis for $\mathcal{U}$, $AV = V\Lambda$, and $A^*V = V\overline{\Lambda}$.

P.14.51 Let $\mathcal{V}$ denote the inner product space of finitely nonzero bilateral complex sequences $\mathbf{a} = (\ldots, a_{-1}, \underline{a_0}, a_1, \ldots)$, in which the underline denotes the zeroth position. The inner product on $\mathcal{V}$ is $\langle \mathbf{a}, \mathbf{b} \rangle = \sum_{n=-\infty}^{\infty} a_n \overline{b_n}$. Let $U \in \mathcal{L}(\mathcal{V})$ be the *right-shift operator* $U(\ldots, a_{-1}, \underline{a_0}, a_1, \ldots) = (\ldots, a_{-2}, \underline{a_{-1}}, a_0, \ldots)$. (a) Compute U^*, the adjoint of U, and show that $U^*U = UU^*$. (b) Show that there is a subspace $\mathcal{U}$ of $\mathcal{V}$ that is invariant under U, but not under U^*. Compare with Theorem 14.9.13.

P.14.52 Suppose that $A \in M_n$ is normal and $\mathcal{U}$ is A-invariant. Let P denote the orthogonal projection onto $\mathcal{U}$. (a) Show that $\operatorname{tr}(AP - PA)^*(AP - PA) = 0$. (b) Conclude that $AP = PA$, which is Theorem 14.9.13.b.

P.14.53 Let $r \in \{1, 2, \ldots, n\}$ and let $A \in M_n$ and $B \in M_r$ be Hermitian. Let $\lambda_1, \lambda_2, \ldots, \lambda_n$ be the eigenvalues of A. Suppose that $\mathbf{x}_1, \mathbf{x}_2, \ldots, \mathbf{x}_r \in \mathbb{C}^n$ are orthonormal and $A\mathbf{x}_i = \lambda_i \mathbf{x}_i$ for each $i = 1, 2, \ldots, r$. Let $X = [\mathbf{x}_1 \ \mathbf{x}_2 \ \ldots \ \mathbf{x}_r]$ and $\Lambda = \operatorname{diag}(\lambda_1, \lambda_2, \ldots, \lambda_r)$. (a) Prove that the n eigenvalues of the Hermitian matrix $A + XBX^*$ are $\lambda_{r+1}, \lambda_{r+2}, \ldots, \lambda_n$ together with the r eigenvalues of $\Lambda + B$. *Hint*: Consider a unitary similarity via $U = [X \ X']$. (b) What does this say if $r = 1$? Compare with Theorem 11.6.1 (Brauer's theorem).

P.14.54 Let $A \in \mathsf{M}_n(\mathbb{C})$ and review Theorem 13.6.1. (a) If A is normal, show that A is similar to A^T via a symmetric unitary similarity matrix. (b) If A is unitary, show that $A = XY$, in which $X, Y \in \mathsf{M}_n(\mathbb{C})$ are unitary and symmetric.

P.14.55 Let $A \in \mathsf{M}_m$ and $B \in \mathsf{M}_n$. If $A \oplus B$ is normal, show that A and B are normal.

P.14.56 Let $A \in \mathsf{M}_m$ and $B \in \mathsf{M}_n$ be normal. (a) Show that $A \otimes B \in \mathsf{M}_{mn}$ is normal. (b) If $A \otimes B$ is normal and neither A nor B is the zero matrix, show that A and B are normal.

14.11 Notes

A real normal matrix with some non-real eigenvalues can be diagonalized by a unitary similarity, but not by a real orthogonal similarity. However, it is real orthogonally similar to a direct sum of a real diagonal matrix and 2×2 real matrices of the form (14.1.7), in which $b \neq 0$. Examples of such matrices are nonzero real skew-symmetric matrices and real orthogonal matrices that are not idempotent. For a discussion of real normal matrices, see [HJ13, Section 2.5].

More than 80 equivalent characterizations of normality are known; Theorem 14.8.1 lists several of them.

14.12 Some Important Concepts

- Definition and alternative characterizations of a normal matrix.

- Spectral theorem for complex normal and real symmetric matrices.

- Commuting normal matrices are simultaneously unitarily diagonalizable.

- Commuting real symmetric matrices are simultaneously real orthogonally diagonalizable.

- Defect from normality.

- Fuglede–Putnam theorem.

- Circulant matrices are a commuting family of normal matrices that are diagonalized simultaneously by the Fourier matrix.

- Symmetric circulant matrices are diagonalized by the Hartley matrix.

- Characterization of Hermitian, skew-Hermitian, unitary, and orthogonal projection matrices as normal matrices with certain spectra.

- Cartesian decomposition.

- Spectral theorem for symmetric normal and symmetric unitary matrices.

- Normal matrices are similar if and only if they are unitarily similar.

- Spectral projections and the spectral resolution of a normal matrix.

15 Positive Semidefinite Matrices

Many interesting mathematical ideas evolve from analogies. If we think of matrices as analogs of complex numbers, then the representation $z = a + bi$ suggests the Cartesian decomposition $A = H + iK$ of a square complex matrix, in which Hermitian matrices play the role of real numbers; see Definition 14.6.6.

Hermitian matrices with nonnegative eigenvalues are natural analogs of nonnegative real numbers. They arise in statistics (correlation matrices and the normal equations for least-squares problems), Lagrangian mechanics (the kinetic energy functional), and quantum mechanics (density matrices). They are the subject of this chapter.

15.1 Positive Semidefinite Matrices

Definition 15.1.1 Let $A \in M_n(\mathbb{F})$.

(a) A is *positive semidefinite* if it is Hermitian and $\langle Ax, x \rangle \geq 0$ for all $x \in \mathbb{F}^n$.

(b) A is *positive definite* if it is Hermitian and $\langle Ax, x \rangle > 0$ for all nonzero $x \in \mathbb{F}^n$.

Whenever we speak of a positive semidefinite matrix, it is always with the understanding that it is Hermitian. A real positive semidefinite matrix must be symmetric.

Definition 15.1.1 has an appealing geometric interpretation for real positive definite matrices. If θ is the angle between the real vectors x and Ax, then the condition

$$0 < \langle Ax, x \rangle = \|Ax\|_2 \|x\|_2 \cos\theta$$

means that $\cos\theta > 0$, that is, $|\theta| < \pi/2$. See Figure 15.1 for an illustration in $\mathbb{R}^2$.

Example 15.1.2 The matrix $A = \begin{bmatrix} 1 & -1 \\ -1 & 1 \end{bmatrix} \in M_2(\mathbb{R})$ is real symmetric. For any $x = \begin{bmatrix} x_1 \\ x_2 \end{bmatrix} \in \mathbb{R}^2$,

$$\langle Ax, x \rangle = (x_1 - x_2)x_1 + (-x_1 + x_2)x_2 = (x_1 - x_2)^2 \geq 0,$$

so A is positive semidefinite. Since $\langle Ax, x \rangle = 0$ for $x = [1\ 1]^T$, A is not positive definite. The eigenvalues of A are the roots of $p_A(z) = z(z - 2) = 0$, which are 0 and 2. The matrix $B = \frac{1}{\sqrt{2}}A$ satisfies $A = B^*B$. Moreover, $C = QB$ satisfies $A = C^*C$ for any real orthogonal $Q \in M_2(\mathbb{R})$.

The following theorem lists several equivalent criteria for a matrix to be positive semidefinite.

Theorem 15.1.3 *Let $A \in M_n(\mathbb{F})$. The following are equivalent:*

(a) *A is positive semidefinite.*

(b) *A is Hermitian and all its eigenvalues are nonnegative.*

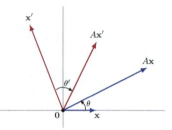

Figure 15.1 If $A \in M_2(\mathbb{R})$ is positive definite, then the angle between any nonzero $\mathbf{x} \in \mathbb{R}^2$ and $A\mathbf{x}$ is less than $\frac{\pi}{2}$.

(c) *There is a $B \in M_n(\mathbb{F})$ such that $A = B^*B$.*

(d) *For some positive integer m, there is a $B \in M_{m \times n}(\mathbb{F})$ such that $A = B^*B$.*

Proof (a) $\Rightarrow$ (b) Since A is positive semidefinite, it is Hermitian. Let $(\lambda, \mathbf{x})$ be an eigenpair of A, in which $\mathbf{x} \in \mathbb{F}^n$ is a unit vector. Then $\lambda = \lambda \langle \mathbf{x}, \mathbf{x} \rangle = \langle \lambda \mathbf{x}, \mathbf{x} \rangle = \langle A\mathbf{x}, \mathbf{x} \rangle \geq 0$.

(b) $\Rightarrow$ (c) Since A is Hermitian, the spectral theorem (Theorem 14.2.2) says that there is a unitary $U \in M_n(\mathbb{F})$ and a real diagonal $\Lambda = \mathrm{diag}(\lambda_1, \lambda_2, \ldots, \lambda_n)$ such that $A = U\Lambda U^*$. Since each λ_i is nonnegative, $D = \mathrm{diag}(\lambda_1^{1/2}, \lambda_2^{1/2}, \ldots, \lambda_n^{1/2}) \in M_n(\mathbb{R})$. Let $B = UDU^* \in M_n(\mathbb{F})$. Then

$$B^*B = (UDU^*)^*(UDU^*) = UD^*U^*UDU^* = UDDU^* = UD^2U^* = U\Lambda U^* = A.$$

(c) $\Rightarrow$ (d) Take $m = n$.

(d) $\Rightarrow$ (a) If $B \in M_{m \times n}(\mathbb{F})$ and $A = B^*B$, then A is Hermitian. If $\mathbf{x} \in \mathbb{F}^n$, then $\langle A\mathbf{x}, \mathbf{x} \rangle = \langle B^*B\mathbf{x}, \mathbf{x} \rangle = \langle B\mathbf{x}, B\mathbf{x} \rangle = \|B\mathbf{x}\|_2^2 \geq 0$, so A is positive semidefinite. $\square$

Example 15.1.4 Suppose that $A \in M_n(\mathbb{R})$ is symmetric and $\langle A\mathbf{x}, \mathbf{x} \rangle \geq 0$ for all $\mathbf{x} \in \mathbb{R}^n$. Since A is Hermitian, the equivalence of (a) and (b) in Theorem 15.1.3 with $\mathbb{F} = \mathbb{R}$ ensures that all its eigenvalues are nonnegative. The same equivalence, this time with $\mathbb{F} = \mathbb{C}$, reveals that $\langle A\mathbf{x}, \mathbf{x} \rangle \geq 0$ for all $\mathbf{x} \in \mathbb{C}^n$. Thus, a real symmetric matrix is positive semidefinite on $\mathbb{R}^n$ if and only if it is positive semidefinite on $\mathbb{C}^n$. For example, if A is the matrix Example 15.1.2, then

$$\langle A\mathbf{x}, \mathbf{x} \rangle = (x_1 - x_2)\overline{x_1} + (-x_1 + x_2)\overline{x_2} = |x_1 - x_2|^2 \geq 0 \quad \text{for } \mathbf{x} = \begin{bmatrix} x_1 \\ x_2 \end{bmatrix} \in \mathbb{C}^2.$$

Example 15.1.5 The matrix $A = \begin{bmatrix} 0 & 1 \\ -1 & 0 \end{bmatrix}$ satisfies $\langle A\mathbf{x}, \mathbf{x} \rangle = x_1 x_2 - x_1 x_2 = 0$ for all $\mathbf{x} = \begin{bmatrix} x_1 \\ x_2 \end{bmatrix} \in \mathbb{R}^2$. However, the real matrix A is not positive semidefinite because it is not symmetric. Indeed, for $\mathbf{x} = [1 \ i]^\mathsf{T}$, we have $\langle A\mathbf{x}, \mathbf{x} \rangle = 2i$.

Example 15.1.6 An orthogonal projection $P \in M_n$ is Hermitian and Theorem 14.6.1.c ensures that $\mathrm{spec}\, P \subseteq \{0, 1\}$. If $a \geq 0$, then aP is Hermitian and $\mathrm{spec}(aP) \subseteq \{0, a\}$. Thus, any real nonnegative scalar multiple of an orthogonal projection is positive semidefinite. In Example 15.1.2, $A = 2\mathbf{v}\mathbf{v}^*$, in which $\mathbf{v} = \frac{1}{\sqrt{2}} \begin{bmatrix} 1 \\ -1 \end{bmatrix}$.

There is a version of Theorem 15.1.3 for positive definite matrices.

Theorem 15.1.7 *Let $A \in M_n(\mathbb{F})$. The following are equivalent:*

(a) *A is positive definite.*

(b1) *A is Hermitian and all its eigenvalues are positive.*

(b2) *A is positive semidefinite and invertible.*

(b3) *A is invertible and A^{-1} is positive definite.*

(c) *There is an invertible $B \in M_n(\mathbb{F})$ such that $A = B^*B$.*

(d) *For some $m \geq n$, there is a $B \in M_{m \times n}(\mathbb{F})$ such that* rank $B = n$ *and* $A = B^*B$.

Proof We indicate how the proof of each implication in Theorem 15.1.3 can be modified to obtain a proof of the corresponding implication in the positive definite case.

(a) $\Rightarrow$ (b1) Observe that $\lambda = \langle Ax, x \rangle > 0$.

(b1) $\Leftrightarrow$ (b2) A matrix with nonnegative eigenvalues is invertible if and only if its eigenvalues are positive.

(b1) $\Leftrightarrow$ (b3) A is Hermitian if and only if A^{-1} is Hermitian. The eigenvalues of A are the reciprocals of the eigenvalues of A^{-1}.

(b1) $\Rightarrow$ (c) Λ has positive diagonal entries, so $B = UDU^* \in M_n(\mathbb{F})$ is invertible and $B^*B = A$.

(c) $\Rightarrow$ (d) Take $m = n$.

(d) $\Rightarrow$ (a) Since rank $B = n$, the rank-nullity theorem ensures that null $B = \{0\}$. If $x \neq 0$, then $Bx \neq 0$ and hence $\langle Ax, x \rangle = \|Bx\|_2^2 > 0$. $\qquad\square$

Theorem 15.1.3.c says that $A \in M_n(\mathbb{F})$ is positive semidefinite if and only if it is the Gram matrix of n vectors in some $\mathbb{F}^m$; see Definition 8.4.14. Theorem 15.1.7.c says that A is positive definite if and only if those vectors are linearly independent.

Theorem 15.1.8 *Let $A \in M_n$ be positive semidefinite.*

(a) tr $A \geq 0$, *with strict inequality if and only if $A \neq 0$.*

(b) det $A \geq 0$, *with strict inequality if and only if A is positive definite.*

Proof Let $\lambda_1, \lambda_2, \ldots, \lambda_n$ be the eigenvalues of A, which are nonnegative. They are positive if A is positive definite.

(a) tr $A = \lambda_1 + \lambda_2 + \cdots + \lambda_n \geq 0$, with equality if and only if each $\lambda_i = 0$. A Hermitian matrix is zero if and only if all its eigenvalues are zero; see P.14.3.

(b) det $A = \lambda_1 \lambda_2 \cdots \lambda_n$ is nonnegative. It is positive if and only if every eigenvalue is positive, which means that A is positive definite. $\qquad\square$

If $A \in M_n(\mathbb{F})$ is positive semidefinite and nonzero, then Example 15.1.2 ensures that there are many $B \in M_n(\mathbb{F})$ such that $A = B^*B$. The following theorem says that they all have the same null space and the same rank.

Theorem 15.1.9 *Suppose that $B \in M_{m \times n}(\mathbb{F})$ and let $A = B^*B$.*

(a) null $A =$ null B.

(b) rank $A =$ rank B.

(c) $A = 0$ *if and only if* $B = 0$.

(d) $\operatorname{col} A = \operatorname{col} B^*$.

(e) *If B is normal, then* $\operatorname{col} A = \operatorname{col} B$.

Proof (a) If $B\mathbf{x} = \mathbf{0}$, then $A\mathbf{x} = B^*B\mathbf{x} = \mathbf{0}$ and hence $\operatorname{null} B \subseteq \operatorname{null} A$. Conversely, if $A\mathbf{x} = \mathbf{0}$, then

$$0 = \langle A\mathbf{x}, \mathbf{x} \rangle = \langle B^*B\mathbf{x}, \mathbf{x} \rangle = \langle B\mathbf{x}, B\mathbf{x} \rangle = \|B\mathbf{x}\|_2^2, \tag{15.1.10}$$

so $B\mathbf{x} = \mathbf{0}$. It follows that $\operatorname{null} A \subseteq \operatorname{null} B$ and hence $\operatorname{null} A = \operatorname{null} B$.

(b) Since A and B both have n columns, (a) and the rank-nullity theorem ensure that $\operatorname{rank} A = n - \operatorname{nullity} A = n - \operatorname{nullity} B = \operatorname{rank} B$.

(c) The assertion follows from (a).

(d) Since A is Hermitian, $\operatorname{col} A = (\operatorname{null} A^*)^\perp = (\operatorname{null} A)^\perp = (\operatorname{null} B)^\perp = \operatorname{col} B^*$; see (8.2.4).

(e) This follows from (d) and Theorem 14.1.9.e.   □

One consequence of the factorization $A = B^*B$ of a positive semidefinite matrix is known as *column inclusion*. Another is the fact that every leading principal submatrix of A is positive semidefinite (positive definite if A is).

Corollary 15.1.11 (Column Inclusion) *Let $A \in \mathsf{M}_n$ be positive semidefinite and write*

$$A = \begin{bmatrix} A_{11} & A_{12} \\ A_{12}^* & A_{22} \end{bmatrix}, \quad \text{in which } A_{11} \in \mathsf{M}_k \text{ and } 1 \leq k \leq n - 1. \tag{15.1.12}$$

Then A_{11} is positive semidefinite, $\det A_{11} \geq 0$, and

$$\operatorname{col} A_{12} \subseteq \operatorname{col} A_{11}. \tag{15.1.13}$$

In particular, there is an $X \in \mathsf{M}_{k \times (n-k)}$ such that $A_{12} = A_{11}X$. Consequently,

$$\operatorname{rank} A_{12} \leq \operatorname{rank} A_{11}. \tag{15.1.14}$$

If A is positive definite, then A_{11} is positive definite and $\det A_{11} > 0$.

Proof Let $B \in \mathsf{M}_n$ be such that $A = B^*B$ (Theorem 15.1.3.c) and partition $B = [B_1 \ B_2]$, in which $B_1 \in \mathsf{M}_{n \times k}$. Then

$$\begin{bmatrix} A_{11} & A_{12} \\ A_{12}^* & A_{22} \end{bmatrix} = A = B^*B = \begin{bmatrix} B_1^* \\ B_2^* \end{bmatrix} \begin{bmatrix} B_1 & B_2 \end{bmatrix} = \begin{bmatrix} B_1^*B_1 & B_1^*B_2 \\ B_2^*B_1 & B_2^*B_2 \end{bmatrix},$$

so $A_{11} = B_1^*B_1$ is positive semidefinite (Theorem 15.1.3.c) and $\operatorname{col} A_{12} = \operatorname{col} B_1^*B_2 \subseteq \operatorname{col} B_1^*$. The preceding theorem ensures that $\operatorname{col} B_1^* = \operatorname{col} B_1^*B_1 = \operatorname{col} A_{11}$, which proves (15.1.13). Thus,

$$\operatorname{rank} A_{12} = \dim \operatorname{col} A_{12} \leq \dim \operatorname{col} A_{11} = \operatorname{rank} A_{11}.$$

If A is positive definite, then Theorem 15.1.7.c ensures that B is invertible. Consequently, $\operatorname{rank} B_1 = k$ and $B_1^*B_1$ is invertible (Theorem 15.1.7.d). The assertions about $\det A_{11}$ follow from Theorem 15.1.8.b.   □

Example 15.1.15 Suppose that $A = [a_{ij}] \in \mathsf{M}_n$ is positive semidefinite and partitioned as in (15.1.12). If $A_{11} = [a_{11}] = 0$, then (15.1.13) ensures that $A_{12} = [a_{12} \ a_{13} \ \ldots \ a_{1n}] = 0$.

Example 15.1.16 Suppose that $A \in M_n$ is positive semidefinite and partitioned as in (15.1.12). If $A_{11} = \begin{bmatrix} 1 & 1 \\ 1 & 1 \end{bmatrix}$, then $\operatorname{col} A_{11} = \operatorname{span}\{[1\ 1]^{\mathsf{T}}\}$. Therefore, each column of A_{12} is a scalar multiple of $[1\ 1]^{\mathsf{T}}$, so $A_{12} \in M_{2 \times (n-2)}$ has equal rows. It follows that the first two rows of A are equal.

Example 15.1.17 The matrix $\begin{bmatrix} 0 & 1 \\ 1 & 0 \end{bmatrix}$ shows that the hypothesis "positive semidefinite" in Corollary 15.1.11 cannot be replaced by "Hermitian."

If $A \in M_n$ is positive definite, Corollary 15.1.11 says that each of its leading principal minors is positive. This necessary condition is also sufficient; see Corollary 15.4.2 and Theorem 18.4.1.

Corollary 15.1.18 *Let $A \in M_n$ be positive semidefinite and let $\mathbf{x} \in \mathbb{C}^n$. Then $\langle A\mathbf{x}, \mathbf{x} \rangle = 0$ if and only if $A\mathbf{x} = \mathbf{0}$.*

Proof If $A\mathbf{x} = \mathbf{0}$, then $\langle A\mathbf{x}, \mathbf{x} \rangle = \langle \mathbf{0}, \mathbf{x} \rangle = 0$. Conversely, if $\langle A\mathbf{x}, \mathbf{x} \rangle = 0$ and $A = B^*B$, then (15.1.10) ensures that $B\mathbf{x} = \mathbf{0}$, which implies that $A\mathbf{x} = B^*B\mathbf{x} = \mathbf{0}$. $\qquad\square$

Example 15.1.19 The Hermitian matrix $D = \operatorname{diag}(1, -1)$ and the nonzero vector $\mathbf{x} = [1\ 1]^{\mathsf{T}}$ satisfy $\langle D\mathbf{x}, \mathbf{x} \rangle = 0$ and $D\mathbf{x} \neq \mathbf{0}$. Thus, the hypothesis "positive semidefinite" cannot be weakened to "Hermitian" in the preceding corollary.

A motivating analogy for the following theorem is that real linear combinations of real numbers are real, and nonnegative linear combinations of nonnegative real numbers are nonnegative.

Theorem 15.1.20 *Let $A, B \in M_n(\mathbb{F})$ and let $a, b \in \mathbb{R}$.*

(a) *If A and B are Hermitian, then $aA + bB$ is Hermitian.*

(b) *If a and b are nonnegative and if A and B are positive semidefinite, then $aA + bB$ is positive semidefinite.*

(c) *If a and b are positive, if A and B are positive semidefinite, and if at least one of A and B is positive definite, then $aA + bB$ is positive definite.*

Proof (a) $(aA + bB)^* = \bar{a}A^* + \bar{b}B^* = aA + bB$.

(b) Part (a) ensures that $aA + bB$ is Hermitian, and

$$\langle (aA + bB)\mathbf{x}, \mathbf{x} \rangle = a\langle A\mathbf{x}, \mathbf{x} \rangle + b\langle B\mathbf{x}, \mathbf{x} \rangle \geq 0 \tag{15.1.21}$$

for every $\mathbf{x} \in \mathbb{F}^n$ because both summands are nonnegative. Consequently, $aA + bB$ is positive semidefinite.

(c) The assertion follows from (15.1.21) because both summands are nonnegative and at least one of them is positive if $\mathbf{x} \neq \mathbf{0}$. $\qquad\square$

The preceding theorem asserts something that is not obvious: the sum of two Hermitian matrices with nonnegative eigenvalues has nonnegative eigenvalues.

Example 15.1.22 The matrices $A = \begin{bmatrix} 1 & -1 \\ -1 & 2 \end{bmatrix}$, $B = \begin{bmatrix} 4 & 3 \\ 3 & 3 \end{bmatrix}$, and $A + B = \begin{bmatrix} 5 & 2 \\ 2 & 5 \end{bmatrix}$ are positive definite. Rounded to two decimal places, $\operatorname{spec} A = \{2.62, 0.38\}$, $\operatorname{spec} B = \{6.54, 0.46\}$, and $\operatorname{spec}(A + B) = \{7, 3\}$.

Example 15.1.23 Consider $A = \begin{bmatrix} 1 & 3 \\ 0 & 1 \end{bmatrix}$, $B = \begin{bmatrix} 1 & 0 \\ 3 & 1 \end{bmatrix}$, and $A + B = \begin{bmatrix} 2 & 3 \\ 3 & 2 \end{bmatrix}$. Then $\operatorname{spec} A = \operatorname{spec} B = \{1\}$, but $\operatorname{spec}(A + B) = \{5, -1\}$. Since A and B are not Hermitian, this does not contradict the preceding theorem.

Here are two ways to make new positive semidefinite matrices from old ones.

Theorem 15.1.24 *Let $A, S \in \mathsf{M}_n$ and $B \in \mathsf{M}_m$, and suppose that A and B are positive semidefinite.*

(a) *$A \oplus B$ is positive semidefinite; it is positive definite if and only if A and B are positive definite.*

(b) *S^*AS is positive semidefinite; it is positive definite if and only if A is positive definite and S is invertible.*

Proof Theorem 15.1.3.c says that there are $P \in \mathsf{M}_n$ and $Q \in \mathsf{M}_m$ such that $A = P^*P$ and $B = Q^*Q$.

(a) Let $C = A \oplus B$. Then $C^* = A^* \oplus B^* = A \oplus B = C$, so C is Hermitian. Compute

$$C = (P^*P) \oplus (Q^*Q) = (P \oplus Q)^*(P \oplus Q).$$

It follows from Theorem 15.1.3.c that C is positive semidefinite. A direct sum is invertible if and only if each summand is invertible, so C is positive definite if and only if both A and B are positive definite.

(b) S^*AS is Hermitian and $S^*AS = S^*P^*PS = (PS)^*(PS)$ is positive semidefinite. Theorem 15.1.7.c says that S^*AS is positive definite if and only if PS is invertible, which happens if and only if S and P are both invertible. Finally, P is invertible if and only if A is positive definite. $\square$

15.2 Schur Complements and Diagonal Dominance

The Schur complement (3.3.9) provides a way to construct new positive semidefinite matrices, and a way to verify that a given Hermitian matrix is positive definite.

Theorem 15.2.1 *Let $A \in \mathsf{M}_n$ be Hermitian and partitioned as*

$$A = \begin{bmatrix} A_{11} & A_{12} \\ A_{12}^* & A_{22} \end{bmatrix},$$

in which $1 \leq k \leq n - 1$ and $A_{11} \in \mathsf{M}_k$ is positive definite.

(a) *A is positive semidefinite if and only if $A_{22} - A_{12}^* A_{11}^{-1} A_{12}$ is positive semidefinite.*

(b) *A is positive definite if and only if $A_{22} - A_{12}^* A_{11}^{-1} A_{12}$ is positive definite.*

Proof Let $S = \begin{bmatrix} I_k & -A_{11}^{-1} A_{12} \\ 0 & I_{n-k} \end{bmatrix}$, which is invertible, and compute

$$S^*AS = \begin{bmatrix} I_k & 0 \\ -A_{12}^* A_{11}^{-1} & I_{n-k} \end{bmatrix} \begin{bmatrix} A_{11} & A_{12} \\ A_{12}^* & A_{22} \end{bmatrix} \begin{bmatrix} I_k & -A_{11}^{-1} A_{12} \\ 0 & I_{n-k} \end{bmatrix} = A_{11} \oplus (A_{22} - A_{12}^* A_{11}^{-1} A_{12}).$$

Theorem 15.1.24 ensures that A is positive semidefinite (respectively, positive definite) if and only if the Schur complement $A/A_{11} = A_{22} - A_{12}^* A_{11}^{-1} A_{12}$ is positive semidefinite (respectively, positive definite). $\square$

Example 15.2.2 Consider $A = \begin{bmatrix} a & \mathbf{x}^* \\ \mathbf{x} & B \end{bmatrix} \in M_n$, in which $a > 0$, $\mathbf{x} \in \mathbb{C}^{n-1}$, and $B \in M_{n-1}$ is Hermitian. The preceding theorem says that A is positive definite if and only if $B - a^{-1}\mathbf{x}\mathbf{x}^*$ is positive definite.

Example 15.2.3 Let $A = \begin{bmatrix} a & b \\ \bar{b} & c \end{bmatrix}$ be Hermitian and suppose that $a > 0$. Theorem 15.2.1 says that A is positive definite if and only if

$$c - a^{-1}b\bar{b} = a^{-1}(ac - |b|^2) = a^{-1}\det A > 0,$$

that is, if and only if $\det A > 0$. Moreover, A is positive semidefinite if and only if $\det A \geq 0$.

A sufficient condition for positive semidefiniteness follows from Gershgorin's disk theorem.

Theorem 15.2.4 *Let $A = [a_{ij}] \in M_n$ be Hermitian and suppose that it has nonnegative diagonal entries.*

(a) *If A is diagonally dominant, then it is positive semidefinite.*

(b) *If A is diagonally dominant and invertible, then it is positive definite.*

(c) *If A is strictly diagonally dominant, then it is positive definite.*

(d) *If A is diagonally dominant, has no zero entries, and $|a_{kk}| > R'_k(A)$ for at least one $k \in \{1, 2, \ldots, n\}$, then A is positive definite.*

Proof (a) Since A has real eigenvalues, Corollary 9.4.15.b ensures that $\mathrm{spec}\, A \subseteq [0, \infty)$. The assertion follows from Theorem 15.1.3.

(b) The assertion follows from (a) and Theorem 15.1.7.b2.

(c) The assertion follows from (b) and Corollary 9.4.15.a.

(d) The assertion follows from Theorem 9.4.20. $\qquad\qquad\square$

Example 15.2.5 The matrix

$$A = \begin{bmatrix} 2 & 1 & 0 \\ 1 & 3 & 1 \\ 0 & 1 & 2 \end{bmatrix} \tag{15.2.6}$$

is Hermitian and strictly diagonally dominant, so it is positive definite; see Figure 15.2a.

Example 15.2.7 The matrix

$$A = \begin{bmatrix} 2 & 1 & 1 \\ 1 & 3 & 1 \\ 1 & 1 & 2 \end{bmatrix}$$

satisfies the criterion in Theorem 15.2.4.d, so it is positive definite. Its eigenvalues are 1 and $3 \pm \sqrt{2}$.

Example 15.2.8 The matrix

$$A = \begin{bmatrix} 5 & 2 & 0 \\ 2 & 3 & 2 \\ 0 & 2 & 5 \end{bmatrix} \tag{15.2.9}$$

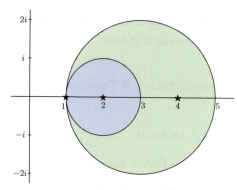

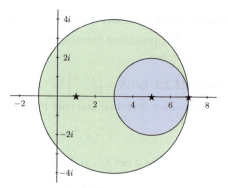

(a) The Hermitian matrix (15.2.6) is strictly di-agonally dominant, so it is positive definite. Its Gershgorin region is contained in the right half plane. Its eigenvalues are 4, 2, and 1.

(b) The Hermitian matrix (15.2.9) is positive definite even though it is not diagonally dom-inant. Its Gershgorin region extends into the left half plane. Its eigenvalues are 7, 5, and 1.

Figure 15.2 Gershgorin regions for the Hermitian matrices (15.2.6) and (15.2.9).

does not satisfy any of the criteria in Theorem 15.2.4; see Figure 15.2b. Theorem 15.2.1 says that A is positive definite if and only if the Schur complement

$$S = \begin{bmatrix} 3 & 2 \\ 2 & 5 \end{bmatrix} - \frac{1}{5} \begin{bmatrix} 2 \\ 0 \end{bmatrix} [2 \ 0] = \frac{1}{5} \begin{bmatrix} 11 & 10 \\ 10 & 25 \end{bmatrix}$$

is positive definite. Since S is strictly diagonally dominant, Theorem 15.2.4 ensures that it (and hence also A) is positive definite. The eigenvalues of A are 1, 5, and 7.

15.3 The Square Root of a Positive Semidefinite Matrix

Two nonnegative real numbers are equal if and only if their squares are equal. An analogous statement is valid for Hermitian matrices with nonnegative eigenvalues, but both conditions (Hermitian as well as nonnegative eigenvalues) are essential. For example, the unequal (but non-Hermitian) matrices $A = \begin{bmatrix} 0 & 1 \\ 0 & 0 \end{bmatrix}$ and $B = \begin{bmatrix} 0 & 0 \\ 1 & 0 \end{bmatrix}$ have nonnegative eigenvalues and $A^2 = B^2$. The unequal Hermitian (but not positive semidefinite) matrices $A = \begin{bmatrix} 1 & 0 \\ 0 & -1 \end{bmatrix}$ and $B = \begin{bmatrix} 0 & 1 \\ 1 & 0 \end{bmatrix}$ also have $A^2 = B^2$.

Lemma 15.3.1 *Let $B, C \in \mathsf{M}_n$ be positive semidefinite diagonal matrices. If $B^2 = C^2$, then $B = C$.*

Proof Let $B = \operatorname{diag}(b_1, b_2, \ldots, b_n)$ and $C = \operatorname{diag}(c_1, c_2, \ldots, c_n)$ have real nonnegative diagonal entries. If $B^2 = C^2$, then $b_i^2 = c_i^2$ for each $i = 1, 2, \ldots, n$. Since $b_i \geq 0$ and $c_i \geq 0$, we have $b_i = c_i$ for each $i = 1, 2, \ldots, n$. $\square$

The square root function on positive semidefinite matrices is a special case of the functional calculus for diagonalizable matrices discussed in Section 10.5. For an alternative approach to uniqueness of the square root, see P.15.18.

Theorem 15.3.2 *Let $A \in \mathsf{M}_n(\mathbb{F})$ be positive semidefinite. There is a unique positive semidefinite $B \in \mathsf{M}_n(\mathbb{F})$ such that $B^2 = A$. Moreover,*

(a) *There is a real polynomial p such that $B = p(A)$.*

(b) *null $A = $ null B and col $A = $ col B, and*

(c) *A is positive definite if and only if B is positive definite.*

Proof (a) Use Theorem 14.2.2 to write $A = U\Lambda U^*$, in which $U \in \mathsf{M}_n(\mathbb{F})$ is unitary and $\Lambda = \mathrm{diag}(\lambda_1, \lambda_2, \ldots, \lambda_n)$ is diagonal. The eigenvalues λ_i are real and nonnegative; they may appear on the diagonal of Λ in any order. We claim that $B = U\Lambda^{1/2}U^*$, in which $\Lambda^{1/2} = \mathrm{diag}(\lambda_1^{1/2}, \lambda_2^{1/2}, \ldots, \lambda_n^{1/2})$. Let p be a real polynomial such that $p(\lambda_i) = \lambda_i^{1/2} \geq 0$ for $i = 1, 2, \ldots, n$. Then $B = Up(\Lambda)U^*$ and

$$B^2 = Up(\Lambda)U^*Up(\Lambda)U^* = Up(\Lambda)^2U^* = U\Lambda U^* = A.$$

Moreover, $B = Up(\Lambda)U^* = p(U\Lambda U^*) = p(A) \in \mathsf{M}_n(\mathbb{F})$; see Theorem 2.6.9.b.

Suppose that $C \in \mathsf{M}_n$ is positive semidefinite and $C^2 = A$. Then $B = p(A) = p(C^2)$ is a polynomial in C, so B commutes with C. Theorem 14.2.11 tells us that B and C are simultaneously unitarily diagonalizable. Let $V \in \mathsf{M}_n$ be unitary and such that $B = VEV^*$ and $C = VFV^*$, in which E and F are diagonal and have nonnegative diagonal entries. Since

$$VE^2V^* = (VEV^*)^2 = B^2 = A = C^2 = (VFV^*)^2 = VF^2V^*,$$

it follows that $E^2 = V^*AV = F^2$. Lemma 15.3.1 ensures that $E = F$ and hence $C = B$.

(b) See Theorem 15.1.9.

(c) Since $B^2 = A$, it follows that A is invertible if and only if B is invertible. All the eigenvalues of an invertible positive semidefinite matrix are positive, so the assertion follows. $\square$

Definition 15.3.3 If $A \in \mathsf{M}_n$ is positive semidefinite, its *positive semidefinite square root* is the unique positive semidefinite matrix $A^{1/2}$ such that $(A^{1/2})^2 = A$. If A is positive definite, we define $A^{-1/2} = (A^{1/2})^{-1}$.

Example 15.3.4 The real symmetric matrix

$$A = \begin{bmatrix} 56 & 10 & -20 \\ 10 & 41 & -10 \\ -20 & -10 & 56 \end{bmatrix} \tag{15.3.5}$$

has a spectral decomposition $A = U\Lambda U^*$, in which $\Lambda = \mathrm{diag}(36, 36, 81)$ and

$$U = \frac{1}{3}\begin{bmatrix} 1 & -2 & -2 \\ 2 & 2 & -1 \\ 2 & -1 & 2 \end{bmatrix}.$$

The eigenvalues of A are positive, so A is positive definite. Its unique positive definite square root is

$$U\Lambda^{1/2}U^* = \frac{1}{9}\begin{bmatrix} 1 & -2 & -2 \\ 2 & 2 & -1 \\ 2 & -1 & 2 \end{bmatrix}\begin{bmatrix} 6 & 0 & 0 \\ 0 & 6 & 0 \\ 0 & 0 & 9 \end{bmatrix}\begin{bmatrix} 1 & 2 & 2 \\ -2 & 2 & -1 \\ -2 & -1 & 2 \end{bmatrix} = \frac{1}{3}\begin{bmatrix} 22 & 2 & -4 \\ 2 & 19 & -2 \\ -4 & -2 & 22 \end{bmatrix}. \tag{15.3.6}$$

The Lagrange interpolation polynomial (2.7.7) such that $p(36) = 6$ and $p(81) = 9$ is $p(x) = \frac{1}{5}(x - 36) - \frac{2}{15}(x - 81) = \frac{1}{15}x + \frac{18}{5}$. A calculation reveals that $p(A) = \frac{1}{15}A + \frac{18}{5}I$ is the matrix $A^{1/2}$ in (15.3.6).

The preceding example computes $A^{1/2}$ in two different ways. The first (via the spectral theorem) requires knowledge of the eigenvalues and eigenvectors of A. The second (via Lagrange interpolation) requires knowledge of spec A; we do not need to know the eigenvalue multiplicities or the eigenvectors.

A single polynomial can take several positive semidefinite matrices to their positive semidefinite square roots. The following corollary shows how this can be done for two matrices.

Corollary 15.3.7 *Let $A \in \mathsf{M}_n$ and $B \in \mathsf{M}_m$ be positive semidefinite. There is a real polynomial p such that $A^{1/2} = p(A)$ and $B^{1/2} = p(B)$.*

Proof Theorem 15.1.24 ensures that $A \oplus B$ and $A^{1/2} \oplus B^{1/2}$ are positive semidefinite. Since $(A^{1/2} \oplus B^{1/2})^2 = A \oplus B$, it follows from the uniqueness assertion in Theorem 15.3.2 that $(A \oplus B)^{1/2} = A^{1/2} \oplus B^{1/2}$. Moreover, there is a polynomial p such that $(A \oplus B)^{1/2} = p(A \oplus B)$, so

$$A^{1/2} \oplus B^{1/2} = (A \oplus B)^{1/2} = p(A \oplus B) = p(A) \oplus p(B),$$

which implies that $A^{1/2} = p(A)$ and $B^{1/2} = p(B)$.   □

The proof of Theorem 15.3.2 contains a recipe to compute the positive semidefinite square root of a positive semidefinite matrix: Diagonalize $A = U\Lambda U^*$, form $\Lambda^{1/2}$, and obtain $A^{1/2} = U\Lambda^{1/2}U^*$. There is a way to bypass the diagonalization for 2×2 positive semidefinite matrices.

Proposition 15.3.8 *Let $A \in \mathsf{M}_2$ be positive semidefinite and nonzero, and let*

$$\tau = (\operatorname{tr} A + 2(\det A)^{1/2})^{1/2}.$$

Then $\tau > 0$ and

$$A^{1/2} = \frac{1}{\tau}(A + (\det A)^{1/2}I). \tag{15.3.9}$$

Proof Theorem 15.1.8 ensures that $\det A \geq 0$ and $\operatorname{tr} A > 0$, so $\tau^2 \geq \operatorname{tr} A > 0$. Theorem 15.1.20 guarantees that (15.3.9) is positive semidefinite since it is a nonnegative linear combination of positive semidefinite matrices; it suffices to show that its square is A. Compute

$$\frac{1}{\tau^2}\left(A + (\det A)^{1/2}I\right)^2 = \frac{1}{\tau^2}\left(A^2 + 2(\det A)^{1/2}A + (\det A)I\right)$$

$$= \frac{1}{\tau^2}\left((\operatorname{tr} A)A - (\det A)I + 2(\det A)^{1/2}A + (\det A)I\right)$$

$$= \frac{1}{\tau^2}\left(\operatorname{tr} A + 2(\det A)^{1/2}\right)A$$

$$= A.$$

We used Theorem 11.2.1 to make the substitution $A^2 = (\operatorname{tr} A)A - (\det A)I$; see (10.1.1).   □

Example 15.3.10 Consider the real diagonally dominant symmetric matrix $A = \begin{bmatrix} 2 & 1 \\ 1 & 1 \end{bmatrix}$. Theorem 15.2.4.d ensures that A is positive definite. To find $A^{1/2}$, Proposition 15.3.8 tells us to compute $\tau = \left(\operatorname{tr} A + 2(\det A)^{1/2}\right)^{1/2} = (3 + 2)^{1/2} = \sqrt{5}$ and $A^{1/2} = \frac{1}{\sqrt{5}}(A + I) = \frac{1}{\sqrt{5}}\begin{bmatrix} 3 & 1 \\ 1 & 2 \end{bmatrix}$. Although this is the only positive semidefinite square root of A, there are other real symmetric square roots. For example, $B = \begin{bmatrix} 1 & 1 \\ 1 & 0 \end{bmatrix}$ satisfies $B^2 = A$. The eigenvalues of B are $\frac{1}{2}(1 \pm \sqrt{5})$, one of which is negative; see Example 10.5.4.

The product of two positive real numbers is positive, but the product of two positive definite matrices need not be positive definite. For example, the product

$$\begin{bmatrix} 2 & 1 \\ 1 & 1 \end{bmatrix}\begin{bmatrix} 1 & 1 \\ 1 & 2 \end{bmatrix} = \begin{bmatrix} 3 & 4 \\ 2 & 3 \end{bmatrix} \qquad (15.3.11)$$

of two positive definite matrices is not Hermitian, so it is not positive definite. However, it has positive eigenvalues $3 \pm 2\sqrt{2}$. The final theorem in this section provides an explanation for this observation.

Theorem 15.3.12 *Let $A, B \in M_n$ and suppose that A is positive definite.*

(a) *If B is Hermitian, then AB is diagonalizable and has real eigenvalues.*

(b) *If B is positive semidefinite, then AB is diagonalizable and has nonnegative real eigenvalues. If, in addition, $B \neq 0$, then AB has at least one positive eigenvalue.*

(c) *If B is positive definite, then AB is diagonalizable and has positive real eigenvalues.*

Proof Consider the similarity

$$A^{-1/2}(AB)A^{1/2} = A^{-1/2}A^{1/2}A^{1/2}BA^{1/2} = A^{1/2}BA^{1/2} = (A^{1/2})^*B(A^{1/2}),$$

in which $C = (A^{1/2})^*B(A^{1/2})$ is Hermitian and therefore diagonalizable. Since AB is similar to C, it is diagonalizable and has the same eigenvalues as C.

(a) The eigenvalues of the Hermitian matrix C are real.

(b) Theorem 15.1.24.b says that C is positive semidefinite, so its eigenvalues are real and nonnegative. If they are all zero, then $C = 0$ since it is diagonalizable. If $C = 0$, then $B = A^{-1/2}CA^{-1/2} = 0$.

(c) AB is invertible and has nonnegative eigenvalues, so they are positive. $\square$

Corollary 15.3.13 *Let $A, B \in M_n$. Suppose that A is positive definite and B is positive semidefinite. Then $\det(A + B) \geq \det A$ with equality if and only if $B = 0$.*

Proof The preceding theorem says that the eigenvalues $\lambda_1, \lambda_2, \ldots, \lambda_n$ of $A^{-1}B$ are nonnegative and $\lambda_1 = \lambda_2 = \cdots = \lambda_n = 0$ if and only if $B = 0$. Compute

$$\det(A + B) = \det\big(A(I + A^{-1}B)\big) = (\det A)\det(I + A^{-1}B) = (\det A)\prod_{i=1}^{n}(1 + \lambda_i) \geq \det A,$$

with equality if and only if $\lambda_1 = \lambda_2 = \cdots = \lambda_n = 0$. $\square$

15.4 The Cholesky Factorization

In Section 4.3, we found that a system of linear equations $A\mathbf{x} = \mathbf{y}$, in which A is invertible, can be solved by backward and forward substitutions if A has an LU factorization. Theorem 4.3.5 says that A has an LU factorization if and only if every leading principal submatrix is invertible, and Corollary 15.1.11 says that this condition is satisfied if A is positive definite. Thus, a system of linear equations $A\mathbf{x} = \mathbf{y}$ can be solved by backward and forward substitution if A is positive definite.

Corollary 15.1.11 ensures that every leading principal minor of a positive definite matrix is positive. The following theorem describes a variant of the LU factorization that is available for positive definite matrices.

Theorem 15.4.1 (Cholesky Factorization) *Let $A \in M_n$ be Hermitian and suppose that every leading principal minor of A is positive. Let $A = LU$ be the LU-factorization of A, in which L is unit lower triangular, $U = [u_{ij}]$ is upper triangular, and both L and U are unique.*

(a) *$D = \mathrm{diag}(u_{11}, u_{22}, \ldots, u_{nn})$ is positive definite.*

(b) *$C = LD^{1/2}$ is the unique lower triangular matrix with positive diagonal entries such that $A = CC^*$.*

(c) *A is positive definite.*

Proof (a) Using the partitioned matrices in (4.3.6), we find that the leading $k \times k$ principal submatrices of A, L, and U satisfy $A_{(k)} = L_{(k)}U_{(k)}$ for $k = 1, 2, \ldots, n$. Then

$$u_{11}u_{22} \cdots u_{kk} = (\det L_{(k)})(\det U_{(k)}) = \det A_{(k)} > 0$$

for each $k = 1, 2, \ldots, n$. An induction shows that each $u_{kk} > 0$.

(b) Theorem 4.3.5 ensures that A has an LU factorization. Since A is Hermitian, we have

$$LU = A = A^* = U^*L^* = U^*D^{-1}DL^* = (D^{-1}U)^*(DL^*),$$

in which $(D^{-1}U)^*$ is unit lower triangular and DL^* is upper triangular. Theorem 4.3.5.b ensures that $U = DL^*$, so $A = LU = LDL^* = (LD^{1/2})(LD^{1/2})^* = CC^*$, as asserted. If $A = BB^*$ and B is lower triangular with positive diagonal entries, then $CC^* = BB^*$ and $B^{-1}C = ((B^{-1}C)^{-1})^*$. Since $B^{-1}C$ is lower triangular and Hermitian, it follows that $B^{-1}C$ is diagonal and has positive diagonal entries. Finally, the relation $B^{-1}C = (B^{-1}C)^{-1}$ ensures that $B^{-1}C = I$, so $B = C$.

(c) Since $A = CC^*$ and C is invertible, A is positive definite; see Theorem 15.1.7.c.   □

The preceding theorem provides a practical necessary and sufficient condition for positive definiteness. For a different approach to this condition, see Theorem 18.4.1.

Corollary 15.4.2 (Sylvester Criterion) *A Hermitian matrix is positive definite if and only if all of its leading principal minors are positive.*

Proof Let $A \in M_n$ be Hermitian. If A is positive definite, then Corollary 15.1.11 ensures that all of its leading principal minors are positive. Conversely, if every principal minor of A is positive, Theorem 15.4.1.c ensures that A is positive definite.   □

Example 15.4.3 The leading principal minors of the 4×4 *min matrix*

$$A = [\min\{i, j\}] = \begin{bmatrix} 1 & 1 & 1 & 1 \\ 1 & 2 & 2 & 2 \\ 1 & 2 & 3 & 3 \\ 1 & 2 & 3 & 4 \end{bmatrix}$$

are 1, 1, 1, and 1. Since A is Hermitian, the preceding corollary ensures that it is positive definite. Using the algorithm illustrated in Example 4.3.9, we compute the Cholesky factorization of A as follows. First,

$$A - \begin{bmatrix} 1 \\ 1 \\ 1 \\ 1 \end{bmatrix} [1\ 1\ 1\ 1] = \begin{bmatrix} 0 & 0 & 0 & 0 \\ 0 & 1 & 1 & 1 \\ 0 & 1 & 2 & 2 \\ 0 & 1 & 2 & 3 \end{bmatrix},$$

then

$$\begin{bmatrix} 0 & 0 & 0 & 0 \\ 0 & 1 & 1 & 1 \\ 0 & 1 & 2 & 2 \\ 0 & 1 & 2 & 3 \end{bmatrix} - \begin{bmatrix} 0 \\ 1 \\ 1 \\ 1 \end{bmatrix} [0\ 1\ 1\ 1] = \begin{bmatrix} 0 & 0 & 0 & 0 \\ 0 & 0 & 0 & 0 \\ 0 & 0 & 1 & 1 \\ 0 & 0 & 1 & 2 \end{bmatrix},$$

and

$$\begin{bmatrix} 0 & 0 & 0 & 0 \\ 0 & 0 & 0 & 0 \\ 0 & 0 & 1 & 1 \\ 0 & 0 & 1 & 2 \end{bmatrix} - \begin{bmatrix} 0 \\ 0 \\ 1 \\ 1 \end{bmatrix} [0\ 0\ 1\ 1] = \begin{bmatrix} 0 & 0 & 0 & 0 \\ 0 & 0 & 0 & 0 \\ 0 & 0 & 0 & 0 \\ 0 & 0 & 0 & 1 \end{bmatrix} = \begin{bmatrix} 0 \\ 0 \\ 0 \\ 1 \end{bmatrix} [0\ 0\ 0\ 1].$$

Now use (3.1.17) to collapse the representation of A as a sum of outer products:

$$A = \begin{bmatrix} 1 & 1 & 1 & 1 \\ 1 & 2 & 2 & 2 \\ 1 & 2 & 3 & 3 \\ 1 & 2 & 3 & 4 \end{bmatrix} = \underbrace{\begin{bmatrix} 1 & 0 & 0 & 0 \\ 1 & 1 & 0 & 0 \\ 1 & 1 & 1 & 0 \\ 1 & 1 & 1 & 1 \end{bmatrix}}_{C} \underbrace{\begin{bmatrix} 1 & 1 & 1 & 1 \\ 0 & 1 & 1 & 1 \\ 0 & 0 & 1 & 1 \\ 0 & 0 & 0 & 1 \end{bmatrix}}_{C^*},$$

which is the Cholesky factorization $A = CC^*$. Since

$$\underbrace{\begin{bmatrix} 1 & -1 & 0 & 0 \\ 0 & 1 & -1 & 0 \\ 0 & 0 & 1 & -1 \\ 0 & 0 & 0 & 1 \end{bmatrix}}_{C^{-*}} \underbrace{\begin{bmatrix} 1 & 1 & 1 & 1 \\ 0 & 1 & 1 & 1 \\ 0 & 0 & 1 & 1 \\ 0 & 0 & 0 & 1 \end{bmatrix}}_{C^*} = \begin{bmatrix} 1 & 0 & 0 & 0 \\ 0 & 1 & 0 & 0 \\ 0 & 0 & 1 & 0 \\ 0 & 0 & 0 & 1 \end{bmatrix},$$

the inverse of C^* is the upper bidiagonal matrix with main diagonal entries 1 and entries -1 on the first superdiagonal. Therefore, the inverse of the 4×4 min matrix is the tridiagonal matrix

$$(CC^*)^{-1} = C^{-*}C^{-1}$$

$$= \begin{bmatrix} 1 & -1 & 0 & 0 \\ 0 & 1 & -1 & 0 \\ 0 & 0 & 1 & -1 \\ 0 & 0 & 0 & 1 \end{bmatrix} \begin{bmatrix} 1 & 0 & 0 & 0 \\ -1 & 1 & 0 & 0 \\ 0 & -1 & 1 & 0 \\ 0 & 0 & -1 & 1 \end{bmatrix} = \begin{bmatrix} 2 & -1 & 0 & 0 \\ -1 & 2 & -1 & 0 \\ 0 & -1 & 2 & -1 \\ 0 & 0 & -1 & 1 \end{bmatrix}.$$

See P.15.35 for the $n \times n$ min matrix.

15.5 Simultaneous Diagonalization of Quadratic Forms

A real quadratic form in n variables $x_1, x_2, \ldots, x_n$ is a real-valued function on $\mathbb{R}^n$ of the form

$$\mathbf{x} \mapsto \langle A\mathbf{x}, \mathbf{x} \rangle = \sum_{i,j=1}^{n} a_{ij} x_i x_j,$$

in which $A \in M_n(\mathbb{R})$ is symmetric and $\mathbf{x} = [x_i] \in \mathbb{R}^n$. The spectral theorem says that there is a real orthogonal $Q \in M_n(\mathbb{R})$ and a real diagonal $\Lambda = \mathrm{diag}(\lambda_1, \lambda_2, \ldots, \lambda_n)$ such that $A = Q\Lambda Q^\mathsf{T}$. If $\mathbf{y} = [y_i] = Q^\mathsf{T}\mathbf{x}$, then

$$\sum_{i,j=1}^{n} a_{ij} x_i x_j = \langle A\mathbf{x}, \mathbf{x}\rangle = \langle Q\Lambda Q^\mathsf{T}\mathbf{x}, \mathbf{x}\rangle = \langle \Lambda Q^\mathsf{T}\mathbf{x}, Q^\mathsf{T}\mathbf{x}\rangle = \langle \Lambda\mathbf{y}, \mathbf{y}\rangle = \sum_{i=1}^{n} \lambda_i y_i^2 \quad (15.5.1)$$

is a weighted sum of squares of the new variables $y_1, y_2, \ldots, y_n$. The weights λ_i in (15.5.1) are the eigenvalues of A.

Example 15.5.2 In the study of central conic sections in plane geometry, there is a real symmetric invertible $A = [a_{ij}] \in M_2(\mathbb{R})$ and a real quadratic form

$$a_{11} x_1^2 + 2a_{12} x_1 x_2 + a_{22} x_2^2 = \langle A\mathbf{x}, \mathbf{x}\rangle = \langle Q\Lambda Q^\mathsf{T}\mathbf{x}, \mathbf{x}\rangle = \langle \Lambda Q^\mathsf{T}\mathbf{x}, Q^\mathsf{T}\mathbf{x}\rangle = \lambda_1 y_1^2 + \lambda_2 y_2^2.$$

The plane curve $\langle A\mathbf{x}, \mathbf{x}\rangle = 1$ is an ellipse if the eigenvalues of A are both positive; it is a hyperbola if one eigenvalue is positive and the other is negative. The change of variables $\mathbf{x} \mapsto Q^\mathsf{T}\mathbf{x}$ is from one orthonormal basis (the standard basis for $\mathbb{R}^2$) to the orthonormal basis formed by the columns of Q^T.

What happens if we have two real quadratic forms $\langle A\mathbf{x}, \mathbf{x}\rangle$ and $\langle B\mathbf{x}, \mathbf{x}\rangle$? For example, the first might represent the kinetic energy of a collection of masses and the second might represent the potential energy of the collection. Or they might represent plane ellipses corresponding to the orbits of two different planets. Can we find a real orthogonal matrix Q such that $Q^\mathsf{T}AQ$ and $Q^\mathsf{T}BQ$ are both diagonal? Not unless A commutes with B, which is a strong requirement; this is Theorem 14.2.11.b. However, if we are willing to change variables to a new basis that is not necessarily orthonormal, then we need not require that A and B commute. It is sufficient that one of them be positive definite, and that is often the case in applications.

Theorem 15.5.3 *Let $A, B \in M_n(\mathbb{F})$ be Hermitian, and suppose that A is positive definite. There is an invertible $S \in M_n(\mathbb{F})$ such that $A = SIS^*$, $B = S\Lambda S^*$, and $\Lambda \in M_n(\mathbb{R})$ is a real diagonal matrix whose diagonal entries are the eigenvalues of $A^{-1}B$.*

Proof The matrix $A^{-1/2}BA^{-1/2} \in M_n(\mathbb{F})$ is Hermitian, so there is a unitary $U \in M_n(\mathbb{F})$ and a real diagonal matrix Λ such that $A^{-1/2}BA^{-1/2} = U\Lambda U^*$. Let $S = A^{1/2}U$. Then $SIS^* = A^{1/2}UU^*A^{1/2} = A^{1/2}A^{1/2} = A$ and $S\Lambda S^* = A^{1/2}U\Lambda U^*A^{1/2} = A^{1/2}A^{-1/2}BA^{-1/2}A^{1/2} = B$. The eigenvalues of $A^{-1/2}BA^{-1/2}$ are real; they are the same as the eigenvalues of Λ and $A^{-1/2}A^{-1/2}B = A^{-1}B$ (Theorems 10.7.2 and 15.3.12.a). $\qquad\square$

Example 15.5.4 Consider the real symmetric matrices

$$A = \begin{bmatrix} 20 & -20 \\ -20 & 40 \end{bmatrix} \quad \text{and} \quad B = \begin{bmatrix} -34 & 52 \\ 52 & -56 \end{bmatrix}. \quad (15.5.5)$$

Theorem 15.2.4.d ensures that A is positive definite. We want an invertible matrix S and a diagonal matrix Λ such that $A = SIS^*$ and $B = S\Lambda S^*$. First, use (15.3.9) to compute $A^{1/2} = \begin{bmatrix} 4 & -2 \\ -2 & 6 \end{bmatrix}$ and $A^{-1/2} = \frac{1}{10}\begin{bmatrix} 3 & 1 \\ 1 & 2 \end{bmatrix}$. Then

$$A^{-1/2}BA^{-1/2} = \frac{1}{2}\begin{bmatrix} -1 & 3 \\ 3 & -1 \end{bmatrix} = U\Lambda U^*, \text{ in which } \Lambda = \begin{bmatrix} 1 & 0 \\ 0 & -2 \end{bmatrix} \text{ and } U = \frac{1}{\sqrt{2}}\begin{bmatrix} 1 & 1 \\ 1 & -1 \end{bmatrix}.$$

The preceding theorem says that if $S = A^{1/2}U = \sqrt{2}\begin{bmatrix} 1 & 3 \\ 2 & -4 \end{bmatrix}$, then $A = SIS^*$ and $B = S\Lambda S^*$. The quadratic forms

$$20x^2 - 40xy + 40y^2 \quad \text{and} \quad -34x^2 + 104xy - 56y^2$$

become $\xi^2 + \eta^2$ and $\xi^2 - 2\eta^2$ with respect to the new variables $\xi = \sqrt{2}x + 2\sqrt{2}y$ and $\eta = 3\sqrt{2}x - 4\sqrt{2}y$.

15.6 The Schur Product Theorem

Definition 15.6.1 Let $A = [a_{ij}]$ and $B = [b_{ij}] \in \mathsf{M}_{m \times n}$. The *Hadamard product* of A and B is $A \circ B = [a_{ij}b_{ij}] \in \mathsf{M}_{m \times n}$.

Example 15.6.2 The Hadamard product of $A = \begin{bmatrix} 1 & 2 & 3 \\ 4 & 5 & 6 \end{bmatrix}$ and $B = \begin{bmatrix} 6 & 5 & 0 \\ 3 & 2 & 1 \end{bmatrix}$ is

$$A \circ B = \begin{bmatrix} 6 & 10 & 0 \\ 12 & 10 & 6 \end{bmatrix}.$$

The Hadamard product is also called the *entrywise product* or the *Schur product*. The following theorems list some of its properties.

Theorem 15.6.3 *Let* $A = [a_{ij}] \in \mathsf{M}_{m \times n}$, *let* $B, C \in \mathsf{M}_{m \times n}$, *and let* $\gamma \in \mathbb{C}$.

(a) $A \circ B = B \circ A$.

(b) $A \circ (\gamma B + C) = (\gamma B + C) \circ A = \gamma(A \circ B) + A \circ C$.

(c) $A \circ E = A$, *in which* $E \in \mathsf{M}_{m \times n}$ *is the all-ones matrix. If* $a_{ij} \neq 0$ *for all* i,j, *then* $A \circ [a_{ij}^{-1}] = E$.

(d) $(A \circ B)^* = A^* \circ B^*$.

(e) *If* $A, B \in \mathsf{M}_n$ *are Hermitian, then* $A \circ B$ *is Hermitian.*

(f) $A \circ I_n = \operatorname{diag}(a_{11}, a_{22}, \ldots, a_{nn})$.

(g) *If* $\Lambda = \operatorname{diag}(\lambda_1, \lambda_2, \ldots, \lambda_n)$, *then* $\Lambda A + A\Lambda = [\lambda_i + \lambda_j] \circ A = A \circ [\lambda_i + \lambda_j]$.

Proof See P.15.62 for (a) – (f). To prove (g), compute $\Lambda A + A\Lambda = [\lambda_i a_{ij} + a_{ij}\lambda_j] = [(\lambda_i + \lambda_j)a_{ij}] = [\lambda_i + \lambda_j] \circ A$ and use (a). $\qquad\square$

Theorem 15.6.4 *Let* $\mathbf{x} = [x_i], \mathbf{y} = [y_i] \in \mathbb{C}^n$ *and let* $A \in \mathsf{M}_n$.

(a) $\mathbf{x} \circ \mathbf{y} = [x_i y_i] \in \mathbb{C}^n$.

(b) $\mathbf{xx}^* \circ \mathbf{yy}^* = (\mathbf{x} \circ \mathbf{y})(\mathbf{x} \circ \mathbf{y})^*$.

(c) *Let* $D = \operatorname{diag}(x_1, x_2, \ldots, x_n)$. *Then* $\mathbf{xx}^* \circ A = DAD^* = A \circ \mathbf{xx}^*$.

Proof See P.15.63 for (a) and (c). To prove (b), compute $\mathbf{xx}^* = [x_i \bar{x}_j] \in \mathsf{M}_n$, $\mathbf{yy}^* = [y_i \bar{y}_j] \in \mathsf{M}_n$, and $\mathbf{xx}^* \circ \mathbf{yy}^* = [x_i \bar{x}_j y_i \bar{y}_j] = [x_i y_i \overline{x_j y_j}] = (\mathbf{x} \circ \mathbf{y})(\mathbf{x} \circ \mathbf{y})^*$. $\qquad\square$

The ordinary matrix product of two positive semidefinite matrices need not be positive semidefinite; see (15.3.11). However, the following theorem shows that the Hadamard product of two positive semidefinite matrices is positive semidefinite; see also P.15.70.

Theorem 15.6.5 (Schur Product Theorem) *If $A, B \in \mathsf{M}_n$ are positive semidefinite, then $A \circ B$ is positive semidefinite.*

Proof Theorem 15.1.3.c ensures that there are $X, Y \in \mathsf{M}_n$ such that $A = XX^*$ and $B = YY^*$. Partition $X = [\mathbf{x}_1 \ \mathbf{x}_2 \ \cdots \ \mathbf{x}_n]$ and $Y = [\mathbf{y}_1 \ \mathbf{y}_2 \ \cdots \ \mathbf{y}_n]$ according to their columns. Then

$$A = XX^* = \sum_{i=1}^{n} \mathbf{x}_i \mathbf{x}_i^* \quad \text{and} \quad B = YY^* = \sum_{i=1}^{n} \mathbf{y}_i \mathbf{y}_i^*. \tag{15.6.6}$$

Therefore,

$$A \circ B = \left(\sum_{i=1}^{n} \mathbf{x}_i \mathbf{x}_i^* \right) \circ \left(\sum_{j=1}^{n} \mathbf{y}_j \mathbf{y}_j^* \right) = \sum_{i,j=1}^{n} \left((\mathbf{x}_i \mathbf{x}_i^*) \circ (\mathbf{y}_j \mathbf{y}_j^*) \right) = \sum_{i,j=1}^{n} (\mathbf{x}_i \circ \mathbf{y}_j)(\mathbf{x}_i \circ \mathbf{y}_j)^*,$$

in which we have used properties of the Hadamard product listed in the preceding two theorems. This shows that $A \circ B$ is a sum of positive semidefinite matrices (Theorem 15.1.3.c), so Theorem 15.1.20 ensures that $A \circ B$ is positive semidefinite. $\qquad\square$

A Hadamard product $A \circ B$ of positive semidefinite matrices can be positive definite even if one factor is not invertible.

Example 15.6.7 Let $A = \begin{bmatrix} 1 & 1 \\ 1 & 2 \end{bmatrix}$ and $B = \begin{bmatrix} 1 & \sqrt{2} \\ \sqrt{2} & 2 \end{bmatrix}$. Then A is positive definite, B is positive semidefinite and not invertible, and $A \circ B = \begin{bmatrix} 1 & \sqrt{2} \\ \sqrt{2} & 4 \end{bmatrix}$ is positive definite.

Corollary 15.6.8 *Let $A, B \in \mathsf{M}_n$ be positive semidefinite.*

(a) *If A is positive definite and every main diagonal entry of B is positive, then $A \circ B$ is positive definite.*

(b) *If A and B are positive definite, then $A \circ B$ is positive definite.*

Proof (a) Let $\lambda_1 \leq \lambda_2 \leq \cdots \leq \lambda_n$ be the eigenvalues of A. Then $\lambda_1 > 0$ since A is positive definite. The matrix $A - \lambda_1 I$ is Hermitian and has eigenvalues $0, \lambda_2 - \lambda_1, \lambda_3 - \lambda_1, \ldots, \lambda_n - \lambda_1$, all of which are nonnegative. Therefore, $A - \lambda_1 I$ is positive semidefinite. The preceding theorem ensures that $(A - \lambda_1 I) \circ B$ is positive semidefinite. Let $\mathbf{x} = [x_i] \in \mathbb{C}^n$ be nonzero, let $B = [b_{ij}] \in \mathsf{M}_n$, let $b = \min_{1 \leq i \leq n} b_{ii}$, and compute

$$0 \leq \langle ((A - \lambda_1 I) \circ B)\mathbf{x}, \mathbf{x} \rangle = \langle ((A \circ B) - \lambda_1 I \circ B)\mathbf{x}, \mathbf{x} \rangle$$
$$= \langle (A \circ B)\mathbf{x}, \mathbf{x} \rangle - \langle (\lambda_1 I \circ B)\mathbf{x}, \mathbf{x} \rangle$$
$$= \langle (A \circ B)\mathbf{x}, \mathbf{x} \rangle - \lambda_1 \langle \operatorname{diag}(b_{11}, b_{22}, \ldots, b_{nn})\mathbf{x}, \mathbf{x} \rangle$$
$$= \langle (A \circ B)\mathbf{x}, \mathbf{x} \rangle - \lambda_1 \sum_{i=1}^{n} b_{ii} |x_i|^2.$$

Therefore,

$$\langle (A \circ B)\mathbf{x}, \mathbf{x} \rangle \geq \lambda_1 \sum_{i=1}^{n} b_{ii} |x_i|^2 \geq \lambda_1 b \sum_{i=1}^{n} |x_i|^2 = \lambda_1 b \|\mathbf{x}\|_2^2 > 0, \tag{15.6.9}$$

so $A \circ B$ is positive definite.

(b) If B is positive definite, then $b_{ii} = \langle B\mathbf{e}_i, \mathbf{e}_i \rangle > 0$ for $i = 1, 2, \ldots, n$, so the assertion follows from (a). $\qquad\square$

The preceding corollary does not exhaust all the ways that $A \circ B$ can be positive definite. In the following example, A and B are positive semidefinite and neither is invertible.

Example 15.6.10 Let

$$A = \begin{bmatrix} 2 & 1 & 1 \\ 1 & 1 & 1 \\ 1 & 1 & 1 \end{bmatrix} \quad \text{and} \quad B = \begin{bmatrix} 2 & 1 & 1 \\ 1 & 1 & 0 \\ 1 & 0 & 1 \end{bmatrix}.$$

To two decimal places, the eigenvalues of A are 3.41, 0.59, and 0; the eigenvalues of B are 3, 1, and 0. Their Hadamard product

$$A \circ B = \begin{bmatrix} 4 & 1 & 1 \\ 1 & 1 & 0 \\ 1 & 0 & 1 \end{bmatrix}$$

is positive definite; to two decimal places, its eigenvalues are 4.56, 1, and 0.44, and its leading principal minors are 4, 3, and 2.

Example 15.6.11 Let $A, B \in M_n$ be positive definite and consider the *Lyapunov equation*

$$AX + XA = B. \tag{15.6.12}$$

Since $\operatorname{spec} A \cap \operatorname{spec}(-A) = \varnothing$, Theorem 11.4.1 ensures that (15.6.12) has a unique solution X. We claim that X is positive definite. Let $A = U\Lambda U^*$ be a spectral decomposition, in which $\Lambda = \operatorname{diag}(\lambda_1, \lambda_2, \ldots, \lambda_n)$ and $U \in M_n$ is unitary. Write (15.6.12) as $U\Lambda U^*X + XU\Lambda U^* = B$, which is equivalent to

$$\Lambda(U^*XU) + (U^*XU)\Lambda = U^*BU. \tag{15.6.13}$$

Use Theorem 15.6.3.g to write (15.6.13) as $[\lambda_i + \lambda_j] \circ (U^*XU) = U^*BU$, which is equivalent to

$$U^*XU = [(\lambda_i + \lambda_j)^{-1}] \circ (U^*BU).$$

Since B is positive definite, U^*BU is positive definite (Theorem 15.1.24.b). The Hermitian matrix $[(\lambda_i + \lambda_j)^{-1}] \in M_n$ has positive main diagonal entries and is positive semidefinite; see P.15.54. Corollary 15.6.8 ensures that $[(\lambda_i + \lambda_j)^{-1}] \circ (U^*BU)$ is positive definite. The solution to (15.6.12) is $X = U([(\lambda_i + \lambda_j)^{-1}] \circ (U^*BU))U^*$. Theorem 15.1.24.b tells us that X is positive definite.

15.7 Problems

P.15.1 Which of $\begin{bmatrix} 1 & 0 \\ 0 & 0 \end{bmatrix}$, $\begin{bmatrix} 1 & 1 \\ 1 & 0 \end{bmatrix}$, $\begin{bmatrix} 1 & 1 \\ 1 & 1 \end{bmatrix}$, $\begin{bmatrix} 2 & 1 \\ 1 & 1 \end{bmatrix}$, and $\begin{bmatrix} 3 & 2 \\ 2 & 1 \end{bmatrix}$ are positive semidefinite? Positive definite?

P.15.2 What is the smallest real number x such that $\begin{bmatrix} 4 & 7 \\ 7 & x \end{bmatrix}$ is positive semidefinite?

P.15.3 Compute the positive definite square root of $\begin{bmatrix} 5 & 4 \\ 4 & 5 \end{bmatrix}$.

P.15.4 (a) Let $\mathbf{a} \in \mathbb{C}^n$ be nonzero. Compute the positive semidefinite square root of $A = \mathbf{a}\mathbf{a}^*$.
(b) Compute the positive semidefinite square root of

$$A = \begin{bmatrix} 1 & 2 & 3 & 4 \\ 2 & 4 & 6 & 8 \\ 3 & 6 & 9 & 12 \\ 4 & 8 & 12 & 16 \end{bmatrix}.$$

P.15.5 Let $A = \begin{bmatrix} 1 & 0 \\ -2 & 1 \end{bmatrix}$. (a) Show that $\langle A\mathbf{x}, \mathbf{x} \rangle \geq 0$ for all $\mathbf{x} \in \mathbb{R}^2$ and $\operatorname{spec} A = \{1\}$. (b) Compute $\langle A\mathbf{x}, \mathbf{x} \rangle$ for $\mathbf{x} = [1 \ i]^\mathsf{T}$. (c) Is A positive semidefinite? Why?

P.15.6 Use Schur complements to show that the matrices A, B, and $A \circ B$ in Example 15.6.10 are positive semidefinite.

P.15.7 Prove that the matrix in (15.3.6) is positive definite and show that its square is the matrix in (15.3.5). Do not use the spectral theorem.

P.15.8 Let $K \in M_n$ be Hermitian. Show that $K = 0$ if and only if $\langle K\mathbf{x}, \mathbf{x} \rangle = 0$ for all $\mathbf{x} \in \mathbb{C}^n$. *Hint*: Consider $\langle K(\mathbf{x} + e^{i\theta}\mathbf{y}), \mathbf{x} + e^{i\theta}\mathbf{y} \rangle$.

P.15.9 Let $A \in M_n$. (a) If A is Hermitian, show that $\langle A\mathbf{x}, \mathbf{x} \rangle$ is real for all $\mathbf{x} \in \mathbb{C}^n$. (b) If $\langle A\mathbf{x}, \mathbf{x} \rangle$ is real for all $\mathbf{x} \in \mathbb{C}^n$, show that A is Hermitian. *Hint*: Consider the Cartesian decomposition $A = H + iK$.

P.15.10 Let $\Lambda \in M_n(\mathbb{R})$ be diagonal. Show that $\Lambda = P - Q$, in which $P, Q \in M_n(\mathbb{R})$ are diagonal and positive semidefinite.

P.15.11 Let $A \in M_n$ be Hermitian. Show that there are commuting positive semidefinite $B, C \in M_n$ such that $A = B - C$.

P.15.12 Let $A \in M_n$ be positive semidefinite and partitioned as in (15.1.12). (a) Show that $\operatorname{rank}[A_{11} \ A_{12}] = \operatorname{rank} A_{11}$. (b) Is this true if A is merely Hermitian?

P.15.13 Let $A \in M_2$ and adopt the notation of Proposition 15.3.8. (a) Show that $\tau = \operatorname{tr} A^{1/2}$. (b) If $\operatorname{spec} A = \{\lambda, \mu\}$ and $\lambda \neq \mu$, use (10.8.1) to show that

$$A^{1/2} = \frac{\sqrt{\lambda} - \sqrt{\mu}}{\lambda - \mu} A + \frac{\lambda\sqrt{\mu} - \mu\sqrt{\lambda}}{\lambda - \mu} I = \frac{1}{\sqrt{\lambda} + \sqrt{\mu}} \left(A + \sqrt{\lambda\mu} I \right)$$

and verify that this is the same matrix as in (15.3.9). (c) What happens if $\lambda = \mu$?

P.15.14 Let $A, B \in M_n$ be positive semidefinite. Show that AB is positive semidefinite if and only if A commutes with B.

P.15.15 Let $A, B \in M_n$ be positive semidefinite. (a) Show that $A = 0$ if and only if $\operatorname{tr} A = 0$. (b) Show that $A + B = 0$ if and only if $A = B = 0$.

P.15.16 Suppose that $A \in M_n$ is nilpotent, $A = H + iK$ is a Cartesian decomposition, and H is positive semidefinite. Prove that $A = 0$.

P.15.17 Let $A, C \in M_n$ be positive semidefinite and suppose that $C^2 = A$. Let $B = A^{1/2}$ and let p be a polynomial such that $B = p(A)$. Provide details for the following outline of a proof that $C = B$: (a) B commutes with C. (b) $(B - C)^* B(B - C) + (B - C)^* C(B - C) = (B^2 - C^2)(B - C) = 0$. (c) $(B - C)B(B - C) = (B - C)C(B - C) = 0$. (d) $B - C = 0$. *Hint*: P.10.12.

P.15.18 Let $A, B \in M_n$ be positive semidefinite and suppose that $A^2 = B^2$. (a) If A and B are invertible, show that $V = A^{-1}B$ is unitary. Why are the eigenvalues of V real and positive? Deduce that $V = I$ and $A = B$. (b) If $A = 0$, show that $B = 0$. *Hint*: P.10.12 or Theorem 15.1.9. (c) Suppose that A and B are not invertible and $A \neq 0$. Let $U = [U_1 \ U_2] \in M_n$ be unitary, in which the columns of U_2 are an orthonormal basis for $\operatorname{null} A$. Show that

$$U^* A U = \begin{bmatrix} U_1^* A U_1 & U_1^* A U_2 \\ U_2^* A U_1 & U_2^* A U_2 \end{bmatrix} = \begin{bmatrix} U_1^* A U_1 & 0 \\ 0 & 0 \end{bmatrix}$$

and let $U^* B U = \begin{bmatrix} U_1^* B U_1 & U_1^* B U_2 \\ U_2^* B U_1 & U_2^* B U_2 \end{bmatrix}$. Why is $B^2 U_2 = 0$? Why is $BU_2 = 0$? Why is $(U^* A U)^2 = (U^* B U)^2$? Deduce from (a) that $A = B$. (d) State a theorem that summarizes what you have proved in (a)–(c).

P.15.19 Let $A, B \in \mathsf{M}_n$ and suppose that A is positive semidefinite. (a) If B is Hermitian, show that AB need not be diagonalizable. *Hint*: Consider $A = \begin{bmatrix} 0 & 0 \\ 0 & 1 \end{bmatrix}$ and $B = \begin{bmatrix} 0 & 1 \\ 1 & 0 \end{bmatrix}$.
(b) If B is positive semidefinite, show that AB has nonnegative real eigenvalues. *Hint*: Consider the eigenvalues and Jordan blocks of $AB = A^{1/2}A^{1/2}B$ and $A^{1/2}BA^{1/2}$.

P.15.20 Let $A \in \mathsf{M}_n$ be Hermitian and let $\mathbf{x} \in \mathbb{C}^n$ be a unit vector. (a) Is it possible for $A\mathbf{x}$ to be nonzero and orthogonal to $\mathbf{x}$? (b) Is that possible if A is positive semidefinite? Discuss.

P.15.21 Let $X, Y \in \mathsf{M}_{n \times r}$ with $n \geq 2$ and rank $X = r \geq 1$. (a) Show that there is a positive definite $A \in \mathsf{M}_n$ such that $AX = Y$ if and only if X^*Y is positive definite. (b) What does this say if $r = 1$? *Hint*: ($\Leftarrow$) Let $A = QR$ be a QR factorization and let $U^* = [Q \ Q_2]$ be unitary. Then $AX = Y$ is equivalent to $BUQ = UYR^{-1}$, in which $B = UAU^*$. Write $UYR^{-1} = \begin{bmatrix} Z \\ W \end{bmatrix}$, in which $Z = Q^*YR^{-1}$ is positive definite. Choose $\lambda > 0$ so that $B = \begin{bmatrix} Z & W^* \\ W & \lambda I_{n-r} \end{bmatrix}$ is positive definite.

P.15.22 Let $A \in \mathsf{M}_n(\mathbb{F})$ be positive definite. Let $(\lambda_1, \mathbf{u}_1)$ and $(\lambda_n, \mathbf{u}_n)$ be eigenpairs of A, in which $\mathbf{u}_1, \mathbf{u}_n$ are orthonormal and λ_1 and λ_n are, respectively, its smallest and largest eigenvalues. (a) Let $s, t \in \mathbb{R}$ with $t > 0$. Show that $s - t \leq \frac{s^2}{4t}$, with equality for $s = 2t$. *Hint*: Start with $(s - 2t)^2 \geq 0$. (b) If $\mathbf{x} \in \mathbb{C}^n$ is nonzero, show that

$$\frac{\mathbf{x}^*A\mathbf{x}}{\|\mathbf{x}\|_2\|A\mathbf{x}\|_2} \geq \frac{2\sqrt{\lambda_1\lambda_n}}{\lambda_1 + \lambda_n}. \tag{15.7.1}$$

Hint: Show that $(\lambda_n I - A)(A - \lambda_1 I)$ is positive semidefinite and deduce that

$$\|A\mathbf{x}\|_2^2 \leq \underbrace{(\lambda_1 + \lambda_n)\mathbf{x}^*A\mathbf{x}}_{s} - \underbrace{\lambda_1\lambda_n\|\mathbf{x}\|_2^2}_{t}.$$

(c) Show that equality holds in (15.7.1) for $\mathbf{x} = \sqrt{\lambda_n}\mathbf{u}_1 + \sqrt{\lambda_1}\mathbf{u}_n$. (d) If $\mathbb{F} = \mathbb{R}$, why is (15.7.1) a stronger statement than the one in the caption for Figure 15.1? (e) Deduce from (15.7.1) that

$$(\mathbf{x}^*A\mathbf{x})(\mathbf{x}^*A^{-1}\mathbf{x}) \leq \frac{(\lambda_1 + \lambda_n)^2}{4\lambda_1\lambda_n}\|\mathbf{x}\|_2^4.$$

This is the *Kantorovich inequality*. *Hint*: Let $\mathbf{x} = A^{-1/2}\mathbf{y}$ in (15.7.1).

P.15.23 If $A \in \mathsf{M}_n$ is positive definite, show that $(A^{-1})^{1/2} = (A^{1/2})^{-1}$.

P.15.24 Let $A = [a_{ij}] \in \mathsf{M}_2$ be Hermitian. (a) Show that A is positive semidefinite if and only if tr $A \geq 0$ and det $A \geq 0$. (b) Show that A is positive definite if and only if tr $A \geq 0$ and det $A > 0$. (c) Show that A is positive definite if and only if $a_{11} > 0$ and det $A > 0$. (d) If $a_{11} \geq 0$ and det $A \geq 0$, show that A need not be positive semidefinite.

P.15.25 Give an example of a Hermitian $A \in \mathsf{M}_3$ for which tr $A > 0$, det $A > 0$, and A is not positive semidefinite.

P.15.26 Let $A \in \mathsf{M}_n$ be Hermitian and not invertible. (a) If its leading principal minors of size $1, 2, \ldots, n - 1$ are positive, show that A is positive semidefinite; this is the Sylvester criterion for positive semidefiniteness. (b) Use this criterion to show that

$$A = \begin{bmatrix} 1 & 1 & 3 \\ 1 & 2 & 4 \\ 3 & 4 & 10 \end{bmatrix}$$

is positive semidefinite. (c) Confirm that A is positive semidefinite by computing a lower triangular matrix C such that $A = CC^*$. Is C invertible?

P.15.27 The 4×4 *max matrix* is

$$M = [\max\{i,j\}] = \begin{bmatrix} 1 & 2 & 3 & 4 \\ 2 & 2 & 3 & 4 \\ 3 & 3 & 3 & 4 \\ 4 & 4 & 4 & 4 \end{bmatrix}.$$

(a) Show that M is not positive definite. (b) The entries of

$$A = \left[\frac{1}{\max\{i,j\}} \right] = \begin{bmatrix} 1 & \frac{1}{2} & \frac{1}{3} & \frac{1}{4} \\ \frac{1}{2} & \frac{1}{2} & \frac{1}{3} & \frac{1}{4} \\ \frac{1}{3} & \frac{1}{3} & \frac{1}{3} & \frac{1}{4} \\ \frac{1}{4} & \frac{1}{4} & \frac{1}{4} & \frac{1}{4} \end{bmatrix}$$

are the reciprocals of the entries of M. Show that the leading principal minors of A are 1, $\left(\frac{1}{2!}\right)^2$, $\left(\frac{1}{3!}\right)^2$, and $\left(\frac{1}{4!}\right)^2$. Why is A positive definite? (c) Use the algorithm illustrated in Example 15.4.3 to show that

$$C = \begin{bmatrix} 1 & 0 & 0 & 0 \\ \frac{1}{2} & \frac{1}{2} & 0 & 0 \\ \frac{1}{3} & \frac{1}{3} & \frac{1}{3} & 0 \\ \frac{1}{4} & \frac{1}{4} & \frac{1}{4} & \frac{1}{4} \end{bmatrix}$$

provides the Cholesky factorization $A = CC^*$. (d) Show that

$$C^{-1} = \begin{bmatrix} 1 & 0 & 0 & 0 \\ -1 & 2 & 0 & 0 \\ 0 & -2 & 3 & 0 \\ 0 & 0 & -3 & 4 \end{bmatrix} \quad \text{and deduce that} \quad A^{-1} = \begin{bmatrix} 2 & -2 & 0 & 0 \\ -2 & 8 & -6 & 0 \\ 0 & -6 & 18 & -12 \\ 0 & 0 & -12 & -16 \end{bmatrix}.$$

P.15.28 Let

$$A = \begin{bmatrix} 5 & 7 & 6 & 5 \\ 7 & 10 & 8 & 7 \\ 6 & 8 & 10 & 9 \\ 5 & 7 & 9 & 10 \end{bmatrix}.$$

(a) Find the LU decomposition of A. (b) Deduce from (a) that A is positive definite and find $\det A$. (c) Why does A^{-1} have integer entries? (d) What are the $(1,1)$ and $(2,4)$ entries of A^{-1}? (e) Use (a) to find the Cholesky factorization $A = CC^*$.

P.15.29 Prove the Cauchy–Schwarz inequality in $\mathbb{F}^n$ by computing $\det A^*A$, in which $A = [\mathbf{x} \ \mathbf{y}]$ and $\mathbf{x}, \mathbf{y} \in \mathbb{F}^n$.

P.15.30 Let $A, B \in \mathsf{M}_n$ be positive definite. Although AB need not be positive definite, show that $\operatorname{tr} AB > 0$ and $\det AB > 0$.

P.15.31 The *Jordan product* of $A, B \in \mathsf{M}_n$ is $A \bullet B = \frac{1}{2}(AB + BA)$. Suppose that A and B are positive definite. (a) Show that $A \bullet B$ is Hermitian and $\operatorname{tr}(A \bullet B) > 0$. (b) Let $A = \begin{bmatrix} 5 & 3 \\ 3 & 5 \end{bmatrix}$, $B_1 = \begin{bmatrix} 8 & 0 \\ 0 & 1 \end{bmatrix}$, and $B_2 = \begin{bmatrix} 10 & 0 \\ 0 & 1 \end{bmatrix}$. Show that A, B_1, B_2, and $A \bullet B_1$ are positive definite, but $A \bullet B_2$ is not. What is $\det(A \bullet B_2)$?

P.15.32 Let $A \in \mathsf{M}_n$ be normal. Show that A is positive semidefinite if and only if $\operatorname{tr} AB$ is real and nonnegative for every positive semidefinite $B \in \mathsf{M}_n$.

P.15.33 Consider $A = \begin{bmatrix} 10 & 0 \\ 0 & 1 \end{bmatrix}$, $B = \begin{bmatrix} 1 & -1 \\ -1 & 2 \end{bmatrix}$, and $C = \begin{bmatrix} 3 & 5 \\ 5 & 10 \end{bmatrix}$. Show that A, B, and C are positive definite, $\det ABC > 0$, and $\operatorname{tr} ABC < 0$. Conclude that the eigenvalues of ABC are not all real and nonnegative.

P.15.34 Let $A, B \in \mathsf{M}_n$ and suppose that A is positive semidefinite. Show that $AB = BA$ if and only if $A^{1/2}B = BA^{1/2}$.

P.15.35 Review Example 15.4.3. (a) Prove by induction or otherwise that the $n \times n$ min matrix has a Cholesky factorization LL^*, in which $L \in \mathsf{M}_n$ is lower triangular and has every entry on and below the diagonal equal to 1. (b) Show that the inverse of L^* is the upper bidiagonal matrix with entries 1 on the diagonal, and entries -1 in the superdiagonal. (c) Conclude that the inverse of the $n \times n$ min matrix is the tridiagonal matrix with entries -1 in the superdiagonal and subdiagonal, and entries 2 on the main diagonal except for the (n, n) entry, which is 1.

P.15.36 Consider the $n \times n$ tridiagonal matrix with entries 2 on the main diagonal and entries -1 in the superdiagonal and subdiagonal. This matrix arises in numerical solution schemes for differential equations. Use the preceding problem to show that it is positive definite.

P.15.37 Let $A = [a_{ij}] \in \mathsf{M}_n$ be positive definite and let $k \in \{1, 2, \ldots, n - 1\}$. Show in the following two ways that the $k \times k$ leading principal submatrix of A is positive definite. (a) Use the Cholesky factorization of A. (b) Consider $\langle A\mathbf{x}, \mathbf{x} \rangle$, in which the last $n - k$ entries of $\mathbf{x}$ are zero.

P.15.38 Let $A \in \mathsf{M}_n$ be positive definite (respectively, positive semidefinite). Show that every principal submatrix of A is positive definite (respectively, positive semidefinite).

P.15.39 Let $A \in \mathsf{M}_n$ be positive semidefinite and partitioned as in (15.1.12). Show that $\operatorname{rank} A_{11} = \operatorname{rank}[A_{11}\ A_{12}]$.

P.15.40 Let a nonzero $A \in \mathsf{M}_n$ be positive semidefinite and let $r = \operatorname{rank} A$. (a) Show that A has an $r \times r$ positive definite principal submatrix. *Hint*: Permute r linearly independent rows to the top and use the preceding problem. (b) Show that

$$\operatorname{rank} A = \max\{k : B \text{ is a } k \times k \text{ principal submatrix of } A \text{ and } \det B \neq 0\}. \quad (15.7.2)$$

Compare with the assertion in (7.6.5) and give an example of a (necessarily not positive semidefinite) matrix for which (15.7.2) is false.

P.15.41 Let $A \in \mathsf{M}_n$ be positive semidefinite. (a) Use the QR factorization $A^{1/2} = QR$ to show that $A = R^*R$. (b) Compare this factorization with the Cholesky factorization in Theorem 15.4.1. (c) If A is positive definite, what can you say about R and the Cholesky factor C?

P.15.42 Let $A = [a_{ij}] \in \mathsf{M}_n$ be positive definite. Let $A = CC^*$ be its Cholesky factorization, in which $C = [c_{ij}]$. Partition $C^* = [\mathbf{x}_1\ \mathbf{x}_2\ \ldots\ \mathbf{x}_n]$ according to its columns. Show that $a_{ii} = \|\mathbf{x}_i\|_2^2 \geq c_{ii}^2$ and deduce that $\det A \leq a_{11}a_{22}\cdots a_{nn}$, with equality if and only if A is diagonal. This is one version of *Hadamard's inequality*.

P.15.43 Partition $A = [\mathbf{a}_1\ \mathbf{a}_2\ \ldots\ \mathbf{a}_n] \in \mathsf{M}_n$ according to its columns. Apply the result in the preceding problem to A^*A and prove that $|\det A| \leq \|\mathbf{a}_1\|_2\|\mathbf{a}_2\|_2\cdots\|\mathbf{a}_n\|_2$, with equality if and only if either A has orthogonal columns or at least one of its columns is zero. This is another version of *Hadamard's inequality*; see (7.6.2).

P.15.44 Describe the conic section $5x^2 - 4xy + 8y^2 = 36$.

P.15.45 Describe the conic section $9x^2 - 4xy + 6y^2 - 10x - 20y = 5$. *Hint*: Rewrite the equation as $\mathbf{x}^\mathsf{T}A\mathbf{x} + \mathbf{k}^\mathsf{T}\mathbf{x} = 5$.

P.15.46 Determine what type (parabola, ellipse,...) of conic section is described by (a) $5x^2 - 7xy + 3y^2 = 47$, and (b) $5x^2 - 8xy + 3y^2 = 47$.

P.15.47 Let $A, B \in M_n$ be Hermitian and suppose that A is invertible. If there is an invertible $S \in M_n$ such that S^*AS and S^*BS are both diagonal, show that $A^{-1}B$ is diagonalizable and has real eigenvalues. This necessary condition for simultaneous diagonalization of two Hermitian matrices is known to be sufficient as well; see [HJ13, Thm. 4.5.17].

P.15.48 Let $A = \begin{bmatrix} 0 & 1 \\ 1 & 0 \end{bmatrix}$. Show that there is no invertible $S \in M_2$ such that S^*AS and S^*BS are both diagonal if (a) $B = \begin{bmatrix} 1 & 0 \\ 0 & -1 \end{bmatrix}$, or (b) if $B = \begin{bmatrix} 1 & 1 \\ 1 & 0 \end{bmatrix}$.

P.15.49 Let $A \in M_n$ and suppose that $A + A^*$ is positive definite. Show that there is an invertible $S \in M_n$ such that S^*AS is diagonal.

P.15.50 Show that a square complex matrix is the product of two positive definite matrices if and only if it is diagonalizable and has positive eigenvalues.

P.15.51 Let $A, B \in M_n$ be Hermitian and suppose that A is positive definite. Show that $A + B$ is positive definite if and only if $\operatorname{spec} A^{-1}B \subseteq (-1, \infty)$.

P.15.52 Let $A \in M_n$ be positive semidefinite and partitioned as in (15.1.12). (a) If A_{11} is not invertible, prove that A is not invertible. (b) If A_{11} is invertible, must A be invertible?

P.15.53 Let V be an $\mathbb{F}$-inner product space and let $\mathbf{u}_1, \mathbf{u}_2, \ldots, \mathbf{u}_n \in V$. (a) Show that the Gram matrix $G = [\langle \mathbf{u}_j, \mathbf{u}_i \rangle] \in M_n(\mathbb{F})$ is positive semidefinite. (b) Show that G is positive definite if and only if $\mathbf{u}_1, \mathbf{u}_2, \ldots, \mathbf{u}_n$ are linearly independent. (c) Show that $[(i + j - 1)^{-1}] \in M_n$ in P.5.24 is positive definite.

P.15.54 Let $\lambda_1, \lambda_2, \ldots, \lambda_n$ be real and positive. (a) Show that

$$\frac{1}{\lambda_i + \lambda_j} = \int_0^\infty e^{-\lambda_i t} e^{-\lambda_j t}\, dt \qquad \text{for } i, j = 1, 2, \ldots, n.$$

(b) Show that the *Cauchy matrix* $[(\lambda_i + \lambda_j)^{-1}] \in M_n$ is positive semidefinite, and it is positive definite if $\lambda_1, \lambda_2, \ldots, \lambda_n$ are distinct. *Hint*: P.2.55.

P.15.55 Let $A = [a_{ij}] \in M_3$ be positive semidefinite and partitioned as in (15.1.12). Suppose that $A_{11} = \begin{bmatrix} 1 & e^{i\theta} \\ e^{-i\theta} & 1 \end{bmatrix}$, in which $\theta \in \mathbb{R}$. If $a_{13} = \frac{1}{2}$, what is a_{23}?

P.15.56 Let $A = [a_{ij}] \in M_n$ be positive semidefinite and partitioned as in (15.1.12). Show that there is a $Y \in M_{k \times (n-k)}$ such that $A_{12} = YA_{22}$. Deduce that $\operatorname{rank} A_{12} \leq \operatorname{rank} A_{22}$.

P.15.57 Let $A = [a_{ij}] \in M_n$ be partitioned as in (15.1.12). (a) Show that

$$\operatorname{rank} A \leq \operatorname{rank}[A_{11}\ A_{12}] + \operatorname{rank}[A_{12}^*\ A_{22}].$$

(b) If A is positive semidefinite, deduce that $\operatorname{rank} A \leq \operatorname{rank} A_{11} + \operatorname{rank} A_{22}$. (c) Show by example that the inequality in (b) need not be correct if A is Hermitian but not positive semidefinite.

P.15.58 Let $A, B \in M_n$ be normal. Use Definition 14.1.1 and Theorem 15.1.9 to show that $\operatorname{rank} AB = \operatorname{rank} BA$. See P.14.40 for a different proof.

P.15.59 Let $H \in M_n$ be Hermitian and suppose that $\operatorname{spec} H \subseteq [-1, 1]$. (a) Show that $I - H^2$ is positive semidefinite. (b) Let $U = H + i(I - H^2)^{1/2}$. Show that U is unitary and $H = \frac{1}{2}(U + U^*)$.

P.15.60 Let $A \in M_n$. Use the Cartesian decomposition and the preceding problem to show that A is a linear combination of four or fewer unitary matrices. See P.17.50 for a related result.

P.15.61 Let $\mathscr{U}_n$ denote the set of unitary matrices in M_n. Show that (a) span $\mathscr{U}_n = \mathsf{M}_n$ and (b) M_n has a basis comprised of unitary matrices. See P.7.26 for a unitary basis for M_2.

P.15.62 Prove (a)–(f) in Theorem 15.6.3.

P.15.63 Prove (a) and (c) in Theorem 15.6.4.

P.15.64 Let $A = [a_{ij}] \in \mathsf{M}_n$ be normal and let $A = U\Lambda U^*$ be a spectral decomposition, in which $\Lambda = \mathrm{diag}(\lambda_1, \lambda_2, \ldots, \lambda_n)$ and $U = [u_{ij}] \in \mathsf{M}_n$ is unitary. Let $\mathbf{a} = [a_{11}\ a_{22}\ \ldots\ a_{nn}]^\mathsf{T}$ and $\boldsymbol{\lambda} = [\lambda_1\ \lambda_2\ \ldots\ \lambda_n]^\mathsf{T}$. (a) Show that

$$\mathbf{a} = (U \circ \overline{U})\boldsymbol{\lambda}. \tag{15.7.3}$$

(b) Show that $U \circ \overline{U}$ is doubly stochastic.

P.15.65 Let $A \in \mathsf{M}_3$ be normal and suppose that 1, i, and $2 + 2i$ are its eigenvalues. (a) Use (15.7.3) to draw a region in the complex plane that must contain the diagonal entries of A. (b) Could $2i$ be a diagonal entry of A? How about -1 or $1 + i$? (c) Why must every main diagonal entry of A have nonnegative real part? (d) How large could the real part of a diagonal entry be?

P.15.66 Let $A = [a_{ij}] \in \mathsf{M}_n$ be normal and let $\mathrm{spec}\,A = \{\lambda_1, \lambda_2, \ldots, \lambda_n\}$. (a) Use (15.7.3) to show that $\min_{1 \leq i \leq n} \mathrm{Re}\,\lambda_i \leq \min_{1 \leq i \leq n} \mathrm{Re}\,a_{ii}$. (b) Illustrate this inequality with a figure that plots the eigenvalues and main diagonal entries in the complex plane. (c) What can you say about $\max \mathrm{Re}\,a_{ii}$ and $\max \mathrm{Re}\,\lambda_i$? *Hint*: Project your figure onto the x-axis.

P.15.67 Let $A = [a_{ij}] \in \mathsf{M}_n$ be Hermitian and let $\lambda_1 \leq \lambda_2 \leq \cdots \leq \lambda_n$ be its eigenvalues. (a) Use (15.7.3) to show that $\lambda_1 \leq \min_{1 \leq i \leq n} a_{ii}$. (b) What can you say about the maximum diagonal entry and λ_n?

P.15.68 Let $A, B \in \mathsf{M}_n$ be positive semidefinite. Let γ be the largest main diagonal entry of B and let λ_n be the largest eigenvalue of A. Show that

$$\langle (A \circ B)\mathbf{x}, \mathbf{x} \rangle \leq \gamma \lambda_n \|\mathbf{x}\|_2^2 \quad \text{for } \mathbf{x} \in \mathbb{C}^n. \tag{15.7.4}$$

Hint: Proceed as in the proof of Corollary 15.6.8.

P.15.69 Let $A, B \in \mathsf{M}_n$. Suppose that A is diagonalizable and $\mathrm{spec}\,A \cap \mathrm{spec}(-A) = \varnothing$. Modify the argument in Example 15.6.11 to show that there is a unique $X \in \mathsf{M}_n$ such that $AX + XA = B$.

P.15.70 Let $k > 2$ and let $A_1, A_2, \ldots, A_k \in \mathsf{M}_n$ be positive semidefinite. (a) Show that $A_1 \circ A_2 \circ \cdots \circ A_k$ is positive semidefinite. (b) If every main diagonal entry of A_i is positive for all $i = 1, 2, \ldots, k$ and if A_j is positive definite for some $j \in \{1, 2, \ldots, k\}$, show that $A_1 \circ A_2 \circ \cdots \circ A_k$ is positive definite.

P.15.71 Let $A = [a_{ij}] \in \mathsf{M}_n$ be positive semidefinite. Show that $B = [e^{a_{ij}}] \in \mathsf{M}_n$ is positive semidefinite. *Hint*: See P.A.15 and represent B as a sum of positive real multiples of Hadamard products.

P.15.72 Let $A, B \in \mathsf{M}_n$ be positive semidefinite. Show that $\mathrm{rank}(A \circ B) \leq (\mathrm{rank}\,A)(\mathrm{rank}\,B)$. *Hint*: Only $\mathrm{rank}\,A$ summands are needed to represent A in (15.6.6).

P.15.73 Let $A, B \in \mathsf{M}_n$. Suppose that A is positive definite and B is positive semidefinite. Show that $\mathrm{rank}(A \circ B) \geq \mathrm{rank}\,B$ and give an example to show that strict inequality is possible. *Hint*: If $r = \mathrm{rank}\,B > 0$, let B_1 be a positive definite $r \times r$ principal submatrix of B (P.15.40) and let A_1 be the corresponding principal submatrix of A. Then $A_1 \circ B_1$ is a principal submatrix of $A \circ B$.

P.15.74 Let $A, B \in M_n$ and suppose that A is positive definite. If B is positive semidefinite and has exactly k nonzero main diagonal entries, show that $\text{rank}(A \circ B) = k$.

P.15.75 Let $A \in M_{m \times n}(\mathbb{F})$. Suppose that $\text{rank}\, A = r \geq 1$ and let the columns of $X \in M_{m \times r}(\mathbb{F})$ be a basis for $\text{col}\, A$. Let $Y = (X^*X)^{-1}X^*A$ and show that $A = XY$. Why is this a full-rank factorization?

P.15.76 Let $A \in M_{m \times n}(\mathbb{F})$. Use Theorem 15.1.9.b and Theorem 2.3.1 to show that $\text{rank}\, A = \text{rank}\, \overline{A}$.

P.15.77 If $A \in M_m$ and $B \in M_n$ are positive semidefinite, show that $A \otimes B$ is positive semidefinite.

P.15.78 Let $A = [a_{ij}] \in M_3$ and $B = [b_{ij}] \in M_3$. Write down the 3×3 principal submatrix of $A \otimes B \in M_9$ whose entries lie in the intersections of its first, fifth, and ninth rows with its first, fifth, and ninth columns. Do you recognize this matrix?

P.15.79 Let $A = [a_{ij}] \in M_n$ and $B = [b_{ij}] \in M_n$. (a) Show that $A \circ B \in M_n$ is a principal submatrix of $A \otimes B \in M_{n^2}$. (b) Deduce Theorem 15.6.5 from (a) and P.15.77.

P.15.80 Let $A, B \in M_n$, $\mathbf{u} \in \mathbb{C}^n$, $\Delta = \text{diag}\, \mathbf{u}$, and $P = [\mathbf{e}_1 \otimes \mathbf{e}_1 \;\; \mathbf{e}_2 \otimes \mathbf{e}_2 \; \ldots \; \mathbf{e}_n \otimes \mathbf{e}_n] \in M_{n^2 \times n}$. Show that (a) $P\mathbf{u} = \text{vec}\, \Delta$, (b) $P^*P = I_n$, and (c) $A \circ B = P^*(A \otimes B)P$.

P.15.81 Maintain the notation of the preceding problem. Suppose that A and B are positive semidefinite, $X, Y \in M_n$, $A = X^*X$, and $B = Y^*Y$. (a) Show that A is positive definite if and only if X is invertible. (b) Show that B has no zero main diagonal entries if and only if Y has no zero columns. (c) Show that $\mathbf{u}^*(A \circ B)\mathbf{u} = \|(X \otimes Y)P\mathbf{u}\|_2^2 = \|\text{vec}\,(Y\Delta X^{\mathsf{T}})\|_2^2$. (d) If A is positive definite and B has no zero diagonal entries, deduce from (c) that $A \circ B$ is positive definite.

15.8 Notes

See [HJ13, Section 3.5] for various ways in which a matrix can be factored as a product of triangular and permutation matrices. For example, each $A \in M_n$ can be factored as $A = PLR$, in which L is lower triangular, R is upper triangular, and P is a permutation matrix.

See [HJ13, Section 4.5] for a discussion of how Theorem 15.5.3 is used to analyze small oscillations of a collection of masses about a stable equilibrium.

In 1899, the French mathematician Jacques Hadamard published a paper about power series $f(z) = \sum_{k=0}^{\infty} a_k z^k$ and $g(z) = \sum_{k=0}^{\infty} b_k z^k$, in which he studied the product $(f \circ g)(z) = \sum_{k=0}^{\infty} a_k b_k z^k$. Because of that paper, Hadamard's name has been linked with term-by-term (or entrywise) products of various kinds. The first systematic study of entrywise products of matrices was published in 1911 by Isaai Schur. For a historical survey of the Hadamard product, see [HJ94, Section 5.0]. For a detailed treatment of Hadamard products and applications, see [HJ94, Chapter 5 and Section 6.3]; see also [HJ13, Section 7.5].

In Example 15.6.10, $\text{rank}\, A = 2$ and every pair of distinct columns of B is linearly independent. For an explanation of why this is sufficient to ensure that $A \circ B$ is positive definite, see [HY20].

The eigenvalues of the tridiagonal matrix in P.15.35 are $4\sin^2((2k+1)\pi/(4n+2))$, for $k = 0, 1, \ldots, n-1$, so the eigenvalues of the $n \times n$ min matrix are $\frac{1}{4}\csc^2((2k+1)\pi/(4n+2))$, for $k = 0, 1, \ldots, n-1$. For a continuous analog of the Cholesky factorization of the min matrix, see P.16.37.

15.9 Some Important Concepts

- Positive definite and semidefinite matrices.

- Characterization of positive semidefinite matrices as Gram matrices.

- Characterization of positive semidefinite matrices as normal matrices with real nonnegative eigenvalues.

- Column inclusion.

- Diagonal dominance and positive definiteness.

- Unique positive semidefinite square root of a positive semidefinite matrix.

- Cholesky factorization of a positive definite matrix.

- Simultaneous diagonalization of a positive definite matrix and a Hermitian matrix.

- Hadamard product and the Schur product theorem.

16 The Singular Value and Polar Decompositions

This chapter is about the singular value decomposition, a matrix factorization that shares some features with the spectral decomposition of a normal matrix. However, every real or complex matrix, normal or not, square or not, has a singular value decomposition. It was discovered in the nineteenth century and was applied in the 1930s to problems in psychometrics. Widespread use of the singular value decomposition in applications began only in the 1960s, when accurate and efficient algorithms were developed for its numerical computation. Today, singular value subroutines are embedded in thousands of computer algorithms to solve problems in data analysis, image compression, chemical physics, least-squares problems, low-dimensional approximation of high-dimensional data, genomics, robotics, and many other fields.

16.1 The Singular Value Decomposition

The spectral theorem for a normal matrix $A \in \mathsf{M}_n$ says that

$$A = U\Lambda U^* = \lambda_1 \mathbf{u}_1 \mathbf{u}_1^* + \lambda_2 \mathbf{u}_2 \mathbf{u}_2^* + \cdots + \lambda_n \mathbf{u}_n \mathbf{u}_n^*,$$

in which $U = [\mathbf{u}_1 \ \mathbf{u}_2 \ \ldots \ \mathbf{u}_n] \in \mathsf{M}_n$ is unitary and $\lambda_1, \lambda_2, \ldots, \lambda_n$ are the eigenvalues of A. If $m \geq n$, the singular value decomposition of $A \in \mathsf{M}_{m \times n}$ is

$$A = V\Sigma W^* = \sigma_1 \mathbf{v}_1 \mathbf{w}_1^* + \sigma_2 \mathbf{v}_2 \mathbf{w}_2^* + \cdots + \sigma_n \mathbf{v}_n \mathbf{w}_n^*,$$

in which $V = [\mathbf{v}_1 \ \mathbf{v}_2 \ \ldots \ \mathbf{v}_m] \in \mathsf{M}_m$ and $W = [\mathbf{w}_1 \ \mathbf{w}_2 \ \ldots \ \mathbf{w}_n] \in \mathsf{M}_n$ are unitary and $\sigma_1 \geq \sigma_2 \geq \cdots \geq \sigma_n \geq 0$ are the decreasingly ordered eigenvalues of the positive semidefinite matrix $(A^*A)^{1/2} \in \mathsf{M}_n$.

The following example illustrates a special case of the singular value decomposition.

Example 16.1.1 Let $A \in \mathsf{M}_n$ be invertible and let $U = A(A^*A)^{-1/2}$. Then $U^*U = (A^*A)^{-1/2}(A^*A)(A^*A)^{-1/2} = I$, so U is unitary. We have $A = U(A^*A)^{1/2}$, which presents A as the product of a unitary matrix and a positive definite matrix. Let $A^*A = W\Sigma^2 W^*$ be a spectral decomposition, in which $W \in \mathsf{M}_n$ is unitary and the diagonal entries of $\Sigma = \mathrm{diag}(\sigma_1, \sigma_2, \ldots, \sigma_n)$ are the positive square roots of the eigenvalues of A^*A, arranged in decreasing order. Then $(A^*A)^{1/2} = W\Sigma W^*$ and $A = U(A^*A)^{1/2} = UW\Sigma W^* = V\Sigma W^*$, in which $V = UW \in \mathsf{M}_n$ is unitary.

If $A \in \mathsf{M}_{m \times n}$ and $\mathrm{rank}\, A = r \geq 1$, Theorem 10.7.2 says that the positive semidefinite matrices $A^*A \in \mathsf{M}_n$ and $AA^* \in \mathsf{M}_m$ have the same nonzero eigenvalues. These eigenvalues are positive, and there are r of them (Theorem 15.1.9). The positive square roots of these eigenvalues have a special name.

Definition 16.1.2 Let $A \in \mathsf{M}_{m \times n}$ and let $q = \min\{m, n\}$. If rank $A = r \geq 1$, let $\sigma_1 \geq \sigma_2 \geq \cdots \geq \sigma_r > 0$ be the decreasingly ordered positive eigenvalues of $(A^*A)^{1/2}$. The *singular values* of A are $\sigma_1, \sigma_2, \ldots, \sigma_r$ and $\sigma_{r+1} = \sigma_{r+2} = \cdots = \sigma_q = 0$. If $A = 0$, then the singular values of A are $\sigma_1 = \sigma_2 = \cdots = \sigma_q = 0$.

The singular values of $A \in \mathsf{M}_n$ are the eigenvalues of $(A^*A)^{1/2}$, which are the same as the eigenvalues of $(AA^*)^{1/2}$.

Example 16.1.3 Consider

$$A = \begin{bmatrix} 0 & 20 & 30 \\ 0 & 0 & 20 \\ 0 & 0 & 0 \end{bmatrix} \quad \text{and} \quad A^* = \begin{bmatrix} 0 & 0 & 0 \\ 20 & 0 & 0 \\ 30 & 20 & 0 \end{bmatrix}.$$

Then spec $A = \{0\}$, rank $A = 2$,

$$A^*A = \begin{bmatrix} 0 & 0 & 0 \\ 0 & 400 & 600 \\ 0 & 600 & 1300 \end{bmatrix}, \quad (A^*A)^{1/2} = \begin{bmatrix} 0 & 0 & 0 \\ 0 & 16 & 12 \\ 0 & 12 & 34 \end{bmatrix},$$

spec $A^*A = \{1600, 100, 0\}$, and spec$(A^*A)^{1/2} = \{40, 10, 0\}$. The singular values of A are $40, 10, 0$ and its eigenvalues are $0, 0, 0$.

Theorem 16.1.4 *Let $A \in \mathsf{M}_{m \times n}$, let $r = $ rank A and $q = \min\{m, n\}$, let $\sigma_1 \geq \sigma_2 \geq \cdots \geq \sigma_q$ be the singular values of A, and let $c \in \mathbb{C}$.*

(a) *If $r \geq 1$, then $\sigma_1^2, \sigma_2^2, \ldots, \sigma_r^2$ are the positive eigenvalues of A^*A and AA^*.*

(b) *$\sum_{i=1}^{q} \sigma_i^2 = \operatorname{tr} A^*A = \operatorname{tr} AA^* = \|A\|_F^2$.*

(c) *$A, A^*, A^\mathsf{T},$ and $\overline{A}$ have the same singular values.*

(d) *The singular values of cA are $|c|\sigma_1, |c|\sigma_2, \ldots, |c|\sigma_q$.*

Proof (a) Theorem 15.3.2 ensures that $\sigma_1^2, \sigma_2^2, \ldots, \sigma_r^2$ are the positive eigenvalues of $A^*A = ((A^*A)^{1/2})^2$. Theorem 10.7.2 says that they are the positive eigenvalues of AA^*.

(b) $\operatorname{tr} A^*A = \|A\|_F^2$ is a definition (see Example 5.4.5), and $\operatorname{tr} A^*A$ is the sum of the eigenvalues of A^*A.

(c) The positive semidefinite matrices A^*A, AA^*, $A^\mathsf{T}\overline{A} = \overline{A^*A}$, and $\overline{A}A^\mathsf{T} = \overline{AA^*}$ have the same nonzero eigenvalues; see Theorem 10.2.9.

(d) The eigenvalues of the matrix $(cA^*)(cA) = |c|^2 A^*A$ are $|c|^2\sigma_1^2, |c|^2\sigma_2^2, \ldots, |c|^2\sigma_q^2$; see Corollary 11.1.4. $\qquad\square$

The symbol Σ in the following theorem is conventional notation for the *singular value matrix*; in this context it has nothing to do with summation.

Theorem 16.1.5 (Singular Value Decomposition) *Let $A \in \mathsf{M}_{m \times n}(\mathbb{F})$ be nonzero and let $r = $ rank A. Let $\sigma_1 \geq \sigma_2 \geq \cdots \geq \sigma_r > 0$ be the positive singular values of A and define*

$$\Sigma_r = \begin{bmatrix} \sigma_1 & & \\ & \ddots & \\ & & \sigma_r \end{bmatrix} \in \mathsf{M}_r(\mathbb{R}).$$

Then there are unitary matrices $V \in \mathsf{M}_m(\mathbb{F})$ and $W \in \mathsf{M}_n(\mathbb{F})$ such that

$$A = V \Sigma W^*, \tag{16.1.6}$$

in which

$$\Sigma = \begin{bmatrix} \Sigma_r & 0_{r \times (n-r)} \\ 0_{(m-r) \times r} & 0_{(m-r) \times (n-r)} \end{bmatrix} \in \mathsf{M}_{m \times n}(\mathbb{R}) \tag{16.1.7}$$

is the same size as A. If $m = n$, then $V, W \in \mathsf{M}_n(\mathbb{F})$ and $\Sigma = \Sigma_r \oplus 0_{n-r}$. If $V = [\mathbf{v}_1 \ \mathbf{v}_2 \ \cdots \ \mathbf{v}_m]$ and $W = [\mathbf{w}_1 \ \mathbf{w}_2 \ \cdots \ \mathbf{w}_n]$, then

$$A = \sigma_1 \mathbf{v}_1 \mathbf{w}_1^* + \sigma_2 \mathbf{v}_2 \mathbf{w}_2^* + \cdots + \sigma_r \mathbf{v}_r \mathbf{w}_r^*. \tag{16.1.8}$$

Proof (a) Let $m \geq n$ and suppose that A has full column rank, that is $r = n$. Then A^*A is positive definite, as is $(A^*A)^{-1/2}$. Let $U = A(A^*A)^{-1/2} \in \mathsf{M}_{m \times n}(\mathbb{F})$. Then $U^*U = (A^*A)^{-1/2}(A^*A)(A^*A)^{-1/2} = I_n$, so U has orthonormal columns. We have $A = U(A^*A)^{1/2}$. Let $A^*A = W\Sigma_n^2 W^*$ be a spectral decomposition, in which $W \in \mathsf{M}_n(\mathbb{F})$ is unitary, $\Sigma_n^2 = \operatorname{diag}(\sigma_1^2, \sigma_2^2, \ldots, \sigma_n^2)$, and $\sigma_1 \geq \sigma_2 \geq \cdots \geq \sigma_n > 0$. Then $(A^*A)^{1/2} = W\Sigma_n W^*$ and $A = U(A^*A)^{1/2} = (UW)\Sigma W^*$. If $m = n$, then $V = UW$ is unitary and $A = V\Sigma W^*$ with $\Sigma = \Sigma_n$. If $m > n$, then $UW \in \mathsf{M}_{m \times n}(\mathbb{F})$ has orthonormal columns and there is an $X \in \mathsf{M}_{m \times (m-n)}(\mathbb{F})$ such that $V = [UW \ X] \in \mathsf{M}_m(\mathbb{F})$ is unitary. Then

$$A = \underbrace{[UW \ X]}_{V} \underbrace{\begin{bmatrix} \Sigma_n \\ 0_{(m-n) \times n} \end{bmatrix}}_{\Sigma} W^* = V\Sigma W^*.$$

(b) Now let $m \geq n$ and suppose that $1 \leq r \leq n - 1$. Let $Z = [Z_1 \ Z_2] \in \mathsf{M}_n(\mathbb{F})$ be unitary, in which the columns of $Z_2 \in \mathsf{M}_{m \times (n-r)}(\mathbb{F})$ are a basis for null $A \subseteq \mathbb{F}^n$. Then $AZ = [AZ_1 \ AZ_2] = [AZ_1 \ 0]$, so $r = \operatorname{rank} A = \operatorname{rank} AZ = \operatorname{rank}[AZ_1 \ 0] = \operatorname{rank} AZ_1$. Thus, $AZ_1 \in \mathsf{M}_{m \times r}(\mathbb{F})$ has full column rank. Then (a) ensures that $AZ_1 = V\Sigma Y^*$, in which $V \in \mathsf{M}_m(\mathbb{F})$ and $Y \in \mathsf{M}_r(\mathbb{F})$ are unitary. Thus,

$$AZ = [V\Sigma Y^* \ 0] = V \begin{bmatrix} \Sigma_r & 0_{r \times (n-r)} \\ 0_{(m-r) \times r} & 0_{(m-r) \times (n-r)} \end{bmatrix} \begin{bmatrix} Y^* & 0 \\ 0 & I_{n-r} \end{bmatrix}$$

and

$$A = V \underbrace{\begin{bmatrix} \Sigma_r & 0_{r \times (n-r)} \\ 0_{(m-r) \times r} & 0_{(m-r) \times (n-r)} \end{bmatrix}}_{\Sigma} \underbrace{\begin{bmatrix} Y^* & 0 \\ 0 & I_{n-r} \end{bmatrix} Z^*}_{W^*} = V\Sigma W^*,$$

in which V and W are unitary.

(c) If $n < m$, apply (a) or (b) to A^* and take the adjoint of the resulting factorization; see P.16.5. The representation (16.1.8) of A as a sum of r outer products follows from using (3.1.17) to expand (16.1.6).

If A is real, the constructions in the proof of the factorization (16.1.6) ensure that the unitary factors V and W are real. □

Definition 16.1.9 A *singular value decomposition* of $A \in \mathsf{M}_{m \times n}(\mathbb{F})$ is a factorization of the form $A = V\Sigma W^*$, in which $V \in \mathsf{M}_m(\mathbb{F})$ and $W \in \mathsf{M}_n(\mathbb{F})$ are unitary and $\Sigma \in \mathsf{M}_{m \times n}(\mathbb{R})$ is the matrix (16.1.7).

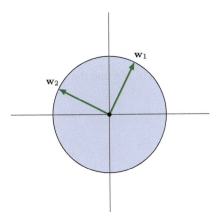

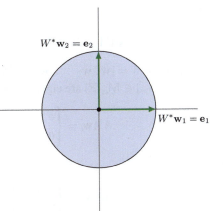

(a) The unit disk and the right singular vectors $\mathbf{w}_1, \mathbf{w}_2$ of A.

(b) Action of W^* (rotation by $\tan^{-1}(-2) \approx -63.4°$) on the vectors in (a).

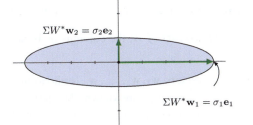

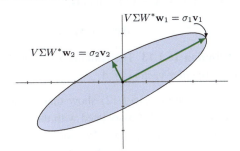

(c) Action of $\Sigma = \mathrm{diag}(4, 1)$ on the vectors in (b).

(d) Action of V (rotation by $\tan^{-1}(\frac{1}{2}) \approx 26.6°$) on the vectors in (c).

Figure 16.1 Singular value decomposition analysis of the matrix A in (16.1.11). It transforms the unit disk into an ellipse whose semiaxes have lengths equal to the singular values of A.

Example 16.1.10 Consider

$$A = \begin{bmatrix} 2 & 3 \\ 0 & 2 \end{bmatrix}, \tag{16.1.11}$$

which has a singular value decomposition

$$\begin{bmatrix} 2 & 3 \\ 0 & 2 \end{bmatrix} = \underbrace{\begin{bmatrix} \frac{2}{\sqrt{5}} & -\frac{1}{\sqrt{5}} \\ \frac{1}{\sqrt{5}} & \frac{2}{\sqrt{5}} \end{bmatrix}}_{V} \underbrace{\begin{bmatrix} 4 & 0 \\ 0 & 1 \end{bmatrix}}_{\Sigma} \underbrace{\begin{bmatrix} \frac{1}{\sqrt{5}} & \frac{2}{\sqrt{5}} \\ -\frac{2}{\sqrt{5}} & \frac{1}{\sqrt{5}} \end{bmatrix}}_{W^*}.$$

Let $V = [\mathbf{v}_1 \ \mathbf{v}_2]$ and $W = [\mathbf{w}_1 \ \mathbf{w}_2]$. Figure 16.1 illustrates how the three factors in the singular value decomposition of A act on the unit disk.

What do the unitary matrices V and W represent? What are their columns and how are they related? The matrices $\Sigma^\mathsf{T}\Sigma = \Sigma_r^2 \oplus 0_{n-r} \in \mathsf{M}_n(\mathbb{R})$ and $\Sigma\Sigma^\mathsf{T} = \Sigma_r^2 \oplus 0_{m-r} \in \mathsf{M}_m(\mathbb{R})$ are diagonal. Thus,

$$A^*AW = W(\Sigma^\mathsf{T}\Sigma)W^*W = W(\Sigma^\mathsf{T}\Sigma) = W(\Sigma_r^2 \oplus 0_{n-r})$$

and

$$AA^*V = (V\Sigma\Sigma^T V^*)V = V(\Sigma\Sigma^T) = V(\Sigma_r^2 \oplus 0_{m-r}),$$

so the columns of $W = [\mathbf{w}_1 \ \mathbf{w}_2 \ \cdots \ \mathbf{w}_n] \in M_n(\mathbb{F})$ are eigenvectors of A^*A and the columns of $V = [\mathbf{v}_1 \ \mathbf{v}_2 \ \cdots \ \mathbf{v}_m] \in M_m(\mathbb{F})$ are eigenvectors of AA^*. We have

$$A^*A\mathbf{w}_i = \begin{cases} \sigma_i^2 \mathbf{w}_i & \text{for } i = 1, 2, \ldots, r, \\ \mathbf{0} & \text{for } i = r+1, r+2, \ldots, n, \end{cases}$$

and

$$AA^*\mathbf{v}_i = \begin{cases} \sigma_i^2 \mathbf{v}_i & \text{for } i = 1, 2, \ldots, r, \\ \mathbf{0} & \text{for } i = r+1, r+2, \ldots, m. \end{cases}$$

The columns of V and W are linked by the identities $AW = V\Sigma W^*W = V\Sigma$ and $A^*V = W\Sigma^T V^*V = W\Sigma^T$, that is,

$$A\mathbf{w}_i = \sigma_i \mathbf{v}_i \quad \text{and} \quad A^*\mathbf{v}_i = \sigma_i \mathbf{w}_i \quad \text{for } i = 1, 2, \ldots, r. \tag{16.1.12}$$

Definition 16.1.13 The columns of V are *left singular vectors of A*; the columns of W are *right singular vectors of A*.

The identities (16.1.12) show that if right singular vectors $\mathbf{w}_1, \mathbf{w}_2, \ldots, \mathbf{w}_r$ (orthonormal eigenvectors of A^*A associated with the eigenvalues $\sigma_1^2, \sigma_2^2, \ldots, \sigma_r^2$) are known, then left singular vectors are $\mathbf{v}_i = \sigma_i^{-1} A\mathbf{w}_i$ for $i = 1, 2, \ldots, r$. It is not necessary to compute a spectral decomposition of AA^* to obtain the left singular vectors. The following example illustrates a four-step algorithm that finds a singular value decomposition of a small matrix.

Example 16.1.14 Let us compute a singular value decomposition of $A = \begin{bmatrix} 3 & 0 \\ 4 & 5 \end{bmatrix}$.

Step 1. $A^*A = \begin{bmatrix} 25 & 20 \\ 20 & 25 \end{bmatrix}$. Its eigenvalues are $\sigma_1^2 = 45$ and $\sigma_2^2 = 5$, so the singular values of A are $\sigma_1 = 3\sqrt{5}$ and $\sigma_2 = \sqrt{5}$.

Step 2. Unit eigenvectors of A^*A are $\mathbf{w}_1 = \frac{1}{\sqrt{2}}\begin{bmatrix} 1 \\ 1 \end{bmatrix}$ and $\mathbf{w}_2 = \frac{1}{\sqrt{2}}\begin{bmatrix} -1 \\ 1 \end{bmatrix}$. These are right singular vectors of A.

Step 3. Use (16.1.12) to compute $\mathbf{v}_1 = \sigma_1^{-1} A\mathbf{w}_1 = \frac{1}{\sqrt{10}}\begin{bmatrix} 1 \\ 3 \end{bmatrix}$ and $\mathbf{v}_2 = \sigma_2^{-1} A\mathbf{w}_2 = \frac{1}{\sqrt{10}}\begin{bmatrix} -3 \\ 1 \end{bmatrix}$. These are left singular vectors of A.

Step 4. $A = V\Sigma W^*$, in which $V = \frac{1}{\sqrt{10}}\begin{bmatrix} 1 & -3 \\ 3 & 1 \end{bmatrix}$, $\Sigma = \begin{bmatrix} 3\sqrt{5} & 0 \\ 0 & \sqrt{5} \end{bmatrix}$, and $W = \frac{1}{\sqrt{2}}\begin{bmatrix} 1 & -1 \\ 1 & 1 \end{bmatrix}$.

Example 16.1.15 Here are two different singular value decompositions of a rank-2 real 3×2 matrix:

$$A = \begin{bmatrix} 4 & 0 \\ -5 & -3 \\ 2 & 6 \end{bmatrix} = \underbrace{\begin{bmatrix} \frac{1}{3} & -\frac{2}{3} & \frac{2}{3} \\ -\frac{2}{3} & \frac{1}{3} & \frac{2}{3} \\ \frac{2}{3} & \frac{2}{3} & \frac{1}{3} \end{bmatrix}}_{V} \underbrace{\begin{bmatrix} 6\sqrt{2} & 0 \\ 0 & 3\sqrt{2} \\ 0 & 0 \end{bmatrix}}_{\Sigma} \underbrace{\begin{bmatrix} \frac{1}{\sqrt{2}} & \frac{1}{\sqrt{2}} \\ -\frac{1}{\sqrt{2}} & \frac{1}{\sqrt{2}} \end{bmatrix}}_{W^*} \tag{16.1.16}$$

$$= \begin{bmatrix} -\frac{1}{3} & -\frac{2}{3} & \frac{2}{3} \\ \frac{2}{3} & \frac{1}{3} & \frac{2}{3} \\ -\frac{2}{3} & \frac{2}{3} & \frac{1}{3} \end{bmatrix} \begin{bmatrix} 6\sqrt{2} & 0 \\ 0 & 3\sqrt{2} \\ 0 & 0 \end{bmatrix} \begin{bmatrix} -\frac{1}{\sqrt{2}} & -\frac{1}{\sqrt{2}} \\ -\frac{1}{\sqrt{2}} & \frac{1}{\sqrt{2}} \end{bmatrix}.$$

In the second factorization, we changed the signs of the entries in the first columns of V and W, but that did not change the product of the factors in the singular value decomposition. The singular value matrix Σ in a singular value decomposition of A is always uniquely determined by A. The two unitary factors are never uniquely determined.

Example 16.1.17 The singular value decomposition (16.1.16) provides the representation

$$
A = 6\sqrt{2}\,\mathbf{v}_1\mathbf{w}_1^* + 3\sqrt{2}\,\mathbf{v}_2\mathbf{w}_2^* = \underbrace{\begin{bmatrix} \frac{1}{3} & -\frac{2}{3} \\ -\frac{2}{3} & \frac{1}{3} \\ \frac{2}{3} & \frac{2}{3} \end{bmatrix}}_{X} \underbrace{\begin{bmatrix} 6\sqrt{2} & 0 \\ 0 & 3\sqrt{2} \end{bmatrix}}_{\Sigma_2} \underbrace{\begin{bmatrix} \frac{1}{\sqrt{2}} & \frac{1}{\sqrt{2}} \\ -\frac{1}{\sqrt{2}} & \frac{1}{\sqrt{2}} \end{bmatrix}}_{Y^*} = X\Sigma_2 Y^*.
$$

This factorization is more compact than (16.1.16) and shows that the third column of V plays no role in representing the entries of A.

The terminology for multiplicities of singular values and equality of singular values is parallel to that of eigenvalues.

Definition 16.1.18 Let $A, B \in \mathsf{M}_{m\times n}$ and suppose that rank $A = r$. The *multiplicity* of a positive singular value σ_i of A is the multiplicity of σ_i^2 as an eigenvalue of A^*A (or of AA^*). The multiplicity of a zero singular value of A (if any) is $\min\{m, n\} - r$. Singular values with multiplicity 1 are *simple*. If every singular value of A is simple, its singular values are *distinct*. We say that A and B have the same singular values if A^*A and B^*B have the same eigenvalues (with the same multiplicities); see Definition 10.2.5.

16.2 The Compact Singular Value Decomposition

If the rank of $A \in \mathsf{M}_{m\times n}(\mathbb{F})$ is smaller than m and n, or if $m \neq n$, then some of the information encoded in the unitary factors of a singular value decomposition of A plays no role in representing the entries of A; see Example 16.1.17. In this section, we show how to organize the singular value decomposition in a compact way that preserves all (and only) the information needed to compute every entry of A. In Section 17.1, we investigate a broader question that arises in image compression: how to truncate the singular value decomposition in a way that permits us to compute an approximation to every entry of A, to any desired accuracy.

Let $A = V\Sigma W^*$ be a singular value decomposition and let $\Sigma_r = \mathrm{diag}(\sigma_1, \sigma_2, \ldots, \sigma_r)$ be the $r \times r$ upper-left corner of Σ. Partition $V = [V_r\ V'] \in \mathsf{M}_m(\mathbb{F})$ and $W = [W_r\ W'] \in \mathsf{M}_n(\mathbb{F})$, in which $V_r \in \mathsf{M}_{m\times r}(\mathbb{F})$, $V' \in \mathsf{M}_{m\times(m-r)}(\mathbb{F})$, $W_r \in \mathsf{M}_{n\times r}(\mathbb{F})$, and $W' \in \mathsf{M}_{n\times(n-r)}(\mathbb{F})$. Then

$$
A = V\Sigma W^* = [V_r\ V'] \begin{bmatrix} \Sigma_r & 0_{r\times(n-r)} \\ 0_{(m-r)\times r} & 0_{(m-r)\times(n-r)} \end{bmatrix} \begin{bmatrix} W_r^* \\ W'^* \end{bmatrix}
$$

$$
= [V_r\ V'] \begin{bmatrix} \Sigma_r W_r^* \\ 0_{(m-r)\times r} \end{bmatrix} = V_r\Sigma_r W_r^* + V'0_{(m-r)\times r} = V_r\Sigma_r W_r^*. \tag{16.2.1}
$$

Unless $r = m = n$, the last $m - r$ columns of V and/or the last $n - r$ columns of W are absent in the representation (16.2.1) of A.

Theorem 16.2.2 (Compact Singular Value Decomposition) *Let* $A \in \mathsf{M}_{m\times n}(\mathbb{F})$ *be nonzero, let* $r = \mathrm{rank}\,A$, *and let* $A = V\Sigma W^*$ *be a singular value decomposition. Partition* $V = [V_r\ V'] \in \mathsf{M}_m(\mathbb{F})$ *and* $W = [W_r\ W'] \in \mathsf{M}_n(\mathbb{F})$, *in which* $V_r = [\mathbf{v}_1\ \mathbf{v}_2\ \cdots\ \mathbf{v}_r] \in \mathsf{M}_{m\times r}(\mathbb{F})$

and $W_r = [\mathbf{w}_1 \ \mathbf{w}_2 \ \ldots \ \mathbf{w}_r] \in M_{n \times r}(\mathbb{F})$, *and let* $\Sigma_r = \operatorname{diag}(\sigma_1, \sigma_2, \ldots, \sigma_r)$. *Then* $\sigma_1 \geq \sigma_2 \geq \cdots \geq \sigma_r > 0$ *and*

$$A = V_r \Sigma_r W_r^*. \tag{16.2.3}$$

Moreover, $(\sigma_i^2, \mathbf{w}_i)$ *is an eigenpair of* A^*A *and* $\mathbf{v}_i = \sigma_i^{-1} A \mathbf{w}_i$ *for each* $i = 1, 2, \ldots, r$.

Proof The presentation (16.2.3) is in (16.2.1). The identity $A^*A = W_r \Sigma_r V_r^* V_r \Sigma_r W_r^* = W_r \Sigma_r^2 W_r^*$ identifies each $(\sigma_i^2, \mathbf{w}_i)$ as an eigenpair of A^*A and (16.2.3) ensures that $A\mathbf{w}_i = V_r \Sigma_r W_r^* \mathbf{w}_i = V_r \Sigma_r \mathbf{e}_i = \sigma_i \mathbf{v}_i$ for each $i = 1, 2, \ldots, r$. $\qquad\square$

Definition 16.2.4 Let $A \in M_{m \times n}(\mathbb{F})$ be nonzero and let $r = \operatorname{rank} A$. A factorization $A = X \Sigma_r Y^*$ is a *compact singular value decomposition* if $X \in M_{m \times r}(\mathbb{F})$ and $Y \in M_{n \times r}(\mathbb{F})$ have orthonormal columns, $\Sigma_r = \operatorname{diag}(\sigma_1, \sigma_2, \ldots, \sigma_r)$, and $\sigma_1 \geq \sigma_2 \geq \cdots \geq \sigma_r > 0$.

Theorem 16.2.2 ensures that every nonzero $A \in M_{m \times n}(\mathbb{F})$ has a compact singular value decomposition in which the matrices X and Y are, respectively, the first $r = \operatorname{rank} A$ columns of the unitary matrices V and W in some singular value decomposition of A. Other choices of X and Y are always possible, so a compact singular value decomposition of A is not unique. However, all of the compact singular value decompositions of A can be obtained from any one of them; we return to this matter in Theorem 16.2.10.

Example 16.2.5 Let us compute a compact singular value decomposition $A = X \Sigma_2 Y^*$ for $A = \begin{bmatrix} 3 & 2 & 2 \\ 2 & 3 & -2 \end{bmatrix}$.

Step 1. The eigenvalues of $AA^* = \begin{bmatrix} 17 & 8 \\ 8 & 17 \end{bmatrix}$ are 25 and 9, which are the same as the positive eigenvalues of A^*A (we work with AA^* here because it is 2×2; note that A^*A is 3×3). The positive singular values of A are $\sigma_1 = 5$ and $\sigma_2 = 3$.

Step 2. Unit eigenvectors of A^*A are $\mathbf{w}_1 = \frac{1}{\sqrt{2}} \begin{bmatrix} 1 \\ 1 \\ 0 \end{bmatrix}$ and $\mathbf{w}_2 = \frac{1}{3\sqrt{2}} \begin{bmatrix} 1 \\ -1 \\ 4 \end{bmatrix}$. These are the columns of Y (right singular vectors of A associated with its positive singular values).

Step 3. Compute $\mathbf{v}_1 = \sigma_1^{-1} A \mathbf{w}_1 = \frac{1}{\sqrt{2}} \begin{bmatrix} 1 \\ 1 \end{bmatrix}$ and $\mathbf{v}_2 = \sigma_2^{-1} A \mathbf{w}_2 = \frac{1}{\sqrt{2}} \begin{bmatrix} 1 \\ -1 \end{bmatrix}$. These are the columns of X (left singular vectors of A associated with its positive singular values).

Step 4. $A = X \Sigma_2 Y^*$, in which $X = \frac{1}{\sqrt{2}} \begin{bmatrix} 1 & 1 \\ 1 & -1 \end{bmatrix}$, $\Sigma_2 = \begin{bmatrix} 5 & 0 \\ 0 & 3 \end{bmatrix}$, and $Y = \frac{1}{3\sqrt{2}} \begin{bmatrix} 3 & 1 \\ 3 & -1 \\ 0 & 4 \end{bmatrix}$.

Theorem 16.2.6 *Let* $A \in M_{m \times n}(\mathbb{F})$, *let* $r = \operatorname{rank} A \geq 1$, *and let* $A = X \Sigma_r Y^*$ *be a compact singular value decomposition (16.2.3). Then*

$$\operatorname{col} A = \operatorname{col} X = (\operatorname{null} A^*)^\perp \tag{16.2.7}$$

and

$$\operatorname{null} A = (\operatorname{col} Y)^\perp = (\operatorname{col} A^*)^\perp. \tag{16.2.8}$$

Proof The representation $A = X \Sigma_r Y^*$ ensures that $\operatorname{col} A \subseteq \operatorname{col} X$. The reverse containment follows from $X = X \Sigma_r I_r \Sigma_r^{-1} = X \Sigma_r Y^* Y \Sigma_r^{-1} = AY \Sigma_r^{-1}$. This shows that $\operatorname{col} X \subseteq \operatorname{col} A$, so $\operatorname{col} X = \operatorname{col} A$. The identity $\operatorname{col} A = (\operatorname{null} A^*)^\perp$ is (8.2.4), and (16.2.8) follows in the same way from applying (16.2.7) to A^* and taking orthogonal complements. $\qquad\square$

The positive definite diagonal factor Σ_r in (16.2.3) is uniquely determined by A; its diagonal entries are the positive singular values of A in decreasing order. However, the factors X and Y

in (16.2.3) are not uniquely determined. For example, we can replace X by XD and Y by YD for any diagonal unitary matrix $D \in \mathsf{M}_r$. If some singular value has multiplicity greater than 1, then other types of nonuniqueness are possible. In Theorem 17.6.1, in Lemma 17.4.1, and in other applications, we need to be able to describe precisely how any two compact singular value decompositions of a given matrix are related. The following lemma is a first step toward that description.

Lemma 16.2.9 *Let $A, B \in \mathsf{M}_n$ be positive definite and let $U, V \in \mathsf{M}_n$ be unitary. If $UA = BV$, then $U = V$.*

Proof We have

$$(UAU^*)^2 = UA^2U^* = UAA^*U^* = (UA)(UA)^* = (BV)(BV)^* = BVV^*B^* = B^2.$$

The uniqueness of the positive semidefinite square root (Theorem 15.3.2) ensures that $UAU^* = B$, or $UA = BU$. Since $UA = BV$ as well, and B is invertible, it follows that $V = U$. □

Theorem 16.2.10 *Let $A \in \mathsf{M}_{m \times n}(\mathbb{F})$ be nonzero, let $r = \operatorname{rank} A$, and let $A = X_1 \Sigma_r Y_1^* = X_2 \Sigma_r Y_2^*$ be compact singular value decompositions.*

(a) *There is a unitary $U \in \mathsf{M}_r(\mathbb{F})$ such that $X_1 = X_2 U$, $Y_1 = Y_2 U$, and $U\Sigma_r = \Sigma_r U$.*

(b) *If A has d distinct positive singular values $s_1 > s_2 > \cdots > s_d > 0$ with respective multiplicities $n_1, n_2, \ldots, n_d$, then $\Sigma_r = s_1 I_{n_1} \oplus s_2 I_{n_2} \oplus \cdots \oplus s_d I_{n_d}$ and U in (a) can be any matrix of the form*

$$U = U_1 \oplus U_2 \oplus \cdots \oplus U_d, \quad \text{in which } U_i \in \mathsf{M}_{n_i} \text{ and } i = 1, 2, \ldots, d;$$

each direct summand U_i is unitary.

Proof (a) Theorem 16.2.6 tells us that $\operatorname{col} X_1 = \operatorname{col} X_2 = \operatorname{col} A$, and Theorem 8.3.9 provides a unitary $U \in \mathsf{M}_r(\mathbb{F})$ such that $X_1 = X_2 U$. Similar reasoning shows that there is a unitary $V \in \mathsf{M}_r(\mathbb{F})$ such that $Y_1 = Y_2 V$. Then $X_2 \Sigma_r Y_2^* = X_1 \Sigma_r Y_1^* = X_2 U \Sigma_r V^* Y_2^*$, so $X_2(\Sigma_r - U\Sigma_r V^*) Y_2^* = 0$. Consequently, $0 = X_2^* X_2 (\Sigma_r - U\Sigma_r V^*) Y_2^* Y_2 = \Sigma_r - U\Sigma_r V^*$, which is equivalent to $U\Sigma_r = \Sigma_r V$. It follows from Lemma 16.2.9 that $U = V$, and hence U commutes with Σ_r.

(b) Since U commutes with Σ_r, a direct sum of distinct scalar matrices, the asserted block diagonal form of U follows from Lemma 3.2.17 and Theorem 7.1.9.b. □

The geometric content of the preceding theorem is that, while the individual left and right singular vectors of A are never uniquely determined, the subspace spanned by the right (respectively, left) singular vectors associated with each distinct singular value of A is uniquely determined. This observation is a familiar fact expressed in a new vocabulary. The subspace of right (respectively, left) singular vectors of A associated with a singular value σ_i is the eigenspace of A^*A (respectively, AA^*) associated with the eigenvalue σ_i^2.

16.3 The Polar Decomposition

In this section, we explore matrix analogs of the polar form $z = |z|e^{i\theta} = e^{i\theta}|z|$ of a nonzero complex number z, in which $e^{i\theta}$ is a complex number with modulus 1. Since unitary matrices are natural analogs of complex numbers with modulus 1, we investigate matrix factorizations of the form $A = PU$ or $A = UQ$, in which P and Q are positive semidefinite and U is unitary. Such factorizations are possible (Theorem 16.3.9), and the factors commute if and only if A is normal.

The following definition is suggested by $|z| = (\bar{z}z)^{1/2}$ for $z \in \mathbb{C}$.

Definition 16.3.1 The *modulus* of $A \in \mathsf{M}_{m \times n}(\mathbb{F})$ is the positive semidefinite matrix $|A| = (A^*A)^{1/2} \in \mathsf{M}_n(\mathbb{F})$.

Example 16.3.2 If $A \in \mathsf{M}_{m \times n}(\mathbb{F})$, then $|A^*| = (AA^*)^{1/2} \in \mathsf{M}_m(\mathbb{F})$.

Example 16.3.3 The modulus of a 1×1 matrix $A = [a]$ is $|A| = [\bar{a}a]^{1/2} = [|a|]$, in which $|a|$ is the modulus of the complex number a.

Example 16.3.4 The modulus of the zero matrix in $\mathsf{M}_{m \times n}$ is 0_n.

Example 16.3.5 The modulus of a column vector $\mathbf{x} \in \mathbb{F}^n$ is $(\mathbf{x}^*\mathbf{x})^{1/2} = \|\mathbf{x}\|_2$, its Euclidean norm.

Example 16.3.6 If $\mathbf{x} \in \mathbb{F}^n$ is nonzero, then the modulus of the row vector $\mathbf{x}^*$ is $(\mathbf{x}\mathbf{x}^*)^{1/2} = \|\mathbf{x}\|_2^{-1}\mathbf{x}\mathbf{x}^* \in \mathsf{M}_n$. To verify this, compute

$$(\|\mathbf{x}\|_2^{-1}\mathbf{x}\mathbf{x}^*)^2 = \|\mathbf{x}\|_2^{-2}\mathbf{x}\mathbf{x}^*\mathbf{x}\mathbf{x}^* = \|\mathbf{x}\|_2^{-2}\|\mathbf{x}\|_2^2\mathbf{x}\mathbf{x}^* = \mathbf{x}\mathbf{x}^*.$$

Example 16.3.7 The moduli of $A = \begin{bmatrix} 0 & 1 \\ 0 & 0 \end{bmatrix}$ and $B = \begin{bmatrix} 0 & 0 \\ 1 & 0 \end{bmatrix}$ are $|A| = \begin{bmatrix} 0 & 0 \\ 0 & 1 \end{bmatrix}$ and $|B| = \begin{bmatrix} 1 & 0 \\ 0 & 0 \end{bmatrix}$. Observe that $|A|\,|B| = \begin{bmatrix} 0 & 0 \\ 0 & 0 \end{bmatrix} \neq \begin{bmatrix} 1 & 0 \\ 0 & 0 \end{bmatrix} = |AB|$, so the modulus function is not multiplicative.

The modulus of a matrix can be expressed as a product of factors extracted from its singular value decomposition.

Lemma 16.3.8 *Let $A \in \mathsf{M}_{m \times n}(\mathbb{F})$ be nonzero, let rank $A = r \geq 1$, and let $A = V\Sigma W^*$ be a singular value decomposition. Let $\sigma_1 \geq \sigma_2 \geq \cdots \geq \sigma_r > 0$ be the positive singular values of A and let $\Sigma_r = \mathrm{diag}(\sigma_1, \sigma_2, \ldots, \sigma_r)$. For any integer $k \geq r$, define $\Sigma_k = \Sigma_r \oplus 0_{k-r} \in \mathsf{M}_k(\mathbb{R})$. Then $|A| = W\Sigma_n W^* \in \mathsf{M}_n(\mathbb{F})$ and $|A^*| = V\Sigma_m V^* \in \mathsf{M}_m(\mathbb{F})$.*

Proof The matrix $\Sigma \in \mathsf{M}_{m \times n}(\mathbb{R})$ is defined in (16.1.7). A computation reveals that $\Sigma^\mathsf{T}\Sigma = \Sigma_r^2 \oplus 0_{n-r}$ and $\Sigma\Sigma^\mathsf{T} = \Sigma_r^2 \oplus 0_{m-r}$, so $(\Sigma^\mathsf{T}\Sigma)^{1/2} = \Sigma_r \oplus 0_{n-r} = \Sigma_n$ and $(\Sigma\Sigma^\mathsf{T})^{1/2} = \Sigma_r \oplus 0_{m-r} = \Sigma_m$. Now compute

$$|A| = (A^*A)^{1/2} = (W\Sigma^\mathsf{T}V^*V\Sigma W^*)^{1/2} = (W\Sigma^\mathsf{T}\Sigma W^*)^{1/2} = W(\Sigma^\mathsf{T}\Sigma)^{1/2}W^* = W\Sigma_n W^*$$

and

$$|A^*| = (AA^*)^{1/2} = (V\Sigma W^*W\Sigma^\mathsf{T}V^*)^{1/2} = (V\Sigma\Sigma^\mathsf{T}V^*)^{1/2} = V(\Sigma\Sigma^\mathsf{T})^{1/2}V^* = V\Sigma_m V^*. \quad \square$$

Theorem 16.3.9 (Polar Decomposition) *Let $A \in \mathsf{M}_{m \times n}(\mathbb{F})$. There is a $U \in \mathsf{M}_{m \times n}(\mathbb{F})$ such that $A = U|A| = |A^*|U$, in which U has orthonormal columns if $m \geq n$ or orthonormal rows if $m \leq n$.*

Proof Let $A = V\Sigma W^*$ be a singular value decomposition with unitary $V \in M_m(\mathbb{F})$ and $W \in M_n(\mathbb{F})$. Suppose that $m \geq n$. Partition $\Sigma = \begin{bmatrix} \Sigma_n \\ 0_{(m-n)\times n} \end{bmatrix}$, in which $\Sigma_n \in M_n(\mathbb{R})$ is diagonal and positive semidefinite. Partition $V = [V_n \ V']$, in which $V_n \in M_{m\times n}(\mathbb{F})$ and $V' \in M_{m\times(m-n)}(\mathbb{F})$. Let $U = V_n W^*$. Then U has orthonormal columns,

$$A = [V_n \ V'] \begin{bmatrix} \Sigma_n \\ 0_{(m-n)\times n} \end{bmatrix} W^* = V_n \Sigma_n W^* = (V_n W^*)(W \Sigma_n W^*) = U|A|,$$

and

$$A = V_n \Sigma_n W^* = (V_n \Sigma_n V_n^*)(V_n W^*) = [V_n \ V'] \begin{bmatrix} \Sigma_n & 0 \\ 0 & 0_{m-n} \end{bmatrix} [V_n \ V']^* U$$

$$= (V\Sigma_m V^*)U = |A^*|U.$$

If $m \leq n$, apply the preceding results to A^*. If $m = n$, then $V_n = V$ and $W_m = W$. $\qquad \square$

Definition 16.3.10 Let $A \in M_{m\times n}(\mathbb{F})$. A factorization $A = U|A|$ as in the preceding theorem is a *right polar decomposition*; $A = |A^*|U$ is a *left polar decomposition* of A.

Example 16.3.11 Using the singular value decomposition in Example 16.1.10 we find that the unitary polar factor of $A = \begin{bmatrix} 2 & 3 \\ 0 & 2 \end{bmatrix}$ is

$$U = VW^* = \begin{bmatrix} \frac{2}{\sqrt{5}} & -\frac{1}{\sqrt{5}} \\ \frac{1}{\sqrt{5}} & \frac{2}{\sqrt{5}} \end{bmatrix} \begin{bmatrix} \frac{1}{\sqrt{5}} & \frac{2}{\sqrt{5}} \\ -\frac{2}{\sqrt{5}} & \frac{1}{\sqrt{5}} \end{bmatrix} = \begin{bmatrix} \frac{4}{5} & \frac{3}{5} \\ -\frac{3}{5} & \frac{4}{5} \end{bmatrix}.$$

Compute $A^*A = \begin{bmatrix} 4 & 6 \\ 6 & 13 \end{bmatrix}$ and $AA^* = \begin{bmatrix} 13 & 6 \\ 6 & 4 \end{bmatrix}$ and use (15.3.9) to obtain $|A| = (A^*A)^{1/2} = \begin{bmatrix} 8/5 & 6/5 \\ 6/5 & 17/5 \end{bmatrix}$ and $|A^*| = (AA^*)^{1/2} = \begin{bmatrix} 17/5 & 6/5 \\ 6/5 & 8/5 \end{bmatrix}$. Right and left polar decompositions of A are

$$\underbrace{\begin{bmatrix} 2 & 3 \\ 0 & 2 \end{bmatrix}}_{} = \underbrace{\begin{bmatrix} \frac{4}{5} & \frac{3}{5} \\ -\frac{3}{5} & \frac{4}{5} \end{bmatrix}}_{U} \underbrace{\begin{bmatrix} \frac{8}{5} & \frac{6}{5} \\ \frac{6}{5} & \frac{17}{5} \end{bmatrix}}_{|A|} = \underbrace{\begin{bmatrix} \frac{17}{5} & \frac{6}{5} \\ \frac{6}{5} & \frac{8}{5} \end{bmatrix}}_{|A^*|} \underbrace{\begin{bmatrix} \frac{4}{5} & \frac{3}{5} \\ -\frac{3}{5} & \frac{4}{5} \end{bmatrix}}_{U}.$$

Figure 16.2 illustrates the actions of $|A|$ and U on the unit disk.

We next investigate the uniqueness of the factors in left and right polar decompositions.

Example 16.3.12 The factorizations

$$A = \begin{bmatrix} 0 & 1 \\ 0 & 0 \end{bmatrix} = \underbrace{\begin{bmatrix} 0 & 1 \\ e^{i\theta} & 0 \end{bmatrix}}_{U} \underbrace{\begin{bmatrix} 0 & 0 \\ 0 & 1 \end{bmatrix}}_{|A|} = \underbrace{\begin{bmatrix} 1 & 0 \\ 0 & 0 \end{bmatrix}}_{|A^*|} \underbrace{\begin{bmatrix} 0 & 1 \\ e^{i\theta} & 0 \end{bmatrix}}_{U}$$

are valid for every $\theta \in [0, 2\pi)$. The unitary factors in polar decompositions of a square matrix need not be unique if the matrix is not invertible.

Theorem 16.3.13 *Let $A \in M_{m\times n}(\mathbb{F})$.*

(a) *If $m \geq n$ and $A = UB$, in which B is positive semidefinite and $U \in M_{m\times n}(\mathbb{F})$ has orthonormal columns, then $B = |A|$. If $\operatorname{rank} A = n$, then $|A|$ is invertible, $U = A|A|^{-1}$, and $\operatorname{col} A = \operatorname{col} U$.*

(b) *If $m \leq n$ and $A = CU$, in which C is positive semidefinite and $U \in M_{m\times n}(\mathbb{F})$ has orthonormal rows, then $C = |A^*|$. If $\operatorname{rank} A = m$, then $|A^*|$ is invertible, $U = |A^*|^{-1}A$, and $\operatorname{row} A = \operatorname{row} U$.*

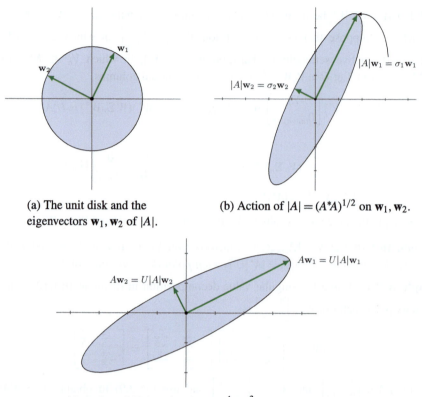

(a) The unit disk and the
eigenvectors $\mathbf{w}_1, \mathbf{w}_2$ of $|A|$.

(b) Action of $|A| = (A^*A)^{1/2}$ on $\mathbf{w}_1, \mathbf{w}_2$.

(c) Action of U (rotation by $\tan^{-1}\left(-\frac{3}{4}\right) \approx -36.9°$) on the
vectors in (b).

Figure 16.2 Right polar decomposition of the matrix A in Example 16.3.11. Notice that (a) and (c) are the same as (a) and (d) in Figure 16.1.

(c) *Let $m = n$ and let $U, V \in \mathsf{M}_n(\mathbb{F})$ be unitary.*

   (i) *If $A = U|A|$, then $A = |A^*|U$.*

   (ii) *If $A = |A^*|V$, then $A = V|A|$.*

   (iii) *If A is invertible and $A = U|A| = |A^*|V$, then $U = V = |A^*|^{-1}A = A|A|^{-1}$.*

Proof (a) If $A = UB$, then $A^*A = BU^*UB = B^2$, so Theorem 15.3.2 ensures that $|A| = (A^*A)^{1/2} = (B^2)^{1/2} = B$. If rank $A = n$, then Theorems 15.1.9 and 15.3.2 tell us that A^*A and $|A|$ are invertible. Consequently, $U = A|A|^{-1}$, $U^*U = |A|^{-1}A^*A|A|^{-1} = |A|^{-1}|A|^2|A|^{-1} = I_n$, and col $U = \operatorname{col}(A|A|^{-1}) = \operatorname{col} A$.

(b) Apply (a) to A^*.

(c.i) Since $A = U|A| = (U|A|U^*)U$, it follows from (b) that $U|A|U^* = |A^*|$.

(c.ii) Since $A = |A^*|V = V(V^*|A^*|V)$, it follows from (a) that $V^*|A^*|V = |A|$.

(c.iii) If $A = U|A|$, then it follows from (c.i) that $A = |A^*|U$. Since $A = |A^*|V$ and $|A^*|$ is invertible, we conclude that $U = V$. One could also invoke Lemma 16.2.9 to obtain this conclusion. $\qquad\square$

Example 16.3.14 Suppose that $A \in \mathsf{M}_{m \times n}$ and rank $A = n$. Let $A = U|A|$ be a polar decomposition, and let $P = A(A^*A)^{-1}A^*$. Then $P = U|A||A|^{-2}|A|U^* = UU^*$. Since the columns of U are an orthonormal basis for col A, Example 8.3.5 ensures that P is the orthogonal projection onto col A. See (8.5.5) for a different presentation of the orthogonal projection onto col A.

Example 16.3.15 Let

$$A = \begin{bmatrix} 1 & 1 \\ 0 & 1 \end{bmatrix}, \tag{16.3.16}$$

and compute $A^*A = \begin{bmatrix} 1 & 1 \\ 1 & 2 \end{bmatrix}$ and $AA^* = \begin{bmatrix} 2 & 1 \\ 1 & 1 \end{bmatrix}$. Proposition 15.3.8 tells us that $|A| = (A^*A)^{1/2} = \begin{bmatrix} 2/\sqrt{5} & 1/\sqrt{5} \\ 1/\sqrt{5} & 3/\sqrt{5} \end{bmatrix}$ and $|A^*| = (AA^*)^{1/2} = \begin{bmatrix} 3/\sqrt{5} & 1/\sqrt{5} \\ 1/\sqrt{5} & 2/\sqrt{5} \end{bmatrix}$. The (unique) unitary factor in the polar decompositions of A is

$$U = |A^*|^{-1}A = \begin{bmatrix} \frac{2}{\sqrt{5}} & \frac{-1}{\sqrt{5}} \\ \frac{-1}{\sqrt{5}} & \frac{3}{\sqrt{5}} \end{bmatrix} \begin{bmatrix} 1 & 1 \\ 0 & 1 \end{bmatrix} = \begin{bmatrix} \frac{2}{\sqrt{5}} & \frac{1}{\sqrt{5}} \\ \frac{-1}{\sqrt{5}} & \frac{2}{\sqrt{5}} \end{bmatrix}. \tag{16.3.17}$$

Then

$$|A^*|U = \begin{bmatrix} \frac{3}{\sqrt{5}} & \frac{1}{\sqrt{5}} \\ \frac{1}{\sqrt{5}} & \frac{2}{\sqrt{5}} \end{bmatrix} \begin{bmatrix} \frac{2}{\sqrt{5}} & \frac{1}{\sqrt{5}} \\ \frac{-1}{\sqrt{5}} & \frac{2}{\sqrt{5}} \end{bmatrix} = \begin{bmatrix} 1 & 1 \\ 0 & 1 \end{bmatrix} = A,$$

$$U|A| = \begin{bmatrix} \frac{2}{\sqrt{5}} & \frac{1}{\sqrt{5}} \\ \frac{-1}{\sqrt{5}} & \frac{2}{\sqrt{5}} \end{bmatrix} \begin{bmatrix} \frac{2}{\sqrt{5}} & \frac{1}{\sqrt{5}} \\ \frac{1}{\sqrt{5}} & \frac{3}{\sqrt{5}} \end{bmatrix} = \begin{bmatrix} 1 & 1 \\ 0 & 1 \end{bmatrix} = A,$$

and

$$|A|U = \begin{bmatrix} \frac{2}{\sqrt{5}} & \frac{1}{\sqrt{5}} \\ \frac{1}{\sqrt{5}} & \frac{3}{\sqrt{5}} \end{bmatrix} \begin{bmatrix} \frac{2}{\sqrt{5}} & \frac{1}{\sqrt{5}} \\ \frac{-1}{\sqrt{5}} & \frac{2}{\sqrt{5}} \end{bmatrix} = \begin{bmatrix} \frac{3}{5} & \frac{4}{5} \\ \frac{-1}{5} & \frac{7}{5} \end{bmatrix} \neq U|A|.$$

The matrix A in the preceding example is not normal. Notice that $|A|U \neq U|A|$, that is, the factors in its polar decomposition do not commute. What can we say about the polar factors of normal matrices?

Theorem 16.3.18 Let $A \in \mathsf{M}_n$ and suppose that $A = U|A|$, in which $U \in \mathsf{M}_n$ is unitary. Then A is normal if and only if $U|A| = |A|U$.

Proof If $U|A| = |A|U$, compute

$$A^*A = |A|^2 = |A|^2UU^* = |A|U|A|U^* = U|A|\,|A|U^* = AA^*,$$

which shows that A is normal. Conversely, if A is normal, then $A^*A = AA^*$ and $|A^*| = (AA^*)^{1/2} = (A^*A)^{1/2} = |A|$. Theorem 16.3.9 says that $A = U|A| = (U|A|U^*)U$. It follows from Theorem 16.3.13.c that $U|A|U^* = |A^*| = |A|$, so $U|A| = |A|U$. □

16.4 Unitary Equivalence and Bidiagonal Matrices

Definition 16.4.1 $A, B \in \mathsf{M}_{m \times n}$ are *unitarily equivalent* if there are unitary $U \in \mathsf{M}_m$ and $V \in \mathsf{M}_n$ such that $A = UBV$.

Unitary equivalence should not be confused with unitary similarity (Definition 7.2.5).

The singular value decomposition $A = V \Sigma W^*$ says that A is unitarily equivalent to Σ, whose singular values are the singular values of A. Unitary equivalence is an equivalence relation (Definition 2.6.11 and P.16.6), for which the following theorem gives a necessary and sufficient condition.

Theorem 16.4.2 $A, B \in M_{m \times n}$ *are unitarily equivalent if and only if they have the same singular values.*

Proof $(\Rightarrow)$ If $A = UBV$, in which $U \in M_m$ and $V \in M_n$ are unitary, then $A^*A = V^*B^*U^*UBV = V^*B^*BV$ is similar to B^*B, so A^*A and B^*B have the same eigenvalues. That is, A and B have the same singular values.
$(\Leftarrow)$ If the singular values of A and B are the same, then they have singular value decompositions $A = V_1 \Sigma W_1^*$ and $B = V_2 \Sigma W_2^*$. Then A is unitarily equivalent to Σ and B is unitarily equivalent to Σ. Since unitary equivalence is an equivalence relation, A and B are unitarily equivalent. $\qquad \square$

The following strategy to compute singular values has proven to be extraordinarily successful: First, reduce to a standard form with many zero entries via unitary equivalences. Then employ a singular-values algorithm that is highly efficient for matrices of the standard form. A special type of upper triangular matrix is the most widely used standard form. The number of its nonzero entries is less than twice the number of its singular values.

Definition 16.4.3 Let $B = [b_{ij}] \in M_{m \times n}$ with $m \geq n$. Then B is *upper bidiagonal* if $b_{ij} = 0$ whenever $j > i + 1, j < i$, or $i > n$, that is,

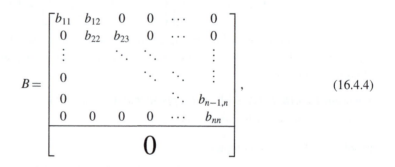

$$B = \begin{bmatrix} b_{11} & b_{12} & 0 & 0 & \cdots & 0 \\ 0 & b_{22} & b_{23} & 0 & \cdots & 0 \\ \vdots & & \ddots & \ddots & & \vdots \\ 0 & & & \ddots & \ddots & \vdots \\ 0 & & & & \ddots & b_{n-1,n} \\ 0 & 0 & 0 & 0 & \cdots & b_{nn} \\ & & & 0 & & \end{bmatrix}, \qquad (16.4.4)$$

in which there can be some zeros on the diagonal and the first superdiagonal.

Theorem 16.4.5 *Let $m \geq n \geq 2$ and let $A \in M_{m \times n}(\mathbb{F})$. There are unitary $U \in M_m(\mathbb{F})$ and $V \in M_n(\mathbb{F})$ such that UAV is upper bidiagonal.*

Proof The proof provides an algorithm to obtain the desired upper bidiagonal matrix via a sequence of left and right multiplications by unitary matrices that does not require computation of any eigenvalues.

Step 1. Partition $A = [\mathbf{a}_1 \ A_1]$, in which $A_1 \in M_{m \times (n-1)}(\mathbb{F})$. If $\mathbf{a}_1 = \mathbf{0}$, go to Step 2. If $\mathbf{a}_1 \neq \mathbf{0}$, let $U_1 \in M_m(\mathbb{F})$ be a unitary matrix whose first column is $\mathbf{a}_1 / \|\mathbf{a}_1\|_2$ (Example 7.1.21).

Then $U_1^*A = [U_1^*\mathbf{a}_1 \ U_1^*A_1] = \begin{bmatrix} \|\mathbf{a}_1\|_2 & \mathbf{r}_1^\mathsf{T} \\ 0_{(m-1)\times 1} & A_2 \end{bmatrix}$, in which $\mathbf{r}_1^\mathsf{T}$ is the first row of $U_1^*A_1$ and $A_2 \in$
$M_{(m-1)\times(n-1)}(\mathbb{F})$. If $\mathbf{r}_1 = \mathbf{0}$, go to Step 3; otherwise go to Step 2.

Step 2. Let $V_1 \in M_{n-1}(\mathbb{F})$ be a unitary matrix whose first column is $\overline{\mathbf{r}_1}/\|\mathbf{r}_1\|_2$. Then

$$U_1^*A(I_{n-1}\oplus V_1) = \begin{bmatrix} \|\mathbf{a}_1\|_2 & \mathbf{r}_1^\mathsf{T}V_1 \\ 0_{(n-1)\times 1} & A_2V_1 \end{bmatrix} = \begin{bmatrix} \|\mathbf{a}_1\|_2 & \|\mathbf{r}_1\|_2 & 0_{1\times(n-2)} \\ 0_{(n-1)\times 1} & \mathbf{a}_2 & A_3 \end{bmatrix},$$

in which $A_2V_1 = [\mathbf{a}_2 \ A_3]$.

Step 3. Repeat Steps 1 and 2, alternating between left and right multiplications to create column zero vectors below the diagonal and row zero vectors to the right of the first superdiagonal. Stop when upper bidiagonal form is achieved. □

The preceding proof is in the same spirit as that of Theorem 7.5.3. In this case, however, we have the freedom to choose successive left and right unitary multipliers independently. In Theorem 16.4.5, we used unitary equivalences; in Theorem 7.5.3 we used unitary similarities.

16.5 Problems

P.16.1 Find singular value decompositions for (a) $\begin{bmatrix} 0 & 1 \\ 1 & 0 \end{bmatrix}$, (b) $\begin{bmatrix} 1 & 0 \\ 0 & -1 \end{bmatrix}$, (c) $\begin{bmatrix} 1 & 1 \\ 1 & 1 \end{bmatrix}$, and
(d) $\begin{bmatrix} 0 & -1 \\ 1 & 0 \end{bmatrix}$.

P.16.2 Let $A \in M_n$ be unitary. Find a singular value decomposition and a polar decomposition for A.

P.16.3 Let $A \in M_n$. Show that A is unitary if and only if all its singular values are equal to 1.

P.16.4 Let $A \in M_n$ be normal and let $A = U\Lambda U^*$ be a spectral decomposition. (a) Find a singular value decomposition for A. What are the singular values of A? (b) Find left and right polar decompositions for A.

P.16.5 Let $A \in M_n$. Show that the adjoint of a singular value decomposition of A is a singular value decomposition of the adjoint of A.

P.16.6 Show that unitary equivalence is an equivalence relation.

P.16.7 Find a singular value decomposition and left/right polar decompositions for

$$A = \begin{bmatrix} 0 & 2 & 0 \\ 0 & 0 & 3 \\ 0 & 0 & 0 \end{bmatrix}.$$

P.16.8 If $A \in M_n$ and $A = V\Sigma W^*$ is a singular value decomposition, find a singular value decomposition for AA^*A.

P.16.9 Let $r \geq 1$. Let $X \in M_{m\times r}$ and $Y \in M_{n\times r}$ have orthonormal columns. Let $\Sigma_r = \text{diag}(\sigma_1, \sigma_2, \ldots, \sigma_r)$ with $\sigma_1 \geq \sigma_2 \geq \cdots \geq \sigma_r > 0$. Show that the diagonal entries of Σ_r are the positive singular values of $X\Sigma_rY^*$.

P.16.10 Let $A \in M_n$. Show that there is a singular value decomposition $A = V\Sigma W^*$ and a spectral decomposition $A = U\Lambda U^*$ such that $U = V = W$ and $\Sigma = \Lambda$ if and only if A is positive semidefinite.

P.16.11 Let $a, b \in \mathbb{C}$ and let $C = \begin{bmatrix} a & -b \\ b & a \end{bmatrix}$. (a) Show that C is normal and its eigenvalues are $a \pm ib$. (b) Show that the singular values of C are $(|a|^2 + |b|^2 \pm 2|\text{Im}\,\overline{a}b|)^{1/2}$ and verify that these are the absolute values of its eigenvalues. (c) Verify that $\|C\|_F^2 = \sigma_1^2 + \sigma_2^2$

and $|\det C| = \sigma_1 \sigma_2$. (d) If a and b are real, show that C is a real scalar multiple of a real orthogonal matrix.

P.16.12 Let $A \in M_m$ and $B \in M_n$. Let $A = V\Sigma_A W^*$ and $B = X\Sigma_B Y^*$ be singular value decompositions. Find a singular value decomposition of $A \oplus B$.

P.16.13 (a) Find a matrix that has equal singular values but distinct eigenvalues. (b) Find a matrix that has equal eigenvalues but distinct singular values.

P.16.14 Let $A \in M_n$ be normal. (a) If A has distinct singular values, show that it has distinct eigenvalues. (b) What can you say about the converse?

P.16.15 Let $A = [a_{ij}] \in M_n$ have rank $r \geq 1$ and let $A = V\Sigma W^*$ be a singular value decomposition in which $\Sigma = \text{diag}(\sigma_1, \sigma_2, \ldots, \sigma_n)$, $V = [v_{ij}]$, and $W = [w_{ij}]$. Show that $a_{ij} = \sum_{k=1}^{r} \sigma_k v_{ik} \overline{w_{jk}}$ for $i, j = 1, 2, \ldots, n$.

P.16.16 (a) Describe an algorithm, similar to the one employed in the proof of Theorem 16.1.5, that begins with a spectral decomposition $AA^* = V\Sigma^2 V^*$ and produces a singular value decomposition of A. (b) Can any matrix V of orthonormal eigenvectors of AA^* be a factor in a singular value decomposition $A = V\Sigma W^*$? (c) Can a singular value decomposition of A be obtained by finding a spectral decomposition $A^*A = W\Sigma^2 W^*$, a spectral decomposition $AA^* = V\Sigma^2 V^*$, and setting $A = V\Sigma W^*$? Discuss. *Hint*: Consider $A = I$.

P.16.17 Let $A \in M_{m \times n}(\mathbb{F})$ with $m \geq n$. Show that there are orthonormal vectors $\mathbf{x}_1, \mathbf{x}_2, \ldots, \mathbf{x}_n \in \mathbb{F}^n$ such that $A\mathbf{x}_1, A\mathbf{x}_2, \ldots, A\mathbf{x}_n$ are orthogonal.

P.16.18 The proof of Theorem 16.2.2 shows how to construct a compact singular value decomposition from a given singular value decomposition. Describe how to construct a singular value decomposition from a given compact singular value decomposition.

P.16.19 Let $\mathbf{x} \in \mathbb{F}^n$ be nonzero. Find singular value decompositions, compact singular value decompositions, and polar decompositions of the $n \times 1$ matrix $\mathbf{x}$ and the $1 \times n$ matrix $\mathbf{x}^*$.

P.16.20 Let $A \in M_{m \times n}(\mathbb{F})$ have rank $r \geq 1$ and let $A = V_1 \Sigma W_1^* = V_2 \Sigma W_2^*$ be singular value decompositions. Suppose that A has d distinct positive singular values $s_1 > s_2 > \cdots > s_d > 0$ and that $\Sigma_r = s_1 I_{n_1} \oplus s_2 I_{n_2} \oplus \cdots \oplus s_d I_{n_d}$. (a) Show that

$$V_1 = V_2(U_1 \oplus U_2 \oplus \cdots \oplus U_d \oplus Z_1) = V_2(U \oplus Z_1), \text{ and}$$
$$W_1 = W_2(U_1 \oplus U_2 \oplus \cdots \oplus U_d \oplus Z_2) = W_2(U \oplus Z_2),$$

in which $U_1 \in M_{n_1}(\mathbb{F})$, $U_2 \in M_{n_2}(\mathbb{F})$, \ldots, $U_d \in M_{n_d}(\mathbb{F})$, $Z_1 \in M_{m-r}(\mathbb{F})$, and $Z_2 \in M_{n-r}(\mathbb{F})$. The summand Z_1 is absent if $r = m$; Z_2 is absent if $r = n$. (b) Why must $U_1, U_2, \ldots, U_d$ be unitary?

P.16.21 Show that Lemma 16.2.9 is false if the hypothesis "A and B are positive definite" is replaced by "A and B are Hermitian and invertible."

P.16.22 Show that Lemma 16.2.9 is false if the hypothesis "U and V are unitary" is replaced by "U and V are invertible."

P.16.23 Let $A \in M_n(\mathbb{F})$, $m > n$, and $B \in M_{m \times n}(\mathbb{F})$. If $\text{rank} B = n$, use a singular value decomposition of B to show that there is a unitary $V \in M_m(\mathbb{F})$ and an invertible $S \in M_n(\mathbb{F})$ such that $BAB^* = V(SAS^* \oplus 0_{m-n})V^*$.

P.16.24 Let $A \in M_n(\mathbb{F})$ and let $A = V\Sigma W^*$ be a singular value decomposition, in which $\Sigma = \text{diag}(\sigma_1, \sigma_2, \ldots, \sigma_n)$. The eigenvalues of the $n \times n$ Hermitian matrix A^*A are $\sigma_1^2, \sigma_2^2, \ldots, \sigma_n^2$. (a) Use P.10.17 to show that the eigenvalues of the $2n \times 2n$ Hermitian matrix $B = \begin{bmatrix} 0 & A \\ A^* & 0 \end{bmatrix}$ are $\pm\sigma_1, \pm\sigma_2, \ldots, \pm\sigma_n$. (b) Compute the block matrix $C = (V \oplus W)^* B(V \oplus W)$. (c) Show that C is permutation similar to

$$\begin{bmatrix} 0 & \sigma_1 \\ \sigma_1 & 0 \end{bmatrix} \oplus \begin{bmatrix} 0 & \sigma_2 \\ \sigma_2 & 0 \end{bmatrix} \oplus \cdots \oplus \begin{bmatrix} 0 & \sigma_n \\ \sigma_n & 0 \end{bmatrix}.$$

Deduce that the eigenvalues of B are $\pm\sigma_1, \pm\sigma_2, \ldots, \pm\sigma_n$.

P.16.25 Let $A \in M_n$ and $a \in \mathbb{R}$. Let $\sigma_1, \sigma_2, \ldots, \sigma_n$ be the singular values of A. Show that the eigenvalues of $B = \begin{bmatrix} A^*A & A^* \\ A & aI_n \end{bmatrix}$ are $\frac{1}{2}(\sigma_i^2 + a \pm \sqrt{(\sigma_i^2 - a)^2 + 4\sigma_i^2})$ for $i = 1, 2, \ldots, n$.

P.16.26 Let $A \in M_2$ be nonzero and let $s = (\|A\|_F^2 + 2|\det A|)^{1/2}$. (a) Show that $|A^*| = \frac{1}{s}(AA^* + |\det A|I)$ and $|A| = \frac{1}{s}(A^*A + |\det A|I)$. (b) Use (a) to verify the entries of $|A|$ and $|A^*|$ in Example 16.3.12.

P.16.27 Let $A, B \in M_n(\mathbb{F})$. (a) Show that $AA^* = BB^*$ if and only if there is a unitary $U \in M_n(\mathbb{F})$ such that $A = BU$. (b) Deduce from (a) that $A^*A = B^*B$ if and only if there is a unitary $U \in M_n(\mathbb{F})$ such that $A = UB$.

P.16.28 Suppose that $A \in M_n$ is not invertible, let $\mathbf{x} \in \mathbb{C}^n$ be a unit vector such that $A^*\mathbf{x} = \mathbf{0}$, let $A = |A^*|U$ be a polar decomposition, and consider the Householder matrix $U_\mathbf{x} = I - 2\mathbf{x}\mathbf{x}^*$. Show that (a) $|A^*|U_\mathbf{x} = |A^*|$, (b) $U_\mathbf{x}U$ is unitary, (c) $A = |A^*|(U_\mathbf{x}U)$ is a polar decomposition, and (d) $U \neq U_\mathbf{x}U$.

P.16.29 Let $A \in M_n(\mathbb{F})$. Prove that the following are equivalent: (a) A is invertible. (b) The unitary factor in a left polar decomposition $A = |A^*|U$ is unique. (c) The unitary factor in a right polar decomposition $A = U|A|$ is unique.

P.16.30 Let $a, b \in \mathbb{C}$ be nonzero scalars and let $\theta \in \mathbb{R}$. Let

$$A = \begin{bmatrix} 0 & a & 0 \\ 0 & 0 & b \\ 0 & 0 & 0 \end{bmatrix}, \quad Q = \begin{bmatrix} 0 & 0 & 0 \\ 0 & |a| & 0 \\ 0 & 0 & |b| \end{bmatrix}, \quad \text{and} \quad U_\theta = \begin{bmatrix} 0 & \frac{a}{|a|} & 0 \\ 0 & 0 & \frac{b}{|b|} \\ e^{i\theta} & 0 & 0 \end{bmatrix}.$$

(a) Show that Q is positive semidefinite, U_θ is unitary, and $A = U_\theta Q$ for all $\theta \in \mathbb{R}$. (b) Find a positive semidefinite P such that $A = PU_\theta$ for all $\theta \in \mathbb{R}$. (c) Describe all values of a and b for which A is normal.

P.16.31 Let $A \in M_n(\mathbb{F})$. Give three proofs that $A(A^*A)^{1/2} = (AA^*)^{1/2}A$ using (a) the singular value decomposition, (b) right and left polar decomposition, and (c) the fact that $(A^*A)^{1/2} = p(A^*A)$ for some polynomial p.

P.16.32 Let $W, S \in M_n$. Suppose that W is unitary, S is invertible (but not necessarily unitary), and SWS^* is unitary. Show that W and SWS^* are unitarily similar. *Hint:* Represent $S = |S^*|U$ as a polar decomposition and use Theorem 16.3.13.

P.16.33 Let $A, B \in M_n$. Theorem 10.7.2 ensures that AB and BA have the same eigenvalues. Do they have the same singular values? Let $A = \begin{bmatrix} 0 & 0 \\ 1 & 0 \end{bmatrix}$ and $B = \begin{bmatrix} 0 & 0 \\ 0 & 2 \end{bmatrix}$. What are the singular values of AB and BA? Which matrix is not normal?

P.16.34 Let $A, B \in M_n$ and suppose that $AB = 0$. (a) Give an example to show that $BA \neq 0$ is possible. Which of the matrices in your example is not normal? (b) If A and B are Hermitian, prove that $BA = 0$. (c) If A and B are normal, use the polar decomposition to show that $|A| |B| = 0$. Deduce that $BA = 0$. (d) If A and B are normal, use P.15.58 to show that $BA = 0$.

P.16.35 Let $A, B \in M_n$ be normal. Use the polar decomposition and Theorem 17.3.12 to show that AB and BA have the same singular values. Where does this argument fail

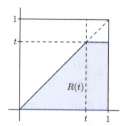

Figure 16.3 The region $R(t)$ in P.16.37.

if either A or B is not normal? *Hint:* Why are the singular values of AB the same as those of $|A|\,|B|$ and $(|A|\,|B|)^*$?

P.16.36 Let $A \in M_{m \times n}(\mathbb{F})$ with $m \geq n$. Let $A = U|A|$ be a polar decomposition. Show that $\|A\mathbf{x}\|_2 = \||A|\mathbf{x}\|_2$ for all $\mathbf{x} \in \mathbb{F}^n$.

P.16.37 Let T be the Volterra operator on $C[0,1]$; see P.6.12, P.9.30, and P.14.29.
(a) Show that

$$(TT^*f)(t) = \int_0^1 \min(t,s)f(s)\,ds \qquad (16.5.1)$$

and

$$(T^*Tf)(t) = \int_0^1 \min(1-t, 1-s)f(s)\,ds.$$

Hint: Express $(TT^*f)(t)$ as a double integral over the region $R(t)$ in Figure 16.3.
(b) Deduce that

$$\langle T^*f, T^*f \rangle = \int_0^1 \int_0^1 \min(t,s)f(s)\overline{f(t)}\,ds\,dt \geq 0 \quad \text{for all } f \in C[0,1].$$

(c) For $n = 0, 1, 2\ldots$, show that $(\pi^{-2}(n + \frac{1}{2})^{-2}, \sin(\pi(n + \frac{1}{2})t))$ is an eigenpair of TT^*. For a discrete analog of (16.5.1), see P.15.35.

P.16.38 Let $A \in M_m$ and $B \in M_n$, and let $A = V\Sigma_A W^*$ and $B = X\Sigma_B Y^*$ be singular value decompositions. (a) Show that $A \otimes B = (V \otimes X)(\Sigma_A \otimes \Sigma_B)(W \otimes Y)^*$. (b) What are the singular values of $A \otimes B$? (c) Deduce from (b) that $\mathrm{rank}(A \otimes B) = (\mathrm{rank}\,A)(\mathrm{rank}\,B)$.

P.16.39 Let $A, B \in M_n$ and let $A = U|A|$ and $B = V|B|$ be polar decompositions. Show that (a) $|A^*| = U|A|U^*$, (b) $U|A|^{1/2} = |A^*|^{1/2}U$, (c) $A = |A^*|^{1/2}U|A|^{1/2}$, and (d)

$$(A + B) \oplus 0_n = \begin{bmatrix} |A^*|^{1/2} & |B^*|^{1/2} \\ 0 & 0 \end{bmatrix} \begin{bmatrix} U & 0 \\ 0 & V \end{bmatrix} \begin{bmatrix} |A|^{1/2} & 0 \\ |B|^{1/2} & 0 \end{bmatrix}.$$

(e) Show that $\mathrm{rank}(A + B) \leq \mathrm{rank}\,A + \mathrm{rank}\,B$.

P.16.40 If $A \in M_n$ and $A = V\Sigma W^*$ is a singular value decomposition, what is a singular value decomposition of $\mathrm{adj}\,A$?

16.6 Notes

For an explicit formula for a 2×2 unitary factor to accompany the positive semidefinite factors in P.16.26, see [HJ13, 7.3.P28].

In operator theory, *unitary equivalence* refers to the notion we have called *unitary similarity*.

Fast and accurate numerical computation of matrix singular values first became possible in 1965 when Gene Golub and William Kahan developed the strategy described in Section 16.4. Their original paper is [GK65].

16.7 Some Important Concepts

- Singular values.
- Singular value decomposition.
- Compact singular value decomposition.
- Uniqueness of the factors in a compact singular value decomposition.
- Modulus of a matrix.
- Left and right polar decompositions.
- Uniqueness of the factors in a polar decomposition.
- Polar factors of a normal matrix.
- Unitary equivalence of a matrix to an upper bidiagonal matrix.

In this chapter, we investigate applications and consequences of the singular value decomposition. For example, it provides a systematic way to approximate a matrix by a matrix of lower rank. It also permits us to define a generalized inverse for matrices that are not invertible (and need not even be square). The singular value decomposition has a pleasant special form for complex symmetric matrices.

The largest singular value is especially important; it turns out to be a norm (the spectral norm) on matrices. We use the spectral norm to study how the solution of a linear system changes if the system is perturbed, and how the eigenvalues of a matrix can change if it is perturbed.

17.1 Singular Values and Approximations

The outer-product representation (3.1.17) ensures that a compact singular value decomposition (16.2.1) of a rank-r matrix $A \in \mathsf{M}_{m \times n}(\mathbb{F})$ can be written as

$$A = V_r \Sigma_r W_r^* = \sum_{i=1}^{r} \sigma_i \mathbf{v}_i \mathbf{w}_i^*, \tag{17.1.1}$$

in which $V_r = [\mathbf{v}_1 \ \mathbf{v}_2 \ \dots \ \mathbf{v}_r] \in \mathsf{M}_{m \times r}(\mathbb{F})$ and $W_r = [\mathbf{w}_1 \ \mathbf{w}_2 \ \dots \ \mathbf{w}_r] \in \mathsf{M}_{n \times r}(\mathbb{F})$. The summands are pairwise orthogonal with respect to the Frobenius inner product since

$$
\begin{aligned}
\langle \sigma_i \mathbf{v}_i \mathbf{w}_i^*, \sigma_j \mathbf{v}_j \mathbf{w}_j^* \rangle_{\mathrm{F}} &= \operatorname{tr}((\sigma_j \mathbf{v}_j \mathbf{w}_j^*)^* (\sigma_i \mathbf{v}_i \mathbf{w}_i^*)) \\
&= \operatorname{tr}(\sigma_i \sigma_j \mathbf{w}_j \mathbf{v}_j^* \mathbf{v}_i \mathbf{w}_i^*) \\
&= \sigma_i \sigma_j \delta_{ij} \operatorname{tr} \mathbf{w}_j \mathbf{w}_i^* = \sigma_i \sigma_j \delta_{ij} \mathbf{w}_i^* \mathbf{w}_j \\
&= \sigma_i \sigma_j \delta_{ij}
\end{aligned}
\tag{17.1.2}
$$

for $i, j = 1, 2, \dots, r$. They are decreasing with respect to the Frobenius norm:

$$\| \sigma_i \mathbf{v}_i \mathbf{w}_i^* \|_{\mathrm{F}}^2 = \sigma_i^2 \geq \sigma_{i+1}^2 = \| \sigma_{i+1} \mathbf{v}_{i+1} \mathbf{w}_{i+1}^* \|_{\mathrm{F}}^2.$$

Orthogonality facilitates computation of the Frobenius norm of sums such as (17.1.1):

$$\| A \|_{\mathrm{F}}^2 = \left\| \sum_{i=1}^{r} \sigma_i \mathbf{v}_i \mathbf{w}_i^* \right\|_{\mathrm{F}}^2 = \sum_{i=1}^{r} \| \sigma_i \mathbf{v}_i \mathbf{w}_i^* \|_{\mathrm{F}}^2 = \sum_{i=1}^{r} \sigma_i^2.$$

Monotonicity suggests that we approximate the rank-r matrix A by a rank-k matrix of the form

$$A_k = \sum_{i=1}^{k} \sigma_i \mathbf{v}_i \mathbf{w}_i^*, \quad \text{in which } k = 1, 2, \dots, r, \tag{17.1.3}$$

obtained by truncating the sum (17.1.1). The orthogonality relations (17.1.2) also facilitate computation of the norm of the residual in this approximation:

$$\|A - A_k\|_F^2 = \left\| \sum_{i=k+1}^{r} \sigma_i \mathbf{v}_i \mathbf{w}_i^* \right\|_F^2 = \sum_{i=k+1}^{r} \sigma_i^2 \|\mathbf{v}_i \mathbf{w}_i^*\|_F^2 = \sum_{i=k+1}^{r} \sigma_i^2. \tag{17.1.4}$$

If m, n, and r are large, and if the singular values of A decrease rapidly, then A_k can be close to A in norm for modest values of k. In order to store or transmit A itself, mn numbers are required; $r(m + n + 1)$ numbers are required to construct A from its compact singular value decomposition; $k(m+n+1)$ numbers are required to construct the approximation A_k from the representation (17.1.3). If k is small compared to m and n, this is a big savings.

With the help of the outer-product representation (3.1.17) again, we can write the truncated sum (17.1.3) as a *truncated singular value decomposition*

$$A_k = V_k \Sigma_k W_k^*, \quad \text{in which } k = 1, 2, \ldots, r.$$

The columns of $V_k \in \mathsf{M}_{m \times k}(\mathbb{F})$ and $W_k \in \mathsf{M}_{n \times k}(\mathbb{F})$ are, respectively, the first k columns of V_r and W_r, respectively, and $\Sigma_k = \operatorname{diag}(\sigma_1, \sigma_2, \ldots, \sigma_k)$.

Example 17.1.5 Figure 17.1 illustrates how truncated singular value decompositions can be used to construct approximations to a familiar image. The Mona Lisa is represented as a real 800×800 matrix of digitized gray scale values. Using only 2.5% of the original data, a recognizable image is recovered; with 15% of the original data an excellent image appears. Figure 17.2 shows that the singular values decay rapidly, which explains the quality of our low-rank approximations.

Original image, Rank: 800   Data used: 2.50%, Rank: 10   Data used: 5.00%, Rank: 20

Data used: 10.00%, Rank: 40   Data used: 15.00%, Rank: 60   Data used: 25.00%, Rank: 100

Figure 17.1 The original Mona Lisa image is represented by a full-rank 800×800 matrix A. The rank-k approximations $A_k = \sum_{i=1}^{k} \sigma_i \mathbf{v}_i \mathbf{w}_i^\mathsf{T}$ are displayed for $k = 10, 20, 40, 60, 100$.

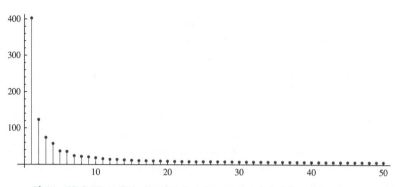

Figure 17.2 Plot of the first 50 singular values of the Mona Lisa image.

17.2 The Spectral and Frobenius Norms

Imagine that some physical or economic process is modeled as a linear transformation $\mathbf{x} \mapsto A\mathbf{x}$, in which $A \in \mathsf{M}_{m\times n}(\mathbb{F})$. We want to know if a likely range of inputs $\mathbf{x}$ could result in outputs $A\mathbf{x}$ that exceed specified limits. Could a certain level of engine power and rate of climb result in vibrations severe enough to tear the wings off a new type of aircraft? Could a proposed program of future bond purchases by the Federal Reserve result in housing price inflation that exceeds 5%? The singular value decomposition can help us bound the size of the outputs of a linear model as a scalar multiple of the size of the inputs.

Let $A \in \mathsf{M}_{m\times n}(\mathbb{F})$ have rank r and let $A = V\Sigma W^*$ be a singular value decomposition. The transformation $\mathbf{x} \mapsto A\mathbf{x}$ is a composition of a unitary transformation $\mathbf{x} \mapsto W^*\mathbf{x}$, scaling along the coordinate axes $W^*\mathbf{x} \mapsto \Sigma(W^*\mathbf{x})$, and another unitary transformation $\Sigma W^*\mathbf{x} \mapsto V(\Sigma W^*\mathbf{x})$. The two unitary transformations preserve Euclidean norms, but the scaling transformation due to the singular value matrix Σ can change the norm of a vector. This qualitative observation is motivation for some careful analysis.

Definition 17.2.1 Let $X \in \mathsf{M}_{m\times n}(\mathbb{F})$, let $q = \min\{m, n\}$, and let $\sigma_1(X) \geq \sigma_2(X) \geq \cdots \geq \sigma_q(X) \geq 0$ be the singular values of X. Define $\sigma_{\max}(X) = \sigma_1(X)$.

Theorem 17.2.2 Let $A \in \mathsf{M}_{m\times n}(\mathbb{F})$, let $q = \min\{m, n\}$, and let $A = V\Sigma W^*$ be a singular value decomposition, in which $V \in \mathsf{M}_m(\mathbb{F})$ and $W = [\mathbf{w}_1\ \mathbf{w}_2\ \ldots\ \mathbf{w}_n] \in \mathsf{M}_n(\mathbb{F})$ are unitary.

(a) $\sigma_{\max}(I_n) = 1$ and $\|I_n\|_{\mathrm{F}} = \sqrt{n}$.

(b) $\|A\mathbf{x}\|_2 \leq \sigma_{\max}(A)$ for all $\mathbf{x} \in \mathbb{F}^n$ such that $\|\mathbf{x}\|_2 = 1$, with equality for $\mathbf{x} = \mathbf{w}_1$.

(c) $\sigma_{\max}(A) = \max_{\|\mathbf{x}\|_2 = 1} \|A\mathbf{x}\|_2 = \|A\mathbf{w}_1\|_2$.

(d) $\sigma_{\max}(A) \leq \|A\|_{\mathrm{F}}$, with equality if and only if $\operatorname{rank} A \leq 1$.

(e) $\|A\mathbf{x}\|_2 \leq \|A\|_{\mathrm{F}}\|\mathbf{x}\|_2$ for all $\mathbf{x} \in \mathbb{F}^n$.

Proof Let $\operatorname{rank} A = r$. Because all the assertions about A are vacuous if $A = 0$, we may assume that $r \geq 1$.

(a) The largest eigenvalue of $I_n^* I_n$ is 1 and $\operatorname{tr} I_n^* I_n = n$.

(b) and (c) Let $\Sigma = \mathrm{diag}(\sigma_1, \sigma_2, \ldots, \sigma_n)$ and $\|\mathbf{x}\|_2 = 1$. Then

$$\|A\mathbf{x}\|_2^2 = \|V\Sigma W^*\mathbf{x}\|_2^2 = \|\Sigma W^*\mathbf{x}\|_2^2 = \left\|[\sigma_i\langle\mathbf{x},\mathbf{w}_i\rangle]\right\|_2^2 = \sum_{i=1}^{r}\sigma_i^2|\langle\mathbf{x},\mathbf{w}_i\rangle|^2$$

$$\leq \sigma_{\max}^2(A)\sum_{i=1}^{r}|\langle\mathbf{x},\mathbf{w}_i\rangle|^2 \leq \sigma_{\max}^2(A)\sum_{i=1}^{n}|\langle\mathbf{x},\mathbf{w}_i\rangle|^2$$

$$= \sigma_{\max}^2(A)\|W\mathbf{x}\|_2^2 = \sigma_{\max}^2(A)\|\mathbf{x}\|_2^2 = \sigma_{\max}^2(A),$$

in which both inequalities are equalities for $\mathbf{x} = \mathbf{w}_1$.

(d) Observe that $\sigma_{\max}^2(A) = \sigma_1^2 \leq \sigma_1^2 + \sigma_2^2 + \cdots + \sigma_r^2 = \|A\|_F^2$ (Theorem 16.1.4) with equality if and only if $\sigma_2^2 = \cdots = \sigma_r^2 = 0$, which occurs if and only if $r \leq 1$.

(e) This follows from (b) and (d). $\qquad\qquad\qquad\qquad\qquad\qquad\qquad\qquad\qquad\qquad\qquad\square$

In addition, the function $\sigma_{\max} : \mathsf{M}_{m\times n} \to [0,\infty)$ has the following properties.

Theorem 17.2.3 *Let $A, B \in \mathsf{M}_{m\times n}(\mathbb{F})$.*

(a) $\sigma_{\max}(A)$ *is real and nonnegative.*

(b) $\sigma_{\max}(A) = 0$ *if and only if $A = 0$.*

(c) $\sigma_{\max}(cA) = |c|\sigma_{\max}(A)$ *for all $c \in \mathbb{F}$.*

(d) $\sigma_{\max}(A + B) \leq \sigma_{\max}(A) + \sigma_{\max}(B)$.

Proof (a) Every singular value is real and nonnegative by definition.

(b) Since singular values are decreasingly ordered, $\sigma_{\max}(A) = 0$ if and only if all $\sigma_i(A) = 0$, which occurs if and only if $\|A\|_F = 0$, that is, $A = 0$.

(c) The largest eigenvalue of $(cA)^*(cA) = |c|^2 A^*A$ is $|c|^2\sigma_1(A)^2$.

(d) Theorem 17.2.2.b ensures that there are unit vectors $\mathbf{x}$, $\mathbf{y}$, and $\mathbf{z}$ such that for any unit vector $\mathbf{u}$ we have $\|A\mathbf{u}\|_2 \leq \|A\mathbf{x}\|_2 = \sigma_{\max}(A)$, $\|B\mathbf{u}\|_2 \leq \|B\mathbf{y}\|_2 = \sigma_{\max}(B)$, and $\|(A+B)\mathbf{u}\|_2 \leq \|(A+B)\mathbf{z}\|_2 = \sigma_{\max}(A+B)$. Then

$$\sigma_{\max}(A+B) = \|(A+B)\mathbf{z}\|_2 \leq \|A\mathbf{z}\|_2 + \|B\mathbf{z}\|_2$$

$$\leq \|A\mathbf{x}\|_2 + \|B\mathbf{y}\|_2 = \sigma_{\max}(A) + \sigma_{\max}(B). \qquad\qquad\square$$

The preceding theorem says that $\sigma_{\max}(\,\cdot\,) : \mathsf{M}_{m\times n}(\mathbb{F}) \to [0,\infty)$ satisfies the four axioms in Definition 5.5.1. Therefore, this function is a *norm* on the $\mathbb{F}$-vector space $\mathsf{M}_{m\times n}(\mathbb{F})$.

Definition 17.2.4 The function $\|\cdot\|_2 : \mathsf{M}_{m\times n} \to [0,\infty)$ defined by $\|A\|_2 = \sigma_{\max}(A)$ is the *spectral norm.*

Which is better, the spectral norm or the Frobenius norm? It depends. If $\mathrm{rank}\,A \geq 2$, then $\|A\|_2 < \|A\|_F$ (Theorem 17.2.2.d), so the spectral norm typically gives tighter upper bounds in an estimate. However, the Frobenius norm of a matrix is likely to be easier to compute than its spectral norm: $\|A\|_F^2$ can be computed via arithmetic operations (see Example 5.4.5), but $\|A\|_2^2$, the largest eigenvalue of A^*A, is typically not so easy to compute.

Example 17.2.5 Consider $A = \begin{bmatrix} 1 & 0 \\ 0 & 0 \end{bmatrix}$, $B = \begin{bmatrix} 0 & 0 \\ 0 & 1 \end{bmatrix}$, $A + B = \begin{bmatrix} 1 & 0 \\ 0 & 1 \end{bmatrix}$, and $A - B = \begin{bmatrix} 1 & 0 \\ 0 & -1 \end{bmatrix}$. Then $\|A+B\|_2^2 + \|A-B\|_2^2 = 2$ and $2\|A\|_2^2 + 2\|B\|_2^2 = 4$. Therefore, the spectral norm is not derived from an inner product; see Theorem 5.4.9.e. However, the Frobenius norm is derived from the inner product $\langle A, B \rangle_F = \operatorname{tr} B^* A$; see Example 5.4.5.

The following theorem says that the Frobenius and spectral norms are *submultiplicative*.

Theorem 17.2.6 *Let $A \in \mathsf{M}_{m \times n}$ and $B \in \mathsf{M}_{n \times k}$.*

(a) $\|AB\|_F \le \|A\|_F \|B\|_F$.

(b) $\|AB\|_2 \le \|A\|_2 \|B\|_2$.

(c) $\|A\mathbf{x}\|_2 \le \|A\|_2 \|\mathbf{x}\|_2$ *for every* $\mathbf{x} \in \mathbb{C}^n$.

Proof (a) Partition $B = [\mathbf{b}_1 \ \mathbf{b}_2 \ \ldots \ \mathbf{b}_k]$ according to its columns. The basic idea in the following computation is that the square of the Frobenius norm of a matrix is not only the sum of the squares of the moduli of its entries, it is also the sum of the squares of the Euclidean norms of its columns. With the help of the upper bound in Theorem 17.2.2.e, compute

$$\|AB\|_F^2 = \left\| [A\mathbf{b}_1 \ A\mathbf{b}_2 \ \ldots \ A\mathbf{b}_k] \right\|_F^2 = \|A\mathbf{b}_1\|_2^2 + \|A\mathbf{b}_2\|_2^2 + \cdots + \|A\mathbf{b}_k\|_2^2$$
$$\le \|A\|_F^2 \|\mathbf{b}_1\|_2^2 + \|A\|_F^2 \|\mathbf{b}_2\|_2^2 + \cdots + \|A\|_F^2 \|\mathbf{b}_k\|_2^2$$
$$= \|A\|_F^2 \left(\|\mathbf{b}_1\|_2^2 + \|\mathbf{b}_2\|_2^2 + \cdots + \|\mathbf{b}_k\|_2^2 \right) = \|A\|_F^2 \|B\|_F^2.$$

(b) For any $\mathbf{x} \in \mathbb{C}^k$, use Theorem 17.2.2.b to compute $\|AB\mathbf{x}\|_2 \le \|A\|_2 \|B\mathbf{x}\|_2 \le \|A\|_2 \|B\|_2 \|\mathbf{x}\|_2$. Let $\mathbf{y}$ be a unit vector such that $\|AB\mathbf{y}\|_2 = \sigma_{\max}(AB)$. Then $\|AB\|_2 = \sigma_{\max}(AB) = \|AB\mathbf{y}\|_2 \le \|A\|_2 \|B\|_2$.

(c) Let $k = 1$ and $B = \mathbf{x}$ in (b). The spectral norm of a vector (an $n \times 1$ matrix) is equal to its Euclidean norm. $\qquad \square$

In the preceding section, we used the Mona Lisa image to illustrate that the truncations A_k in (17.1.3) can provide useful approximations to a rank-r matrix A, even if k is only about 10% of r. Is this the best we can do? Does some other rank-r matrix provide an even better approximation to A? The following theorem says that, with respect to the spectral norm, the answer is no.

Theorem 17.2.7 *Let $A \in \mathsf{M}_{m \times n}(\mathbb{F})$ have rank $r \ge 2$ and let $A = V_r \Sigma_r W_r^*$ be the compact singular value decomposition (16.2.1). Let $1 \le k \le r - 1$, let $B \in \mathsf{M}_{m \times n}(\mathbb{F})$ have rank k, and let $A_k = \sum_{i=1}^k \sigma_i \mathbf{v}_i \mathbf{w}_i^*$. Then $\|A - B\|_2 \ge \|A - A_k\|_2 = \sigma_{k+1}$.*

Proof If we write $\Sigma_r = \Sigma_k \oplus S_{r-k}$, in which $S_{r-k} = \operatorname{diag}(\sigma_{k+1}, \sigma_{k+2}, \ldots, \sigma_r)$, then σ_{k+1} is the largest singular value of S_{r-k}. Compute

$$A - A_k = V_r \begin{bmatrix} \Sigma_k & 0 \\ 0 & S_{r-k} \end{bmatrix} W_r^* - V_r \begin{bmatrix} \Sigma_k & 0 \\ 0 & 0_{r-k} \end{bmatrix} W_r^* = V_r \begin{bmatrix} 0 & 0 \\ 0 & S_{r-k} \end{bmatrix} W_r^*,$$

which is a compact singular value decomposition. Therefore, $\|A - A_k\|_2 = \sigma_{k+1}$.

The rank-nullity theorem (Theorem 4.1.1) ensures that nullity $B = n - k$. Partition $W_r = [\mathbf{w}_1 \ \mathbf{w}_2 \ \ldots \ \mathbf{w}_r]$ and let $\beta = \mathbf{w}_1, \mathbf{w}_2, \ldots, \mathbf{w}_{k+1}$. The subspace intersection theorem (Theorem 4.1.11) ensures that

$$\dim(\operatorname{null} B \cap \operatorname{span} \beta) \ge (n - k) + (k + 1) - n = 1.$$

Let $\mathbf{x}$ be a unit vector in null $B \cap$ span β. Then $B\mathbf{x} = 0$ and $\mathbf{x}$ is orthogonal to $\mathbf{w}_{k+2}, \mathbf{w}_{k+3},$ $\ldots, \mathbf{w}_r$. Theorem 17.2.2.c says that

$$\|A - B\|_2^2 \geq \|(A - B)\mathbf{x}\|_2^2 = \|A\mathbf{x} - B\mathbf{x}\|_2^2 = \|A\mathbf{x}\|_2^2 = \left\| \sum_{i=1}^{r} \sigma_i \mathbf{v}_i \langle \mathbf{x}, \mathbf{w}_i \rangle \right\|_2^2$$

$$= \left\| \sum_{i=1}^{k+1} \sigma_i \mathbf{v}_i \langle \mathbf{x}, \mathbf{w}_i \rangle \right\|_2^2 = \sum_{i=1}^{k+1} \sigma_i^2 |\langle \mathbf{x}, \mathbf{w}_i \rangle|_2^2 \geq \sigma_{k+1}^2 \sum_{i=1}^{k+1} |\langle \mathbf{x}, \mathbf{w}_i \rangle|_2^2 = \sigma_{k+1}^2.$$

In this calculation, we use the facts that $\mathbf{v}_1, \mathbf{v}_2, \ldots, \mathbf{v}_{k+1}$ are orthonormal and the unit vector $\mathbf{x}$ is in a subspace for which β is an orthonormal basis. $\qquad \square$

Example 17.2.8 Let us compute the best rank-1 approximation to

$$A = \begin{bmatrix} 1 & 1 \\ 5 & -1 \\ -1 & 5 \end{bmatrix}.$$

The eigenvalues of $A^*A = \begin{bmatrix} 29 & -9 \\ -9 & 27 \end{bmatrix}$ are $\sigma_1^2 = 36$ and $\sigma_2^2 = 18$. An eigenvector corresponding to σ_1^2 is $\mathbf{w}_1 = [\frac{-1}{\sqrt{2}} \ \frac{1}{\sqrt{2}}]^\mathsf{T}$; this is a right singular vector of A. The corresponding left singular vector is $\mathbf{v}_1 = \frac{1}{\sigma_1} A \mathbf{w}_1 = [0 \ \frac{-1}{\sqrt{2}} \ \frac{1}{\sqrt{2}}]^\mathsf{T}$, so the best rank-1 approximation to A is

$$A_1 = \sigma_1 \mathbf{v}_1 \mathbf{w}_1^* = 6 \begin{bmatrix} 0 \\ \frac{-1}{\sqrt{2}} \\ \frac{1}{\sqrt{2}} \end{bmatrix} \begin{bmatrix} \frac{-1}{\sqrt{2}} & \frac{1}{\sqrt{2}} \end{bmatrix} = \begin{bmatrix} 0 & 0 \\ 3 & -3 \\ -3 & 3 \end{bmatrix}.$$

As a check, compute

$$\begin{bmatrix} 1 & 1 \\ 5 & -1 \\ -1 & 5 \end{bmatrix} - \begin{bmatrix} 0 & 0 \\ 3 & -3 \\ -3 & 3 \end{bmatrix} = \begin{bmatrix} 1 & 1 \\ 2 & 2 \\ 2 & 2 \end{bmatrix} = \begin{bmatrix} 1 \\ 2 \\ 2 \end{bmatrix} \begin{bmatrix} 1 & 1 \end{bmatrix},$$

whose spectral norm is $\sigma_2 = 3\sqrt{2}$; see P.17.4.

17.3 Singular Values and Eigenvalues

Many interesting inequalities involve singular values and eigenvalues; we discuss just a few of them. In this section, it is convenient to index the eigenvalues of $A \in \mathsf{M}_n$ in such a way that $|\lambda_1(A)| = \rho(A)$ (the spectral radius) and their moduli are decreasingly ordered:

$$|\lambda_1(A)| \geq |\lambda_2(A)| \geq \cdots \geq |\lambda_n(A)|.$$

For brevity, we sometimes write $|\lambda_1| \geq |\lambda_2| \geq \cdots \geq |\lambda_n|$.

If A is invertible, then all of its eigenvalues are nonzero. Corollary 11.1.6 ensures that the eigenvalues of A^{-1} are the reciprocals of the eigenvalues of A. Therefore, the decreasingly ordered moduli of the eigenvalues of A^{-1} are

$$|\lambda_n(A)^{-1}| \geq |\lambda_{n-1}(A)^{-1}| \geq \cdots \geq |\lambda_1(A)^{-1}|.$$

Consequently, $\rho(A^{-1}) = |\lambda_n(A)^{-1}|$. There is an analogous identity for singular values.

Definition 17.3.1 Let $\sigma_1(A) \geq \sigma_2(A) \geq \cdots \geq \sigma_n(A) \geq 0$ be the singular values of $A \in \mathsf{M}_n$ and define $\sigma_{\min}(A) = \sigma_n(A)$.

Theorem 17.3.2 *Let* $A \in \mathsf{M}_n$.

(a) *If A is invertible, then* $\|A^{-1}\|_2 = 1/\sigma_{\min}(A)$.

(b) $|\lambda_1(A)\lambda_2(A) \cdots \lambda_n(A)| = |\det A| = \sigma_1(A)\sigma_2(A) \cdots \sigma_n(A)$.

(c) $\rho(A) \leq \|A\|_2$.

(d) $\sigma_{\min}(A) \leq |\lambda_n(A)|$.

(e) $\lambda_n(A) = 0$ *if and only if* $\sigma_{\min}(A) = 0$.

Proof (a) If $A = V\Sigma W^*$ is a singular value decomposition, then $A^{-1} = W\Sigma^{-1}V^*$ and the largest diagonal entry in Σ^{-1} is $1/\sigma_{\min}(A)$. This is the spectral norm of A^{-1}.

(b) Corollary 11.1.3 ensures that $\lambda_1\lambda_2 \cdots \lambda_n = \det A$. Let $A = V\Sigma W^*$ be a singular value decomposition and compute

$$|\det A| = |\det V\Sigma W^*| = |(\det V)(\det \Sigma)(\overline{\det W})| = |\det V|\,(\det \Sigma)\,|\det W| = \sigma_1\sigma_2 \cdots \sigma_n.$$

(c) Let $(\lambda_1, \mathbf{x})$ be an eigenpair of A with $\|\mathbf{x}\|_2 = 1$ and use Theorem 17.2.2.b to compute

$$\rho(A) = |\lambda_1| = \|\lambda_1\mathbf{x}\|_2 = \|A\mathbf{x}\|_2 \leq \|A\|_2\|\mathbf{x}\|_2 = \|A\|_2. \tag{17.3.3}$$

(d) If A is invertible, apply (17.3.3) to A^{-1}:

$$\frac{1}{|\lambda_n(A)|} = |\lambda_1(A^{-1})| = \rho(A^{-1}) \leq \sigma_{\max}(A^{-1}) = \frac{1}{\sigma_{\min}(A)} = \|A^{-1}\|_2,$$

so $\sigma_{\min}(A) \leq |\lambda_n(A)|$. If A is not invertible, then $\det A = 0$, so $\sigma_{\min}(A) = \lambda_n(A) = 0$.

(e) If $\lambda_n(A) = 0$, then (d) implies that $\sigma_{\min}(A) = 0$. If $\sigma_{\min}(A) = 0$, then rank $A < n$, so the rank-nullity theorem implies that nullity $A \geq 1$. Consequently, $0 \in \operatorname{spec} A$ and hence $\lambda_n(A) = 0$. $\square$

The singular values and eigenvalues of a matrix are typically different, but the special cases of normal and positive semidefinite matrices are noteworthy.

Theorem 17.3.4 *Let* $A = [a_{ij}] \in \mathsf{M}_n$.

(a) $|\lambda_1(A)|^2 + |\lambda_2(A)|^2 + \cdots + |\lambda_n(A)|^2 \leq \sigma_1^2(A) + \sigma_2^2(A) + \cdots + \sigma_n^2(A)$.

(b) *A is normal if and only if* $|\lambda_i(A)| = \sigma_i(A)$ *for each* $i = 1, 2, \ldots, n$.

(c) *A is positive semidefinite if and only if* $\lambda_i(A) = \sigma_i(A)$ *for each* $i = 1, 2, \ldots, n$.

Proof (a) Schur's inequality (14.3.2) says that $\sum_{i=1}^n |\lambda_i|^2 \leq \operatorname{tr} A^*A$, and $\operatorname{tr} A^*A = \sum_{i=1}^n \sigma_i^2$.

(b) If $|\lambda_i| = \sigma_i$ for each $i = 1, 2, \ldots, n$, then

$$|\lambda_1|^2 + |\lambda_2|^2 + \cdots + |\lambda_n|^2 = \sigma_1^2 + \sigma_2^2 + \cdots + \sigma_n^2 = \operatorname{tr} A^*A,$$

so Theorem 14.3.1 ensures that A is normal. Conversely, if A is normal, the spectral theorem says that $A = U\Lambda U^*$, in which $\Lambda = \operatorname{diag}(\lambda_1, \lambda_2 \ldots, \lambda_n)$ and U is unitary. Then $A^*A = U\overline{\Lambda}U^*U\Lambda U^* = U\operatorname{diag}(|\lambda_1|^2, |\lambda_2|^2, \ldots, |\lambda_n|^2)U^*$, so the squared singular values of A are $|\lambda_1|^2, |\lambda_2|^2, \ldots, |\lambda_n|^2$.

(c) It follows from (b) that $\lambda_i = \sigma_i$ for each $i = 1, 2, \ldots, n$ if and only if A is normal and has real nonnegative eigenvalues. This is the case if and only if A is Hermitian and has nonnegative eigenvalues (Theorem 14.6.1), that is, if and only if A is positive semidefinite (Theorem 15.1.3). $\qquad\square$

Example 17.3.5 Since the singular values of a normal matrix are the moduli of its eigenvalues, the spectral norm of a normal matrix A equals its spectral radius:

$$\|A\|_2 = \rho(A) \text{ if } A \in \mathsf{M}_n \text{ is normal.} \tag{17.3.6}$$

In particular, if $A = \operatorname{diag}(\lambda_1, \lambda_2, \ldots, \lambda_n)$, then $\|A\|_2 = \max_{1 \le i \le n} |\lambda_i| = \rho(A)$.

Example 17.3.7 Since $\|A\|_2 \ge \rho(A)$ for every $A \in \mathsf{M}_n$ (Theorem 17.3.2.c), it follows from (17.3.6) that A cannot be normal if $\|A\|_2 > \rho(A)$. For example, $B = \begin{bmatrix} 2 & 3 \\ 0 & 2 \end{bmatrix}$ has singular values 4 and 1, and $\rho(B) = 2$, so $\|B\|_2 > \rho(B)$ and hence B is not normal. However, if $\|A\|_2 = \rho(A)$, it does not follow that A is normal. For example, consider the non-normal matrix

$$C = [4] \oplus \begin{bmatrix} 2 & 3 \\ 0 & 2 \end{bmatrix}. \tag{17.3.8}$$

We have $\operatorname{spec} C = \{4, 2\}$, $\operatorname{spec} C^*C = \{16, 1\}$, and $\|C\|_2 = \rho(C)$.

The singular values of a matrix can be any nonnegative real numbers, but the singular values of unitary matrices are severely constrained.

Theorem 17.3.9 *A square matrix A is a scalar multiple of a unitary matrix if and only if* $\sigma_{\max}(A) = \sigma_{\min}(A)$.

Proof If $A = cU$ for some scalar c and a unitary matrix U, then $A^*A = |c|^2 U^*U = |c|^2 I$, so every singular value of A is $|c|$. Conversely, let $A = V\Sigma W^*$ be a singular value decomposition of $A \in \mathsf{M}_n$. The hypothesis is that $\Sigma = sI$ for some $s \ge 0$. Then $A = sVW^*$ is a scalar multiple of the unitary matrix VW^*. $\qquad\square$

Similarity preserves eigenvalues, but it need not preserve singular values. The following example illustrates this point.

Example 17.3.10 Theorem 12.5.3 ensures that

$$A = \begin{bmatrix} 1 & 1 \\ 0 & 1 \end{bmatrix} \quad \text{and} \quad B = \begin{bmatrix} 1 & 2 \\ 0 & 1 \end{bmatrix} \tag{17.3.11}$$

are similar. A computation reveals that $A^*A = \begin{bmatrix} 1 & 1 \\ 1 & 2 \end{bmatrix}$ and $B^*B = \begin{bmatrix} 1 & 2 \\ 2 & 5 \end{bmatrix}$. Since $\operatorname{tr} A^*A \ne \operatorname{tr} B^*B$, the eigenvalues of A^*A and B^*B are not the same. Consequently, A and B do not have the same singular values.

The following theorem reminds us that unitary equivalence preserves singular values. Unitary similarity preserves eigenvalues and singular values.

Theorem 17.3.12 *Let $A \in \mathsf{M}_{m \times n}$, and let $U \in \mathsf{M}_m$ and $V \in \mathsf{M}_n$ be unitary.*

(a) *A and UAV have the same singular values. In particular, these matrices have the same spectral and Frobenius norms.*

(b) *If $m = n$, then A and UAU^* have the same eigenvalues and the same singular values.*

Proof (a) See Theorem 16.4.2. The spectral and Frobenius norms of a matrix are determined by its singular values, so UAV and A have the same spectral and Frobenius norms.

(b) The matrices A and UAU^* are unitarily similar, so they have the same eigenvalues. Part (a) ensures that they have the same singular values. □

How is the sum of the eigenvalues of a square matrix related to the sum of its singular values?

Theorem 17.3.13 *If $A \in M_n$, then $|\operatorname{tr} A| \leq \operatorname{tr} |A|$, that is,*

$$\left| \sum_{k=1}^{n} \lambda_k(A) \right| \leq \sum_{k=1}^{n} \sigma_k(A). \tag{17.3.14}$$

Proof Let $A = [a_{ij}] = V \Sigma W^*$ be a singular value decomposition, in which $V = [v_{ij}]$ and $W = [w_{ij}]$ are unitary, and $\Sigma = \operatorname{diag}(\sigma_1, \sigma_2, \ldots, \sigma_n)$. The assertion that $\operatorname{tr} |A| = \sum_{i=1}^{n} \sigma_i$ follows from Definitions 16.1.2 and 16.3.1. In the following argument, we use the fact that the columns of V and W are unit vectors. A calculation reveals that

$$a_{ii} = \sum_{k=1}^{n} \sigma_k v_{ik} \overline{w_{ik}} \quad \text{for } i = 1, 2, \ldots, n,$$

and

$$\operatorname{tr} A = \sum_{i=1}^{n} a_{ii} = \sum_{i=1}^{n} \sum_{k=1}^{n} \sigma_k v_{ik} \overline{w_{ik}} = \sum_{k=1}^{n} \sigma_k \sum_{i=1}^{n} v_{ik} \overline{w_{ik}} = \sum_{k=1}^{n} \sigma_k \langle \mathbf{v}_k, \mathbf{w}_k \rangle.$$

Now use the triangle and Cauchy–Schwarz inequalities to compute

$$|\operatorname{tr} A| = \left| \sum_{k=1}^{n} \sigma_k \langle \mathbf{v}_k, \mathbf{w}_k \rangle \right| \leq \sum_{k=1}^{n} \sigma_k |\langle \mathbf{v}_k, \mathbf{w}_k \rangle| \leq \sum_{k=1}^{n} \sigma_k \|\mathbf{v}_k\|_2 \|\mathbf{w}_k\|_2 = \sum_{k=1}^{n} \sigma_k(A). \quad \square$$

Example 17.3.15 For $A = \begin{bmatrix} 2 & 3 \\ 0 & 2 \end{bmatrix}$, we have $|\operatorname{tr} A| = 4 < 5 = 4 + 1 = \sigma_1(A) + \sigma_2(A) = \operatorname{tr} |A|$.

We now use the invariance principles in Theorem 17.3.12 to modify (17.3.14) in two different ways.

Theorem 17.3.16 *Let $A = [a_{ij}] \in M_n$. Then*

$$|\lambda_1(A)| + |\lambda_2(A)| + \cdots + |\lambda_n(A)| \leq \sigma_1(A) + \sigma_2(A) + \cdots + \sigma_n(A) \tag{17.3.17}$$

and

$$|a_{11}| + |a_{22}| + \cdots + |a_{nn}| \leq \sigma_1(A) + \sigma_2(A) + \cdots + \sigma_n(A). \tag{17.3.18}$$

Proof To obtain the inequality in (17.3.17), let $U \in M_n$ be unitary and such that $U^*AU = T$ is upper triangular and has diagonal entries $\lambda_1, \lambda_2, \ldots, \lambda_n$; this is Theorem 11.1.1. Then let $\theta_1, \theta_2, \ldots, \theta_n \in \mathbb{R}$ be such that $\lambda_k = e^{i\theta_k}|\lambda_k|$ for $k = 1, 2, \ldots, n$. Let $D = \operatorname{diag}(e^{-i\theta_1}, e^{-i\theta_2}, \ldots, e^{-i\theta_n})$, which is unitary. The matrix DT has main diagonal entries $|\lambda_1|, |\lambda_2|, \ldots, |\lambda_n|$. Because $DT = DU^*AU$, Theorem 17.3.12.a ensures that DT and A have the same singular values. Now apply (17.3.14) to DT:

$$\sum_{k=1}^{n} |\lambda_k| = \operatorname{tr} DT = |\operatorname{tr} DT| \leq \sum_{k=1}^{n} \sigma_k(DT) = \sum_{k=1}^{n} \sigma_k(T) = \sum_{k=1}^{n} \sigma_k(A).$$

Now let $\theta_1, \theta_2, \ldots, \theta_n \in \mathbb{R}$ be such that $a_{kk} = e^{i\theta_k}|a_{kk}|$ for $k = 1, 2, \ldots, n$. Let $E = \text{diag}(e^{-i\theta_1}, e^{-i\theta_2}, \ldots, e^{-i\theta_n})$, which is unitary. Then EA has main diagonal entries $|a_{11}|, |a_{22}|, \ldots, |a_{nn}|$ and the same singular values as A, so an application of (17.3.14) to ET gives the desired inequality:

$$\sum_{k=1}^{n} |a_{kk}| = \text{tr}\, EA = |\text{tr}\, EA| \leq \sum_{k=1}^{n} \sigma_k(EA) = \sum_{k=1}^{n} \sigma_k(A). \qquad \square$$

17.4 The Pseudoinverse

With the help of the singular value decomposition of $A \in \mathsf{M}_{m \times n}$, we can define a matrix $A^\dagger \in \mathsf{M}_{m \times n}$ that has many properties of an inverse. The following lemma is the first step.

Lemma 17.4.1 *Let $A \in \mathsf{M}_{m \times n}(\mathbb{F})$, let $r = \text{rank}\, A \geq 1$, and let $A = X_1 \Sigma_r Y_1^* = X_2 \Sigma_r Y_2^*$ be compact singular value decompositions. Then $Y_1 \Sigma_r^{-1} X_1^* = Y_2 \Sigma_r^{-1} X_2^*$.*

Proof Theorem 16.2.10 ensures that there is a unitary $U \in \mathsf{M}_r$ such that $X_1 = X_2 U$, $Y_1 = Y_2 U$, and $U\Sigma_r = \Sigma_r U$. Then

$$\Sigma_r^{-1} U = \Sigma_r^{-1}(U\Sigma_r)\Sigma_r^{-1} = \Sigma_r^{-1}(\Sigma_r U)\Sigma_r^{-1} = U\Sigma_r^{-1}.$$

It follows that

$$Y_1 \Sigma_r^{-1} X_1^* = Y_2 U \Sigma_r^{-1} U^* X_2^* = Y_2 \Sigma_r^{-1} U U^* X_2^* = Y_2 \Sigma_r^{-1} X_2^*. \qquad \square$$

Definition 17.4.2 Let $A \in \mathsf{M}_{m \times n}(\mathbb{F})$. If $r = \text{rank}\, A \geq 1$ and $A = X\Sigma_r Y^*$ is a compact singular value decomposition, then $A^\dagger = Y\Sigma_r^{-1} X^* \in \mathsf{M}_{n \times m}(\mathbb{F})$ is the *pseudoinverse* (or *Moore–Penrose pseudoinverse*) of A. If $A = 0_{m \times n}$, define $A^\dagger = 0_{n \times m}$.

Although there are compact singular value decompositions $A = X_1 \Sigma_r Y_1^* = X_2 \Sigma_r Y_2^*$ in which $X_1 \neq X_2$ or $Y_1 \neq Y_2$, Lemma 17.4.1 ensures that $Y_1 \Sigma_r^{-1} X_1^* = Y_2 \Sigma_r^{-1} X_2^*$. Therefore, $A^\dagger$ is well-defined and it is correct to speak of the pseudoinverse of A.

Example 17.4.3 The compact singular value decomposition $A = \begin{bmatrix} 0 & 1 \\ 0 & 0 \end{bmatrix} = \begin{bmatrix} 1 \\ 0 \end{bmatrix} [1] [0 \quad 1]$ permits us to compute $A^\dagger = \begin{bmatrix} 0 \\ 1 \end{bmatrix} [1] [1 \quad 0] = \begin{bmatrix} 0 & 0 \\ 1 & 0 \end{bmatrix}$. Observe that $AA^\dagger = \begin{bmatrix} 1 & 0 \\ 0 & 0 \end{bmatrix}$ and $A^\dagger A = \begin{bmatrix} 0 & 0 \\ 0 & 1 \end{bmatrix}$ are orthogonal projections, $AA^\dagger A = \begin{bmatrix} 0 & 1 \\ 0 & 0 \end{bmatrix} = A$, and $A^\dagger AA^\dagger = \begin{bmatrix} 0 & 0 \\ 1 & 0 \end{bmatrix} = A^\dagger$; see P.17.21.

Example 17.4.4 A compact singular value decomposition of a nonzero $\mathbf{x} \in \mathsf{M}_{n \times 1}$ is $\mathbf{x} = (\mathbf{x}/\|\mathbf{x}\|)\,[\|\mathbf{x}\|]\,[1]$. Therefore, $\mathbf{x}^\dagger = [1]\,[\|\mathbf{x}\|^{-1}]\,(\mathbf{x}/\|\mathbf{x}\|)^* = \mathbf{x}^*/\|\mathbf{x}\|^2$. Observe that $\mathbf{x}\mathbf{x}^\dagger = \|\mathbf{x}\|_2^{-2}\mathbf{x}\mathbf{x}^*$ and $\mathbf{x}^\dagger\mathbf{x} = \|\mathbf{x}\|_2^{-2}\mathbf{x}^*\mathbf{x} = 1$ are orthogonal projections, $\mathbf{x}\mathbf{x}^\dagger\mathbf{x} = \mathbf{x}$, and $\mathbf{x}^\dagger\mathbf{x}\mathbf{x}^\dagger = \mathbf{x}^\dagger$. This is a special case of the Penrose identities (17.8.1).

Example 17.4.5 Suppose that $n > r > 1$. Let $B \in \mathsf{M}_r$ be invertible and let $B = V\Sigma W^*$ be a singular value decomposition. Then $\begin{bmatrix} B & 0 \\ 0 & 0_{n-r} \end{bmatrix} = \begin{bmatrix} V \\ 0 \end{bmatrix} \Sigma \begin{bmatrix} V \\ 0 \end{bmatrix}^*$ is a compact singular value decomposition of $B \oplus 0_{n-r}$. Therefore, its pseudoinverse is

$$\begin{bmatrix} B & 0 \\ 0 & 0_{n-r} \end{bmatrix}^\dagger = \begin{bmatrix} V \\ 0 \end{bmatrix} \Sigma^{-1} \begin{bmatrix} W \\ 0 \end{bmatrix}^* = \begin{bmatrix} V\Sigma^{-1}W^* & 0 \\ 0 & 0_{n-r} \end{bmatrix} = \begin{bmatrix} B^{-1} & 0 \\ 0 & 0_{n-r} \end{bmatrix}.$$

Example 17.4.6 If $A \in \mathsf{M}_{m \times n}(\mathbb{F})$ and rank $A = n$ (full column rank), we claim that

$$A^\dagger = (A^*A)^{-1}A^*. \qquad (17.4.7)$$

Let $A = V \Sigma W^* = [V_n \; V'] \begin{bmatrix} \Sigma_n \\ 0 \end{bmatrix} W^*$ be a singular value decomposition. Then $A = V_n \Sigma_n W^*$ is a compact singular value decomposition, $A^* = W \Sigma_n V_n^*$, and $A^*A = W \Sigma_n V_n^* V_n \Sigma_n W^* = W \Sigma_n^2 W^*$. Thus, $(A^*A)^{-1}A^* = W \Sigma_n^{-2} W^* W \Sigma_n V_n^* = W \Sigma_n^{-1} V_n^* = A^\dagger$.

Example 17.4.8 If $A \in \mathsf{M}_{m \times n}(\mathbb{F})$ and rank $A = m$ (full row rank), we claim that

$$A^\dagger = A^*(AA^*)^{-1}. \qquad (17.4.9)$$

Write $A = V \Sigma_m W_m^*$, as in the preceding example, and compute $AA^* = V \Sigma_m^2 V^*$, so $A^*(AA^*)^{-1} = W_m \Sigma_m V^* V \Sigma_m^{-2} V^* = W_m \Sigma_m^{-1} V^* = A^\dagger$.

The following theorem formalizes some of the observations in the preceding two examples.

Theorem 17.4.10 *Let* $A \in \mathsf{M}_{m \times n}(\mathbb{F})$ *be nonzero.*

(a) $\operatorname{col} A^\dagger = \operatorname{col} A^*$ *and* $\operatorname{null} A^\dagger = \operatorname{null} A^*$.

(b) $AA^\dagger \in \mathsf{M}_m(\mathbb{F})$ *is the orthogonal projection of* $\mathbb{F}^m$ *onto* $\operatorname{col} A$ *and* $I - AA^\dagger$ *is the orthogonal projection of* $\mathbb{F}^m$ *onto* $\operatorname{null} A^*$.

(c) $A^\dagger A \in \mathsf{M}_n(\mathbb{F})$ *is the orthogonal projection of* $\mathbb{F}^n$ *onto* $\operatorname{col} A^*$ *and* $I - A^\dagger A$ *is the orthogonal projection of* $\mathbb{F}^n$ *onto* $\operatorname{null} A$.

(d) $AA^\dagger A = A$.

(e) $A^\dagger AA^\dagger = A^\dagger$.

(f) *If* $m = n$ *and* A *is invertible, then* $A^\dagger = A^{-1}$.

(g) $(A^\dagger)^\dagger = A$.

(h) *If* $V \in \mathsf{M}_m$ *and* $W \in \mathsf{M}_n$ *are unitary, then* $(VAW)^\dagger = W^* A^\dagger V^*$.

Proof Let $r = \operatorname{rank} A$ and let $A = X \Sigma_r Y^*$ be a compact singular value decomposition. Then $\Sigma_r \in \mathsf{M}_r$ is invertible, and X and Y have orthonormal columns, so they have full column rank.

(a) We have $A^* = Y \Sigma_r X^*$, so $\operatorname{col} A^* = \operatorname{col} Y = \operatorname{col} A^\dagger$ and $\operatorname{null} A^* = (\operatorname{col} X)^\perp = \operatorname{null} A^\dagger$.

(b) Compute $AA^\dagger = X \Sigma_r Y^* Y \Sigma_r^{-1} X^* = X \Sigma_r \Sigma_r^{-1} X^* = XX^*$. Then invoke (16.2.7) and Example 8.3.5.

(c) Compute $A^\dagger A = Y \Sigma_r^{-1} X^* X \Sigma_r Y^* = Y \Sigma_r^{-1} \Sigma_r Y^* = YY^*$. Then invoke (16.2.8) and Example 8.3.5.

(d) This follows from (b).

(e) This follows from (c) and (a).

(f) If $m = n$ and A is invertible, then rank $A = n$, $\Sigma_n \in \mathsf{M}_n$ is invertible, and $X, Y \in \mathsf{M}_n$ are unitary. Consequently, $A^{-1} = (X \Sigma_n Y^*)^{-1} = Y \Sigma_n^{-1} X^* = A^\dagger$.

(g) Compute $(Y \Sigma_r^{-1} X^*)^\dagger = X \Sigma_r Y^* = A$.

(h) Since $VAW = (VX) \Sigma_r (W^*Y)^*$ is a compact singular value decomposition, $(VAW)^\dagger = (W^*Y) \Sigma_r^{-1} (VX)^* = W^* (Y \Sigma_r^{-1} X^*) V^* = W^* A^\dagger V^*$. $\qquad \square$

If $A \in \mathsf{M}_{m \times n}(\mathbb{F})$ has full column rank, then Theorem 17.4.10.b and Example 17.4.6 ensure that

$$A(A^*A)^{-1}A^*$$

is the projection of $\mathbb{F}^m$ onto col A. If $n = 1$ and $A = \mathbf{u} \in \mathbb{F}^m$ is nonzero, this becomes

$$\mathbf{u}(\mathbf{u}^*\mathbf{u})^{-1}\mathbf{u}^* = \frac{\mathbf{u}\mathbf{u}^*}{\|\mathbf{u}\|_2^2},$$

which is the projection of $\mathbb{F}^m$ onto span$\{\mathbf{u}\}$; see Example 8.3.8.

The pseudoinverse shares some properties of the inverse and adjoint. For example, if A is invertible, then $A^\dagger = A^{-1}$; if A is unitary, then $A^\dagger = A^*$. Theorem 17.4.10.h exhibits an important case in which $(AB)^\dagger = B^\dagger A^\dagger$, but this is not always true; see P.17.30. The following theorem involves an identity for the pseudoinverse that is analogous to the definition of normality.

Theorem 17.4.11 *Let $A \in \mathsf{M}_n$ and suppose that $1 \leq r = \operatorname{rank} A < n$. The following are equivalent:*

(a) $AA^\dagger = A^\dagger A$.

(b) *There is a unitary $U \in \mathsf{M}_n$ and an invertible $B \in \mathsf{M}_r$ such that $A = U \begin{bmatrix} B & 0 \\ 0 & 0_{n-r} \end{bmatrix} U^*$.*

(c) *There is a polynomial p such that $A^\dagger = p(A)$.*

Proof (a) $\Rightarrow$ (b) Let $A = X\Sigma_r Y^*$ be a compact singular value decomposition, so $A^\dagger = Y\Sigma_r^{-1}X^*$. Then

$$AA^\dagger = X\Sigma_r Y^* Y\Sigma_r^{-1}X^* = X\Sigma_r\Sigma_r^{-1}X^* = XX^*$$

and

$$A^\dagger A = Y\Sigma_r^{-1}X^* X\Sigma_r Y^* = Y\Sigma_r^{-1}\Sigma_r Y^* = YY^*.$$

The hypothesis is equivalent to the identity $XI_rX^* = YI_rY^*$, which exhibits two compact singular value decompositions of the same matrix. Theorem 16.2.10.a ensures that there is a unitary $Z \in \mathsf{M}_r$ such that $Y = XZ$. Let $U = [X\ X_2] \in \mathsf{M}_n$ be unitary, and let $B = \Sigma_r Z^*$, which is invertible. Now compute

$$A = X\Sigma_r Y^* = [X\ X_2]\begin{bmatrix} \Sigma_r & 0 \\ 0 & 0 \end{bmatrix}\begin{bmatrix} Y^* \\ X_2^* \end{bmatrix} = U\begin{bmatrix} \Sigma_r & 0 \\ 0 & 0 \end{bmatrix}\begin{bmatrix} Z^*X^* \\ X_2^* \end{bmatrix}$$

$$= U\begin{bmatrix} \Sigma_r Z & 0 \\ 0 & 0 \end{bmatrix}\begin{bmatrix} X^* \\ X_2^* \end{bmatrix} = U\begin{bmatrix} B & 0 \\ 0 & 0 \end{bmatrix}U^*,$$

which is a factorization of the asserted form.

(b) $\Rightarrow$ (c) Theorem 17.4.10.h and Example 17.4.5 tell us that the pseudoinverse of $A = U(B \oplus 0_{n-r})U^*$ is $U(B^{-1} \oplus 0_{n-r})U^*$. Corollary 11.2.11 ensures that there is a polynomial f such that $B^{-1} = f(B)$. Let $p(z) = f(z) + f(0)(zf(z) - 1)$. Then $p(B) = B^{-1} + f(0)(BB^{-1} - I) = B^{-1}$ and $p(0_{n-r}) = f(0)I_{n-r} + f(0)(0_{n-r}f(0_{n-r}) - I_{n-r}) = 0_{n-r}$. Therefore,

$$p(A) = U\begin{bmatrix} p(B) & 0 \\ 0 & p(0_{n-r}) \end{bmatrix}U^* = U\begin{bmatrix} B^{-1} & 0 \\ 0 & 0_{n-r} \end{bmatrix}U^* = A^\dagger.$$

(c) $\Rightarrow$ (a) Any polynomial in A commutes with A. $\square$

A matrix that satisfies any of the equivalent conditions in the preceding theorem is an *EP matrix*. See P.14.42 for two additional equivalent characterizations of EP matrices.

In Section 8.2, we studied consistent linear systems of the form

$$A\mathbf{x} = \mathbf{y}, \qquad \text{in which } A \in \mathsf{M}_{m \times n}(\mathbb{F}) \text{ and } \mathbf{y} \in \mathbb{F}^m. \tag{17.4.12}$$

If (17.4.12) has more than one solution, we may wish to find a solution that has minimum Euclidean norm; see Theorem 8.2.6. The following theorem shows that the pseudoinverse identifies the minimum-norm solution, which is the only solution in $\operatorname{col} A^*$.

Theorem 17.4.13 Let $A \in \mathsf{M}_{m \times n}(\mathbb{F})$ and $\mathbf{y} \in \mathbb{F}^n$. Suppose that $A\mathbf{x} = \mathbf{y}$ is consistent and let $\mathbf{x}_0 = A^\dagger \mathbf{y}$.

(a) $\mathbf{x}_0 \in \operatorname{col} A^*$ and $A\mathbf{x}_0 = \mathbf{y}$.

(b) If $\mathbf{x} \in \mathbb{F}^n$ and $A\mathbf{x} = \mathbf{y}$, then $\|\mathbf{x}\|_2 \geq \|\mathbf{x}_0\|_2$, with equality if and only if $\mathbf{x} \in \operatorname{col} A^*$.

(c) If $\mathbf{x} \in \operatorname{col} A^*$ and $A\mathbf{x} = \mathbf{y}$, then $\mathbf{x} = \mathbf{x}_0$.

Proof (a) Since there is an $\mathbf{x} \in \mathbb{F}^n$ such that $A\mathbf{x} = \mathbf{y}$, parts (a) and (d) of Theorem 17.4.10 ensure that $\mathbf{x}_0 \in \operatorname{col} A^*$ and $A\mathbf{x}_0 = AA^\dagger \mathbf{y} = AA^\dagger A\mathbf{x} = A\mathbf{x} = \mathbf{y}$.

(b) Use Theorem 17.4.10.c and (8.2.4) to compute

$$\|\mathbf{x}\|_2^2 = \|A^\dagger A\mathbf{x} + (I - A^\dagger A)\mathbf{x}\|_2^2 = \|A^\dagger A\mathbf{x}\|_2^2 + \|(I - A^\dagger A)\mathbf{x}\|_2^2$$
$$= \|A^\dagger \mathbf{y}\|_2^2 + \|(I - A^\dagger A)\mathbf{x}\|_2^2 = \|\mathbf{x}_0\|_2^2 + \|(I - A^\dagger A)\mathbf{x}\|_2^2 \geq \|\mathbf{x}_0\|_2^2.$$

The preceding inequality is an equality if and only if $(I - A^\dagger A)\mathbf{x} = \mathbf{0}$, which occurs if and only if $\mathbf{x} \in \operatorname{col} A^*$ (Theorem 17.4.10.c).

(c) If $\mathbf{x} \in \operatorname{col} A^*$, then $\mathbf{x} = A^\dagger A\mathbf{x}$. If, in addition, $A\mathbf{x} = \mathbf{y}$, then

$$\|\mathbf{x} - \mathbf{x}_0\|_2 = \|A^\dagger A\mathbf{x} - A^\dagger \mathbf{y}\|_2 = \|A^\dagger A\mathbf{x} - A^\dagger A\mathbf{x}\|_2 = 0. \qquad \square$$

Example 17.4.14 We can use the pseudoinverse to obtain the unique minimum-norm solution to the consistent linear system $A\mathbf{x} = \mathbf{y}$ in Example 8.2.8. We have

$$A = \begin{bmatrix} 1 & 2 & 3 \\ 4 & 5 & 6 \\ 7 & 8 & 9 \end{bmatrix} \quad \text{and} \quad \mathbf{y} = \begin{bmatrix} 3 \\ 3 \\ 3 \end{bmatrix}.$$

A computation using the pivot-column decomposition and P.17.25 reveals that

$$A^\dagger \mathbf{y} = \begin{bmatrix} -\frac{23}{36} & -\frac{1}{6} & \frac{11}{36} \\ -\frac{1}{18} & 0 & \frac{1}{18} \\ \frac{19}{36} & \frac{1}{6} & -\frac{7}{36} \end{bmatrix} \begin{bmatrix} 3 \\ 3 \\ 3 \end{bmatrix} = \begin{bmatrix} -\frac{3}{2} \\ 0 \\ \frac{3}{2} \end{bmatrix},$$

which is the solution obtained by different means in Example 8.2.8.

The pseudoinverse also helps us analyze least-squares problems. If a system of the form (17.4.12) is inconsistent, then there is no $\mathbf{x}$ such that $A\mathbf{x} = \mathbf{y}$. However, we may wish to find an $\mathbf{x}_0$ such that $\|A\mathbf{x}_0 - \mathbf{y}\|_2 \leq \|A\mathbf{x} - \mathbf{y}\|_2$ for all $\mathbf{x} \in \mathbb{F}^n$; see Section 8.5. Among all such vectors, we may also wish to identify one that has minimum Euclidean norm. The pseudoinverse identifies a vector that has both of these minimality properties.

Theorem 17.4.15 Let $A \in \mathsf{M}_{m \times n}(\mathbb{F})$, let $\mathbf{y} \in \mathbb{F}^m$, and let $\mathbf{x}_0 = A^\dagger \mathbf{y}$.

(a) For every $\mathbf{x} \in \mathbb{F}^n$, we have $\|A\mathbf{x}_0 - \mathbf{y}\|_2 \leq \|A\mathbf{x} - \mathbf{y}\|_2$, with equality if and only if $\mathbf{x} - \mathbf{x}_0 \in \operatorname{null} A$.

(b) If $\mathbf{x} \in \mathbb{F}^n$ and $\|A\mathbf{x} - \mathbf{y}\|_2 = \|A\mathbf{x}_0 - \mathbf{y}\|_2$, then $\|\mathbf{x}\|_2 \geq \|\mathbf{x}_0\|_2$, with equality if and only if $\mathbf{x} = \mathbf{x}_0$.

Proof (a) Since $A\mathbf{x} - AA^\dagger\mathbf{y} \in \operatorname{col} A$ and $(I - AA^\dagger)\mathbf{y} \in (\operatorname{col} A)^\perp$, the Pythagorean theorem ensures that

$$\|A\mathbf{x} - \mathbf{y}\|^2 = \|A\mathbf{x} - AA^\dagger\mathbf{y} + AA^\dagger\mathbf{y} - \mathbf{y}\|_2^2 = \|A\mathbf{x} - AA^\dagger\mathbf{y}\|_2^2 + \|AA^\dagger\mathbf{y} - \mathbf{y}\|_2^2$$
$$= \|A(\mathbf{x} - \mathbf{x}_0)\|_2^2 + \|A\mathbf{x}_0 - \mathbf{y}\|_2^2 \geq \|A\mathbf{x}_0 - \mathbf{y}\|_2^2,$$

with equality if and only if $A(\mathbf{x} - \mathbf{x}_0) = \mathbf{0}$, that is, $\mathbf{x} - \mathbf{x}_0 \in \operatorname{null} A$.

(b) If $\|A\mathbf{x} - \mathbf{y}\|_2 = \|A\mathbf{x}_0 - \mathbf{y}\|_2$, then (a) ensures that $\mathbf{x} = \mathbf{x}_0 + \mathbf{w}$, in which $\mathbf{x}_0 = A^\dagger\mathbf{y} \in \operatorname{col} A^*$ and $\mathbf{w} \in \operatorname{null} A = (\operatorname{col} A^*)^\perp$. Therefore, $\|\mathbf{x}\|_2^2 = \|\mathbf{x}_0 + \mathbf{w}\|_2^2 = \|\mathbf{x}_0\|_2^2 + \|\mathbf{w}\|_2^2 \geq \|\mathbf{x}_0\|_2^2$, with equality if and only if $\mathbf{w} = \mathbf{0}$ if and only if $\mathbf{x} = \mathbf{x}_0$. $\qquad\square$

Example 17.4.16 We can use the pseudoinverse to obtain the unique least-squares solution to the inconsistent linear system $A\mathbf{x} = \mathbf{y}$ in Example 8.5.8. We have

$$A = \begin{bmatrix} 0 & 1 \\ 1 & 1 \\ 2 & 1 \\ 3 & 1 \\ 4 & 1 \end{bmatrix} \quad \text{and} \quad \mathbf{y} = \begin{bmatrix} 1 \\ 1 \\ 3 \\ 3 \\ 4 \end{bmatrix}.$$

Since A has full column rank, we can use (17.4.7) to evaluate the pseudoinverse:

$$A^\dagger = (A^*A)^{-1}A^* = \begin{bmatrix} 30 & 10 \\ 10 & 5 \end{bmatrix}^{-1} A^* = \frac{1}{10}\begin{bmatrix} 1 & -2 \\ -2 & 6 \end{bmatrix}\begin{bmatrix} 0 & 1 & 2 & 3 & 4 \\ 1 & 1 & 1 & 1 & 1 \end{bmatrix}$$
$$= \frac{1}{10}\begin{bmatrix} -2 & -1 & 0 & 1 & 2 \\ 6 & 4 & 2 & 0 & -2 \end{bmatrix}.$$

The solution is

$$A^\dagger\mathbf{y} = \frac{1}{10}\begin{bmatrix} -2 & -1 & 0 & 1 & 2 \\ 6 & 4 & 2 & 0 & -2 \end{bmatrix}\begin{bmatrix} 1 \\ 1 \\ 3 \\ 3 \\ 4 \end{bmatrix} = \begin{bmatrix} \frac{4}{5} \\ \frac{4}{5} \end{bmatrix},$$

which is the solution obtained by other means in Example 8.5.8.

Here is a summary of the preceding theorems.

(a) $\mathbf{x}_0 = A^\dagger\mathbf{y}$ is a least-squares solution of the linear system $A\mathbf{x} = \mathbf{y}$ (which need not be consistent). Among all least-squares solutions, $\mathbf{x}_0$ is the only one with smallest norm.

(b) If $A\mathbf{x} = \mathbf{y}$ is consistent, then $\mathbf{x}_0 = A^\dagger\mathbf{y}$ is a solution. Among all solutions, $\mathbf{x}_0$ has the smallest norm, and it is the only one in $\operatorname{col} A^*$.

The final example in this section illustrates why the pseudoinverse joins Cramer's rule, the Jordan canonical form, and the determinant on the list of important theoretical tools that are not recommended for use in numerical algorithms.

Example 17.4.17 Let $z \in \mathbb{C}$ and let $A(z) = \begin{bmatrix} 1 & 0 \\ 0 & z \end{bmatrix}$. Then

$$A^{\dagger}(z) = \begin{cases} \begin{bmatrix} 1 & 0 \\ 0 & z^{-1} \end{bmatrix} & \text{if } z \neq 0, \\[2ex] \begin{bmatrix} 1 & 0 \\ 0 & 0 \end{bmatrix} & \text{if } z = 0. \end{cases}$$

For small $|z|$, we have $\|A(z) - A(0)\|_2 = |z|$ (small), but $\|A^{\dagger}(z) - A^{\dagger}(0)\|_2 = |z|^{-1}$ (not small).

17.5 The Spectral Condition Number

Definition 17.5.1 Let $A \in \mathsf{M}_n$. If A is invertible, its *spectral condition number* is $\kappa_2(A) = \|A\|_2 \|A^{-1}\|_2$. If A is not invertible, its spectral condition number is undefined.

Theorem 17.3.2.a and the definition of the spectral norm permit us to express the spectral condition number of an invertible A as a ratio of singular values:

$$\kappa_2(A) = \frac{\sigma_{\max}(A)}{\sigma_{\min}(A)}. \tag{17.5.2}$$

Theorem 17.5.3 *Let $A, B \in \mathsf{M}_n$ be invertible.*

(a) $\kappa_2(AB) \leq \kappa_2(A)\kappa_2(B)$.

(b) $\kappa_2(A) \geq 1$, *with equality if and only if A is a nonzero scalar multiple of a unitary matrix.*

(c) *If c is a nonzero scalar, then $\kappa_2(cA) = \kappa_2(A)$.*

(d) $\kappa_2(A) = \kappa_2(A^*) = \kappa_2(\overline{A}) = \kappa_2(A^{\mathsf{T}}) = \kappa_2(A^{-1})$.

(e) $\kappa_2(A^*A) = \kappa_2(AA^*) = \kappa_2(A)^2$.

(f) *If $U, V \in \mathsf{M}_n$ are unitary, then $\kappa_2(UAV) = \kappa_2(A)$.*

Proof (a) Use Theorem 17.2.6.b and compute

$$\kappa_2(AB) = \|AB\|_2 \|B^{-1}A^{-1}\|_2 \leq \|A\|_2 \|B\|_2 \|B^{-1}\|_2 \|A^{-1}\|_2 = \kappa_2(A)\kappa_2(B).$$

(b) It follows from (a) and Theorem 17.2.2.a that

$$1 = \|I_n\|_2 = \|AA^{-1}\|_2 \leq \|A\|_2 \|A^{-1}\|_2 = \kappa_2(A),$$

with equality if and only if $\sigma_{\max}(A) = \sigma_{\min}(A)$ if and only if A is a nonzero scalar multiple of a unitary matrix (Theorem 17.3.9).

(c) Compute

$$\kappa_2(cA) = \|cA\|_2 \|(cA)^{-1}\|_2 = \|cA\|_2 \|c^{-1}A^{-1}\|_2$$
$$= |c| \|A\|_2 |c^{-1}| \|A^{-1}\|_2 = \|A\|_2 \|A^{-1}\|_2 = \kappa_2(A).$$

(d) Observe that $A, A^*, \overline{A}$, and A^{T} all have the same singular values; see Theorem 16.1.4. Also, $\kappa_2(A^{-1}) = \|A^{-1}\|_2 \|A\|_2 = \kappa_2(A)$.

(e) We have $\kappa_2(A^*A) = \sigma_{\max}(A^*A)/\sigma_{\min}(A^*A) = \sigma_{\max}^2(A)/\sigma_{\min}^2(A) = \kappa_2(A)^2 = \kappa_2(A^*)^2 = \kappa_2(AA^*)$.

(f) The singular values of A and UAV are the same, so their spectral condition numbers are the same. $\square$

The spectral condition number of an invertible matrix plays a key role in understanding how the solution of a linear system and computation of eigenvalues can be affected by errors in the data or in the computation.

Let $A \in M_n(\mathbb{F})$ be invertible, let $\mathbf{y} \in \mathbb{F}^n$ be nonzero, and let $\mathbf{x} \in \mathbb{F}^n$ be the unique (necessarily nonzero) solution of the linear system $A\mathbf{x} = \mathbf{y}$. Then $\|\mathbf{y}\|_2 = \|A\mathbf{x}\|_2 \leq \|A\|_2 \|\mathbf{x}\|_2$ (see Theorem 17.2.6), which we rewrite as

$$\frac{1}{\|\mathbf{x}\|_2} \leq \frac{\|A\|_2}{\|\mathbf{y}\|_2}. \tag{17.5.4}$$

What happens to the solution if we perturb $\mathbf{y} \rightarrow \mathbf{y} + \mathbf{y}'$, in which $\|\mathbf{y}'\|_2$ is typically small? We have a new system

$$A(\mathbf{x} + \mathbf{x}') = \mathbf{y} + \mathbf{y}'. \tag{17.5.5}$$

Since $A\mathbf{x} = \mathbf{y}$, we have $A\mathbf{x} + A\mathbf{x}' = \mathbf{y} + A\mathbf{x}' = \mathbf{y} + \mathbf{y}'$, so the resulting perturbation $\mathbf{x}'$ to $\mathbf{x}$ is the unique solution to $A\mathbf{x}' = \mathbf{y}'$. We have $\mathbf{x}' = A^{-1}\mathbf{y}'$, and therefore

$$\|\mathbf{x}'\|_2 = \|A^{-1}\mathbf{y}'\|_2 \leq \|A^{-1}\|_2 \|\mathbf{y}'\|_2. \tag{17.5.6}$$

Now combine (17.5.4) and (17.5.6) to obtain

$$\frac{\|\mathbf{x}'\|_2}{\|\mathbf{x}\|_2} \leq \frac{\|A\|_2 \|A^{-1}\|_2 \|\mathbf{y}'\|_2}{\|\mathbf{y}\|_2} = \kappa_2(A)\frac{\|\mathbf{y}'\|_2}{\|\mathbf{y}\|_2},$$

which provides an upper bound

$$\frac{\|\mathbf{x}'\|_2}{\|\mathbf{x}\|_2} \leq \kappa_2(A)\frac{\|\mathbf{y}'\|_2}{\|\mathbf{y}\|_2} \tag{17.5.7}$$

on the relative error $\|\mathbf{x}'\|_2/\|\mathbf{x}\|_2$ in the solution of the perturbed linear system (17.5.5).

If $\kappa_2(A)$ is not large, the relative error in the solution of $A\mathbf{x} = \mathbf{y}$ cannot be much worse than the relative error in the data. In this case, A is *well conditioned*. For example, if A is a nonzero scalar multiple of a unitary matrix, then $\kappa_2(A) = 1$ and A is *perfectly conditioned*. In this case, the relative error in the solution is at most the relative error in the data. If $\kappa_2(A)$ is large, however, the relative error in the solution could be much larger than the relative error in the data. In this case, A is *ill conditioned*.

Example 17.5.8 Consider $A\mathbf{x} = \mathbf{y}$, in which $A = \begin{bmatrix} 1 & 2 \\ 2 & 4.001 \end{bmatrix}$ and $\mathbf{y} = \begin{bmatrix} 4 \\ 8.001 \end{bmatrix}$. It has the unique solution $\mathbf{x} = \begin{bmatrix} 2 \\ 1 \end{bmatrix}$. Let $\mathbf{y}' = \begin{bmatrix} 0.001 \\ -0.002 \end{bmatrix}$. The system $A\mathbf{x}' = \mathbf{y}'$ has the unique solution $\mathbf{x}' = \begin{bmatrix} 8.001 \\ -4.000 \end{bmatrix}$. The relative errors in the data and in the solution are

$$\frac{\|\mathbf{y}'\|_2}{\|\mathbf{y}\|_2} = 2.5001 \times 10^{-4} \quad \text{and} \quad \frac{\|\mathbf{x}'\|_2}{\|\mathbf{x}\|_2} = 4.0004.$$

The spectral condition number of A is $\kappa_2(A) = 2.5008 \times 10^4$ and the upper bound in (17.5.7) is

$$\frac{\|\mathbf{x}'\|_2}{\|\mathbf{x}\|_2} = 4.0004 \leq 6.2523 = \kappa_2(A)\frac{\|\mathbf{y}'\|_2}{\|\mathbf{y}\|_2}.$$

Because A is ill conditioned, a small relative change in the data can (and in this case does) cause a large relative change in the solution.

Conditioning issues were one reason we expressed reservations in Section 8.5 about finding a least-squares solution of an inconsistent linear system $A\mathbf{x} = \mathbf{y}$ by solving the normal equations

$$A^*A\mathbf{x} = A^*\mathbf{y}. \tag{17.5.9}$$

If A is square and invertible (but not a scalar multiple of a unitary matrix), the matrix A^*A in the normal equations (17.5.9) is more ill conditioned than A because $\kappa_2(A) > 1$ and $\kappa_2(A^*A) = \kappa_2(A)^2$. However, if one uses a QR factorization of A, then Theorem 17.5.3.f ensures that the equivalent linear system $R\mathbf{x} = Q^*\mathbf{y}$ (see (8.5.10)) is as well conditioned as A:

$$\kappa_2(R) = \kappa_2(Q^*A) = \kappa_2(A).$$

Interpretation of results from eigenvalue computations must take into account uncertainties in the input data (perhaps it arises from physical measurements) as well as roundoff errors due to finite precision arithmetic. One way to model uncertainty in the results is to imagine that the computation has been done with perfect precision, but for a slightly different matrix. The following discussion adopts that point of view, and gives a framework to assess the quality of computed eigenvalues.

Lemma 17.5.10 *Let $A \in \mathsf{M}_n$. If $\|A\|_2 < 1$, then $I + A$ is invertible.*

Proof Theorem 17.3.2.c ensures that every $\lambda \in \operatorname{spec} A$ satisfies $|\lambda| \leq \|A\|_2 < 1$. Therefore, $-1 \notin \operatorname{spec} A$ and $0 \notin \operatorname{spec}(I + A)$, so $I + A$ is invertible. $\qquad\square$

The following theorem relies on the preceding lemma and on the fact that the spectral norm of a diagonal matrix is the largest of the moduli of its diagonal entries; see Example 17.3.5. We introduce a perturbation $A \to A + E$, in which $\|E\|_2$ is typically small.

Theorem 17.5.11 (Bauer–Fike) *Let $A \in \mathsf{M}_n$ be diagonalizable and let $A = S\Lambda S^{-1}$, in which $S \in \mathsf{M}_n$ is invertible and $\Lambda = \operatorname{diag}(\lambda_1, \lambda_2, \ldots, \lambda_n)$. Let $E \in \mathsf{M}_n$. If $\lambda \in \operatorname{spec}(A + E)$, then there is some $i \in \{1, 2, \ldots, n\}$ such that*

$$|\lambda_i - \lambda| \leq \kappa_2(S)\|E\|_2. \tag{17.5.12}$$

Proof If λ is an eigenvalue of A, then $\lambda = \lambda_i$ for some $i \in \{1, 2, \ldots, n\}$ and (17.5.12) is satisfied. Now assume that λ is not an eigenvalue of A, so $\Lambda - \lambda I$ is invertible. Since λ is an eigenvalue of $A + E$, we know that $A + E - \lambda I$ is not invertible. Therefore,

$$I + (\Lambda - \lambda I)^{-1}S^{-1}ES = (\Lambda - \lambda I)^{-1}(\Lambda - \lambda I + S^{-1}ES)$$
$$= (\Lambda - \lambda I)^{-1}(S^{-1}AS - \lambda S^{-1}S + S^{-1}ES)$$
$$= (\Lambda - \lambda I)^{-1}S^{-1}(A + E - \lambda I)S$$

is not invertible. It follows from the preceding lemma and Theorem 17.2.6.b that

$$1 \leq \|(\Lambda - \lambda I)^{-1}S^{-1}ES\|_2 \leq \|(\Lambda - \lambda I)^{-1}\|_2\|S^{-1}ES\|_2$$
$$= \max_{1 \leq i \leq n} |\lambda_i - \lambda|^{-1}\|S^{-1}ES\|_2 = \frac{\|S^{-1}ES\|_2}{\min_{1 \leq i \leq n} |\lambda_i - \lambda|}$$
$$\leq \frac{\|S^{-1}\|_2\|E\|_2\|S\|_2}{\min_{1 \leq i \leq n} |\lambda_i - \lambda|} = \kappa_2(S)\frac{\|E\|_2}{\min_{1 \leq i \leq n} |\lambda_i - \lambda|}.$$

This inequality implies (17.5.12). $\qquad\square$

If A is Hermitian or normal, then the matrix of eigenvectors S may be chosen to be unitary. In this case, $\kappa_2(S) = 1$ and the error in the computed eigenvalue cannot be greater than the spectral norm of E. However, if our best choice of S has a large spectral condition number, we should be skeptical about the quality of a computed eigenvalue. For example, the latter situation could occur if A has distinct eigenvalues and two of its eigenvectors are nearly collinear.

Example 17.5.13 The matrix $A = \begin{bmatrix} 1 & 1000 \\ 0 & 2 \end{bmatrix}$ has distinct eigenvalues 1 and 2, so it is diagonalizable. Perturb $A \mapsto A + E$, in which $E = \begin{bmatrix} 0 & 0 \\ 0.01 & 0 \end{bmatrix}$. Then $\|E\|_2 = 0.01$ and $A + E = \begin{bmatrix} 1 & 1000 \\ 0.01 & 2 \end{bmatrix}$ has approximate eigenvalues -1.702 and 4.702. Even though E has small norm, neither eigenvalue of $A + E$ is a good approximation to an eigenvalue of A. A matrix of eigenvectors of A is $S = \begin{bmatrix} 1 & 1 \\ 0 & 0.001 \end{bmatrix}$, for which $\kappa_2(S) \approx 2000$. For the eigenvalue $\lambda \approx 4.702$ of $A + E$, the bound in (17.5.12) is

$$2.702 \approx |2 - \lambda| \le \kappa_2(S)\|E\|_2 \approx 20,$$

so our computations are consistent with the Bauer–Fike bound.

17.6 Complex Symmetric Matrices

If $A \in \mathsf{M}_n$ and $A = A^\mathsf{T}$, then A is *complex symmetric*. Real symmetric matrices are normal, but non-real symmetric matrices need not be normal; they need not even be diagonalizable (Example 17.6.4).

Theorem 17.6.1 (Autonne) *Let $A \in \mathsf{M}_n$ be complex symmetric, let* $\operatorname{rank} A = r \ge 1$, *let* $\sigma_1 \ge \sigma_2 \ge \cdots \ge \sigma_r > 0$ *be the positive singular values of A, and let* $\Sigma_r = \operatorname{diag}(\sigma_1, \sigma_2, \ldots, \sigma_r)$. *There is a unitary $U \in \mathsf{M}_n$ such that*

$$A = U\Sigma U^\mathsf{T} \quad \text{and} \quad \Sigma = \Sigma_r \oplus 0_{n-r}. \tag{17.6.2}$$

Proof Let $A = V\Sigma W^*$ be a singular value decomposition, let

$$B = W^\mathsf{T} A W = W^\mathsf{T}(V\Sigma W^*)W = (W^\mathsf{T} V)\Sigma,$$

and let $X = W^\mathsf{T} V$. Then B is complex symmetric, X is unitary, and $B = X\Sigma$ is a polar decomposition. Since $B^*B = \Sigma X^*X\Sigma = \Sigma^2$ is real and $B = B^\mathsf{T}$, we have $B^*B = \overline{B^*B} = B^\mathsf{T}\overline{B} = BB^*$, that is, B is normal. Partition $X = \begin{bmatrix} X_{11} & X_{12} \\ X_{21} & X_{22} \end{bmatrix}$, in which $X_{11} \in \mathsf{M}_r$. The normality of B and Theorem 16.3.18 ensure that $\Sigma = \Sigma_r \oplus 0_{n-r}$ commutes with X, and Lemma 3.2.17.b tells us that $X_{12} = 0$ and $X_{21} = 0$. Therefore,

$$B = X\Sigma = X_{11}\Sigma_r \oplus X_{22}0_{n-r} = X_{11}\Sigma_r \oplus 0_{n-r},$$

in which X_{11} is unitary (Theorem 7.1.9.b) and commutes with Σ_r. The symmetry of B tells us that $X_{11}\Sigma_r \oplus 0_{n-r} = B = B^\mathsf{T} = \Sigma_r X_{11}^\mathsf{T} \oplus 0_{n-r}$. Since $X_{11}\Sigma_r$ and $\Sigma_r X_{11}^\mathsf{T}$ are right and left polar decompositions of an invertible matrix, Theorem 16.3.13.c.iii says that $X_{11} = X_{11}^\mathsf{T}$, that is, X_{11} is unitary and symmetric. Lemma 14.6.10 ensures that there is a symmetric unitary $Y \in \mathsf{M}_r$ such that $X_{11} = Y^2$ and Y is a polynomial in X_{11}. Since X_{11} commutes with Σ_r, it follows that Y commutes with Σ_r (Theorem 2.6.9). Therefore,

$$B = X_{11}\Sigma_r \oplus 0_{n-r} = Y^2\Sigma_r \oplus 0_{n-r} = Y\Sigma_r Y \oplus 0_{n-r} = (Y \oplus I_r)(\Sigma_r \oplus 0_{n-r})(Y \oplus I_r) = Z\Sigma_r Z,$$

in which $Z = Y \oplus I_{n-r}$ is unitary and symmetric. Consequently,

$$A = \overline{W}BW^* = \overline{W}Z\Sigma_r ZW^* = (\overline{W}Z)\Sigma_r(\overline{W}Z^{\mathsf{T}})^{\mathsf{T}} = (\overline{W}Z)\Sigma_r(\overline{W}Z)^{\mathsf{T}} = U\Sigma U^{\mathsf{T}},$$

in which $U = \overline{W}Z$ is unitary. $\qquad\square$

Example 17.6.3 Consider the 1×1 symmetric matrix $[-1]$. Then $[-1] = [i][1][i]$ and $[i]$ is unitary. There is no real number c such that $[-1] = [c][1][c] = [c^2]$. Thus, if A is real symmetric, there need not be a real orthogonal U such that $A = U\Sigma U^{\mathsf{T}}$ is a singular value decomposition. However, Autonne's theorem says that this factorization is always possible with a complex unitary U.

Example 17.6.4 Here is a singular value decomposition of the type guaranteed by Autonne's theorem:

$$A = \begin{bmatrix} 1 & i \\ i & -1 \end{bmatrix} = \underbrace{\begin{bmatrix} \frac{1}{\sqrt{2}} & -\frac{1}{\sqrt{2}} \\ \frac{i}{\sqrt{2}} & \frac{i}{\sqrt{2}} \end{bmatrix}}_{U} \underbrace{\begin{bmatrix} 2 & 0 \\ 0 & 0 \end{bmatrix}}_{\Sigma} \underbrace{\begin{bmatrix} \frac{1}{\sqrt{2}} & \frac{i}{\sqrt{2}} \\ -\frac{1}{\sqrt{2}} & \frac{i}{\sqrt{2}} \end{bmatrix}}_{U^{\mathsf{T}}}.$$

Notice that $A^2 = 0$, so both eigenvalues of A are zero (see Theorem 9.3.3). If A were diagonalizable, it would be similar to the zero matrix, which it is not. Consequently, A is a complex symmetric matrix that is not diagonalizable.

The special singular value decomposition (17.6.2) of a complex symmetric matrix implies a special compact singular value decomposition.

Corollary 17.6.5 Let $A \in \mathsf{M}_n$ be symmetric and let $\operatorname{rank} A = r \geq 1$. Let $A = U\Sigma U^{\mathsf{T}}$ be a singular value decomposition, in which U is unitary, $\Sigma = \Sigma_r \oplus 0_{n-r}$, and Σ_r is invertible. Write $U = [U_r\ U']$, in which $U_r \in \mathsf{M}_{n\times r}$.

(a) $A = U_r\Sigma_r U_r^{\mathsf{T}}$.

(b) If $W_r, V_r \in \mathsf{M}_{n\times r}$ have orthonormal columns and $A = W_r\Sigma_r W_r^{\mathsf{T}} = V_r\Sigma_r V_r^{\mathsf{T}}$, then there is a real orthogonal $Q \in \mathsf{M}_r$ such that $V_r = W_r Q$ and Q commutes with Σ_r.

Proof (a) With the given partitions of U and Σ, we have $A = U\Sigma U^{\mathsf{T}} = U_r\Sigma_r U_r^{\mathsf{T}}$, which is a compact singular value decomposition.

(b) The hypothesis is that W_r and V_r have orthonormal columns and $A = W_r\Sigma_r\overline{W}_r^* = V_r\Sigma_r\overline{V}_r^*$. Theorem 16.2.10.a ensures that there is a unitary $Q \in \mathsf{M}_r$ that commutes with Σ_r and satisfies $V_r = W_r Q$ and $\overline{V}_r = \overline{W}_r Q$. The identities $Q = W_r^* V_r$ and $Q = \overline{W}_r^*\overline{V}_r$ imply that $Q = \overline{Q}$, so Q is real orthogonal. $\qquad\square$

17.7 Idempotent Matrices

Orthogonal projections and the non-Hermitian matrix $A = \begin{bmatrix} 1 & 2 \\ 0 & 0 \end{bmatrix}$ are idempotent ($A^2 = A$). In the following theorem, we use the singular value decomposition to find a standard form to which any idempotent matrix is unitarily similar.

Theorem 17.7.1 *Suppose that $A \in \mathsf{M}_n(\mathbb{F})$ is idempotent and has rank $r \geq 1$. If A has any singular values that are greater than 1, denote them by $\sigma_1 \geq \sigma_2 \geq \cdots \geq \sigma_k > 1$. There is a unitary $U \in \mathsf{M}_n(\mathbb{F})$ such that*

$$U^*AU = \left(\begin{bmatrix} 1 & \sqrt{\sigma_1^2 - 1} \\ 0 & 0 \end{bmatrix} \oplus \cdots \oplus \begin{bmatrix} 1 & \sqrt{\sigma_k^2 - 1} \\ 0 & 0 \end{bmatrix} \right) \oplus I_{r-k} \oplus 0_{n-r-k}. \tag{17.7.2}$$

One or more of the direct summands may be absent. The first is present only if $k \geq 1$; the second is present only if $r > k$; the third is present only if $n > r + k$.

Proof Example 11.4.12 shows that A is unitarily similar to

$$\begin{bmatrix} I_r & X \\ 0 & 0_{n-r} \end{bmatrix}, \quad \text{in which } X \in \mathsf{M}_{r \times (n-r)}. \tag{17.7.3}$$

If $X = 0$, then A is unitarily similar to $I_r \oplus 0_{n-r}$, so A is an orthogonal projection.

Suppose that rank $X = p \geq 1$ and let $\tau_1 \geq \tau_2 \geq \cdots \geq \tau_p > 0$ be the positive singular values of X. Since A is unitarily similar to (17.7.3), AA^* is unitarily similar to

$$\begin{bmatrix} I_r & X \\ 0 & 0_{n-r} \end{bmatrix} \begin{bmatrix} I_r & 0 \\ X^* & 0_{n-r} \end{bmatrix} = (I_r + XX^*) \oplus 0_{n-r}.$$

Therefore, the eigenvalues of AA^* are $1 + \tau_1^2, 1 + \tau_2^2, \ldots, 1 + \tau_p^2$, together with 1 (with multiplicity $r - p$) and 0 (with multiplicity $n - r$). Since A has k singular values greater than 1, it follows that $p = k$. Moreover,

$$\sigma_i^2 = \tau_i^2 + 1 \quad \text{and} \quad \tau_i = \sqrt{\sigma_i^2 - 1} \quad \text{for } i = 1, 2, \ldots, k.$$

Let $X = V\Sigma W^*$ be a singular value decomposition, in which $V \in \mathsf{M}_r$ and $W \in \mathsf{M}_{n-r}$ are unitary. Let $\Sigma_k = \mathrm{diag}(\tau_1, \tau_2, \ldots, \tau_k)$ be the $k \times k$ upper-left corner of Σ. Consider the unitary similarity

$$\begin{bmatrix} V & 0 \\ 0 & W \end{bmatrix}^* \begin{bmatrix} I_r & X \\ 0 & 0_{n-r} \end{bmatrix} \begin{bmatrix} V & 0 \\ 0 & W \end{bmatrix} = \begin{bmatrix} V^*V & V^*XW \\ 0 & 0_{n-r} \end{bmatrix} = \begin{bmatrix} I_r & \Sigma \\ 0 & 0_{n-r} \end{bmatrix}$$
$$= \begin{bmatrix} I_p & 0 & \Sigma_k & 0 \\ 0 & I_{r-k} & 0 & 0 \\ 0_k & 0 & 0_k & 0 \\ 0 & 0 & 0 & 0_{n-r-k} \end{bmatrix}. \tag{17.7.4}$$

Then (17.7.4) is permutation similar to

$$\begin{bmatrix} I_k & \Sigma_k & 0 & 0 \\ 0_k & 0_k & 0 & 0 \\ 0 & 0 & I_{r-k} & 0 \\ 0 & 0 & 0 & 0_{n-r-k} \end{bmatrix} = \begin{bmatrix} I_k & \Sigma_k \\ 0_k & 0_k \end{bmatrix} \oplus I_{r-k} \oplus 0_{n-r-k}. \tag{17.7.5}$$

The 2×2 block matrix in (17.7.5) is permutation similar to

$$\begin{bmatrix} 1 & \tau_1 \\ 0 & 0 \end{bmatrix} \oplus \cdots \oplus \begin{bmatrix} 1 & \tau_k \\ 0 & 0 \end{bmatrix} = \begin{bmatrix} 1 & \sqrt{\sigma_1^2 - 1} \\ 0 & 0 \end{bmatrix} \oplus \cdots \oplus \begin{bmatrix} 1 & \sqrt{\sigma_k^2 - 1} \\ 0 & 0 \end{bmatrix};$$

see (7.3.8). Therefore, A is unitarily similar to the direct sum (17.7.2). If A is real, Corollary 14.6.12 ensures that it is real orthogonally similar to (17.7.2). $\square$

17.8 Problems

P.17.1 Let $A = \begin{bmatrix} 2 & 3 \\ 0 & 2 \end{bmatrix}$. Compute the eigenvalues, spectral radius, singular values, determinant, spectral norm, and Frobenius norm of A and A^{-1}.

P.17.2 Compute the spectral and Frobenius norms of $A = \begin{bmatrix} 1 & 1 \\ -1 & 1 \end{bmatrix}$, $B = \begin{bmatrix} 1 & 1 \\ 0 & 1 \end{bmatrix}$, and $C = \begin{bmatrix} 1 & 1 \\ 1 & 1 \end{bmatrix}$. Conclude that (a) replacing an entry of a matrix by zero can increase or decrease its spectral norm, and (b) changing the sign of an entry can increase or decrease its spectral norm. (c) What can you say about the Frobenius norm?

P.17.3 Let $\mathbf{x} \in \mathbb{F}^n$. Show that the Frobenius and spectral norms of the $n \times 1$ matrix $\mathbf{x}$ are equal to $\|\mathbf{x}\|_2$.

P.17.4 Let $\mathbf{x} \in \mathbb{F}^m$ and $\mathbf{y} \in \mathbb{F}^n$. Show that the Frobenius and spectral norms of the $m \times n$ matrix $\mathbf{xy}^*$ are equal to $\|\mathbf{x}\|_2\|\mathbf{y}\|_2$.

P.17.5 Consider the matrices V and W in Example 7.1.14. Denote by A the matrix obtained by replacing the last column of V by a column of zeros. Denote by B the matrix obtained by replacing the first row of W with a row of zeros. What are the singular values of A and B?

P.17.6 What non-invertible matrix is closest to $A = \mathrm{diag}(1, 2, 3, 4)$ with respect to the spectral norm? Why?

P.17.7 If $A = \sum_{i=1}^r \sigma_i \mathbf{v}_i \mathbf{w}_i^*$ as in (17.1.1), explain why $A^\dagger \mathbf{x} = \sum_{i=1}^r \frac{\langle \mathbf{x}, \mathbf{v}_i \rangle}{\sigma_i} \mathbf{w}_i$ for all $\mathbf{x} \in \mathbb{C}^n$.

P.17.8 Let the singular values of $A \in \mathsf{M}_2$ be $\sigma_1 \geq \sigma_2 \geq 0$. (a) Show that

$$\{\sigma_1, \sigma_2\} = \left\{ \frac{1}{2} \left(\sqrt{\|A\|_F^2 + 2|\det A|} \pm \sqrt{\|A\|_F^2 - 2|\det A|} \right) \right\}.$$

(b) Use this formula to find the spectral norms of A and B in (17.3.11).

P.17.9 Let $A, B \in \mathsf{M}_n$. (a) Show that $\rho(AB) = \rho(BA) \leq \min\{\|AB\|_2, \|BA\|_2\}$. (b) Exhibit $A, B \in \mathsf{M}_2$ such that $\|AB\|_2 \neq \|BA\|_2$.

P.17.10 Let $A, B \in \mathsf{M}_n$ and suppose that AB is normal. Show that $\|AB\|_2 \leq \|BA\|_2$. *Hint:* Theorem 17.3.2 and (17.3.6).

P.17.11 Let $A = [\mathbf{a}_1\ \mathbf{a}_2\ \ldots\ \mathbf{a}_n] \in \mathsf{M}_{m \times n}$. Show that $\|\mathbf{a}_i\|_2 \leq \|A\|_2$ for each $i = 1, 2, \ldots, n$.

P.17.12 Let $A, B \in \mathsf{M}_n$. Show that $\|AB\|_F \leq \|A\|_2 \|B\|_F$.

P.17.13 Let $A \in \mathsf{M}_{m \times n}$. Show that $\|A\|_F = \|A^*\|_F$ and $\|A\|_2 = \|A^*\|_2$.

P.17.14 Let $A, B \in \mathsf{M}_{m \times n}$. (a) Prove that $\|A \circ B\|_F \leq \|A\|_F \|B\|_F$. This is an analog of Theorem 17.2.6.a. (b) Verify the inequality in (a) for the matrices in Example 15.6.2.

P.17.15 Let $A, B \in \mathsf{M}_n$. Suppose that A is positive definite and B is complex symmetric. Use Theorem 17.6.1 to show that there is an invertible $S \in \mathsf{M}_n$ and a diagonal $\Sigma \in \mathsf{M}_n$ with nonnegative diagonal entries such that $A = SIS^*$ and $B = S\Sigma S^\mathsf{T}$. *Hint:* $A^{-1/2} B \overline{A}^{-1/2} = U\Sigma U^\mathsf{T}$ is complex symmetric. Consider $S = A^{1/2} U$.

P.17.16 Let $A \in \mathsf{M}_n$, let $\mathbf{x} \in \mathbb{C}^n$, and consider the complex quadratic form $q_A(\mathbf{x}) = \mathbf{x}^\mathsf{T} A \mathbf{x}$. (a) Show that $q_A(\mathbf{x}) = q_{\frac{1}{2}(A + A^\mathsf{T})}(\mathbf{x})$, so we may assume that A is complex symmetric. (b) If A is symmetric and has singular values $\sigma_1, \sigma_2, \ldots, \sigma_n$, show that there is a change of variables $\mathbf{x} \mapsto U\mathbf{x} = \mathbf{y} = [y_i]$ such that U is unitary and $q_A(\mathbf{y}) = \sigma_1 y_1^2 + \sigma_2 y_2^2 + \cdots + \sigma_n y_n^2$.

P.17.17 If $A \in \mathsf{M}_n$ is complex symmetric and rank $A = r$, show that it has a full-rank factorization of the form $A = BB^\mathsf{T}$ with $B \in \mathsf{M}_{n \times r}$.

P.17.18 Let $A \in \mathsf{M}_n$ be complex symmetric and have distinct singular values. Suppose that $A = U\Sigma U^\mathsf{T}$ and $A = V\Sigma V^\mathsf{T}$ are singular value decompositions of A. Show that

$U = VD$, in which $D = \text{diag}(d_1, d_2, \ldots, d_n)$ and $d_i = \pm 1$ for each $i = 1, 2, \ldots, n-1$. If A is invertible, then $d_n = \pm 1$. If A is not invertible, show that d_n can be any complex number with modulus 1.

P.17.19 Let $A \in M_n$. Show that A is complex symmetric if and only if it has polar decompositions $A = |A|^\mathsf{T} U = U|A|$, in which the unitary factor U is complex symmetric.

P.17.20 Let $A \in M_{m \times n}$ and let c be a nonzero scalar. Show that $(A^\mathsf{T})^\dagger = (A^\dagger)^\mathsf{T}$, $(A^*)^\dagger = (A^\dagger)^*$, $(\overline{A})^\dagger = \overline{A^\dagger}$, and $(cA)^\dagger = c^{-1}A^\dagger$.

P.17.21 Let $A \in M_{m \times n}(\mathbb{F})$ and let $B, C \in M_{n \times m}(\mathbb{F})$. The *Penrose identities* are

$$(AB)^* = AB, \quad (BA)^* = BA, \quad ABA = A, \quad \text{and} \quad BAB = B. \qquad (17.8.1)$$

(a) Verify that $B = A^\dagger$ satisfies these identities. (b) If $(AC)^* = AC$, $(CA)^* = CA$, $ACA = A$, and $CAC = C$, show that $B = C$ by justifying every step in the following calculation:

$$B = B(AB)^* = BB^*A^* = B(AB)^*(AC)^*$$
$$= BAC = (BA)^*(CA)^*C = A^*C^*C = (CA)^*C = C.$$

(c) Conclude that $B = A^\dagger$ if and only if B satisfies the identities (17.8.1).

P.17.22 Let $A \in M_{m \times n}(\mathbb{F})$ have rank $r \geq 1$ and let $A = RS$ be a full-rank factorization; see Theorem 2.3.7. Show that R^*AS^* is invertible. Prove that

$$A^\dagger = S^*(R^*AS^*)^{-1}R^* \qquad (17.8.2)$$

by verifying that $S^*(R^*AS^*)^{-1}R^*$ satisfies the identities (17.8.1). If A has low rank, this can be an efficient way to calculate the pseudoinverse of A.

P.17.23 (a) Use (17.8.2) to compute the pseudoinverse of $A = \begin{bmatrix} 1 & 1 \\ 0 & 0 \end{bmatrix}$. (b) Use Theorem 17.4.11 to show that A is not an EP matrix.

P.17.24 Let $A \in M_m$ and $B \in M_n$. (a) Use (17.8.1) to show that $(A \oplus B)^\dagger = A^\dagger \oplus B^\dagger$. (b) Can you prove this without using the Penrose identities?

P.17.25 Let $A \in M_{m \times n}$ be nonzero and suppose that $A = XY$ is a full-rank factorization. Show that $A^\dagger = Y^\dagger X^\dagger$.

P.17.26 Suppose that $A \in M_n$ is invertible. Use (17.8.1) and (17.8.2) to show that $A^\dagger = A^{-1}$.

P.17.27 Let $\mathbf{x} \in \mathbb{F}^m$ and $\mathbf{y} \in \mathbb{F}^n$ be nonzero and let $A = \mathbf{x}\mathbf{y}^*$. Use (17.8.2) to show that $A^\dagger = \|\mathbf{x}\|_2^{-2}\|\mathbf{y}\|_2^{-2}A^*$. What is $A^\dagger$ if A is an all-ones matrix?

P.17.28 Let $\Lambda = \text{diag}(\lambda_1, \lambda_2, \ldots, \lambda_n) \in M_n$. Show that $\Lambda^\dagger = [\lambda_1]^\dagger \oplus [\lambda_2]^\dagger \oplus \cdots \oplus [\lambda_n]^\dagger$.

P.17.29 If $A \in M_n$ is normal and $A = U\Lambda U^*$ is a spectral decomposition, show that $A^\dagger = U\Lambda^\dagger U^*$.

P.17.30 Let $A = \begin{bmatrix} 1 & 0 \\ 0 & 2 \end{bmatrix}$ and $B = \begin{bmatrix} 1 & 1 \\ 1 & 1 \end{bmatrix}$. Show that $(AB)^\dagger \neq B^\dagger A^\dagger$.

P.17.31 Show that the matrix A in Example 17.4.14 is an EP matrix. Is A normal?

P.17.32 If $A \in M_n$ is a circulant matrix, show that its pseudoinverse is a circulant matrix. *Hint*: A is normal. Use Theorem 17.4.11.

P.17.33 Let $A \in M_n$ be invertible. Find a noninvertible $B \in M_n$ such that $\|A - B\|_F = \|A - B\|_2 = \sigma_{\min}(A)$.

P.17.34 If $A \in M_n$ is complex symmetric, prove that $A^\dagger$ is complex symmetric.

P.17.35 Let $A \in M_n$ be idempotent. (a) Show that $I - A$ is idempotent. (b) Show that A and $I - A$ have the same singular values that are greater than 1 (if any). (c) If $A \neq 0$ and $A \neq I$, show that the largest singular value of A equals the largest singular value of $I - A$.

P.17.36 Let $A, B \in M_n$ be idempotent. Show that A and B are unitarily similar if and only if they have the same singular values.

P.17.37 Let $A \in M_n$ be idempotent. Show in two ways that $\|A\|_2 \leq 1$ if and only if A is an orthogonal projection. (a) Show that $\|Ax - A^*Ax\|_2 = 0$ for all $\mathbf{x} \in \mathbb{C}^n$, and (b) use Theorem 17.7.1.

P.17.38 Show that the idempotent matrix in Example 8.3.16 is unitarily similar to $\begin{bmatrix} 1 & 3 \\ 0 & 0 \end{bmatrix}$, which is a matrix of the form (8.7.2).

P.17.39 Let $A \in M_n$ have rank $r \geq 1$, let $\sigma_1 \geq \sigma_2 \geq \cdots \geq \sigma_r > 0$ be the positive singular values of A, and suppose that A is nilpotent of index two. (a) Prove that A is unitarily similar to

$$\begin{bmatrix} 0 & \sigma_1 \\ 0 & 0 \end{bmatrix} \oplus \cdots \oplus \begin{bmatrix} 0 & \sigma_r \\ 0 & 0 \end{bmatrix} \oplus 0_{n-2r}. \tag{17.8.3}$$

Hint: Adapt the proof of Theorem 17.7.1. (b) If A is real, show that it is real orthogonally similar to (17.8.3).

P.17.40 Let $A, B \in M_n$ and suppose that $A^2 = 0 = B^2$. Show that A and B are unitarily similar if and only if they have the same singular values.

P.17.41 Let $\tau > 0$. (a) Show that $C = \begin{bmatrix} 1 & \tau \\ 0 & -1 \end{bmatrix}$ satisfies $C^2 = I$. (b) Compute the singular values of C, show that they are reciprocals, and explain why one of them is greater than 1.

P.17.42 Let $A \in M_n$ and suppose that A is an involution ($A^2 = I$). Let $\sigma_1 \geq \sigma_2 \geq \cdots \geq \sigma_k > 1$ be the singular values of A that are greater than 1, if any. (a) Show that $\operatorname{spec} A \subseteq \{1, -1\}$. (b) Show that the singular values of A are $\sigma_1, \sigma_1^{-1}, \sigma_2, \sigma_2^{-1}, \ldots, \sigma_k, \sigma_k^{-1}$, together with 1 (with multiplicity $n - 2k$). *Hint*: $(AA^*)^{-1} = A^*A$.

P.17.43 Let $A \in M_n$. (a) Show that A is idempotent if and only if $2A - I$ is an involution. (b) Show that A is an involution if and only if $\frac{1}{2}(A + I)$ is idempotent.

P.17.44 Let $A \in M_n(\mathbb{F})$, let p be the multiplicity of 1 as an eigenvalue of A, and let k be the number of singular values of A that are greater than 1. Show that A is an involution if and only if there is a unitary $U \in M_n(\mathbb{F})$ and positive real numbers $\tau_1, \tau_2, \ldots, \tau_k$ such that

$$U^*AU = \begin{bmatrix} 1 & \tau_1 \\ 0 & -1 \end{bmatrix} \oplus \cdots \oplus \begin{bmatrix} 1 & \tau_k \\ 0 & -1 \end{bmatrix} \oplus I_{p-k} \oplus (-I_{n-p-k}). \tag{17.8.4}$$

Hint: Use the preceding problem and (17.7.2).

P.17.45 Suppose that $A, B \in M_n$ are involutions. Show that A and B are unitarily similar if and only if they have the same singular values, and $+1$ is an eigenvalue with the same multiplicity for each of them.

P.17.46 Let $A \in M_{m \times n}$. Show that $\|A\|_2 \leq 1$ if and only if $I_n - A^*A$ is positive semidefinite. A matrix that satisfies either of these conditions is a *contraction*.

P.17.47 Let $A \in M_n$ be normal and let $A = U\Lambda U^*$ be a spectral decomposition. (a) Show that $H(A) = UH(\Lambda)U^*$; see Definition 14.6.6. (b) If A is a contraction, show that $H(A) = I$ if and only if $A = I$.

P.17.48 Let $U, V \in M_n$ be unitary and let $C = \frac{1}{2}(U + V)$. (a) Show that C is a contraction. (b) Show that C is unitary if and only if $U = V$.

P.17.49 Let $C \in M_n$ be a contraction. Show that there are unitary $U, V \in M_n$ such that $C = \frac{1}{2}(U + V)$. *Hint*: If $0 \leq \sigma \leq 1$ and $s_{\pm} = \sigma \pm i\sqrt{1 - \sigma^2}$, then $\sigma = \frac{1}{2}(s_+ + s_-)$ and $|s_{\pm}| = 1$.

P.17.50 Deduce from the preceding problem that every square complex matrix is a linear combination of at most two unitary matrices. See P.15.60 for a related result.

P.17.51 Use P.17.50 to show that M_n has a basis consisting of unitary matrices. See P.15.61 for a different proof of this result.

P.17.52 Let $A, B \in M_n$ and let p be a polynomial. (a) Show that $Bp(AB) = p(BA)B$. (b) If A is a contraction, use (a) to show that

$$A^*(I - AA^*)^{1/2} = (I - A^*A)^{1/2}A^*. \qquad (17.8.5)$$

P.17.53 Let $A \in M_n$. Use the singular value decomposition to prove (17.8.5).

P.17.54 Let $A \in M_n$. Show that A is a contraction if and only if for some $m \geq n$, there is a block matrix of the form $U = \begin{bmatrix} A & B \\ C & D \end{bmatrix} \in M_m$ that is unitary. Thus, every contraction is a principal submatrix of some unitary matrix. *Hint*: Start with $B = (I - AA^*)^{1/2}$.

P.17.55 Let $B \in M_n$ and suppose that $1 \notin \operatorname{spec} B$. (a) Show that $(I - B)^{-1} + (I - B^*)^{-1} - I = (I - B)^{-1}(I - BB^*)(I - B^*)^{-1}$. (b) Deduce from (a) that B is a contraction if and only if $(I - B)^{-1} + (I - B^*)^{-1} - I$ is positive semidefinite.

P.17.56 Let $A \in M_n$, $B = \begin{bmatrix} 0 & A \\ 0 & 0 \end{bmatrix} \in M_{2n}$, and $C = \begin{bmatrix} I & A \\ A^* & I \end{bmatrix} \in M_{2n}$. (a) Show that $\|A\|_2 = \|B\|_2$. (b) Use the preceding problem to show that A is a contraction if and only if C is positive semidefinite. *Hint*: $1 \notin \operatorname{spec} B$. (c) Use the singular value decomposition to prove the assertion in (b).

P.17.57 Show that the spectral norm of a nonzero idempotent matrix can be any number in the real interval $[1, \infty)$, but it cannot be any number in the real interval $[0, 1)$.

P.17.58 Under the same assumptions used to derive the upper bound (17.5.7), prove the lower bound

$$\frac{1}{\kappa_2(A)} \frac{\|\mathbf{y}'\|_2}{\|\mathbf{y}\|_2} \leq \frac{\|\mathbf{x}'\|_2}{\|\mathbf{x}\|_2}$$

for the relative error in a solution of the perturbed linear system (17.5.5).

P.17.59 Define the *Frobenius condition number* of an invertible matrix $A \in M_n$ by $\kappa_F(A) = \|A\|_F \|A^{-1}\|_F$. Let $B \in M_n$ be invertible. Prove the following: (a) $\kappa_F(AB) \leq \kappa_F(A)\kappa_F(B)$. (b) $\kappa_F(A) \geq \sqrt{n}$. (c) $\kappa_F(cA) = \kappa_F(A)$ for any nonzero scalar c. (d) $\kappa_F(A) = \kappa_F(A^*) = \kappa_F(\overline{A}) = \kappa_F(A^\mathsf{T})$.

P.17.60 Prove the upper bound

$$\frac{\|\mathbf{x}'\|_2}{\|\mathbf{x}\|_2} \leq \kappa_F(A) \frac{\|\mathbf{y}'\|_2}{\|\mathbf{y}\|_2} \qquad (17.8.6)$$

for the relative error in the linear system analyzed in Section 17.5. If A is unitary, compare the bounds (17.5.7) and (17.8.6) and discuss.

P.17.61 Let $A = [A_{ij}] \in M_{2n}$ be a 2×2 block matrix with each $A_{ij} \in M_n$. Let $B = [\|A_{ij}\|_2] \in M_2$. Let $\mathbf{x} \in \mathbb{C}^{2n}$ be a unit vector, partitioned as $\mathbf{x} = \begin{bmatrix} \mathbf{x}_1 \\ \mathbf{x}_2 \end{bmatrix}$, in which $\mathbf{x}_1, \mathbf{x}_2 \in \mathbb{C}^n$. Provide details for the computation

$$\|A\mathbf{x}\|_2 = \left(\|A_{11}\mathbf{x}_1 + A_{12}\mathbf{x}_2\|_2^2 + \| \cdots \|_2^2 \right)^{1/2}$$

$$\leq \left(\left(\|A_{11}\|_2 \|\mathbf{x}_1\|_2 + \|A_{12}\|_2 \|\mathbf{x}_2\|_2 \right)^2 + (\cdots)^2 \right)^{1/2}$$

$$= \left\| B \begin{bmatrix} \|\mathbf{x}_1\|_2 \\ \|\mathbf{x}_2\|_2 \end{bmatrix} \right\|_2 \leq \max\{\|B\mathbf{y}\|_2 : \mathbf{y} \in \mathbb{C}^2, \|\mathbf{y}\|_2 = 1\}.$$

Conclude that $\|A\|_2 \leq \|B\|_2$.

P.17.62   Let $A = [A_{ij}] \in M_{kn}$ be a $k \times k$ block matrix with each $A_{ij} \in M_n$. Let $B = \left[\|A_{ij}\|_2 \right] \in M_k$. Show that $\|A\|_2 \le \|B\|_2$.

P.17.63   Let $A = [a_{ij}] \in M_n$ and let $B = \left[|a_{ij}| \right]$. Show that $\|A\|_2 \le \|B\|_2$.

17.9 Notes

The truncation (17.1.3) is a best rank-k approximation to A with respect to the Frobenius norm as well as the spectral norm (Theorem 17.2.7). This best approximation is unique if and only if $\sigma_{k-1} > \sigma_k$. See [HJ13, Section 7.4.2].

The structure of the example (17.3.8) is not an accident. If $A \in M_n$ is not a scalar matrix and $\rho(A) = \|A\|_2$, then A is unitarily similar to a matrix of the form $\lambda I \oplus B$, in which $\lambda \in \operatorname{spec} A$ and $|\lambda| = \rho(A)$, $\rho(B) < \rho(A)$, and $\|B\|_2 \le \rho(A)$.

Error bounds for eigenvalues or solutions of linear systems have been studied by numerical analysts since the 1950s. For more information and references to an extensive literature, see [GVL13].

17.10 Some Important Concepts

- Approximate a matrix with a truncated singular value decomposition.

- The largest singular value is a submultiplicative norm on M_n: the spectral norm.

- The Frobenius norm is submultiplicative on M_n.

- Unitary similarity, unitary equivalence, eigenvalues, and singular values.

- Pseudoinverse of a matrix; minimum-norm and least-squares solutions of a linear system.

- Spectral condition number.

- Error bounds for solutions of linear systems.

- Error bounds for eigenvalue computations (Bauer–Fike theorem).

- Singular value decomposition of a complex symmetric matrix (Autonne theorem).

- Unitary similarity of idempotent matrices.

If a Hermitian matrix is perturbed by adding a rank-1 Hermitian matrix, or by bordering to obtain a larger Hermitian matrix, the eigenvalues of the respective matrices are related by interlacing inequalities. We use subspace intersections (Theorem 4.1.11) to study eigenvalue interlacing and the related inequalities of Weyl (Theorem 18.7.1). We discuss applications of eigenvalue interlacing, including Sylvester's principal-minor criterion for positive definiteness, singular value interlacing between a matrix and a submatrix, and majorization inequalities between the eigenvalues and diagonal entries of a Hermitian matrix.

Sylvester's inertia theorem (1852) says that although the transformation $A \mapsto SAS^*$ can change all the nonzero eigenvalues of a Hermitian matrix, it cannot change the number of positive eigenvalues or the number of negative ones. We use subspace intersections to prove Sylvester's theorem, and then use the polar decomposition to prove a generalization of the inertia theorem for normal matrices.

18.1 The Rayleigh Quotient

The eigenvalues of a Hermitian $A \in M_n$ are real. In this chapter, we arrange them in increasing order

$$\lambda_1(A) \le \lambda_2(A) \le \cdots \le \lambda_n(A).$$

We abbreviate $\lambda_i(A)$ to λ_i if only one matrix is under discussion. The foundation upon which many eigenvalue results stand is that if $\mathbf{x} \in \mathbb{C}^n$ is nonzero, then the *Rayleigh quotient*

$$\frac{\langle A\mathbf{x}, \mathbf{x} \rangle}{\langle \mathbf{x}, \mathbf{x} \rangle} = \frac{\mathbf{x}^* A \mathbf{x}}{\mathbf{x}^* \mathbf{x}}$$

is a lower bound for λ_n and an upper bound for λ_1. This principle was discovered by John William Strutt, the third Baron Rayleigh and 1904 Nobel laureate for his discovery of argon.

Theorem 18.1.1 (Rayleigh) *Let $A \in M_n$ be Hermitian, with eigenvalues $\lambda_1 \le \lambda_2 \le \cdots \le \lambda_n$ and corresponding orthonormal eigenvectors $\mathbf{u}_1, \mathbf{u}_2, \ldots, \mathbf{u}_n \in \mathbb{C}^n$. Let $1 \le p \le q \le n$. If $\mathbf{x} \in \mathrm{span}\{\mathbf{u}_p, \mathbf{u}_{p+1}, \ldots, \mathbf{u}_q\}$ is a unit vector, then*

$$\lambda_p \le \langle A\mathbf{x}, \mathbf{x} \rangle \le \lambda_q. \tag{18.1.2}$$

Proof The list $\beta = \mathbf{u}_p, \mathbf{u}_{p+1}, \ldots, \mathbf{u}_q$ is an orthonormal basis for $\mathrm{span}\,\beta$. If $\mathbf{x} \in \mathrm{span}\,\beta$ is a unit vector, Theorem 6.2.5 ensures that $\mathbf{x} = c_p \mathbf{u}_p + c_{p+1}\mathbf{u}_{p+1} + \cdots + c_q \mathbf{u}_q$, in which each $c_i = \langle \mathbf{x}, \mathbf{u}_i \rangle$ and

$$\sum_{i=p}^{q} |c_i|^2 = \|\mathbf{x}\|_2^2 = 1. \tag{18.1.3}$$

The ordering of the eigenvalues and (18.1.3) ensure that

$$\lambda_p = \sum_{i=p}^{q} \lambda_p |c_i|^2 \leq \sum_{i=p}^{q} \lambda_i |c_i|^2 \leq \sum_{i=p}^{q} \lambda_q |c_i|^2 = \lambda_q. \tag{18.1.4}$$

Since

$$\langle A\mathbf{x}, \mathbf{x} \rangle = \left\langle \sum_{i=p}^{q} c_i A\mathbf{u}_i, \mathbf{x} \right\rangle = \left\langle \sum_{i=p}^{q} c_i \lambda_i \mathbf{u}_i, \mathbf{x} \right\rangle = \sum_{i=p}^{q} \lambda_i c_i \langle \mathbf{u}_i, \mathbf{x} \rangle = \sum_{i=p}^{q} \lambda_i c_i \overline{c_i} = \sum_{i=p}^{q} \lambda_i |c_i|^2,$$

the central term in (18.1.4) is $\langle A\mathbf{x}, \mathbf{x} \rangle$, which verifies (18.1.2). $\qquad\square$

The upper bound in (18.1.2) is attained by $\mathbf{x} = \mathbf{u}_q$ and the lower bound is attained by $\mathbf{x} = \mathbf{u}_p$. Therefore,

$$\min_{\substack{\mathbf{x} \in \mathcal{U} \\ \|\mathbf{x}\|_2 = 1}} \langle A\mathbf{x}, \mathbf{x} \rangle = \lambda_p \qquad \text{and} \qquad \max_{\substack{\mathbf{x} \in \mathcal{U} \\ \|\mathbf{x}\|_2 = 1}} \langle A\mathbf{x}, \mathbf{x} \rangle = \lambda_q, \tag{18.1.5}$$

in which $\mathcal{U} = \mathrm{span}\{\mathbf{u}_p, \mathbf{u}_{p+1}, \ldots, \mathbf{u}_q\}$. If $p = 1$ and $q = n$, then $\mathcal{U} = \mathbb{C}^n$ and (18.1.5) becomes

$$\min_{\|\mathbf{x}\|_2 = 1} \langle A\mathbf{x}, \mathbf{x} \rangle = \lambda_1 \qquad \text{and} \qquad \max_{\|\mathbf{x}\|_2 = 1} \langle A\mathbf{x}, \mathbf{x} \rangle = \lambda_n.$$

Consequently,

$$\lambda_1 \leq \langle A\mathbf{x}, \mathbf{x} \rangle \leq \lambda_n \tag{18.1.6}$$

for any unit vector $\mathbf{x}$. For example, if $\mathbf{x} = \mathbf{e}_i$, then $\langle A\mathbf{e}_i, \mathbf{e}_i \rangle = a_{ii}$ is a diagonal entry of A and (18.1.6) tells us that

$$\lambda_1 \leq a_{ii} \leq \lambda_n \qquad \text{for } i = 1, 2, \ldots, n. \tag{18.1.7}$$

Example 18.1.8 Let

$$A = \begin{bmatrix} 6 & 3 & 0 & 3 \\ 3 & -2 & 5 & 2 \\ 0 & 5 & -2 & 3 \\ 3 & 2 & 3 & 4 \end{bmatrix}.$$

The inequalities (18.1.7) tell us that $\lambda_1 \leq -2$ and $\lambda_4 \geq 6$. Additional bounds can be obtained from (18.1.6) with any unit vector $\mathbf{x}$. For example, with $\mathbf{x} = \frac{1}{2}[1 \ -1 \ 1 \ -1]^{\mathsf{T}}$ we obtain $\langle A\mathbf{x}, \mathbf{x} \rangle = -4.5$. With $\mathbf{x} = \frac{1}{2}[1 \ 1 \ 1 \ 1]^{\mathsf{T}}$ we obtain $\langle A\mathbf{x}, \mathbf{x} \rangle = 9.5$, so $\lambda_1 \leq -4.5 < 9.5 \leq \lambda_4$. Rounded to two decimal places, the actual values are $\lambda_1 = -7.46$ and $\lambda_4 = 10.29$.

18.2 Eigenvalue Interlacing for Sums of Hermitian Matrices

Example 18.2.1 Consider the Hermitian matrices

$$A = \begin{bmatrix} 6 & 3 & 0 & 3 \\ 3 & -2 & 5 & 2 \\ 0 & 5 & -2 & 3 \\ 3 & 2 & 3 & 4 \end{bmatrix} \qquad \text{and} \qquad E = \begin{bmatrix} 4 & -2 & 0 & 2 \\ -2 & 1 & 0 & -1 \\ 0 & 0 & 0 & 0 \\ 2 & -1 & 0 & 1 \end{bmatrix} = 6\mathbf{u}\mathbf{u}^*,$$

in which $\mathbf{u} = \frac{1}{\sqrt{6}}[2 \; -1 \; 0 \; 1]^{\mathsf{T}}$. Rounded to two decimal places, the respective eigenvalues are

$$A : -7.46, \; -0.05, \; 3.22, \; 10.29,$$
$$E : 0, \; 0, \; 0, \; 6,$$
$$A + E : -6.86, \; 0.74, \; 4.58, \; 13.54, \quad \text{and}$$
$$A - E : -8.92, \; -1.99, \; 2.21, \; 8.70.$$

The increasingly ordered eigenvalues of A and $A \pm E$ satisfy

$$\lambda_1(A) < \lambda_1(A+E) < \lambda_2(A) < \lambda_2(A+E) < \lambda_3(A) < \lambda_3(A+E) < \lambda_4(A) < \lambda_4(A+E)$$

and

$$\lambda_1(A-E) < \lambda_1(A) < \lambda_2(A-E) < \lambda_2(A) < \lambda_3(A-E) < \lambda_3(A) < \lambda_4(A-E) < \lambda_4(A).$$

Adding the positive semidefinite matrix E to A increases each of its eigenvalues, subject to the upper bounds $\lambda_i(A+E) < \lambda_{i+1}(A)$ for $i = 1, 2, 3$. Subtracting E from A decreases each of its eigenvalues, subject to the lower bounds $\lambda_i(A-E) > \lambda_{i-1}(A)$ for $i = 2, 3, 4$. The eigenvalues of $A \pm E$ *interlace* the eigenvalues of A. Figure 18.1 illustrates the eigenvalue interlacings in this example, which are not accidental.

Theorem 18.2.2 *Let $n \geq 2$ and let $A, E \in \mathsf{M}_n$ be Hermitian. Suppose that E is positive semidefinite and has rank 1. The increasingly ordered eigenvalues of A and $A \pm E$ satisfy*

$$\lambda_i(A) \leq \lambda_i(A+E) \leq \lambda_{i+1}(A) \leq \lambda_n(A+E) \quad \text{for } i = 1, 2, \ldots, n-1, \quad (18.2.3)$$

and

$$\lambda_i(A-E) \leq \lambda_i(A) \leq \lambda_{i+1}(A-E) \leq \lambda_n(A) \quad \text{for } i = 1, 2, \ldots, n-1. \quad (18.2.4)$$

Proof Let $\mathbf{u}_1, \mathbf{u}_2, \ldots, \mathbf{u}_n$ and $\mathbf{v}_1, \mathbf{v}_2, \ldots, \mathbf{v}_n$ be orthonormal eigenvectors corresponding to the increasingly ordered eigenvalues of A and $A + E$, respectively. Let $i \in \{1, 2, \ldots, n-1\}$, $\mathcal{U} = \mathrm{span}\{\mathbf{u}_1, \mathbf{u}_2, \ldots, \mathbf{u}_{i+1}\}$, $\mathcal{V} = \mathrm{span}\{\mathbf{v}_i, \mathbf{v}_{i+1}, \ldots, \mathbf{v}_n\}$, and $\mathcal{W} = \mathrm{null}\, E$. Then $\dim \mathcal{U} = i+1$, $\dim \mathcal{V} = n - i + 1$, and $\dim \mathcal{W} = n - \mathrm{rank}\, E = n - 1$. Because $\dim \mathcal{U} + \dim \mathcal{V} = n + 2$, Theorem 4.1.11 ensures that $\dim(\mathcal{U} \cap \mathcal{V}) \geq 2$. Since

$$\dim(\mathcal{U} \cap \mathcal{V}) + \dim \mathcal{W} \geq 2 + (n-1) = n + 1,$$

the same theorem says that there is a unit vector $\mathbf{x} \in \mathcal{U} \cap \mathcal{V} \cap \mathcal{W}$. Use Theorem 18.1.1 to compute

$$\lambda_i(A+E) \leq \langle (A+E)\mathbf{x}, \mathbf{x} \rangle \qquad \text{since } \mathbf{x} \in \mathcal{V}$$
$$= \langle A\mathbf{x}, \mathbf{x} \rangle + \langle E\mathbf{x}, \mathbf{x} \rangle = \langle A\mathbf{x}, \mathbf{x} \rangle \qquad \text{since } \mathbf{x} \in \mathcal{W}$$
$$\leq \lambda_{i+1}(A) \qquad \text{since } \mathbf{x} \in \mathcal{U}.$$

This shows that $\lambda_i(A+E) \leq \lambda_{i+1}(A)$ for $i = 1, 2, \ldots, n-1$.

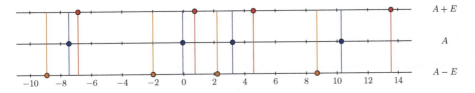

Figure 18.1 Eigenvalues of A and $A \pm E$ in Example 18.2.1; rank $E = 1$.

Let $i \in \{1, 2, \ldots, n\}$, $\mathcal{U} = \text{span}\{\mathbf{u}_i, \mathbf{u}_{i+1}, \ldots, \mathbf{u}_n\}$, and $\mathcal{V} = \text{span}\{\mathbf{v}_1, \mathbf{v}_2, \ldots, \mathbf{v}_i\}$. Then $\dim \mathcal{U} = n - i + 1$ and $\dim \mathcal{V} = i$, so $\dim \mathcal{U} + \dim \mathcal{V} = n + 1$. Theorem 4.1.11 ensures that there is a unit vector $\mathbf{x} \in \mathcal{U} \cap \mathcal{V}$. Use Theorem 18.1.1, and the fact that E is positive semidefinite, to compute

$$
\begin{aligned}
\lambda_i(A) &\leq \langle A\mathbf{x}, \mathbf{x} \rangle && \text{since } \mathbf{x} \in \mathcal{U} \\
&\leq \langle A\mathbf{x}, \mathbf{x} \rangle + \langle E\mathbf{x}, \mathbf{x} \rangle = \langle (A + E)\mathbf{x}, \mathbf{x} \rangle && \text{since } \langle E\mathbf{x}, \mathbf{x} \rangle \geq 0 \\
&\leq \lambda_i(A + E) && \text{since } \mathbf{x} \in \mathcal{V}.
\end{aligned}
$$

This shows that $\lambda_i(A) \leq \lambda_i(A + E)$ for $i = 1, 2, \ldots, n$.

We have now proved the interlacing inequalities (18.2.3). Replace A with $A - E$ and replace $A + E$ with A in (18.2.3) to obtain (18.2.4). □

The preceding theorem provides bounds on the eigenvalues of a Hermitian matrix $A \in \mathsf{M}_n$ that is perturbed by adding or subtracting a rank-1 positive semidefinite Hermitian matrix E. No matter how large the entries of E are, the eigenvalues $\lambda_2(A \pm E), \lambda_3(A \pm E), \ldots, \lambda_{n-1}(A \pm E)$ are confined to intervals between a pair of adjacent eigenvalues of A. Since (14.2.4) ensures that a rank-r Hermitian matrix H is a real linear combination of r rank-1 Hermitian matrices, we can apply Theorem 18.2.2 r times to find bounds for the eigenvalues of $A + H$; see P.18.6.

18.3 Eigenvalue Interlacing for Bordered Hermitian Matrices

We now study another context in which eigenvalue interlacing occurs.

Example 18.3.1 Let

$$
A = \begin{bmatrix} 4 & -2 & 1 & 3 & 1 \\ -2 & 6 & 4 & -1 & -3 \\ 1 & 4 & 4 & -1 & -2 \\ 3 & -1 & -1 & 2 & -2 \\ 1 & -3 & -2 & -2 & -2 \end{bmatrix} = \begin{bmatrix} B & \mathbf{y} \\ \mathbf{y}^* & c \end{bmatrix},
$$

in which

$$
B = \begin{bmatrix} 4 & -2 & 1 & 3 \\ -2 & 6 & 4 & -1 \\ 1 & 4 & 4 & -1 \\ 3 & -1 & -1 & 2 \end{bmatrix}, \qquad \mathbf{y} = \begin{bmatrix} 1 \\ -3 \\ -2 \\ -2 \end{bmatrix}, \qquad \text{and} \quad c = -2.
$$

Rounded to two decimal places, the eigenvalues of A and B are

$$
A: -4.38, -0.60, 2.15, 6.15, 10.68 \qquad \text{and} \qquad B: -1.18, 1.38, 5.91, 9.89.
$$

The increasingly ordered eigenvalues of A and B interlace:

$$
\lambda_1(A) < \lambda_1(B) < \lambda_2(A) < \lambda_2(B) < \lambda_3(A) < \lambda_3(B) < \lambda_4(A) < \lambda_4(B) < \lambda_5(A);
$$

see the first two lines in Figure 18.2.

Theorem 18.3.2 *Let $B \in \mathsf{M}_n$ be Hermitian, $\mathbf{y} \in \mathbb{C}^n$, $c \in \mathbb{R}$, and*

$$
A = \begin{bmatrix} B & \mathbf{y} \\ \mathbf{y}^* & c \end{bmatrix} \in \mathsf{M}_{n+1}.
$$

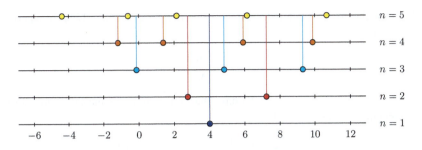

Figure 18.2 Eigenvalues of the $n \times n$ leading principal submatrices of A in Example 18.3.1 for $n = 1, 2, 3, 4, 5$; $n = 5$ corresponds to A and $n = 4$ to B.

The increasingly ordered eigenvalues of A and B satisfy

$$\lambda_i(A) \leq \lambda_i(B) \leq \lambda_{i+1}(A) \quad \textit{for } i = 1, 2, \ldots, n. \tag{18.3.3}$$

Proof We proceed in parallel to the proof of Theorem 18.2.2. Let $\mathbf{u}_1, \mathbf{u}_2, \ldots, \mathbf{u}_{n+1} \in \mathbb{C}^{n+1}$ and $\mathbf{v}_1, \mathbf{v}_2, \ldots, \mathbf{v}_n \in \mathbb{C}^n$ be orthonormal eigenvectors corresponding to the increasingly ordered eigenvalues of A and B, respectively, and let $\mathbf{w}_i = \begin{bmatrix} \mathbf{v}_i \\ 0 \end{bmatrix} \in \mathbb{C}^{n+1}$ for $i = 1, 2, \ldots, n$.

Let $i \in \{1, 2, \ldots, n\}$, $\mathcal{U} = \mathrm{span}\{\mathbf{u}_i, \mathbf{u}_{i+1}, \ldots, \mathbf{u}_{n+1}\} \subseteq \mathbb{C}^{n+1}$, $\mathcal{V} = \mathrm{span}\{\mathbf{v}_1, \mathbf{v}_2, \ldots, \mathbf{v}_i\}$, and $\mathcal{W} = \mathrm{span}\{\mathbf{w}_1, \mathbf{w}_2, \ldots, \mathbf{w}_i\} \subseteq \mathbb{C}^{n+1}$. Then $\dim \mathcal{U} = n - i + 2$ and $\dim \mathcal{V} = \dim \mathcal{W} = i$. Since $\mathcal{U}$ and $\mathcal{W}$ are subspaces of $\mathbb{C}^{n+1}$ and $\dim \mathcal{U} + \dim \mathcal{W} = n + 2$, there is a unit vector $\mathbf{x} = \begin{bmatrix} \mathbf{v} \\ 0 \end{bmatrix} \in \mathcal{U} \cap \mathcal{W}$, in which $\mathbf{v} \in \mathcal{V}$. Therefore,

$$\begin{aligned} \lambda_i(A) &\leq \langle A\mathbf{x}, \mathbf{x} \rangle && \text{since } \mathbf{x} \in \mathcal{U} \\ &= \langle B\mathbf{v}, \mathbf{v} \rangle && \text{since } \mathbf{x} \in \mathcal{W} \\ &\leq \lambda_i(B) && \text{since } \mathbf{v} \in \mathcal{V}. \end{aligned}$$

Let $i \in \{1, 2, \ldots, n\}$, $\mathcal{U} = \mathrm{span}\{\mathbf{u}_1, \mathbf{u}_2, \ldots, \mathbf{u}_{i+1}\}$, $\mathcal{V} = \mathrm{span}\{\mathbf{v}_i, \mathbf{v}_{i+1}, \ldots, \mathbf{v}_n\}$, and $\mathcal{W} = \mathrm{span}\{\mathbf{w}_i, \mathbf{w}_{i+1}, \ldots, \mathbf{w}_n\}$. Then $\dim \mathcal{U} = i + 1$ and $\dim \mathcal{V} = \dim \mathcal{W} = n - i + 1$, so $\dim \mathcal{U} + \dim \mathcal{W} = n + 2$. Thus, there is a unit vector $\mathbf{x} = \begin{bmatrix} \mathbf{v} \\ 0 \end{bmatrix} \in \mathcal{U} \cap \mathcal{W}$, in which $\mathbf{v} \in \mathcal{V}$. Then

$$\begin{aligned} \lambda_i(B) &\leq \langle B\mathbf{v}, \mathbf{v} \rangle && \text{since } \mathbf{x} \in \mathcal{V} \\ &= \langle A\mathbf{x}, \mathbf{x} \rangle && \text{since } \mathbf{x} \in \mathcal{W} \\ &\leq \lambda_{i+1}(A) && \text{since } \mathbf{v} \in \mathcal{U}. \end{aligned} \qquad \square$$

The first corollary of the preceding theorem provides interlacing inequalities for singular values. In the proof, we have to deal with conflicting conventions: singular values are indexed in decreasing order, while eigenvalues are indexed in increasing order.

Corollary 18.3.4 *Let $n \geq 2$ and $A \in \mathsf{M}_n$. Delete a row or column of A and denote the resulting matrix by B, which is in $\mathsf{M}_{(n-1) \times n}$ or in $\mathsf{M}_{n \times (n-1)}$. The decreasingly ordered singular values of A and B satisfy*

$$\sigma_i(A) \geq \sigma_i(B) \geq \sigma_{i+1}(A) \qquad \textit{for } i = 1, 2, \ldots, n-1. \tag{18.3.5}$$

Proof Suppose that a row is deleted and denote that row by $\mathbf{y}^* \in \mathsf{M}_{1 \times n}$. Let P be a permutation matrix such that

$$PA = \begin{bmatrix} B \\ \mathbf{y}^* \end{bmatrix}.$$

Theorem 17.3.12.a ensures that A and PA have the same singular values, the squares of which are the eigenvalues of the bordered matrix

$$(PA)(PA)^* = \begin{bmatrix} B \\ \mathbf{y}^* \end{bmatrix} [B^* \ \mathbf{y}] = \begin{bmatrix} BB^* & B\mathbf{y} \\ \mathbf{y}^*B^* & \mathbf{y}^*\mathbf{y} \end{bmatrix}.$$

Arranged in increasing order, the eigenvalues of AA^* are $\sigma_n^2(A) \leq \sigma_{n-1}^2(A) \leq \cdots \leq \sigma_1^2(A)$, that is

$$\lambda_i(AA^*) = \sigma_{n-i+1}^2(A) \qquad \text{for } i = 1, 2, \ldots, n.$$

The increasingly ordered eigenvalues of BB^* are $\sigma_{n-1}^2(B) \leq \sigma_{n-2}^2(B) \leq \cdots \leq \sigma_1^2(B)$, that is,

$$\lambda_i(BB^*) = \sigma_{n-i}^2(B) \qquad \text{for } i = 1, 2, \ldots, n - 1.$$

Now invoke (18.3.3), which says that the increasingly ordered eigenvalues of BB^* interlace the increasingly ordered eigenvalues of AA^*, that is,

$$\sigma_{n-i+1}^2(A) \leq \sigma_{n-i}^2(B) \leq \sigma_{n-i}^2(A) \qquad \text{for } i = 1, 2, \ldots, n - 1.$$

Change variables in the indices to obtain

$$\sigma_i^2(A) \geq \sigma_i^2(B) \geq \sigma_{i+1}^2(A) \qquad \text{for } i = 1, 2, \ldots, n - 1,$$

which is equivalent to (18.3.5). If a column of A is deleted, apply the preceding result to A^*. □

Example 18.3.6 Consider

$$A = \begin{bmatrix} 1 & 2 & 3 & 4 \\ 8 & 7 & 6 & 5 \\ 10 & 11 & 9 & 8 \\ 12 & 13 & 15 & 14 \end{bmatrix} \quad \text{and} \quad B = \begin{bmatrix} 1 & 2 & 3 & 4 \\ 8 & 7 & 6 & 5 \\ 12 & 13 & 15 & 14 \end{bmatrix}.$$

Rounded to two decimal places, the singular values of A and B are 35.86, 4.13, 1.00, 0.48 and 30.42, 3.51, 0.51, respectively. The interlacing inequalities (18.3.5) are satisfied.

The next corollary generalizes Theorem 18.3.2 to address leading principal submatrices of any size.

Corollary 18.3.7 *Let $B \in \mathsf{M}_n$ and $C \in \mathsf{M}_m$ be Hermitian, let $Y \in \mathsf{M}_{n \times m}$, and let*

$$A_m = \begin{bmatrix} B & Y \\ Y^* & C \end{bmatrix} \in \mathsf{M}_{n+m}.$$

The increasingly ordered eigenvalues of A_m and B satisfy

$$\lambda_i(A_m) \leq \lambda_i(B) \leq \lambda_{i+m}(A_m) \qquad \text{for } i = 1, 2, \ldots, n. \tag{18.3.8}$$

Proof We proceed by induction on m. Theorem 18.3.2 establishes the base case $m = 1$. For the induction step, let $m \geq 1$ and suppose that the inequalities (18.3.8) are valid. Write A_{m+1} as

$$A_{m+1} = \begin{bmatrix} A_m & \mathbf{y} \\ \mathbf{y}^* & c \end{bmatrix} \in \mathsf{M}_{n+m+1}.$$

Theorem 18.3.2 ensures that the increasingly ordered eigenvalues of A_m and A_{m+1} satisfy

$$\lambda_i(A_{m+1}) \leq \lambda_i(A_m) \leq \lambda_{i+1}(A_{m+1}) \qquad \text{for } i = 1, 2, \ldots, n+m. \tag{18.3.9}$$

The induction hypothesis and (18.3.9) ensure that

$$\lambda_i(A_{m+1}) \leq \lambda_i(A_m) \leq \lambda_i(B) \qquad \text{for } i = 1, 2, \ldots, n, \tag{18.3.10}$$

as well as

$$\lambda_i(B) \leq \lambda_{i+m}(A_m) \leq \lambda_{i+m+1}(A_{m+1}) \qquad \text{for } i = 1, 2, \ldots, n. \tag{18.3.11}$$

Now combine (18.3.10) and (18.3.11), and conclude that

$$\lambda_i(A_{m+1}) \leq \lambda_i(B) \leq \lambda_{i+m+1}(A_{m+1}) \qquad \text{for } i = 1, 2, \ldots, n.$$

This completes the induction. $\qquad\square$

Example 18.3.12 Figure 18.2 illustrates the interlacing inequalities (18.3.8) for the 5×5 matrix A in Example 18.3.1 and $m = 5, 4, 3, 2,$ and 1.

18.4 Sylvester's Criterion

The criterion for positive definiteness in the following theorem has already been proved using a Cholesky factorization (Corollary 15.4.2); we now prove it using eigenvalue interlacing.

Theorem 18.4.1 (Sylvester Criterion) *A Hermitian matrix is positive definite if and only if all of its leading principal minors are positive.*

Proof Let $A \in \mathsf{M}_n$ be Hermitian.
($\Rightarrow$) Suppose that $A \in \mathsf{M}_n$ is positive definite and partition it as $A = \begin{bmatrix} A_{11} & A_{12} \\ A_{12}^* & A_{22} \end{bmatrix}$, in which $A_{11} \in \mathsf{M}_k$. If $\mathbf{y} \in \mathbb{C}^k$ is nonzero and $\mathbf{x} = \begin{bmatrix} \mathbf{y} \\ \mathbf{0} \end{bmatrix} \in \mathbb{C}^n$, then $\mathbf{x} \neq \mathbf{0}$ and $0 < \langle A\mathbf{x}, \mathbf{x} \rangle = \langle A_{11}\mathbf{y}, \mathbf{y} \rangle$. We conclude that each leading $k \times k$ principal submatrix of A is positive definite, and consequently its determinant, a leading principal minor of A, is positive (Theorem 15.1.8.b).
($\Leftarrow$) We proceed by induction on n. For the base case $n = 1$, the hypothesis is that $\det A = a_{11} > 0$, so there is nothing to prove. For the induction step, let $n \geq 1$ and assume that an $n \times n$ Hermitian matrix is positive definite if all of its leading principal minors are positive. Let $A \in \mathsf{M}_{n+1}$ be Hermitian and suppose that all of its leading principal minors are positive. Partition A as

$$A = \begin{bmatrix} B & \mathbf{y} \\ \mathbf{y}^* & a \end{bmatrix}, \qquad \text{in which } B \in \mathsf{M}_n.$$

Then $\det A > 0$ and each leading principal minor of B is positive. The induction hypothesis ensures that B is positive definite, so its eigenvalues are positive. Theorem 18.3.2 tells us that the eigenvalues of B interlace those of A, that is,

$$\lambda_1(A) \leq \lambda_1(B) \leq \lambda_2(A) \leq \cdots \leq \lambda_n(B) \leq \lambda_{n+1}(A).$$

Since $\lambda_1(B) > 0$, it follows that $\lambda_2(A), \lambda_3(A), \ldots, \lambda_{n+1}(A)$ are positive. We also know that

$$\lambda_1(A)\big(\lambda_2(A)\lambda_3(A)\cdots\lambda_{n+1}(A)\big) = \det A > 0,$$

so $\lambda_1(A) > 0$ and A is positive definite because its eigenvalues are positive. This completes the induction. $\qquad\square$

Example 18.4.2 Consider the Hermitian matrix

$$A = \begin{bmatrix} 2 & 2 & 2 \\ 2 & 3 & 3 \\ 2 & 3 & 4 \end{bmatrix}.$$

Its leading principal minors are 2, 2, and 2, so Sylvester's criterion ensures that it is positive definite. It is not diagonally dominant, so the criteria in Theorem 15.2.4 cannot be used to show that A is positive definite.

18.5 Diagonal Entries and Eigenvalues of Hermitian Matrices

If $A \in M_n$ has real diagonal entries, there is a permutation matrix $P \in M_n$ such that the diagonal entries of PAP^T are in increasing order; see the discussion following Definition 7.3.3. The matrices A and PAP^T are similar, so they have the same eigenvalues.

Definition 18.5.1 The *increasingly ordered diagonal entries* of a Hermitian matrix $A = [a_{ij}] \in M_n$ are the real numbers $d_1 \leq d_2 \leq \cdots \leq d_n$ defined by

$$d_1 = \min\{a_{11}, a_{22}, \ldots, a_{nn}\}$$
$$d_j = \min\{a_{11}, a_{22}, \ldots, a_{nn}\}\backslash\{d_1, d_2, \ldots, d_{j-1}\} \qquad \text{for } j = 2, 3, \ldots, n.$$

Theorem 18.5.2 (Schur) *Let $A = [a_{ij}] \in M_n$ be Hermitian. Let $\lambda_1 \leq \lambda_2 \leq \cdots \leq \lambda_n$ be its increasingly ordered eigenvalues and let $d_1 \leq d_2 \leq \cdots \leq d_n$ be its increasingly ordered diagonal entries. Then*

$$\sum_{i=1}^{k} \lambda_i \leq \sum_{i=1}^{k} d_i \qquad \text{for } k = 1, 2, \ldots, n, \tag{18.5.3}$$

with equality for $k = n$.

Proof A permutation similarity of A does not change its eigenvalues, so we may assume that $a_{ii} = d_i$, that is, $a_{11} \leq a_{22} \leq \cdots \leq a_{nn}$. Let $k \in \{1, 2, \ldots, n\}$ and partition $A = \begin{bmatrix} A_{11} & A_{12} \\ A_{12}^* & A_{22} \end{bmatrix}$, in which $A_{11} \in M_k$. The inequalities (18.3.8) ensure that the increasingly ordered eigenvalues of A and its principal submatrix A_{11} satisfy $\lambda_i(A) \leq \lambda_i(A_{11})$ for $i = 1, 2, \ldots, k$. Therefore,

$$\sum_{i=1}^{k} \lambda_i(A) \leq \sum_{i=1}^{k} \lambda_i(A_{11}) = \operatorname{tr} A_{11} = \sum_{i=1}^{k} a_{ii} = \sum_{i=1}^{k} d_i,$$

which proves (18.5.3). If $k = n$, then both sides of (18.5.3) equal $\operatorname{tr} A$. $\qquad\square$

Inequalities of the form (18.5.3) are *majorization inequalities*. See P.18.37 for a different, but equivalent, version of the majorization inequalities.

Example 18.5.4 Let A be the matrix in Example 18.3.1. To two decimal places, quantities in the inequalities (18.5.3) are

k	1	2	3	4	5
$\sum_{i=1}^{k} \lambda_i$	-4.38	-4.98	-2.83	3.32	14
$\sum_{i=1}^{k} a_i$	-2	0	4	8	14

18.6 *Congruence and Inertia of Hermitian Matrices

In Section 15.5, the transformation $B \longmapsto SBS^*$ arose in our discussion of simultaneous diagonalization of two quadratic forms. We now study the properties of this transformation, with initial emphasis on its effect on the eigenvalues of a Hermitian matrix.

Definition 18.6.1 Let $A, B \in \mathsf{M}_n$. Then A is *congruent to B if there is an invertible $S \in \mathsf{M}_n$ such that $A = SBS^*$.

Example 18.6.2 If two matrices are unitarily similar, then they are *congruent. In particular, real orthogonal similarity and permutation similarity are *congruences.

Like similarity and unitary equivalence, *congruence is an equivalence relation; see P.18.3.
Suppose that $A \in \mathsf{M}_n$ is Hermitian and $S \in \mathsf{M}_n$ is invertible. Then SAS^* is Hermitian and rank $A = \operatorname{rank} SAS^*$ (Theorem 4.2.1). Therefore, if A is not invertible, the multiplicities of 0 as an eigenvalue of A and SAS^* are the same (Theorem 10.4.14). What other properties do A and SAS^* have in common? As a first step toward an answer, we introduce a type of diagonal matrix to which each Hermitian matrix is *congruent.

Definition 18.6.3 A *real inertia matrix* is a diagonal matrix whose entries are in $\{-1, 0, 1\}$.

Theorem 18.6.4 *Let $A \in \mathsf{M}_n$ be Hermitian. Then A is *congruent to a real inertia matrix D, in which the number of $+1$ diagonal entries in D is the number of positive eigenvalues of A. The number of -1 diagonal entries in D is the number of negative eigenvalues of A.*

Proof Let $A = U\Lambda U^*$ be a spectral decomposition, in which U is unitary, $\Lambda = \operatorname{diag}(\lambda_1, \lambda_2, \ldots, \lambda_n)$, and the eigenvalues $\lambda_1, \lambda_2, \ldots, \lambda_n$ are in any desired order. Let $G = \operatorname{diag}(g_1, g_2, \ldots, g_n)$, in which

$$g_i = \begin{cases} |\lambda_i|^{-1/2} & \text{if } \lambda_i \neq 0, \\ 1 & \text{if } \lambda_i = 0. \end{cases}$$

Then

$$g_i \lambda_i g_i = \begin{cases} \dfrac{\lambda_i}{|\lambda_i|} = \operatorname{sgn} \lambda_i & \text{if } \lambda_i \neq 0, \\ 0 & \text{if } \lambda_i = 0. \end{cases}$$

If $S = GU^*$, then $SAS^* = GU^*(U\Lambda U^*)UG = G\Lambda G$ is a real inertia matrix that is *congruent to A and has the asserted numbers of positive, negative, and zero diagonal entries. $\qquad\square$

Lemma 18.6.5 *If two real inertia matrices of the same size have the same numbers of* $+1$ *and* -1 *diagonal entries, then they are* *congruent.

Proof Let $D_1, D_2 \in \mathsf{M}_n$ be real inertia matrices. The hypotheses ensure that D_1 and D_2 have the same eigenvalues, and Theorem 14.7.1.d tells us that they are permutation similar. Since permutation similarity is a *congruence, D_1 and D_2 are *congruent. □

Example 18.6.6 The real inertia matrices $D_1 = \mathrm{diag}(1, -1, 1)$ and $D_2 = \mathrm{diag}(-1, 1, 1)$ are *congruent via the permutation matrix

$$P = \begin{bmatrix} 0 & 1 & 0 \\ 1 & 0 & 0 \\ 0 & 0 & 1 \end{bmatrix}.$$

The following result is the key to a proof of the converse of Lemma 18.6.5.

Lemma 18.6.7 *Let* $A \in \mathsf{M}_n$ *be Hermitian, let* k *be a positive integer, and let* $\mathcal{U}$ *be a* k-*dimensional subspace of* $\mathbb{C}^n$. *If* $\langle A\mathbf{x}, \mathbf{x} \rangle > 0$ *for every nonzero* $\mathbf{x} \in \mathcal{U}$, *then* A *has at least* k *positive eigenvalues.*

Proof Let $\lambda_1 \leq \lambda_2 \leq \cdots \leq \lambda_n$ be the increasingly ordered eigenvalues of A, with corresponding orthonormal eigenvectors $\mathbf{u}_1, \mathbf{u}_2, \ldots, \mathbf{u}_n \in \mathbb{C}^n$, and let $\mathcal{V} = \mathrm{span}\{\mathbf{u}_1, \mathbf{u}_2, \ldots, \mathbf{u}_{n-k+1}\}$. Since

$$\dim \mathcal{U} + \dim \mathcal{V} = k + (n - k + 1) = n + 1,$$

Theorem 4.1.11 ensures that there is a nonzero $\mathbf{x} \in \mathcal{U} \cap \mathcal{V}$. Then $\langle A\mathbf{x}, \mathbf{x} \rangle > 0$ because $\mathbf{x} \in \mathcal{U}$. Since $\mathbf{x} \in \mathcal{V}$, Theorem 18.1.1 tells us that $\langle A\mathbf{x}, \mathbf{x} \rangle \leq \lambda_{n-k+1}$. Therefore,

$$0 < \langle A\mathbf{x}, \mathbf{x} \rangle \leq \lambda_{n-k+1} \leq \lambda_{n-k+2} \leq \cdots \leq \lambda_n,$$

that is, the k largest eigenvalues of A are positive. □

The following theorem is the main result of this section.

Theorem 18.6.8 (Sylvester's Inertia Theorem) *Two Hermitian matrices of the same size are* *congruent if and only if they have the same number of positive eigenvalues and the same number of negative eigenvalues.

Proof Let $A, B \in \mathsf{M}_n$ be Hermitian. If A and B have the same number of positive eigenvalues and the same number of negative eigenvalues, then they have the same numbers of zero eigenvalues. Theorem 18.6.4 ensures that A and B are *congruent to real inertia matrices that have the same eigenvalues, and Lemma 18.6.5 ensures that those real inertia matrices are *congruent. It follows from the transitivity of *congruence that A and B are *congruent.

For the converse, suppose that $A = S^*BS$ and $S \in \mathsf{M}_n$ is invertible. Since A and B are diagonalizable and $\mathrm{rank}\, A = \mathrm{rank}\, B$, they have the same number of zero eigenvalues and the same number of nonzero eigenvalues (Theorem 10.4.14). If they have the same number of positive eigenvalues, then they also have the same number of negative eigenvalues.

If A has no positive eigenvalues, then every eigenvalue of A is negative or zero. Therefore, $-A$ is positive semidefinite (Theorem 15.1.3), in which case $-B = S^*(-A)S$ is also positive semidefinite (Theorem 15.1.24.b). It follows (Theorem 15.1.3 again) that every eigenvalue of $-B$ is nonnegative. Therefore, B has no positive eigenvalues.

Let $\lambda_1 \leq \lambda_2 \leq \cdots \leq \lambda_n$ be the increasingly ordered eigenvalues of A, with corresponding orthonormal eigenvectors $\mathbf{u}_1, \mathbf{u}_2, \ldots, \mathbf{u}_n \in \mathbb{C}^n$. Suppose that A has exactly k positive eigenvalues for some $k \in \{1, 2, \ldots, n\}$. Then

$$0 < \lambda_{n-k+1} \leq \lambda_{n-k+2} \leq \cdots \leq \lambda_n$$

and $\mathcal{U} = \mathrm{span}\{\mathbf{u}_{n-k+1}, \mathbf{u}_{n-k+2}, \ldots, \mathbf{u}_n\}$ has dimension k. If $\mathbf{x} \in \mathcal{U}$ is nonzero, then $\langle A\mathbf{x}, \mathbf{x}\rangle \geq \lambda_{n-k+1}\|\mathbf{x}\|_2^2 > 0$ (Theorem 18.1.1). The set $S\mathcal{U} = \{S\mathbf{x} : \mathbf{x} \in \mathcal{U}\}$ is a subspace of $\mathbb{C}^n$ (see Example 1.3.12). The vectors $S\mathbf{u}_{n-k+1}, S\mathbf{u}_{n-k+2}, \ldots, S\mathbf{u}_n$ span $S\mathcal{U}$, and Example 4.2.3 shows that they are linearly independent. Therefore, $\dim S\mathcal{U} = k$. If $\mathbf{y} \in S\mathcal{U}$ is nonzero, then there is a nonzero $\mathbf{x} \in \mathcal{U}$ such that $\mathbf{y} = S\mathbf{x}$. Then $\langle B\mathbf{y}, \mathbf{y}\rangle = (S\mathbf{x})^* B(S\mathbf{x}) = \mathbf{x}^* S^* BS\mathbf{x} = \langle A\mathbf{x}, \mathbf{x}\rangle > 0$. Lemma 18.6.7 ensures that B has at least k positive eigenvalues, that is, B has at least as many positive eigenvalues as A. Now interchange the roles of A and B in the preceding argument, and conclude that A has at least as many positive eigenvalues as B. Therefore, A and B have the same number of positive eigenvalues. $\square$

Example 18.6.9 Consider

$$A = \begin{bmatrix} -2 & -4 & -6 \\ -4 & -5 & 0 \\ -6 & 0 & 34 \end{bmatrix} \quad \text{and} \quad L = \begin{bmatrix} 1 & 0 & 0 \\ -2 & 1 & 0 \\ 5 & -4 & 1 \end{bmatrix}.$$

Then

$$LA = \begin{bmatrix} -2 & -4 & -6 \\ 0 & 3 & 12 \\ 0 & 0 & 4 \end{bmatrix} \quad \text{and} \quad LAL^* = \begin{bmatrix} -2 & 0 & 0 \\ 0 & 3 & 0 \\ 0 & 0 & 4 \end{bmatrix},$$

so A is *congruent to the real inertia matrix $\mathrm{diag}(-1, 1, 1)$. Theorem 18.6.8 tells us that A has two positive eigenvalues and one negative eigenvalue. Rounded to two decimal places, the eigenvalues of A are -8.07, 0.09, and 34.98. See P.18.31 for an elaboration of this example.

18.7 Weyl's Inequalities

Interlacing inequalities such as those in Theorem 18.2.2 are special cases of inequalities for the eigenvalues of two Hermitian matrices and their sum.

Theorem 18.7.1 (Weyl) *Let* $A, B \in \mathsf{M}_n$ *be Hermitian. For each* $i = 1, 2, \ldots, n$, *the increasingly ordered eigenvalues of* A, B, *and* $A + B$ *satisfy*

$$\lambda_i(A + B) \leq \lambda_{i+j}(A) + \lambda_{n-j}(B) \qquad \text{for } j = 0, 1, \ldots, n - i, \tag{18.7.2}$$

and

$$\lambda_{i-j+1}(A) + \lambda_j(B) \leq \lambda_i(A + B) \qquad \text{for } j = 1, 2, \ldots, i. \tag{18.7.3}$$

Proof Let $\mathbf{u}_1, \mathbf{u}_2, \ldots, \mathbf{u}_n$, $\mathbf{v}_1, \mathbf{v}_2, \ldots, \mathbf{v}_n$, and $\mathbf{w}_1, \mathbf{w}_2, \ldots, \mathbf{w}_n$, be orthonormal eigenvectors of A, $A + B$, and B, respectively, corresponding to their increasingly ordered eigenvalues. Let $i \in \{1, 2, \ldots, n\}$ and $j \in \{0, 1, \ldots, n - i\}$. Let $\mathcal{U} = \mathrm{span}\{\mathbf{u}_1, \mathbf{u}_2, \ldots, \mathbf{u}_{i+j}\}$, $\mathcal{V} = \mathrm{span}\{\mathbf{v}_i, \mathbf{v}_{i+1}, \ldots, \mathbf{v}_n\}$, and $\mathcal{W} = \mathrm{span}\{\mathbf{w}_1, \mathbf{w}_2, \ldots, \mathbf{w}_{n-j}\}$. Then

$$\dim \mathcal{U} + \dim \mathcal{V} = (i + j) + (n - i + 1) = n + j + 1,$$

so Theorem 4.1.11 ensures that $\dim(\mathcal{U} \cap \mathcal{V}) \geq j + 1$. Since

$$\dim(\mathcal{U} \cap \mathcal{V}) + \dim \mathcal{W} \geq (j+1) + (n-j) = n+1,$$

Theorem 4.1.11 tells us that there is a unit vector $\mathbf{x} \in \mathcal{U} \cap \mathcal{V} \cap \mathcal{W}$. Now compute

$$\begin{aligned}
\lambda_i(A+B) &\leq \langle (A+B)\mathbf{x}, \mathbf{x} \rangle = \langle A\mathbf{x}, \mathbf{x} \rangle + \langle B\mathbf{x}, \mathbf{x} \rangle && \text{since } \mathbf{x} \in \mathcal{V} \\
&\leq \lambda_{i+j}(A) + \langle B\mathbf{x}, \mathbf{x} \rangle && \text{since } \mathbf{x} \in \mathcal{U} \\
&\leq \lambda_{i+j}(A) + \lambda_{n-j}(B) && \text{since } \mathbf{x} \in \mathcal{W}.
\end{aligned}$$

This verifies the inequalities (18.7.2).

Now consider the Hermitian matrices $-A$, $-(A+B)$, and $-B$. The increasingly ordered eigenvalues of $-A$ are $-\lambda_n(A) \leq -\lambda_{n-1}(A) \leq \cdots \leq -\lambda_1(A)$, that is, $\lambda_i(-A) = -\lambda_{n-i+1}(A)$ for each $i = 1, 2, \ldots, n$. The inequalities (18.7.2) involve the increasingly ordered eigenvalues

$$\begin{aligned}
\lambda_i(-(A+B)) &= -\lambda_{n-i+1}(A+B), \\
\lambda_{i+j}(-A) &= -\lambda_{n-i-j+1}(A), \quad \text{and} \\
\lambda_{n-j}(-B) &= -\lambda_{j+1}(B).
\end{aligned}$$

We have

$$-\lambda_{n-i+1}(A+B) \leq -\lambda_{n-i-j+1}(A) - \lambda_{j+1}(B) \qquad \text{for } j = 0, 1, \ldots, n-i,$$

that is,

$$\lambda_{n-i+1}(A+B) \geq \lambda_{n-i-j+1}(A) + \lambda_{j+1}(B) \qquad \text{for } j = 0, 1, \ldots, n-i.$$

After a change of variables in the indices we obtain

$$\lambda_i(A+B) \geq \lambda_{i-j+1}(A) + \lambda_j(B) \qquad \text{for } j = 1, 2, \ldots, i.$$

This verifies the inequalities (18.7.3). $\square$

Example 18.7.4 Weyl's inequalities imply the interlacing inequalities (18.2.3). Adopt the notation of Theorem 18.2.2, let $B = E$, and observe that $\lambda_i(B) = 0$ for $i = 1, 2, \ldots, n-1$. If we invoke (18.7.2) and let $j = 1$, we obtain

$$\lambda_i(A+E) \leq \lambda_{i+1}(A) + \lambda_{n-1}(E) = \lambda_{i+1}(A) \quad \text{for } i = 1, 2, \ldots, n-1. \tag{18.7.5}$$

If we invoke (18.7.3) and let $j = 1$, we obtain

$$\lambda_i(A+E) \geq \lambda_i(A) + \lambda_1(E) = \lambda_i(A). \tag{18.7.6}$$

Example 18.7.7 Weyl's inequalities provide some new information about the eigenvalues of A and $A + E$, in which E is positive semidefinite and has rank 1. If we invoke (18.7.2), let $j = 0$, and observe that $\lambda_n(E) = \|E\|_2$ (the spectral norm of E), we obtain

$$\lambda_i(A+E) \leq \lambda_i(A) + \|E\|_2 \quad \text{for } i = 1, 2, \ldots, n. \tag{18.7.8}$$

The bounds (18.7.5), (18.7.6), and (18.7.8) tell us that

$$0 \leq \lambda_i(A+E) - \lambda_i(A) \leq \min\left\{\|E\|_2, \lambda_{i+1}(A) - \lambda_i(A)\right\} \quad \text{for } i = 1, 2, \ldots, n-1.$$

Each eigenvalue $\lambda_i(A+E)$ differs from $\lambda_i(A)$ by no more than $\|E\|_2$ for each $i = 1, 2, \ldots, n$. This gives valuable information if E has small norm. However, no matter how large the norm of E is, $\lambda_i(A+E) \leq \lambda_{i+1}(A)$ for each $i = 1, 2, \ldots, n-1$.

18.8 *Congruence and Inertia of Normal Matrices

A version of Sylvester's inertia theorem is valid for normal matrices, and the generalization illuminates the role of the numbers of positive and negative eigenvalues in the theory of *congruence for Hermitian matrices. We begin with a generalization of real inertia matrices.

Definition 18.8.1 An *inertia matrix* is a complex diagonal matrix whose nonzero diagonal entries (if any) have modulus 1.

If an inertia matrix is nonzero, it is permutation similar to the direct sum of a diagonal unitary matrix and a zero matrix. An inertia matrix with real entries is a real inertia matrix (Definition 18.6.3).

Theorem 18.8.2 *Let $A \in M_n$ be normal. Then A is *congruent to an inertia matrix D, in which for each $\theta \in [0, 2\pi)$, the number of diagonal entries $e^{i\theta}$ in D equals the number of eigenvalues of A that lie on the ray $\{\rho e^{i\theta} : \rho > 0\}$ in the complex plane.*

Proof Let $A = U\Lambda U^*$ be a spectral decomposition, in which U is unitary and $\Lambda = \operatorname{diag}(\lambda_1, \lambda_2, \ldots, \lambda_n)$. Each nonzero eigenvalue has the polar form $\lambda_j = |\lambda_j| e^{i\theta_j}$ for a unique $\theta_j \in [0, 2\pi)$. Let $G = \operatorname{diag}(g_1, g_2, \ldots, g_n)$, in which

$$g_j = \begin{cases} |\lambda_j|^{-1/2} & \text{if } \lambda_j \neq 0, \\ 1 & \text{if } \lambda_j = 0. \end{cases}$$

Then

$$g_j \lambda_j g_j = \begin{cases} \frac{\lambda_j}{|\lambda_j|} = e^{i\theta_j} & \text{if } \lambda_j \neq 0, \\ 0 & \text{if } \lambda_j = 0. \end{cases}$$

If $S = GU^*$, then $SAS^* = GU^*(U\Lambda U^*)UG = G\Lambda G$ is an inertia matrix that is *congruent to A and has the asserted number of diagonal entries $e^{i\theta}$. □

Figure 18.3 indicates the locations of the nonzero eigenvalues of a normal matrix A; the points on the unit circle indicate the nonzero eigenvalues of an inertia matrix that is *congruent to A.

The preceding theorem tells us that each normal matrix A is *congruent to an inertia matrix D that is determined by the eigenvalues of A; this is an analog of Theorem 18.6.4. We claim that D is uniquely determined, up to permutation of its diagonal entries. The following lemma is the key to our proof of this claim. It is motivated by the observation that a nonzero inertia matrix is permutation similar to the direct sum of a diagonal unitary matrix and a zero matrix (which may be absent). To make productive use of this observation, we need to understand the relationship between two *congruent unitary matrices.

Lemma 18.8.3 *Two unitary matrices are *congruent if and only if they are similar.*

Proof Let $V, W \in M_n$ be unitary. If they are similar, they have the same eigenvalues, so Theorem 14.7.1.b ensures that they are unitarily similar. Unitary similarity is a *congruence.

Conversely, if V and W are *congruent, then there is an invertible $S \in M_n$ such that $V = SWS^*$. Let $S = PU$ be a left polar decomposition, in which P is positive definite and U is unitary. Then $V = SWS^* = PUWU^*P$, so

$$P^{-1}V = (UWU^*)P, \tag{18.8.4}$$

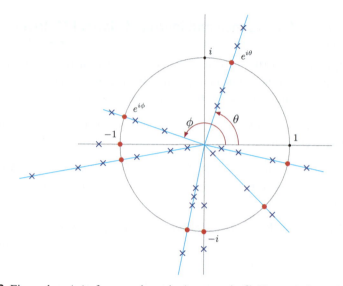

Figure 18.3 Eigenvalues ($\times$) of a normal matrix A on rays in $\mathbb{C}$. Nonzero eigenvalues ($\bullet$) of an inertia matrix D that is *congruent to A. The respective multiplicities of $e^{i\theta}$, $e^{i\phi}$, -1, and $-i$ as eigenvalues of D are 4, 3, 1, and 2.

in which P and P^{-1} are positive definite and V and UWU^* are unitary. Let $A = P^{-1}V$. The left side of (18.8.4) is a left polar decomposition of the invertible matrix A, while the right side of (18.8.4) is a right polar decomposition of A. Theorem 16.3.13.c.iii ensures that $V = UWU^*$, that is, V and W are (unitarily) similar. □

Theorem 18.8.5 *Two inertia matrices are *congruent if and only if they are permutation similar.*

Proof Let $D_1, D_2 \in \mathsf{M}_n$ be inertia matrices. If they are permutation similar, they are *congruent since a permutation similarity is a *congruence.

Conversely, suppose that D_1 and D_2 are *congruent and let $r = \operatorname{rank} D_1 = \operatorname{rank} D_2$ (Theorem 4.2.1). If $r = 0$, then $D_1 = D_2 = 0$. If $r = n$, then D_1 and D_2 are unitary and the preceding lemma ensures that they are similar. Theorem 14.7.1.d tells us that they are permutation similar. Now suppose that $1 \le r \le n-1$. After permutation similarities (*congruences), if necessary, we may assume that $D_1 = V_1 \oplus 0_{n-r}$ and $D_2 = V_2 \oplus 0_{n-r}$, in which $V_1, V_2 \in \mathsf{M}_r$ are diagonal unitary matrices. Let $S \in \mathsf{M}_n$ be invertible and such that $SD_1S^* = D_2$. Partition $S = [S_{ij}]$ conformally with D_1 and D_2, so $S_{11} \in \mathsf{M}_r$. Then

$$\begin{bmatrix} V_2 & 0 \\ 0 & 0 \end{bmatrix} = D_2 = SD_1S^* = \begin{bmatrix} S_{11} & S_{12} \\ S_{21} & S_{22} \end{bmatrix} \begin{bmatrix} V_1 & 0 \\ 0 & 0 \end{bmatrix} \begin{bmatrix} S_{11}^* & S_{21}^* \\ S_{12}^* & S_{22}^* \end{bmatrix} = \begin{bmatrix} S_{11}V_1S_{11}^* & S_{11}V_1S_{21}^* \\ S_{21}V_1S_{11}^* & S_{21}V_1S_{21}^* \end{bmatrix},$$

which tells us that $S_{11}V_1S_{11}^* = V_2$. Since V_2 is invertible, we conclude that S_{11} is invertible and V_1 is *congruent to V_2. The preceding lemma ensures that V_1 and V_2 are similar, from which it follows that D_1 and D_2 are similar. Theorem 14.7.1.d tells us that they are permutation similar. □

Theorem 18.8.6 (Inertia Theorem for Normal Matrices) *Two normal matrices of the same size are *congruent if and only if, for each $\theta \in [0, 2\pi)$, they have the same number of eigenvalues on the ray $\{\rho e^{i\theta} : \rho > 0\}$ in the complex plane.*

Proof Let $A, B \in M_n$ be normal. Let D_1 and D_2 be inertia matrices that are *congruent to A and B, respectively, and are constructed according to the recipe in Theorem 18.8.2.

If A is *congruent to B, the transitivity of *congruence ensures that D_1 is *congruent to D_2. Theorem 18.8.5 tells us that D_1 and D_2 are permutation similar, so they have the same eigenvalues (diagonal entries) with the same multiplicities. The construction in Theorem 18.8.2 ensures that A and B have the same number of eigenvalues on the ray $\{\rho e^{i\theta} : \rho > 0\}$ for each $\theta \in [0, 2\pi)$.

Conversely, the hypotheses on the eigenvalues of A and B ensure that the diagonal entries of D_1 can be obtained by permuting the diagonal entries of D_2, so D_1 and D_2 are permutation similar and *congruent. The transitivity of *congruence again ensures that A and B are *congruent. □

Sylvester's inertia theorem is a corollary of the inertia theorem for normal matrices.

Corollary 18.8.7 *Two Hermitian matrices of the same size are* *congruent if and only if they have the same number of positive eigenvalues and the same number of negative eigenvalues.*

Proof If a Hermitian matrix has any nonzero eigenvalues, they are real and hence are on the rays $\{\rho e^{i\theta} : \rho > 0\}$ with $\theta = 0$ or $\theta = \pi$. □

A square matrix is Hermitian if and only if it is *congruent to a real inertia matrix. Although every normal matrix is *congruent to an inertia matrix, a matrix can be *congruent to an inertia matrix without being normal; see P.18.46. Moreover, not every matrix is *congruent to an inertia matrix; see P.18.50.

18.9 Problems

P.18.1 Let A be the matrix in Example 18.1.8. Use (18.1.6) and $\mathbf{x} = \frac{1}{\sqrt{2}}(\mathbf{e}_i \pm \mathbf{e}_j)$ with four different choices of i and j to obtain upper bounds on $\lambda_1(A)$ and lower bounds on $\lambda_4(A)$. What are your best bounds?

P.18.2 Use Sylvester's criterion to show that

$$A = \begin{bmatrix} 2 & 2 & 3 \\ 2 & 5 & 5 \\ 3 & 5 & 7 \end{bmatrix}$$

is positive definite.

P.18.3 Show that *congruence is an equivalence relation on M_n.

P.18.4 Let

$$A = \begin{bmatrix} 0 & 0 & 0 & 1 \\ 0 & 0 & 0 & 2 \\ 0 & 0 & 0 & 3 \\ 1 & 2 & 3 & 4 \end{bmatrix}.$$

(a) Use Theorem 18.3.2 to show that 0 has multiplicity at least two as an eigenvalue of A. (b) Show that rank $A = 2$ and use this fact to show that 0 has multiplicity two as an eigenvalue of A. Why must A have one positive eigenvalue and one negative eigenvalue?

P.18.5 Let $A, E \in M_4$ be Hermitian. Suppose that E is positive semidefinite and has rank 2. Show that $\lambda_1(A + E) \le \lambda_3(A)$, $\lambda_2(A + E) \le \lambda_4(A)$, and $\lambda_i(A) \le \lambda_i(A + E)$ for $i = 1, 2, 3, 4$.

P.18.6 Let $A, E \in M_n$ be Hermitian. Suppose that E is positive semidefinite and rank $r \ge 1$. Show that $\lambda_i(A) \le \lambda_i(A + E)$ for $i = 1, 2, \ldots, n$ and $\lambda_i(A + E) \le \lambda_{i+r}(A)$ for $i = 1, 2, \ldots, n - r$.

P.18.7 Let $A \in M_n$ be Hermitian and let $\lambda_1 \le \lambda_2 \le \cdots \le \lambda_n$ be its increasingly ordered eigenvalues. Show that

$$\min_{\substack{\mathbf{x} \in \mathbb{C}^n \\ \mathbf{x} \neq \mathbf{0}}} \frac{\langle A\mathbf{x}, \mathbf{x} \rangle}{\langle \mathbf{x}, \mathbf{x} \rangle} = \lambda_1 \quad \text{and} \quad \max_{\substack{\mathbf{x} \in \mathbb{C}^n \\ \mathbf{x} \neq \mathbf{0}}} \frac{\langle A\mathbf{x}, \mathbf{x} \rangle}{\langle \mathbf{x}, \mathbf{x} \rangle} = \lambda_n.$$

P.18.8 Let $A \in M_n$ be Hermitian, with increasingly ordered eigenvalues $\lambda_1 \le \lambda_2 \le \cdots \le \lambda_n$ and corresponding orthonormal eigenvectors $\mathbf{u}_1, \mathbf{u}_2, \ldots, \mathbf{u}_n \in \mathbb{C}^n$. Let $i_1, i_2, \ldots, i_k$ be integers such that $1 \le i_1 < i_2 < \cdots < i_k \le n$. If $\mathbf{x} \in \text{span}\{\mathbf{u}_{i_1}, \mathbf{u}_{i_2}, \ldots, \mathbf{u}_{i_k}\}$ is a unit vector, show that $\lambda_{i_1} \le \langle A\mathbf{x}, \mathbf{x} \rangle \le \lambda_{i_k}$.

P.18.9 Adopt the notation of Theorem 18.1.1. (a) Show that $\{\langle A\mathbf{x}, \mathbf{x} \rangle : \mathbf{x} \in \mathbb{C}^n, \|\mathbf{x}\|_2 = 1\} = [\lambda_1, \lambda_n] \subseteq \mathbb{R}$. (b) If A is real symmetric, show that $\{\langle A\mathbf{x}, \mathbf{x} \rangle : \mathbf{x} \in \mathbb{R}^n, \|\mathbf{x}\|_2 = 1\} = [\lambda_1, \lambda_n] \subseteq \mathbb{R}$.

P.18.10 Suppose that the eigenvalues $\lambda_1, \lambda_2, \ldots, \lambda_n$ of a normal matrix $A \in M_n$ are the vertices of a convex polygon $\mathscr{W}(A)$ in the complex plane; see Figure 18.4 for an illustration with $n = 5$. Show that $\mathscr{W}(A) = \{\langle A\mathbf{x}, \mathbf{x} \rangle : \mathbf{x} \in \mathbb{C}^n \text{ and } \|\mathbf{x}\|_2 = 1\}$. The set $\mathscr{W}(A)$ is the *numerical range* (also known as the *field of values*) of A.

P.18.11 Sketch the sets (a) $\{\langle J_2\mathbf{x}, \mathbf{x} \rangle : \mathbf{x} \in \mathbb{R}^n \text{ and } \|\mathbf{x}\|_2 = 1\}$, and (b) $\mathscr{W}(J_2) = \{\langle A\mathbf{x}, \mathbf{x} \rangle : \mathbf{x} \in \mathbb{C}^n \text{ and } \|\mathbf{x}\|_2 = 1\}$.

P.18.12 If $A, B \in M_n$ are unitarily similar, show that $\mathscr{W}(A) = \mathscr{W}(B)$, that is, A and B have the same numerical range.

P.18.13 Let $A = \text{diag}(1, 2, 4)$ and $B = \text{diag}(1, 3, 4)$. Show that $\mathscr{W}(A) = \mathscr{W}(B)$ but A is not unitarily similar to B.

P.18.14 If $1 \le m \le n - 1$ and $B \in M_m$ is a principal submatrix of $A \in M_n$, show that $\mathscr{W}(B) \subseteq \mathscr{W}(A)$.

P.18.15 Let $A \in M_n$. Show that $\text{spec } A \subseteq \mathscr{W}(A)$.

P.18.16 Let $A, B \in M_n$. Show that $\text{spec}(A + B) \subseteq \mathscr{W}(A) + \mathscr{W}(B)$.

P.18.17 If $A \in M_n$ is a scalar matrix, what is $\mathscr{W}(A)$?

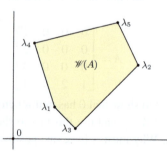

Figure 18.4 Numerical range of a 5×5 normal matrix A with $\text{spec } A = \{\lambda_1, \lambda_2, \lambda_3, \lambda_4, \lambda_5\}$.

P.18.18 Let $A \in \mathsf{M}_n$. (a) Show that rank $A = 1$ if and only if there are nonzero $\mathbf{u}, \mathbf{v} \in \mathbb{C}^n$ such that $A = \mathbf{u}\mathbf{v}^*$. (b) Show that A is Hermitian and rank $A = 1$ if and only if there is a nonzero $\mathbf{u} \in \mathbb{C}^n$ such that $A = \mathbf{u}\mathbf{u}^*$ or $A = -\mathbf{u}\mathbf{u}^*$.

P.18.19 Adopt the notation in Theorem 18.1.1. If $\mathbf{x} \in \mathcal{U}$, show that (a) $\langle A\mathbf{x}, \mathbf{x} \rangle = \lambda_p$ if and only if $A\mathbf{x} = \lambda_p \mathbf{x}$, and (b) $\langle A\mathbf{x}, \mathbf{x} \rangle = \lambda_q$ if and only if $A\mathbf{x} = \lambda_q \mathbf{x}$.

P.18.20 Let $A = [a_{ij}] \in \mathsf{M}_n$ be Hermitian with eigenvalues $\lambda_1 \leq \lambda_2 \leq \cdots \leq \lambda_n$. If $i \in \{1, 2, \ldots, n\}$ is such that $a_{ii} = \lambda_1$ or $a_{ii} = \lambda_n$, show that $a_{ij} = a_{ji} = 0$ for all $j \neq i$. *Hint*: Suppose that A is positive definite and $a_{11} = \lambda_n(A)$ (otherwise, consider a permutation similarity of $\pm A + \mu I$). Partition $A = \begin{bmatrix} \lambda_n(A) & \mathbf{x}^* \\ \mathbf{x} & B \end{bmatrix}$. Use Theorem 14.3.1 and (18.3.3) to show that $\|A\|_F^2 = \lambda_n(A)^2 + 2\|\mathbf{x}\|_2^2 + \|B\|_F^2 \geq \lambda_n(A)^2 + 2\|\mathbf{x}\|_2^2 + \sum_{i=1}^{n-1} \lambda_i(A)^2 = \|A\|_F^2 + 2\|\mathbf{x}\|_2^2$.

P.18.21 Let $A = [a_{ij}] \in \mathsf{M}_4$ be Hermitian and suppose that spec $A = \{1, 2, 3, 4\}$. If $a_{11} = 1$, $a_{22} = 2$, and $a_{44} = 4$, what are the other entries? Why?

P.18.22 Adopt the notation in Theorem 18.2.2. If $\lambda \in$ spec A has multiplicity 2 or greater, show that $\lambda \in$ spec$(A + E)$.

P.18.23 Give an example of a 2×2 Hermitian matrix that has nonnegative leading principal minors but is not positive semidefinite. Which of its principal minors is negative?

P.18.24 Suppose that $A \in \mathsf{M}_n$ is Hermitian and let $B \in \mathsf{M}_{n-1}$ be its $(n-1) \times (n-1)$ leading principal submatrix. (a) If B is positive semidefinite and rank $A =$ rank B, show that A is positive semidefinite. (b) If the first $n-1$ principal minors of A are positive and $\det A = 0$, show that A is positive semidefinite.

P.18.25 Show that

$$A = \begin{bmatrix} 2 & 0 & 0 & 0 & 1 & 1 \\ 0 & 2 & 0 & 0 & 1 & 1 \\ 0 & 0 & 3 & 1 & 1 & 1 \\ 0 & 0 & 1 & 3 & 1 & 1 \\ 1 & 1 & 1 & 1 & 5 & 1 \\ 1 & 1 & 1 & 1 & 1 & 5 \end{bmatrix}$$

is positive definite.

P.18.26 Suppose that $0, 0, \ldots, 0, 1$ are the eigenvalues of a Hermitian $A = [a_{ij}] \in \mathsf{M}_n$. Show that $0 \leq a_{ii} \leq 1$ for each $i = 1, 2, \ldots, n$.

P.18.27 If $10, 25, 26, 39, 50$ are the eigenvalues of a Hermitian $A \in \mathsf{M}_5$, show that A cannot have main diagonal entries $14, 16, 30, 44, 46$.

P.18.28 Deduce the inequalities (18.1.7) from the majorization inequalities (18.5.3).

P.18.29 Deduce from Theorem 18.6.8 that two real inertia matrices are *congruent if and only if they are similar.

P.18.30 Let $D = \mathrm{diag}(1, -1) \in \mathsf{M}_2$. Is D *congruent to $-D$? Is I_2 *congruent to $-I_2$? Why?

P.18.31 Let A and L be the matrices in Example 18.6.9. Let

$$L_1 = \begin{bmatrix} 1 & 0 & 0 \\ -2 & 1 & 0 \\ -3 & 0 & 1 \end{bmatrix} \quad \text{and} \quad L_2 = \begin{bmatrix} 1 & 0 & 0 \\ 0 & 1 & 0 \\ 0 & -4 & 1 \end{bmatrix}.$$

(a) Show that $L = L_2 L_1$. (b) Compute $L_1 A$, and then compute $L_2(L_1 A) = LA$. Explain why L_1 and L_2 encode a sequence of elementary row operations that reduce A to upper triangular form. (c) Now compute $((LA)L_1^*)L_2^*$ and explain what happens. (d) To determine the signs of the eigenvalues of A, why does it suffice to examine only the main diagonal entries of LA? (e) When it works, the reduction algorithm in this problem identifies the inertia matrix of a Hermitian matrix without computing any eigenvalues. Does it work for $A = \begin{bmatrix} 0 & 1 \\ 1 & 0 \end{bmatrix}$? What is the inertia matrix of A?

P.18.32 If $H \in \mathsf{M}_n$ is Hermitian, define

$$i_+(H) = \text{the number of positive eigenvalues of } H,$$
$$i_-(H) = \text{the number of negative eigenvalues of } H, \quad \text{and}$$
$$i_0(H) = \text{the number of zero eigenvalues of } H.$$

Let $A = [A_{ij}] \in \mathsf{M}_n$ be a Hermitian 2×2 block matrix, in which $A_{11} \in \mathsf{M}_k$ is invertible. Define $S = \begin{bmatrix} I_k & 0 \\ -A_{12}^* A_{11}^{-1} & I_{n-k} \end{bmatrix} \in \mathsf{M}_n$. (a) Show that $SAS^* = A_{11} \oplus A/A_{11}$, in which $A/A_{11} = A_{22} - A_{12}^* A_{11}^{-1} A_{12}$ is the Schur complement of A_{11} in A. (b) Prove *Haynsworth's inertia theorem*:

$$i_+(A) = i_+(A_{11}) + i_+(A/A_{11}),$$
$$i_-(A) = i_-(A_{11}) + i_-(A/A_{11}), \quad \text{and}$$
$$i_0(A) = i_0(A/A_{11}).$$

(c) Prove that A is positive definite if and only if both A_{11} and A/A_{11} are positive definite.

P.18.33 Let $A, B \in \mathsf{M}_n$ be Hermitian. (a) Use Weyl's inequalities to show that the increasingly ordered eigenvalues of A and $A + B$ satisfy

$$\lambda_i(A) + \lambda_1(B) \leq \lambda_i(A + B) \leq \lambda_i(A) + \lambda_n(B) \qquad \text{for } i = 1, 2, \ldots, n.$$

(b) Show that $|\lambda_i(A + B) - \lambda_i(A)| \leq \|B\|_2$ for each $i = 1, 2, \ldots, n$. (c) Compare the bounds in (b) with the Bauer–Fike bounds in Theorem 17.5.11. Discuss.

P.18.34 Let $A, B \in \mathsf{M}_n$ be Hermitian. (a) If B has exactly p positive eigenvalues and $1 \leq p \leq n - 1$, use Weyl's inequalities to show that the increasingly ordered eigenvalues of A and $A + B$ satisfy $\lambda_i(A + B) \leq \lambda_{i+p}(A)$ for $i = 1, 2, \ldots, n - p$. (b) If B has exactly q negative eigenvalues and $1 \leq q \leq n - 1$, show that $\lambda_{i-q}(A) \leq \lambda_i(A + B)$ for $i = q + 1, q + 2, \ldots, n$.

P.18.35 Let $A, B \in \mathsf{M}_n$ be Hermitian and suppose that B is positive semidefinite. Show that the increasingly ordered eigenvalues of A and $A + B$ satisfy $\lambda_i(A) \leq \lambda_i(A + B)$ for $i = 1, 2, \ldots, n$. This is the *monotonicity theorem*. What can you say if B is positive definite?

P.18.36 Consider the non-Hermitian matrices $A = \begin{bmatrix} 0 & 0 \\ 1 & 0 \end{bmatrix}$ and $B = \begin{bmatrix} 0 & 1 \\ 0 & 0 \end{bmatrix}$. Show that the eigenvalues of A, B, and $A + B$ do not satisfy Weyl's inequalities.

P.18.37 Adopt the notation of Theorem 18.5.2. Show that the "bottom-up" majorization inequalities (18.5.3) are equivalent to the "top-down" inequalities

$$a_n + a_{n-1} + \cdots + a_{n-k+1} \leq \lambda_n + \lambda_{n-1} + \cdots + \lambda_{n-k+1} \qquad \text{for } k = 1, 2, \ldots, n,$$

with equality for $k = n$.

P.18.38 Let $A \in \mathsf{M}_n$, let $\sigma_1 \geq \sigma_2 \geq \cdots \geq \sigma_n$ be its singular values, and let $c_1 \geq c_2 \geq \cdots \geq c_n$ be the decreasingly ordered Euclidean norms of the columns of A. (a) Show that

$$c_n^2 + c_{n-1}^2 + \cdots + c_{n-k+1}^2 \geq \sigma_n^2 + \sigma_{n-1}^2 + \cdots + \sigma_{n-k+1}^2 \qquad (18.9.1)$$

for $k = 1, 2, \ldots, n$, with equality for $k = n$. (b) If A has some columns or rows with small Euclidean norm, why must it have some small singular values? (c) If A is normal, what does (18.9.1) tell you about its eigenvalues?

P.18.39 Let $\lambda, c \in \mathbb{R}$, $\mathbf{y} \in \mathbb{C}^n$, and $A = \begin{bmatrix} \lambda I_n & \mathbf{y} \\ \mathbf{y}^* & c \end{bmatrix} \in M_{n+1}$. (a) Use Theorem 18.3.2 to show that λ is an eigenvalue of A with multiplicity at least $n - 1$. (b) Use the Cauchy expansion (3.3.16) to compute p_A and determine the eigenvalues of A.

P.18.40 Adopt the notation of Definition 11.3.10. (a) Show that the companion matrix of $f = z^n + c_{n-1}z^{n-1} + \cdots + c_1 z + c_0$ can be written as

$$C_f = \begin{bmatrix} \mathbf{0}^\mathsf{T} & -c_0 \\ I_{n-1} & \mathbf{y} \end{bmatrix} \in M_n, \quad \text{in which } \mathbf{y} = \begin{bmatrix} -c_1 \\ -c_2 \\ \vdots \\ -c_{n-1} \end{bmatrix}.$$

(b) Show that $C_f^* C_f = \begin{bmatrix} I_{n-1} & \mathbf{y} \\ \mathbf{y}^* & c \end{bmatrix}$, in which $c = |c_0|^2 + |c_1|^2 + \cdots + |c_n|^2$. (c) Show that the singular values $\sigma_1 \geq \sigma_2 \geq \cdots \geq \sigma_n$ of C_f satisfy $\sigma_1^2 = \frac{1}{2}(c + 1 + \sqrt{(c+1)^2 - 4|c_0|^2})$, $\sigma_2 = \sigma_3 = \cdots = \sigma_{n-1} = 1$, and $\sigma_n^2 = \frac{1}{2}(c + 1 - \sqrt{(c+1)^2 - 4|c_0|^2})$.

P.18.41 Let $A \in M_{n+1}$ be Hermitian and let $i \in \{1, 2, \ldots, n+1\}$. Let B denote the $n \times n$ matrix obtained by deleting the ith row and ith column of A. Show that the eigenvalues of A and B satisfy the interlacing inequalities (18.3.3).

P.18.42 Let $A \in M_n$ and let B denote the matrix obtained by deleting either (a) two rows, (b) two columns, or (c) a row and a column from A. What is the analog of (18.3.5) in this case?

P.18.43 Let $A = [a_{ij}] \in M_n(\mathbb{R})$ be tridiagonal and suppose that $a_{i,i+1}a_{i+1,i} > 0$ for each $i = 1, 2, \ldots, n - 1$. Show that: (a) There is a real diagonal matrix D with positive diagonal entries such that DAD^{-1} is symmetric. (b) A is diagonalizable. (c) $\operatorname{spec} A \subset \mathbb{R}$. (d) Each eigenvalue of A has geometric multiplicity 1. *Hint*: Why is $\operatorname{rank}(A - \lambda I) \geq n - 1$ for every $\lambda \in \mathbb{C}$? (e) A has distinct eigenvalues. (f) If B is obtained from A by deleting any row and the corresponding column, then the eigenvalues of B are real, distinct, and interlace the eigenvalues of A.

P.18.44 Let $A, B \in M_n$ be positive semidefinite. Use (15.7.4) and (18.1.7) to show that $\|A \circ B\|_2 \leq \|A\|_2 \|B\|_2$.

P.18.45 Let $A, B \in M_n$. (a) Prove that $\|A \circ B\|_2 \leq \|A\|_2 \|B\|_2$. This is an analog of Theorem 17.2.6.b. *Hint*: Use P.15.79, (18.3.5), and P.16.38 (b) Verify the inequality in (a) for the matrices in Example 15.6.2.

P.18.46 A normal matrix is *congruent to an inertia matrix, but a matrix can be *congruent to an inertia matrix without being normal. Let $S = \begin{bmatrix} 1 & 1 \\ 0 & -1 \end{bmatrix}$ and $D = \begin{bmatrix} 1 & 0 \\ 0 & i \end{bmatrix}$. Compute $A = SDS^*$ and show that A is not normal.

P.18.47 Consider the matrix A defined in the preceding problem. Compute $A^{-*}A$ and show that it is similar to a unitary matrix. Why is this not surprising?

P.18.48 If $A \in M_n$ is invertible, let $B = A^{-*}A$. (a) Show that B is similar to B^{-*}. (b) What can you say about the Jordan canonical form of B? (c) Let A be the matrix defined in P.18.50. Verify that B is similar to B^{-*}.

P.18.49 Let $U, V \in M_n$ be unitary. (a) If U and V are *congruent, show that they are unitarily similar. (b) If U and V are similar, show that they are *congruent. *Hint*: See Theorem 18.8.6.

P.18.50 Suppose that $A \in M_n$ is invertible and *congruent to an inertia matrix D. (a) Show that $A^{-*}A$ is similar to the unitary matrix D^2. (b) Show that $A = \begin{bmatrix} 0 & 1 \\ 2 & 0 \end{bmatrix}$ is not *congruent to an inertia matrix.

P.18.51 Let $\lambda_1, \lambda_2, \ldots, \lambda_n$ be the eigenvalues of $A \in M_n$, and let $A = H(A) + iK(A)$ be its Cartesian decomposition. (a) If A is normal, show that $\mathrm{Re}\,\lambda_1, \mathrm{Re}\,\lambda_2, \ldots, \mathrm{Re}\,\lambda_n$ are the eigenvalues of $H(A)$, and $\mathrm{Im}\,\lambda_1, \mathrm{Im}\,\lambda_2, \ldots, \mathrm{Im}\,\lambda_n$ are the eigenvalues of $K(A)$. Conclude that the number of nonreal eigenvalues of A in the upper half plane $\{z \in \mathbb{C} : \mathrm{Im}\,z > 0\}$ equals the number of positive eigenvalues of $K(A)$. (b) Consider $A = J_2(0)$ (the nilpotent Jordan block of size 2) and show that the conclusion in (a) need not be true if A is not normal.

P.18.52 Let $A \in M_{27}$ be the normal matrix whose eigenvalues are plotted in Figure 18.3, let $S \in M_{27}$ be invertible, and suppose that $B = SAS^*$ is normal. The relationship between the eigenvalues of A and B is described in Theorem 18.8.6; in this problem we take a different approach and exploit the fact that the rotations $e^{i\alpha}A$ and $e^{i\alpha}B$ are *congruent for any real α. Let $e^{i\alpha}A = H(e^{i\alpha}A) + iK(e^{i\alpha}A)$ and $e^{i\alpha}B = H(e^{i\alpha}B) + iK(e^{i\alpha}B)$ be Cartesian decompositions. Observe that $K(e^{i\alpha}B) = SK(e^{i\alpha}A)S^*$, so $K(e^{i\alpha}A)$ and $K(e^{i\alpha}B)$ have the same number of positive eigenvalues (Theorem 18.6.8). Figure 18.3 indicates that A has exactly 7 nonreal eigenvalues in the upper half plane. (a) Why do $K(A)$ and $K(B)$ each have exactly 7 positive eigenvalues? (b) Deduce from (a) and the preceding problem that B has exactly 7 nonreal eigenvalues in the upper half plane. (c) If ε is sufficiently small and positive, why does $e^{-i\varepsilon}A$ have exactly 8 eigenvalues in the upper half plane? (d) Why do $K(e^{-i\varepsilon}A)$ and $K(e^{-i\varepsilon}B)$ each have exactly 8 positive eigenvalues? (e) Deduce from (d) that $e^{-i\varepsilon}B$ has exactly 8 eigenvalues in the upper half plane and therefore B has exactly one eigenvalue on the negative real axis (the same as A). (f) Modify this line of argument to show that B has exactly 3 eigenvalues on the ray $\{re^{i\phi} : r > 0\}$. (g) What can you say in general?

P.18.53 If a Hermitian matrix is bordered and the resulting matrix is not Hermitian, then there need not be any eigenvalue interlacing. Let $\lambda \in \mathbb{R}$ and let $A = \begin{bmatrix} \lambda & y \\ 1 & c \end{bmatrix} \in M_2$. (a) For suitable choices of y and c, show that $p_A(z)$ can be any monic polynomial of degree 2. (b) What can you say about the eigenvalues of A?

P.18.54 (a) If $z \in \mathbb{C}$ is nonzero, show that $\mathrm{Re}\,z$ and $\mathrm{Re}\,(z^{-1})$ have the same sign. *Hint:* Consider $z^{-1}(z + \bar{z})\bar{z}^{-1}$. (b) If $A \in M_n$ is invertible, show that $H(A)$ and $H(A^{-1})$ have the same inertia matrix; see Definition 14.6.6.

18.10 Notes

Some authors use a *decreasing* order $\lambda_1 \geq \lambda_2 \geq \cdots \geq \lambda_n$ for the eigenvalues of a Hermitian matrix. This convention has the advantage that the eigenvalues of a positive semidefinite matrix are its singular values, with the same indexing. It has the disadvantage that the indices and eigenvalues are not ordered in the same way as real numbers.

P.18.23 shows that a Hermitian matrix whose leading principal submatrices have nonnegative determinants need not be positive semidefinite. However, if every principal submatrix of a Hermitian matrix has nonnegative determinant, then the matrix is positive semidefinite; see [HJ13, Theorem 7.2.5.a].

The inequalities (18.2.3), (18.3.3), and (18.5.3) characterize the structures associated with them. For example, if real numbers λ_i and μ_i satisfy $\lambda_i \leq \mu_i \leq \lambda_{i+1} \leq \mu_n$ for $i = 1, 2, \ldots, n-1$, then there is a Hermitian $A \in \mathsf{M}_n$ and a positive semidefinite rank-1 matrix $E \in \mathsf{M}_n$ such that $\lambda_1, \lambda_2, \ldots, \lambda_n$ are the eigenvalues of A and $\mu_1, \mu_2, \ldots, \mu_n$ are the eigenvalues of $A + E$. If $\mu_1 \leq \lambda_1 \leq \mu_2 \leq \cdots \leq \lambda_n \leq \mu_{n+1}$, then there is a Hermitian $A \in \mathsf{M}_n$ such that $\lambda_1, \lambda_2, \ldots, \lambda_n$ are the eigenvalues of A and $\mu_1, \mu_2, \ldots, \mu_{n+1}$ are the eigenvalues of a Hermitian matrix that borders it. If $\mu_1 + \mu_2 + \cdots + \mu_k \leq \lambda_1 + \lambda_2 + \cdots + \lambda_k$ for $k = 1, 2, \ldots, n$, with equality for $k = n$, then there is a Hermitian $A \in \mathsf{M}_n$ whose eigenvalues and main diagonal entries are $\lambda_1, \lambda_2, \ldots, \lambda_n$ and $\mu_1, \mu_2, \ldots, \mu_n$, respectively. Proofs of these and other inverse eigenvalue problems are in [HJ13, Section 4.3].

Theorem 18.6.8 was proved for real symmetric matrices in 1852 by James Joseph Sylvester, who wrote "my view of the physical meaning of quantity of matter inclines me, upon the ground of analogy, to give [it] the name of the Law of Inertia for Quadratic Forms." Sylvester's theorem was generalized to normal matrices (Theorem 18.8.6) in 2001, and to all square complex matrices in 2006. For more information about *congruence, see [HJ13, Section 4.5].

Matrices $A, B \in \mathsf{M}_n$ are *congruent* if there is an invertible $S \in \mathsf{M}_n$ such that $A = SBS^\mathsf{T}$. Congruence is an equivalence relation on M_n, but many of its properties are quite different from those of *congruence. For example, two complex symmetric matrices of the same size are congruent if and only if they have the same rank. Each complex symmetric matrix is unitarily congruent to a diagonal matrix with real nonnegative diagonal entries. Although $-A$ need not be *congruent to A, $-A$ is always congruent to A. However, A^T is always both congruent and *congruent to A. See [HJ13, Section 4.5] for more information about congruence.

Equality occurs in the Weyl inequalities (18.7.2) if and only if there is a nonzero $\mathbf{x}$ such that $A\mathbf{x} = \lambda_{i+j}(A)\mathbf{x}$, $B\mathbf{x} = \lambda_{n-j}(B)\mathbf{x}$, and $(A + B)\mathbf{x} = \lambda_i(A + B)\mathbf{x}$. For a discussion of this and other cases of equality for the inequalities in this chapter, see [HJ13, Section 4.3].

If $A \in \mathsf{M}_n$, then $\mathscr{W}(A)$ (its numerical range) is a convex subset of $\mathbb{C}$. For a proof and more information about the numerical range see [HJ94, Chapter 1].

18.11 Some Important Concepts

- The Rayleigh quotient and eigenvalues of a Hermitian matrix.

- Eigenvalue interlacing for a Hermitian matrix and a rank-1 additive perturbation.

- Weyl's inequalities for eigenvalues of a sum of two Hermitian matrices.

- Eigenvalue interlacing for a bordered Hermitian matrix.

- Eigenvalue interlacing for a principal submatrix of a Hermitian matrix.

- Singular value interlacing if a row or column is deleted from a matrix.

- Sylvester's principal-minor criterion for a positive definite matrix.

- Majorization inequalities for the eigenvalues and diagonal entries of a Hermitian matrix.

- Sylvester's inertia theorem about *congruence of Hermitian matrices and its generalization to normal matrices.

Norms and Matrix Norms

The Euclidean norm plays a role in many topics discussed in this book, but other norms arise in current applications of matrix mathematics. For example, the ℓ_1 norm (absolute sum norm)

$$\|\mathbf{x}\|_1 = |x_1| + |x_2| + \cdots + |x_n|$$

of $\mathbf{x} = [x_i] \in \mathbb{R}^n$ is used in the *mean absolute deviation*

$$\frac{1}{n}(|x_1 - \mu| + |x_2 - \mu| + \cdots + |x_n - \mu|) = \frac{1}{n}\|\mathbf{x} - \mu\mathbf{e}\|_1, \quad \text{in which } \mu = \frac{1}{n}\mathbf{e}^{\mathsf{T}}\mathbf{x}.$$

This is a measure of scatter often used in robust statistical modeling instead of the *root mean square deviation*

$$\frac{1}{\sqrt{n}}\left(|x_1 - \mu|^2 + |x_2 - \mu|^2 + \cdots + |x_n - \mu|^2\right)^{1/2} = \frac{1}{\sqrt{n}}\|\mathbf{x} - \mu\mathbf{e}\|_2,$$

which uses the Euclidean norm. The ℓ_1 norm is employed in the basis-pursuit algorithm for compressive sensing, which has achieved impressive success in medical imaging, astronomy, facial recognition, and reflection seismology. The ℓ_∞ norm (max norm),

$$\|\mathbf{x}\|_\infty = \max\{|x_1|, |x_2|, \ldots, |x_n|\},$$

is used in robotics, where strict control of each individual entry of $\mathbf{x}$ is required.

In this chapter, we introduce new examples of norms, with special attention to submultiplicative norms on matrices. These norms are well adapted to applications involving power series of matrices and iterative numerical algorithms. We use them to prove a formula for the spectral radius that is the key to a fundamental theorem on positive matrices in the next chapter.

19.1 Norms of Vectors

The following example introduces a family of norms that includes the absolute sum, Euclidean, and max norms. Infinite-dimensional versions of this family appear in functional analysis and Fourier series.

Example 19.1.1 For $p \geq 1$, the ℓ_p-norm on $\mathbb{F}^n$ is

$$\|\mathbf{x}\|_p = (|x_1|^p + |x_2|^p + \cdots + |x_n|^p)^{1/p}. \tag{19.1.2}$$

It is evident that $\|\mathbf{x}\|_p \geq 0$ for all $\mathbf{x} \in \mathbb{F}^n$ (nonnegativity); $\|\mathbf{x}\|_p = 0$ if and only if $\mathbf{x} = \mathbf{0}$ (positivity) and $\|c\mathbf{x}\|_p = |c|\|\mathbf{x}\|_p$ for all $c \in \mathbb{F}$ and all $\mathbf{x} \in \mathbb{F}^n$ (homogeneity). In this book, we use the ℓ_p-norm for only three values of p, and we have already verified the triangle inequality for them: $p = 1$ (Example 5.5.2); $p = 2$ (Corollary 5.4.19), and $p = \infty$ (Example 5.5.3). See P.19.14 for a proof of the triangle inequality for other values of p.

A second family of norms also includes the absolute sum and max norms. It plays an important role in the theory of unitarily invariant matrix norms and arises in order and rank statistics.

Example 19.1.3 If $\mathbf{x} = [x_i] \in \mathbb{F}^n$, let $i_1, i_2, \ldots, i_n$ be distinct elements of $\{1, 2, \ldots, n\}$ such that $|x_{i_1}| \geq |x_{i_2}| \geq \cdots \geq |x_{i_n}|$. For each $k = 1, 2, \ldots, n$ the *k-norm* of $\mathbf{x}$ is

$$\|\mathbf{x}\|_{[k]} = |x_{i_1}| + |x_{i_2}| + \cdots + |x_{i_k}|.$$

If $k = 1$, the k-norm is the max norm; if $k = n$, it is the absolute sum norm.

As functions of a vector $\mathbf{x}$, the ℓ_p-norms and the k-norms depend only on the absolute values of the entries of $\mathbf{x}$. Some norms in the following family differ in this regard.

Example 19.1.4 Let $\| \cdot \|$ be a norm on $\mathbb{F}^n$ and let $S \in \mathsf{M}_n(\mathbb{F})$ be invertible. For $\mathbf{x} \in \mathbb{F}^n$, define

$$\|\mathbf{x}\|_S = \|S\mathbf{x}\|.$$

Then $\|\mathbf{x}\|_S = \|S\mathbf{x}\| \geq 0$ and $\|S\mathbf{x}\| = 0$ if and only if $S\mathbf{x} = \mathbf{0}$, which happens if and only if $\mathbf{x} = \mathbf{0}$ since S is invertible. In addition, $\|c\mathbf{x}\|_S = \|S(c\mathbf{x})\| = |c|\|S\mathbf{x}\| = |c|\|\mathbf{x}\|_S$ and

$$\|\mathbf{x} + \mathbf{y}\|_S = \|S\mathbf{x} + S\mathbf{y}\| \leq \|S\mathbf{x}\| + \|S\mathbf{y}\| = \|\mathbf{x}\|_S + \|\mathbf{y}\|_S.$$

Thus, $\| \cdot \|_S$ is a norm on $\mathbb{F}^n$. For example, consider the ℓ_1-norm and $S = \begin{bmatrix} 1 & -1 \\ 0 & 1 \end{bmatrix}$. Then $\|\mathbf{x}\|_S = |x_1 - x_2| + |x_2|$. Notice that

$$\left\| \begin{bmatrix} 1 \\ 1 \end{bmatrix} \right\|_S = 1, \quad \left\| \begin{bmatrix} 0 \\ 1 \end{bmatrix} \right\|_S = 2, \quad \text{and} \quad \left\| \begin{bmatrix} -1 \\ 1 \end{bmatrix} \right\|_S = 3.$$

Unlike the norms in the preceding two examples, this norm is not a function of the absolute values of the entries of $\mathbf{x}$.

19.2 Norms of Matrices

Definition 19.2.1 A norm $\| \cdot \|$ on the $\mathbb{F}$-vector space $\mathsf{M}_n(\mathbb{F})$ is a *matrix norm* if it is submultiplicative: $\|AB\| \leq \|A\| \, \|B\|$ for all $A, B \in \mathsf{M}_n(\mathbb{F})$.

Lemma 19.2.2 *If $\| \cdot \|$ is a matrix norm on M_n, then $\|A^k\| \leq \|A\|^k$ for all $A \in \mathsf{M}_n$ and $k = 1, 2, \ldots$.*

Proof We proceed by induction. Let S_k be the statement $\|A^k\| \leq \|A\|^k$. In the base case $k = 1$ there is nothing to prove. Suppose that $k \geq 2$ and S_k is true. Submultiplicativity and the induction hypothesis ensure that $\|A^{k+1}\| = \|A^k A\| \leq \|A^k\| \, \|A\| \leq \|A\|^k \|A\| = \|A\|^{k+1}$, which completes the induction. $\qquad \square$

A norm $\| \cdot \|$ on $\mathbb{F}^{n^2}$ is associated with the norm $N(\cdot)$ on $\mathsf{M}_n(\mathbb{F})$ defined by $N(A) = \|\text{vec}\,A\|$; see Definition 3.4.15. Our first example of a norm of this type is a matrix norm.

Example 19.2.3 The ℓ_1 *norm* on M_n is defined for each $A = [a_{ij}] \in \mathsf{M}_n$ by

$$\|A\|_1 = \|\text{vec}\,A\|_1 = \sum_{i,j=1}^{n} |a_{ij}|. \tag{19.2.4}$$

To verify that it is a matrix norm, let $B = [b_{ij}] \in \mathsf{M}_n$ and compute

$$\|AB\|_1 = \sum_{i,j=1}^{n} \left| \sum_{k=1}^{n} a_{ik} b_{kj} \right| \leq \sum_{i,j,k=1}^{n} |a_{ik} b_{kj}| \leq \sum_{i,j,k,\ell=1}^{n} |a_{ik}| \, |b_{\ell j}|$$

$$= \left(\sum_{i,k=1}^{n} |a_{ik}| \right) \left(\sum_{j,\ell=1}^{n} |b_{\ell j}| \right) = \|A\|_1 \|B\|_1.$$

Our second example is a familiar matrix norm; see Example 5.4.5.

Example 19.2.5 The *Frobenius norm* on $\mathsf{M}_{m \times n}$ is the Euclidean norm of $\mathrm{vec}\, A$:

$$\|A\|_{\mathrm{F}} = \|\mathrm{vec}\, A\|_2 = \left(\mathrm{tr}\, A^* A \right)^{1/2} = \left(\sum_{i,j} |a_{ij}|^2 \right)^{1/2}.$$

In Theorem 17.2.6, we showed that the Frobenius norm is a matrix norm on M_n.

If we have a sequence of $m \times n$ matrices $A_k = [a_{ij}^{(k)}]$ for $k = 1, 2, \ldots$, then $\lim_{k \to \infty} A_k = B = [b_{ij}]$ means that $\lim_{k \to \infty} a_{ij}^{(k)} = b_{ij}$ for each i, j; see Definition 13.2.1. Since $|a_{ij}^{(k)}| \leq \|A_k\|_{\mathrm{F}}$, it follows that $\lim_{k \to \infty} A_k = 0$ if and only if $\lim_{k \to \infty} \|A_k\|_{\mathrm{F}} = 0$.

Our third example is not a matrix norm, but a positive multiple of it is a matrix norm.

Example 19.2.6 The *max norm* on M_n is

$$\|A\|_\infty = \|\mathrm{vec}\, A\|_\infty = \max_{1 \leq i,j \leq n} |a_{ij}|.$$

Consider $A = \begin{bmatrix} 1 & 1 \\ 0 & 1 \end{bmatrix}$ and $A^2 = \begin{bmatrix} 1 & 2 \\ 0 & 1 \end{bmatrix}$. Then

$$\|A^2\|_\infty = 2 \nleq 1 = \|A\|_\infty^2,$$

so $\| \cdot \|_\infty$ is not a matrix norm. However, the *n-max norm*

$$N_{n\text{-max}}(\cdot) = n\| \cdot \|_\infty \tag{19.2.7}$$

is a matrix norm on M_n:

$$N_{n\text{-max}}(AB) = n \max_{1 \leq i,j \leq n} \left| \sum_{k=1}^{n} a_{ik} b_{kj} \right| \leq n \max_{1 \leq i,j \leq n} \sum_{k=1}^{n} |a_{ik} b_{kj}|$$

$$\leq n \sum_{k=1}^{n} \|A\|_\infty \|B\|_\infty = (n\|A\|_\infty)(n\|B\|_\infty)$$

$$= N_{n\text{-max}}(A) N_{n\text{-max}}(B).$$

The following example is associated with the max norm in a different way.

Example 19.2.8 Partition $A = [a_{ij}] = [\mathbf{a}_1 \; \mathbf{a}_2 \; \ldots \; \mathbf{a}_n] \in \mathsf{M}_n$ according to its columns and define the *column max norm* on M_n by

$$N_{\mathrm{col\ max}}(A) = \sum_{j=1}^{n} \|\mathbf{a}_j\|_\infty. \tag{19.2.9}$$

One can check that $N_{\text{col max}}(\cdot)$ is a norm on M_n (see P.19.5). The following computation with $B = [b_{ij}] = [\mathbf{b}_1\ \mathbf{b}_2\ \ldots\ \mathbf{b}_n] \in \mathsf{M}_n$ verifies submultiplicativity:

$$
\begin{aligned}
N_{\text{col max}}(AB) &= \sum_{j=1}^{n} \|A\mathbf{b}_j\|_\infty = \sum_{j=1}^{n} \left\| \sum_{k=1}^{n} \mathbf{a}_k b_{kj} \right\|_\infty \leq \sum_{j=1}^{n}\sum_{k=1}^{n} \|\mathbf{a}_k b_{kj}\|_\infty \\
&= \sum_{j=1}^{n}\sum_{k=1}^{n} \|\mathbf{a}_k\|_\infty |b_{kj}| \leq \sum_{j=1}^{n}\sum_{k=1}^{n} \|\mathbf{a}_k\|_\infty \|\mathbf{b}_j\|_\infty \\
&= \left(\sum_{k=1}^{n} \|\mathbf{a}_k\|_\infty \right)\left(\sum_{j=1}^{n} \|\mathbf{b}_j\|_\infty \right) = N_{\text{col max}}(A) N_{\text{col max}}(B).
\end{aligned}
$$

In Example 19.1.4, we showed how to modify a given norm on $\mathbb{F}^n$ to create new norms on $\mathbb{F}^n$. A similar construction modifies a given matrix norm to create new matrix norms.

Example 19.2.10 Let $\|\cdot\|$ be a matrix norm on $\mathsf{M}_n(\mathbb{F})$ and let $S \in \mathsf{M}_n(\mathbb{F})$ be invertible. For each $A \in \mathsf{M}_n(\mathbb{F})$ let

$$
N_S(A) = \|SAS^{-1}\|. \tag{19.2.11}
$$

Then $N_S(A) \geq 0$ with equality only if $SAS^{-1} = 0$, that is, if and only if $A = 0$. In addition, $N_S(cA) = \|ScAS^{-1}\| = |c|\,\|SAS^{-1}\| = |c|N_S(A)$ and

$$
N_S(A + B) = \|SAS^{-1} + SBS^{-1}\| \leq \|SAS^{-1}\| + \|SBS^{-1}\| = N_S(A) + N_S(B).
$$

Finally, $N_S(AB) = \|(SAS^{-1})(SBS^{-1})\| \leq \|SAS^{-1}\|\,\|SBS^{-1}\| = N_S(A)N_S(B)$, so $N_S(\cdot)$ is a matrix norm.

19.3 Induced Matrix Norms

In Theorem 17.2.2, we proved that the largest singular value function $\sigma_{\max}(\cdot) : \mathsf{M}_n(\mathbb{F}) \to \mathbb{R}$ has the following properties:

(a) $\|A\mathbf{x}\|_2 \leq \sigma_{\max}(A)$ for every $\mathbf{x} \in \mathbb{F}^n$ such that $\|\mathbf{x}\|_2 = 1$, and

(b) A has a right singular vector $\mathbf{w}$ such that $\|\mathbf{w}\|_2 = 1$ and $\|A\mathbf{w}\|_2 = \sigma_{\max}(A)$.

In Theorem 17.2.3, we used (a) and (b), and some properties of $\sigma_{\max}(\cdot)$, to show that $\sigma_{\max}(\cdot)$ is a matrix norm. In this section, we use the following theorem to create matrix norms from norms on $\mathbb{F}^n$.

Theorem 19.3.1 *Let $\|\cdot\|$ be a norm on $\mathbb{F}^n$ and suppose that $N(\cdot) : \mathsf{M}_n(\mathbb{F}) \to \mathbb{R}$ is a function such that:*

(a) *$\|A\mathbf{x}\| \leq N(A)$ for every $A \in \mathsf{M}_n(\mathbb{F})$ and every $\mathbf{x} \in \mathbb{F}^n$ such that $\|\mathbf{x}\| = 1$.*

(b) *For each $A \in \mathsf{M}_n(\mathbb{F})$ there is some $\mathbf{y} \in \mathbb{F}^n$ such that $\|\mathbf{y}\| = 1$ and $\|A\mathbf{y}\| = N(A)$.*

Then $N(\cdot)$ is a matrix norm on $\mathsf{M}_n(\mathbb{F})$.

Proof We must verify the four properties in Definition 5.5.1 and submultiplicativity. (Nonnegativity) Let $\mathbf{x} \in \mathbb{F}^n$ be a unit vector, that is, $\|\mathbf{x}\| = 1$. Then $N(A) \geq \|A\mathbf{x}\| \geq 0$ for every $A \in \mathsf{M}_n(\mathbb{F})$.

(Positivity) If $N(A) = 0$, then $0 = N(A) \geq \|A\mathbf{x}\| \geq 0$, and hence $A\mathbf{x} = \mathbf{0}$ for every unit vector $\mathbf{x}$, so $A = 0$. If $A = 0$ and $\mathbf{y}$ is a unit vector such that $\|A\mathbf{y}\| = N(A)$, then $0 = \|A\mathbf{y}\| = N(A)$.

(Homogeneity) If $c = 0$ or $A = 0$, then $N(cA) = N(0) = 0 = |c|N(A)$. If both $c \in \mathbb{F}$ and A are nonzero, let $\mathbf{x}$ and $\mathbf{y}$ be unit vectors such that $\|A\mathbf{x}\| = N(A)$ and $\|(cA)\mathbf{y}\| = N(cA)$. Then

$$N(cA) = \|(cA)\mathbf{y}\| = |c|\,\|A\mathbf{y}\|$$
$$\leq |c|N(A) \tag{19.3.2}$$
$$= |c|\,\|A\mathbf{x}\| = \|A(c\mathbf{x})\| \leq N(cA).$$

Consequently, (19.3.2) is an equality, so $N(cA) = |c|N(A)$.

(Triangle inequality) Let $\mathbf{z}$ be a unit vector such that $\|(A + B)\mathbf{z}\| = N(A + B)$. Then

$$N(A + B) = \|(A + B)\mathbf{z}\| = \|A\mathbf{z} + B\mathbf{z}\| \leq \|A\mathbf{z}\| + \|B\mathbf{z}\| \leq N(A) + N(B).$$

(Submultiplicativity) Let $\mathbf{z}$ be a unit vector such that $\|(AB)\mathbf{z}\| = N(AB)$. If $B\mathbf{z} = \mathbf{0}$, then $AB\mathbf{z} = \mathbf{0}$, $N(AB) = \|(AB)\mathbf{z}\| = \|\mathbf{0}\| = 0$, and hence $0 = N(AB) \leq N(A)N(B)$. If $B\mathbf{z} \neq \mathbf{0}$, then

$$N(AB) = \|(AB)\mathbf{z}\| = \left\| A\frac{B\mathbf{z}}{\|B\mathbf{z}\|} \right\| \|B\mathbf{z}\| \leq N(A)\,N(B). \qquad \square$$

Definition 19.3.3 If $\|\cdot\|$ is a norm on $\mathbb{F}^n$ and $N(\cdot) : \mathsf{M}_n(\mathbb{F}) \to \mathbb{R}$ satisfies (a) and (b) in Theorem 19.3.1, then $N(\cdot)$ is an *induced matrix norm*. It is *induced* by the norm $\|\cdot\|$ on $\mathbb{F}^n$.

In the next example, we revisit the spectral norm and use the preceding theorem to streamline the proof of Theorem 17.2.3.

Example 19.3.4 Let $A \in \mathsf{M}_n(\mathbb{F})$ and let $A = V\Sigma W^*$ be a singular value decomposition. Then $V \in \mathsf{M}_n(\mathbb{F})$ and $W = [\mathbf{w}_1\ \mathbf{w}_2\ \ldots\ \mathbf{w}_n] \in \mathsf{M}_n(\mathbb{F})$ are unitary, $\Sigma = \mathrm{diag}(\sigma_1, \sigma_2 \ldots, \sigma_n)$, and $\sigma_{\max}(A) = \sigma_1 \geq \sigma_2 \geq \cdots \geq \sigma_n \geq 0$. For any $\mathbf{x} \in \mathbb{F}^n$ such that $\|\mathbf{x}\|_2 = 1$, we have

$$\|A\mathbf{x}\|_2^2 = \|V\Sigma W^*\mathbf{x}\|_2^2 = \|\Sigma W^*\mathbf{x}\|_2^2 = \sum_{i=1}^{n} \sigma_i^2 |\langle \mathbf{x}, \mathbf{w}_i \rangle|^2$$
$$\leq \sum_{i=1}^{n} \sigma_1^2 |\langle \mathbf{x}, \mathbf{w}_i \rangle|^2 = \sigma_1^2 \sum_{i=1}^{n} |\langle \mathbf{x}, \mathbf{w}_i \rangle|^2 = \sigma_1^2 \|W^*\mathbf{x}\|_2^2$$
$$= \sigma_1^2 \|\mathbf{x}\|_2^2 = \sigma_{\max}^2(A),$$

with equality if $\mathbf{x} = W\mathbf{e}_1$. The preceding theorem ensures that the *spectral norm* $\|A\|_2 = \sigma_{\max}(A)$ is a matrix norm on $\mathsf{M}_n(\mathbb{F})$; it is induced by the Euclidean norm on $\mathbb{F}^n$.

Example 19.3.5 What matrix norm on $\mathsf{M}_n(\mathbb{F})$ is induced by the ℓ_1 vector norm on $\mathbb{F}^n$? Partition $A = [\mathbf{a}_1\ \mathbf{a}_2\ \ldots\ \mathbf{a}_n] \in \mathsf{M}_n(\mathbb{F})$ according to its columns and let $\mathbf{x} = [x_i] \in \mathbb{F}^n$ be a unit vector with respect to the ℓ_1 norm, that is, $\|\mathbf{x}\|_1 = |x_1| + |x_2| + \cdots + |x_n| = 1$. Then

$$\|A\mathbf{x}\|_1 = \left\| \sum_{j=1}^{n} x_j \mathbf{a}_j \right\|_1 \leq \sum_{j=1}^{n} \|x_j \mathbf{a}_j\|_1 = \sum_{j=1}^{n} |x_j|\,\|\mathbf{a}_j\|_1$$
$$\leq \sum_{j=1}^{n} |x_j| \max_{1 \leq k \leq n} \|\mathbf{a}_k\|_1 = \|\mathbf{x}\|_1 \max_{1 \leq k \leq n} \|\mathbf{a}_k\|_1 = \max_{1 \leq k \leq n} \|\mathbf{a}_k\|_1.$$

If m is an index such that $\|\mathbf{a}_m\|_1 = \max_{1 \leq k \leq n} \|\mathbf{a}_k\|_1$, then $\|A\mathbf{e}_m\|_1 = \|\mathbf{a}_m\|_1 = \max_{1 \leq k \leq n} \|\mathbf{a}_k\|_1$.

Theorem 19.3.1 ensures that the *maximum absolute column-sum norm*

$$N_1(A) = \max_{1 \leq j \leq n} \|\mathbf{a}_j\|_1 \tag{19.3.6}$$

is a matrix norm on $\mathsf{M}_n(\mathbb{F})$. It is induced by the ℓ_1 norm on $\mathbb{F}^n$.

Example 19.3.7 What matrix norm on $\mathsf{M}_n(\mathbb{F})$ is induced by the ℓ_∞ vector norm on $\mathbb{F}^n$? Let $A = [a_{ij}] \in \mathsf{M}_n(\mathbb{F})$, partition $A^\mathsf{T} = [\mathbf{r}_1 \; \mathbf{r}_2 \; \ldots \; \mathbf{r}_n]$ according to its columns, and let $\mathbf{x} = [x_i] \in \mathbb{F}^n$ be a unit vector with respect to the ℓ_∞ norm, that is, $\|\mathbf{x}\|_\infty = \max_{1 \le i \le n} |x_i| = 1$. Then

$$\|A\mathbf{x}\|_\infty = \max_{1 \le i \le n} \left| \sum_{j=1}^{n} a_{ij} x_j \right| \le \max_{1 \le i \le n} \sum_{j=1}^{n} |a_{ij}| \, |x_j|$$

$$\le \max_{1 \le i \le n} \sum_{j=1}^{n} |a_{ij}| \max_{1 \le k \le n} |x_k| = \max_{1 \le i \le n} \sum_{j=1}^{n} |a_{ij}| \, \|\mathbf{x}\|_\infty = \max_{1 \le i \le n} \|\mathbf{r}_i\|_1.$$

Let m be an index such that $\|\mathbf{r}_m\|_1 = \max_{1 \le i \le n} \|\mathbf{r}_i\|_1$ and define $\mathbf{y} = [y_i] \in \mathbb{F}^n$ by

$$y_j = \begin{cases} \dfrac{\overline{a_{mj}}}{|a_{mj}|} & \text{if } a_{mj} \ne 0, \\ 1 & \text{if } a_{mj} = 0. \end{cases}$$

Then $\|\mathbf{y}\|_\infty = 1$ and

$$\max_{1 \le i \le n} \|\mathbf{r}_i\|_1 \ge \|A\mathbf{y}\|_\infty = \max_{1 \le i \le n} \left| \sum_{j=1}^{n} a_{ij} y_j \right| \ge \left| \sum_{j=1}^{n} a_{mj} y_j \right| \tag{19.3.8}$$

$$= \sum_{j=1}^{n} |a_{mj}| = \|\mathbf{r}_m\|_1 = \max_{1 \le i \le n} \|\mathbf{r}_i\|_1,$$

which means that the inequalities in (19.3.8) are equalities. In particular, $\|A\mathbf{y}\|_\infty = \max_{1 \le i \le n} \|\mathbf{r}_i\|_1$. Theorem 19.3.1 ensures that the *maximum absolute row-sum norm*

$$N_\infty(A) = \max_{1 \le k \le n} \|\mathbf{r}_k\|_1 \tag{19.3.9}$$

is a matrix norm on $\mathsf{M}_n(\mathbb{F})$. It is induced by the ℓ_∞ norm on $\mathbb{F}^n$.

19.4 Matrix Norms and the Spectral Radius

Iterative methods are used widely in numerical algorithms, and it is often the case that an interative method converges if and only if the spectral radius (Definition 13.2.6) of a particular matrix A is less than one. The following theorem says that this is the case if some matrix norm of A is less than one; any matrix norm will do.

Theorem 19.4.1 *Let $\| \cdot \|$ be a matrix norm on M_n and let $A \in \mathsf{M}_n$. Then $\rho(A) \le \|A\|$.*

Proof Let $(\lambda, \mathbf{x})$ be an eigenpair of A such that $|\lambda| = \rho(A)$ and let $X = [\mathbf{x} \; \mathbf{x} \; \ldots \; \mathbf{x}] \in \mathsf{M}_n$. Then $AX = [A\mathbf{x} \; A\mathbf{x} \; \ldots \; A\mathbf{x}] = [\lambda \mathbf{x} \; \lambda \mathbf{x} \; \ldots \; \lambda \mathbf{x}] = \lambda X$, and hence

$$\rho(A)\|X\| = |\lambda| \, \|X\| = \|\lambda X\| = \|AX\| \le \|A\| \, \|X\|.$$

Since X (and hence also $\|X\|$) is nonzero, it follows that $\rho(A) \le \|A\|$. $\qquad \square$

Example 19.4.2 Let

$$A = \begin{bmatrix} 1 & 2 & 3 \\ 4 & 5 & 6 \\ 7 & 8 & 10 \end{bmatrix}. \tag{19.4.3}$$

To two decimal places, the eigenvalues of A are 16.71, 0.20, and -0.91, so $\rho(A) = 16.71$. The four matrix norms in Section 19.2 give the following upper bounds for $\rho(A)$: the ℓ_1 norm is $\|A\|_1 = 46$; the n-max norm is $N_{n\text{-max}}(A) = 30$; the column max norm is $N_{\text{col max}}(A) = 25$; and the Frobenius norm is $\|A\|_F = 17.44$. The three induced matrix norms in Section 19.3 give additional upper bounds: the spectral norm is $\|A\|_2 = 17.41$; the maximum absolute column-sum norm is $N_1(A) = 19$; and the maximum absolute row-sum norm is $N_\infty(A) = 25$.

No single matrix norm gives the best upper bound for the spectral radius in all cases, and in practice one is limited to using matrix norms that are feasible to compute. For example, the maximum absolute column and row sum norms can be computed with pencil and paper, the square root in the Frobenius norm begs for a calculator, and the spectral norm requires special software.

Any matrix norm provides an upper bound on the spectral radius, and it is possible for the upper bound to be an equality. For example, the matrix A in (19.4.3) is diagonalizable, so there is an invertible $S \in \mathsf{M}_3$ such that $A = S\Lambda S^{-1}$ and Λ is diagonal. Each row of Λ has only one nonzero entry (one of the eigenvalues of A), so $N_\infty(\Lambda) = \rho(A)$. Example 19.2.10 ensures that $X \mapsto N_\infty(S^{-1}XS)$ is a matrix norm on M_n such that $N_\infty(S^{-1}AS) = N_\infty(\Lambda) = \rho(A)$.

In general, what can be said about a matrix A if there is a matrix norm $\|\cdot\|$ such that $\rho(A) = \|A\|$? Inequalities between matrix norms play a role in answering this question. As examples of the kinds of inequalities we have in mind, consider the maximum absolute row-sum norm and the Frobenius norm. For any $A = [a_{ij}] \in \mathsf{M}_n$, a computation using the Cauchy–Schwarz inequality provides an upper bound for the maximum absolute row-sum norm:

$$N_\infty(A) = \max_{1 \leq i \leq n} \sum_{j=1}^{n} |a_{ij}| \leq \max_{1 \leq i \leq n} \left(\left(\sum_{j=1}^{n} |a_{ij}|^2 \right)^{1/2} \left(\sum_{j=1}^{n} 1 \right)^{1/2} \right)$$

$$\leq \sqrt{n} \left(\sum_{i,j=1}^{n} |a_{ij}|^2 \right)^{1/2} = \sqrt{n} \|A\|_F, \qquad (19.4.4)$$

with equality for $A = \mathbf{e}_1\mathbf{e}^T$, in which $\mathbf{e} \in \mathbb{F}^n$ is the all-ones vector. To obtain a lower bound, we compute

$$\|A\|_F^2 = \sum_{i=1}^{n} \sum_{j=1}^{n} |a_{ij}|^2 \leq \sum_{i=1}^{n} \left(\sum_{j=1}^{n} |a_{ij}| \right)^2 \leq \sum_{i=1}^{n} \left(\max_{1 \leq k \leq n} \sum_{j=1}^{n} |a_{kj}| \right)^2 = nN_\infty(A)^2, \quad (19.4.5)$$

with equality for $A = \mathbf{e}\mathbf{e}_1^T$. The following theorem addresses inequalities between other pairs of matrix norms.

Theorem 19.4.6 *The inequalities in Table 19.1 are valid. In addition, suppose that $S \in \mathsf{M}_n(\mathbb{F})$ is invertible, $\|\cdot\|$ is a matrix norm, $\|A\| \leq c\|A\|_F$, and $\|A\|_F \leq d\|A\|$ for all $A \in \mathsf{M}_n(\mathbb{F})$. The function $A \mapsto N_S(A) = \|SAS^{-1}\|$ is a matrix norm such that*

$$N_S(A) \leq \left(c\|S\|\,\|S^{-1}\| \right) \|A\|_F \qquad \text{and} \qquad \|A\|_F \leq \left(d\|S^{-1}\|_F\|S\|_F \right) N_S(A) \qquad (19.4.7)$$

for all $A \in \mathsf{M}_n(\mathbb{F})$.

Proof The verifications are similar to the computations in (19.4.4) and (19.4.5); see P.19.27. Singular values are useful if the spectral norm is involved. For example,

$$\|A\|_F^2 = \sum_{i=1}^{n} \sigma_i^2(A) \leq \sum_{i=1}^{n} \sigma_{\max}^2(A) = n\|A\|_2^2,$$

Table 19.1 Inequalities between matrix norms on $\mathsf{M}_n(\mathbb{F})$.

ℓ_1 matrix norm	$\|A\|_1 \leq n\|A\|_F$	$\|A\|_F \leq \|A\|_1$
n-max norm	$N_{n\text{-max}}(A) \leq n\|A\|_F$	$\|A\|_F \leq N_{n\text{-max}}(A)$
column max norm	$N_{\text{col max}}(A) \leq \sqrt{n}\|A\|_F$	$\|A\|_F \leq \sqrt{n}N_{\text{col max}}(A)$
spectral norm	$\|A\|_2 \leq \|A\|_F$	$\|A\|_F \leq \sqrt{n}\|A\|_2$
maximum absolute column sum	$N_1(A) \leq \sqrt{n}\|A\|_F$	$\|A\|_F \leq \sqrt{n}N_1(A)$
maximum absolute row sum	$N_\infty(A) \leq \sqrt{n}\|A\|_F$	$\|A\|_F \leq \sqrt{n}N_\infty(A)$

with equality if A is unitary, and

$$\|A\|_2^2 = \sigma_{\max}^2(A) \leq \sum_{i=1}^n \sigma_i^2(A) = \|A\|_F^2,$$

with equality if $A = \mathbf{e}_1\mathbf{e}_1^\mathsf{T}$.

Example 19.2.10 shows that $N_S(\cdot)$ is a matrix norm. The inequalities in (19.4.7) can be verified as follows:

$$N_S(A) = \|SAS^{-1}\| \leq (\|S\| \, \|S^{-1}\|)\|A\| \leq (c\|S\| \, \|S^{-1}\|)\|A\|_F$$

and

$$\|A\|_F = \|S^{-1}(SAS^{-1})S\|_F \leq \|S^{-1}\|_F \|S\|_F \|SAS^{-1}\|_F \leq (d\|S^{-1}\|_F\|S\|_F)N_S(A). \qquad \square$$

We now return to the question of what can be said about a matrix A if there is a matrix norm $\|\cdot\|$ such that $\rho(A) = \|A\|$. The answer involves the Jordan structures associated with the maximum-modulus eigenvalues of A.

Theorem 19.4.8 *Let $A \in \mathsf{M}_n$. There is a matrix norm $\|\cdot\|$ on M_n and a $c > 0$ such that $\rho(A) = \|A\|$ and*

$$\|X\|_F \leq c\|X\| \quad \text{for all } X \in \mathsf{M}_n \tag{19.4.9}$$

if and only if every maximum-modulus eigenvalue of A is semisimple.

Proof If $A = 0$, the Frobenius norm satisfies $\|A\|_F = \rho(A)$ and (19.4.9). Moreover, the only eigenvalue of A is semisimple. Now assume that $A \neq 0$.

($\Rightarrow$) Since $\rho(A) = \|A\| \neq 0$, let $B = \|A\|^{-1}A = \rho(A)^{-1}A$ and write $B^k = [b_{ij}^{(k)}]$. Then $\rho(B) = \|B\| = 1$ and

$$|b_{ij}^{(k)}| \leq \|B^k\|_F \leq c\|B^k\| \leq c\|B\|^k = c$$

for each $i,j = 1, 2, \ldots, n$ and all $k = 1, 2, \ldots$ This says that B is power bounded, and Theorem 13.2.10 ensures that each of its maximum-modulus eigenvalues is semisimple.

($\Leftarrow$) Suppose that some maximum-modulus eigenvalue of A is not semisimple. If there is a matrix norm $\|\cdot\|$ and a $c > 0$ that satisfy (19.4.9) and $\rho(A) = \|A\|$, then $\rho(A) = \|A\| \neq 0$ and $B = \|A\|^{-1}A = \rho(A)^{-1}A$ has an eigenvalue with modulus one that is not semisimple. Theorem 13.2.10 says that B is not power bounded, so $\|B^k\|_F$ is unbounded as $k \to \infty$. However, $\|B\| = 1$ and (19.4.9) ensures that $\|B^k\|_F \leq c\|B^k\| \leq c\|B\|^k = c$ for all $k = 1, 2, \ldots$. This contradiction shows that a matrix norm cannot satisfy both (19.4.9) and $\rho(A) = \|A\|$. $\qquad \square$

If some maximum-modulus eigenvalue of $A \in \mathsf{M}_n$ is not semisimple, then $\rho(A) < \|A\|$ for every matrix norm $\|\cdot\|$ on M_n that satisfies (19.4.9). Nevertheless, the following theorem shows that for each $\varepsilon > 0$ there is a matrix norm $\|\cdot\|$ that satisfies (19.4.9) and $\|A\| < \rho(A)+\varepsilon$.

Theorem 19.4.10 *Let $A \in \mathsf{M}_n$ and let $N_\infty(\cdot)$ be the maximum absolute row-sum matrix norm.*

(a) *For each $\varepsilon > 0$, there is an invertible $S \in \mathsf{M}_n$ such that $\rho(A) \leq N_\infty(S^{-1}AS) \leq \rho(A) + \varepsilon$.*

(b) *If every maximum modulus eigenvalue of A is semisimple, then there is an invertible $S \in \mathsf{M}_n$ such that $\rho(A) = N_\infty(S^{-1}AS)$.*

Proof Let $\mu_1, \mu_2, \ldots, \mu_d$ be the distinct eigenvalues of A.

(a) Let (12.5.1) be the Jordan canonical form of A. Let $S \in \mathsf{M}_n$ be invertible and such that $A = SJ_\varepsilon S^{-1}$, in which J_ε is the direct sum in (12.5.4). Then $N_\infty(S^{-1}AS) = N_\infty(J_\varepsilon) \leq \max_{1 \leq k \leq d}(|\mu_k| + \varepsilon) = \rho(A) + \varepsilon$.

(b) If every maximum-modulus eigenvalue of A is semisimple, then every Jordan block associated with a maximum-modulus eigenvalue is 1×1. Choose $\varepsilon > 0$ such that

$$\max\{|\mu_i| + \varepsilon < \rho(A) : i = 1, 2, \ldots, d \text{ and } |\mu_i| < \rho(A)\},$$

and then choose S as in (a). Then $N_\infty(S^{-1}AS) = N_\infty(J_\varepsilon) = \rho(A)$. □

We now turn to a celebrated relationship between matrix norms and the spectral radius.

Theorem 19.4.11 (Gelfand formula) *Let $\|\cdot\|$ be a matrix norm on M_n such that*

$$\|\cdot\| \leq d\|\cdot\|_F \text{ for some } d > 0. \tag{19.4.12}$$

Then

$$\rho(A) = \lim_{k \to \infty} \|A^k\|^{1/k} \tag{19.4.13}$$

for every $A \in \mathsf{M}_n$.

Proof Theorem 19.4.1 ensures that $\rho(A)^k = \rho(A^k) \leq \|A^k\|$, so $\rho(A) \leq \|A^k\|^{1/k}$ for all $k = 1, 2, \ldots$. If $\varepsilon > 0$ and $B = (\rho(A) + \varepsilon)^{-1}A$, then $\rho(B) = \rho(A)(\rho(A) + \varepsilon)^{-1} < 1$. Theorem 13.2.7 says that B is convergent, so $\lim_{k \to \infty} \|B^k\|_F = 0$. Since $\|B^k\| \leq d\|B^k\|_F$, there is some N (it depends on ε, d, and A) such that $\|B^k\| = \|(\rho(A) + \varepsilon)^{-k}A^k\| \leq 1$ for every $k \geq N$. This means that $\|A^k\| \leq (\rho(A) + \varepsilon)^k$ and $\|A^k\|^{1/k} \leq \rho(A) + \varepsilon$ for every $k \geq N$. Since $\varepsilon > 0$ is arbitrary and $\rho(A) \leq \|A^k\|^{1/k} \leq \rho(A) + \varepsilon$ for every $k \geq N$, we conclude that $\lim_{k \to \infty} \|A^k\|^{1/k}$ exists and equals $\rho(A)$. □

Every matrix norm introduced in this chapter satisfies (19.4.9) and (19.4.12), with explicit values for c and d. Therefore, Theorem 19.4.8 and the Gelfand formula (19.4.13) are valid for all of these matrix norms. A much stronger result is known: Every pair of norms on M_n (however constructed, matrix norm or not) satisfies inequalities of the form (19.4.9) and (19.4.12); see the Notes at the end of this chapter.

19.5 The Point Jacobi Iterative Method

Many matrix problems that arise in applications are solved using computer programs that take an initial approximation and refine it iteratively until some stopping criterion is satisfied. The iterative algorithm discussed in this section has been used to solve large sparse systems of linear equations for which direct methods such as Gaussian elimination are impractical.

Suppose that we wish to solve a linear system $A\mathbf{x} = \mathbf{b}$, in which $A = [a_{ij}] \in \mathsf{M}_n$ is an invertible matrix whose diagonal entries are all nonzero. Let $D = \mathrm{diag}(a_{11}, a_{22}, \dots, a_{nn})$. The *point Jacobi iterative method* reformulates the equation $A\mathbf{x} = \mathbf{b}$ as $(D + (A - D)\mathbf{x} = \mathbf{b}$, or $D\mathbf{x} = (D - A)\mathbf{x} + \mathbf{b}$. Then

$$\mathbf{x} = D^{-1}(D - A)\mathbf{x} + D^{-1}\mathbf{b} = B\mathbf{x} + D^{-1}\mathbf{b},$$

in which

$$B = D^{-1}(D - A) = \left[\delta_{ij} - \frac{a_{ij}}{a_{ii}} \right] \in \mathsf{M}_n \qquad (19.5.1)$$

is the *point Jacobi matrix*.

The point Jacobi iterative method takes an initial vector $\mathbf{x}^{(0)}$ and performs the iteration

$$\mathbf{x}^{(k)} = B\mathbf{x}^{(k-1)} + D^{-1}\mathbf{b} \qquad \text{for each } k = 1, 2, \dots. \qquad (19.5.2)$$

After k steps, the difference between the solution $\mathbf{x}$ and the iterate $\mathbf{x}^{(k)}$ is

$$
\begin{aligned}
\mathbf{x} - \mathbf{x}^{(k)} &= (B\mathbf{x} + D^{-1}\mathbf{b}) - \left(B\mathbf{x}^{(k-1)} + D^{-1}\mathbf{b} \right) \\
&= B\left(\mathbf{x} - \mathbf{x}^{(k-1)} \right) = B^2\left(\mathbf{x} - \mathbf{x}^{(k-2)} \right) = \cdots \\
&= B^k\left(\mathbf{x} - \mathbf{x}^{(0)} \right).
\end{aligned}
$$

To ensure that $\lim_{k\to\infty} \mathbf{x}^{(k)} = \mathbf{x}$ for every initial vector $\mathbf{x}^{(0)}$, we must require that B is convergent, that is, $\lim_{k\to\infty} B^k = 0$. Theorem 13.2.7 tells us that the spectral radius of B must be less than one. This occurs if some matrix norm of B (any one will do) is less than one.

If we use the maximum absolute row-sum norm, then the criterion is

$$N_\infty(B) = \max_{1 \le i \le n} \sum_{j \ne i} \frac{|a_{ij}|}{|a_{ii}|} < 1,$$

which is equivalent to $\sum_{j \ne i} |a_{ij}| < |a_{ii}|$ for each $i = 1, 2, \dots, n$. Thus, the point Jacobi iterative method converges to the solution of $A\mathbf{x} = \mathbf{b}$ if A is strictly diagonally dominant.

Example 19.5.3 Consider the following matrix A and its point Jacobi matrix B:

$$
A = \begin{bmatrix} 4 & 2 & -2 \\ 0 & 4 & 2 \\ 1 & 0 & 4 \end{bmatrix} \quad \text{and} \quad B = \begin{bmatrix} 0 & -\frac{1}{2} & \frac{1}{2} \\ 0 & 0 & -\frac{1}{2} \\ -\frac{1}{4} & 0 & 0 \end{bmatrix}.
$$

Does point Jacobi iteration converge in this case? We may become increasingly doubtful as we compute $N_\infty(B) = 1$, $N_1(B) = 1$, $N_{\text{col max}}(B) = \frac{5}{4}$, $N_{n\text{-max}}(B) = \frac{3}{2}$, and $\|B\|_1 = \frac{7}{4}$, all of which are greater than 1. But convergence is ensured because $\|B\|_F^2 = \frac{13}{16} < 1$. The spectral radius of B is approximately 0.46.

The solution of $Ax = b = [2\ 14\ 13]^\mathsf{T}$ is $x = [1\ 2\ 3]^\mathsf{T}$. Starting with $x^{(0)} = [1\ 1\ 1]^\mathsf{T}$, the point Jacobi iterates are

$$x^{(1)} \approx \begin{bmatrix} 0.5 \\ 3.0 \\ 3.0 \end{bmatrix}, \qquad x^{(5)} \approx \begin{bmatrix} 1.0235 \\ 2.0078 \\ 2.9766 \end{bmatrix}, \qquad x^{(10)} \approx \begin{bmatrix} 1.0005 \\ 2.0001 \\ 2.9995 \end{bmatrix}.$$

The point Jacobi iterative method can converge even if $\|B\| \geq 1$ for every matrix norm in Table 19.1; see P.19.26. The essential condition is that $\rho(B) < 1$.

19.6 The Power Method for a Dominant Eigenpair

The iterative algorithm discussed in this section is widely believed to be in daily use to find a dominant eigenpair of the Google matrix, which provides the PageRank of websites; see Example 11.6.3.

In Section 13.4, we learned that the eigenvalue $\lambda = 1$ of a positive stochastic matrix is not only a simple eigenvalue equal to the spectral radius, it is the only eigenvalue of maximum modulus. In the next chapter, we learn that the spectral radius of any positive matrix (stochastic or not) is a simple eigenvalue that is the only eigenvalue of maximum modulus.

Definition 19.6.1 An eigenvalue λ of $A \in \mathsf{M}_n$ is *dominant* if $|\lambda| = \rho(A)$, λ is a simple eigenvalue, and no other eigenvalue of A has modulus equal to $\rho(A)$. An eigenpair (λ, x) of A is a *dominant eigenpair* if λ is the dominant eigenvalue of A.

Suppose that $n \geq 2$ and $A \in \mathsf{M}_n$ has the dominant eigenvalue λ. Then there is an invertible $S \in \mathsf{M}_n$ such that

$$A = S \begin{bmatrix} \lambda & 0^\mathsf{T} \\ 0 & J \end{bmatrix} S^{-1},$$

in which $J \in \mathsf{M}_{n-1}$ is a Jordan matrix and $\rho(J) < |\lambda|$. The first column of S is an eigenvector of A associated with the dominant eigenvalue λ. Partition $S = [x\ X]$ and $S^{-*} = [y\ Y]$, in which $X, Y \in \mathsf{M}_{n \times (n-1)}$. Since we can replace S by cS and S^{-1} by $c^{-1}S^{-1}$ for any nonzero scalar c, we may assume that $\|x\|_2 = 1$. Then (λ, x) is a dominant eigenpair of A and

$$I = S^{-1}S = \begin{bmatrix} y^* \\ Y^* \end{bmatrix} [x\ X] = \begin{bmatrix} y^*x & y^*X \\ Y^*x & Y^*X \end{bmatrix} = \begin{bmatrix} 1 & 0^\mathsf{T} \\ 0 & I_{n-1} \end{bmatrix}. \tag{19.6.2}$$

The powers of A are

$$A^k = S \begin{bmatrix} \lambda^k & 0^\mathsf{T} \\ 0 & J^k \end{bmatrix} S^{-1} = [x\ X] \begin{bmatrix} \lambda^k & 0^\mathsf{T} \\ 0 & J^k \end{bmatrix} \begin{bmatrix} y^* \\ Y^* \end{bmatrix} = \lambda^k xy^* + XJ^kY^* \quad \text{for } k = 1, 2, \ldots. \tag{19.6.3}$$

The *power method* to find a dominant eigenpair begins by choosing an initial unit vector $x^{(0)}$. Since the columns of S are a basis for $\mathbb{C}^n$, we have $x^{(0)} = ax + Xb$ for some $a \in \mathbb{C}$ and some $b \in \mathbb{C}^{n-1}$. It turns out to be essential to assume that $a \neq 0$, so if we have any *a priori* information about x we should choose $x^{(0)}$ to be as good an approximation to x as we can muster. Absent any information about x, one is advised to choose $x^{(0)}$ at random. Using (19.6.3) and the identities in (19.6.2), we find that

$$
\begin{aligned}
A^k \mathbf{x}^{(0)} &= (\lambda^k \mathbf{x}\mathbf{y}^* + XJ^k Y^*)\mathbf{x}^{(0)} \\
&= \lambda^k (\mathbf{y}^* \mathbf{x}^{(0)})\mathbf{x} + XJ^k Y^* \mathbf{x}^{(0)} \\
&= \lambda^k \big(\mathbf{y}^*(a\mathbf{x} + X\mathbf{b})\mathbf{x} + XJ^k(aY^*\mathbf{x} + Y^*X\mathbf{b})\big) \\
&= \lambda^k (a\mathbf{y}^*\mathbf{x} + \mathbf{y}^*X\mathbf{b})\mathbf{x} + XJ^k(aY^*\mathbf{x} + Y^*X\mathbf{b}) \\
&= \lambda^k (a + 0)\mathbf{x} + XJ^k(\mathbf{0} + \mathbf{b}) \\
&= \lambda^k a\mathbf{x} + XJ^k\mathbf{b} \\
&= \lambda^k (a\mathbf{x} + X(\lambda^{-1}J)^k\mathbf{b}) \\
&= \lambda^k (a\mathbf{x} + XB^k\mathbf{b}) \qquad \text{for } k = 1, 2, \ldots,
\end{aligned}
$$

in which $B = \lambda^{-1}J$ and $\rho(B) < 1$, so $\lim_{k\to\infty} B^k = 0$. If we write $\lambda = |\lambda| e^{i\theta}$, in which $0 \le \theta < 2\pi$, then the normalized iterates are

$$
\begin{aligned}
\mathbf{z}_k &= \frac{A^k \mathbf{x}^{(0)}}{\|A^k \mathbf{x}^{(0)}\|_2} = \frac{\lambda^k \left(a\mathbf{x} + XB^k\mathbf{b}\right)}{\|\lambda^k \left(a\mathbf{x} + XB^k\mathbf{b}\right)\|_2} \\
&= \frac{\lambda^k}{|\lambda^k|} \frac{a\mathbf{x} + XB^k\mathbf{b}}{\|a\mathbf{x} + XB^k\mathbf{b}\|_2} \\
&= \frac{e^{ik\theta} a\mathbf{x} + X(e^{i\theta}B)^k\mathbf{b}}{\|a\mathbf{x} + XB^k\mathbf{b}\|_2}.
\end{aligned}
\tag{19.6.4}
$$

In the denominator, we have

$$
\lim_{k\to\infty} \|a\mathbf{x} + XB^k\mathbf{b}\|_2 = \|a\mathbf{x}\|_2 = |a|,
$$

which is why it is essential to assume at the beginning that $a \ne 0$. In the numerator, we have $\lim_{k\to\infty} X(e^{i\theta}B)^k\mathbf{b} = \mathbf{0}$. If we write $a = e^{i\phi}|a|$, in which $0 \le \phi < 2\pi$, then the normalized iterates (19.6.4) are

$$
\mathbf{z}_k = e^{i(\phi + k\theta)}\mathbf{x} + \mathbf{d}_k,
\tag{19.6.5}
$$

in which $\mathbf{d}_1, \mathbf{d}_2, \ldots$ is a bounded sequence of vectors such that $\lim_{k\to\infty} \|\mathbf{d}_k\|_2 = 0$. Now compute the Rayleigh quotients

$$
\begin{aligned}
r_k &= \frac{\mathbf{z}_k^* A \mathbf{z}_k}{\mathbf{z}_k^* \mathbf{z}_k} = \frac{(e^{i(\phi+k\theta)}\mathbf{x} + \mathbf{d}_k)^* A(e^{i(\phi+k\theta)}\mathbf{x} + \mathbf{d}_k)}{\|e^{i(\phi+k\theta)}\mathbf{x} + \mathbf{d}_k\|_2^2} \\
&= \frac{(e^{i(\phi+k\theta)}\mathbf{x} + \mathbf{d}_k)^* (\lambda e^{i(\phi+k\theta)}\mathbf{x} + A\mathbf{d}_k)}{\|e^{i(\phi+k\theta)}\mathbf{x} + \mathbf{d}_k\|_2^2} \\
&= \frac{\lambda + e^{-i(\phi+k\theta)}\mathbf{x}^* A\mathbf{d}_k + \lambda e^{i(\phi+k\theta)}\mathbf{d}_k^*\mathbf{x} + \mathbf{d}_k^* A\mathbf{d}_k}{1 + 2\,\mathrm{Re}\,(e^{i(\phi+k\theta)}\mathbf{d}_k^*\mathbf{x}) + \|\mathbf{d}_k\|_2^2} \\
&= \frac{\lambda + \delta_k}{1 + \eta_k},
\end{aligned}
$$

in which the scalar sequences δ_k and η_k tend to zero as $k \to \infty$ because $\mathbf{d}_k \to \mathbf{0}$. The conclusion is that for each sufficiently large k, $(r_k, \mathbf{z}_k)$ is an approximate dominant eigenpair of A, and $r_k \to \lambda$. If λ is not real and positive, then $\theta \ne 0$ and we cannot expect the sequence $\mathbf{z}_k$ to converge. Nevertheless, for sufficiently large k, (19.6.5) ensures that $\mathbf{z}_k$ is a unit vector that is close to a unit eigenvector of A associated with the dominant eigenvalue λ.

In practice, one normalizes at each iteration to avoid overflow or underflow; this is the *power method with scaling*: First choose a unit vector $\mathbf{x}^{(0)}$ at random (or with some knowledge of a dominant eigenvector, if available). Then for each $k = 1, 2, \ldots$, compute

$\mathbf{y}^{(k)} = A\mathbf{x}^{(k-1)}$, $\mathbf{x}^{(k)} = \mathbf{y}^{(k)}/\|\mathbf{y}^{(k)}\|_2$, and the Rayleigh quotients $r_k = (\mathbf{x}^{(k)})^*A\mathbf{x}^{(k)}$. Stop when empirical convergence is observed.

Example 19.6.6 Here are the first four iterates and Rayleigh quotients for the power method and the matrix A in (19.4.3). The dominant eigenvalue of A is approximately 16.7075. The initial vector $\mathbf{x}^{(0)} = [1\ 1\ 1]^\mathsf{T}/\sqrt{3}$ yields

$$\mathbf{x}^{(1)} \approx \begin{bmatrix} 0.2016 \\ 0.5039 \\ 0.8399 \end{bmatrix}, \quad \mathbf{x}^{(2)} \approx \begin{bmatrix} 0.2247 \\ 0.5040 \\ 0.8340 \end{bmatrix}, \quad \mathbf{x}^{(3)} \approx \begin{bmatrix} 0.2234 \\ 0.5039 \\ 0.8343 \end{bmatrix}, \quad \mathbf{x}^{(4)} \approx \begin{bmatrix} 0.2235 \\ 0.5039 \\ 0.8343 \end{bmatrix},$$

and

$$r_1 \approx 16.5926, \qquad r_2 \approx 16.7133, \qquad r_3 \approx 16.7072, \qquad r_4 \approx 16.7075.$$

Despite its name, the power method does not compute powers of A (n^3 scalar multiplications at each step); it computes matrix-vector products (n^2 scalar multiplications at each step).

19.7 Problems

P.19.1 Let $\mathbf{x} = [-1\ 1\ -2\ 3]^\mathsf{T}$. Compute its ℓ_1, ℓ_2, and ℓ_∞ norms; its ℓ_p norm for $p = 3$; and its k-norms for $k = 1, 2, 3, 4$.

P.19.2 Let $A = \begin{bmatrix} 1 & 2 \\ 3 & 4 \end{bmatrix}$. Compute $\|A\|_1$, $\|A\|_\mathrm{F}$, $\|A\|_\infty$, $N_{n\text{-max}}(A)$, $N_{\text{col max}}(A)$, $\|A\|_2^2$, $N_1(A)$, $N_\infty(A)$, and $\rho(A)$.

P.19.3 (a) Does the point Jacobi iterative method converge for the matrix in the preceding problem? Why? (b) Does it converge for $A = \begin{bmatrix} 2 & 1 \\ 3 & 4 \end{bmatrix}$? Why?

P.19.4 Show that the k-norms $\|\cdot\|_{[k]}$ on $\mathbb{F}^n$ satisfy the four axioms for a norm.

P.19.5 (a) Show that $N_{\text{col max}}(\cdot)$ is a norm on M_n. (b) Show that $N_{\text{col max}}(A) \le N_{n\text{-max}}(A)$ for all $A \in \mathsf{M}_n$. (c) Find a nonzero A such that $N_{\text{col max}}(A) = N_{n\text{-max}}(A)$.

P.19.6 If $\|\cdot\|$ is a norm on $\mathbb{F}^n$ and $c > 0$, show that $c\|\cdot\|$ is a norm on $\mathbb{F}^n$.

P.19.7 Let $\|\cdot\|$ be a matrix norm and let $A = \begin{bmatrix} 0 & c \\ 0 & 0 \end{bmatrix}$. Although $\rho(A) \le \|A\|$ for all $c \in \mathbb{R}$, show that $\|A\| - \rho(A)$ can be arbitrarily large.

P.19.8 Let $\|\cdot\|$ be a matrix norm. (a) If $c \ge 1$, show that $c\|\cdot\|$ is a matrix norm. (b) Give an example to show that if $0 < c < 1$, then $c\|\cdot\|$ is a norm, but it need not be a matrix norm. (c) Give an example to show that $\frac{1}{2}\|\cdot\|$ can be a matrix norm.

P.19.9 (a) Find two matrix norms $\|\cdot\|$ on M_n such that $\|A\| = \|A^*\|$ for all $A \in \mathsf{M}_n$. (b) Find a matrix norm on M_n such that $\|A\| \ne \|A^*\|$ for some $A \in \mathsf{M}_n$.

P.19.10 If $\|\cdot\|$ is a matrix norm on M_n, define $\|\cdot\|_*$ by $\|A\|_* = \|A^*\|$. (a) Prove that $\|\cdot\|_*$ is a matrix norm. (b) What matrix norm do you get if you apply this construction to $N_{\text{col max}}(\cdot)$, the column max norm?

P.19.11 (a) Show that the ℓ_p-norms and the k-norms are *permutation-invariant norms*. That is, for any $\mathbf{x} \in \mathbb{F}^n$ and any permutation matrix $P \in \mathsf{M}_n$, the vectors $P\mathbf{x}$ and $\mathbf{x}$ have the same norm. (b) Give an example of a norm on $\mathbb{F}^n$ that is not permutation invariant.

P.19.12 Let $\|\cdot\|_p$ be the function defined in (19.1.2). (a) If $0 < p < 1$, show that $\|\cdot\|_p$ satisfies three of the four properties in Definition 5.5.1. (b) Let $\mathbf{x} = [1\ 2]^\mathsf{T}$ and $\mathbf{y} = [2\ 1]^\mathsf{T}$. Compute $\|\mathbf{x}+\mathbf{y}\|_{2/3}$, $\|\mathbf{x}\|_{2/3}$, and $\|\mathbf{y}\|_{2/3}$. What do you conclude? (c) Examine Figure 19.1. Is $\{\mathbf{x} \in \mathbb{R}^2 : \|\mathbf{x}\|_{2/3} \le 1\}$ convex? (d) Is $\{\mathbf{x} \in \mathbb{R}^2 : \|\mathbf{x}\|_3 \le 1\}$ convex?

P.19.13 Show that $\|\mathbf{x}\|_\infty = \lim_{p \to \infty} \|\mathbf{x}\|_p$ for each $\mathbf{x} \in \mathbb{C}^n$; see Example 19.1.1.

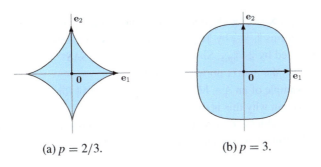

(a) $p = 2/3$.   (b) $p = 3$.

Figure 19.1 $\{\mathbf{x} \in \mathbb{R}^2 : \|\mathbf{x}\|_p \leq 1\}$.

P.19.14 The *Minkowski inequality* is the triangle inequality for the ℓ_p norms (19.1.2). Let $\mathbf{x} = [x_i] \in \mathbb{C}^n$ and $\mathbf{y} = [y_i] \in \mathbb{C}^n$. Let $a > 0$, $b > 0$, and $p > 1$. Define

$$f(t) = t^{1-p}a^p + (1-t)^{1-p}b^p \qquad \text{for } t \in (0, 1).$$

Provide details for the following steps in a proof of the Minkowski inequality for $p \in (1, \infty)$. (a) Show that $f''(t) > 0$ for $t \in (0, 1)$. (b) Show that $f'(\frac{a}{a+b}) = 0$ and $\min_{t \in (0,1)} f(t) = (a+b)^p$. (c) If x_i and y_i are nonzero, show that

$$|x_i + y_i|^p \leq (|x_i| + |y_i|)^p \leq t^{1-p}|x_i|^p + (1-t)^{1-p}|y_i|^p \qquad \text{for all } t \in (0, 1). \quad (19.7.1)$$

(d) Show that (19.7.1) is valid even if $x_i = 0$ or $y_i = 0$. (e) Show that

$$\|\mathbf{x} + \mathbf{y}\|_p^p \leq t^{1-p}\|\mathbf{x}\|_p^p + (1-t)^{1-p}\|\mathbf{y}\|_p^p \qquad \text{for all } t \in (0, 1).$$

(f) Deduce that $\|\mathbf{x} + \mathbf{y}\|_p^p \leq (\|\mathbf{x}\|_p + \|\mathbf{y}\|_p)^p$ for all $\mathbf{x}, \mathbf{y} \in \mathbb{C}^n$ and conclude that $\|\mathbf{x} + \mathbf{y}\|_p \leq \|\mathbf{x}\|_p + \|\mathbf{y}\|_p$ for all $\mathbf{x}, \mathbf{y} \in \mathbb{C}^n$, which is the Minkowski inequality.

P.19.15 For any $\mathbf{x} \in \mathbb{F}^n$, show that: (a) $\|\mathbf{x}\|_1 \leq \sqrt{n}\|\mathbf{x}\|_2$ and $\|\mathbf{x}\|_2 \leq \|\mathbf{x}\|_1$; (b) $\|\mathbf{x}\|_\infty \leq \|\mathbf{x}\|_2$ and $\|\mathbf{x}\|_2 \leq \sqrt{n}\|\mathbf{x}\|_\infty$; and (c) $\|\mathbf{x}\|_1 \leq n\|\mathbf{x}\|_\infty$ and $\|\mathbf{x}\|_\infty \leq \|\mathbf{x}\|_1$. For each of the inequalities in (a), (b), and (c), exhibit a nonzero vector for which the asserted bound is attained.

P.19.16 If $N(\cdot)$ is an induced matrix norm, show that $N(I) = 1$.

P.19.17 Show that none of the matrix norms $\|\cdot\|_1$, $\|\cdot\|_F$, $N_{n\text{-max}}(\cdot)$, and $N_{\text{col max}}(\cdot)$ on M_n (see Section 19.2) are induced by norms on $\mathbb{C}^n$ if $n \geq 2$.

P.19.18 If $A \in \mathsf{M}_n$ is not invertible, show that $N(I - A) \geq 1$ for every matrix norm $N(\cdot)$.

P.19.19 Let $\|\cdot\|$ be a matrix norm on $\mathsf{M}_n(\mathbb{F})$. If $A \in \mathsf{M}_n(\mathbb{F})$ is invertible and $\lambda \in \operatorname{spec} A$, show that $|\lambda| \geq 1/\|A^{-1}\|$.

P.19.20 If $N(\cdot)$ is the norm on $\mathsf{M}_n(\mathbb{F})$ induced by the norm $\|\cdot\|$ on $\mathbb{F}^n$, show that

$$N(A) = \max_{\|\mathbf{x}\| = 1} \|A\mathbf{x}\| = \max_{\mathbf{x} \neq 0} \frac{\|A\mathbf{x}\|}{\|\mathbf{x}\|}. \qquad (19.7.2)$$

Use this characterization to show that $N(\cdot)$ is a matrix norm. *Hint*: Built into the hypothesis is the assumption that $N(\cdot)$ satisfies the conditions (a) and (b) in Theorem 19.3.1.

P.19.21 Let $\|\cdot\|_a$ and $\|\cdot\|_b$ be norms on an $\mathbb{F}$-vector space $\mathcal{V}$ and let $N(\cdot) = \max\{\|\cdot\|_a, \|\cdot\|_b\}$. (a) Show that $N(\cdot)$ is a norm on $\mathcal{V}$. (b) If $\mathcal{V} = \mathsf{M}_n(\mathbb{F})$ and $\|\cdot\|_a$ and $\|\cdot\|_b$ are matrix norms, show that $N(\cdot)$ is a matrix norm.

P.19.22 Let $p \in \{1, 2, \infty\}$, let $\| \cdot \|_p$ denote the ℓ_p norm on $\mathbb{C}^n$, let $S \in \mathsf{M}_n$ be invertible, and define the norms $\| \cdot \|_{p,S}$ by $\|\mathbf{x}\|_{p,S} = \|S\mathbf{x}\|_p$. Let $N_{p,S}(\cdot)$ denote the matrix norm on M_n that is induced by $\| \cdot \|_{p,S}$. Show that $N_{p,S}(A) = N_p(SAS^{-1})$ for all $A \in \mathsf{M}_n$, in which $N_p(\cdot)$ denotes the matrix norm induced by $\| \cdot \|_p$.

P.19.23 Give an example of an $A \in \mathsf{M}_2$ such that $\rho(A) < \|A\|$ for every matrix norm $\| \cdot \|$ on M_2 and explain why this is the case.

P.19.24 Let $A \in \mathsf{M}_n$ and let $\lambda \in \operatorname{spec} A$. Show that λ is semisimple if and only if the Jordan canonical form of A has the following property: If $J_k(\lambda)$ is a Jordan block of A, then $k = 1$.

P.19.25 Use Theorem 9.4.2 to show (without invoking anything about norms) that the point Jacobi matrix B in (19.5.1) has spectral radius less than one if A is strictly diagonally dominant.

P.19.26 Let

$$A = \begin{bmatrix} 1 & 20 & -20 \\ 10 & 1 & 10 \\ 20 & 20 & 1 \end{bmatrix}.$$

(a) Is A invertible? (b) Is A strictly diagonally dominant? (c) Compute the point Jacobi matrix B associated with A. (d) Compute the values of $\|B\|$ for every matrix norm in Table 19.1 except the spectral norm. What do these values tell you about $\rho(B)$? (e) Compute $\rho(B)$. *Hint*: One approach is to find the pivot column decomposition $B = PR$ (Theorem 1.7.5) and then find the eigenvalues of RP (Theorem 10.7.2). (f) Why is the point Jacobi iterative method guaranteed to converge to the solution of $A\mathbf{x} = \mathbf{b}$ for any given $\mathbf{b}$? (g) Compute several Jacobi iterates (19.5.2) and show that they converge to $\mathbf{x}$ in just three steps, that is, $\mathbf{x} = \mathbf{x}^{(3)} = B^2\mathbf{b} + B\mathbf{b} + \mathbf{b}$. Why does this happen?

P.19.27 Verify four of the eight inequalities in Table 19.1 that have not been verified in the text. In each case, exhibit a nonzero matrix for which the inequality is an equality. This shows that each inequality is best possible.

P.19.28 Use Table 19.1 to show that $\|A\|_1 \leq n^{3/2}\|A\|_2$ and $N_\infty(A) \leq nN_1(A)$.

P.19.29 Show that the bound $\|A\|_F \leq \sqrt{n}\|A\|_2$ in Table 19.1 can be improved to $\|A\|_F \leq \sqrt{\operatorname{rank} A}\,\|A\|_2$.

P.19.30 Let $A \in \mathsf{M}_n(\mathbb{F})$ be positive definite and define $f : \mathbb{F}^n \to \mathbb{R}$ by $f(\mathbf{x}) = (\mathbf{x}^*A\mathbf{x})^{1/2}$. Show that $f(\cdot)$ is a norm on $\mathbb{F}^n$.

P.19.31 If $n \geq 2$ and λ is a dominant eigenvalue of $A \in \mathsf{M}_n$, show that $|\lambda| > 0$. What happens if $n = 1$?

P.19.32 Let $A \in \mathsf{M}_2(\mathbb{R})$. (a) If A does not have a real eigenvalue, prove that it does not have a dominant eigenvalue. (b) If A has a dominant eigenvalue, prove that both of its eigenvalues are real.

19.8 Notes

With tools from real analysis, it can be shown that if $\| \cdot \|_a$ and $\| \cdot \|_b$ are any norms on $\mathbb{F}^n$, then there are positive constants c and d such that $c\|\mathbf{x}\|_a \leq \|\mathbf{x}\|_b \leq d\|\mathbf{x}\|_a$ for all $\mathbf{x} \in \mathbb{F}^n$. Consequently, the conditions (19.4.9) and (19.4.12) are satisfied for every matrix norm on M_n, so they can be omitted from Theorems 19.4.8 and 19.4.11, respectively. It can also be shown that if $\| \cdot \|$ is any norm on M_n, then there is a scalar $c \geq 1$ such that $c\| \cdot \|$ is a matrix norm. Consequently, the Gelfand formula (19.4.13) is valid for any norm on M_n. Finally, it can be shown that if $\| \cdot \|$ is any norm on $\mathbb{F}^n$, then there is a $\mathbf{y} \in \mathbb{F}^n$ such that $\|\mathbf{y}\| = 1$ and $\|A\mathbf{x}\| \leq \|A\mathbf{y}\|$ for all $\mathbf{x} \in \mathbb{F}^n$ such that $\|\mathbf{x}\| = 1$. This means that $\| \cdot \|$ induces a matrix norm on M_n via the recipe in (19.7.2). For details about these and other results about norms that require tools from analysis, see [HJ13, Chapter 5].

19.9 Some Important Concepts

- Matrix norms: axioms and examples.

- Induced matrix norms.

- The spectral radius is dominated by every matrix norm.

- A is convergent if there is a matrix norm $\| \cdot \|$ such that $\|A\| < 1$.

- The spectral radius of a given matrix can be arbitrarily closely approximated by matrix norms.

- The Gelfand formula for the spectral radius.

- The Jacobi iterative method for solving certain linear systems.

- The power method for finding a dominant eigenpair.

Positive and Nonnegative Matrices

This chapter is about some remarkable properties of positive matrices, by which we mean square matrices with real positive entries. Positive matrices are found in economic models, genetics, biology, team rankings, network analysis, Google's PageRank, and city planning. The spectral radius of any matrix is the absolute value of an eigenvalue, but for a positive matrix the spectral radius itself is an eigenvalue, and it is positive and dominant. It is associated with a positive eigenvector, whose ordered entries have been used for ranking sports teams, priority setting, and resource allocation in multicriteria decision making. Since the spectral radius is a dominant eigenvalue, an associated positive eigenvector can be computed by the power method.

Some properties of positive matrices are shared by nonnegative matrices that satisfy certain auxiliary conditions. One condition that we investigate in this chapter is that some positive power has no zero entries.

20.1 Nonnegative Matrices

We begin with some general observations about nonnegative matrices.

Definition 20.1.1 Let $A = [a_{ij}] \in M_{m \times n}(\mathbb{F})$ and $B = [b_{ij}] \in M_{m \times n}(\mathbb{F})$.

(a) In this chapter only, we define $|A| = [|a_{ij}|] \in M_{m \times n}(\mathbb{R})$ (entrywise absolute values).

(b) If $\mathbb{F} = \mathbb{R}$, we write $A > B$ if $a_{ij} > b_{ij}$ for all i, j, and $A \geq B$ if $a_{ij} \geq b_{ij}$ for all i, j; the reversed inequalities are defined in a similar fashion.

(c) A is *positive* if its entries are all real and positive; it is *nonnegative* if its entries are all real and nonnegative.

The following theorem lists some facts about inequalities and nonnegative matrices that we use in this chapter. The triangle inequality is our principal tool, and we make use of the maximum absolute row-sum norm $N_\infty(\,\cdot\,)$.

Theorem 20.1.2 Let $A = [a_{ij}] \in M_n$, $B = [b_{ij}] \in M_n$, and $\mathbf{x} = [x_i] \in \mathbb{C}^n$.

(a) $|A\mathbf{x}| \leq |A|\,|\mathbf{x}|$.

(b) $|AB| \leq |A|\,|B|$. In particular, if $A \geq 0$ and $B \geq 0$, then $AB \geq 0$.

(c) $|A^k| \leq |A|^k$ for $k = 1, 2, \ldots$.

(d) If $A \geq 0$ and $\mathbf{x} \geq \mathbf{0}$, then $A\mathbf{x} \geq \mathbf{0}$.

(e) If $A > 0$, $\mathbf{x} \geq \mathbf{0}$, and $\mathbf{x} \neq \mathbf{0}$, then $A\mathbf{x} > \mathbf{0}$.

(f) *If $0 \leq A \leq B$, then $0 \leq A^k \leq B^k$ for $k = 1, 2, \ldots$.*

(g) *If $|A| \leq B$, then $N_\infty(A) = N_\infty(|A|) \leq N_\infty(B)$.*

Proof (a) For $i = 1, 2, \ldots$, the triangle inequality ensures that

$$(|A\mathbf{x}|)_i = \left| \sum_{j=1}^n a_{ij} x_j \right| \leq \sum_{j=1}^n |a_{ij} x_j| = \sum_{j=1}^n |a_{ij}| \, |x_j| = (|A| \, |\mathbf{x}|)_i.$$

(b) Partition $B = [\mathbf{b}_1 \; \mathbf{b}_2 \; \ldots \; \mathbf{b}_n]$ according to its columns and apply (a):

$$|AB| = \left[|A\mathbf{b}_1| \; |A\mathbf{b}_2| \; \ldots \; |A\mathbf{b}_n| \right] \leq \left[|A||\mathbf{b}_1| \quad |A||\mathbf{b}_2| \quad \ldots \quad |A||\mathbf{b}_n| \right] = |A| \, |B|.$$

If A and B are real and nonnegative, then $0 \leq |AB| \leq |A| \, |B| = AB$.

(c) This follows from (b) and induction.

(d) $(A\mathbf{x})_i = \sum_{j=1}^n a_{ij} x_j \geq 0$ for each $i = 1, 2, \ldots, n$.

(e) This is Lemma 13.4.2.

(f) We proceed by induction. Let S_k be the statement $0 \leq A^k \leq B^k$. In the base case $k = 1$, there is nothing to prove. If $k \geq 1$ and S_k is true, then $B^k - A^k \geq 0$, so the columns of $B^k - A^k$ are nonnegative. Then (d) ensures that $B^{k+1} - BA^k = B(B^k - A^k) \geq 0$. Write this as $B^{k+1} \geq BA^k = (B - A)A^k + A^{k+1} \geq A^{k+1}$, which shows that S_{k+1} is true. This completes the induction.

(g) $N_\infty(A) = N_\infty(|A|) = \max_{1 \leq i \leq n} \sum_{j=1}^n |a_{ij}| \leq \max_{1 \leq i \leq n} \sum_{j=1}^n |b_{ij}| = N_\infty(B)$. $\square$

The inequality in (g) is valid for the maximum absolute column-sum norm, the Frobenius norm, the ℓ_1 norm, and any norm that depends only on the absolute values of the entries of A.

The spectral radius of a nonnegative matrix cannot decrease if any entry is increased. We use the Gelfand formula for the spectral radius to prove this fundamental fact.

Theorem 20.1.3 *Let $A, B \in \mathsf{M}_n$ with B nonnegative. If $|A| \leq B$, then $\rho(A) \leq \rho(|A|) \leq \rho(B)$.*

Proof The preceding theorem ensures that $|A^k| \leq |A|^k \leq B^k$ and $N_\infty(A^k) = N_\infty(|A^k|) \leq N_\infty(|A|^k) \leq N_\infty(B^k)$ for $k = 1, 2, \ldots$. Therefore, $N_\infty(A^k)^{1/k} \leq N_\infty(|A|^k)^{1/k} \leq N_\infty(B^k)^{1/k}$ for $k = 1, 2, \ldots$. The assertion follows from Theorem 19.4.11. $\square$

Example 20.1.4 The preceding theorem says that the spectral radius of a complex matrix cannot decrease if every entry is replaced by its absolute value. However, $A = \begin{bmatrix} 1 & -1 \\ -1 & 1 \end{bmatrix}$ and $B = \begin{bmatrix} 1 & 1 \\ -1 & 1 \end{bmatrix}$ have $\rho(A) = 2 > \sqrt{2} = \rho(B)$, so the spectral radius can decrease if a single entry is replaced by its absolute value.

A nonzero nonnegative matrix can have a zero spectral radius; any nilpotent Jordan block $J_k(0)$ provides an example. However, a positive matrix has a positive spectral radius.

Corollary 20.1.5 *Let $A = [a_{ij}] \in \mathsf{M}_n$ be nonnegative.*

(a) *If B is a principal submatrix of A, then $\rho(B) \leq \rho(A)$.*

(b) $\max_{1 \leq i \leq n} a_{ii} \leq \rho(A)$.

(c) *If A has a positive diagonal entry, then $\rho(A) > 0$.*

(d) *If $A > 0$, then $\rho(A) > 0$.*

Proof (a) Suppose that B is $m \times m$. If $m = n$ there is nothing to prove, so assume that $1 \leq m \leq n - 1$. After a permutation similarity of A, we may assume that B is a leading $m \times m$ principal submatrix of A. Then $0 \leq B \oplus 0_{n-m} \leq A$ and $\rho(B \oplus 0_{n-m}) = \rho(B)$, so Theorem 20.1.3 ensures that $\rho(B) \leq \rho(A)$.

(b) This is the case $m = 1$ in (a).

(c) This follows from (b).

(d) This follows from (c).   □

The maximum absolute row sum of a matrix is an upper bound for its spectral radius, but the minimum absolute row sum of matrix need not be a lower bound for its spectral radius. For example, $A = \begin{bmatrix} 1 & 1 \\ -1 & 1 \end{bmatrix}$ has $\rho(A) = \sqrt{2}$ and the minimum absolute row sum of A is 2. However, the minimum row sum of a nonnegative matrix is a lower bound for its spectral radius.

Theorem 20.1.6 *If $A = [a_{ij}] \in M_n$ is nonnegative, then*

$$\min_{1 \leq i \leq n} \sum_{j=1}^{n} a_{ij} \leq \rho(A) \leq \max_{1 \leq i \leq n} \sum_{j=1}^{n} a_{ij}. \tag{20.1.7}$$

Proof Since $\max_{1 \leq i \leq n} \sum_{j=1}^{n} a_{ij} = N_\infty(A)$, the upper bound follows from Theorem 19.4.1. Let $\mu = \min_{1 \leq i \leq n} \sum_{j=1}^{n} a_{ij}$. If $\mu = 0$, there is nothing to prove. If $\mu > 0$, then every row sum of A is positive and not less than μ. Define $B = [b_{ij}]$ by

$$b_{ij} = \frac{\mu a_{ij}}{\sum_{k=1}^{n} a_{ik}} \qquad \text{for } i, j = 1, 2, \ldots, n.$$

Then $0 \leq b_{ij} \leq a_{ij}$ for all $i, j = 1, 2, \ldots, n$, so Theorem 20.1.3 ensures that $\rho(B) \leq \rho(A)$. Notice that $B\mathbf{e} = \mu\mathbf{e}$, in which $\mathbf{e} \in \mathbb{R}^n$ is the all-ones vector. Every row sum of B equals μ, which is an eigenvalue of B. Therefore, $\mu = N_\infty(B) \geq \rho(B)$, which implies that $\rho(B) = \mu$ since $\mu \in \operatorname{spec} B$. Therefore, $\mu \leq \rho(A)$ as asserted.   □

If $A\mathbf{x} = \lambda\mathbf{x}$, then Theorem 20.1.2.a says that $|\lambda|\,|\mathbf{x}| \leq |A|\,|\mathbf{x}|$. This is an example of an inequality of the form $b\mathbf{x} \leq A\mathbf{x} \leq c\mathbf{x}$, which arises frequently in the study of nonnegative matrices. The following theorem shows how to infer bounds on the spectral radius from such inequalities.

Theorem 20.1.8 *Let $A = [a_{ij}] \in M_n$ be nonnegative and let $\mathbf{x} = [x_i] \in \mathbb{R}^n$ be nonnegative and nonzero.*

(a) *Let $c > 0$. If $A\mathbf{x} \geq c\mathbf{x}$, then $\rho(A) \geq c$; if $A\mathbf{x} > c\mathbf{x}$, then $\rho(A) > c$.*

(b) *Let $b > 0$ and suppose that $\mathbf{x} > \mathbf{0}$. If $A\mathbf{x} \leq b\mathbf{x}$, then $\rho(A) \leq b$; if $A\mathbf{x} < b\mathbf{x}$, then $\rho(A) < b$.*

Proof (a) We proceed by induction. Let S_k be the statement $A^k\mathbf{x} \geq c^k\mathbf{x}$ for $k = 1, 2, \ldots$. In the base case $k = 1$ there is nothing to prove. If S_k is true for some $k \geq 1$, then $A^k\mathbf{x} - c^k\mathbf{x} \geq \mathbf{0}$. Theorem 20.1.2.d ensures that

$$A(A^k\mathbf{x} - c^k\mathbf{x}) = A^{k+1}\mathbf{x} - c^k A\mathbf{x} \geq \mathbf{0}.$$

Therefore, $A^{k+1}\mathbf{x} \geq c^k A\mathbf{x} \geq c^{k+1}\mathbf{x}$, so S_{k+1} is true. This completes the induction. Then $(c^{-1}A)^k\mathbf{x} \geq \mathbf{x}$ for all $k = 1, 2, \ldots$. Since $\mathbf{x} \neq \mathbf{0}$, this means that $c^{-1}A$ is not convergent.

Theorem 13.2.7 ensures that $\rho(c^{-1}A) \geq 1$, so $\rho(A) \geq c$. Now suppose that $A\mathbf{x} > c\mathbf{x}$. For each i such that $x_i \neq 0$, we have

$$\frac{(A\mathbf{x})_i}{x_i} > c, \quad \text{so} \quad \mu = \min\left\{ \frac{(A\mathbf{x})_i}{x_i} : 1 \leq i \leq n \text{ and } x_i \neq 0 \right\} > c.$$

Since $A\mathbf{x} \geq \mu\mathbf{x}$, we have shown that $\rho(A) \geq \mu > c$.

(b) If $A\mathbf{x} \leq b\mathbf{x}$, induction shows that $A^k\mathbf{x} \leq b^k\mathbf{x}$ for $k = 1, 2, \ldots$. Then $(b^{-1}A)^k\mathbf{x} \leq \mathbf{x}$ for all $k = 1, 2, \ldots$. Since $\mathbf{x} > \mathbf{0}$, it follows that $b^{-1}A$ is power bounded. Theorem 13.2.7 ensures that $\rho(b^{-1}A) \leq 1$, so $\rho(A) \leq b$. If $A\mathbf{x} < b\mathbf{x}$, then $\mu = \max_{x_i \neq 0} \frac{(A\mathbf{x})_i}{x_i} < b$. We have $A\mathbf{x} \leq \mu\mathbf{x}$, so $\rho(A) \leq \mu < b$. $\qquad\square$

Some nonnegative matrices have no positive eigenvector. For example, nonzero scalar multiples of $\mathbf{e}_1$ are the only eigenvectors of a nilpotent Jordan block $J_k(0)$. However, if a nonzero nonnegative matrix A has a positive eigenvector, then its associated eigenvalue is $\rho(A)$ and A is diagonally similar to a positive scalar multiple of a stochastic matrix.

Corollary 20.1.9 *Let* $A = [a_{ij}] \in \mathsf{M}_n(\mathbb{R})$ *be nonnegative and nonzero. Suppose that* $(\lambda, \mathbf{x})$ *is an eigenpair of* A *in which* $\mathbf{x} = [x_i] > \mathbf{0}$, *and let* $D = \mathrm{diag}(x_1, x_2, \ldots, x_n)$.

(a) $\lambda = \rho(A) > 0$.

(b) $M = D^{-1}\left(\rho(A)^{-1}A\right)D$ *is row stochastic.*

(c) *Every maximum-modulus eigenvalue of* A *is semisimple.*

(d) *If* $\mathbf{y} \in \mathbb{R}^n$ *is nonzero and* $A^\mathsf{T}\mathbf{y} \geq \rho(A)\mathbf{y}$, *then* $(\rho(A), \mathbf{y})$ *is an eigenpair of* A^T.

Proof (a) If the kth row of A is nonzero, then $(A\mathbf{x})_k > 0$, so $\lambda = (A\mathbf{x})_k / x_k > 0$. Thus, $\lambda > 0$ and $\lambda\mathbf{x} \leq A\mathbf{x} \leq \lambda\mathbf{x}$. The preceding theorem ensures that $\lambda \leq \rho(A) \leq \lambda$, so $\lambda = \rho(A) > 0$.

(b) Let $\mathbf{e} \in \mathbb{R}^n$ be the all-ones vector. Then

$$M\mathbf{e} = D^{-1}\left(\rho(A)^{-1}A\right)D\mathbf{e} = \rho(A)^{-1}D^{-1}A\mathbf{x} = \rho(A)^{-1}D^{-1}\rho(A)\mathbf{x} = D^{-1}\mathbf{x} = \mathbf{e},$$

so M is row stochastic.

(c) Theorem 13.4.6 says that $\rho(M) = 1$ and every maximum-modulus eigenvalue of M is semisimple. Since A is similar to $\rho(A)M$, it follows that every maximum-modulus eigenvalue of A is semisimple.

(d) We know that $A^\mathsf{T}\mathbf{y} - \rho(A)\mathbf{y} \geq \mathbf{0}$ and $\mathbf{y} \neq \mathbf{0}$. If $A^\mathsf{T}\mathbf{y} - \rho(A)\mathbf{y} \neq \mathbf{0}$, then the positivity of $\mathbf{x}$ ensures that $\mathbf{x}^\mathsf{T}(A^\mathsf{T}\mathbf{y} - \rho(A)\mathbf{y}) > 0$. However,

$$\mathbf{x}^\mathsf{T}(A^\mathsf{T}\mathbf{y} - \rho(A)\mathbf{y}) = (A\mathbf{x})^\mathsf{T}\mathbf{y} - \rho(A)\mathbf{x}^\mathsf{T}\mathbf{y} = (\rho(A)\mathbf{x})^\mathsf{T}\mathbf{y} - \rho(A)\mathbf{x}^\mathsf{T}\mathbf{y}$$
$$= \rho(A)\mathbf{x}^\mathsf{T}\mathbf{y} - \rho(A)\mathbf{x}^\mathsf{T}\mathbf{y} = \mathbf{0},$$

which is a contradiction. Therefore, $A^\mathsf{T}\mathbf{y} - \rho(A)\mathbf{y} = \mathbf{0}$, which means that $(\rho(A), \mathbf{y})$ is an eigenpair of A^T. $\qquad\square$

20.2 Positive Matrices

The fundamental fact about a positive square matrix is that it has a positive eigenvector.

Theorem 20.2.1 *Let $A \in M_n(\mathbb{R})$ be positive and let $(\lambda, \mathbf{x})$ be an eigenpair of A such that $|\lambda| = \rho(A)$.*

(a) $\rho(A) > 0$.

(b) $A|\mathbf{x}| = \rho(A)|\mathbf{x}|$, *so* $\rho(A) \in \operatorname{spec} A$.

(c) $|\mathbf{x}| > \mathbf{0}$.

(d) $\mathbf{x} = e^{i\theta} |\mathbf{x}|$ *for some* $\theta \in [0, 2\pi)$.

Proof (a) This is Corollary 20.1.5.d.

(b) Since $\mathbf{x} \neq \mathbf{0}$, we have $A|\mathbf{x}| > \mathbf{0}$ (Theorem 20.1.2.e). We also have

$$\rho(A)|\mathbf{x}| = |\lambda \mathbf{x}| = |A\mathbf{x}| \leq A|\mathbf{x}|, \qquad (20.2.2)$$

so $\mathbf{z} = A|\mathbf{x}| - \rho(A)|\mathbf{x}| \geq \mathbf{0}$. If $\mathbf{z} \neq \mathbf{0}$, then $A\mathbf{z} > \mathbf{0}$, that is, $A(A|\mathbf{x}|) - \rho(A)(A|\mathbf{x}|) > \mathbf{0}$. Theorem 20.1.8.a ensures that $\rho(A) > \rho(A)$, which is a contradiction. We conclude that $\mathbf{z} = \mathbf{0}$, which proves (b).

(c) Since $A|\mathbf{x}| = \rho(A)|\mathbf{x}|$, $\rho(A) > 0$, and $A|\mathbf{x}| > \mathbf{0}$, it follows that $|\mathbf{x}| > \mathbf{0}$.

(d) The inequality (20.2.2) is an equality, so we have equality in the triangle inequality

$$\left| \sum_{j=1}^{n} a_{1j}x_j \right| \leq \sum_{j=1}^{n} a_{1j}|x_j|.$$

Since each $a_{1j} > 0$, and each $x_j \neq 0$, it follows that all of the complex numbers $x_1, x_2, \ldots, x_n$ lie on the same ray from the origin in the complex plane; see (A.3.11). This shows that $\mathbf{x} = e^{i\theta}|\mathbf{x}|$ for some $\theta \in [0, 2\pi)$. $\qquad\square$

A positive eigenvector for a positive matrix A can be constructed as follows: Let $(\lambda, \mathbf{x})$ be any eigenpair with $|\lambda| = \rho(A)$. Then $|\mathbf{x}|$ is a positive eigenvector. This recipe sounds highly nonunique, but it is not. There is only one such λ, and it is the spectral radius of A. Furthermore, its eigenspace $\mathcal{E}_{\rho(A)}(A)$ is one dimensional, so positive scalar multiples of $|\mathbf{x}|$ are the only positive vectors in $\mathcal{E}_{\rho(A)}(A)$. The scalar factor is determined by the condition $\mathbf{e}^\mathsf{T}|\mathbf{x}| = 1$.

We can make a similar argument starting with an eigenpair $(\lambda, \mathbf{y})$ of A^T, in which $|\lambda| = \rho(A)$. The only such λ is $\rho(A)$ itself, and $|\mathbf{y}|$ is a positive eigenvector that is unique up to a positive scalar multiple, which is determined by the condition $|\mathbf{y}|^\mathsf{T}|\mathbf{x}| = 1$.

Theorem 20.2.3 (Perron) *Let $A \in M_n(\mathbb{R})$ be positive.*

(a) $\rho(A) > 0$.

(b) *There is a unique* $\mathbf{x} = [x_i] \in \mathbb{R}^n$ *such that* $A\mathbf{x} = \rho(A)\mathbf{x}$ *and* $\mathbf{e}^\mathsf{T}\mathbf{x} = x_1 + x_2 + \cdots + x_n = 1$; *moreover,* $\mathbf{x} > \mathbf{0}$.

(c) $\rho(A)$ *is a simple eigenvalue of* A.

(d) $|\lambda| < \rho(A)$ *if* $\lambda \in \operatorname{spec} A$ *and* $\lambda \neq \rho(A)$.

(e) *There is a unique* $\mathbf{y} \in \mathbb{R}^n$ *such that* $A^\mathsf{T}\mathbf{y} = \rho(A)\mathbf{y}$ *and* $\mathbf{y}^\mathsf{T}\mathbf{x} = 1$. *Moreover,* $\mathbf{y} > \mathbf{0}$.

(f) $\lim_{k \to \infty} (\rho(A)^{-1}A)^k = \mathbf{x}\mathbf{y}^\mathsf{T}$.

Proof The preceding theorem ensures that A has an eigenpair $(\rho(A), \mathbf{x})$ such that $\rho(A) > 0$ and $\mathbf{x} > \mathbf{0}$. We may assume that $\mathbf{x}$ is scaled so that $\mathbf{e}^\mathsf{T}\mathbf{x} = 1$. Let $D = \mathrm{diag}(x_1, x_2, \ldots, x_n)$. Corollary 20.1.9.b says that $A = \rho(A)DMD^{-1}$, in which M is row stochastic. The diagonal similarity ensures that M is positive. Theorem 13.4.8 says that $\lambda = 1$ is the only maximum-modulus eigenvalue of M, and its algebraic and geometric multiplicities are both 1. Therefore, $\rho(A)$ is the only maximum-modulus eigenvalue of A, and it is a simple eigenvalue. Theorem 13.4.11 says that there is a unique $\mathbf{z} \in \mathbb{R}^n$ such that $M^\mathsf{T}\mathbf{z} = \mathbf{z}$ and $\mathbf{e}^\mathsf{T}\mathbf{z} = 1$; moreover, $\mathbf{z} > \mathbf{0}$. If we let $\mathbf{y} = D^{-1}\mathbf{z}$, then $\mathbf{y} > \mathbf{0}$ and

$$\mathbf{y} = D^{-1}\mathbf{z} = D^{-1}M^\mathsf{T}\mathbf{z} = D^{-1}(\rho(A)^{-1}DA^\mathsf{T}D^{-1})\mathbf{z} = (\rho(A)^{-1}A^\mathsf{T})D^{-1}\mathbf{z} = \rho(A)^{-1}A^\mathsf{T}\mathbf{y}.$$

Therefore, $A^\mathsf{T}\mathbf{y} = \rho(A)\mathbf{y}$ and $\mathbf{y}^\mathsf{T}\mathbf{x} = \mathbf{z}^\mathsf{T}D^{-1}\mathbf{x} = \mathbf{z}^\mathsf{T}\mathbf{e} = 1$. Finally, Theorem 13.4.11 ensures that

$$(\rho(A)^{-1}A)^k = (DMD^{-1})^k = DM^kD^{-1} \quad \rightarrow \quad D(\mathbf{e}\mathbf{z}^\mathsf{T})D^{-1} = (D\mathbf{e})(D^{-1}\mathbf{z})^\mathsf{T} = \mathbf{x}\mathbf{y}^\mathsf{T}$$

as $k \rightarrow \infty$. $\qquad\qquad\qquad\qquad\qquad\qquad\qquad\qquad\qquad\qquad\qquad\qquad\qquad\qquad\qquad\qquad$ $\square$

Definition 20.2.4 The unique positive scaled eigenvector characterized in (b) of the preceding theorem is the *(right) Perron vector* of A; $\rho(A)$ is the *Perron root* of A. The unique scaled eigenvector of A^T characterized in (e) is the *left Perron vector* of A.

20.3 Primitive Matrices

If A is nonnegative and has some zero entries, then it is possible that $A^p > 0$ for some $p > 1$, but the zeros have to be in the right places.

Example 20.3.1 If we define

$$W = \begin{bmatrix} 0 & 1 & 0 \\ 0 & 0 & 1 \\ 1 & 1 & 0 \end{bmatrix}, \quad \text{then} \quad W^4 = \begin{bmatrix} 0 & 1 & 1 \\ 1 & 1 & 1 \\ 1 & 2 & 1 \end{bmatrix} \quad \text{and} \quad W^5 = \begin{bmatrix} 1 & 1 & 1 \\ 1 & 2 & 1 \\ 1 & 2 & 2 \end{bmatrix}.$$

The majority of the entries of W are zero, but $W^5 > 0$.

If $A = \begin{bmatrix} 1 & 1 \\ 0 & 1 \end{bmatrix}$, then $A^p = \begin{bmatrix} 1 & p \\ 0 & 1 \end{bmatrix}$ for $p = 1, 2, \ldots$. The majority of the entries of A are non-zero, but A^p is not positive for any $p \geq 1$. Finally, consider $B = \begin{bmatrix} 0 & 1 \\ 1 & 1 \end{bmatrix}$ and $B^2 = \begin{bmatrix} 1 & 1 \\ 1 & 2 \end{bmatrix} > 0$. Observe that A and B have just one zero entry each, but in different places.

We now show that if a nonnegative matrix has a positive power, then it has a positive eigenvector and enjoys the other conclusions of Theorem 20.2.1.

Theorem 20.3.2 *Let $A \in M_n(\mathbb{R})$ be nonnegative and suppose that $A^p > 0$ for some positive integer p. Let $(\lambda, \mathbf{x})$ be an eigenpair of A such that $|\lambda| = \rho(A)$.*

(a) $\rho(A) > 0$.

(b) $A|\mathbf{x}| = \rho(A)|\mathbf{x}|$, *so* $\rho(A) \in \mathrm{spec}\,A$.

(c) $|\mathbf{x}| > \mathbf{0}$.

(d) $\mathbf{x} = e^{i\theta}|\mathbf{x}|$ *for some* $\theta \in [0, 2\pi)$.

Proof (a) Apply Theorem 20.2.1.a to A^p and conclude that $\rho(A^p) = \rho(A)^p > 0$, so $\rho(A) > 0$.
(c) and (d) We have $A^p \mathbf{x} = \lambda^p \mathbf{x}$, $\mathbf{x} \neq \mathbf{0}$, and $|\lambda^p| = \rho(A^p)$, so parts (b), (c), and (d) of Theorem 20.2.1 ensure that $A^p |\mathbf{x}| = \rho(A)^p |\mathbf{x}|$, $|\mathbf{x}| > 0$, and $\mathbf{x} = e^{i\theta} |\mathbf{x}|$ for some $\theta \in [0, 2\pi)$.

(b) The identity $A\mathbf{x} = \lambda\mathbf{x}$ implies that $A|\mathbf{x}| \geq \rho(A)|\mathbf{x}|$, and an induction shows that

$$A^p |\mathbf{x}| \geq \rho(A)A^{p-1}|\mathbf{x}| \geq \rho(A)^2 A^{p-2}|\mathbf{x}| \geq \cdots \geq \rho(A)^{p-1}A|\mathbf{x}| \geq \rho(A)^p |\mathbf{x}|. \quad (20.3.3)$$

However, $A^p |\mathbf{x}| = \lambda^p |\mathbf{x}| = \rho(A)^p |\mathbf{x}|$, so every inequality in (20.3.3) is an equality. In particular, $\rho(A)^{p-1}A|\mathbf{x}| = \rho(A)^p |\mathbf{x}|$, which implies (b). $\qquad\square$

Definition 20.3.4 A square nonnegative matrix A is *primitive* if A^p is positive for some $p \geq 1$.

Theorem 20.3.5 *Let $A \in \mathsf{M}_n(\mathbb{R})$ be primitive.*

(a) $\rho(A) > 0$.

(b) *There is a unique $\mathbf{x} = [x_i] \in \mathbb{R}^n$ such that $A\mathbf{x} = \rho(A)\mathbf{x}$ and $\mathbf{e}^\mathsf{T}\mathbf{x} = x_1 + x_2 + \cdots + x_n = 1$; moreover, $\mathbf{x} > \mathbf{0}$.*

(c) $\rho(A)$ *is a simple eigenvalue of A.*

(d) $|\lambda| < \rho(A)$ *if $\lambda \in \operatorname{spec} A$ and $\lambda \neq \rho(A)$.*

(e) *There is a unique $\mathbf{y} \in \mathbb{R}^n$ such that $A^\mathsf{T}\mathbf{y} = \rho(A)\mathbf{y}$ and $\mathbf{y}^\mathsf{T}\mathbf{x} = 1$. Moreover, $\mathbf{y} > \mathbf{0}$.*

(f) $\lim_{k\to\infty} (\rho(A)^{-1}A)^k = \mathbf{x}\mathbf{y}^\mathsf{T}$.

Proof The preceding theorem ensures that A has an eigenpair $(\rho(A), \mathbf{x})$ such that $\rho(A) > 0$ and $\mathbf{x} > \mathbf{0}$. We may assume that $\mathbf{x}$ is scaled so that $\mathbf{e}^\mathsf{T}\mathbf{x} = 1$. Let $D = \operatorname{diag}(x_1, x_2, \ldots, x_n)$. Corollary 20.1.9.b says that $A = \rho(A)DMD^{-1}$, in which M is row stochastic. The diagonal similarity ensures that M is primitive, and the assertions follow from Theorem 13.4.15. $\qquad\square$

The spectral radius of a primitive matrix is its *Perron root*, and the positive vectors $\mathbf{x}$ and $\mathbf{y}$ in the preceding theorem are its *right and left Perron vectors*, respectively.

20.4 The Power Method for Primitive Matrices

The power method works well for positive or primitive matrices because the Perron root and the right and left Perron vectors are all real and positive.

Let $n \geq 2$ and let $A \in \mathsf{M}_n(\mathbb{R})$ be positive or primitive. Let ρ be the Perron root of A and let $\mathbf{x}$ and $\mathbf{y}$ be its right and left Perron vectors, respectively. Following the development of the power method in Section 19.6, we let

$$A = S \begin{bmatrix} \rho & \mathbf{0}^\mathsf{T} \\ \mathbf{0} & J \end{bmatrix} S^{-1}$$

be the Jordan canonical form of A, in which $S = [\mathbf{x} \ X]$, $S^{-*} = [\mathbf{y} \ Y]$, $\|\mathbf{x}\|_1 = 1$, and $\mathbf{y}^\mathsf{T}\mathbf{x} = 1$. Since the right Perron vector is a unit vector with respect to the ℓ_1 norm, it is convenient to use this norm instead of the Euclidean norm in the algorithm. In (19.6.3), we found that the powers of A are

$$A^k = S \begin{bmatrix} \rho^k & \mathbf{0}^\mathsf{T} \\ \mathbf{0} & J^k \end{bmatrix} S^{-1} = \rho^k \mathbf{x}\mathbf{y}^\mathsf{T} + XJ^k Y^* \quad \text{for } k = 1, 2, \ldots.$$

Choose a positive initial vector $\mathbf{x}^{(0)}$ and write it as $\mathbf{x}^{(0)} = a\mathbf{x} + X\mathbf{b}$. Then $\mathbf{y}^\mathsf{T}\mathbf{x}^{(0)} > 0$ and the identity (19.6.2) ensures that

$$\mathbf{y}^\mathsf{T}\mathbf{x}^{(0)} = \mathbf{y}^\mathsf{T}(a\mathbf{x} + X\mathbf{b}) = a(\mathbf{y}^\mathsf{T}\mathbf{x}) + (\mathbf{y}^\mathsf{T}X)\mathbf{b} = a > 0$$

and

$$Y^*\mathbf{x}^{(0)} = Y^*(a\mathbf{x} + X\mathbf{b}) = aY^*\mathbf{x} + Y^*X\mathbf{b} = \mathbf{b}.$$

The power method iterates are

$$
\begin{aligned}
A^k\mathbf{x}^{(0)} &= (\rho^k\mathbf{x}\mathbf{y}^\mathsf{T} + XJ^kY^*)\mathbf{x}^{(0)} \\
&= \rho^k\mathbf{x}(\mathbf{y}^\mathsf{T}\mathbf{x}^{(0)}) + XJ^k(Y^*\mathbf{x}^{(0)}) \\
&= \rho^k a\mathbf{x} + XJ^k\mathbf{b} \\
&= \rho^k(a\mathbf{x} + X(\rho^{-1}J)^k\mathbf{b}) \\
&= \rho^k(a\mathbf{x} + XB^k\mathbf{b})
\end{aligned}
$$

(20.4.1)

for $k = 1, 2, \ldots$. Since $B = \rho^{-1}J$ and $\rho(B) < 1$, we have $\lim_{k\to\infty} B^k = 0$. The normalized iterates are

$$
\begin{aligned}
\mathbf{z}_k &= \frac{A^k\mathbf{x}^{(0)}}{\|A^k\mathbf{x}^{(0)}\|_1} = \frac{\rho^k(a\mathbf{x} + XB^k\mathbf{b})}{\|\rho^k(a\mathbf{x} + XB^k\mathbf{b})\|_1} = \frac{\rho^k(a\mathbf{x} + XB^k\mathbf{b})}{\rho^k\|a\mathbf{x} + XB^k\mathbf{b}\|_1} \\
&= \frac{a\mathbf{x} + XB^k\mathbf{b}}{\|a\mathbf{x} + XB^k\mathbf{b}\|_1} \quad\to\quad \frac{a\mathbf{x}}{\|a\mathbf{x}\|_1} = \frac{a\mathbf{x}}{a\|\mathbf{x}\|_1} = \mathbf{x} \quad \text{as } k \to \infty.
\end{aligned}
$$

The conclusion is that the sequence of normalized iterates converges to the Perron vector.

In the power method with scaling, choose a positive initial vector $\mathbf{x}^{(0)}$. For each $k = 1, 2, \ldots$, compute $\mathbf{w}^{(k)} = A\mathbf{x}^{(k-1)}$, $\|\mathbf{w}^{(k)}\|_1$, and $\mathbf{x}^{(k)} = \mathbf{w}^{(k)}/\|\mathbf{w}^{(k)}\|_1$. Then $\mathbf{x}^{(k)} \to \mathbf{x}$ and $\|\mathbf{w}^{(k)}\|_1 \to \rho(A)$ as $k \to \infty$. Stop when empirical convergence is observed.

Example 20.4.2 Let A be the matrix in (19.4.3), for which $\rho(A) \approx 16.708$. The initial vector $\mathbf{x}^{(0)} = [1\ 1\ 1]^\mathsf{T}$ yields

$$
\mathbf{x}^{(1)} \approx \begin{bmatrix} 0.13043 \\ 0.32609 \\ 0.54348 \end{bmatrix}, \quad
\mathbf{x}^{(2)} \approx \begin{bmatrix} 0.14378 \\ 0.32254 \\ 0.53368 \end{bmatrix}, \quad
\mathbf{x}^{(3)} \approx \begin{bmatrix} 0.14308 \\ 0.32268 \\ 0.53424 \end{bmatrix}, \quad
\mathbf{x}^{(4)} \approx \begin{bmatrix} 0.14312 \\ 0.32267 \\ 0.53421 \end{bmatrix},
$$

and

$$\|\mathbf{w}^{(1)}\|_1 \approx 15.333, \quad \|\mathbf{w}^{(2)}\|_1 \approx 16.783, \quad \|\mathbf{w}^{(3)}\|_1 \approx 16.703, \quad \|\mathbf{w}^{(4)}\|_1 \approx 16.708.$$

20.5 Problems

P.20.1 What is the spectral radius of $A = \begin{bmatrix} 1 & 1 \\ -1 & -1 \end{bmatrix}$? Show that the hypothesis of nonnegativity cannot be omitted from Corollary 20.1.5.

P.20.2 If $A \in M_n(\mathbb{R})$ is nonnegative and $A^p > 0$, show that $A^k > 0$ for all $k \geq p$.

P.20.3 Let $A \in M_2(\mathbb{R})$ be positive. Show that both of its eigenvalues are real.

P.20.4 If $A = [a_{ij}] \in M_n(\mathbb{R})$ is positive and $\mathbf{x} = [x_i]$ is its Perron vector, show that $\rho(A) = \sum_{i,j=1}^{n} a_{ij}x_j$.

P.20.5 Let $a, b \in (0, 1)$ and let $A = \begin{bmatrix} 1-a & b \\ a & 1-b \end{bmatrix}$. Find the Perron root and the left and right Perron vectors of A.

P.20.6 The rate at which the power method converges to a dominant eigenpair of a positive matrix $A \in M_n(\mathbb{R})$ depends on the spectral radius of the matrix B in (20.4.1). (a) Show that $\rho(B) = |\lambda_{n-1}(A)|/\rho(A)$, in which $\lambda_{n-1}(A)$ is the second-largest-modulus eigenvalue of $A = [a_{ij}]$. It is known that

$$\frac{|\lambda_{n-1}(A)|}{\rho(A)} \leq \frac{1 - \kappa^2}{1 + \kappa^2}, \tag{20.5.1}$$

in which $\kappa = (\min_{1 \leq i,j \leq n} a_{ij})/(\max_{1 \leq i,j \leq n} a_{ij})$. (b) Compute the values of both sides of the inequality in (20.5.1) for the matrix A in Example 20.4.2, whose eigenvalues are approximately 16.708, −0.906, and 0.198. The value of the upper bound in (20.5.1) suggests that convergence of the power method might be rather slow for A, but we found that excellent accuracy was achieved in only four iterations. In practice, the upper bound (20.5.1) is often overly conservative.

P.20.7 If $A \in M_n(\mathbb{R})$ is primitive, show that $\rho(A) = \lim_{k \to \infty} (\operatorname{tr} A^k)^{1/k}$.

P.20.8 For each $n = 2, 3, \ldots$, define the *Wielandt matrix*

$$W_n = \begin{bmatrix} \mathbf{0} & I_{n-1} \\ 1 & \mathbf{e}_1^{\mathsf{T}} \end{bmatrix}. \tag{20.5.2}$$

(a) What is W_2? W_2^2? (b) Show that W_3 is the matrix W in Example 20.3.1. (c) Show that W_4^k is not positive for $k = 1, 2, \ldots, 9$, but $W_4^{10} > 0$. *Hint:* Consider P.20.2 and compute $W_4^2, (W_4^2)^2, (W_4^4)^2, W_4 W_4^8, W_4^2 W_4^8$.

P.20.9 If $A, B, C, D \in M_n(\mathbb{R})$, $A \geq B \geq 0$, and $C \geq D \geq 0$, show that $AC \geq BD \geq 0$.

P.20.10 Let $A, B \in M_n(\mathbb{R})$ be nonnegative and suppose that $A \geq B \geq 0$. Prove that $\|A\|_2 \geq \|B\|_2$.

P.20.11 Let $A \in M_n(\mathbb{R})$ be positive. (a) Show that its largest singular value $\sigma_{\max}(A)$ is simple and has positive left and right singular vectors. (b) If $\mathbf{z} \in \mathbb{R}^n$ is positive and $A\mathbf{z} = \rho(A)\mathbf{z} = A^{\mathsf{T}}\mathbf{z}$, show that $\sigma_{\max}(A) = \rho(A)$. (c) Let

$$D = \begin{bmatrix} 16 & 3 & 2 & 13 \\ 5 & 10 & 11 & 8 \\ 9 & 6 & 7 & 12 \\ 4 & 15 & 14 & 1 \end{bmatrix} \tag{20.5.3}$$

and let $\rho(D), \lambda_2, \lambda_3, \lambda_4$ be the eigenvalues of D. What are the values of $\rho(D)$ and $\sigma_{\max}(D)$? Why? (d) Find left and right singular vectors associated with $\sigma_{\max}$. (e) Let $B = D - \frac{\rho(D)}{4}\mathbf{e}\mathbf{e}^{\mathsf{T}}$. Show that the eigenvalues of B are $0, \lambda_2, \lambda_3, \lambda_4$. *Hint:* Theorem 11.6.1. (f) Show that $K_4 B K_4 = -B$, in which K_4 is the reversal matrix (7.1.11). (g) If $\lambda \in \operatorname{spec} B$, why is $-\lambda \in \operatorname{spec} B$? (h) Why must at least one of $\lambda_2, \lambda_3, \lambda_4$ be 0? Conclude that D is not invertible.

P.20.12 Let $A \in M_n(\mathbb{R})$ be positive and let $\mathbf{x} = [x_i] \in \mathbb{R}^n$ be its Perron vector. If either of the inequalities in (20.1.7) is an equality, prove that $\mathbf{x} = \frac{1}{n}\mathbf{e}$, every row sum of A equals $\rho(A)$, and both inequalities in (20.1.7) are equalities.

P.20.13 Let $A \in M_n(\mathbb{R})$ be positive and symmetric, and suppose that it has exactly one positive eigenvalue. (a) Prove that each principal submatrix of A has exactly one positive eigenvalue. (b) Prove that $a_{ij} \geq \sqrt{a_{ii}a_{jj}}$ for all $i, j = 1, 2, \ldots, n$. *Hint:* Consider a principal submatrix of size $n-1$ and use interlacing. What is its positive eigenvalue?

P.20.14 Let $A = \begin{bmatrix} 1 & 1 \\ 0 & 1 \end{bmatrix}$ and $B = \begin{bmatrix} 1 & 2 \\ 0 & 1 \end{bmatrix}$. Show that $0 \leq A \leq B$, $A \neq B$, and $\rho(A) = \rho(B)$.

P.20.15 Let $A = \begin{bmatrix} 1 & 1 \\ 1 & 1 \end{bmatrix}$ and $B = \begin{bmatrix} 1 & 2 \\ 1 & 1 \end{bmatrix}$. Show that $0 < A \le B$, $A \ne B$, and $\rho(A) < \rho(B)$.

P.20.16 Let $A, B \in M_n(\mathbb{R})$. Suppose that $0 < A \le B$ and $A \ne B$. Show that $\rho(A) < \rho(B)$. *Hint*: Let **x** be the right Perron vector of B and let **y** be the left Perron vector of A. Compare $\mathbf{y}^\mathsf{T} A \mathbf{x}$ and $\mathbf{y}^\mathsf{T} B \mathbf{x}$.

P.20.17 Let $A, B \in M_n(\mathbb{R})$ be nonnegative. Suppose that $0 \le A \le B$ and $A \ne B$. If $B > 0$, show that $\rho(A) < \rho(B)$. *Hint*: Add $\varepsilon > 0$ to any zero entries of A.

P.20.18 Let $A \in M_n(\mathbb{R})$ be positive and let $(\lambda, \mathbf{z})$ be an eigenpair of A, in which $\mathbf{z} \ge \mathbf{0}$. Use the principle of biorthogonality (P.10.29) to show that λ is the Perron root of A. Why is **z** positive?

P.20.19 Let $A, B \in M_n(\mathbb{R})$ be positive. If A commutes with B, show that the right (respectively, left) Perron vector of A is the right (respectively, left) Perron vector of B.

20.6 Notes

Oskar Perron published Theorem 20.2.3 in 1907. Five years later, Ferdinand Georg Frobenius extended Perron's theorem to nonnegative matrices $A \in M_n(\mathbb{R})$ that are *irreducible*, that is, A is not permutation similar to a block upper triangular matrix $\begin{bmatrix} B & C \\ 0 & D \end{bmatrix}$, in which B and D are square. An equivalent characterization of irreducibility is $(I+A)^{n-1} > 0$. The results discussed in this chapter are often referred to as *Perron–Frobenius theory*.

Using methods of analysis, one can show that if $A \in M_n(\mathbb{R})$ is nonnegative, then it has an eigenpair $(\rho(A), \mathbf{x})$ in which $\mathbf{x} \ge \mathbf{0}$. For a proof, and for more information about nonnegative, positive, primitive, and irreducible matrices, see [HJ13, Chapter 8].

If $A \in M_n(\mathbb{R})$ is nonnegative and we want to decide if it is primitive, how many powers A^p do we have to compute before concluding that A is not primitive? In 1950, Helmut Wielandt proved that A is primitive if and only if $A^{n^2 - 2n + 2} > 0$. Thus, if we have found a $p \ge n^2 - 2n + 2$

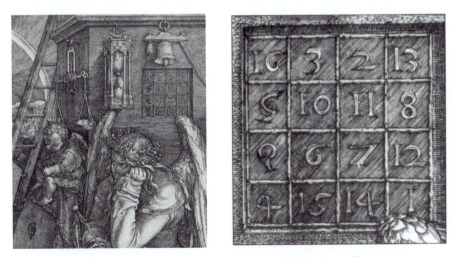

(a) Magic square at upper right. (b) Dürer's magic square.

Figure 20.1 Details from Dürer's *Melencolia*.

such that A^p is not positive, then A is not primitive. For each $n = 2, 3, \ldots$, the matrix W_n in (20.5.2) has $W_n^p > 0$ for every integer $p \geq n^2 - 2n + 2$ but not for any $p \leq n^2 - 2n + 1$.

The matrix (20.5.3) appears in Albrecht Dürer's 1514 engraving *Melencolia*; see Figure 20.1 and the entries in positions (4,2) and (4,3) of Figure 20.1.b. The sums of the entries in the rows, columns, diagonals, corners, middle square, and quadrants of (20.5.3) are each equal to its spectral radius (34), which equals its spectral norm. In fact, there are 86 different combinations of four entries that add up to the spectral radius. Dürer's matrix is an extraordinary example of a *magic square*. Its eigenvalues are 34, 8, −8, and 0.

20.7 Some Important Concepts

- Spectral radius inequalities for nonnegative matrices.

- Primitive matrices.

- The spectral radius of a primitive matrix is a positive dominant eigenvalue; it has a positive associated eigenvector.

- Powers of a primitive matrix A with unit spectral radius converge to a rank-1 matrix that is the outer product of the left and right Perron vectors of A.

- The power method for a primitive matrix converges.

APPENDIX A
Complex Numbers

A.1 Complex Numbers

A *complex number* is an expression of the form $a + bi$, in which $a, b \in \mathbb{R}$ (the set of real numbers) and i is a symbol with the property $i^2 = -1$. The set of all complex numbers is denoted by $\mathbb{C}$. The real numbers a and b are the *real* and *imaginary parts*, respectively, of the complex number $z = a + bi$. We write Re $z = a$ and Im $z = b$ (note that Im z is a real number). The number $i = 0 + 1i$ is the *imaginary unit*.

We identify a complex number $z = a + bi$ with the ordered pair $(a, b) \in \mathbb{R}^2$. In this context, we refer to $\mathbb{C}$ as the *complex plane*; see Figure A.1a. We regard $\mathbb{R}$ as a subset of $\mathbb{C}$ by identifying the real number a with the complex number $a + 0i = (a, 0)$. If b is a real number, we identify the *purely imaginary* number bi with $(0, b)$. The number $0 = 0 + 0i$ is simultaneously real and imaginary; it is the *origin*.

Definition A.1.1 Let $z = a + bi$ and $w = c + di$ be complex numbers. The *sum $z + w$* and *product zw* of z and w are defined by

$$z + w = (a + c) + (b + d)i, \tag{A.1.2}$$

and

$$zw = (ac - bd) + (ad + bc)i, \tag{A.1.3}$$

respectively.

If we write $z = (a, b)$ and $w = (c, d)$, then (A.1.2) is vector addition in $\mathbb{R}^2$; see Figure A.1b. Therefore, $\mathbb{C}$ can be thought of as $\mathbb{R}^2$, with the additional operation (A.1.3) of complex multiplication.

Example A.1.4 Let $z = 1 + 2i$ and $w = 3 + 4i$. Then $z + w = 4 + 6i$ and $zw = -5 + 10i$.

Although the definition of complex multiplication may appear unmotivated, there are good algebraic and geometric reasons to define the product of complex numbers in this manner. If we manipulate the imaginary unit i in the same manner as we manipulate real numbers, we find that

$$zw = (a + bi)(c + di) = ac + bic + adi + bidi$$
$$= ac + bdi^2 + i(ad + bc) = (ac - bd) + i(ad + bc)$$

since $i^2 = -1$. Thus, the definition (A.1.3) is a consequence of the assumption that complex arithmetic obeys the same algebraic rules as real arithmetic. This requires proof. The computation

$$(a + bi)(c + di) = (ac - bd) + (ad + bc)i = (ca - db) + (da + cb)i = (c + di)(a + bi)$$

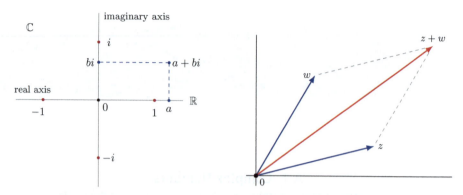

(a) The complex number $a + bi$ is identified with the point whose Cartesian coordinates are (a, b).

(b) Addition in the complex plane corresponds to vector addition in $\mathbb{R}^2$.

Figure A.1 Complex numbers as points and vectors in a plane.

shows that complex multiplication is *commutative*: $zw = wz$ for all $z, w \in \mathbb{C}$. To verify directly that the complex numbers satisfy the associative and distributive rules requires more work. Fortunately, a linear algebraic device saves us much effort and provides valuable geometric insight.

Consider

$$\mathcal{C} = \left\{ \begin{bmatrix} a & -b \\ b & a \end{bmatrix} : a, b \in \mathbb{R} \right\} \subseteq \mathsf{M}_2(\mathbb{R})$$

and observe that the sum

$$\begin{bmatrix} a & -b \\ b & a \end{bmatrix} + \begin{bmatrix} c & -d \\ d & c \end{bmatrix} = \begin{bmatrix} a+c & -(b+d) \\ b+d & a+c \end{bmatrix} \tag{A.1.5}$$

and product

$$\begin{bmatrix} a & -b \\ b & a \end{bmatrix} \begin{bmatrix} c & -d \\ d & c \end{bmatrix} = \begin{bmatrix} ac - bd & -(ad + bc) \\ ad + bc & ac - bd \end{bmatrix} \tag{A.1.6}$$

of matrices from $\mathcal{C}$ also belongs to $\mathcal{C}$. Thus, $\mathcal{C}$ is closed under matrix addition and matrix multiplication.

For each complex number $z = a + bi$, define $M_z \in \mathcal{C}$ by

$$M_z = \begin{bmatrix} a & -b \\ b & a \end{bmatrix}. \tag{A.1.7}$$

Then (A.1.5) and (A.1.6) are equivalent to

$$M_z + M_w = M_{z+w} \qquad \text{and} \qquad M_z M_w = M_{zw}, \tag{A.1.8}$$

in which $z = a + bi$ and $w = c + di$. In particular,

$$M_1 = \begin{bmatrix} 1 & 0 \\ 0 & 1 \end{bmatrix}, \qquad M_i = \begin{bmatrix} 0 & -1 \\ 1 & 0 \end{bmatrix},$$

and $(M_i)^2 = M_{-1}$, which reflects the fact that $i^2 = -1$.

The map $z \mapsto M_z$ is a one-to-one correspondence between $\mathbb{C}$ and $\mathcal{C}$ that respects complex addition and multiplication. Anything that we can prove about addition and multiplication in $\mathcal{C}$ translates into a corresponding statement about $\mathbb{C}$ and vice versa. For example, complex arithmetic is associative and distributive because matrix arithmetic is associative and distributive. For a direct verification of associativity of multiplication, see P.A.9.

One can think of $\mathcal{C}$ as a copy of the complex numbers that is contained in $M_2(\mathbb{R})$. Complex addition and multiplication correspond to matrix addition and matrix multiplication, respectively, in $\mathcal{C}$.

Every nonzero complex number $z = a + bi$ has a multiplicative inverse, denoted by z^{-1}, which has the property that $zz^{-1} = z^{-1}z = 1$. The existence of multiplicative inverses is not obvious from the definition (A.1.3), although (A.1.8) helps make this clear. The inverse of z should be the complex number corresponding to

$$(M_z)^{-1} = \frac{1}{\det M_z} \begin{bmatrix} a & b \\ -b & a \end{bmatrix} = \begin{bmatrix} \frac{a}{a^2+b^2} & \frac{b}{a^2+b^2} \\ \frac{-b}{a^2+b^2} & \frac{a}{a^2+b^2} \end{bmatrix}.$$

This leads to

$$z^{-1} = \frac{a}{a^2 + b^2} - \frac{b}{a^2 + b^2}i, \tag{A.1.9}$$

which can be verified by direct computation.

Example A.1.10 If $z = 1 + 2i$, then $z^{-1} = \frac{1}{5} - \frac{2}{5}i$.

The linear transformation on $\mathbb{R}^2$ induced by M_z acts as follows:

$$\begin{bmatrix} a & -b \\ b & a \end{bmatrix} \begin{bmatrix} c \\ d \end{bmatrix} = \begin{bmatrix} ac - bd \\ ad + bc \end{bmatrix}.$$

Thus, multiplication by z maps (c, d), which represents $w = c + di$, to $(ac - bd, ad + bc)$, which represents zw. Figure A.2 illustrates how multiplication by a nonzero $z = a + bi$ rotates the complex plane about the origin through an angle θ such that $\cos \theta = a/\sqrt{a^2 + b^2}$; it scales by the positive factor $\sqrt{a^2 + b^2}$.

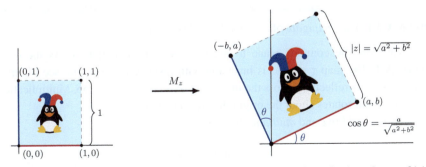

Figure A.2 Multiplication by a nonzero complex number $z = a + bi$ scales by a factor of $|z| = \sqrt{a^2 + b^2}$ and induces a rotation of the complex plane about the origin through an angle θ such that $\cos \theta = a/\sqrt{a^2 + b^2}$.

A.2 Modulus and Argument

Definition A.2.1 Let $z = a + bi$ be a complex number. The *modulus* of z is the nonnegative real number defined by $|z| = \sqrt{a^2 + b^2}$. The *argument* of z, denoted $\arg z$, is any angle θ that satisfies

$$\cos \theta = \frac{\operatorname{Re} z}{|z|} = \frac{a}{\sqrt{a^2 + b^2}} \quad \text{if } z \neq 0;$$

$\arg 0$ is undefined; see Figure A.3.

Example A.2.2 If $z = 1 + \sqrt{3}i$, then $|z| = 2$, $\cos \theta = \frac{1}{2}$, and $\arg z = \pi/3$.

The argument of a nonzero complex number is determined only up to an additive multiple of 2π. For example, $-\frac{\pi}{2}$, $\frac{3\pi}{2}$, and $-\frac{5\pi}{2}$ are arguments of $z = -i$. The modulus of z is also called the *absolute value* or *norm* of z. Since $|a + 0i| = \sqrt{a^2} = |a|$, the modulus function is an extension to $\mathbb{C}$ of the absolute value function on $\mathbb{R}$. We have $|z| = 0$ if and only if $z = 0$.

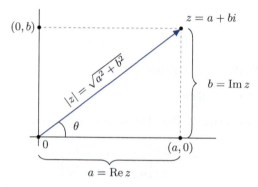

Figure A.3 The modulus $|z|$ and argument θ of a nonzero complex number; $\cos \theta = (\operatorname{Re} z)/|z|$.

A.3 Conjugation and Complex Arithmetic

Definition A.3.1 The *conjugate* of $z = a + bi \in \mathbb{C}$ is $\bar{z} = a - bi$.

As a map from the complex plane to itself, $z \mapsto \bar{z}$ is reflection across the real axis; see Figure A.4. The map $z \mapsto -z$ is inversion with respect to the origin; see Figure A.5a. The map $z \mapsto -\bar{z}$ is reflection across the imaginary axis; see Figure A.5b. Multiplication by $\bar{z}$ stretches by $|z|$ and rotates about the origin through an angle of $-\arg z$.

Conjugation is *additive* and *multiplicative*, that is,

$$\overline{z + w} = \bar{z} + \bar{w}$$

and

$$\overline{zw} = \bar{z}\,\bar{w}, \tag{A.3.2}$$

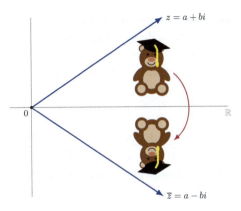

Figure A.4 Complex conjugation reflects the complex plane across the real axis.

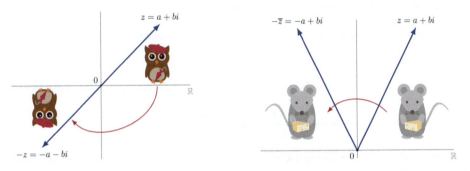

(a) The map $z \mapsto -z$ rotates by π around the origin.

(b) The map $z \mapsto -\bar{z}$ is reflection across the imaginary axis.

Figure A.5 Rotations and reflections in the complex plane.

respectively. To verify (A.3.2), let $z = a + bi$, let $w = c + di$, and compute

$$\overline{zw} = \overline{(ac - bd) + i(ad + bc)} = (ac - bd) - i(ad + bc)$$
$$= \big(ac - (-b)(-d)\big) + i\big(a(-d) + b(-c)\big)$$
$$= (a - bi)(c - di)$$
$$= \bar{z}\,\bar{w}.$$

Another proof of (A.3.2) follows from (A.1.8) and the fact that $M_{\bar{z}} = M_z^{\mathsf{T}}$. Since

$$M_{\overline{zw}} = (M_{zw})^{\mathsf{T}} = (M_z M_w)^{\mathsf{T}} = M_w^{\mathsf{T}} M_z^{\mathsf{T}} = M_{\bar{w}} M_{\bar{z}} = M_{\bar{z}} M_{\bar{w}},$$

we conclude that $\overline{zw} = \bar{z}\,\bar{w}$

The important identity

$$|z|^2 = z\bar{z} \tag{A.3.3}$$

relates the modulus to conjugation. It can be verified by direct computation:

$$z\bar{z} = (a + bi)(a - bi) = a^2 + b^2 + i(ab - ba) = a^2 + b^2 = |z|^2.$$

For a geometric explanation of (A.3.3), observe that multiplication by z and $\bar{z}$ rotate by $\arg z$ and $-\arg z$, respectively, while both scale by a factor of $|z|$. Multiplication by $z\bar{z}$ therefore induces no net rotation and scales by a factor of $|z|^2$.

We can use (A.3.3) to simplify quotients of complex numbers. If $z \neq 0$, then

$$\frac{w}{z} = \frac{w\bar{z}}{z\bar{z}} = \frac{w\bar{z}}{|z|^2}. \tag{A.3.4}$$

Example A.3.5 $\dfrac{1+i}{1+\sqrt{3}i} = \dfrac{1+i}{1+\sqrt{3}i} \cdot \dfrac{1-\sqrt{3}i}{1-\sqrt{3}i} = \dfrac{1+\sqrt{3}}{4} + \dfrac{1-\sqrt{3}}{4}i.$

If we let $w = 1$ in (A.3.4), we obtain an alternative version of (A.1.9) for the multiplicative inverse of a complex number:

$$z^{-1} = \frac{\bar{z}}{|z|^2} = \frac{a-bi}{a^2+b^2}, \quad \text{in which } z \neq 0.$$

Example A.3.6 $\dfrac{1}{1+i} = \dfrac{1}{1+i} \cdot \dfrac{1-i}{1-i} = \dfrac{1-i}{1-i^2} = \dfrac{1}{2} - \dfrac{1}{2}i.$

Like its real counterpart, the complex absolute value is *multiplicative*, that is,

$$|zw| = |z||w| \tag{A.3.7}$$

for all $z, w \in \mathbb{C}$. This is more subtle than it appears, for (A.3.7) is equivalent to the identity

$$\underbrace{(ac-bd)^2 + (ad+bc)^2}_{|zw|^2} = \underbrace{(a^2+b^2)}_{|z|^2}\underbrace{(c^2+d^2)}_{|w|^2}.$$

One can verify (A.3.7) by using (A.3.3): $|zw|^2 = (zw)(\overline{zw}) = (z\bar{z})(w\bar{w}) = |z|^2|w|^2$. We can also establish (A.3.7) with determinants and the fact that $\det M_z = a^2 + b^2 = |z|^2$:

$$|zw|^2 = \det M_{zw} = \det(M_z M_w) = (\det M_z)(\det M_w) = |z|^2|w|^2.$$

An important consequence of (A.3.7) is that $zw = 0$ if and only if $z = 0$ or $w = 0$. If $zw = 0$, then $0 = |zw| = |z||w|$, so at least one of $|z|$ or $|w|$ is zero, which implies that $z = 0$ or $w = 0$.

The real and imaginary parts of $z = a + bi$ are:

$$a = \operatorname{Re} z = \frac{1}{2}(z+\bar{z}) \qquad \text{and} \qquad b = \operatorname{Im} z = \frac{1}{2i}(z-\bar{z}). \tag{A.3.8}$$

Moreover, z is real if and only if $\bar{z} = z$, and z is purely imaginary if and only if $\bar{z} = -z$. Observe that

$$|\operatorname{Re} z| \leq |z| \qquad \text{and} \qquad |\operatorname{Im} z| \leq |z| \tag{A.3.9}$$

since $|z|$ is the length of the hypotenuse of a right triangle with sides of length $|\operatorname{Re} z|$ and $|\operatorname{Im} z|$; see Figure A.3. Alternatively, if $a, b \in \mathbb{R}$, then $a^2 \leq a^2 + b^2$, so $|a| = \sqrt{a^2} \leq \sqrt{a^2+b^2}$, with equality if and only if $b^2 = 0$. This shows that $|\operatorname{Re} z| \leq |z|$, with equality if and only if z is real. The same reasoning shows that $|\operatorname{Im} z| \leq |z|$, with equality if and only if z is purely imaginary.

An inequality stronger than (A.3.9) is the *triangle inequality*,

$$|z+w| \leq |z| + |w| \quad \text{for } z, w \in \mathbb{C}. \tag{A.3.10}$$

If we think of z and w as vectors in $\mathbb{R}^2$, then the triangle inequality asserts that the length of any one side of a triangle is at most the sum of the lengths of the other two sides;

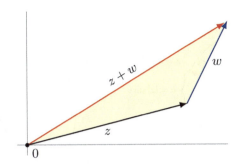

Figure A.6 The triangle inequality $|z + w| \le |z| + |w|$.

see Figure A.6. Here is an algebraic proof of the triangle inequality that demonstrates some techniques to deal with complex numbers:

$$
\begin{aligned}
\left(|z| + |w|\right)^2 - |z + w|^2 &= \left(|z|^2 + 2|z||w| + |w|^2\right) - (z + w)(\bar{z} + \bar{w}) \\
&= |z|^2 + 2|z||w| + |w|^2 - |z|^2 - |w|^2 - z\bar{w} - \bar{z}w \\
&= 2|z||\bar{w}| - (z\bar{w} + \bar{z}w) \\
&= 2|z\bar{w}| - 2\operatorname{Re}(z\bar{w}) \\
&= 2\left(|z\bar{w}| - \operatorname{Re}(z\bar{w})\right) \\
&\ge 0.
\end{aligned}
$$

We invoke (A.3.9) for the final inequality. The proof shows that $|z + w| \le |z| + |w|$, with equality if and only if $|z\bar{w}| = \operatorname{Re}(z\bar{w})$. Our discussion of (A.3.10) shows that $\operatorname{Re}(z\bar{w}) = |z\bar{w}|$ if and only if $z\bar{w}$ is real and nonnegative. If $z\bar{w} = 0$, then either $z = 0$ or $w = 0$. If $z\bar{w} > 0$, then

$$
w = \frac{wz\bar{w}}{z\bar{w}} = \frac{zw\bar{w}}{z\bar{w}} = z\frac{|w|^2}{z\bar{w}}.
$$

Thus, (A.3.10) is an equality if and only if one of z or w is a real nonnegative multiple of the other. This is, $z = 0$, or $w = 0$, or z and w are nonzero and lie on the same ray from the origin.

An induction argument generalizes (A.3.10):

$$
|z_1 + z_2 + \cdots + z_n| \le |z_1| + |z_2| + \cdots + |z_n| \quad \text{for } z_1, z_2, \ldots, z_n \in \mathbb{C}, \tag{A.3.11}
$$

with equality if and only if all of the nonzero summands z_i lie on the same ray from the origin.

A.4 Polar Form of a Complex Number

Polar coordinates yield insights into complex multiplication. For $z = a + bi$, let $a = r\cos\theta$ and $b = r\sin\theta$, in which $r = |z|$ and $\theta = \arg z$. Thus,

$$
a + bi = r(\cos\theta + i\sin\theta),
$$

in which $r \ge 0$; see Figure A.7a. Then $\cos\theta + i\sin\theta$ lies on the *unit circle* $\{z \in \mathbb{C} : |z| = 1\}$ since $|\cos\theta + i\sin\theta| = \sqrt{\cos^2\theta + \sin^2\theta} = 1$.

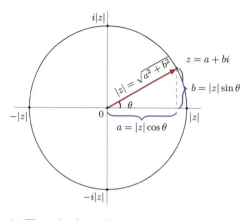

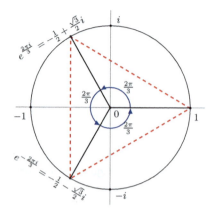

(a) The polar form of a complex number.

(b) The cube roots of unity are the vertices of an equilateral triangle inscribed in the unit circle.

Figure A.7 Complex numbers and roots of unity in polar form.

Writing z in its *polar form* $z = r(\cos\theta + i\sin\theta)$ separates the two distinct transformations induced by multiplication by z:

- Multiplication by $r \geq 0$ scales by a factor r.

- Multiplication by $\cos\theta + i\sin\theta$ rotates about the origin through an angle θ.

This suggests a rule for complex multiplication: to multiply complex numbers, add the arguments and multiply the lengths. A short computation shows why this is true. The product of $z = r_1(\cos\theta_1 + i\sin\theta_1)$ and $w = r_2(\cos\theta_2 + i\sin\theta_2)$ is

$$zw = r_1 r_2 (\cos\theta_1 + i\sin\theta_1)(\cos\theta_2 + i\sin\theta_2)$$
$$= r_1 r_2 [(\cos\theta_1 \cos\theta_2 - \sin\theta_1 \sin\theta_2) + i(\cos\theta_1 \sin\theta_2 + \sin\theta_1 \cos\theta_2)]$$
$$= r_1 r_2 [\cos(\theta_1 + \theta_2) + i\sin(\theta_1 + \theta_2)].$$

Thus, $|zw| = r_1 r_2 = |z||w|$ and $\arg zw = \theta_1 + \theta_2 = \arg z + \arg w$.

If we let $z = r(\cos\theta + i\sin\theta)$ with $r > 0$ and apply the preceding rule repeatedly, we obtain *de Moivre's formula*

$$z^n = r^n(\cos n\theta + i\sin n\theta) \quad \text{for } n \in \mathbb{Z}. \tag{A.4.1}$$

Example A.4.2 Let $r = 1$ and $n = 2$ in (A.4.1): $(\cos\theta + i\sin\theta)^2 = \cos 2\theta + i\sin 2\theta$. Expand the left side of the preceding equation and obtain

$$(\cos^2\theta - \sin^2\theta) + (2\sin\theta\cos\theta)i = \cos 2\theta + i\sin 2\theta. \tag{A.4.3}$$

Comparison of the real and imaginary parts of (A.4.3) gives the *double angle formulas*

$$\cos 2\theta = \cos^2\theta - \sin^2\theta \qquad \text{and} \qquad \sin 2\theta = 2\sin\theta\cos\theta.$$

Definition A.4.4 For $n = 1, 2, \ldots$, the solutions to $z^n = 1$ are the *nth roots of unity*.

Example A.4.5 The cube roots of unity can be computed with de Moivre's formula. If $z^3 = 1$, then (A.4.1) ensures that $r^3 = 1$ and $\cos 3\theta + i\sin 3\theta = 1$. Since $r \geq 0$, it follows that $r = 1$.

The restrictions on θ tell us that $3\theta = 2\pi k$ for some integer k. Since $\theta = \frac{2\pi k}{3}$, there are only three distinct angles $\theta \in [0, 2\pi)$ that can arise this way, namely 0, $\frac{2\pi}{3}$, and $\frac{4\pi}{3}$. Therefore, the cube roots of unity are

$$\cos 0 + i \sin 0 = 1,$$

$$\cos \frac{2\pi}{3} + i \sin \frac{2\pi}{3} = -\frac{1}{2} + i\frac{\sqrt{3}}{2}, \quad \text{and}$$

$$\cos \frac{4\pi}{3} + i \sin \frac{4\pi}{3} = -\frac{1}{2} - i\frac{\sqrt{3}}{2}.$$

These points form the vertices of an equilateral triangle; see Figure A.7b.

For a nonzero $z = r(\cos \theta + i \sin \theta)$, there are precisely n distinct nth roots of z:

$$r^{1/n} \left[\cos\left(\frac{\theta + 2\pi k}{n}\right) + i \sin\left(\frac{\theta + 2\pi k}{n}\right) \right] \quad \text{for } k = 0, 1, \ldots, n-1,$$

in which $r^{1/n}$ denotes the positive nth root of the positive real number r.

Example A.4.6 The cube roots of $1 + i$ are $2^{1/6}\left(\cos(\frac{\pi}{12} + \frac{2\pi k}{3}) + i\sin(\frac{\pi}{12} + \frac{2\pi k}{3})\right)$ for $k = 0, 1, 2$.

A.5 The Complex Exponential Function

Definition A.5.1 Let $z = x + iy$, in which x and y are real. The *complex exponential* function e^z is defined by

$$e^z = e^{x+iy} = e^x(\cos y + i \sin y). \tag{A.5.2}$$

Example A.5.3 $e^{4+7i} = e^4(\cos 7 + i \sin 7)$.

The complex exponential function satisfies the addition formula

$$e^{z+w} = e^z e^w \tag{A.5.4}$$

for all complex z, w. A special case of (A.5.2) is *Euler's formula*

$$e^{iy} = \cos y + i \sin y \quad \text{for } y \in \mathbb{R}. \tag{A.5.5}$$

Although the definition (A.5.2) may appear mysterious, it can be justified through the use of power series; see P.A.15.

Example A.5.6 Euler's formula implies many trigonometric identities. For example, equate real and imaginary parts in

$$\cos(\theta_1 + \theta_2) + i \sin(\theta_1 + \theta_2) = e^{i(\theta_1 + \theta_2)} = e^{i\theta_1} e^{i\theta_2}$$
$$= (\cos \theta_1 + i \sin \theta_1)(\cos \theta_2 + i \sin \theta_2)$$
$$= (\cos \theta_1 \cos \theta_2 - \sin \theta_1 \sin \theta_2)$$
$$+ i(\cos \theta_1 \sin \theta_2 + \sin \theta_1 \cos \theta_2)$$

and obtain the addition formulas for sine and cosine:

$$\cos(\theta_1 + \theta_2) = (\cos \theta_1 \cos \theta_2 - \sin \theta_1 \sin \theta_2) \quad \text{and}$$
$$\sin(\theta_1 + \theta_2) = \cos \theta_1 \sin \theta_2 + \sin \theta_1 \cos \theta_2.$$

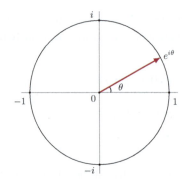

Figure A.8 The unit circle in the complex plane.

Euler's formula (A.5.5) implies that

$$|e^{i\theta}|^2 = (\cos\theta + i\sin\theta)(\cos\theta - i\sin\theta) = \cos^2\theta + \sin^2\theta = 1.$$

Thus, $\{e^{i\theta} : -\pi < \theta \le \pi\}$ is the unit circle in the complex plane; see Figure A.8.
   The polar form of a complex number is

$$z = r(\cos\theta + i\sin\theta) = re^{i\theta},$$

and de Moivre's formula is $z^n = r^n e^{in\theta}$.

Example A.5.7 The cube of $1 + i = \sqrt{2}e^{\pi i/4}$ is $2^{3/2}e^{3\pi i/4} = 2^{3/2}(\cos\frac{3\pi}{4} + i\sin\frac{3\pi}{4})$.

A.6 Problems

P.A.1 Show that: (a) $(1 + i)^4 = -4$. (b) $(1 - i)^{-1} - (1 + i)^{-1} = i$. (c) $(i - 1)^{-4} = -1/4$.
(d) $10(1 + 3i)^{-1} = 1 - 3i$. (e) $(\sqrt{3} + i)^3 = 8i$.

P.A.2 Evaluate the following expressions. Write your answer as $a + bi$, in which a and b
are real. (a) $(1 + i)(2 + 3i)$. (b) $\frac{2+3i}{1+i}$. (c) $\left(\frac{(2+i)^2}{4-3i}\right)^2$. (d) $e^{i\alpha}\overline{e^{i\beta}}$, in which α and β are
real. (e) $\overline{(1 - i)}\overline{(2 + 2i)}$. (f) $|2 - i|^3$.

P.A.3 Let $z = 3 + 4i$. Plot and label the following points in the complex plane: z, $-z$, $\bar{z}$, $-\bar{z}$,
$1/z$, $1/\bar{z}$, and $-1/\bar{z}$.

P.A.4 Compute $(1 + i)^3$ without using the polar form and compare with Example A.5.7.

P.A.5 Write $z = \sqrt{3} - i$ in polar form $z = re^{i\theta}$.

P.A.6 Write the following in polar form: (a) $1 + i$, (b) $(1 + i)^2$, and (c) $(1 + i)^4$.

P.A.7 What are the square roots of $1 + i$? Draw a picture that shows their locations.

P.A.8 What are the cube roots of $8i$? Draw a picture that shows their locations.

P.A.9 Verify that complex multiplication is associative, that is, show that

$$[(a + bi)(c + di)](e + fi) = (a + bi)[(c + di)(e + fi)]$$

for all $a, b, c, d, e, f \in \mathbb{R}$.

P.A.10 Show that $|z| = 1$ if and only if $z \ne 0$ and $1/z = \bar{z}$.

P.A.11 If $z \ne 0$, show that $\overline{1/z} = 1/\bar{z}$.

P.A.12 (a) Show that $\text{Re}(z^{-1}) > 0$ if and only if $\text{Re}\,z > 0$. (b) What about $\text{Im}(z^{-1}) > 0$?

P.A.13 Let $z = a + bi$ and $w = c + di$, in which $a, b, c, d \in \mathbb{R}$. Let $\mathbf{u} = [a\ b]^{\mathsf{T}}$ and $\mathbf{v} = [c\ d]^{\mathsf{T}}$. Show that the inequality $|\operatorname{Re}(z\overline{w})| \leq |z\overline{w}|$ (see (A.3.9)) is equivalent to the Cauchy–Schwarz inequality $|\langle \mathbf{u}, \mathbf{v} \rangle| \leq \|\mathbf{u}\|\|\mathbf{v}\|$ in $\mathbb{R}^2$ (see (5.4.15)).

P.A.14 Show that $|z + w|^2 - |z - w|^2 = 4\operatorname{Re}(z\overline{w})$ and explain why this identity is equivalent to the polarization identity (5.4.24) in $\mathbb{R}^2$.

P.A.15 Derive Euler's formula (A.5.5) by substituting $z = iy$ into the power series expansion $e^z = \sum_{n=0}^{\infty} \frac{z^n}{n!}$ for the exponential function. *Hint*: $\cos z = \sum_{n=0}^{\infty} (-1)^n \frac{z^{2n}}{(2n)!}$ and $\sin z = \sum_{n=0}^{\infty} (-1)^n \frac{z^{2n+1}}{(2n+1)!}$.

P.A.16 Let a, b, c be distinct complex numbers. Prove that a, b, c are the vertices of an equilateral triangle if and only if $a^2 + b^2 + c^2 = ab + bc + ca$. *Hint*: Consider the relationship among $b - a$, $c - b$, and $a - c$.

P.A.17 Suppose that $|a| = |b| = |c| \neq 0$ and $a + b + c = 0$. What can you conclude about a, b, c?

P.A.18 Prove that

$$1 + 2\sum_{n=1}^{N} \cos nx = \frac{\sin((N + \frac{1}{2})x)}{\sin \frac{x}{2}}$$

for $x \neq 0$ by using the formula $1 + z + \cdots + z^{n-1} = \frac{1 - z^n}{1 - z}$ for the sum of a finite geometric series, in which $z \neq 1$; see P.D.5.

P.A.19 Use (A.5.2) to prove the addition formula (A.5.4).

P.A.20 If $\theta \in \mathbb{R}$, show that $\cos \theta = \frac{1}{2}(e^{i\theta} + e^{-i\theta})$ and $\sin \theta = \frac{1}{2i}(e^{i\theta} - e^{-i\theta})$.

APPENDIX B
Polynomials

B.1 Polynomials

A *complex polynomial of degree* $k \geq 0$ is a function of the form

$$p(z) = c_k z^k + c_{k-1} z^{k-1} + \cdots + c_1 z + c_0, \quad \text{in which } c_0, c_1, \ldots, c_k \in \mathbb{C} \text{ and } c_k \neq 0. \quad \text{(B.1.1)}$$

The scalars $c_0, c_1, \ldots, c_k$ are the *coefficients* of p. We denote the degree of p by $\deg p$. If all of its coefficients are real, then p is a *real polynomial*. A real polynomial is a special type of complex polynomial. The *zero polynomial* is the function $p(z) = 0$; by convention, the degree of the zero polynomial is $-\infty$. If $c_k = 1$ in (B.1.1), then p is a *monic* polynomial. Polynomials of degree 1 or greater are *nonconstant polynomials*; polynomials of degree less than one are *constant polynomials*.

The word "polynomial," when used without qualification, means "complex polynomial." For p and c_k as in (B.1.1), the polynomial $p(z)/c_k$ is monic.

If p and q are polynomials and c is a scalar, then cp, $p + q$, and pq are polynomials. For example, if $c = 5, p(z) = z^2 + 1$, and $q(z) = z - 3$, then $cp(z) = 5z^2 + 5, p(z) + q(z) = z^2 + z - 2$, and $p(z)q(z) = z^3 - 3z^2 + z - 3$.

B.2 The Division Algorithm

The following polynomial version of long division with remainder is the *division algorithm*. If f and g are polynomials such that $1 \leq \deg g \leq \deg f$, then there are unique polynomials q and r such that $f = gq + r$ and $\deg r < \deg g$. The polynomials f, g, q, r are the *dividend, divisor, quotient*, and *remainder*, respectively. For example, if

$$f(z) = 2z^4 + z^3 - z^2 + 1 \quad \text{and} \quad g(z) = z^2 - 1,$$

then

$$
\begin{array}{r}
2z^2 + z + 1 \\
z^2 - 1 \overline{\smash{\big)}\ 2z^4 + z^3 - z^2 \qquad\quad + 1} \\
\underline{2z^4 \qquad\quad - 2z^2} \\
z^3 + z^2 \\
\underline{z^3 \qquad\quad - z} \\
z^2 + z \\
\underline{z^2 \qquad\quad - 1} \\
z + 2
\end{array}
$$

and hence $q(z) = 2z^2 + z + 1$ and $r(z) = z + 2$.

B.3 Zeros and Roots

Let λ be a complex number and let p be a polynomial. Then λ is a *zero of* p (alternatively, λ is a *root of the equation* $p(z) = 0$) if $p(\lambda) = 0$.

A real polynomial might have no real zeros. For example, $p(z) = z^2 + 1$ has no real zeros, but $\pm i$ are non-real zeros of p. One of the reasons for using complex scalars in linear algebra is the following result, which can be proved with methods from complex analysis or topology.

Theorem B.3.1 (Fundamental Theorem of Algebra) *Every nonconstant polynomial has a zero in* $\mathbb{C}$.

If $\deg p \geq 1$ and $\lambda \in \mathbb{C}$, then the division algorithm ensures that $p(z) = (z - \lambda)q(z) + r$, in which r is a constant polynomial. If λ is a zero of p, then $0 = p(\lambda) = 0 + r$, so $r = 0$ and $p(z) = (z - \lambda)q(z)$, in which case we have *factored* out the zero λ from p.

This process can be repeated. For the polynomial (B.1.1), we obtain

$$p(z) = c_k(z - \lambda_1)(z - \lambda_2) \cdots (z - \lambda_k), \tag{B.3.2}$$

in which the (not necessarily distinct) complex numbers

$$\lambda_1, \lambda_2, \ldots, \lambda_k \tag{B.3.3}$$

are the zeros of p. If $\mu_1, \mu_2, \ldots, \mu_d$ are the *distinct* elements of the list (B.3.3), we can write (B.3.2) as

$$p(z) = c_k(z - \mu_1)^{n_1}(z - \mu_2)^{n_2} \cdots (z - \mu_d)^{n_d}, \tag{B.3.4}$$

in which $\mu_i \neq \mu_j$ for $i \neq j$. The exponent n_i in (B.3.4), the number of times that μ_i appears in the list (B.3.3), is the *multiplicity* of the zero μ_i of p. Thus, $n_1 + n_2 + \cdots + n_d = n = \deg p$. For example, $p(z) = 3(z - 4)(z - 7)(z - 7) = 3(z - 4)(z - 7)^2$ has $\mu_1 = 4$, $n_1 = 1$, $\mu_2 = 7$, and $n_2 = 2$.

B.4 Identity Theorems for Polynomials

Let f and g be polynomials and suppose that $\deg f = n \geq 1$.

(a) The sum of the multiplicities of the zeros of f is n.

(b) f has at most n distinct zeros.

(c) If $\deg g \leq n$ and g has at least $n + 1$ distinct zeros, then g is the zero polynomial.

(d) If $g(z) = 0$ for infinitely many distinct values of z, then g is the zero polynomial.

(e) If $\deg g \leq n$, if $z_1, z_2, \ldots, z_{n+1}$ are distinct complex numbers, and if $f(z_i) = g(z_i)$ for each $i = 1, 2, \ldots, n + 1$, then $f = g$.

(f) If fg is the zero polynomial, then g is the zero polynomial (since $\deg f \geq 1$).

B.5 Polynomials and Matrices

For a polynomial $p(z) = c_k z^k + c_{k-1} z^{k-1} + \cdots + c_1 z + c_0$ and $A \in \mathsf{M}_n$, we define

$$p(A) = c_k A^k + c_{k-1} A^{k-1} + \cdots + c_1 A + c_0 I.$$

Consider $A = \begin{bmatrix} 1 & 2 \\ 3 & 4 \end{bmatrix}$ and $p(z) = z^2 - 2z + 1$. Then

$$p(A) = \begin{bmatrix} 1 & 2 \\ 3 & 4 \end{bmatrix}\begin{bmatrix} 1 & 2 \\ 3 & 4 \end{bmatrix} - 2\begin{bmatrix} 1 & 2 \\ 3 & 4 \end{bmatrix} + \begin{bmatrix} 1 & 0 \\ 0 & 1 \end{bmatrix} = \begin{bmatrix} 7 & 10 \\ 15 & 22 \end{bmatrix} - \begin{bmatrix} 2 & 4 \\ 6 & 8 \end{bmatrix}$$
$$+ \begin{bmatrix} 1 & 0 \\ 0 & 1 \end{bmatrix} = \begin{bmatrix} 6 & 6 \\ 9 & 15 \end{bmatrix}.$$

The associativity of matrix multiplication (see P.3.24) ensures that the same result is obtained using the factorization $p(z) = (z - 1)^2$:

$$p(A) = (A - I)^2 = \begin{bmatrix} 0 & 2 \\ 3 & 3 \end{bmatrix}\begin{bmatrix} 0 & 2 \\ 3 & 3 \end{bmatrix} = \begin{bmatrix} 6 & 6 \\ 9 & 15 \end{bmatrix}.$$

If p and q are polynomials and $A \in M_n$, then

$$p(A) + q(A) = (p + q)(A)$$

and

$$p(A)q(A) = (pq)(A) = (qp)(A) = q(A)p(A).$$

In particular, A commutes with $p(A)$ for any polynomial p. For $\Lambda = \mathrm{diag}(\lambda_1, \lambda_2, \ldots, \lambda_n)$, we have

$$p(\Lambda) = \mathrm{diag}(p(\lambda_1), p(\lambda_2), \ldots, p(\lambda_n)). \tag{B.5.1}$$

If $A = [a_{ij}]$ is a triangular matrix (upper or lower), then the diagonal entries of $p(A)$ are $p(a_{11}), p(a_{22}), \ldots, p(a_{nn})$; see P.C.20.

B.6 Problems

P.B.1 Let $p(z) = z^3 + z^2 + z + 1$ and $g(z) = z^2 + 2z$. Compute $q(z)$ and $r(z)$ in the identity $p = gq + r$.

P.B.2 If p is a real polynomial, show that $p(\lambda) = 0$ if and only if $p(\bar{\lambda}) = 0$.

P.B.3 Show that a real polynomial can be factored into real linear factors and real quadratic factors that have no real zeros.

P.B.4 Show that every real polynomial of odd degree has a real zero. *Hint*: Use the intermediate value theorem from calculus.

P.B.5 Adopt the notation of Appendix B.2. Let $p(z) = 3z^2 + 2z - 1$, let λ be a scalar, and let $g(z) = z - \lambda$. The division algorithm says that $p = gq + r$, in which $\deg r \leq 1$. (a) Carry out the polynomial long division to obtain $q(z) = 3z + 3\lambda + 2$ and $r = 3\lambda^2 + 2\lambda - 1$. (b) Explain why $z - \lambda$ divides $p(z)$ with zero remainder if and only if $p(\lambda) = 0$.

P.B.6 Let p be a polynomial of degree $n \geq 2$ and let $\lambda \in \mathbb{C}$. Let q be the unique polynomial of degree $n - 1$ such that $p(z) = (z - \lambda)q(z) + r$ (the division algorithm), in which $r \in \mathbb{C}$. (a) Show that $r = p(\lambda)$ and $q(\lambda) = p'(\lambda)$. (b) Explain why $p(z) = (z - \lambda)q(z)$ if and only if $p(\lambda) = 0$. (c) What can you say if $p(\lambda) = p'(\lambda) = 0$?

P.B.7 Let $p(z) = c_k z^k + c_{k-1}z^{k-1} + \cdots + c_1 z + c_0$, in which $k \geq 1$, each c_i is a nonnegative integer, and $c_k \geq 1$. Prove the following statements: (a) $p(t + 2) = c_k t^k + d_{k-1}t^{k-1} + \cdots + d_1 t + d_0$, in which each d_i is a nonnegative integer and $d_0 \geq 2^k$. (b) $p(nd_0 + 2)$

is divisible by d_0 for each $n = 1, 2, \ldots$. (c) $p(n)$ is not a prime for infinitely many positive integers n. This was proved by C. Goldbach in 1752.

P.B.8   Let $h(z)$ be a polynomial and suppose that $z(z - 1)h(z) = 0$ for all $z \in [0, 1]$. Prove that h is the zero polynomial.

P.B.9   If c is a nonzero scalar and p, q are nonzero polynomials, show that (a) $\deg(cp) = \deg p$, (b) $\deg(p + q) \leq \max\{\deg p, \deg q\}$, and (c) $\deg(pq) = \deg p + \deg q$.

P.B.10 What happens in P.B.9 if p is the zero polynomial?

P.B.11 Prove the uniqueness assertion of the division algorithm. That is, if f and g are polynomials such that $1 \leq \deg g \leq \deg f$ and if $q_1, q_2, r_1,$ and r_2 are polynomials such that $\deg r_1 < \deg g$, $\deg r_2 < \deg g$, and $f = gq_1 + r_1 = gq_2 + r_2$, then $q_1 = q_2$ and $r_1 = r_2$.

P.B.12 Give an example of a nonconstant function $f : \mathbb{R} \to \mathbb{R}$ such that $f(t) = 0$ for infinitely many distinct values of t. Is f a polynomial?

APPENDIX C
Basic Linear Algebra

C.1 Functions and Sets

A *set* is a collection of distinct elements. Let $\mathscr{X}$ and $\mathscr{Y}$ be sets. The notation $f\colon \mathscr{X} \to \mathscr{Y}$ indicates that f is a *function* whose *domain* is $\mathscr{X}$ and *codomain* is $\mathscr{Y}$. That is, f assigns a definite value $f(x) \in \mathscr{Y}$ to each $x \in \mathscr{X}$. A function may assign the same value to two different elements in its domain, that is, $x_1 \neq x_2$ and $f(x_1) = f(x_2)$ is possible. But $x_1 = x_2$ and $f(x_1) \neq f(x_2)$ is not permitted.

The *range* of $f\colon \mathscr{X} \to \mathscr{Y}$ is

$$\operatorname{ran} f = \{f(x)\colon x \in \mathscr{X}\} = \{y \in \mathscr{Y}\colon y = f(x) \text{ for some } x \in \mathscr{X}\},$$

which is a subset of $\mathscr{Y}$. A function $f\colon \mathscr{X} \to \mathscr{Y}$ is *onto* if $\operatorname{ran} f = \mathscr{Y}$, that is, if the range and codomain of f are equal. A function $f\colon \mathscr{X} \to \mathscr{Y}$ is *one to one* if $f(x_1) = f(x_2)$ implies that $x_1 = x_2$. Equivalently, f is one to one if $x_1 \neq x_2$ implies that $f(x_1) \neq f(x_2)$; see Figure C.1.

Suppose that $f\colon \mathscr{X} \to \mathscr{Y}$ is onto and one to one. Then for each $y \in \mathscr{Y}$ there is a unique $x \in \mathscr{X}$ such that $f(x) = y$. Define $g\colon \mathscr{Y} \to \mathscr{X}$ as follows: $g(y) = x$ if $f(x) = y$. The function g is the *inverse* of f and we write $g = f^{-1}$. Notice that $f^{-1}(f(x)) = x$ for every $x \in \mathscr{X}$ and $f(f^{-1}(y)) = y$ for every $y \in \mathscr{Y}$.

We say that $x_1, x_2, \ldots, x_k$ are *distinct* if $i, j \in \{1, 2, \ldots, k\}$ and $i \neq j$ implies that $x_i \neq x_j$. For example, the distinct elements in the list $1, 1, 2, 3, 3, 4, 5, 5$ are $1, 2, 3, 4, 5$.

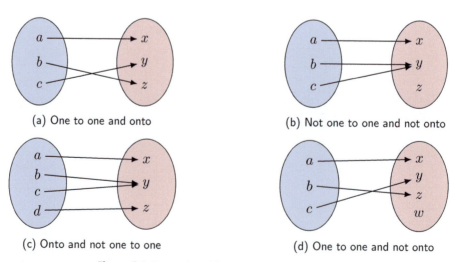

(a) One to one and onto

(b) Not one to one and not onto

(c) Onto and not one to one

(d) One to one and not onto

Figure C.1 Properties of functions: one to one and onto.

C.2 Matrices

We denote the field of real numbers by $\mathbb{R}$ and the field of complex numbers by $\mathbb{C}$. Real or complex numbers are called *scalars*. The only scalars that we consider are complex numbers, which we sometimes restrict to being real. See Appendix A for a discussion of complex numbers.

An $m \times n$ *matrix* is a rectangular array

$$A = [a_{ij}] = \begin{bmatrix} a_{11} & a_{12} & \cdots & a_{1n} \\ a_{21} & a_{22} & \cdots & a_{2n} \\ \vdots & \vdots & \ddots & \vdots \\ a_{m1} & a_{m2} & \cdots & a_{mn} \end{bmatrix} \tag{C.2.1}$$

of real or complex numbers; it has *size* $m \times n$. We identify 1×1 matrices with scalars. The (i, j) *entry* of A is a_{ij}. Two matrices are *equal* if they have the same size and if their corresponding entries are equal. An $n \times n$ matrix is a *square matrix*; we say that it has *size* n. The set of all $m \times n$ matrices with complex entries is denoted by $\mathsf{M}_{m \times n}(\mathbb{C})$, or by $\mathsf{M}_n(\mathbb{C})$ if $m = n$. For convenience, we write $\mathsf{M}_n(\mathbb{C}) = \mathsf{M}_n$ and $\mathsf{M}_{m \times n}(\mathbb{C}) = \mathsf{M}_{m \times n}$. The set of $m \times n$ matrices with real entries is denoted by $\mathsf{M}_{m \times n}(\mathbb{R})$, or by $\mathsf{M}_n(\mathbb{R})$ if $m = n$. Matrices in $\mathsf{M}_{m \times 1}$ are *column vectors*; matrices in $\mathsf{M}_{1 \times n}$ are *row vectors*.

Rows and Columns

For each $j = 1, 2, \ldots, n$, the jth *column* of the matrix (C.2.1) is the $m \times 1$ matrix

$$\mathbf{c}_j = \begin{bmatrix} a_{1j} \\ a_{2j} \\ \vdots \\ a_{mj} \end{bmatrix}.$$

For each $i = 1, 2, \ldots, m$, the ith *row* of the matrix in (C.2.1) is the $1 \times n$ matrix that we denote by (see p. 437 for the transpose of a matrix)

$$\mathbf{r}_i^\mathsf{T} = [a_{i1} \ a_{i2} \ \ldots \ a_{in}].$$

It is often convenient to write the matrix (C.2.1) as a $1 \times n$ array of columns

$$A = [\mathbf{c}_1 \ \mathbf{c}_2 \ \ldots \ \mathbf{c}_n] \tag{C.2.2}$$

or an $m \times 1$ array of rows

$$A = \begin{bmatrix} \mathbf{r}_1^\mathsf{T} \\ \mathbf{r}_2^\mathsf{T} \\ \vdots \\ \mathbf{r}_m^\mathsf{T} \end{bmatrix}. \tag{C.2.3}$$

Addition and Scalar Multiplication

If $A = [a_{ij}]$ and $B = [b_{ij}]$ are $m \times n$ matrices, then $A + B$ is the $m \times n$ matrix whose (i, j) entry is $a_{ij} + b_{ij}$. If $A \in \mathsf{M}_{m \times n}$ and c is a scalar, then $cA = [ca_{ij}]$ is the $m \times n$ matrix obtained by multiplying each entry of A by c. A *zero matrix* is an $m \times n$ matrix whose entries are all zero. Such a matrix is denoted by 0, although subscripts can be attached to indicate its size. Let $A, B \in M_{m \times n}$ and let c, d be scalars.

(a) $A + B = B + A$.

(d) $c(A + B) = cA + cB$.

(b) $A + (B + C) = (A + B) + C$.

(e) $c(dA) = (cd)A = d(cA)$.

(c) $A + 0 = A = 0 + A$.

(f) $(c + d)A = cA + dA$.

Multiplication

If $A = [a_{ij}] \in M_{m \times r}$ and $B = [b_{ij}] \in M_{r \times n}$, then the (i, j) entry of the *product* $AB = [c_{ij}] \in M_{m \times n}$ is

$$c_{ij} = \sum_{k=1}^{r} a_{ik} b_{kj}. \tag{C.2.4}$$

This sum involves entries in the ith row of A and the jth column of B. The number of columns of A must equal the number of rows of B. If we write $B = [\mathbf{b}_1 \ \mathbf{b}_2 \ \dots \ \mathbf{b}_n]$ as a $1 \times n$ array of its columns, then (C.2.4) says that

$$AB = [A\mathbf{b}_1 \ A\mathbf{b}_2 \ \dots \ A\mathbf{b}_n].$$

See Chapter 3 for other interpretations of matrix multiplication.

Let A, B, and C be matrices of appropriate sizes and let c be a scalar.

(a) $A(BC) = (AB)C$.

(c) $(A + B)C = AC + BC$.

(b) $A(B + C) = AB + AC$.

(d) $(cA)B = c(AB) = A(cB)$.

Be careful: $AB = AC$ does not imply that $B = C$; see P.C.7.

Commuting Matrices

We say that $A, B \in M_n$ *commute* if $AB = BA$. Some pairs of matrices in M_n do not commute. For example,

$$\begin{bmatrix} 0 & 1 \\ 0 & 0 \end{bmatrix} \begin{bmatrix} 0 & 0 \\ 0 & 1 \end{bmatrix} = \begin{bmatrix} 0 & 1 \\ 0 & 0 \end{bmatrix}, \quad \text{but} \quad \begin{bmatrix} 0 & 0 \\ 0 & 1 \end{bmatrix} \begin{bmatrix} 0 & 1 \\ 0 & 0 \end{bmatrix} = \begin{bmatrix} 0 & 0 \\ 0 & 0 \end{bmatrix}.$$

Identity Matrices

The matrix

$$I_n = \begin{bmatrix} 1 & 0 & 0 & \cdots & 0 \\ 0 & 1 & 0 & \cdots & 0 \\ 0 & 0 & 1 & \cdots & 0 \\ \vdots & \vdots & \vdots & \ddots & \vdots \\ 0 & 0 & 0 & \cdots & 1 \end{bmatrix} \in M_n$$

is the $n \times n$ *identity matrix*. That is, $I_n = [\delta_{ij}]$, in which

$$\delta_{ij} = \begin{cases} 1 & \text{if } i = j, \\ 0 & \text{if } i \neq j, \end{cases} \tag{C.2.5}$$

is the *Kronecker delta*. If the size is clear from context, we write I with no subscript. For every $A \in M_{m \times n}$, we have $AI_n = A = I_m A$.

Triangular Matrices

We say that $A = [a_{ij}] \in \mathbf{M}_{m \times n}$ is: *upper triangular* if $a_{ij} = 0$ for all $i > j$; *strictly upper triangular* if $a_{ij} = 0$ for all $i \geq j$; *lower triangular* if $a_{ij} = 0$ for all $i < j$; *strictly lower triangular* if $a_{ij} = 0$ for all $i \leq j$; *unit lower triangular* if it is lower triangular and $a_{ii} = 1$ for all i; *triangular* if it is either upper triangular or lower triangular. See Figure C.2.

$$
\begin{bmatrix}
a_{11} & a_{12} & a_{13} & a_{14} & a_{15} \\
0 & a_{22} & a_{23} & a_{24} & a_{25} \\
0 & 0 & a_{33} & a_{34} & a_{35} \\
0 & 0 & 0 & a_{44} & a_{45} \\
0 & 0 & 0 & 0 & a_{55}
\end{bmatrix}
\qquad
\begin{bmatrix}
0 & a_{12} & a_{13} & a_{14} & a_{15} \\
0 & 0 & a_{23} & a_{24} & a_{25} \\
0 & 0 & 0 & a_{34} & a_{35} \\
0 & 0 & 0 & 0 & a_{45}
\end{bmatrix}
\qquad
\begin{bmatrix}
1 & 0 & 0 & 0 \\
a_{21} & 1 & 0 & 0 \\
a_{31} & a_{32} & 1 & 0 \\
a_{41} & a_{42} & a_{43} & 1 \\
a_{51} & a_{52} & a_{53} & a_{54}
\end{bmatrix}
$$

(a) Upper triangular (b) Strictly upper triangular (c) Unit lower triangular

Figure C.2 Triangular matrices can be square or non-square.

Diagonal Matrices

We say that $A = [a_{ij}] \in \mathbf{M}_{m \times n}$ is *diagonal* if $a_{ij} = 0$ whenever $i \neq j$; see Figure C.3. That is, any nonzero entry of A must lie on the *main diagonal* of A, which consists of the *diagonal entries* a_{ij} with $i = j$; the entries a_{ij} with $i \neq j$ are the *off-diagonal entries* of A. The notation $\mathrm{diag}(\lambda_1, \lambda_2, \ldots, \lambda_n)$ denotes the $n \times n$ diagonal matrix whose diagonal entries are $\lambda_1, \lambda_2, \ldots, \lambda_n$, in that order. Two square diagonal matrices of the same size commute. A *scalar matrix* is a square diagonal matrix of the form $\mathrm{diag}(c, c, \ldots, c) = cI$ for some scalar c.

$$
\begin{bmatrix}
a_{11} & 0 & 0 & 0 & 0 \\
0 & a_{22} & 0 & 0 & 0 \\
0 & 0 & a_{33} & 0 & 0 \\
0 & 0 & 0 & a_{44} & 0 \\
0 & 0 & 0 & 0 & a_{55}
\end{bmatrix}
\qquad
\begin{bmatrix}
a_{11} & 0 & 0 & 0 & 0 \\
0 & a_{22} & 0 & 0 & 0 \\
0 & 0 & a_{33} & 0 & 0 \\
0 & 0 & 0 & a_{44} & 0
\end{bmatrix}
\qquad
\begin{bmatrix}
a_{11} & 0 & 0 & 0 \\
0 & a_{22} & 0 & 0 \\
0 & 0 & a_{33} & 0 \\
0 & 0 & 0 & a_{44} \\
0 & 0 & 0 & 0
\end{bmatrix}
$$

(a) Diagonal, $m = n$ (b) Diagonal, $m < n$ (c) Diagonal, $m > n$

Figure C.3 Diagonal matrices can be square or non-square.

Superdiagonals and Subdiagonals

The (first) *superdiagonal* of $A = [a_{ij}] \in \mathbf{M}_{m \times n}$ contains the entries a_{ij} with $j - i = 1$. The kth superdiagonal contains the entries a_{ij} with $j - i = k \geq 1$. The kth *subdiagonal* contains the entries a_{ij} with $i - j = k \geq 1$; see Figure C.4.

$$
\begin{bmatrix}
a_{11} & a_{12} & a_{13} & a_{14} & a_{15} \\
a_{21} & a_{22} & a_{23} & a_{24} & a_{25} \\
a_{31} & a_{32} & a_{33} & a_{34} & a_{35} \\
a_{41} & a_{42} & a_{43} & a_{44} & a_{45} \\
a_{51} & a_{52} & a_{53} & a_{54} & a_{55}
\end{bmatrix}
\quad
\begin{bmatrix}
a_{11} & a_{12} & a_{13} & a_{14} & a_{15} \\
a_{21} & a_{22} & a_{23} & a_{24} & a_{25} \\
a_{31} & a_{32} & a_{33} & a_{34} & a_{35} \\
a_{41} & a_{42} & a_{43} & a_{44} & a_{45}
\end{bmatrix}
\quad
\begin{bmatrix}
a_{11} & a_{12} & a_{13} & a_{14} \\
a_{21} & a_{22} & a_{23} & a_{24} \\
a_{31} & a_{32} & a_{33} & a_{34} \\
a_{41} & a_{42} & a_{43} & a_{44} \\
a_{51} & a_{52} & a_{53} & a_{54}
\end{bmatrix}
$$

(a) First superdiagonal (b) Second superdiagonal (c) First subdiagonal

Figure C.4 Superdiagonals and subdiagonals.

Tridiagonal and Bidiagonal Matrices

A matrix $A = [a_{ij}] \in M_{m \times n}$ is *tridiagonal* if $a_{ij} = 0$ whenever $|i - j| \geq 2$. A tridiagonal matrix is *bidiagonal* if either its subdiagonal or its superdiagonal contains only zeros; see Figure C.5.

$$\begin{bmatrix} a_{11} & a_{12} & 0 & 0 & 0 \\ a_{21} & a_{22} & a_{23} & 0 & 0 \\ 0 & a_{32} & a_{33} & a_{34} & 0 \\ 0 & 0 & a_{43} & a_{44} & a_{45} \\ 0 & 0 & 0 & a_{54} & a_{55} \end{bmatrix} \qquad \begin{bmatrix} a_{11} & 0 & 0 & 0 & 0 \\ a_{21} & a_{22} & 0 & 0 & 0 \\ 0 & a_{32} & a_{33} & 0 & 0 \\ 0 & 0 & a_{43} & a_{44} & 0 \end{bmatrix} \qquad \begin{bmatrix} a_{11} & a_{12} & 0 & 0 & 0 \\ 0 & a_{22} & a_{23} & 0 & 0 \\ 0 & 0 & a_{33} & a_{34} & 0 \\ 0 & 0 & 0 & a_{44} & a_{45} \\ 0 & 0 & 0 & 0 & a_{55} \\ 0 & 0 & 0 & 0 & 0 \end{bmatrix}$$

(a) Tridiagonal   (b) Bidiagonal   (c) Bidiagonal

Figure C.5 Bidiagonal and tridiagonal matrices.

Submatrices

A *submatrix* of $A \in M_{m \times n}$ is a matrix whose entries lie in the intersections of specified rows and columns of A. A $k \times k$ *principal submatrix* of A is a submatrix whose entries lie in the intersections of rows $i_1, i_2, \ldots, i_k$ and columns $i_1, i_2, \ldots, i_k$ of A, for some indices $i_1 < i_2 < \cdots < i_k$. The $k \times k$ *leading principal submatrix* of A is the submatrix $A_{(k)}$ whose entries lie in the intersections of rows $1, 2, \ldots, k$ and columns $1, 2, \ldots, k$. The $k \times k$ *trailing principal submatrix* of A is the submatrix whose entries lie in the intersections of rows $m - k + 1$, $m - k + 2, \ldots, m$ and columns $n - k + 1, n - k + 2, \ldots, n$; see Figure C.6.

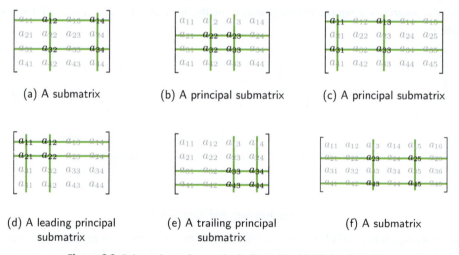

(a) A submatrix   (b) A principal submatrix   (c) A principal submatrix

(d) A leading principal submatrix   (e) A trailing principal submatrix   (f) A submatrix

Figure C.6 Submatrices of a matrix, indicated by highlighted entries.

Inverses and Powers

We say that $A \in M_n$ is *invertible* if there exists a $B \in M_n$ such that

$$AB = I_n = BA. \tag{C.2.6}$$

Such a B is an *inverse* of A. If A has no inverse, then A is *noninvertible* (also referred to as *singular*). Either of the equalities in (C.2.6) implies the other. That is, if $A, B \in M_n$, then $AB = I$ if and only if $BA = I$; see Theorem 2.2.13 and Example 4.1.5.

Not every square matrix has an inverse. However, a matrix has at most one inverse; see Corollary 2.2.14. As a consequence, if A is invertible, we speak of *the* inverse of A, rather than *an* inverse of A. If A is invertible, then the inverse of A is denoted by A^{-1}. It satisfies $AA^{-1} = I = A^{-1}A$. The 2×2 case is noteworthy: if $ad - bc \neq 0$, then

$$\begin{bmatrix} a & b \\ c & d \end{bmatrix}^{-1} = \frac{1}{ad - bc} \begin{bmatrix} d & -b \\ -c & a \end{bmatrix}. \tag{C.2.7}$$

For $A \in \mathsf{M}_n$, define

$$A^0 = I \quad \text{and} \quad A^k = \underbrace{AA \cdots A}_{k \text{ times}}. \tag{C.2.8}$$

If A is invertible, we define $A^{-k} = (A^{-1})^k$ for $k = 1, 2, \ldots.$ Let A and B be matrices of appropriate sizes, let j, k be integers, and let c be a scalar.

(a) $A^j A^k = A^{j+k} = A^k A^j$.

(b) $(A^{-1})^{-1} = A$.

(c) $(A^j)^{-1} = A^{-j}$.

(d) If $c \neq 0$, then $(cA)^{-1} = c^{-1}A^{-1}$.

(e) $(AB)^{-1} = B^{-1}A^{-1}$.

Transpose

The *transpose* of $A = [a_{ij}] \in \mathsf{M}_{m \times n}$ is the matrix $A^\mathsf{T} \in \mathsf{M}_{n \times m}$ whose (i, j) entry is a_{ji}. Let A and B be matrices of appropriate sizes and let c be a scalar.

(a) $(A^\mathsf{T})^\mathsf{T} = A$.

(b) $(A \pm B)^\mathsf{T} = A^\mathsf{T} \pm B^\mathsf{T}$.

(c) $(cA)^\mathsf{T} = cA^\mathsf{T}$.

(d) $(AB)^\mathsf{T} = B^\mathsf{T}A^\mathsf{T}$.

(e) If A is invertible, then $(A^\mathsf{T})^{-1} = (A^{-1})^\mathsf{T}$. We write $(A^{-1})^\mathsf{T} = A^{-\mathsf{T}}$.

Conjugate

The *conjugate* of $A \in \mathsf{M}_{m \times n}$ is the matrix $\overline{A} \in \mathsf{M}_{m \times n}$ whose (i, j) entry is $\overline{a_{ij}}$, the complex conjugate of a_{ij}. Thus,

$$\overline{(\overline{A})} = A, \qquad \overline{A + B} = \overline{A} + \overline{B}, \qquad \text{and} \qquad \overline{AB} = \overline{A}\,\overline{B}.$$

If A has only real entries, then $A = \overline{A}$. The conjugate $\overline{\mathbf{v}}$ of $\mathbf{v} \in \mathbb{C}^m$ follows the rules for $\mathsf{M}_{m \times 1}$.

Conjugate Transpose

The *conjugate transpose* of $A \in \mathsf{M}_{m \times n}$ is the matrix $A^* = \overline{A^\mathsf{T}} = (\overline{A})^\mathsf{T} \in \mathsf{M}_{n \times m}$ whose (i, j) entry is $\overline{a_{ji}}$. If A has only real entries, then $A^* = A^\mathsf{T}$. The conjugate transpose of a matrix is also known as its *adjoint*. Let A and B be matrices of appropriate sizes and let c be a scalar.

(a) $(A^*)^* = A$.

(b) $(A \pm B)^* = A^* \pm B^*$.

(c) $(cA)^* = \overline{c}A^*$.

(d) $(AB)^* = B^*A^*$.

(e) If A is invertible, then $(A^*)^{-1} = (A^{-1})^*$. We write $(A^{-1})^* = A^{-*}$.

Special Types of Matrices

Let $A \in \mathsf{M}_n$.

(a) If $A^* = A$, then A is *Hermitian*; if $A^* = -A$, then A is *skew Hermitian*.

(b) If $A^\mathsf{T} = A$, then A is *symmetric*; if $A^\mathsf{T} = -A$, then A is *skew symmetric*.

(c) If $A^*A = I$, then A is *unitary*; if A is real and $A^\mathsf{T}A = I$, then A is *real orthogonal*.

(d) If $A^*A = AA^*$, then A is *normal*.

(e) If $A^2 = I$, then A is an *involution*.

(f) If $A^2 = A$, then A is *idempotent*.

(g) If $A^k = 0$ for some positive integer k, then A is *nilpotent*.

(h) If $A^\mathsf{T}A = I$, then A is *orthogonal*.

Vandermonde Matrices

The matrix $V(z_1, z_2, \ldots, z_n) = [z_i^{n-j}] \in \mathsf{M}_n$ is a *Vandermonde matrix*. For $n = 1, 2, 3$,

$$V(z_1) = [1], \qquad V(z_1, z_2) = \begin{bmatrix} z_1 & 1 \\ z_2 & 1 \end{bmatrix}, \quad \text{and} \quad V(z_1, z_2, z_3) = \begin{bmatrix} z_1^2 & z_1 & 1 \\ z_2^2 & z_2 & 1 \\ z_3^2 & z_3 & 1 \end{bmatrix}.$$

Vandermonde matrices sometimes appears transposed, with its columns reversed, or with its rows reversed; see (2.8.1). A Vandermonde matrix $V(z_1, z_2, \ldots, z_n)$ is invertible if and only if $z_1, z_2, \ldots, z_n$ are distinct; see P.2.53.

Trace

The *trace* of $A = [a_{ij}] \in \mathsf{M}_n$ is the sum of the diagonal entries of A:

$$\operatorname{tr} A = \sum_{i=1}^{n} a_{ii}.$$

Let A and B be matrices of appropriate sizes and let c be a scalar.

(a) $\operatorname{tr}(cA \pm B) = c \operatorname{tr} A \pm \operatorname{tr} B$.   (d) $\operatorname{tr} A^* = \overline{\operatorname{tr} A}$.

(b) $\operatorname{tr} A^\mathsf{T} = \operatorname{tr} A$.   (e) $\operatorname{tr} AB = \operatorname{tr} BA$.

(c) $\operatorname{tr} \overline{A} = \overline{\operatorname{tr} A}$.

In (e), AB and BA could have different sizes. If $A = [a_{ij}] \in \mathsf{M}_{m \times n}$ and $B = [b_{ij}] \in \mathsf{M}_{n \times m}$, let $AB = [c_{ij}] \in \mathsf{M}_m$ and $BA = [d_{ij}] \in \mathsf{M}_n$. Then

$$\operatorname{tr} AB = \sum_{i=1}^{m} c_{ii} = \sum_{i=1}^{m}\sum_{j=1}^{n} a_{ij}b_{ji} = \sum_{j=1}^{n}\sum_{i=1}^{m} b_{ji}a_{ij} = \sum_{j=1}^{n} d_{jj} = \operatorname{tr} BA. \qquad (\text{C.2.9})$$

For example, if $\mathbf{x} = [x_i], \mathbf{y} = [y_i] \in \mathbb{C}^n$, then $\mathbf{x}\mathbf{y}^\mathsf{T} = [x_i y_j] \in \mathsf{M}_n$, $\mathbf{y}^\mathsf{T}\mathbf{x}$ is a scalar (a 1×1 matrix), and $\operatorname{tr} \mathbf{x}\mathbf{y}^\mathsf{T} = \operatorname{tr} \mathbf{y}^\mathsf{T}\mathbf{x} = \mathbf{y}^\mathsf{T}\mathbf{x} = x_1 y_1 + x_2 y_2 + \cdots + x_n y_n$. Be careful: $\operatorname{tr} ABC$ need not equal $\operatorname{tr} CBA$ or $\operatorname{tr} ACB$. However, (C.2.9) ensures that $\operatorname{tr} ABC = \operatorname{tr} CAB = \operatorname{tr} BCA$.

C.3 Systems of Linear Equations

An $m \times n$ *system of linear equations* (a *linear system*) is a list of linear equations of the form

$$
\begin{aligned}
a_{11}x_1 &+ a_{12}x_2 &+ \cdots &+ a_{1n}x_n &= b_1, \\
a_{21}x_1 &+ a_{22}x_2 &+ \cdots &+ a_{2n}x_n &= b_2, \\
\vdots\quad &\quad \vdots\quad &\ddots\quad &\quad\vdots &\quad\vdots \\
a_{m1}x_1 &+ a_{m2}x_2 &+ \cdots &+ a_{mn}x_n &= b_m.
\end{aligned}
\tag{C.3.1}
$$

It involves m *linear equations* in the n *variables* (or *unknowns*) $x_1, x_2, \ldots, x_n$. The scalars a_{ij} are the *coefficients* of the system (C.3.1); the scalars b_i are the *constant terms*.

By a *solution* to (C.3.1) we mean a list of scalars $x_1, x_2, \ldots, x_n$ that satisfy the m equations in (C.3.1). A system of equations that has no solution is *inconsistent*. If a system has at least one solution, it is *consistent*. There are exactly three possibilities for a system of linear equations: it has no solution, exactly one solution, or infinitely many solutions.

Homogeneous Systems

The system (C.3.1) is *homogeneous* if $b_1 = b_2 = \cdots = b_m = 0$. Every homogeneous system has the *trivial solution* $x_1 = x_2 = \cdots = x_n = 0$. If there are other solutions, they are *nontrivial solutions*. There are only two possibilities for a homogeneous system: it has infinitely many solutions, or it has only the trivial solution. A homogeneous linear system with more unknowns than equations has infinitely many solutions.

Matrix Representation of a Linear System

The linear system (C.3.1) is often written as

$$
A\mathbf{x} = \mathbf{b}, \tag{C.3.2}
$$

in which

$$
A = \begin{bmatrix} a_{11} & a_{12} & \cdots & a_{1n} \\ a_{21} & a_{22} & \cdots & a_{2n} \\ \vdots & \vdots & \ddots & \vdots \\ a_{m1} & a_{m2} & \cdots & a_{mn} \end{bmatrix}, \quad \mathbf{x} = \begin{bmatrix} x_1 \\ x_2 \\ \vdots \\ x_n \end{bmatrix}, \quad \text{and} \quad \mathbf{b} = \begin{bmatrix} b_1 \\ b_2 \\ \vdots \\ b_m \end{bmatrix}. \tag{C.3.3}
$$

The *coefficient matrix* $A = [a_{ij}] \in \mathsf{M}_{m\times n}$ of the system has m rows and n columns if the corresponding system of equations (C.3.1) has m equations in n unknowns. The matrices $\mathbf{x}$ and $\mathbf{b}$ are $n \times 1$ and $m \times 1$, respectively. Matrices such as $\mathbf{x}$ and $\mathbf{b}$ are *column vectors*. We sometimes denote $\mathsf{M}_{n\times 1}(\mathbb{C})$ by $\mathbb{C}^n$ and $\mathsf{M}_{n\times 1}(\mathbb{R})$ by $\mathbb{R}^n$. When we need to identify the entries of a column vector in a line of text, we often write $\mathbf{x} = [x_1 \ x_2 \ \ldots \ x_n]^{\mathsf{T}}$ instead of the tall vertical matrix in (C.3.3). The ith entry of $\mathbf{x}$ is $(\mathbf{x})_i = x_i$.

An $m \times n$ homogeneous linear system can be written as $A\mathbf{x} = \mathbf{0}_m$, in which $A \in \mathsf{M}_{m\times n}$ and $\mathbf{0}_m$ is the $m \times 1$ column vector whose entries are all zero. We say that $\mathbf{0}_m$ is a *zero vector* and write $\mathbf{0}$ with no subscript if the size is clear from context. Since $A\mathbf{0}_n = \mathbf{0}_m$, a homogeneous system always has the trivial solution.

If $A \in \mathsf{M}_n$ is invertible, then (C.3.2) has the unique solution $\mathbf{x} = A^{-1}\mathbf{b}$.

If $\mathbf{x}$ and $\mathbf{x}'$ satisfy (C.3.2), then $A\mathbf{x} = \mathbf{b}$, $A\mathbf{x}' = \mathbf{b}$, and $A(\mathbf{x} - \mathbf{x}') = \mathbf{b} - \mathbf{b} = \mathbf{0}$. Thus, every solution to (C.3.2) can be written as $\mathbf{x} + \mathbf{v}$, in which $A\mathbf{x} = \mathbf{b}$ and $A\mathbf{v} = \mathbf{0}$.

Reduced Row Echelon Form

Three elementary operations can be used to solve a system (C.3.1) of linear equations:

(I) Multiply an equation by a nonzero constant.

(II) Interchange two equations.

(III) Add a multiple of one equation to another.

One can represent the system (C.3.1) as an *augmented matrix*

$$[A \ \mathbf{b}] = \begin{bmatrix} a_{11} & a_{12} & \cdots & a_{1n} & b_1 \\ a_{21} & a_{22} & \cdots & a_{2n} & b_2 \\ \vdots & \vdots & \ddots & \vdots & \vdots \\ a_{m1} & a_{m2} & \cdots & a_{mn} & b_m \end{bmatrix} \tag{C.3.4}$$

and perform *elementary row operations* on (C.3.4) that correspond to the three permissible algebraic operations on the system (C.3.1):

(I) Multiply a row by a nonzero constant.

(II) Interchange two rows.

(III) Add a multiple of one row to another.

Each of these operations is reversible.

The three types of elementary row operations can be used to *row reduce* the augmented matrix (C.3.4) to a simple form from which the solutions to (C.3.1) can be obtained by inspection. A matrix is in *reduced row echelon form* if it satisfies the following:

(a) Rows that consist entirely of zero entries are grouped together at the bottom of the matrix.

(b) If a row does not consist entirely of zero entries, then the first nonzero entry in that row is a one (a *leading one*).

(c) A leading one in a higher row must occur further to the left than a leading one in a lower row.

(d) Every column that contains a leading one must have zero entries everywhere else.

Each matrix can be row reduced to a unique matrix that is in reduced row echelon form; that is, each $A \in \mathbf{M}_{m \times n}$ can be factored as $A = SE$, in which S is invertible (not unique) and E is in reduced row echelon form (unique).

The number of leading ones in the reduced row echelon form of a matrix equals its rank. Other characterizations of the rank of a matrix are discussed in Chapters 2 and 3.

Elementary Matrices

An $n \times n$ matrix is an *elementary matrix* if it can be obtained from I_n by performing an elementary row operation. Every elementary matrix is invertible; its inverse is the elementary matrix that corresponds to reversing the original row operation. Multiplication of a matrix on the left by an elementary matrix performs an elementary row operation on that matrix, so elementary matrices encode the elementary row operations. Here are some 2×2 examples:

(I) Multiply a row by a nonzero constant:

$$\begin{bmatrix} k & 0 \\ 0 & 1 \end{bmatrix} \begin{bmatrix} a_{11} & a_{12} \\ a_{21} & a_{22} \end{bmatrix} = \begin{bmatrix} ka_{11} & ka_{12} \\ a_{21} & a_{22} \end{bmatrix}.$$

The inverse of this elementary matrix is $\begin{bmatrix} k^{-1} & 0 \\ 0 & 1 \end{bmatrix}$.

(II) Interchange two rows:

$$\begin{bmatrix} 0 & 1 \\ 1 & 0 \end{bmatrix} \begin{bmatrix} a_{11} & a_{12} \\ a_{21} & a_{22} \end{bmatrix} = \begin{bmatrix} a_{21} & a_{22} \\ a_{11} & a_{12} \end{bmatrix}.$$

This elementary matrix is its own inverse.

(III) Add a nonzero multiple of one row to another:

$$\begin{bmatrix} 1 & k \\ 0 & 1 \end{bmatrix} \begin{bmatrix} a_{11} & a_{12} \\ a_{21} & a_{22} \end{bmatrix} = \begin{bmatrix} a_{11} + ka_{21} & a_{12} + ka_{22} \\ a_{21} & a_{22} \end{bmatrix}$$

and

$$\begin{bmatrix} 1 & 0 \\ k & 1 \end{bmatrix} \begin{bmatrix} a_{11} & a_{12} \\ a_{21} & a_{22} \end{bmatrix} = \begin{bmatrix} a_{11} & a_{12} \\ ka_{11} + a_{21} & ka_{12} + a_{22} \end{bmatrix}.$$

The inverses of these elementary matrices are $\begin{bmatrix} 1 & -k \\ 0 & 1 \end{bmatrix}$ and $\begin{bmatrix} 1 & 0 \\ -k & 1 \end{bmatrix}$, respectively.

Multiplication of a matrix on the right by an elementary matrix performs column operations. A matrix is invertible if and only if it is a product of elementary matrices.

C.4 Determinants

The determinant function $\det: M_n(\mathbb{C}) \to \mathbb{C}$ is of great theoretical importance, but of limited numerical use. Computation of determinants of large matrices should be avoided in applications. See [MM18, Chapter 6] for a detailed exposition of determinants.

Laplace Expansions

The determinant of an $n \times n$ matrix $A = [a_{ij}] \in M_n$ can be defined and computed inductively as a certain sum of determinants of $(n-1) \times (n-1)$ matrices. If $n = 1$, define $\det A = a_{11}$. If $n \geq 2$, let A_{ij} denote the $(n-1) \times (n-1)$ submatrix obtained by deleting the ith row and jth column of A. Then

$$\det A = \sum_{k=1}^{n} (-1)^{i+k} a_{ik} \det A_{ik} = \sum_{k=1}^{n} (-1)^{k+j} a_{kj} \det A_{kj} \quad \text{for } i, j = 1, 2, \ldots, n. \quad \text{(C.4.1)}$$

The first sum is the *Laplace expansion by minors along row i* and the second is the *Laplace expansion by minors along column j*. The quantity $\det A_{ij}$ is the (i, j) *minor of A*; the signed product $(-1)^{i+j} \det A_{ij}$ is the (i, j) *cofactor of A*.

The sign that accompanies a minor $\det A_{ij}$ in a Laplace expansion (C.4.1) is the sign that appears in position (i, j) of the checkerboard pattern

$$
\begin{bmatrix}
+ & - & + & - & + & \cdots \\
- & + & - & + & - & \cdots \\
+ & - & + & - & + & \cdots \\
- & + & - & + & - & \cdots \\
+ & - & + & - & + & \cdots \\
\vdots & \vdots & \vdots & \vdots & \vdots & \ddots
\end{bmatrix}.
$$

For example, here is how the submatrix A_{23} is obtained by striking out the second row and third column of a 4×4 matrix:

$$
A = \begin{bmatrix}
a_{11} & a_{12} & a_{13} & a_{14} \\
a_{21} & a_{22} & a_{23} & a_{24} \\
a_{31} & a_{32} & a_{33} & a_{34} \\
a_{41} & a_{42} & a_{43} & a_{44}
\end{bmatrix}
\quad \text{and} \quad
A_{23} = \begin{bmatrix}
a_{11} & a_{12} & a_{14} \\
a_{31} & a_{32} & a_{34} \\
a_{41} & a_{42} & a_{44}
\end{bmatrix}.
$$

The $(2, 3)$ cofactor of A is $(-1)^{2+3} \det A_{23} = -\det A_{23}$.

If $n = 2$, the Laplace expansion of the determinant by minors along the first row is

$$
\det \begin{bmatrix} a_{11} & a_{12} \\ a_{21} & a_{22} \end{bmatrix} = a_{11} \det[a_{22}] - a_{12} \det[a_{21}] = a_{11}a_{22} - a_{12}a_{21}.
$$

The Laplace expansion by minors along the second column is

$$
\det \begin{bmatrix} a_{11} & a_{12} \\ a_{21} & a_{22} \end{bmatrix} = -a_{12} \det[a_{21}] + a_{22} \det[a_{11}] = a_{11}a_{22} - a_{12}a_{21}.
$$

If $n = 3$, the Laplace expansion of the determinant by minors along the first row is

$$
\det \begin{bmatrix} a_{11} & a_{12} & a_{13} \\ a_{21} & a_{22} & a_{23} \\ a_{31} & a_{32} & a_{33} \end{bmatrix} = a_{11} \det \begin{bmatrix} a_{22} & a_{23} \\ a_{32} & a_{33} \end{bmatrix} - a_{12} \det \begin{bmatrix} a_{21} & a_{23} \\ a_{31} & a_{33} \end{bmatrix} + a_{13} \det \begin{bmatrix} a_{21} & a_{22} \\ a_{31} & a_{32} \end{bmatrix}
$$

$$
= a_{11}a_{22}a_{33} + a_{12}a_{23}a_{31} + a_{13}a_{32}a_{21} - a_{11}a_{23}a_{32}
$$

$$
- a_{22}a_{13}a_{31} - a_{33}a_{12}a_{21}.
$$

The inductive definition of $\det A$ via Laplace expansion ensures that (a) $\det A$ is a sum of $n!$ distinct signed products of matrix entries; (b) $a_{11}a_{22} \cdots a_{nn}$ is a summand; and (c) each summand contains as a factor exactly one entry from each row and column of A.

Determinants and the Inverse

If $A \in M_n$, the *adjugate* of A is the $n \times n$ matrix

$$
\operatorname{adj} A = [(-1)^{i+j} \det A_{ji}] \in M_n, \tag{C.4.2}
$$

which is the transpose of the matrix of cofactors of A. The matrices A and $\operatorname{adj} A$ satisfy

$$
A(\operatorname{adj} A) = (\operatorname{adj} A)A = (\det A)I. \tag{C.4.3}
$$

If A is invertible, then

$$
A^{-1} = (\det A)^{-1} \operatorname{adj} A. \tag{C.4.4}
$$

Consequently, if A has integer entries and $\det A = 1$, then A^{-1} has integer entries.

Properties of Determinants

Let $A, B \in M_n$ and let c be a scalar.

(a) $\det I = 1$.

(b) $\det A \neq 0$ if and only if A is invertible.

(c) $\det AB = (\det A)(\det B)$; this is the *product rule* for determinants.

(d) $\det AB = \det BA$.

(e) $\det(cA) = c^n \det A$.

(f) $\det \overline{A} = \overline{\det A}$.

(g) $\det A^{\mathsf{T}} = \det A$.

(h) $\det A^* = \overline{\det A}$.

(i) If A is invertible, then $\det(A^{-1}) = (\det A)^{-1}$.

(j) If $A = [a_{ij}] \in M_n$ is upper or lower triangular, then $\det A = a_{11}a_{22} \cdots a_{nn}$.

(k) $\det A \in \mathbb{R}$ if $A \in M_n(\mathbb{R})$.

(l) $\det A$ is a linear function of any one column (or any one row) of A, with the other columns (or rows) held fixed; see Theorem 3.1.25.

(m) $\det A = 0$ if A has two equal columns or two equal rows.

Be careful: $\det(A + B)$ need not equal $\det A + \det B$.

Determinants and Row Reduction

The determinant of an $n \times n$ matrix A can be computed with row reduction and the following properties:

(I) If B is obtained by multiplying each entry of a row of A by a scalar c, then $\det B = c \det A$.

(II) If B is obtained by interchanging two different rows of A, then $\det B = -\det A$.

(III) If B is obtained from A by adding a scalar multiple of a row to a different row, then $\det B = \det A$.

Because $\det A = \det A^{\mathsf{T}}$, column operations have analogous properties.

Permutations and Determinants

A *permutation* of the list $1, 2, \ldots, n$ is a one-to-one function $\sigma : \{1, 2, \ldots, n\} \to \{1, 2, \ldots, n\}$. A permutation induces a reordering of $1, 2, \ldots, n$. For example, $\sigma(1) = 2$, $\sigma(2) = 1$, and $\sigma(3) = 3$ defines a permutation of $1, 2, 3$. There are $n!$ distinct permutations of the list $1, 2, \ldots, n$.

A permutation $\tau : \{1, 2, \ldots, n\} \to \{1, 2, \ldots, n\}$ that interchanges precisely two elements of $1, 2, \ldots, n$ and leaves all others fixed is a *transposition*. Each permutation of $1, 2, \ldots, n$ can be written as a composition of transpositions in many different ways. However, the parity (even or odd) of the number of transpositions involved depends only upon the permutation. We say that a permutation σ is *even* or *odd* depending upon whether an even or odd number of transpositions is required to represent σ. The *sign* of σ is

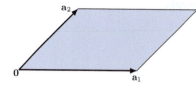 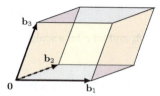

(a) Parallelogram determined by $\mathbf{a}_1, \mathbf{a}_2$ $\in \mathbb{R}^2$.

(b) Parallelepiped determined by $\mathbf{b}_1, \mathbf{b}_2, \mathbf{b}_3$ $\in \mathbb{R}^3$.

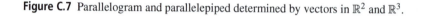

Figure C.7 Parallelogram and parallelepiped determined by vectors in $\mathbb{R}^2$ and $\mathbb{R}^3$.

$$\operatorname{sgn}\sigma = \begin{cases} 1 & \text{if } \sigma \text{ is even,} \\ -1 & \text{if } \sigma \text{ is odd.} \end{cases}$$

The determinant of $A = [a_{ij}] \in \mathsf{M}_n$ can be written

$$\det A = \sum_{\sigma}\left(\operatorname{sgn}\sigma \prod_{i=1}^{n} a_{i\sigma(i)}\right),$$

in which the sum is over all $n!$ permutations of $1, 2, \ldots, n$.

Determinants, Area, and Volume

If $A = [\mathbf{a}_1\ \mathbf{a}_2] \in \mathsf{M}_2(\mathbb{R})$, then $|\det A|$ is the area of the parallelogram (possibly degenerate) determined by $\mathbf{a}_1$ and $\mathbf{a}_2$ (its vertices are at $\mathbf{0}$, $\mathbf{a}_1$, $\mathbf{a}_2$, and $\mathbf{a}_1 + \mathbf{a}_2$); see Figure C.7.a. If $B = [\mathbf{b}_1\ \mathbf{b}_2\ \mathbf{b}_3] \in \mathsf{M}_3(\mathbb{R})$, then $|\det B|$ is the volume of the parallelepiped (possibly degenerate) determined by $\mathbf{b}_1$, $\mathbf{b}_2$, and $\mathbf{b}_3$; see Figure C.7.b.

C.5 Problems

P.C.1 Let $f\colon \{1, 2, \ldots, n\} \to \{1, 2, \ldots, n\}$ be a function. Show that the following are equivalent: (a) f is one to one. (b) f is onto. (c) f is a permutation of $1, 2, \ldots, n$.

P.C.2 Show that: (a) The diagonal entries of a Hermitian matrix are real. (b) The diagonal entries of a skew-Hermitian matrix are purely imaginary. (c) The diagonal entries of a skew-symmetric matrix are zero.

P.C.3 If $A \in \mathsf{M}_n(\mathbb{F})$ is invertible, show that A^T is invertible and $(A^{-1})^\mathsf{T}$ is its inverse.

P.C.4 Let $A = \operatorname{diag}(1, 2)$. If $X \in \mathsf{M}_2$ commutes with A, what can you say about X? For a generalization, see Lemma 3.2.17.

P.C.5 (a) Verify the identity (C.4.3) for a 2×2 matrix. (b) Show that the identity (C.2.7) is (C.4.4).

P.C.6 Deduce (C.4.4) from the identity (C.4.3).

P.C.7 Let $A = \begin{bmatrix} 0 & 1 \\ 0 & 0 \end{bmatrix}$, $B = \begin{bmatrix} 4 & 3 \\ 1 & 2 \end{bmatrix}$, and $C = \begin{bmatrix} 3 & 4 \\ 1 & 2 \end{bmatrix}$. Show that $AB = AC$ even though $B \neq C$.

P.C.8 Let $A = \begin{bmatrix} 1 & i \\ i & -1 \end{bmatrix}$. (a) Show that $A^2 = 0$ even though $A \neq 0$. (b) Find a 3×3 matrix whose square is nonzero and whose cube is zero.

P.C.9 Let $A \in \mathsf{M}_n$. Show that A is idempotent if and only if $I - A$ is idempotent.

P.C.10 Let $A \in \mathsf{M}_n$ be idempotent. Show that A is invertible if and only if $A = I$.

P.C.11 Let $A, B \in \mathsf{M}_n$ be idempotent. Show that $\mathrm{tr}((A - B)^3) = \mathrm{tr}(A - B)$.

P.C.12 (a) Exhibit 2×2 matrices A, B, C such that $\mathrm{tr}\, ABC \neq \mathrm{tr}\, BAC$. (b) Using your matrices in (a), verify that $\mathrm{tr}\, ABC = \mathrm{tr}\, BCA = \mathrm{tr}\, CAB$.

P.C.13 Compute the determinants of

$$\begin{bmatrix} 5 & \sqrt{3}+i \\ \sqrt{3}-i & 4 \end{bmatrix}, \qquad \begin{bmatrix} 1 & 2 & 3 \\ 2 & 2 & 3 \\ 3 & 3 & 3 \end{bmatrix}, \qquad \text{and} \qquad \begin{bmatrix} 1 & 1 & 1 \\ 1 & 2 & 2 \\ 1 & 2 & 3 \end{bmatrix}.$$

P.C.14 Let

$$A = \begin{bmatrix} 1 & 2 \\ 3 & 4 \end{bmatrix}, \quad B = \begin{bmatrix} 8 & 5 \\ 7 & 6 \end{bmatrix}, \quad B_1 = \begin{bmatrix} 3 & 4 \\ 7 & 6 \end{bmatrix},$$

$$B_2 = \begin{bmatrix} 5 & 1 \\ 7 & 6 \end{bmatrix}, \quad B_3 = \begin{bmatrix} 8 & 2 \\ 7 & 4 \end{bmatrix}, \quad B_4 = \begin{bmatrix} 8 & 3 \\ 7 & 2 \end{bmatrix}.$$

(a) Compute AB, BA, $\det A$, $\det B$, $\det AB$, and $\det BA$. (b) Is $AB = BA$? (c) How are $\det AB$, $\det BA$, $\det A$, and $\det B$ related? Is this an accident? (d) Compute $A + B$ and $\det(A + B)$. Is $\det(A + B) = \det A + \det B$? (e) Compute $\det B_1$, $\det B_2$, $\det B_3$, and $\det B_4$. (f) How are $\det B_1 + \det B_2$, $\det B_3 + \det B_4$, and $\det B$ related? Is this an accident?

P.C.15 So-called *sparse matrices* are often encountered in real-world datasets; they have many zero entries and may have some special structure that can be exploited in computations. Consider

$$A = \begin{bmatrix} 1 & 0 & 0 & 0 & 0 \\ 2 & 1 & 3 & 4 & 5 \\ 6 & 0 & 1 & 0 & 0 \\ 7 & 0 & 8 & 1 & 9 \\ 10 & 0 & 11 & 0 & 1 \end{bmatrix}.$$

(a) Compute $\det A$ by using strategic choices of rows and columns to compute Laplace expansions. (b) The sparsity pattern of this matrix forces its inverse to have a zero entry in every position in which A has a zero entry. Use (C.4.2) and (C.4.4) to verify this assertion for four zero entries of your choice, one in each row of A that contains a zero entry.

P.C.16 (a) For each $n = 1, 2, \ldots$, what is the smallest positive integer N such that no $n \times n$ matrix with N zero entries is invertible? (b) Exhibit an invertible $n \times n$ matrix with $N - 1$ zero entries.

P.C.17 Let

$$A = \begin{bmatrix} 2 & 2 & -2 & 0 \\ 1 & 3 & 3 & -1 \\ 3 & 3 & -3 & 2 \end{bmatrix} \qquad \text{and} \qquad B = \begin{bmatrix} 0 & -1 & -3 & -1 \\ -2 & 1 & 0 & 2 \\ 2 & -3 & 1 & 3 \end{bmatrix}.$$

Does there exist an invertible $S \in \mathsf{M}_3$ such that $A = SB$? *Hint*: Consider the reduced row echelon forms of A and B.

P.C.18   Use row reduction to compute rank A and rank A^{T} for

$$A = \begin{bmatrix} 1 & 1 & -3 & -2 \\ -1 & 3 & -3 & 0 \\ 4 & -1 & -1 & -2 \end{bmatrix}.$$

Verify that rank $A = $ rank A^{T}, as predicted by Theorem 2.3.1.

P.C.19   Show that $\begin{bmatrix} \sqrt{2} & -i \\ i & \sqrt{2} \end{bmatrix}$ is orthogonal but not unitary. Why must a real orthogonal matrix be unitary?

P.C.20   Let $A = [a_{ij}] \in \mathsf{M}_n$ be triangular (upper or lower) and let p be a polynomial. Show that the diagonal entries of $p(A)$ are $p(a_{11}), p(a_{22}), \ldots, p(a_{nn})$.

D.1 Mathematical Induction

Suppose that $S_1, S_2, \ldots$ are mathematical statements. The *principle of mathematical induction* asserts that if the statements:

(a) "S_1 is true," and

(b) "If S_n is true, then S_{n+1} is true,"

are true, then S_n is true for all $n \geq 1$. The statement "S_n is true" is the *induction hypothesis*.

If we have proved the *base case* (a), then S_1 is true. If we have also proved the *inductive step* (b), then the truth of S_1 implies the truth of S_2, which implies the truth of S_3, and so forth; see Figure D.1.

For example, let S_n be the statement

$$1 + 2 + \cdots + n = \frac{n(n+1)}{2}.$$

We use mathematical induction to prove that S_n is true for $n = 1, 2, \ldots$. Since $1 = \frac{1 \cdot 2}{2}$, the base case S_1 is true. To show that S_n implies S_{n+1} we must show that the induction hypothesis

$$1 + 2 + \cdots + n = \frac{n(n+1)}{2} \tag{D.1.1}$$

implies that

$$1 + 2 + \cdots + n + (n+1) = \frac{(n+1)((n+1)+1)}{2}. \tag{D.1.2}$$

(a) After you prove the base case, you know that S_1 is true.

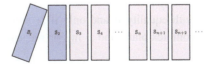

(b) After you prove the inductive step, you know that S_1 implies S_2, so S_2 is true ...

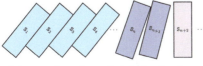

(c) ... and you know that S_n implies S_{n+1}.

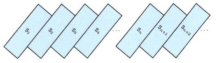

(d) Mathematical induction ensures that S_n is true for all $n = 1, 2, \ldots$.

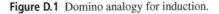

Figure D.1 Domino analogy for induction.

Add $n+1$ to both sides of (D.1.1) and obtain

$$(1+2+\cdots+n)+(n+1) = \frac{n(n+1)}{2}+(n+1)$$
$$= \frac{n(n+1)+2(n+1)}{2}$$
$$= \frac{(n+1)(n+2)}{2}$$
$$= \frac{(n+1)((n+1)+1)}{2},$$

which is (D.1.2). Therefore, S_{n+1} is true if S_n is true. The principle of mathematical induction ensures that S_n is true for all $n = 1, 2, \ldots$.

A statement S_n can be true for many initial values of n without being true for all n. For example, the polynomial $p(n) = n^2 + n + 41$ has the property that $p(n)$ is prime for $n = 1, 2, \ldots, 39$. However, the streak ends with $n = 40$ since $p(40) = 41^2$ is not prime. The polynomial $p(n) = n^6 + 1091$ has the property that $p(n)$ is not prime for $n = 1, 2, \ldots, 3905$, but $p(3906)$ is prime.

D.2 Problems

P.D.1 Use mathematical induction to prove that $1^2 + 2^2 + \cdots + n^2 = \frac{n(n+1)(2n+1)}{6}$ for $n = 1, 2, \ldots$.

P.D.2 Use mathematical induction to prove that $1^3 + 2^3 + \cdots + n^3 = \left(\frac{n(n+1)}{2}\right)^2$ for $n = 1, 2, \ldots$.

P.D.3 Let $A \in M_n$ be invertible. Use mathematical induction to prove that $(A^{-1})^k = (A^k)^{-1}$ for all integers k.

P.D.4 Use mathematical induction to prove Binet's formula (10.5.6) for the Fibonacci numbers.

P.D.5 Use mathematical induction to prove that

$$1 + z + z^2 + \cdots + z^{n-1} = \frac{1 - z^n}{1 - z}$$

for complex $z \neq 1$ and all positive integers n.

P.D.6 Prove A.3.11 by mathematical induction. *Hint*: If z_1, z_2, z_3 are nonzero, then

$$|z_1 + z_2 + z_3| \leq |z_1| + |z_2 + z_3| \leq |z_1| + |z_2| + |z_3|,$$

with equality if and only if z_2 and z_3 lie on the same ray through the origin in the complex plane and z_1 lies on the same ray as $z_2 + z_3$.

References

[Bha05]   Rajendra Bhatia, *Fourier Series*, Classroom Resource Materials Series, Mathematical Association of America, Washington, DC, 2005. Reprint of the 1993 edition [Hindustan Book Agency, New Delhi].

[ĆJ15]   Branko Ćurgus and Robert I. Jewett, Somewhat stochastic matrices, *Amer. Math. Monthly* **122** (2015), no. 1, 36–42.

[Dav63]   Philip J. Davis, *Interpolation and Approximation*, Blaisdell Publishing Co., Ginn and Co., New York, NY, Toronto, London, 1963; Dover Books on Mathematics, 2014.

[DF04]   David S. Dummit and Richard M. Foote, *Abstract Algebra*, 3rd ed., John Wiley & Sons, Hoboken, NJ, 2004.

[Dur12]   Peter Duren, *Invitation to Classical Analysis*, Pure and Applied Undergraduate Texts, Vol. 17, American Mathematical Society, Providence, RI, 2012.

[GK65]   Gene H. Golub and William Kahan, Calculating the singular values and pseudo-inverse of a matrix, *SIAM J. Numer. Anal.* **2** (1965), 205–224.

[GVL13]   Gene H. Golub and Charles F. Van Loan, *Matrix Computations*, 4th ed., Johns Hopkins Studies in the Mathematical Sciences, Johns Hopkins University Press, Baltimore, MD, 2013.

[HJ94]   Roger A. Horn and Charles R. Johnson, *Topics in Matrix Analysis*, Cambridge University Press, Cambridge, 1994. Corrected reprint of the 1991 original.

[HJ13]   Roger A. Horn and Charles R. Johnson, *Matrix Analysis*, 2nd ed., Cambridge University Press, Cambridge, 2013.

[HY20]   Roger A. Horn and Zai Yang, Rank of a Hadamard product, *Linear Algebra Appl.* **591** (2020), 87–98.

[JS96]   Charles R. Johnson and Erik A. Schreiner, The relationship between *AB* and *BA*, *Amer. Math. Monthly* **103** (1996), no. 7, 578–582.

[MM18]   Elizabeth S. Meckes and Mark W. Meckes, *Linear Algebra*, Cambridge University Press, Cambridge, 2018.

[Pra14]   Viktor V. Prasolov, *Problems and Theorems in Linear Algebra*, Translations of Mathematical Monographs, Vol. 134, American Mathematical Society, Providence, RI, 1994.

Index

positive semidefinite matrix, 309
symmetric unitary matrix, 289
standard basis, *see* basis, standard
standard inner product, *see* inner product, standard
stochastic matrix
    column, 262
    definition, 262
    doubly, 262
    eigenvalues with modulus 1, 263
    Hadamard product of a unitary matrix and its
        conjugate, 323
    Jordan canonical form, 263
    limit theorem, 264, 266
    limit theorem if doubly stochastic, 273
    normal, 298
    positive, 263
    positive entries, 264
    power bounded, 262
    row, 262
    spectral radius, 263
strict diagonal dominance
    definition, 192
    implies invertibility, 193
    sufficient condition for positive definiteness, 307
strictly lower triangular matrix, *see* matrix, strictly
        lower triangular
strictly upper triangular matrix, *see* matrix, strictly
        upper triangular
Strutt, John William, 369
subdiagonal, 435
submatrix
    and rank, 157, 321
    definition, 436
    eigenvalue interlacing, 374
    invertible and rank, 157
    leading principal, 436
    principal, 436
    singular value interlacing, 373
    trailing principal, 436
submultiplicative
    definition, 348, 391
    matrix norms are, 391
subspace
    always nonempty, 5
    criterion, 5
    definition, 5
    direct sum, 12
    direct sum of orthogonal complements, 159
    intersection, 11, 77, 89
    invariant, *see* invariant subspace
    is the column space of some matrix, 75
    is the null space of some matrix, 75
    span of union, 12
    subset need not be a, 7
    sum, 12
    union, 11
    zero, 5
sum of subspaces, *see* subspace, sum
superdiagonal, 435

Sylvester
    criterion for positive definiteness, 312, 375
    criterion for positive semidefiniteness, 319, 388
    determinant identity, 72, 154
    inertia theorem, 378, 383
    James Joseph, 389
    law of nullity, 88
    theorem on linear matrix equations, 228
symmetric matrix
    complex, *see* complex symmetric matrix
    definition, 438
    product of two, 269
    real, *see* real symmetric matrix
    real vs. complex, 298
symmetric normal matrix
    real and imaginary parts, 298
    spectral theorem, 289
symmetric relation, 43
symmetric unitary matrix
    spectral theorem, 289
    square root, 289
system of linear equations, *see* linear system

tensor product, *see* Kronecker product
Toeplitz matrix
    commutes with Jordan block, 275
    definition, 275
trace
    and singular values, 352
    definition, 438
    idempotent matrix, 445
    is a linear transformation, 88
    preserved by similarity, 41
    sum of eigenvalues, 206, 222
    $\operatorname{tr} AB = \operatorname{tr} BA$, 438
    zero, *see* Shoda's theorem
trailing principal submatrix, *see* submatrix, trailing
        principal
transitive relation, 43
transpose
    definition, 437
    is idempotent, 437
    of a block matrix, 63
    of inverses, 437
    reverses products, 437
    similarity of a matrix and its, 250, 269
transposition, *see* permutation, transposition
triangle inequality
    complex numbers, 422
    derived norm, 105
    Euclidean length, 97
    in the plane, 95
    norm, 106
triangular matrix
    definition, 435
    diagonal entries of $p(A)$, 430
    products, 64
tridiagonal matrix
    definition, 436